20

Springer Series on Fluorescence

Methods and Applications

Series Editor: Bruno Pedras

Springer Series on Fluorescence

Series Editor: Bruno Pedras

Recently Published and Forthcoming Volumes

Fluorescence Spectroscopy and Microscopy in Biology
Volume Editors: Radek Šachl and Mariana Amaro
Vol. 20, 2023

Modern Applications of Lanthanide Luminescence
Volume Editor: Ana de Bettencourt-Dias
Vol. 19, 2023

Fluorescence in Industry
Volume Editor: Bruno Pedras
Vol. 18, 2019

Perspectives on Fluorescence
A Tribute to Gregorio Weber
Volume Editor: David M. Jameson
Vol. 17, 2016

Fluorescence Studies of Polymer Containing Systems
Volume Editor: Karel rocházka
Vol. 16, 2016

Advanced Photon Counting
Volume Editors: Peter Kapusta, Michael Wahl and Rainer Erdmann
Vol. 15, 2015

Far-Field Optical Nanoscopy
Volume Editors: Philip Tinnefeld, Christian Eggeling and Stefan W. Hell
Vol. 14, 2015

Fluorescent Methods to Study Biological Membranes
Volume Editors: Y. Mély and G. Duportail
Vol. 13, 2013

Fluorescent Proteins II
Application of Fluorescent Protein Technology
Volume Editor: G. Jung
Vol. 12, 2012

Fluorescent Proteins I
From Understanding to Design
Volume Editor: G. Jung
Vol. 11, 2012

Advanced Fluorescence Reporters in Chemistry and Biology III
Applications in Sensing and Imaging
Volume Editor: A.P. Demchenko
Vol. 10, 2011

Advanced Fluorescence Reporters in Chemistry and Biology II
Molecular Constructions, Polymers and Nanoparticles
Volume Editor: A.P. Demchenko
Vol. 9, 2010

Advanced Fluorescence Reporters in Chemistry and Biology I
Fundamentals and Molecular Design
Volume Editor: A.P. Demchenko
Vol. 8, 2010

Lanthanide Luminescence
Photophysical, Analytical and Biological Aspects
Volume Editors: P. Hänninen and H. Härmä
Vol. 7, 2011

Standardization and Quality Assurance in Fluorescence Measurements II
Bioanalytical and Biomedical Applications
Volume Editor: Resch-Genger, U.
Vol. 6, 2008

Standardization and Quality Assurance in Fluorescence Measurements I
Techniques
Volume Editor: U. Resch-Genger
Vol. 5, 2008

Fluorescence of Supermolecules, Polymeres, and Nanosystems
Volume Editor: M.N. Berberan-Santos
Vol. 4, 2007

Fluorescence Spectroscopy in Biology
Volume Editor: M. Hof
Vol. 3, 2004

Fluorescence Spectroscopy, Imaging and Probes
Volume Editor: R. Kraayenhof
Vol. 2, 2002

Fluorescence Spectroscopy and Microscopy in Biology

Volume Editors:
Radek Šachl
Mariana Amaro

With contributions by

Abhinav · D. Y. K. Aik · M. Amaro · K. Bacia · V. Birkedal · A. Bodén · C. Bodner · M. Brameshuber · B. Chmelová · A. Coutinho · J. Damborsky · T. Dellmann · C. Díaz-García · M. S. Dietz · J. Ebenhan · F. Fernandes · V. Fidler · A. J. Garcia Saéz · A. M. Giudici · M. Heilemann · M. Hof · H. Hofmann · J. Humpolíčková · B. Israels L. B.-Å. Johansson · P. Kapusta · A. Kostina · L. M. Lund · R. Macháň · R. Muir · F. Pennacchietti · Z. Prokop · M. L. Renart · O. Rolinski · R. Šachl · M. J. Sarmento · F. Schneider · E. Sezgin · K. Stříšovský · J. Sykora · I. Testa · J. Widengren · T. Wohland · W. Zheng

Springer

Volume Editors
Radek Šachl
J. Heyrovský Institute of Physical Chemistry
Prague, Czech Republic

Mariana Amaro
J. Heyrovský Institute of Physical Chemistry
Prague, Czech Republic

ISSN 1617-1306 ISSN 1865-1313 (electronic)
Springer Series on Fluorescence
ISBN 978-3-031-30361-6 ISBN 978-3-031-30362-3 (eBook)
https://doi.org/10.1007/978-3-031-30362-3

This Springer imprint is published by the registered company Springer Nature Switzerland AG
The registered company address is: Gewerbestrasse 11, 6330 Cham, Switzerland

Series Editor

Bruno Pedras
iBB—Institute for Bioengineering and Biosciences and Associate Laboratory i4HB—Institute for Health and Bioeconomy, Instituto Superior Técnico, Universidade de Lisboa, Av. Rovisco Pais
Lisbon, Portugal

The field of fluorescence continues to grow steadily, both in fundamental aspects and applications in a highly interdisciplinary areas (analytical, physical and organic chemistry, molecular sciences, biology, biomedicine and medical research). The *Springer Series on Fluorescence* aims at drawing together both papers on fundamental research including recent advances in fluorescence methods and techniques as well those on practice-oriented research and pioneering applications.

Aims and Scope

Fluorescence spectroscopy, fluorescence imaging and fluorescent probes are indispensible tools in numerous fields of modern medicine and science, including molecular biology, biophysics, biochemistry, clinical diagnosis and analytical and environmental chemistry. Applications stretch from spectroscopy and sensor technology to microscopy and imaging, to single molecule detection, to the development of novel fluorescent probes, and to proteomics and genomics. The *Springer Series on Fluorescence* aims at publishing state-of-the-art articles that can serve as invaluable tools for both practitioners and researchers being active in this highly interdisciplinary field. The carefully edited collection of papers in each volume will give continuous inspiration for new research and will point to exciting new trends.

Preface

The fields of fluorescence spectroscopy and especially fluorescence microscopy have experienced enormous development since the first volume of the springer series on fluorescence was published, over 15 years ago. Representing such progress is the attribution of The Nobel Prize in Chemistry for the development of super-resolved fluorescence microscopy to our colleagues William E. Moerner, Stefan W. Hell, and Eric Betzig. The publicly most highly recognised award in science brought fluorescence not only to the minds of the public but also into the imagination of a broad variety of researchers. The continuous technological advances that now enable us to detect and monitor single molecules with high spatial and temporal resolution have made fluorescence methods even more popular, namely in biology and its related fields.

This book is a timely consolidation of the latest developments in fluorescence reviewed by experts in the field. It intends to deliver a comprehensive view of significant and fascinating advancements in fluorescence and of their applications in membrane and protein biophysics. The reader will be guided through both basic and advanced fluorescence spectroscopy and microscopy approaches with focus on their application potential. Our wish was to create a book that will be a useful tool for everyone interested in understanding and using fluorescence spectroscopy and microscopy. The book is intended for both complete beginners who want to quickly orient themselves in the large number of existing fluorescent methods and advanced readers who are interested in particular methods and their proper use.

We would like to kindly acknowledge everyone who has contributed to making this book possible. Of those, we would like to highlight all our colleagues who have been gracious and benevolently helped in the book's revision to make it into its best version. We especially would like to thank all the authors who so willingly wrote beautiful chapters to share their invaluable knowledge with the current and next generations of inquiring minds.

Prague, Czech Republic

Radek Šachl
Mariana Amaro

Contents

Part III Applications of Fluorescence Spectroscopy and Microscopy to Biological Membranes

Part IV Applications of Fluorescence Spectroscopy to Protein Studies

Part I
Fluorescence Spectroscopy: Basics and Advanced Approaches

Choosing the Right Fluorescent Probe

Maria J. Sarmento and Fábio Fernandes

Contents

Abstract Fluorescence microscopy and spectroscopy are by now used routinely in any laboratory working on the field of basic and applied biological sciences. A wide and expanding library of small organic fluorophores with radically different properties has been made available, offering great flexibility to the user of fluorescence-based methods. Beyond small organic fluorophores, the development of fluorescent proteins allowed for the introduction of genetically encoded fluorescence tagging, a novel tool that quickly revolutionized cellular imaging and cell biology.

M. J. Sarmento (✉)
Instituto de Medicina Molecular, Faculdade de Medicina, Universidade de Lisboa, Lisbon, Portugal
e-mail: maria.sarmento@medicina.ulisboa.pt

F. Fernandes
iBB – Institute for Bioengineering and Biosciences, Instituto Superior Técnico, Universidade de Lisboa, Lisbon, Portugal

Associate Laboratory i4HB – Institute for Health and Bioeconomy, IST, Universidade de Lisboa, Lisbon, Portugal

Department of Bioengineering, Instituto Superior Técnico, Universidade de Lisboa, Lisbon, Portugal
e-mail: fernandesf@tecnico.ulisboa.pt

R. Šachl, M. Amaro (eds.), *Fluorescence Spectroscopy and Microscopy in Biology*, Springer Ser Fluoresc (2023) 20: 3–52, https://doi.org/10.1007/4243_2022_30,
Published online: 3 January 2023

Still, a considerable fraction of casual users of fluorescence tools do not follow rational considerations when selecting a fluorescent probe for their intended application, often relying on trial and error alone, which inevitably leads to a decrease in data quality and limits the potential of fluorescence-based methods. This chapter aims to present an overview of the most important considerations to be made when selecting a fluorescent reporter. Fluorophore properties and their importance for fluorescence assays will be discussed. A list of different fluorophores and a summary of their properties will also be presented as a tool to assist in the process of choosing the right fluorescence probe. Finally, specific applications such as super-resolution microscopy require very specific fluorophores, and these will be discussed separately.

Keywords Fluorescence · Fluorescent proteins · Labelling · Organic probes · Super-resolution microscopy

The popularity of fluorescence methods in biology is largely associated with their extraordinary sensitivity and flexibility. Nevertheless, the quality of fluorescence-based assays is still fundamentally limited by factors such as number of detected photons, so that the choice of adequate fluorescent probe is pivotal in dictating the quality of fluorescence data.

Organic fluorescent compounds typically present aromatic or conjugated double bonds. Lipids and nucleic acids do not generally present fluorescence, while proteins commonly exhibit fluorescence in the UV range, and in some cases also in the visible range of the spectra [1]. Intrinsic fluorophores are the ones that occur naturally, while extrinsic fluorophores are added to a sample.

1 Intrinsic and Extrinsic Fluorescent Probes

Intrinsic fluorescence or autofluorescence results from the presence of molecules that are naturally fluorescent [2, 3], such as NADH, flavins and porphyrins. These molecules have been used at times as indicators of particular cellular processes [4, 5], but their sparse abundance and poor photophysical properties strictly limit their application. Autofluorescence can also arise from the presence of the aromatic amino acids tryptophan (Trp), tyrosine (Tyr) and phenylalanine (Phe). Particularly, Trp has been extensively used in seminal works on protein conformation and interactions due to its ability to report on the local microenvironment, either through variation in fluorescence intensity and/or the spectral shift of the emission peak [6–9]. However, the study of a particular protein within the cellular environment is impossible, as a multitude of different proteins present one or more tryptophan residues. Bulk Trp fluorescence has still been proven useful to distinguish, for example, oesophageal cancer cells from their healthy counterparts [10], and to monitor bacteria inactivation upon UV light exposure [11]. It is important to mention that molecules such as collagen and elastin also present intrinsic fluorescence, and so the extracellular matrix can also contribute to the overall autofluorescence of the samples [3].

To circumvent these issues, over the past decades extrinsic fluorescence has been extensively explored through the development of a wide variety of fluorophores with improved photophysical properties and capable of very high molecular selectivity [12]. These fluorophores comprise three main classes: organic dyes [13–15], fluorescent proteins [16, 17] and inorganic probes. The latter group, however, that includes quantum dots [18–23], metal complexes [24–26], lanthanide complexes/nanoparticles [27–30] and carbon dots/nanoparticles [31–34], still presents some strong drawbacks that prevent their extensive application in living cells when compared to the other two. Quantum dots, for example, can exhibit significant toxicity mainly due to their heavy metal core and the formation of free radicals upon excitation [35]. Moreover, the need for specific coating to assure selectivity renders quantum dots potentially even larger than a fluorescent protein, hindering their ability to enter the cells through diffusion [35, 36]. Their internalization is then dependent on the physiological internalization processes of the cell or the coupling with molecules able to facilitate the cellular uptake. In this review, we will thus focus on organic probes and fluorescent proteins (FPs), but thorough revisions of the properties and applications of inorganic dyes can be found elsewhere [37].

In general, any extrinsic probe used for live-cell imaging should fulfil defined pre-requisites, also depending on the particular application/technique. Targeting of the molecule or structure of interest must be as specific as possible. Lack of selectivity, even in a small extent, can lead to misinterpretation of the data, even more so in highly sensitive techniques such as single-molecule fluorescence and super-resolution microscopy. The detectability of the probe in the cellular context is also crucial. The photon yield of the probe must be sufficient to allow proper detection of the target molecule at its physiological levels. However, this should not be accomplished through the unreasonable increase of the probe's concentration since it will possibly interfere with the studied intracellular process. The dye should also allow for the use of a low excitation power to prevent any cellular photodamage. Furthermore, recent years have witnessed an increased demand for far-red to near-infrared emitting dyes [38–41]. This class of dyes work in a spectral range where biological samples are actually transparent, thus avoiding autofluorescence, besides reducing part of the spectral overcrowding in multi-channel imaging.

2 Organic Dyes

Arguably the most popular small organic fluorophore, fluorescein, a xanthene-based dye (Fig. 1), was synthesized in the nineteenth century by Baeyer [42] and remains ubiquitous for standard applications in fluorescence spectroscopy and microscopy. Since then, a multitude of different fluorophores and derivatives with different properties have been made available to researchers in biological sciences

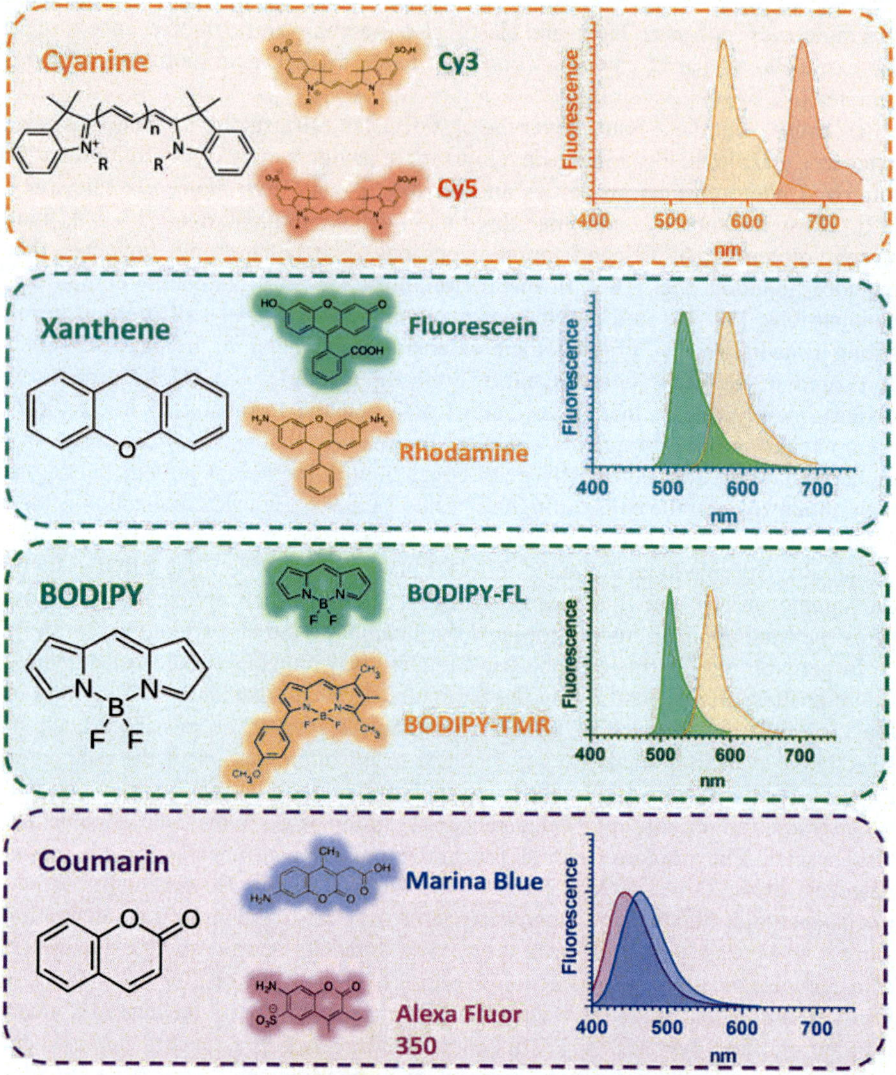

Fig. 1 Popular classes of extrinsic fluorescent probes for life sciences applications. Examples of each class are presented together with the corresponding fluorescence emission spectra

(Table 1). These are by now used in countless applications, from the study of very specific biological processes, such as protein folding and cellular distribution, macromolecule interactions, endocytosis and DNA replication, just to name a few, to the development of different sensors, including for virus detection, and the determination of biological microenvironment properties as membrane potential, viscosity and redox potential [1]. This functional versatility is intimately related with

Table 1 Properties and applications of some of the most popular organic probes used in life sciences

Probe	$\lambda_{max,abs}$ (nm)	$\lambda_{max,em}$ (nm)	τ (ns)	Applications and notes	References
Alexa Fluor 350	346	442	–	UV-excitable dye. Small fluorophore Brighter than AMCA	[43]
Alexa Fluor 488	494	517	4.1	Popular bright green fluorophore	[44, 45]
Alexa Fluor 594	590	617	4.0	Significantly brighter than Texas Red	[44, 46]
Alexa Fluor 633	632	647	3.2	Far-red fluorescence	[43, 47]
Alexa Fluor 647	651	672	1	Far-red fluorescence	[47]
AMCA	349	448	4	Commonly used UV-excitable dye Small fluorophore	[43, 48]
ATTO 488	500	525	4.2	Popular bright green fluorophore	[49]
ATTO 647 N	644	669	2.4	Highly photostable. Very hydrophobic and cationic. Used in high-resolution microscopy	[50]
ATTO 655	633	684	1.8–3.7	Far-red fluorescence. Quenched by Trp and Guanosine. Lifetime sensitive to DNA binding	[51, 52]
ATTO 680	680	700	1.7	Far-red fluorescence. Quenched by Trp and Guanosine	[52]
BODIPY 630/650	625	640	3.9	Narrow emission band. Small Stokes shift	[43]
BODIPY 650/665	646	660		Longest-wavelength BODIPY dye. Narrow emission band. Small Stokes shift	[43]
BODIPY FL	505	513	5.7	Narrow emission band. Small Stokes shift Insensitive to pH (unlike fluorescein) Relatively long lifetime	[53]
BODIPY-TMR	542	574	5.1	Narrow emission band, photoresistant	[43, 54]
Cy3	548	562	0.3	Commonly used hydrophobic yellow cyanine. Used for labelling of oligonucleotides and antibodies. Sulfo-Cy3 presents higher water solubility	[47, 55]
Cy3B	558	572	2.8	Rigidized cyanine with higher fluorescence quantum yield and lifetime than Cy3	[47, 55]

(continued)

Table 1 (continued)

Probe	$\lambda_{max,abs}$ (nm)	$\lambda_{max,em}$ (nm)	τ (ns)	Applications and notes	References
Cy5	646	663	0.98	Commonly used hydrophobic far-red cyanine. Used for labelling of oligonucleotides and antibodies. Sulfo-Cy5 presents higher water solubility	[56]
Cy5.5	675	694	1	Near-IR cyanine. Used for labelling of oligonucleotides and antibodies. Sulfo-Cy5.5 presents considerably higher water solubility	[47]
DiD/DiIC$_{18}$(5)	644	665	~1	Lipophilic dialkylcarbocyanine dye for membrane labelling. Allows for long-term cell tracking	[43]
DiI/DiIC$_{18}$(3)	549	564	1–1.5[a]	Lipophilic dialkylcarbocyanine dye for membrane labelling. Allow for long-term cell tracking	[43, 57]
DPH (MeOH)	350	428	5.0[b]	Membrane viscosity probe	[58, 59]
EDANS	336	520	8.9	Protein labelling, environment sensing, long lifetime	[43, 60]
FITC	494	518	3.7	Commonly used fluorescein derivative for covalent labelling of biomolecules Prone to photobleaching	[61]
Fluorescein (dianion, pH above 6.4)	491	515	4.2	Fluorescence spectroscopy and microscopy in general	[62–64]
Hoechst 33342 (ds DNA)	350	456	2.2	Nuclei imaging, flow cytometry, detection of DNA	[47]
Hoechst 33342 (no DNA)	336	471	0.35	Nuclei imaging, flow cytometry, detection of DNA	[47]
Laurdan (in lipid membranes)	~355	440–490	4–5.9	Measurement of lipid membrane hydration and order	[65]
NBD[c]	478	545	1–10	FRAP, protein labelling, environment sensing	[66–68]
Nile Red (EtOH)	554	629	3.65	Lipophilic stain. Useful for labelling intracellular lipid droplets. Almost non-fluorescent in water	[69, 70]
Pyrene (in cyclohexane)	338	375	100–400	Very long lifetime. Forms excimers which allow for detection of conformational changes of macromolecules or subunit assembly	[43]
Rhodamine 110	505	534	4.0	Frequently used in substrates for protease activity assays	[47]
Rhodamine 6G	526	556	3.9	Commonly used cationic dye. Very photostable	[55]

Rhodamine B	557	578	1.43	Commonly used hydrophobic dye	[55]
SR101	578	605	–	Frequently used for selective astrocyte labelling in neurophysiological experiments	[43, 71]
TRITC	554	573	2.2	Commonly used bright orange dye for labelling of biomolecules	[44, 45]
Texas Red	596	620	4.2	Sulfonyl chloride derivative of SR101 used for conjugation of proteins and oligonucleotides	[44, 47]
TMA-DPH (MeOH)	356	451	0.27	Membrane viscosity probe	[72]
YOYO-1(bound to DNA)	490	507	2.3	Useful for detection of nucleic acids (quantum yield of DNA bound molecule is far superior to the one of free dye)	[47]
YOYO-1(no DNA)	457	549	2.1	–	[47]
YOYO-3	612	631	–	Membrane impermeable DNA-binding dye. Useful for monitoring cell viability	[43]

Unless specified, values are shown for the probes in water

AMCA 7-amino-4-methylcoumarin, *DiD/DiI* 1,1′-dioctadecyl-3,3,3′,3′-tetramethylindodicarbocyanine, *DPH* 1,6-diphenyl-1,3,5-hexatriene, *EDANS* 5-(2-aminoethylamino)-1-naphthalenesulfonic acid, *FITC* fluorescein-5-isothiocyanate, *NBD* 4-nitrobenz-2-oxa-1,3-diazole, *SR101* sulforhodamine 101, *TRITC* tetramethylrhodamine, *TMA-DPH* 1-(4-trimethylammoniumphenyl)-6-phenyl-1,3,5-hexatriene p-toluenesulfonate

[a] Lifetime values measured in live cells

[b] Measured in ethanol

[c] Emission properties of NBD are highly sensitive to solvent polarity, and the quantum yield in water approaches 0. Hence the spectral properties depicted are for the fluorophore in DMSO

some unique features that overcome some of the limitations of FPs (as described in Sect. 3) [73]. Possibly the most significant of these features is their much smaller size, limiting the impact of the dye on biological functions and properties of the target molecules [74]. Historically, another main advantage of organic probes over FPs is their superior photophysical properties (e.g. higher photostability), resulting in a higher photon yield [75] and, consequently, higher quality data. However, this gap has been decreasing over the years due to the development of better performing FPs. They also outshine FPs in terms of labelling versatility. First, organic probes offer the possibility of labelling proteins at various positions, not being limited to the terminal residues. Second, labelling is not restricted to proteins, and thus any molecular structure from DNA to lipids can in principle be tagged through bioorthogonal chemistry.

Despite all the advantages, there are still some limitations that preclude the use of these molecules in particular experiments. For example, synthetic dyes are not genetically encoded, and thus the degree of labelling of a particular target might interfere with the acquisition of quantitative data. Furthermore, when it comes to labelling intracellular structures, the probes must have some solubility in aqueous media but should also be cell-permeant. However, membrane permeability depends on properties such as size, lipophilicity and charge [76], and so several probes have to be loaded into the cell through alternative methods (e.g. microinjection) [77]. An additional problem that frequently arises from loading cells with fluorescent synthetic probes is the non-specific adsorption of these molecules to different cell structures, leading to significant background fluorescence. In this case, extensive washing cycles are often required, which in turn may be incompatible with particular applications involving time-tracking of fast biological functions. This limitation however has been overcome through the development of the so-called smart (or fluorogenic) probes [73, 78]. These probes are initially quenched, emitting very weak or no fluorescence. Upon interaction with the target molecule, a chemical or conformational alteration renders the probe fluorescent at specific wavelengths. One of the most popular examples of smart probes are the DNA-binding cyanines TOTO and YOYO. Both probes are non-fluorescent in aqueous solution. Intercalation with DNA forces the molecules into a fluorescent planar form [79, 80].

Four classes of organic fluorophores have become particularly popular as extrinsic fluorescent probes for life sciences applications. These are cyanines, xanthenes, BODIPY and coumarins (Fig. 1). Although a comprehensive description of all available synthetic fluorophores, their applications and experimental considerations is out of the scope of this chapter, we will present valuable information to help users to choose the appropriate organic dye and experimental design that would ultimately fit their needs. Fundamental photophysical properties will be introduced and their relevance for a successful fluorescence-based experiment will be discussed. The properties of selected examples of commercially available organic dyes will also be presented (Sect. 2.1 and Table 1). Then, several labelling strategies are also described (Sect. 2.2). Finally, due to their intrinsically different properties and experimental requirements, a concise description of membrane probes and their main applications is also included (Sect. 2.3).

2.1 Ideal Properties of Fluorescent Probes

In a fluorescence experiment, the fluorophore will cycle between the ground and excited states, and fluorescence is obtained upon radiative return to the ground state (Fig. 2).

Typically, a single fluorophore can undergo several thousand repeats of this cycle before destruction [1], and this is largely responsible for the extraordinary sensitivity of fluorescence methods. Importantly, due to the presence of multiple processes contributing to the dissipation of energy from the excited state (Fig. 2a), the energy of the fluorescence photon is smaller than the energy of excitation. This shift to a longer wavelength is known as the Stokes shift (Fig. 2b) and is critical to the sensitivity of fluorescence-based methods, as blocking of the excitation light allows for selective detection of fluorescence light [81].

The most important properties of the fluorophore dictate the pattern of the cycle between ground and excited states, and we will discuss them individually in the next sections.

2.1.1 Molecular Brightness

The molecular brightness (B) of a fluorophore, defined as the number of photons per second for a single molecule, is not an intrinsic property of the molecule, as it depends on the excitation intensity, as well as on light collection and detection

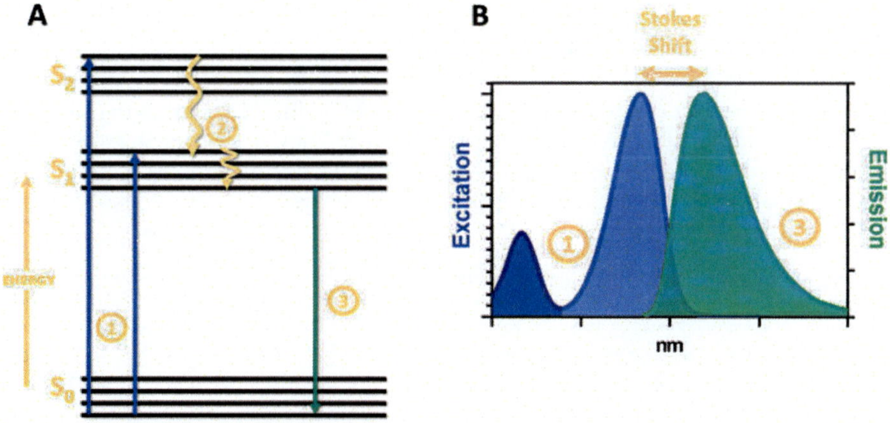

Fig. 2 Excitation and emission of a fluorophore. (**a**) Jablonski diagram depicting the excitation of a fluorophore through absorbance of light (1), creating an excited electronic singlet state (S1 or S2), followed by dissipation of energy through internal conversion, vibrational or solvent relaxation (2). Fluorescence emission takes place upon return to the ground state (3). (**b**) Fluorescence excitation and emission spectra of a fluorophore

efficiency by the instrument during acquisition [1]. Nevertheless, it is proportional to the product of the molar extinction coefficient (ε) at the excitation wavelength and the fluorescence quantum yield (QY) of the molecule [1]:

$$B \approx \varepsilon \times \mathrm{QY} \tag{1}$$

In most applications, the number of photons acquired dictates the quality of recovered data. In many experiments, the impact of a low brightness value can be compensated by increasing the concentration of the fluorophore in the sample or by increasing the excitation intensity. However, when this is not possible, as is often the case for fluorescence microscopy applications, a very high B value is desirable in order to avoid the use of very high-intensity excitation light. The use of high-intensity excitation light can lead to increased background signal, contamination of detection by excitation light and decreased fluorophore stability (see below). The higher the value of B is, the lower the excitation intensity can be to achieve sufficient fluorescence photon counts.

Low brightness of labelled molecules, as defined in Eq. 1, can be certainly compensated by increasing the stoichiometry of dyes to macromolecule (D:M). However, the drawback of such strategy is that the structure or pattern of molecular interactions of the labelled molecule can be negatively impacted. In the case of labelled antibodies, increasing the degree of labelling (DOL, mean number of fluorophores per antibody) leads to sharp decreases in antibody affinity [82, 83]. Additionally, in the case of specific labelling strategies, an increase in D:M stoichiometry could be impossible. For these reasons, the use of probes offering adequately high values of ε and QY is strongly recommended.

2.1.2 Extinction Coefficient

The molar extinction coefficient of a molecule (in $M^{-1}\ cm^{-1}$) describes the capacity for light absorption at a given wavelength. The cross-section for light absorption (σ) of a fluorophore describes the photon capture area of a molecule and is calculated from ε using [1]:

$$\sigma = 3.82 \times 10^{-5} \varepsilon \left(\text{in Å}^2\right) \tag{2}$$

In the case of fluorescein, which presents a considerable ε value of 75,500 $M^{-1}\ cm^{-1}$ at 490 nm [62] (Fig. 3), the cross-section value for light absorption is 2.88 Å^2, which is significantly smaller than its molecular dimensions. In this way, only a small fraction of light encountered by the fluorophore is effectively absorbed by it.

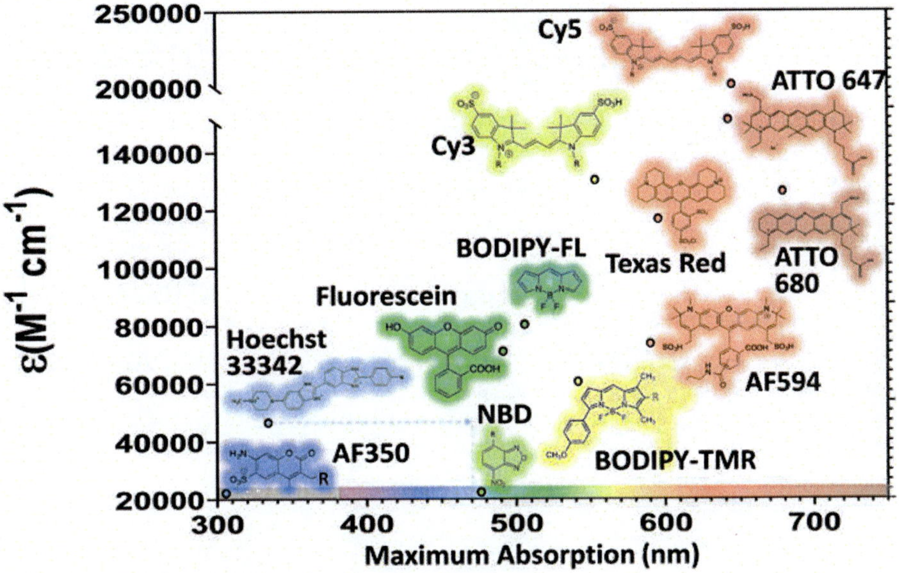

Fig. 3 Extinction coefficients (ε) and absorption maximum (λ_{Abs}^{max}) for a selection of commonly used fluorophores. Exact values of ε and λ_{Abs}^{max} correspond to the positions of open circles closer to each fluorophore. Colours associated with each structure reflect the wavelength of maximum fluorescence emission according to the colour spectrum shown in the bottom. The Stokes shift for Hoechst 33342 (unbound) is represented by an arrow. Cy3 and Cy5 are shown in their sulfonated forms. Values taken from [43, 44, 56, 60, 63, 64, 66, 84–87]

2.1.3 Quantum Yield

Fluorescence quantum yield (QY) is defined as the ratio of the number of photons emitted to the number of photons absorbed. Since the fluorophore molecule can also return to the ground state from the excited state through nonradiative processes, the value of QY is always smaller than 1. Taking once more fluorescein as an example, its quantum yield is 0.93 at 490 nm [62] (Fig. 4). This large QY value is largely responsible for the popularity of fluorescein as a fluorophore. Values of quantum yield are generally listed at the wavelength of maximum absorption.

The presence of efficient electron donors in the vicinity of the fluorophore has been shown to lead to marked decreases in quantum yield. The amino acid tryptophan and to a lesser degree tyrosine, methionine and histidine have been shown to decrease the quantum yield of a large array of fluorophores by different combinations of static and dynamic quenching [88, 89]. On the other hand, carbocyanines, such as Cy3, Cy5 (Fig. 1) or Alexa Fluor 647 (AF647) (Figs. 3 and 4), have been shown to be less susceptible to intramolecular quenching by these amino acids [89]. Similar quenching of fluorophores has been reported for labelled oligonucleotides, as specific nucleotides, namely guanosine, induce significant quenching of several fluorophores in their immediate vicinity [89]. Oxazine derivatives such as

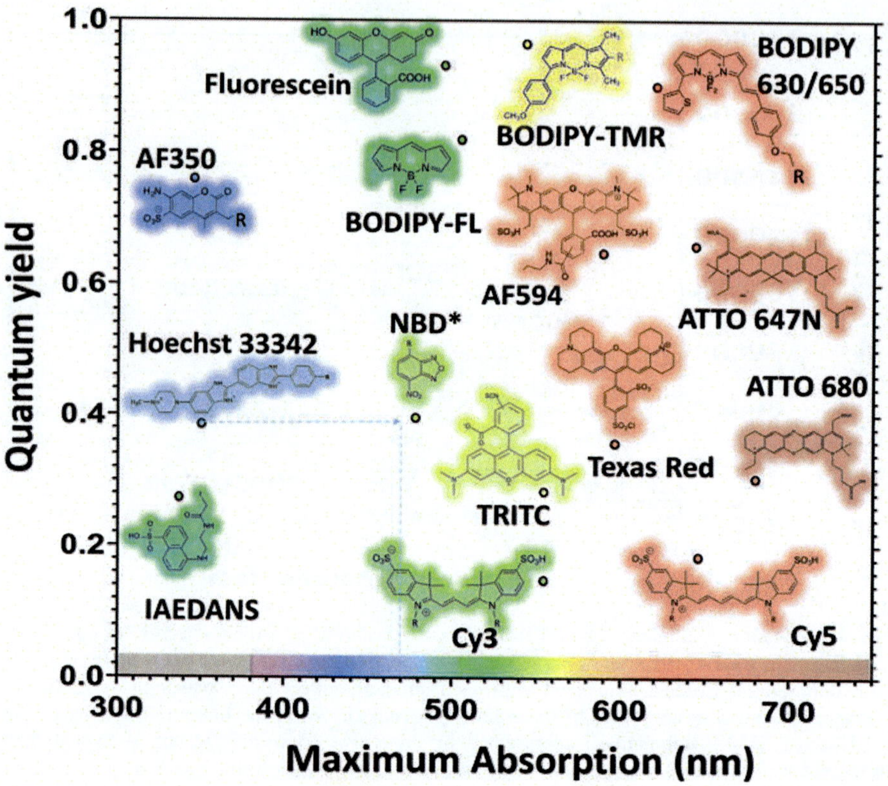

Fig. 4 Fluorescence quantum yield (QY) and absorption maximum (λ_{Abs}^{max}) for a selection of commonly used fluorophores. Exact values of quantum yield and λ_{Abs}^{max} correspond to the positions of open circles closer to each fluorophore. Colours associated with each structure reflect the wavelength of maximum fluorescence emission according to the colour spectrum shown in the bottom. The Stokes shift for Hoechst 33342 (unbound) is represented by an arrow. *Properties of NBD are for the fluorophore in DMSO, as its quantum yield approaches 0 in water. Cy3 and Cy5 are shown in their sulfonated forms. Values taken from [43, 44, 56, 60, 63, 64, 66, 84–87]

ATTO 655 and ATTO 680 (Figs. 2 and 3) are particularly susceptible to these effects, and this has been used to detect peptide–antibody interactions with labelled peptides [90].

2.1.4 Photostability

One of the most important parameters to be considered when choosing a fluorescent probe is its photoresistance, especially when the goal is its use for imaging applications. In fact, one of the most limiting factors in defining fluorescence intensity in fluorescence microscopy is the irreversible photochemical destruction of the fluorophore during continuous illumination. The rate of this process, called photobleaching, determines the number of photons a fluorophore can emit before its destruction [1]. In practice, photobleaching results in a decrease in fluorescence emission overtime during acquisition, as larger fractions of fluorophores are photodamaged.

The average number of photons a fluorophore can emit before its destruction can be estimated based on its quantum yield and photobleaching rates [1]. Fluorophores usually used in fluorescence microscopy applications emit from 10^4 to 10^6 photons during their useful lifetime, while other molecules can only emit around 1,000 photons before irreversible destruction [1, 91]. Returning to the example of fluorescein, which exhibits weak photostability, this molecule can emit between 30,000 and 40,000 photons in water before photobleaching, and only a minor fraction of these will be collected during any experiment [44].

Although the photophysics and photochemistry associated with photobleaching are in general poorly understood, interactions between the fluorophore and molecular oxygen are thought to be one of the main causes for photoinactivation [92]. Upon transition of the fluorophore to long-lived triplet states (intersystem crossing), the molecule resides in the excited state for considerably longer periods of time (ms instead of ns), becoming considerably more susceptible to irreversible chemical reactions with oxygen and other molecules. Photobleaching can be partially reduced by oxygen depletion through the use of antifade agents composed of reactive oxygen scavengers or through the bubbling of N_2. Adding to the complexity of the phenomenon, oxygen is quencher of the triplet state and thus minimizes the lifetime of the reactive excited molecule. One possible consequence of oxygen depletion can be a drastic increase of the lifetime of the molecule in the dark triplet state, reducing the overall fluorescence as well [93]. In fact, other antifade agents make use of triplet-state quenching, reducing the reactivity of the dye.

Hence, particularly for fluorescence microscopy applications, fluorophores that exhibit strong photostability should be ideally chosen. While no rules exist for predicting the photostability of a dye [1], certain trends are observed, as within the same family of compounds, fluorophore elongation and flexibility are typically accompanied by less photostability [94]. The increased photostability of several Alexa dyes was obtained in part from rigidifying modifications of their structure [95]. Dyes with low degrees of triplet-state formation have also been associated with higher photostability [96].

As rhodamine-based dyes are frequently highly photostable, a large selection of these molecules is currently available from different companies with excellent properties. Several Alexa, DyLight, ATTO, HiLyte and Janelia dyes belong to this class and offer bleaching rates several fold lower than older generation dyes, such as fluorescein [81, 95, 97–99]. Improved photostability relative to fluorescein is also observed in boron–dipyrromethene BODIPY dyes [100]. The dyes presented above typically exhibit fluorescence between 450 and 700 nm, and when fluorophores emitting at shorter wavelength are required, coumarins are often used. However, coumarin derivatives are generally less photostable than rhodamine-based fluorophores. Examples of coumarin derivatives are Alexa Fluor 350 and

430, Marina Blue (Fig. 1), as well as ATTO 390, 425 and 465. Cyanine dyes such as from the Cy class, while extremely popular, are also not as photostable as most rhodamine-based derivatives, due to their susceptibility towards photooxidation [101]. On the other hand, very high photostability was also obtained with carbopyronines (e.g. ATTO 647) and oxazine derivatives (e.g. ATTO 655, ATTO 680 and ATTO 700) [102].

While photobleaching is almost always undesirable, in some cases, such as in fluorescence recovery after photobleaching (FRAP) or fluorescence loss in photobleaching (FLIP) experiments, photobleaching can actually be employed to study molecular diffusion rates [103, 104].

2.1.5 Aggregation and Solubility

An increase in labelling stoichiometry often does not necessarily lead to linear increases of fluorescence intensity, as probe self-quenching can reduce the quantum yield of conjugated fluorophores [83]. In some cases, significant intramolecular self-quenching can take place even at moderate ratios of dye to labelled molecule. As an example, in non-specific covalent labelling of proteins with Cy5, functionalization is favoured in the vicinity of already protein-bound Cy5 labels, leading to the formation of clusters of Cy5 within the protein [105]. As a result, a dramatic reduction of Cy5 quantum yield due to static self-quenching is observed even at rather low Cy5/protein ratios, causing multiple labelled proteins to exhibit even lower fluorescence than single-labelled proteins. This behaviour is due to the tendency of Cy5 for dimerization, giving rise to non-fluorescent complexes. This is less problematic for carbocyanines such as Alexa Fluor 647 and 4S-Cy5.5 which carry four sulfonate groups, as their presence increases repulsion between the dyes during conjugation [105, 106].

Dye aggregation is in fact a problem for most of the older generation of fluorophores, limiting the fluorescence of conjugates to coumarins, xanthenes and cyanines [95, 107]. The Alexa Fluor dyes correspond to sulfonated derivatives of fluorophores from these classes. The presence of sulfonated groups allows for reduced interactions during labelling and increased quantum yield of conjugates [95]. Additionally, sulfonation increases solubility of the dyes, allowing for conjugation in aqueous media, without the need for organic solvents.

2.1.6 Fluorescence Emission Spectra

The fluorescence emission spectra of commercially available fluorophores span the near ultraviolet (UV), visible (VIS) and near-infrared ranges (IR) of the electromagnetic spectrum. The choice of fluorophore to be used in a given fluorescence spectral range must of course take into consideration the detection restrictions of the equipment to be employed, such as limitations of filter combinations. The dramatic reduction in photon detection efficiency within typical photomultiplier detectors above 700 nm [2] can also limit the range of useful probes to molecules emitting in the UV-VIS range. Additionally, experiments with biological samples often have to deal with the additional complexity created by the presence of autofluorescence, which is particularly strong in the UV range. These limitations regarding measurements in the UV and infrared are responsible for the popularity of fluorophores emitting visible light between 380 and 750 nm (Figs. 2 and 3).

Experiments with multiple fluorophores rely on accurate isolation of individual signals. This can be challenging in imaging applications heavily dependent on optical filters, as emission spectral overlap of multiple fluorophores can lead to contamination of the fluorescence signal from one molecule in the detection channel of another one (bleedthrough). In equipment with high spectral resolution, spectral acquisition can allow for separation of the signal from two fluorophores emitting in the same range of energies through the use of techniques such as spectral unmixing [38]. However, bleedthrough can be minimized using simply a judicious selection of probes guaranteeing minimal spectral overlap.

Fluorescent probes exhibiting narrow fluorescence emission spectra facilitate the procedure by increasing the spectral separation between coexisting dyes. While the use of narrow bandpass optical filters solves this issue, it also leads to recovery of only a fraction of the total fluorescence from a fluorophore with broad emission spectra. Organic dyes often display narrow fluorescence emission bandwidths, but there are many exceptions. While rhodamine-derived Alexa Fluor dyes and BODIPY derivatives show particularly narrow emission bandwidths, probes from the cyanine class are known for presenting broader emission [25, 39, 40].

One advantage of dyes with broad absorption and emission spectra is their greater versatility since they can be excited with different illumination sources and detected with different optical filter combinations.

Finally, employment of probes in the IR has some advantages. Simultaneous measurements with probes emitting in the UV-VIS range are possible with negligible bleedthrough. This is particularly important when simultaneous measurements with fluorescent proteins are desired, due to their broad emission spectra.

2.1.7 Stokes Shift

The separation between excitation and emission maxima of a fluorophore (Stokes shift) is critical for an efficient detection of fluorescence signal. When the fluorophore exhibits a larger Stokes shift, contamination of the detection with scattered excitation light is less likely and wider bandpass filters can be used for the detection of the fluorescence signal, improving sensitivity of the measurement. Hence, fluorophores with a larger Stokes shift are generally preferred. Large Stokes shift are observed in dyes with asymmetric structure and electron distribution [108]. However, most of the popular fluorophore classes already mentioned (xanthene, BODIPY and cyanine derivatives) present a rather small Stokes shift [108].

Another consequence of a short Stokes shift is the increase in overlap between excitation and emission spectra. As a result, fluorophores with a small Stokes shift are susceptible to homo-FRET events. Unlike typical FRET, where the donor is a chemically distinct molecule from the acceptor, in homo-FRET the donor and acceptor share the same structure. This is useful for detection of homo-oligomerization or clustering. BODIPY derivatives present particularly small Stokes shifts [53] and have been intensively employed in homo-FRET studies [109, 110].

2.1.8 Fluorescence Lifetime

The average time a molecule resides in the excited state following excitation is called fluorescence lifetime (τ). A very high lifetime is disadvantageous for imaging applications as it will correspond to a shorter number of excitation-emission cycles, reducing the number of photons acquired, and thus the fluorescence intensity. Most popular probes exhibit fluorescence lifetimes of 1–6 ns (Table 1). Notable exceptions are cyanine derivatives such as Cy5, which present fluorescence lifetimes lower than 1 ns [56], while pyrene has a characteristic long lifetime of hundreds of ns [43].

Fluorescence lifetime is independent of fluorophore concentration, allowing for highly robust measurements [1]. Analysis of protein interactions in living cells has benefited considerably from lifetime measurements, as FRET imaging by fluorescence lifetime does not require the extensive calibrations and image processing mandatory for intensity-based methods [104].

For some probes, this quantity is heavily dependent on the local environment of the fluorophore and can be used to sense its distribution or conformational changes of labelled macromolecules. Hoechst 33342 or ATTO 655 can be used to detect the presence of double-stranded DNA, as binding of the dyes to the double strand increases the lifetime of the excited state of these molecules (Table 1) [47, 51]. A DNA-binding BODIPY derivative was shown to detect protein–DNA interactions through fluorescence lifetime changes [111]. A similar BODIPY-based molecule with high mitochondrial affinity was employed to quantify mitochondrial membrane viscosity as its fluorescence lifetime increased concomitantly with reduced mobility in the membrane [112].

Fluorescence lifetime measurements can be carried out with a pulsed (time-domain) or modulated excitation sources [1]. When working with pulsed excitation, one must be aware that the fluorescence lifetime to be measured should be significantly shorter than the repetition rate of the laser. In Ti:Sapphire lasers, this value is limited to ~80 MHz, corresponding to a 12.5 ns interval between two pulses. In these conditions, in a sample with a fluorophore presenting a lifetime significantly longer than 3 ns, the sample will still have excited fluorophores immediately before a new pulse, generating an incomplete decay and an erroneous lifetime estimation [1]. For this reason, it is advised that the dye chosen for the fluorescence lifetime measurement should present lifetimes at least four times shorter than the laser repetition rate [113].

2.1.9 pH Sensitivity

As a rule, when choosing a fluorophore for a specific application which does not involve estimating pH, the fluorophore to be chosen should be pH insensitive. Although most commercially available probes are indeed pH independent, there are several notable exceptions, such as fluorescein itself. In fact, fluorescein signal exhibits a strong pH dependence, resulting from the phenolic pK_a of 6.4 [114]. Near neutral pH, the monoanionic form (ε = 29,000–32,600 M^{-1} cm^{-1}, QY = 0.36–0.37) exists in equilibrium with the dianion (ε = 76,900–87,600 M^{-1} cm^{-1}, QY = 0.92–0.95) [26]. Due to the photophysical differences between the two species, the fluorescence intensity of fluorescein is strongly dependent on pH, and when analysing data with fluorescein or fluorescein labelled conjugates, this must be considered. The proximity of the pK_a of fluorescein to the pH of most biological systems is detrimental for multiple applications. For this reason, several derivatives have been introduced, presenting modified pK_a values. Oregon Green, for example, presents a pK_a of 4.8, eliminating pH sensitivity in most biological systems [115].

On the other hand, the pH dependence of fluorescein also allows for the use of the fluorophore as a pH sensor. A fluorescent derivative of fluorescein introduced by Roger Tsien in 1982, 7′-bis-(2-carboxyethyl)-carboxyfluorescein (BCECF), presents a pK_a closer to 7 (6.98), more appropriate for sensing of intracellular pH [116, 117]. The additional carboxylations in this molecule decrease the membrane permeability of the dye significantly, reducing membrane leakage. Cellular loading of the dye can be carried out through the addition of acetoxymethyl (AM) ester groups to the fluorophore, which increase the permeability of the molecule but are hydrolysed by intracellular esterases, leading to intracellular accumulation. BCECF-AM is still one of the most popular intracellular pH sensors. When the goal is to measure pH within acidic compartments, such as lysosomes, then dyes with lower pK_a values can be employed.

2.1.10 Overview of Fluorescent Probe Classes

Coumarins

This family of compounds includes molecules containing the 2H-chromen-2-one motif and is arguably the largest among the small organic dye categories. Coumarins are usually very photostable, with high quantum yield and a very large Stokes shift [13]. Their blue emission makes them a good alternative for multi-colour imaging, being an easy combination for green-to-IR dyes. This excitation at short wavelengths can also be a drawback when imaging cells since it overlaps with autofluorescence. Coumarins also have limited brightness due to weak absorption (e.g. AF350 in Fig. 3) [13].

In any case, these fluorophores have been extensively used in life sciences during the past decades. In a recent example, π-extended fluorescent coumarin (PC6S) was used to image lipid droplets through fluorescence lifetime imaging (FLIM), both in living cells and in the tissues of living mice [118]. Other alternatives, including caged derivatives with activation in the visible and near-IR spectral range, have also been explored. For example, diazocoumarin derivatives have been used to label protein in living cells in a selective and fluorogenic manner, upon photoactivation (uncaging) with visible or near-IR illumination [119].

Cyanines

In this category one can find the brightest fluorophores among organic dyes. Although cyanines usually present low QY ($\leq$0.25), they also have very high extinction coefficients (e.g. Cy5 in Figs. 3 and 4; Table 1) [13]. Cyanine derivatives contain conjugated polymethine chains with quaternary nitrogens in their chemical structure. Adjusting the functional groups and the length of the conjugated chains, the photophysical properties of the fluorophores can be largely tuned.

However, the use of cyanines can be limited by significant photobleaching, as previously discussed, and a small Stokes shift (possibly important for super-resolution microscopy approaches) [120, 121]. Additionally, Cy5 and Cy7 derivatives are easily oxidized in the presence of O_2 and O_3, preventing their use in experiments that require long measurement sessions [122]. The development of new derivatives with enhanced QY and increased oxidation resistance has mitigated these drawbacks.

Cyanines have been used in a great number of different applications. They even gained more relevance with the development of single-molecule localization microscopy (SMLM) techniques (see Sect. 4 below). More recently, cyanine probes have been used, for example, to detect and label mercury [123], DNA [124] and RNA [125] in living cells.

Fluorescein and Rhodamine-Based Dyes

As already mentioned, both fluorescein and rhodamines are xanthene dyes that have been extensively used since they were first synthesized [126, 127]. The reason of this wide popularity is associated mainly with their very high brightness and the fact that both excitation and emission wavelengths fall well within the visible spectral range. As previously described, with a pK_a value of 6.4, fluorescein spectral properties depend strongly on the environment pH (between 5 and 9). They can also present limited photostability [13, 122].

Over the years, numerous fluorescein and rhodamine derivatives have been developed, targeted at countless applications. Among the most commonly used fluorescein derivatives [128, 129] are fluorescein isothiocyanate (FITC), carboxyfluorescein, 5/6-carboxyfluorescein succinimidyl ester, fluorescein amidite (FAM) and fluorescein di-acetate. Rhodamine 6G, 110, 123 and rhodamine B are possibly within the most extensively used rhodamines, together with rhodamine-derived Alexa Fluor dyes. When used for live-cell imaging, properties like permeability, localization and aggregation state will strongly depend on the structural characteristics of each derivative.

BODIPYs

Boron dipyrromethene compounds or BODIPYs are very popular fluorophores due to their enhanced photophysical properties [130]. They present high quantum yields (e.g. BODIPY-TMR in Fig. 4), high brightness, sharp absorption and emission spectra and a very small Stokes shift (Table 1). Possibly the most significant limitation to the use of BODIPY dyes is their susceptibility to oxidation [13].

BODIPY variants included in this category cover a great portion of the visible spectrum, from green (BODIPY FL) to red (BODIPY 650/665). This versatility has been extensively explored through the development of BODIPY conjugates of different biomolecules, including proteins, lipids, lipopolysaccharides and nucleotides, among others. They have also been employed in live-cell imaging (see [131] for a comprehensive review). For example, BODIPYs have been recently used in the detection of hydrogen sulphide and lysosome tracking [132].

2.2 Methods for Fluorescent Protein Labelling

The choice of organic probe must always be performed hand-in-hand with the selection of the appropriate labelling strategy. The latter depends on the available fluorophore derivatives, the target molecule and the experimental requirements. Here, some of the most commonly used strategies for labelling proteins upon purification or within fixed or living cells are discussed (Fig. 5). The extensive description of these and other methodologies can be found in very thorough reviews elsewhere [133].

Covalent Binding to Natural-Occurring Amino Acids

The most common strategy to label purified proteins is to use fluorophore derivatives that directly react with amine groups or cysteine residues (Fig. 5) [134].

By using amine-reactive fluorophore conjugates, such as isothiocyanates or succinimidyl-esters, target proteins will be labelled in either lysine residues or the N-terminal amine. Due to the high frequency of lysine residues in proteins, labelling often results in the binding of several fluorophores at different positions, which can ultimately alter protein function and its possible interactions. Since lysine amine groups have significantly different *pKa* values than the terminal amine (*pKa* 10–11 versus *pKa* 7, respectively), it is possible to preferentially label the N-terminal [135]. Although this solves the issue of multiple labelling, it restricts the fluorophore position at the amino terminal.

Cysteine residues, on the other hand, are much less frequent in proteins and thus offer greater flexibility regarding the position of the fluorophore. In this case, maleimide fluorophore conjugates are used due to their high specificity for the cysteine thiol group [133]. In addition, the conjugation reactions are typically fast at mild pH and temperature.

Labelling amine groups or cysteine residues in intact cells or cell extracts is generally not feasible since it is not possible to reach labelling specificity under these

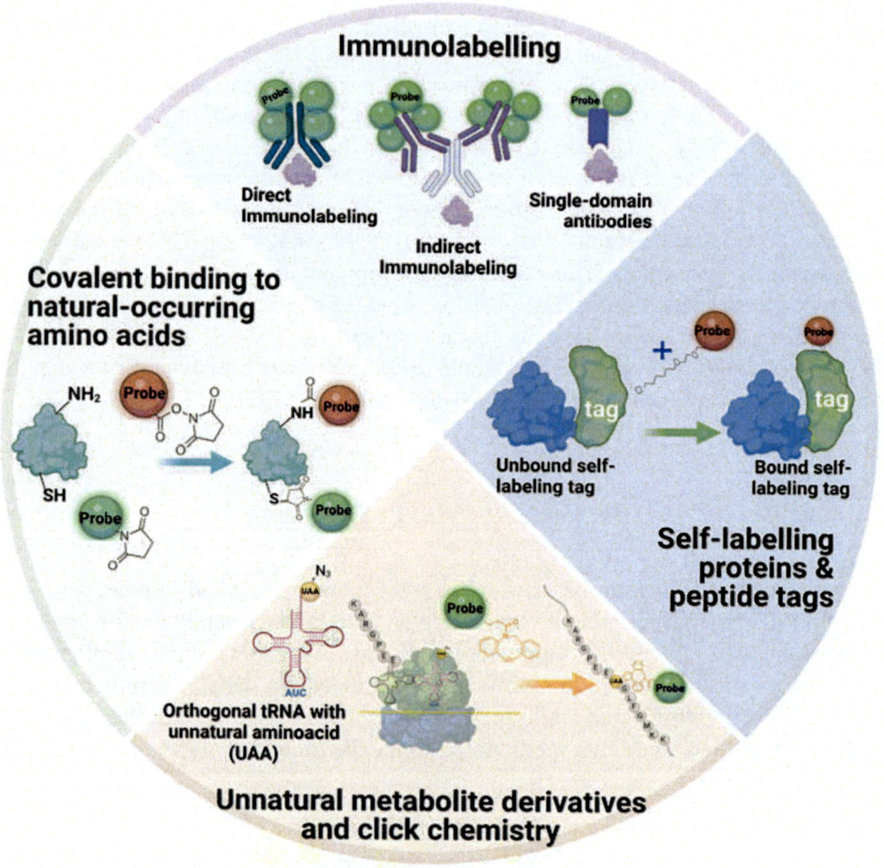

Fig. 5 Summary depiction of some of the most commonly used (or most promising) protein labelling methodologies. The examples depicted for covalent binding to natural-occurring amino acids are related to cysteine and protein amine labelling with maleimide and succinimidyl ester derivatives, respectively. The example portrayed for self-labelling proteins and peptide tags illustrates self-labelling by the HaloTag. The method shown below describes incorporation of an azide-containing unnatural amino acid (UAA) and site-specific labelling of the modified protein with Cu^{2+}-free click-chemistry. Figure created with BioRender (BioRender.com)

conditions. For this reason, these methods are more often employed for labelling of purified proteins.

Immunolabelling

To achieve high labelling specificity within cell samples, the classical approach is to use immunofluorescence [136]. In general, an externally added primary antibody specifically binds the target molecule/structure with high affinity. Subsequently, one or more secondary antibodies (conjugated with a fluorescent probe) form a stable complex with the primary antibody (indirect immunolabelling), allowing the specific detection of the target (Fig. 5). Direct immunolabelling with one labelled specific antibody is also possible. However, antibodies are large molecules of ~150 kDa and thus their use implies a couple of limitations [137]. First, they cannot cross the plasma membrane, limiting their application to fixed, and thus, dead cells. This prevents the study of dynamic processes, providing only a snapshot of the target molecule at the time of fixation. Also, it is important to keep in mind that the fixation protocol itself might interfere with the observed cell morphology and/or the properties of the molecule of interest. Second, the size of the antibodies and that of the antibody complex (primary + secondary) makes it so that the detection system is often much larger than the target molecule or structure [137]. In this case, determination of the target's size becomes extremely difficult and, for example, molecules that are well far apart might look closer or even interacting when immunolabelled. Importantly, this might have serious implications in single-molecule localization microscopy techniques, with the location of the fluorophore possibly being clearly distinct from that of the target molecule.

More recently, these limitations have been overcome by the development of single-domain antibodies (sdAbs) or 'nanobodies' [138–140]. sdAbs are much smaller (~15 kDa) than traditional antibodies, highly stable and soluble in many environments. Depending on the experimental requirements, sdAbs can even be functionalized, for example, to cross the plasma membrane and allowing live-cell imaging. Their smaller size also makes them much better suited to help evaluate the dimensions of biological structures, as opposed to the traditional antibodies.

Self-labelling Proteins & Peptide Tags

Possibly the most widely used strategy to label proteins with synthetic probes inside living cells is the use of small self-labelling proteins or tags (<40 kDa) [141]. In this approach, the target protein is expressed as a fusion construct containing a polypeptide or protein designed to react with a fluorescent probe (or other moieties such as biotin) via a specific biorthogonal reaction. This ensures the covalent binding of virtually any fluorescent probe to the protein of interest as long as the proper biorthogonal functionalization of the probe is available and compatible with the particular self-labelling protein to be used. The versatility of this method is also increased by the number of tags that have already been developed (derived from different proteins): HaloTag (derived from the haloalkane dehalogenase) (Fig. 5) [142], SNAP-tag (derived from human O6-alkylguanine DNA alkyltransferase) [143, 144], CLIP-tag (derived from SNAP-tag) [145], DHFR-tag (derived from dihydrofolate reductase) [146], BL-tag (derived from b-lactamase) [147], PYP-tag (derived from the photoactive yellow protein in purple bacteria) [148] and RA-tag (derived from de novo designed retroaldolase) [149].

Ideally, the tags should be as small as possible to avoid interfering with the target protein, should present low substrate promiscuity, fast labelling kinetics and thermodynamic stability. The main advantages of this approach are associated with the use of synthetic dyes (described above) and their superior photophysical performance, and yet being genetically encoded, thus offering a better control over the degree of labelling. Moreover, it can be used for pulse-chase assays, contrary to FPs

[150]. The disadvantages, however, also stem from the fact that the tags are genetically encoded. First, this means that the labelling is restricted to the N- or C-termini, preventing any site-specific tagging into internal regions of the protein of interest [74]. Second, the tags are still proteins with diameters that can reach 3–4 nm, comparable to that of FPs [35, 74]. The labelling can thus affect the folding, location and function of the native target protein.

Alternatively, a tetracysteine peptide can be genetically introduced into the protein of interest (not necessarily at the termini), instead of the self-labelling protein [151]. The peptide sequence, Cys-Cys-X-X-Cys-Cys (where X is any amino acid apart from cysteine), is then able to specifically react with membrane-permeable biarsenical dyes. The standard biarsenical dyes in use are FlAsH-EDT$_2$ (fluorescein arsenical hairpin binder) and ReAsH (resorufin arsenical hairpin binder) that emit in the green and red regions of the spectrum, respectively [35]. Although the tetracysteine peptide shares most of the advantages with the self-labelling proteins, the major point of differentiation is their size. In this case, the much smaller size of the moiety is less likely to interfere with the conformation and function of the protein of interest. Nevertheless, the use of biarsenical dyes can still perturb the protein's native conditions, can lead to considerable levels of background fluorescence and might induce some cytotoxicity.

Unnatural Metabolite Derivatives & Click-Chemistry

More recently, specific molecular labelling within the cellular environment has been accomplished through the development of the so-called click-chemistry (Fig. 5). Contrary to genetically encoded reporters, it can be used to label not only proteins but also other biomolecules such as DNA, lipids and glycans. This labelling strategy is a two-step process. First, cells must be cultured in the presence of non-natural metabolites with a 'clickable' moiety [12]. For example, nucleotides BrdU (5-bromouridine) and EdU (5-ethynyl-2′-deoxyuridine) have been used to tag nascent RNA and DNA, respectively [152]. Then, fluorescent labelling is accomplished through a biorthogonal 'click' reaction with probe derivatives containing the corresponding 'click' feature (e.g. azide or alkenyl groups). In the case of EdU, azide Alexa Fluor derivatives are commonly used. The most typical biorthogonal reaction is named azide-alkyne cycloaddition, requires copper [Cu(I)] as catalyst and has the main advantages of being fast, regioselective and presenting high yields [153, 154]. However, due to copper cytotoxicity, other alternatives have also been widely employed [155]. These copper-free reactions include the azide-alkyne cycloaddition between an azide and a strained cyclooctyne [156] and the Staudinger ligation between an azide and a phosphine [157]. Although the copper-free character of these reactions makes them uniquely suited for in vivo labelling, they tend to be less efficient, meaning that the reaction time might need to be increased to assure full conjugation of the fluorescent probe derivative with the unnatural metabolite [158]. In any case, when planning an experiment, the choice of reaction will greatly depend on the system, the target biomolecule and the technique to be employed [12]. Moreover, copper toxicity, alterations of membrane permeability and possible background fluorescence must also be considered.

When considering labelling proteins using 'click-chemistry', a promising strategy has recently emerged that combines genetically encoded site-specific labelling with the major advantages of using small synthetic dyes. This is accomplished through the use of unnatural amino acids (UAAs) that can either be conjugated with an organic probe via a biorthogonal reaction (Fig. 5) or be already incorporated as a fluorescent amino acid derivative [159, 160]. UAAs, also called the expanded genetic code, are molecules that are not found in native proteins, but were rather discovered in nature or chemically synthesized [161–163]. Examples include selenocysteine (Sec) and pyrrolysine (Pyl), also referred to as the 21st and 22nd amino acids [164]. In principle, many other UAAs can (and some have been) incorporated into target proteins by using a nonsense codon and the UAA respective orthogonal pair [165]. This strategy confers large flexibility when it comes to the labelling position within the target protein, with the advantage of keeping the labelling ratio under control (since it is genetically encoded). Also, due to their much smaller size, UAAs overcome most issues related to the interference with the native protein, as discussed above for self-labelling proteins. Yet, incorporation of UAAs (and their fluorescent labelling) is still challenging [74, 166, 167] and there is still some work to be done to make it the method of choice for live-cell protein labelling.

2.3 *Membrane Probes*

Fluorescence microscopy and spectroscopy are invaluable tools for the imaging and characterization of cellular membranes or membrane model systems. Membrane staining can be achieved through labelling of specific targets (e.g. glycoproteins and glycolipids with wheat germ agglutinin (WGA) conjugates [168]), with derivatized membrane components (e.g. labelled cholera toxin-A or BODIPY-labelled phospholipids, or through simple partitioning of lipophilic dyes [1]. This section will focus solely on probes of the latter class.

Lipophilic staining of cellular membranes is often employed in cell imaging applications. Lipophilic carbocyanines with hydrocarbon tails, such as DiO (green fluorescence), DiI (red fluorescence), DiD (far-red fluorescence) or DiR (near-infrared fluorescence), are particularly popular for this application [169, 170]. Styryl dyes such as FM4–64 or FM1–43 also belong to the class of lipophilic stains. Unlike lipophilic carbocyanines, these dyes are not lipophilic enough to translocate across the lipid bilayer and become anchored to the outer leaflet of the plasma membrane, with minimal translocation to the cytoplasmic leaflet [171]. This property leads to their popularity as tools for evaluation of intracellular vesicle trafficking [171, 172]. FM4–64 is generally preferred to FM1–43 as it is brighter and more photostable. FM4–64 also allows for simultaneous imaging with green emitting dyes and fluorescent proteins as it exhibits red-shifted fluorescence [171]. All these dyes benefit greatly from the fact that their quantum yield is close to zero in water, thus generating better staining contrast.

Beyond the simple staining of lipid membranes, lipophilic fluorescent membrane probes can provide important information on membrane structure. The time scale of fluorescence is adequate to the analysis of fluorophore dynamics during the excited state and solvent-dependent excited-state relaxation [173, 174]. In turn, information about these processes allows us to evaluate membrane viscosity and lipid packing.

Membrane viscosity has been shown to regulate the activity of several membrane proteins and is linked to multiple physiological processes within the cell [175]. Lateral diffusion can be quantified through measurement of the diffusion coefficient of fluorescent membrane components using fluorescence recovery after photobleaching (FRAP) [176] or fluorescence correlation spectroscopy (FCS) [177]. On the other hand, local dynamics within the membrane can be evaluated through the use of fluorescent microviscosity probes.

DPH (1,6-diphenyl-1,3,5-hexatriene) and its trimethylamino-derivative, TMA-DPH are part of a class of membrane viscosity probes whose fluorescent polarization/anisotropy is strongly dependent on local microviscosity [178, 179]. The DPH fluorophore only absorbs light polarized along its long axis and the polarization of emitted photons has the same orientation [180]. In this way, rotation of the fluorophore (during its excited state) along its long axis has no impact on the polarization of fluorescence, while rotation along the two perpendicular axes induces depolarization [179]. As acyl-chain fluctuations dictate rotation of the fluorophore during its excited state, fluorescence anisotropy values of these dyes reflect local membrane viscosity. Due to the presence of the trimethylamino-derivative, TMA-DPH becomes anchored to the lipid–water interface and reflects viscosity in a slightly more shallow region of the lipid bilayer [178]. Another useful membrane probe sensitive to membrane viscosity is trans-parinaric acid (*t*-PnA) [181]. The fluorescence lifetime of *t*-PnA is strongly dependent on membrane dynamics, and the appearance of a very long lifetime (>30 ns) is used as a fingerprint for the presence of gel phase membrane domains [179, 182, 183]. Finally, a class of molecules known as molecular rotors, which include 9-(dicyanovinyl)-julolidine (DCVJ), 9-(2-carboxy-2-cyano)vinyl julolidine (CCVJ) and BODIPY-C_{12} present fluorophores with the ability of twisting along a single bond. This movement is hindered within ordered membranes influencing fluorescence properties such as quantum yield or fluorescence lifetime [184, 185].

Polarity sensitive probes such as 2-dimethylamino-6-1auroylnaphthalene (Laurdan) can also be used to evaluate membrane packing, as these dyes exhibit strong spectral changes upon changes in membrane ordering [65, 186]. One advantage of Laurdan and similar probes is that membrane order can be estimated from simple ratiometric measurements reflecting spectral shifts, facilitating their implementation in imaging applications.

3 Fluorescent Proteins

Fluorescent proteins (FPs), more precisely the green fluorescent protein (GFP) was discovered in 1962 [187]. Since then, the use of FPs in life sciences has been at the centre of a great and unparalleled revolution in the way we 'see' cellular events at the molecular level. Unsurprisingly, 'the discovery and development of the green fluorescent protein' earned Osamu Shimomura, Martin Chalfie and Roger Tsien the Nobel Prize in Chemistry in 2008, unequivocally stating the importance of such finding for modern science and medicine.

GFP is a 27 kDa protein from the jellyfish *Aequorea victoria* that folds into a β-barrel structure where three sequential amino acids (Ser-Tyr-Gly in positions 65–67) form the protein chromophore upon cyclization, oxidation and dehydration steps, without the need for enzymes or cofactors [16, 188]. The main reason for GFP popularity from the start is associated with the advantage of being genetically encoded, thus bypassing many problems associated with the use of synthetic probes for protein labelling. Namely, their use avoids mis(or multi)labelling of proteins, issues with background fluorescence due to unspecific binding and is also completely compatible with live-cell (and tissue) imaging. Moreover, engineering a fusion construct is relatively easy at the experimental level, being routinely performed in many laboratories.

Although GFP is still frequently used, many other variants with different and improved photophysical properties are now widely available (Table 2). These variants were either discovered in other organisms such as sea anemone [216], copepod [216] and lancelet [217], or developed through the modification of the original protein barrel structure by mutagenesis. Many modifications in the core chromophore structure were performed [206], as well as changes in the side chains of nearby amino acids [16, 218]. The result is a large variety of FPs of different colours (from violet to far-red) covering almost entirely the visible spectrum. Moreover, the development of FPs with large (>100 nm) Stokes shift [219–222], and the possibility of generating tandem FP versions apart from the monomeric ones [223, 224] further increase the number of available choices. Yet, attempts are continuously being made to develop different FPs with, among other features, improved brightness, with other excitation/emission wavelengths and enhanced pH resistance (when compared with the original proteins). The engineering of red fluorescent proteins (RFP) has also been the focus of numerous attempts since they would present a few additional advantages. As already described for organic probes, fluorescence in the red to near-infrared region of the spectrum ensures a more efficient tissue penetration, decreases any issues related to cell autofluorescence and, additionally, relieves some of the spectral crowding (and consequent channel bleedthrough) usually encountered when designing multi-colour imaging experiments. DsRed from *Discosoma* was the first RFP to be isolated [225]. However, this protein shows severe limitations that precluded its generalized use. DsRed assumes a GFP-like intermediate form during maturation and a fraction of the protein exhibits excitation in the green spectral region. Furthermore, it requires over 30 h of

Table 2 Overview of the most commonly used fluorescent proteins, their oligomerization state and main photophysical properties

Protein	Oligomerization	ε (M^{-1} cm^{-1})	$\lambda_{max,abs}$ (nm)	$\lambda_{max,em}$ (nm)	QY	Brightness ($\times 10^3$ M^{-1} cm^{-1})	References
TagBFP	Monomer	52,000	402	457	0.63	33	[189]
mTurquoise	Monomer	30,000	434	474	0.93	28	[190]
MCerulean3	Monomer	40,000	433	475	0.87	35	[191]
EGFP	Weak dimer	55,900	488	507	0.6	34	[192]
rs-EGFP	Monomer	47,000	493	510	0.36	17	[193]
Skylan -NS	Monomer	133,770	499	511	0.59	79	[194]
T-Sapphire	Weak dimer	44,000	399	511	0.6	26	[195]
mGreenLantern	Monomeric	102,000	503	514	0.72	73	[196]
mNeonGreen	Monomer	116,000	506	517	0.8	93	[197]
PA-GFP	Monomer	17,400	504	517	0.79	14	[198]
mClover3	Monomer	109,000	506	518	0.78	85	[199]
rsFastLime	Tetramer	39,094	496	518	0.77	30	[200]
mAmetrine	Monomer	45,000	406	526	0.58	26	[201]
EYFP	Weak dimer	67,000	513	527	0.67	45	[202]
mVenus	Monomer	104,000	515	527	0.64	67	[203]
mCitrine	Monomer	94,000	516	529	0.74	70	[204]
mOrange	Monomer	71,000	548	562	0.69	49	[205]
tdTomato	Tandem dimer	138,000	554	581	0.69	95	[205]
DsRed	Tetramer	75,000	558	583	0.68	51	[206]
rsTagRFP	Monomer	36,800	567	585	0.11	4	[207]
mApple	Monomer	75,000	568	592	0.49	37	[208]
mRuby3	Monomer	128,000	558	592	0.45	58	[199]

mScarlet	Monomer	100,000	569	594	0.7	70	[209]
PamCherry	Monomer	18,000	564	595	0.46	8	[210]
PATagRFP	Monomer	66,000	562	595	0.38	25	[211]
mCherry	Monomer	72,000	587	610	0.22	16	[205]
mKeima	Monomer	14,400	440	620	0.24	3	[212]
PamKate	Monomer	25,000	586	628	0.18	5	[213]
mKate2	Monomer	62,500	588	633	0.4	25	[214]
mNeptune	Monomer	67,000	600	650	0.2	13	[215]

incubation time at 37 °C to achieve steady-state levels and, since it is an obligate tetramer, it might force tetramerization of any target protein [226–228]. For these reasons, additional efforts have been carried out to bioengineer new and better performing RFP. Yet, despite considerable advances, the photophysical properties of RFPs still fall short when compared with the performance of other FPs.

FPs have been employed in the study of an immense number of cellular functions and processes. These have been extensively reviewed elsewhere [229–231]. However, it is worth noting that FPs are particularly well suited for the development of biosensors, allowing the imaging and quantification of diverse cellular processes [232]. Briefly, there are mainly four classes of FP-containing biosensors: (1) FPs with environment-dependent photophysical properties (e.g. brightness or emission wavelength) that detect changes in pH [233], Cl^- [234], metal ions [235] and redox potential [236], among others; (2) FP-target constructs where the photophysical properties of the FP are affected by conformational changes in the target protein and that have been used, for example, to detect hydrogen peroxide [237] and membrane potential variations [238]; (3) Sensors in which detection relies on differences in the FRET efficiency between two spectrally distinct FPs [239]. FRET sensors were already applied to the detection of Ca^{2+} [240], cyclic nucleotides [241] and others; (4) FP-target constructs used as translocation sensors [242]. In other words, when linked to specific target proteins, FPs can report on environmental changes in intercellular compartments.

When expressing FP-containing fusion constructs, a fine balance must be accomplished. The expression level should not be so high that would interfere with cell functioning but should at the same time be high enough to allow a proper fluorescence signal for cell imaging purposes [12]. It is also important to take into account the maturation speed of the chromophore and the possible aggregation of the construct [243]. In extreme cases, this can affect the folding of the FP and consequently the structure of the chromophore itself, thus compromising the fluorescence output of the protein [244]. Depending on the experimental design, optimization of the nucleotide sequence for a particular model organism might also be required [245, 246], as well as the fine-tuning of the length and flexibility of the linker peptide between the FP and the target protein [247, 248].

Despite the wide-ranging use of FPs in cell biology and imaging, this technology still poses some limitations [249]. The most well-known is arguably their size (2–5 nm), which means that in some studies the size of the FP is comparable to that of the target protein, thus preventing, for example, the precise localization of the protein of interest. For the same reason, there is always a chance that the FP not only interferes with the subcellular localization of the target but also impairs the protein's biological function [250]. This is even more important when the initial FP variants are considered since they show a significant tendency to dimerize [251, 252]. This however has been mitigated or completely eliminated in newer FPs through the replacement of hydrophobic amino acids at the surface for positively charged amino acids [204]. Among the disadvantages are also the limitation of labelling at the protein terminals, loss of FP fluorescence upon fixation [253] or brightness dependence on the expression and maturation temperature for some FPs [254]. Moreover, at the single-molecule level, FPs are known to blink at all timescales. Despite being associated with triplet and radical states, blinking of some variants has been related to proton transfer and conformational dynamics, involving different states of the chromophore [255]. Still, in many applications blinking is not a problem, but should be considered at the experiment design stage.

Apart from the issues discussed above, selection of an FP for an experiment should also involve the evaluation of the same properties previously discussed for organic dyes. Some of these are summarized in Table 2.

4 Fluorescent Probes for Super-Resolution Microscopy

Super-resolution microscopy surpasses the diffraction barrier by precluding the simultaneous signalling of adjacent fluorophores. In general terms, this is accomplished by transiently or permanently transferring (switching) fluorophores between two distinguishable states with different spectral, temporal or other detectable response to illumination. The choice of proper fluorophore is thus a requirement to achieve the best possible resolution. Although all the selection criteria described in the sections above still apply here, photophysical parameters such as brightness, photostability and switching kinetics gain additional relevance, depending on the chosen imaging technique. Super-resolution approaches have been thoroughly reviewed elsewhere [256–258].

In single-molecule localization microscopy (SMLM) [259], for example, only a subset of sparse fluorophores gets activated at any given time, allowing the accurate determination of the fluorophore's position with sub-diffraction precision. By repeating the process several times, exciting stochastically different molecules, it is therefore possible to reconstruct an image combining all the determined positions. Choosing photoactivatable fluorophores for SMLM should then consider the fluorescence contrast between *on* and *off* states, the turn-*on* halftime, the photobleaching halftime and the photon yield of individual molecules before bleaching [218, 260, 261]. Desirably, the photon yield should be high not only to ensure accurate position determination but also to allow the use of lower laser power. In case of reversible *on/off* switching, it is also important to take into account the number of photons detected per switching event, the *on/off* duty cycle (fraction of time a fluorophore spends in the fluorescent versus non-fluorescent state), the fatigue resistance (number of switching cycles required to bleach 50% of the initial fluorescence) and the fluorophores' survival fraction after subjected to relatively severe conditions [260]. To ensure the required sparse activation of fluorophores, the *off*-switching (or bleaching) rate must be much larger than the *on*-rate, resulting in a low duty cycle between 10^{-4} and 10^{-6} [262, 263].

In a very distinct approach, stimulated emission depletion (STED) microscopy overcomes the diffraction limit by reversibly silencing fluorophores at predefined positions within the diffraction-limited point-spread function (PSF) [264]. The excitation light is combined with a high-intensity doughnut-shaped laser that depletes

Table 3 Summary of the photoswitchable fluorescent proteins most commonly used for super-resolution microscopy, their oligomerization state, colour transition and photophysical properties of both states

Protein	Oligomerization	Transition[a]	ε (M^{-1} cm^{-1})	$\lambda_{max,abs}$ (nm)	$\lambda_{max,em}$ (nm)	QY[b]	References
Dendra2	Monomer	G/R	45,000/35,000	490/553	507/573	0.5/0.55	[277]
Kaede	Tetramer	G/R	98,800/60,400	508/572	518/580	0.88/0.33	[278]
mClavGR2	Monomer	G/R	19,000/32,000	488/566	504/583	0.77/0.53	[279]
mEos3.2	Monomer	G/R	63,400/32,000	507/572	516/580	0.84/0.55	[280]
mIrisFP	Monomer	G/R	47,000/33,000	486/546	516/578	0.54/0.59	[281]
mMaple3	Monomer	G/R	15,760/23,970	491/568	506/583	0.37/0.52	[263]
NijiFP	Monomer	G/O	41,100/42,000	469/526	507/569	0.64/0.65	[282]
PS-CFP2	Monomer	C/G	43,000/47,000	400/490	468/511	0.2/0.23	[283]
PSmOrange	Monomer	FR/O	32,700/113,300	634/548	662/565	0.28/0.51	[284]

[a] *G/R* green-to-red, *G/O* green-to-orange, *C/G* cyan-to-green, *FR/O* far-red-to-orange
[b] Quantum yield

fluorescence in specific regions (doughnut-shaped periphery) while leaving a central focal spot active. The acquisition is performed by scanning the sample with the two aligned beams and, by making so, only the non-silenced fluorophores in the central complementary regions emit light. As a consequence of this setup, fluorophores for STED microscopy must be quite photostable and thus exhibit high resistance to photobleaching [265–268]. Additional criteria for fluorophore selection include: (1) the emission spectrum of the fluorophore must be compatible with the available STED laser to be used [269]; (2) spectral overlap between stimulated emission and excited-state absorption should be avoided [265]; (3) large Stokes shift probes should be preferred to prevent direct excitation of the fluorophore by the STED beam and also to assist in multi-colour imaging [265, 270, 271]. However, most of the available large Stokes shift probes still do not present the required high brightness and enhanced photostability [265].

Although this description of SMLM and STED is not meant to be extensive, it is clear from these examples that the choice of super-resolution technique determines the fluorophore to be used. As for conventional imaging, many different types of fluorophores were already applied to one or several super-resolution approaches, including small organic probes [73], fluorescent proteins [272], quantum dots [273] and nanoparticles [274], among others. Moreover, fluorophores that use reversible ligand binding [275] or quenching [276] as a switching mechanism instead of photoswitching have also been employed. In the next sections, the most common classes of switching organic probes and FPs (Table 3) are briefly introduced, including some recent examples of their application. Detailed information on their mechanism and performance can be found elsewhere [218, 260, 285].

4.1 Synthetic Probes

Activator-Reporter Dye Pair

This system consists of a 'reporter' probe that switches between a fluorescent and a dark state and whose photoactivation is facilitated by an 'activator' fluorophore placed within 1–2 nm distance [218]. The 'reporter' dye is usually kept in a non-fluorescent state, and the random activation of non-overlapping fluorophores is accomplished by the use of an activation laser that excites the 'activator' probe. The 'activator' can then induce its 'reporter' neighbours to become fluorescent. The final image is consequently reconstructed from emission signal of the 'reporter'.

The first system to be described was the Cy3-Cy5 combination, used by Xiaowei Zhuang to develop stochastic optical reconstruction microscopy (STORM) [286]. Since then, several combinations of 'activator' dyes (e.g. Cy2 or Alexa Fluor 405) and 'reporters' (e.g. Cy5.5, Cy7 or Alexa Fluor 647) were employed [287, 288].

Activator-Free Dyes
Some dyes are stochastically activated in the absence of an activator fluorophore and under continuous laser illumination [218, 289]. However, they frequently require significantly higher laser power. The switching behaviour also entails a specific chemical environment. For this reason, the imaging buffers usually include 'non-common' components such as triplet quenchers, oxygen scavengers, reducing agents and others [290–293].

These dyes were first applied by Sauer and co-workers to perform direct STORM [97, 289]. In this pivotal work, the authors used cyanine dyes (without 'activators') that could be photoswitched under reducing conditions [289]. Recently, a comprehensive study compared the performance of 26 commercially available probes in a STORM setup and identified the best-performing dyes in different colour ranges: Atto488 for the green spectral region, Cy3B for the yellow, Alexa Fluor 647 for the red and DyLight750 in the near IR [260].

Spontaneously Blinking Dyes
Contrary to the probes from the categories described above, spontaneously blinking dyes do not need activators or incident light to become active. Instead, they spontaneously blink in the absence of light in a stochastic and reversible manner [218]. Spontaneously blinking dyes often switch between an 'open' fluorescent state and a 'closed' dark state. One such example is the intramolecular spirocyclization that occurs in hydroxymethyl (HM) rhodamines, from which other related derivatives have been developed. Hydroxymethyl-Si-rhodamine (HM-SiR) [294] and HEtetTFER [295], for example, have been applied to SMLM by Yasuteru Urano and his team, allowing the use of reduced illumination intensities and still presenting a high photon yield. In a different approach, Urano's team also described a different spontaneous blinking mechanism that relies on the reversible ground-state nucleophilic attack of intracellular glutathione (GSH) upon a xanthene fluorophore [296]. This allowed the observation of microtubules and mitochondria within living cells by SMLM. Although not many options are available so far, the recent developments in this area suggest an increase in the use of these dyes in the coming years.

Photochromic Dyes
Fluorophores that switch reversibly between an *on* and *off* state in a light-assisted manner are called photochromic [218]. These include different rhodamines, diarylethenes and photoswitchable cyanines [297]. Rhodamine lactams, for example, are colourless compounds that become fluorescent through the transient cleavage of the lactam bond upon far-UV illumination [298]. The fluorescent molecule then thermally reverts back to its non-emissive state. This reversible photoswitching mechanism is well suited for SMLM and efforts have been made to extend it to other rhodamines with different photophysical properties. Moreover, the addition of different substituents to the rhodamine core has also led to the development of photochromic dyes with activation light that is not in the far-UV spectral range. Phthalimide groups were shown to allow for near-UV (or 2-photon) activation [299] while stilbene groups even made it possible to use visible light [300, 301]. One of

such examples are the 'turn-on mode' fluorescent diarylethenes (fDAEs) recently developed by the Hell group. With yellow light activation, fDAEs were successfully applied to the visualization of cells' vimentin filaments by SMLM [302].

Photoactivatable or 'Caged' Dyes

In contrast to the photochromic probes, 'caged' dyes are irreversibly activated through a photochemical reaction [218]. These molecules combine fluorophores with photolabile (or caging) moieties located at appropriate positions that block them in a non-emissive state. Upon illumination, the protective groups are either removed or modified and the fluorescence emission of the dye is greatly increased [137, 262]. The use of caged dyes thus results in great *on*/*off* contrast and photon yield [218], making them ideal for SMLM techniques such as photoactivated localization microscopy (PALM) [303] and DNA points accumulation in nanoscale topography (DNA-PAINT) [304]. The photoactivation quantum yield can however be a critical element to consider.

Among the most common caging moieties are the o-nitrobenzyl derivatives, azides, coumarinyl groups and 2-diazoketones [305]. Different fluorophores have also been successfully caged, such as Q-rhodamine [303], rhodamine 110 [306], Si-Q-rhodamine [99], carbofluorescein [307], and several others.

4.2 *Fluorescent Proteins*

Photoactivatable (PA-FPs)

This category includes proteins that are irreversibly switched from a non-fluorescent *off*-state to a fluorescent *on*-state upon illumination. The development of PA-FPs goes hand-in-hand with the establishment of SMLM techniques. Indeed, PA-GFP (developed by Patterson and Lippincott-Schwartz [198]) was used in the very first papers on PALM [308, 309]. Due to its photophysical properties, high activation quantum yield and their monomeric state, PA-GFP is still one of the most used and dependable PA-FPs. Since then, several other variants were developed, including the ones within the orange/red spectral range, such as PAmRFP, PAmCherry, PATagRFP and PAmKate.

Nevertheless, the use of PA-FPs makes it hard to detect cells expressing the protein, since before illumination they remain in a non-fluorescent state [218]. This drawback is largely overcome by the use of photoswitchable dyes instead (see below).

Photoswitchable (PS-FPs)

Photoswitchable (or photoconvertible) FPs rely on the irreversible conversion from one fluorescent *on*-state to another of a different colour [285]. Depending on the specific colour transition, PS-FPs can be categorized as (1) GFP-like cyan-to-green proteins (PS-CFP2), (2) Kaede-like green-to-orange/red proteins (Dendra2, mEos3.2, mClavGR2, mMaple3) or (3) DsRed-like green-to-far red proteins (PSmOrange). This spectral coverage together with the photoswitching capability makes them not only ideal for simple SMLM experiments but also allows their use in multi-colour super-resolution imaging.

Although extensively applied, it is worth noting that even the brightest PS-FP remains much dimmer than some organic probes, with photon yields that can be one order of magnitude lower (e.g. comparing Eos with the Cy3-Cy5 pair) [287, 310, 311].

Photochromic (PC-FPs)
As their small-molecule counterparts, photochromic FPs are able to reversibly transition between a dark *off*-state and a fluorescent *on*-state. However, their main advantage over photochromic organic probes is the stability and reliability of their switching system, meaning that they are much more resistant to switching fatigue [262]. This also explains the use of these proteins in non-linear super-resolution imaging approaches such as non-linear structured-illumination microscopy (NL-SIM) [312] or reversible saturable optical fluorescence transition (RESOLFT) [313]. Examples of PC-FPs include rs-EGFP, rsFastlime, rsTagRFP or Skylan-NS.

Photoactivatable/Photoconvertible FPs
This category includes proteins with complex working mechanisms that combine the photoswitching behaviour of PC-FPs and PS-FPs. In other words, proteins such as mIrisFP [281] and NijiFP [282] can be irreversibly photoconverted between two *on*-states (green-to-red transition) and then feature a reversible photoswitching behaviour from both species to a dark *off*-state. Due to their complicated mechanism, these are the only FP variants included in this category.

5 Perspectives

Researchers in the field of biological sciences have now available an enormous abundance of different fluorescent probes to choose from. These present remarkably diverse properties so that specific experiments can make use of ideally suited labels. There are however still some limitations that are expected to be addressed. When it comes to organic dyes, significant difficulties still limit the ability to specifically label target molecules within a living cell or tissue. The next years will undoubtedly bring new and creative strategies combining biorthogonal chemistry with the use of unnatural metabolites and caged dyes, to achieve the required labelling specificity in a live-cell compatible manner. Regarding FPs, researchers are still paving the way to improve their photophysical properties and overall performance. Although they show the ultimate labelling specificity, there is still room for improvement in terms of FP photostability and brightness, to reduce the performance gap that remains when comparing FPs with synthetic dyes. In any case, whether it is organic dyes or FPs, the development of photostable, highly performing far red to near-IR probes is arguably one of the biggest challenges of the next years.

Overall, the available fluorescent probes used for life sciences cover already a very wide spectrum of applications. Nonetheless, the design of new probes must continue to walk hand-in-hand with the development of new techniques and approaches, as were the case of super-resolution microscopy and the development of photoswitchable fluorescent probes.

References

1. Lakowicz JR (2006) Principles of fluorescence spectroscopy. Springer, Boston
2. Andersson H, Baechi T, Hoechl M, Richter C (1998) Autofluorescence of living cells. J Microsc 191:1–7
3. Monici M (2005) Cell and tissue autofluorescence research and diagnostic applications. Biotechnol Annu Rev 11:227–256
4. Brandes R, Bers DM (1996) Increased work in cardiac trabeculae causes decreased mitochondrial NADH fluorescence followed by slow recovery. Biophys J 71:1024–1035
5. Masters BR, Chance B (1999) Redox confocal imaging: intrinsic fluorescent probes of cellular metabolism. In: Fluorescent and luminescent probes for biological activity. Elsevier, pp 361–374
6. Hellmann N, Schneider D (2019) Hands on: using tryptophan fluorescence spectroscopy to study protein structure. In: Protein supersecondary structures. Humana Press, New York, pp 379–401
7. Vivian JT, Callis PR (2001) Mechanisms of tryptophan fluorescence shifts in proteins. Biophys J 80:2093–2109
8. Royer CA (2006) Probing protein folding and conformational transitions with fluorescence. Chem Rev 106:1769–1784
9. Ghisaidoobe ABT, Chung SJ (2014) Intrinsic tryptophan fluorescence in the detection and analysis of proteins: a focus on förster resonance energy transfer techniques. Int J Mol Sci 15: 22518–22538
10. Banerjee B, Graves LR, Utzinger U (2012) Tryptophan fluorescence of cells and tissue in esophageal carcinoma. IEEE Sensors J 12:3273–3274
11. Li R, Dhankhar D, Chen J, Cesario TC, Rentzepis PM (2019) A tryptophan synchronous and normal fluorescence study on bacteria inactivation mechanism. Proc Natl Acad Sci 116: 18822–18826
12. Galas L, Gallavardin T, Bénard M, Lehner A, Schapman D, Lebon A, Komuro H, Lerouge P, Leleu S, Franck X (2018) "Probe, sample, and instrument (PSI)": the hat-trick for fluorescence live cell imaging. Chemosensors 6:40
13. Fu Y, Finney NS (2018) Small-molecule fluorescent probes and their design. RSC Adv 8: 29051–29061
14. Tian X, Murfin LC, Wu L, Lewis SE, James TD (2021) Fluorescent small organic probes for biosensing. Chem Sci 12:3406–3426
15. Huang Y, Zhang Y, Huo F, Wen Y, Yin C (2020) Design strategy and bioimaging of small organic molecule multicolor fluorescent probes. Sci China Chem 63(12):1742–1755
16. Chudakov DM, Matz MV, Lukyanov S, Lukyanov KA (2010) Fluorescent proteins and their applications in imaging living cells and tissues. Physiol Rev 90:1103–1163
17. Rodriguez EA, Campbell RE, Lin JY, Lin MZ, Miyawaki A, Palmer AE, Shu X, Zhang J, Tsien RY (2017) The growing and glowing toolbox of fluorescent and photoactive proteins. Trends Biochem Sci 42:111
18. Abbasi E, Kafshdooz T, Bakhtiary M, Nikzamir N, Nikzamir N, Nikzamir M, Mohammadian M, Akbarzadeh A (2016) Biomedical and biological applications of quantum dots. Artif Cells, Nanomed Biotechnol 44:885–891
19. Hatipoglu MK, Kelestemur S, Culha M (2016) Synthesis and biological applications of quantum dots. Springer, Cham, pp 505–534

20. Petryayeva E, Algar WR, Medintz IL (2013) Quantum dots in bioanalysis: a review of applications across various platforms for fluorescence spectroscopy and imaging. Appl Spectrosc 67:215–252
21. Zhou J, Yang Y, Zhang CY (2015) Toward biocompatible semiconductor quantum dots: from biosynthesis and bioconjugation to biomedical application. Chem Rev 115:11669–11717
22. Bilan R, Fleury F, Nabiev I, Sukhanova A (2015) Quantum dot surface chemistry and functionalization for cell targeting and imaging. Bioconjug Chem 26:609–624
23. Alivisatos AP, Gu W, Larabell C (2005) Quantum dots as cellular probes. Annu Rev Biomed Eng 7:55–76
24. Connell TU, James JL, White AR, Donnelly PS (2015) Protein labelling with versatile phosphorescent metal complexes for live cell luminescence imaging. Chem – A Eur J 21: 14146–14155
25. Fernández-Moreira V, Thorp-Greenwood FL, Coogan MP (2009) Application of d6 transition metal complexes in fluorescence cell imaging. Chem Commun 46:186–202
26. Tzubery A, Melamed-Book N, Tshuva EY (2018) Fluorescent antitumor titanium(IV) salen complexes for cell imaging. Dalton Trans 47:3669–3673
27. Cotruvo JA (2019) The chemistry of lanthanides in biology: recent discoveries, emerging principles, and technological applications. ACS Cent Sci 5:1496–1506
28. Cho U, Chen JK (2020) Lanthanide-based optical probes of biological systems. Cell Chem Biol 27:921–936
29. Mathieu E, Sipos A, Demeyere E, Phipps D, Sakaveli D, Borbas KE (2018) Lanthanide-based tools for the investigation of cellular environments. Chem Commun 54:10021–10035
30. Andraud C, Maury O (2009) Lanthanide complexes for nonlinear optics: from fundamental aspects to applications. Eur J Inorg Chem:4357–4371
31. Ge G, Li L, Wang D, Chen M, Zeng Z, Xiong W, Wu X, Guo C (2021) Carbon dots: synthesis, properties and biomedical applications. J Mater Chem B 9:6553–6575
32. Unnikrishnan B, Wu RS, Wei SC, Huang CC, Chang HT (2020) Fluorescent carbon dots for selective labeling of subcellular organelles. ACS Omega 5:11248–11261
33. Lesani P, Hazeera A, Hadi M, Lu Z, Palomba S, New EJ, Zreiqat H (2021) Design principles and biological applications of red-emissive two-photon carbon dots. Commun Mater 2(1): 1–12
34. Qu D, Wang X, Bao Y, Sun Z (2020) Recent advance of carbon dots in bio-related applications. J Phys Mater 3:022003
35. Jensen EC (2012) Use of fluorescent probes: their effect on cell biology and limitations. Anat Rec Adv Integr Anat Evol Biol 295:2031–2036
36. Wallace PK, Muirhead KA (2007) Cell tracking 2007: a proliferation of probes and applications. Immunol Investig 36:527–561
37. Chen M, He X, Wang K, He D, Yang X, Shi H (2014) Inorganic fluorescent nanoprobes for cellular and subcellular imaging. TrAC – Trends Anal Chem 58:120–129
38. Haque A, Faizi MSH, Rather JA, Khan MS (2017) Next generation NIR fluorophores for tumor imaging and fluorescence-guided surgery: a review. Bioorg Med Chem 25:2017–2034
39. Wan M, Zhu Y, Zou J (2020) Novel near-infrared fluorescent probe for live cell imaging. Exp Ther Med 19:1213
40. Tsubono Y, Kawamoto Y, Hidaka T, Pandian GN, Hashiya K, Bando T, Sugiyama H (2020) A near-infrared fluorogenic pyrrole-imidazole polyamide probe for live-cell imaging of telomeres. J Am Chem Soc 142:17356–17363
41. Liu X, Zhang QY, Wang F, Jiang JH (2019) A near infrared fluorescent probe for the detection and imaging of prolyl aminopeptidase activity in living cells. Analyst 144:5980–5985
42. Baeyer A (1871) Ueber eine neue Klasse von Farbstoffen. Ber Dtsch Chem Ges 4:555–558
43. Johnson I, Spence M (2010) Molecular probes handbook, a guide to fluorescent probes and labeling technologies.11th edn. Life Technologies, New York
44. Tsien RY, Waggoner A (1995) Fluorophores for confocal microscopy. In: Pawley JB (ed) Handbook of biological confocal microscopy. Springer, Boston, pp 267–279

45. Nanguneri S, Flottmann B, Herrmannsdörfer F, Kuner T, Heilemann M (2014) Single-molecule super-resolution imaging by tryptophan-quenching-induced photoswitching of phalloidin-fluorophore conjugates. Microsc Res Tech 77:510–516
46. Haenni D, Zosel F, Reymond L, Nettels D, Schuler B (2013) Intramolecular distances and dynamics from the combined photon statistics of single-molecule FRET and photoinduced electron transfer. J Phys Chem B 117:13015–13028
47. ISS 2021 lifetime data of selected fluorophores
48. Nakata E, Koshi Y, Koga E, Katayama Y, Hamachi I (2005) Double-modification of lectin using two distinct chemistries for fluorescent ratiometric sensing and imaging saccharides in test tube or in cell. J Am Chem Soc 127:13253–13261
49. Urban NT, Foreman MR, Hell SW, Sivan Y (2018) Nanoparticle-assisted STED nanoscopy with gold nanospheres. ACS Photonics 5:2574–2583
50. Adinolfi B, Pellegrino M, Tombelli S, Trono C, Giannetti A, Domenici C, Varchi G, Sotgiu G, Ballestri M, Baldini F (2018) Polymeric nanoparticles promote endocytosis of a survivin molecular beacon: localization and fate of nanoparticles and beacon in human A549 cells. Life Sci 215:106–112
51. Vogelsang J, Cordes T, Forthmann C, Steinhauer C, Tinnefeid P (2009) Controlling the fluorescence of ordinary oxazine dyes for single-molecule switching and superresolution microscopy. Proc Natl Acad Sci U S A 106:8107–8112
52. Ultzinger U (2011) Spectra database hosted at the University of Arizona
53. Kang HC, Haugland RP, Fisher PJ, Prendergast FG (1989) Spectral properties of 4-sulfonato-3,3′,5,5′-tetramethyl-2,2′-pyrronlethen-1,1′-borondifluoride complex (Bodipy), its sodium salt, and protein derivatives. In: Salzman GC (ed) New technologies in cytometry. SPIE, p 68
54. Moens PDJ, Bagatolli LA (2007) Profilin binding to sub-micellar concentrations of phosphatidylinositol (4,5) bisphosphate and phosphatidylinositol (3,4,5) trisphosphate. Biochim Biophys Acta 1768:439–449
55. Sauer M, Hofkens J, Enderlein J (2011) Fluorophores and fluorescent labels. In: Handbook of fluorescence spectroscopy and imaging. Wiley, Weinheim, pp 31–60
56. Texier I, Goutayer M, Da Silva A, Guyon L, Djaker N, Josserand V, Neumann E, Bibette J, Vinet F (2009) Cyanine-loaded lipid nanoparticles for improved in vivo fluorescence imaging. J Biomed Opt 14:054005
57. Lee IH, Saha S, Polley A, Huang H, Mayor S, Rao M, Groves JT (2015) Live cell plasma membranes do not exhibit a miscibility phase transition over a wide range of temperatures. J Phys Chem B 119:4450–4459
58. Georgiev R, Christova D, Todorova L, Georgieva B, Vasileva M, Novakov C, Babeva T (2018) Triblock copolymer micelles as templates for preparation of mesoporous niobia thin films. J Phys Conf Ser 992:012037
59. Gracetto AC, Batistela VR, Caetano W, De Oliveira HPM, Santos WG, Cavalheiro CCS, Hioka N (2010) Unusual 1,6-diphenyl-1,3,5-hexatriene (DPH) spectrophotometric behavior in water/ethanol and water/DMSO mixtures. J Braz Chem Soc 21:1497–1502
60. Hudson EN, Weber G (1973) Synthesis and characterization of two fluorescent sulfhydryl reagents. Biochemistry 12:4154–4161
61. Nunnally BK, He H, Li LC, Tucker SA, McGown LB (1997) Characterization of visible dyes for four-decay fluorescence detection in dna sequencing. Anal Chem 69:2392–2397
62. Strickler SJ, Berg RA (1962) Relationship between absorption intensity and fluorescence lifetime of molecules. J Chem Phys 37:814–822
63. Martin MM, Lindqvist L (1975) The pH dependence of fluorescein fluorescence. J Lumin 10: 381–390
64. Magde D, Rojas GE, Seybold PG (1999) Solvent dependence of the fluorescence lifetimes of xanthene dyes. Photochem Photobiol 70:737–744
65. Parasassi T, Gratton E (1995) Membrane lipid domains and dynamics as detected by Laurdan fluorescence. J Fluoresc 5:59–69

66. Fery-Forgues S, Fayet JP, Lopez A (1993) Drastic changes in the fluorescence properties of NBD probes with the polarity of the medium: involvement of a TICT state? J Photochem Photobiol A Chem 70:229–243
67. Lin S, Struve WS (1991) Time-resolved fluorescence of nitrobenzoxadiazole-aminohexanoic-acid: effect of intermolecular hydrogen-bonding on non-radiative decay. Photochem Photobiol 54:361–365
68. Amaro M, Filipe HAL, Prates Ramalho JP, Hof M, Loura LMS (2016) Fluorescence of nitrobenzoxadiazole (NBD)-labeled lipids in model membranes is connected not to lipid mobility but to probe location. Phys Chem Chem Phys 18:7042–7054
69. Tajalli H, Gilani AG, Zakerhamidi MS, Tajalli P (2008) The photophysical properties of Nile red and Nile blue in ordered anisotropic media. Dyes Pigments 78:15–24
70. Cser A, Nagy K, Biczók L (2002) Fluorescence lifetime of Nile Red as a probe for the hydrogen bonding strength with its microenvironment. Chem Phys Lett 360:473–478
71. Nimmerjahn A, Kirchhoff F, Kerr JND, Helmchen F (2004) Sulforhodamine 101 as a specific marker of astroglia in the neocortex in vivo. Nat Methods 1:31–37
72. Prendergast FG, Callahan PJ, Haugland RP (1981) 1-[4-(Trimethylamino)phenyl]-6-phenylhexa-1,3,5-triene: synthesis, fluorescence properties, and use as a fluorescence probe of lipid bilayers. Biochemistry 20:7333–7338
73. Kozma E, Kele P (2019) Fluorogenic probes for super-resolution microscopy. Org Biomol Chem 17:215–233
74. Wang L, Frei MS, Salim A, Johnsson K (2019) Small-molecule fluorescent probes for live-cell super-resolution microscopy. J Am Chem Soc 141:2770–2781
75. Lavis LD, Raines RT (2008) Bright ideas for chemical biology. ACS Chem Biol 3:142–155
76. Yang NJ, Hinner MJ (2015) Getting across the cell membrane: an overview for small molecules, peptides, and proteins. Methods Mol Biol 1266:29–53
77. Mobbs P, Becker D, Williamson R, Bate M, Warner A (1999) Techniques for dye injection and cell labelling. In: Microelectrode techniques. The Plymouth workshop handbook. The Company of Biologists Ltd, Cambridge, pp 361–387
78. Ji X, Ji K, Chittavong V, Aghoghovbia RE, Zhu M, Wang B (2017) Click and fluoresce: a bioorthogonally activated smart probe for wash-free fluorescent labeling of biomolecules. J Org Chem 82:1471–1476
79. Larsson A, Carlsson C, Jonsson M, Albinsson B (1994) Characterization of the binding of the fluorescent dyes YO and YOYO to DNA by polarized light spectroscopy. J Am Chem Soc 116:8459–8465
80. Hirons GT, Fawcett JJ, Crissman HA (1994) TOTO and YOYO: new very bright fluorochromes for DNA content analyses by flow cytometry. Cytometry 15:129–140
81. Lichtman JW, Conchello JA (2005) Fluorescence microscopy. Nat Methods 2:910–919
82. Vira S, Mekhedov E, Humphrey G, Blank PS (2010) Fluorescent-labeled antibodies: balancing functionality and degree of labeling. Anal Biochem 402:146–150
83. Szabó Á, Szendi-Szatmári T, Ujlaky-Nagy L, Rádi I, Vereb G, Szöllősi J, Nagy P (2018) The effect of fluorophore conjugation on antibody affinity and the photophysical properties of dyes. Biophys J 114:688–700
84. Cosa G, Focsaneanu K-S, McLean JRN, McNamee JP, Scaiano JC (2001) Photophysical properties of fluorescent DNA-dyes bound to single- and double-stranded DNA in aqueous buffered solution. Photochem Photobiol 73:585
85. Mullins JM (2010) Fluorochromes: properties and characteristics. Methods Mol Biol 588: 123–134
86. Li Q, Seeger S (2011) Multidonor deep-UV FRET study of protein – ligand binding and its potential to obtain structure information. J Phys Chem B 115:13643–13649
87. Berlman I (1971) Handbook of fluorescence spectra of aromatic molecules.2nd edn. Elsevier
88. Chen H, Ahsan SS, Santiago-Berrios MB, Abruña HD, Webb WW (2010) Mechanisms of quenching of alexa fluorophores by natural amino acids. J Am Chem Soc 132:7244–7245

89. Marmé N, Knemeyer JP, Sauer M, Wolfrum J (2003) Inter- and intramolecular fluorescence quenching of organic dyes by tryptophan. Bioconjug Chem 14:1133–1139
90. Neuweiler H, Schulz A, Vaiana AC, Smith JC, Kaul S, Wolfrum J, Sauer M (2002) Detection of individual p53-autoantibodies by using quenched peptide-based molecular probes. Angew Chem Int Ed 41:4769–4773
91. Eggeling C, Widengren J, Rigler R, Seidel CAM (1998) Photobleaching of fluorescent dyes under conditions used for single-molecule detection: evidence of two-step photolysis. Anal Chem 70:2651–2659
92. Diaspro A, Chirico G, Usai C, Ramoino P, Dobrucki J (2006) Photobleaching. In: Handbook of biological confocal microscopy. Springer, Boston, pp 690–702
93. Vogelsang J, Kasper R, Steinhauer C, Person B, Heilemann M, Sauer M, Tinnefeld P (2008) A reducing and oxidizing system minimizes photobleaching and blinking of fluorescent dyes. Angew Chem Int Ed 47:5465–5469
94. Demchenko AP (2020) Photobleaching of organic fluorophores: quantitative characterization, mechanisms, protection. Methods Appl Fluoresc 8:022001
95. Panchuk-Voloshina N, Haugland RP, Bishop-Stewart J, Bhalgat MK, Millard PJ, Mao F, Leung WY, Haugland RP (1999) Alexa dyes, a series of new fluorescent dyes that yield exceptionally bright, photostable conjugates. J Histochem Cytochem 47:1179–1188
96. Mitronova GY, Belov VN, Bossi ML, Wurm CA, Meyer L, Medda R, Moneron G, Bretschneider S, Eggeling C, Jakobs S, Hell SW (2010) New fluorinated rhodamines for optical microscopy and nanoscopy. Chem – A Eur J 16:4477–4488
97. Heilemann M, Van De Linde S, Mukherjee A, Sauer M (2009) Super-resolution imaging with small organic fluorophores. Angew Chem Int Ed 48:6903–6908
98. Hnedzko D, McGee DW, Rozners E (2016) Synthesis and properties of peptide nucleic acid labeled at the N-terminus with HiLyte Fluor 488 fluorescent dye. Bioorg Med Chem 24:4199–4205
99. Grimm JB, English BP, Chen J, Slaughter JP, Zhang Z, Revyakin A, Patel R, Macklin JJ, Normanno D, Singer RH, Lionnet T, Lavis LD (2015) A general method to improve fluorophores for live-cell and single-molecule microscopy. Nat Methods 12:244–250
100. Hinkeldey B, Schmitt A, Jung G (2008) Comparative photostability studies of BODIPY and fluorescein dyes by using fluorescence correlation spectroscopy. ChemPhysChem 9:2019–2027
101. Gorka AP, Schnermann MJ (2016) Harnessing cyanine photooxidation: from slowing photobleaching to near-IR uncaging. Curr Opin Chem Biol 33:117–125
102. Zheng Q, Jockusch S, Zhou Z, Blanchard SC (2014) The contribution of reactive oxygen species to the photobleaching of organic fluorophores. Photochem Photobiol 90:448–454
103. Yguerabide J, Schmidt JA, Yguerabide EE (1982) Lateral mobility in membranes as detected by fluorescence recovery after photobleaching. Biophys J 40:69–75
104. Ishikawa-Ankerhold HC, Ankerhold R, Drummen GPC (2012) Advanced fluorescence microscopy techniques–FRAP, FLIP, FLAP, FRET and FLIM. Molecules 17:4047–4132
105. Gruber HJ, Hahn CD, Kada G, Riener CK, Harms GS, Ahrer W, Dax TG, Knaus HG (2000) Anomalous fluorescence enhancement of Cy3 and Cy3.5 versus anomalous fluorescence loss of Cy5 and Cy7 upon covalent linking to IgG and noncovalent binding to avidin. Bioconjug Chem 11:696–704
106. Gebhardt C, Lehmann M, Reif MM, Zacharias M, Gemmecker G, Cordes T (2021) Molecular and spectroscopic characterization of green and red cyanine fluorophores from the Alexa Fluor and AF series. ChemPhysChem 22:1566–1583
107. Valdes-Aguilera O, Neckers DC (1989) Aggregation phenomena in xanthene dyes. Acc Chem Res 22:171–177
108. Demchenko AP (2009) Introduction to fluorescence sensing. Springer, Dordrecht
109. Bergström F, Mikhalyov I, Hägglöf P, Wortmann R, Ny T, Johansson LBÅ (2002) Dimers of dipyrrometheneboron difluoride (BODIPY) with light spectroscopic applications in chemistry and biology. J Am Chem Soc 124:196–204

110. Monteiro ME, Sarmento MJ, Fernandes F (2014) Role of calcium in membrane interactions by PI(4,5)P2 – binding proteins. Biochem Soc Trans 42:1441–1446
111. Dziuba D, Jurkiewicz P, Cebecauer M, Hof M, Hocek M (2016) A rotational BODIPY nucleotide: an environment-sensitive fluorescence-lifetime probe for DNA interactions and applications in live-cell microscopy. Angew Chem 128:182–186
112. Song X, Li N, Wang C, Xiao Y (2017) Targetable and fixable rotor for quantifying mitochondrial viscosity of living cells by fluorescence lifetime imaging. J Mater Chem B 5:360–368
113. Leung RWK, Yeh S-CA, Fang Q (2011) Effects of incomplete decay in fluorescence lifetime estimation. Biomed Opt Express 2:2517–2531
114. Lavis LD, Rutkoski TJ, Raines RT (2007) Tuning the pK a of fluorescein to optimize binding assays environment. The fluorescence of the dye varies likewise. Life Sci 79:6775–6782
115. Haugland RP (2005) The molecular probes handbook: a guide to fluorescent probes and labeling technologies. Invitrogen Corp, Karlsbad
116. Paradiso AM, Tsien RY, Machen TE (1984) Na+-H+ exchange in gastric glands as measured with a cytoplasmic-trapped, fluorescent pH indicator. Proc Natl Acad Sci 81:7436–7440
117. Rink TI, Tsien RY, Pozzan T (1982) Cytoplasmic pH and free Mg2+ in lymphocytes. J Cell Biol 95:189–196
118. Yoshihara T, Maruyama R, Shiozaki S, Yamamoto K, Kato SI, Nakamura Y, Tobita S (2020) Visualization of lipid droplets in living cells and fatty livers of mice based on the fluorescence of π-extended coumarin using fluorescence lifetime imaging microscopy. Anal Chem 92: 4996–5003
119. Yang D, Dai SY (2020) A visible and near-infrared light activatable diazocoumarin probe for fluorogenic protein labeling in living cells. J Am Chem Soc 142:17156–17166
120. Sun W, Guo S, Hu C, Fan J, Peng X (2016) Recent development of chemosensors based on cyanine platforms. Chem Rev 116:7768–7817
121. Shindy HA (2017) Fundamentals in the chemistry of cyanine dyes: a review. Dyes Pigments 145:505–513
122. Zheng Q, Lavis LD (2017) Development of photostable fluorophores for molecular imaging. Curr Opin Chem Biol 39:32–38
123. Li G, Guan Y, Ye F, Liu SH, Yin J (2020) Cyanine-based fluorescent indicator for mercury ion and bioimaging application in living cells. Spectrochim Acta A Mol Biomol Spectrosc 239: 118465
124. Uno K, Sugimoto N, Sato Y (2021) N-aryl pyrido cyanine derivatives are nuclear and organelle DNA markers for two-photon and super-resolution imaging. Nat Commun 12(1): 1–9
125. Aristova D, Kosach V, Chernii S, Slominsky Y, Balanda A, Filonenko V, Yarmoluk S, Rotaru A, Özkan HG, Mokhir A, Kovalska V (2021) Monomethine cyanine probes for visualization of cellular RNA by fluorescence microscopy. Methods Appl Fluoresc 9:045002
126. Beija M, Afonso CAM, Martinho JMG (2009) Synthesis and applications of rhodamine derivatives as fluorescent probes. Chem Soc Rev 38:2410–2433
127. Duan Y, Liu M, Sun W, Wang M, Liu S, Li Q (2009) Recent progress on synthesis of fluorescein probes. Mini Rev Org Chem 6:35–43
128. Rajasekar M (2021) Recent development in fluorescein derivatives. J Mol Struct 1224:129085
129. Le Guern F, Mussard V, Gaucher A, Rottman M, Prim D (2020) Fluorescein derivatives as fluorescent probes for ph monitoring along recent biological applications. Int J Mol Sci 21: 1–23
130. Loudet A, Burgess K (2007) BODIPY dyes and their derivatives: syntheses and spectroscopic properties. Chem Rev 107:4891–4932
131. Martynov VI, Pakhomov AA (2021) BODIPY derivatives as fluorescent reporters of molecular activities in living cells. Russ Chem Rev 90:1213–1262
132. Yue J, Tao Y, Zhang J, Wang H, Wang N, Zhao W (2021) BODIPY-based fluorescent probe for fast detection of hydrogen sulfide and lysosome-targeting applications in living cells. Chem – An Asian J 16:850–855

133. Toseland CP (2013) Fluorescent labeling and modification of proteins. J Chem Biol 6:85–95
134. Obermaier C, Griebel A, Westermeier R (2015) Principles of protein labeling techniques. Methods Mol Biol 1295:153–165
135. Sereda TJ, Mant CT, Quinn AM, Hodges RS (1993) Effect of the alpha-amino group on peptide retention behaviour in reversed-phase chromatography. Determination of the pK (a) values of the alpha-amino group of 19 different N-terminal amino acid residues. J Chromatogr 646:17–30
136. Giepmans BNG, Adams SR, Ellisman MH, Tsien RY (2006) The fluorescent toolbox for assessing protein location and function. Science 312:217–224
137. Fernández-Suárez M, Ting AY (2008) Fluorescent probes for super-resolution imaging in living cells. Nat Rev Mol Cell Biol 9:929–943
138. Klein A, Hank S, Raulf A, Joest EF, Tissen F, Heilemann M, Wieneke R, Tampé R (2018) Live-cell labeling of endogenous proteins with nanometer precision by transduced nanobodies. Chem Sci 9:7835
139. Herce HD, Schumacher D, Schneider AFL, Ludwig AK, Mann FA, Fillies M, Kasper MA, Reinke S, Krause E, Leonhardt H, Cardoso MC, Hackenberger CPR (2017) Cell-permeable nanobodies for targeted immunolabelling and antigen manipulation in living cells. Nat Chem 9(8):762–771
140. Traenkle B, Rothbauer U (2017) Under the microscope: single-domain antibodies for live-cell imaging and super-resolution microscopy. Front Immunol 8:1030
141. Hoelzel CA, Zhang X (2020) Visualizing and manipulating biological processes by using HaloTag and SNAP-tag technologies. ChemBioChem 21:1935–1946
142. Los GV, Encell LP, McDougall MG, Hartzell DD, Karassina N, Zimprich C, Wood MG, Learish R, Ohana RF, Urh M, Simpson D, Mendez J, Zimmerman K, Otto P, Vidugiris G, Zhu J, Darzins A, Klaubert DH, Bulleit RF, Wood KV (2008) HaloTag: a novel protein labeling technology for cell imaging and protein analysis. ACS Chem Biol 3:373–382
143. Keppler A, Gendreizig S, Gronemeyer T, Pick H, Vogel H, Johnsson K (2002) A general method for the covalent labeling of fusion proteins with small molecules in vivo. Nat Biotechnol 21(1):86–89
144. Juillerat A, Gronemeyer T, Keppler A, Gendreizig S, Pick H, Vogel H, Johnsson K (2003) Directed evolution of O6-alkylguanine-DNA alkyltransferase for efficient labeling of fusion proteins with small molecules in vivo. Chem Biol 10:313–317
145. Gautier A, Juillerat A, Heinis C, Corrêa IR, Kindermann M, Beaufils F, Johnsson K (2008) An engineered protein tag for multiprotein labeling in living cells. Chem Biol 15:128–136
146. Miller LW, Cai Y, Sheetz MP, Cornish VW (2005) In vivo protein labeling with trimethoprim conjugates: a flexible chemical tag. Nat Methods 2:255–257
147. Mizukami S, Watanabe S, Hori Y, Kikuchi K (2009) Covalent protein labeling based on noncatalytic β-lactamase and a designed FRET substrate. J Am Chem Soc 131:5016–5017
148. Hori Y, Ueno H, Mizukami S, Kikuchi K (2009) Photoactive yellow protein-based protein labeling system with turn-on fluorescence intensity. J Am Chem Soc 131:16610–16611
149. Liu Y, Zhang X, Tan YL, Bhabha G, Ekiert DC, Kipnis Y, Bjelic S, Baker D, Kelly JW (2014) De novo-designed enzymes as small-molecule-regulated fluorescence imaging tags and fluorescent reporters. J Am Chem Soc 136:13102–13105
150. Liu Y, Miao K, Li Y, Fares M, Chen S, Zhang X (2018) A HaloTag-based multicolor fluorogenic sensor visualizes and quantifies proteome stress in live cells using solvatochromic and molecular rotor-based fluorophores. Biochemistry 57:4663–4674
151. Adams SR (2008) Tags and probes for chemical biology: the biarsenical-tetracysteine protein tag: chemistry and biological applications. In: Chemical biology: from small molecules to systems biology and drug design, vol 1–3. Wiley, pp 427–457
152. Sarmento MJ, Oneto M, Pelicci S, Pesce L, Scipioni L, Faretta M, Furia L, Dellino GI, Pelicci PG, Bianchini P, Diaspro A, Lanzanò L (2018) Exploiting the tunability of stimulated emission depletion microscopy for super-resolution imaging of nuclear structures. Nat Commun 9:3415

153. Liang L, Astruc D (2011) The copper(I)-catalyzed alkyne-azide cycloaddition (CuAAC) "click" reaction and its applications. An overview. Coord Chem Rev 255:2933–2945
154. Presolski SI, Hong VP, Finn MG (2011) Copper-catalyzed azide–alkyne click chemistry for bioconjugation. Curr Protoc Chem Biol 3:153–162
155. Gutmann M, Memmel E, Braun AC, Seibel J, Meinel L, Lühmann T (2016) Biocompatible azide-alkyne "click" reactions for surface decoration of glyco-engineered cells. ChemBioChem 17:866–875
156. Baskin JM, Prescher JA, Laughlin ST, Agard NJ, Chang PV, Miller IA, Lo A, Codelli JA, Bertozzi CR (2007) Copper-free click chemistry for dynamic in vivo imaging. Proc Natl Acad Sci 104:16793–16797
157. Saxon E, Bertozzi CR (2000) Cell surface engineering by a modified Staudinger reaction. Science 287:2007–2010
158. Li L, Zhang Z (2016) Development and applications of the copper-catalyzed azide-alkyne cycloaddition (CuAAC) as a bioorthogonal reaction. Molecules 21:1393
159. Laxman P, Ansari S, Gaus K, Goyette J (2021) The benefits of unnatural amino acid incorporation as protein labels for single molecule localization microscopy. Front Chem 9:161
160. Lee KJ, Kang D, Park HS (2019) Site-specific labeling of proteins using unnatural amino acids. Mol Cells 42:386–396
161. Saal KA, Richter F, Rehling P, Rizzoli SO (2018) Combined use of unnatural amino acids enables dual-color super-resolution imaging of proteins via cick chemistry. ACS Nano 12: 12247–12254
162. Brown W, Liu J, Deiters A (2018) Genetic code expansion in animals. ACS Chem Biol 13: 2375–2386
163. Wiltschi B (2016) Incorporation of non-canonical amino acids into proteins in yeast. Fungal Genet Biol 89:137–156
164. Ambrogelly A, Palioura S, Söll D (2007) Natural expansion of the genetic code. Nat Chem Biol 3(1):29–35
165. Jakob L, Gust A, Grohmann D (2019) Evaluation and optimisation of unnatural amino acid incorporation and bioorthogonal bioconjugation for site-specific fluorescent labelling of proteins expressed in mammalian cells. Biochem Biophys Rep 17:1–9
166. Lang K, Chin JW (2014) Cellular incorporation of unnatural amino acids and bioorthogonal labeling of proteins. Chem Rev 114:4764–4806
167. Chin JW (2014) Expanding and reprogramming the genetic code of cells and animals. Annu Rev Biochem 83:379–408
168. Chazotte B (2011) Labeling membrane glycoproteins or glycolipids with fluorescent wheat germ agglutinin. Cold Spring Harb Protoc:6
169. Honig MG, Hume RI (1986) Fluorescent carbocyanine dyes allow living neurons of identified origin to be studied in long-term cultures. J Cell Biol 103:171–187
170. Kleusch C, Hersch N, Hoffmann B, Merkel R, Csiszár A (2012) Fluorescent lipids: functional parts of fusogenic liposomes and tools for cell membrane labeling and visualization. Molecules 17:1055–1073
171. Bolte S, Talbot C, Boutte Y, Catrice O, Read ND, Satiat-Jeunemaitre B (2004) FM-dyes as experimental probes for dissecting vesicle trafficking in living plant cells. J Microsc 214:159–173
172. Betz WJ, Bewick GS (1993) Optical monitoring of transmitter release and synaptic vesicle recycling at the frog neuromuscular junction. J Physiol 460:287–309
173. Sýkora J, Kapusta P, Fidler V, Hof M (2002) On what time scale does solvent relaxation in phospholipid bilayers happen? Langmuir 18:571–574
174. Loura LMS, Prates Ramalho JP (2009) Fluorescent membrane probes' behavior in lipid bilayers: insights from molecular dynamics simulations. Biophys Rev 1:141–148
175. Lee AG (2004) How lipids affect the activities of integral membrane proteins. Biochim Biophys Acta 1666:62–87

176. Reits EAJ, Neefjes JJ (2001) From fixed to FRAP: measuring protein mobility and activity in living cells. Nat Cell Biol 3:E145
177. Macháň R, Hof M (2010) Lipid diffusion in planar membranes investigated by fluorescence correlation spectroscopy. Biochim Biophys Acta Biomembr 1798:1377–1391
178. do Canto AMTM, Robalo JR, Santos PD, Carvalho AJP, Ramalho JPP, Loura LMS (2016) Diphenylhexatriene membrane probes DPH and TMA-DPH: a comparative molecular dynamics simulation study. Biochim Biophys Acta Biomembr 1858:2647–2661
179. Lentz BR (1989) Membrane "fluidity" as detected by diphenylhexatriene probes. Chem Phys Lipids 50:171–190
180. Lentz BR (1993) Use of fluorescent probes to monitor molecular order and motions within liposome bilayers. Chem Phys Lipids 64:99–116
181. Sklar LA, Hudson BS, Simoni RD (1977) Conjugated polyene fatty acids as fluorescent probes: synthetic phospholipid membrane studies. Biochemistry 16:819–828
182. Nyholm TKM, Lindroos D, Westerlund B, Slotte JP (2011) Construction of a DOPC/PSM/cholesterol phase diagram based on the fluorescence properties of trans-parinaric acid. Langmuir 27:8339–8350
183. de Almeida RFM, Fedorov A, Prieto M (2003) Sphingomyelin/phosphatidylcholine/cholesterol phase diagram: boundaries and composition of lipid rafts. Biophys J 85:2406–2416
184. Haidekker MA, Theodorakis EA (2007) Molecular rotors – fluorescent biosensors for viscosity and flow. Org Biomol Chem 5:1669–1678
185. Kuimova MK (2012) Mapping viscosity in cells using molecular rotors. Phys Chem Chem Phys 14:12671–12686
186. Amaro M, Reina F, Hof M, Eggeling C, Sezgin E (2017) Laurdan and Di-4-ANEPPDHQ probe different properties of the membrane. J Phys D Appl Phys 50:134004
187. Shimomura O, Johnson FH, Saiga Y (1962) Extraction, purification and properties of aequorin, a bioluminescent protein from the luminous hydromedusan, Aequorea. J Cell Comp Physiol 59:223–239
188. Heim R, Prasher DC, Tsien RY (1994) Wavelength mutations and posttranslational autoxidation of green fluorescent protein. Proc Natl Acad Sci U S A 91:12501
189. Subach OM, Gundorov IS, Yoshimura M, Subach FV, Zhang J, Grüenwald D, Souslova EA, Chudakov DM, Verkhusha VV (2008) Conversion of red fluorescent protein into a bright blue probe. Chem Biol 15:1116–1124
190. Goedhart J, Von Stetten D, Noirclerc-Savoye M, Lelimousin M, Joosen L, Hink MA, Van Weeren L, Gadella TWJ, Royant A (2012) Structure-guided evolution of cyan fluorescent proteins towards a quantum yield of 93%. Nat Commun 3:751
191. Markwardt ML, Kremers G-J, Kraft CA, Ray K, Cranfill PJC, Wilson KA, Day RN, Wachter RM, Davidson MW, Rizzo MA (2011) An improved cerulean fluorescent protein with enhanced brightness and reduced reversible photoswitching. PLoS One 6:e17896
192. Cormack BP, Valdivia RH, Falkow S (1996) FACS-optimized mutants of the green fluorescent protein (GFP). Gene 173:33–38
193. Grotjohann T, Testa I, Leutenegger M, Bock H, Urban NT, Lavoie-Cardinal F, Willig KI, Eggeling C, Jakobs S, Hell SW (2011) Diffraction-unlimited all-optical imaging and writing with a photochromic GFP. Nature 478:204–208
194. Zhang X, Zhang M, Li D, He W, Peng J, Betzig E, Xu P (2016) Highly photostable, reversibly photoswitchable fluorescent protein with high contrast ratio for live-cell superresolution microscopy. Proc Natl Acad Sci U S A 113:10364–10369
195. Zapata-Hommer O, Griesbeck O (2003) Efficiently folding and circularly permuted variants of the sapphire mutant of GFP. BMC Biotechnol 3:5
196. Campbell BC, Nabel EM, Murdock MH, Lao-Peregrin C, Tsoulfas P, Blackmore MG, Lee FS, Liston C, Morishita H, Petsko GA (2020) mGreenLantern: a bright monomeric fluorescent protein with rapid expression and cell filling properties for neuronal imaging. Proc Natl Acad Sci U S A 117:30710–30721

197. Shaner NC, Lambert GG, Chammas A, Ni Y, Cranfill PJ, Baird MA, Sell BR, Allen JR, Day RN, Israelsson M, Davidson MW, Wang J (2013) A bright monomeric green fluorescent protein derived from Branchiostoma lanceolatum. Nat Methods 10:407–409
198. Patterson GH, Lippincott-Schwartz J (2002) A photoactivatable GFP for selective photolabeling of proteins and cells. Science 297:1873–1877
199. Bajar BT, Wang ES, Lam AJ, Kim BB, Jacobs CL, Howe ES, Davidson MW, Lin MZ, Chu J (2016) Improving brightness and photostability of green and red fluorescent proteins for live cell imaging and FRET reporting. Sci Rep 6:1–12
200. Stiel AC, Trowitzsch S, Weber G, Andresen M, Eggeling C, Hell SW, Jakobs S, Wahl MC (2007) 1.8 Å bright-state structure of the reversibly switchable fluorescent protein Dronpa guides the generation of fast switching variants. Biochem J 402:35–42
201. Ai HW, Hazelwood KL, Davidson MW, Campbell RE (2008) Fluorescent protein FRET pairs for ratiometric imaging of dual biosensors. Nat Methods 5:401–403
202. Ormö M, Cubitt AB, Kallio K, Gross LA, Tsien RY, Remington SJ (1996) Crystal structure of the Aequorea victoria green fluorescent protein. Science 273:1392–1395
203. Kremers GJ, Goedhart J, Van Munster EB, Gadella TWJ (2006) Cyan and yellow super fluorescent proteins with improved brightness, protein folding, and FRET förster radius. Biochemistry 45:6570–6580
204. Zacharias DA, Violin JD, Newton AC, Tsien RY (2002) Partitioning of lipid-modified monomeric GFPs into membrane microdomains of live cells. Science 296:913–916
205. Shaner NC, Campbell RE, Steinbach PA, Giepmans BNG, Palmer AE, Tsien RY (2004) Improved monomeric red, orange and yellow fluorescent proteins derived from Discosoma sp. red fluorescent protein. Nat Biotechnol 22:1567–1572
206. Matz MV, Fradkov AF, Labas YA, Savitsky AP, Zaraisky AG, Markelov ML, Lukyanov SA (1999) Fluorescent proteins from nonbioluminescent Anthozoa species. Nat Biotechnol 17: 969–973
207. Pletnev S, Subach FV, Dauter Z, Wlodawer A, Verkhusha VV (2012) A structural basis for reversible photoswitching of absorbance spectra in red fluorescent protein rsTagRFP. J Mol Biol 417:144–151
208. Shaner NC, Lin MZ, McKeown MR, Steinbach PA, Hazelwood KL, Davidson MW, Tsien RY (2008) Improving the photostability of bright monomeric orange and red fluorescent proteins. Nat Methods 5:545–551
209. Bindels DS, Haarbosch L, Van Weeren L, Postma M, Wiese KE, Mastop M, Aumonier S, Gotthard G, Royant A, Hink MA, Gadella TWJ (2016) MScarlet: a bright monomeric red fluorescent protein for cellular imaging. Nat Methods 14:53–56
210. Subach FV, Patterson GH, Manley S, Gillette JM, Lippincott-Schwartz J, Verkhusha VV (2009) Photoactivatable mCherry for high-resolution two-color fluorescence microscopy. Nat Methods 6:153–159
211. Subach FV, Patterson GH, Renz M, Lippincott-Schwartz J, Verkhusha VV (2010) Bright monomeric photoactivatable red fluorescent protein for two-color super-resolution sptPALM of live cells. J Am Chem Soc 132:6481–6491
212. Kogure T, Karasawa S, Araki T, Saito K, Kinjo M, Miyawaki A (2006) A fluorescent variant of a protein from the stony coral Montipora facilitates dual-color single-laser fluorescence cross-correlation spectroscopy. Nat Biotechnol 24:577–581
213. Gunewardene MS, Subach FV, Gould TJ, Penoncello GP, Gudheti MV, Verkhusha VV, Hess ST (2011) Superresolution imaging of multiple fluorescent proteins with highly overlapping emission spectra in living cells. Biophys J 101:1522–1528
214. Shcherbo D, Murphy CS, Ermakova GV, Solovieva EA, Chepurnykh TV, Shcheglov AS, Verkhusha VV, Pletnev VZ, Hazelwood KL, Roche PM, Lukyanov S, Zaraisky AG, Davidson MW, Chudakov DM (2009) Far-red fluorescent tags for protein imaging in living tissues. Biochem J 418:567–574
215. Lin MZ, McKeown MR, Ng HL, Aguilera TA, Shaner NC, Campbell RE, Adams SR, Gross LA, Ma W, Alber T, Tsien RY (2009) Autofluorescent proteins with excitation in the optical window for intravital imaging in mammals. Chem Biol 16:1169–1179

216. Shagin DA, Barsova EV, Yanushevich YG, Fradkov AF, Lukyanov KA, Labas YA, Semenova TN, Ugalde JA, Meyers A, Nunez JM, Widder EA, Lukyanov SA, Matz MV (2004) GFP-like proteins as ubiquitous metazoan superfamily: evolution of functional features and structural complexity. Mol Biol Evol 21:841–850
217. Deheyn DD, Kubokawa K, Mccarthy JK, Murakami A, Porrachia M, Rouse GW, Holland ND (2007) Endogenous green fluorescent protein (GFP) in amphioxus. Biol Bull 213:95–100
218. Jradi FM, Lavis LD (2019) Chemistry of photosensitive fuorophores for single-molecule localization microscopy. ACS Chem Biol 14:1077–1090
219. Santos EM, Sheng W, Esmatpour Salmani R, Tahmasebi Nick S, Ghanbarpour A, Gholami H, Vasileiou C, Geiger JH, Borhan B (2021) Design of large Stokes shift fluorescent proteins based on excited state proton transfer of an engineered photobase. J Am Chem Soc 143: 15091–15102
220. Zhao B, Ding W, Tan Z, Tang Q, Zhao K (2019) A large stokes shift fluorescent protein constructed from the fusion of red fluorescent mCherry and far-red fuorescent BDFP1.6. ChemBioChem 20:1167–1173
221. Piatkevich KD, Hulit J, Subach OM, Wu B, Abdulla A, Segall JE, Verkhusha VV (2010) Monomeric red fluorescent proteins with a large Stokes shift. Proc Natl Acad Sci 107:5369–5374
222. Piatkevich KD, Malashkevich VN, Morozova KS, Nemkovich NA, Almo SC, Verkhusha VV (2013) Extended stokes shift in fluorescent proteins: chromophore-protein interactions in a near-infrared TagRFP675 variant. Sci Rep 3:1–11
223. Khmelinskii A, Keller PJ, Bartosik A, Meurer M, Barry JD, Mardin BR, Kaufmann A, Trautmann S, Wachsmuth M, Pereira G, Huber W, Schiebel E, Knop M (2012) Tandem fluorescent protein timers for in vivo analysis of protein dynamics. Nat Biotechnol 30(7): 708–714
224. Shcherbo D, Merzlyak EM, Chepurnykh TV, Fradkov AF, Ermakova GV, Solovieva EA, Lukyanov KA, Bogdanova EA, Zaraisky AG, Lukyanov S, Chudakov DM (2007) Bright far-red fluorescent protein for whole-body imaging. Nat Methods 4:741–746
225. Baird GS, Zacharias DA, Tsien RY (2000) Biochemistry, mutagenesis, and oligomerization of DsRed, a red fluorescent protein from coral. Proc Natl Acad Sci 97:11984–11989
226. Zhang J, Campbell RE, Ting AY, Tsien RY (2002) Creating new fluorescent probes for cell biology. Nat Rev Mol Cell Biol 3:906–918
227. Zacharias DA (2002) Sticky caveats in an otherwise glowing report: oligomerizing fluorescent proteins and their use in cell biology. Sci STKE 2002(131):pe23
228. Tsien RY (1999) Rosy dawn for fluorescent proteins. Nat Biotechnol 17(10):956–957
229. Mishin AS, Belousov VV, Solntsev KM, Lukyanov KA (2015) Novel uses of fluorescent proteins. Curr Opin Chem Biol 27:1–9
230. Stepanenko OV, Stepanenko OV, Shcherbakova DM, Kuznetsova IM, Turoverov KK, Verkhusha VV (2011) Modern fluorescent proteins: from chromophore formation to novel intracellular applications. BioTechniques 51:313–327
231. Davidson MW, Campbell RE (2009) Engineered fluorescent proteins: innovations and applications. Nat Methods 6(10):713–717
232. Kim H, Ju J, Lee HN, Chun H, Seong J (2021) Genetically encoded biosensors based on fluorescent proteins. Sensors (Switzerland) 21:1–18
233. Burgstaller S, Bischof H, Gensch T, Stryeck S, Gottschalk B, Ramadani-Muja J, Eroglu E, Rost R, Balfanz S, Baumann A, Waldeck-Weiermair M, Hay JC, Madl T, Graier WF, Malli R (2019) PH-lemon, a fluorescent protein-based pH reporter for acidic compartments. ACS Sensors 4:883–891
234. Tutol JN, Peng W, Dodani SC (2019) Discovery and characterization of a naturally occurring, turn-on yellow fluorescent protein sensor for chloride. Biochemistry 58:31–35
235. Yu X, Strub MP, Barnard TJ, Noinaj N, Piszczek G, Buchanan SK, Taraska JW (2014) An engineered palette of metal ion quenchable fluorescent proteins. PLoS One 9:e95808

236. Hanson GT, Aggeler R, Oglesbee D, Cannon M, Capaldi RA, Tsien RY, Remington SJ (2004) Investigating mitochondrial redox potential with redox-sensitive green fluorescent protein indicators. J Biol Chem 279:13044–13053
237. Pak VV, Ezeriņa D, Lyublinskaya OG, Pedre B, Tyurin-Kuzmin PA, Mishina NM, Thauvin M, Young D, Wahni K, Martínez Gache SA, Demidovich AD, Ermakova YG, Maslova YD, Shokhina AG, Eroglu E, Bilan DS, Bogeski I, Michel T, Vriz S, Messens J, Belousov VV (2020) Ultrasensitive genetically encoded indicator for hydrogen peroxide identifies roles for the oxidant in cell migration and mitochondrial function. Cell Metab 31: 642–653.e6
238. Baker BJ, Mutoh H, Dimitrov D, Akemann W, Perron A, Iwamoto Y, Jin L, Cohen LB, Isacoff EY, Pieribone VA, Hughes T, Knöpfel T (2008) Genetically encoded fluorescent sensors of membrane potential. Brain Cell Biol 36:53–67
239. Hochreiter B, Garcia AP, Schmid JA (2015) Fluorescent proteins as genetically encoded FRET biosensors in life sciences. Sensors (Basel) 15:26281–26314
240. Miyawaki A, Llopis J, Heim R, Michael McCaffery J, Adams JA, Ikura M, Tsien RY (1997) Fluorescent indicators for Ca2+ based on green fluorescent proteins and calmodulin. Nature 388:882–887
241. Calamera G, Li D, Ulsund AH, Kim JJ, Neely OC, Moltzau LR, Bjørnerem M, Paterson D, Kim C, Levy FO, Andressen KW (2019) FRET-based cyclic GMP biosensors measure low cGMP concentrations in cardiomyocytes and neurons. Commun Biol 2(1):1–12
242. Maryu G, Matsuda M, Aoki K (2016) Multiplexed fluorescence imaging of ERK and akt activities and cell-cycle progression. Cell Struct Funct 41:81–92
243. Cranfill PJ, Sell BR, Baird MA, Allen JR, Lavagnino Z, De Gruiter HM, Kremers GJ, Davidson MW, Ustione A, Piston DW (2016) Quantitative assessment of fluorescent proteins. Nat Methods 13(7):557–562
244. Kremers GJ, Gilbert SG, Cranfill PJ, Davidson MW, Piston DW (2011) Fluorescent proteins at a glance. J Cell Sci 124:157–160
245. Kaishima M, Ishii J, Matsuno T, Fukuda N, Kondo A (2016) Expression of varied GFPs in Saccharomyces cerevisiae: codon optimization yields stronger than expected expression and fluorescence intensity. Sci Rep 6(1):1–15
246. Van Genechten W, Demuyser L, Dedecker P, Van Dijck P (2020) Presenting a codon-optimized palette of fluorescent proteins for use in Candida albicans. Sci Rep 10:1–9
247. Chen X, Zaro JL, Shen WC (2013) Fusion protein linkers: property, design and functionality. Adv Drug Deliv Rev 65:1357–1369
248. Li G, Huang Z, Zhang C, Dong BJ, Guo RH, Yue HW, Yan LT, Xing XH (2016) Construction of a linker library with widely controllable flexibility for fusion protein design. Appl Microbiol Biotechnol 100:215–225
249. Crivat G, Taraska JW (2012) Imaging proteins inside cells with fluorescent tags. Trends Biotechnol 30:8
250. Prescher JA, Bertozzi CR (2005) Chemistry in living systems. Nat Chem Biol 1(1):13–21
251. Ward WW, Prentice HJ, Roth AF, Cody CW, Reeves SC (1982) Spectral perturbations of the Aequorea green-fluorescent protein. Photochem Photobiol 35:803–808
252. Yang F, Moss LG, Phillips GN (1996) The molecular structure of green fluorescent protein. Nat Biotechnol 14(10):1246–1251
253. Swenson ES, Price JG, Brazelton T, Krause DS (2007) Limitations of green fluorescent protein as a cell lineage marker. Stem Cells 25:2593–2600
254. Wiedenmann J, Oswald F, Nienhaus GU (2009) Fluorescent proteins for live cell imaging: opportunities, limitations, and challenges. IUBMB Life 61:1029–1042
255. Ha T, Tinnefeld P (2012) Photophysics of fluorescent probes for single-molecule biophysics and super-resolution imaging. Annu Rev Phys Chem 63:595–617
256. Yang Z, Samanta S, Yan W, Yu B, Qu J (2021) Super-resolution microscopy for biological imaging. In: Optical imaging in human disease and biological research, pp 23–43

257. Schermelleh L, Ferrand A, Huser T, Eggeling C, Sauer M, Biehlmaier O, Drummen GPC (2019) Super-resolution microscopy demystified. Nat Cell Biol 21(1):72–84
258. Valli J, Garcia-Burgos A, Rooney LM, de Melo e Oliveira BV, Duncan RR, Rickman C (2021) Seeing beyond the limit: a guide to choosing the right super-resolution microscopy technique. J Biol Chem 297(1)
259. Lelek M, Gyparaki MT, Beliu G, Schueder F, Griffié J, Manley S, Jungmann R, Sauer M, Lakadamyali M, Zimmer C (2021) Single-molecule localization microscopy. Nat Rev Methods Prim 1(1):1–27
260. Dempsey GT, Vaughan JC, Chen KH, Bates M, Zhuang X (2011) Evaluation of fluorophores for optimal performance in localization-based super-resolution imaging. Nat Methods 8:1027–1040
261. Bates M, Huang B, Zhuang X (2008) Super-resolution microscopy by nanoscale localization of photo-switchable fluorescent probes. Curr Opin Chem Biol 12:505–514
262. Minoshima M, Kikuchi K (2017) Photostable and photoswitching fluorescent dyes for super-resolution imaging. J Biol Inorg Chem 22:639–652
263. Wang S, Moffitt JR, Dempsey GT, Xie XS, Zhuang X (2014) Characterization and development of photoactivatable fluorescent proteins for single-molecule-based superresolution imaging. Proc Natl Acad Sci U S A 111:8452–8457
264. Wu Z, Xu X, Xi P (2021) Stimulated emission depletion microscopy for biological imaging in four dimensions: a review. Microsc Res Tech 84:1947–1958
265. Sednev MV, Belov VN, Hell SW (2015) Fluorescent dyes with large Stokes shifts for super-resolution optical microscopy of biological objects: a review. Methods Appl Fluoresc 3(4): 042004
266. Jeong S, Widengren J, Lee JC (2022) Fluorescent probes for sted optical nanoscopy. Nanomaterials 12:21
267. Yang X, Yang Z, Wu Z, He Y, Shan C, Chai P, Ma C, Tian M, Teng J, Jin D, Yan W, Das P, Qu J, Xi P (2020) Mitochondrial dynamics quantitatively revealed by STED nanoscopy with an enhanced squaraine variant probe. Nat Commun 11(1):1–9
268. Lukinavičius G, Reymond L, D'Este E, Masharina A, Göttfert F, Ta H, Güther A, Fournier M, Rizzo S, Waldmann H, Blaukopf C, Sommer C, Gerlich DW, Arndt HD, Hell SW, Johnsson K (2014) Fluorogenic probes for live-cell imaging of the cytoskeleton. Nat Methods 11:731–733
269. Xu Y, Xu R, Wang Z, Zhou Y, Shen Q, Ji W, Dang D, Meng L, Tang BZ (2021) Recent advances in luminescent materials for super-resolution imaging via stimulated emission depletion nanoscopy. Chem Soc Rev 50:667–690
270. Sreedharan S, Gill MR, Garcia E, Saeed HK, Robinson D, Byrne A, Cadby A, Keyes TE, Smythe C, Pellett P, Bernardino De La Serna J, Thomas JA (2017) Multimodal super-resolution optical microscopy using a transition-metal-based probe provides unprecedented capabilities for imaging both nuclear chromatin and mitochondria. J Am Chem Soc 139: 15907–15913
271. Butkevich AN, Lukinavičius G, D'Este E, Hell SW (2017) Cell-permeant large Stokes shift dyes for transfection-free multicolor nanoscopy. J Am Chem Soc 139:12378–12381
272. Nienhaus K, Ulrich Nienhaus G (2014) Fluorescent proteins for live-cell imaging with super-resolution. Chem Soc Rev 43:1088–1106
273. Feld LG, Shynkarenko Y, Krieg F, Rainò G, Kovalenko MV (2021) Perovskite quantum dots for super-resolution optical microscopy: where strong photoluminescence blinking matters. Adv Opt Mater 9:2100620
274. Jin D, Xi P, Wang B, Zhang L, Enderlein J, Van Oijen AM (2018) Nanoparticles for super-resolution microscopy and single-molecule tracking. Nat Methods 15:415–423
275. Sharonov A, Hochstrasser RM (2006) Wide-field subdiffraction imaging by accumulated binding of diffusing probes. Proc Natl Acad Sci 103:18911–18916
276. Schwering M, Kiel A, Kurz A, Lymperopoulos K, Sprödefeld A, Krämer R, Herten DP (2011) Far-field nanoscopy with reversible chemical reactions. Angew Chem Int Ed 50:2940–2945

277. Gurskaya NG, Verkhusha VV, Shcheglov AS, Staroverov DB, Chepurnykh TV, Fradkov AF, Lukyanov S, Lukyanov KA (2006) Engineering of a monomeric green-to-red photoactivatable fluorescent protein induced by blue light. Nat Biotechnol 24:461–465
278. Ando R, Hama H, Yamamoto-Hino M, Mizuno H, Miyawaki A (2002) An optical marker based on the UV-induced green-to-red photoconversion of a fluorescent protein. Proc Natl Acad Sci U S A 99:12651–12656
279. Hoi H, Shaner NC, Davidson MW, Cairo CW, Wang J, Campbell RE (2010) A monomeric photoconvertible fluorescent protein for imaging of dynamic protein localization. J Mol Biol 401:776–791
280. Zhang M, Chang H, Zhang Y, Yu J, Wu L, Ji W, Chen J, Liu B, Lu J, Liu Y, Zhang J, Xu P, Xu T (2012) Rational design of true monomeric and bright photoactivatable fluorescent proteins. Nat Methods 9:727–729
281. Fuchs J, Böhme S, Oswald F, Hedde PN, Krause M, Wiedenmann J, Nienhaus GU (2010) A photoactivatable marker protein for pulse-chase imaging with superresolution. Nat Methods 7: 627–630
282. Adam V, Moeyaert B, David CC, Mizuno H, Lelimousin M, Dedecker P, Ando R, Miyawaki A, Michiels J, Engelborghs Y, Hofkens J (2011) Rational design of photoconvertible and biphotochromic fluorescent proteins for advanced microscopy applications. Chem Biol 18:1241–1251
283. Xia J, Kim SHH, Macmilllan S, Truant R (2006) Practical three color live cell imaging by widefield microscopy. Biol Proced Online 8:63–68
284. Subach OM, Patterson GH, Ting LM, Wang Y, Condeelis JS, Verkhusha VV (2011) A photoswitchable orange-to-far-red fluorescent protein, PSmOrange. Nat Methods 8:771–780
285. Chozinski TJ, Gagnon LA, Vaughan JC, Puchner EM, Huang B, Gaub HE, Just W (2014) Twinkle, twinkle little star: photoswitchable fluorophores for super-resolution imaging. FEBS Lett 588:3603–3612
286. Rust MJ, Bates M, Zhuang X (2006) Stochastic optical reconstruction microscopy (STORM) provides sub-diffraction-limit image resolution. Nat Methods 3:793
287. Bates M, Huang B, Dempsey GT, Zhuang X (2007) Multicolor super-resolution imaging with photo-switchable fluorescent probes. Science 317:1749
288. Tam J, Cordier GA, Borbely JS, Álvarez ÁS, Lakadamyali M (2014) Cross-talk-free multicolor storm imaging using a single fluorophore. PLoS One 9:e101772
289. Heilemann M, Margeat E, Kasper R, Sauer M, Tinnefeld P (2005) Carbocyanine dyes as efficient reversible single-molecule optical switch. J Am Chem Soc 127:3801–3806
290. Vaughan JC, Dempsey GT, Sun E, Zhuang X (2013) Phosphine quenching of cyanine dyes as a versatile tool for fluorescence microscopy. J Am Chem Soc 135:1197–1200
291. Vaughan JC, Jia S, Zhuang X (2012) Ultrabright photoactivatable fluorophores created by reductive caging. Nat Methods 9:1181–1184
292. Lehmann M, Gottschalk B, Puchkov D, Schmieder P, Schwagerus S, Hackenberger CPR, Haucke V, Schmoranzer J (2015) Multicolor caged dSTORM resolves the ultrastructure of synaptic vesicles in the brain. Angew Chem Int Ed 54:13230–13235
293. Goossen-Schmidt NC, Schnieder M, Hüve J, Klingauf J (2020) Switching behaviour of dSTORM dyes in glycerol-containing buffer. Sci Rep 10(1):1–8
294. Uno SN, Kamiya M, Yoshihara T, Sugawara K, Okabe K, Tarhan MC, Fujita H, Funatsu T, Okada Y, Tobita S, Urano Y (2014) A spontaneously blinking fluorophore based on intramolecular spirocyclization for live-cell super-resolution imaging. Nat Chem 6:681–689
295. Uno SN, Kamiya M, Morozumi A, Urano Y (2017) A green-light-emitting, spontaneously blinking fluorophore based on intramolecular spirocyclization for dual-colour super-resolution imaging. Chem Commun 54:102–105
296. Morozumi A, Kamiya M, Uno SN, Umezawa K, Kojima R, Yoshihara T, Tobita S, Urano Y (2020) Spontaneously blinking fluorophores based on nucleophilic addition/dissociation of intracellular glutathione for live-cell super-resolution imaging. J Am Chem Soc 142: 9625–9633

297. Thiel Z, Rivera-Fuentes P (2018) Photochemically active dyes for super-resolution microscopy. Chimia (Aarau) 72:764–770
298. Li B, Haris U, Aljowni M, Nakatsuka A, Patel SK, Lippert AR (2021) Tuning the photophysical properties of spirolactam rhodamine photoswitches. Isr J Chem 61:244–252
299. Fölling J, Belov V, Kunetsky R, Medda R, Schönle A, Egner A, Eggeling C, Bossi M, Hell SW (2007) Photochromic rhodamines provide nanoscopy with optical sectioning. Angew Chem Int Ed Engl 46:6266–6270
300. Bossi M, Föiling J, Belov VN, Boyarskiy VP, Medda R, Egner A, Eggeling C, Schönle A, Hell SW (2008) Multicolor far-field fluorescence nanoscopy through isolated detection of distinct molecular species. Nano Lett 8:2463–2468
301. Belov VN, Bossi ML, Fölling J, Boyarskiy VP, Hell SW (2009) Rhodamine spiroamides for multicolor single-molecule switching fluorescent nanoscopy. Chem – A Eur J 15:10762–10776
302. Uno K, Aktalay A, Bossi ML, Irie M, Belov VN, Hell SW (2021) Turn-on mode diarylethenes for bioconjugation and fluorescence microscopy of cellular structures. Proc Natl Acad Sci U S A 118:2100165118
303. Hauke S, Von Appen A, Quidwai T, Ries J, Wombacher R (2016) Specific protein labeling with caged fluorophores for dual-color imaging and super-resolution microscopy in living cells. Chem Sci 8:559–566
304. Jang S, Kim M, Shim SH (2020) Reductively caged, photoactivatable DNA-PAINT for high-throughput super-resolution microscopy. Angew Chem Int Ed 59:11758–11762
305. Klán P, Šolomek T, Bochet CG, Blanc A, Givens R, Rubina M, Popik V, Kostikov A, Wirz J (2013) Photoremovable protecting groups in chemistry and biology: reaction mechanisms and efficacy. Chem Rev 113:119–191
306. Banala S, Maurel D, Manley S, Johnsson K (2012) A caged, localizable rhodamine derivative for superresolution microscopy. ACS Chem Biol 7:289–293
307. Butkevich AN, Weber M, Cereceda Delgado AR, Ostersehlt LM, D'Este E, Hell SW (2021) Photoactivatable fluorescent dyes with hydrophilic caging groups and their use in multicolor nanoscopy. J Am Chem Soc 143:18388–18393
308. Betzig E, Patterson GH, Sougrat R, Lindwasser OW, Olenych S, Bonifacino JS, Davidson MW, Lippincott-Schwartz J, Hess HF (2006) Imaging intracellular fluorescent proteins at nanometer resolution. Science 313:1642–1645
309. Hess ST, Girirajan TPK, Mason MD (2006) Ultra-high resolution imaging by fluorescence photoactivation localization microscopy. Biophys J 91:4258–4272
310. Bates M, Blosser TR, Zhuang X (2005) Short-range spectroscopic ruler based on a single-molecule optical switch. Phys Rev Lett 94:108101
311. Shroff H, Galbraith CG, Galbraith JA, White H, Gillette J, Olenych S, Davidson MW, Betzig E (2007) Dual-color superresolution imaging of genetically expressed probes within individual adhesion complexes. Proc Natl Acad Sci 104:20308–20313
312. Li D, Shao L, Chen BC, Zhang X, Zhang M, Moses B, Milkie DE, Beach JR, Hammer JA, Pasham M, Kirchhausen T, Baird MA, Davidson MW, Xu P, Betzig E (2015) Extended-resolution structured illumination imaging of endocytic and cytoskeletal dynamics. Science 349(6251):aab3500
313. Hofmann M, Eggeling C, Jakobs S, Hell SW (2005) Breaking the diffraction barrier in fluorescence microscopy at low light intensities by using reversibly photoswitchable proteins. Proc Natl Acad Sci U S A 102:17565–17569

Fluorescence Kinetics and Time-Resolved Measurement

Vlastimil Fidler and Peter Kapusta

Contents

Abstract The goal of this contribution is threefold:

1. Highlight the richness of fluorescence observables and parameters that enable fluorescent molecules (or biomacromolecule's fluorescent groups) to be important reporters for biological studies
2. Provide a brief overview of current instrumental techniques and methods available for fluorescence spectroscopy studies from steady-state versions to high time resolution in bulk as well as in single-molecule studies

V. Fidler
Brown University, Providence, RI, USA

Faculty of Biomedical Engineering CTU in Prague, Prague, Czech Republic
e-mail: vlastimil_fidler@brown.edu

P. Kapusta (✉)
J. Heyrovsky Institute of Physical Chemistry of the Czech Academy of Sciences, Prague, Czech Republic
e-mail: kapusta@jh-inst.cas.cz

R. Šachl, M. Amaro (eds.), *Fluorescence Spectroscopy and Microscopy in Biology*, Springer Ser Fluoresc (2023) 20: 53–86, https://doi.org/10.1007/4243_2022_31,
Published online: 25 November 2022

3. Based on the previous knowledge (1) and (2), to provide a "helping hand" to molecular biologists and biochemists when they are to decide about fluorescence spectroscopy application for a given research task

Keywords Anisotropy · Fluorescence decay · Nonlinear time-resolved techniques · Photon counting and timing · Polarization · Time-resolved fluorescence

1 Basics and a Bit of History

A mysterious light emitted by some materials – solid or soft matter, inorganic or organic, biological tissue, or living organisms – has been known and attracting attention since ancient times. Already Aristotle knew about the luminescence of dead fish and flesh as well as that of fungi. Emanation of light is described in Chinese, Japanese, and Indian mythologies and ancient histories. Natural examples of light-emitting glow worms, luminescent wood, and even live creatures such as fire-flies were also known already then.

We could consider the year 1603 to be the beginning of modern luminescent material study, though unintended. In that year, the alchemist Vincenzo Cascariolo ground natural mineral barite ($BaSO_4$) found near Bologna and heated it under reducing conditions in the hope of obtaining gold. Instead, he noticed persistent luminescence of this material, which has since then been called Bolognian stone. It is not known which dopants were responsible for its luminescence, but Bologna stone became the first scientifically documented material exhibiting a persistent luminescence. In 1640, Fortunius Licetus published *Litheosphorus Sive de Lapide Bononiensi*, devoted to Bologna stone luminescence.

The term "luminescence" (as opposed to incandescence) was first used in 1888 by Eilhardt Wiedemann who classified six kinds of it according to the method of excitation. This classification – photoluminescence, electroluminescence, thermoluminescence, chemiluminescence, triboluminescence, and crystalloluminescence – has been used to this day.

In 1845, Fredrick W. Herschel discovered that UV light can excite a quinine solution (e.g., tonic water) to emit blue light. This kind of luminescence came to be known as fluorescence.

A comprehensive history of luminescence is described, e.g., in *A History of Luminescence from the Earliest Times Until 1900* by E. Newton Harvey [1].

1.1 Nobel Prizes Related to Fluorescence

Over the last 100 years, fluorescence has gained in importance and has become an important tool in biophysics. This is evidenced by Nobel prizes awarded for research related to fluorescence. First of all, there was an effort to increase time resolution in order to follow the dynamics of chemical processes in real time. For their fundamental studies "of extremely fast chemical reactions, effected by disturbing the equilibrium by means of very short pulses of energy," Manfred Eigen, Ronald George Wreyford Norrish, and George Porter were awarded the Nobel Prize in Chemistry in 1967 [2–4].

They established nanosecond flash photolysis as a very effective way to study intermediates of fast photo-initiated processes (then on the ns time scale). Soon afterward – in 1974 – G. Porter, E.S. Reid, and C. J. Tredwell implemented mode-locked Nd^{3+} glass laser with a CS_2 cell for optical sampling and published fluorescence lifetimes of fluorescein (3.6 ns), eosin (900 ps), and erythrosine (110 ps) – the first such reproducible experimental data on the picosecond (ps) time scale [5].

This breakthrough into the ps region provided a method to follow faster and faster photo-initiated processes in real time. New femtosecond (fs) non/linear spectroscopic techniques were developed, e.g., by Graham Fleming and his group, in order to elucidate key features of natural photosynthesis. In 1999, a Nobel Prize for Chemistry was awarded to Ahmed H. Zewail "for his studies of the transition states of chemical reactions using femtosecond spectroscopy" [6].

The synergistic effects of new instrumental laser techniques yielding new, formerly inaccessible, experimental data and the development of quantum theory calculations launched a new field of femto-chemistry and molecular biology, where a molecule in a new role – such as a fluorophore acting as reporter in a biomolecular superstructure – could be fully utilized. Subsequently, a Nobel Prize for Chemistry was awarded to Roger Y. Tsien, Osamu Shimomura, and Martin Chalfie "for the discovery and development of the green fluorescent protein" in 2008 [7].

When the study of fs processes in real time – the "ultimate time scale" for chemistry – became feasible, new research fields emerged: single-molecule study and spectroscopic techniques with high spatial resolution – i.e., combination of advanced time-resolved (TR) spectroscopy with microscopy. Once again, a Nobel Prize for Chemistry was awarded for significant advancement in this new field to Eric Betzig, Stefan W. Hell, and William E. Moerner "for the development of super-resolved fluorescence microscopy" in 2014 [8].

1.2 Fluorescence as a Tool of Growing Importance

1.2.1 Jabłoński Diagram and Stokes Shift

Alexander Jabłoński used the Franck-Condon principle to explain the main features of luminescence from molecular liquid solutions. A simple diagram he developed in 1933 makes it possible to explain the main spectral and kinetic features of this emission. Subsequently, Jabłoński, together with Kasha, Lewis, and Terenin, proved a singlet and triplet character of the energy levels involved in the process of

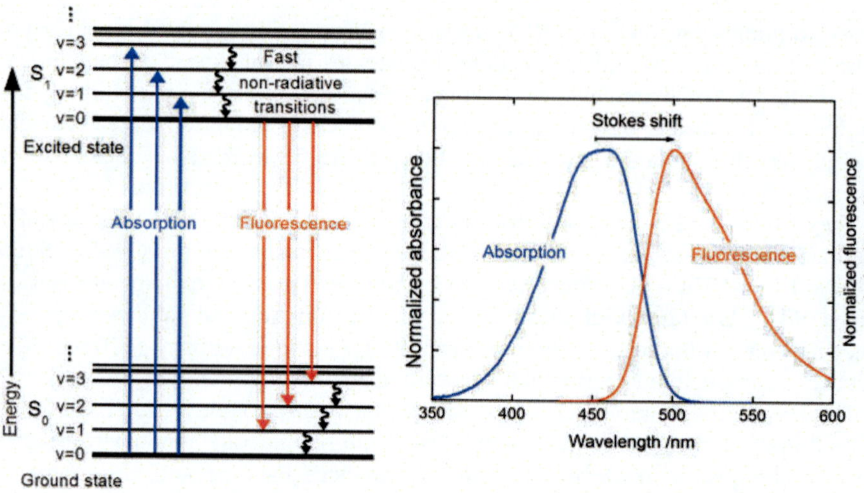

Fig. 1 Jabłoński diagram and its relation to typical absorption and emission spectra of fluorophores. S_0 is the lowest electronic state (ground state) with vibrational levels $v = 0, 1, 2, \ldots$ S_1 is the lowest singlet excited electronic state

emission, and the specific terms fluorescence, phosphorescence, and delayed fluorescence were established (Fig. 1).

This textbook-style energy level diagram illustrates the most typical photophysical processes during excitation and de-excitation of fluorophores utilized in biology. At normal laboratory (or physiological) temperature in a biological sample, practically all fluorophores are in their vibrationally relaxed ($v = 0$), lowest electronic ground state, S_0. Absorption of a photon carrying enough energy to transition the molecule to a higher electronically excited state causes vibrational excitation, too. The excess vibrational energy is dissipated due to very fast vibrational relaxation, typically by molecular collisions. (These are non-radiative transitions.) By emission of a photon (i.e., fluorescence), the molecule loses most of the absorbed energy and returns to a vibrationally excited level of its electronic ground state. Again, fast vibrational relaxation follows, re-forming the vibrationally relaxed, electronic ground state. However, in fluorescence spectroscopy, the observables are the energy of absorbed and emitted photons, more precisely, the distribution of those energies: transition probabilities of absorbing and emitting photons with certain energies. In a Jabłoński diagram, the lengths of straight (non-wavy) arrows are proportional to the absorbed or emitted photon energy. Thus, a typical fluorescence emission spectrum is red-shifted (shifted to lower energies) in comparison to the absorption spectrum. The resulting spectral shift, called the Stokes shift, is one of the important practical consequences of non-radiative relaxations involved in this sequence of events. Vibrational relaxation is not the only source of Stokes shift. In fluid media, solvent relaxation around an excited state (in general, relaxation of its surrounding molecular environment) can lower its energy and contributes significantly to Stokes shifts [9, 10].

1.2.2 Fluorescence as a Reporter of Molecular Properties and Interactions

The fluorescence of a molecule is the light emitted spontaneously due to transition from an excited singlet state (usually S_1) to various vibrational levels of the electronic ground state, i.e., $S_{1,0} \rightarrow S_{0,\nu}$. For many years, just the main parameters of fluorescence, i.e., the spectral distribution of the emission intensity (i.e., emission spectrum) and fluorescence quantum yield, were studied together with study of symmetry relating to the excitation and emission spectra and Stokes shift. This line of research was later followed by studies of fluorescence lifetimes.

Initially fluorescence spectroscopy was used for systematic study of organic molecules and biomolecules with the aim to obtain new information about the molecule itself – e.g., energies of its electronic states and transition rate constants. The rapid development of instrumental techniques – both steady-state and time-resolved spectrofluorometers – helped advance the field to study interactions between the fluorescent molecule and its molecular surroundings and on the ways these interactions influence many observable properties of fluorescence. In such studies, the fluorescent molecule reports on its environment, rather than being the main target of study. This shift of emphasis promoted increased attention to the parameters of excitation and of emitted fluorescence that could carry any information on the interactions between a fluorescent molecule and any kind of molecular surroundings – whether it was a solvent or other subunits of a complex biomolecular structure where the fluorescent subunit constitutes just its part.

1.3 Basic Characteristics of Excitation and Emission

Current advanced experimental techniques allow control and/or analysis of many parameters of the excitation and the emission during a fluorescence study. The information that is sought determines which parameters need to be controlled in a particular study. Currently, the following experimental parameters and modes can be investigated.

Excitation
Intensity I_{exc} at the given wavelength
Excitation spectral profile $I_{exc} = I_{exc}(\lambda)$
Polarization (often linear, sometimes circular, sometimes undefined)
CW or pulsed, constant or modulated
Pulse time profile and pulse repetition rate
Single-photon excitation (this is the most common modality)
Two-photon excitation and entangled two-photon excitation
Excitation by structured multi-mode light
Excitation beam geometry, mode structure

(continued)

Excited sample volume – bulk (e.g., in a cuvette) or confined to a diffraction-limited spot (volume)
Fixed position or scanning
Fluorescence
For a fixed excitation intensity and excitation wavelength:
Total fluorescence intensity within a given emission wavelength region
Total anisotropy
Emission spectrum: $I_{em} = I_{em}(\lambda)$
Emission polarization or anisotropy spectrum
For variable excitation wavelength at constant excitation intensity
Excitation spectrum at λ_{em} = constant; emission intensity is a function of excitation wavelength, $I_{em} = I_{em}(\lambda_{exc})$
Fluorescence quantum yield
Excitation polarization or anisotropy spectrum
Under pulsed or modulated excitation
Fluorescence intensity decay
Fluorescence anisotropy decay (depolarization)
Various fluorescence decay parameters, e.g., multi-exponential, energy transfer model
With spatial resolution
Emission intensity map
Emission spectral map
Intensity decay (lifetime) map, FLIM

Just looking at the above list of parameters to be taken into account when a fluorescence measurement is considered, it is clear that the instrumentation required could be either simple or very *sophisticated* – depending on which fluorescence parameter is expected to provide the new information we seek. Today, the fluorescent molecule (or a fluorescent subunit of a molecular structure) is mostly used as a reporter of intra- and/or inter-molecular interactions which means that the fluorescence kinetic parameters are typically the most important. Consequently, sophisticated instrumental setups with high-time resolution are proliferating even to traditional biochemistry and molecular biology laboratories.

1.4 Kinetics of Fluorescence

The most important molecular fluorescence feature (parameter) – after steady-state fluorescence excitation and emission spectra were measured – is the fluorescence decay kinetics. In the simplest case, we assume that excited state relaxation is governed by mutually independent, spontaneous processes and each of them can be characterized by a pseudo-first-order rate constant *k*. The following simplified Jabłoński diagram (Fig. 2) labels individual relaxation processes with relevant rate constants.

Fig. 2 Rate constants in Jabłoński diagram. k_F – fluorescence rate constant, k_{NR} – sum of various non-radiative decay processes, typically internal conversion and intersystem crossing, k_Q – quenching rate constant, accounting for unavoidable quenching mechanisms induced by the molecular environment of the fluorophore

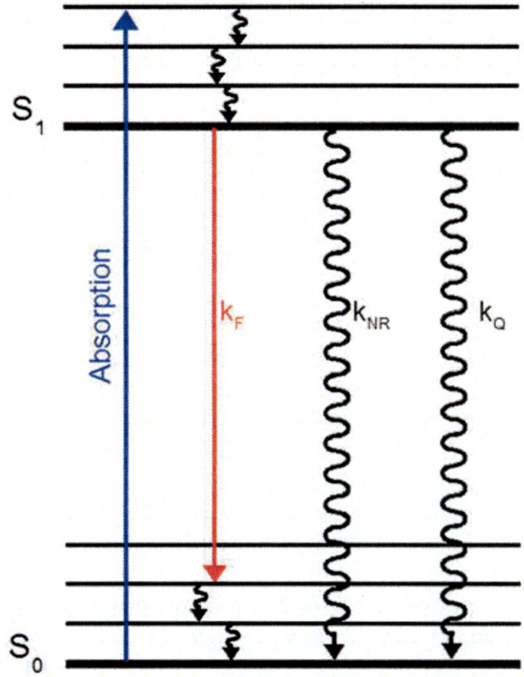

If the assumption of spontaneity holds, then the fluorescence decay is exponential and the one parameter to be extracted from such a measurement is the rate constant of this decay. In practice, this means measurement of the fluorescence lifetime.

Spontaneous fluorescence decay is an example of a first-order process. A fundamental molecular parameter, unique for each fluorophore, is the fluorescence (radiative) rate constant, k_F. If fluorescence were the only process depopulating the relaxed excited S_1 state, then the measured excited state lifetime would be simply $\tau_F = 1/k_F$ and the intensity decay *I(t)* would be described by a simple exponential function of form:

$$I(t) = I(0) \cdot e^{-k_F/t} = I(0) \cdot e^{-t/\tau_F}$$

where *I*(0) means the initial intensity. In this particular case, τ_F would be called the radiative lifetime.

However, as the Jabłoński diagram in Fig. 2 shows, non-radiative processes compete with fluorescence to cause decay of the excited singlet state population. These parallel running processes depopulating the excited state produce an overall decay rate that is the sum of all active rate constants. Let us denote this sum by:

$$k_{\mathrm{Decay}} = k_F + k_{\mathrm{NR}} + k_Q$$

The observable lifetime is then:

$$\tau_{\mathrm{Decay}} = 1/(k_F + k_{\mathrm{NR}} + k_Q)$$

and the experimentally observable decay curve takes the following form:

$$I(t) = I(0) \cdot e^{-k_{\mathrm{Decay}}/t} = I(0) \cdot e^{-t/\tau_{\mathrm{Decay}}}$$

It is important to realize that only the *radiative* rate constant, k_F (and thus τ_F), is unique to a particular fluorophore, and it is impossible to determine this fundamental molecular parameter directly. In fluorescence decay curve measurements, also known as lifetime measurements, τ_{Decay} is the observable parameter. Both k_{NR} and k_Q usually depend on the molecular environment of the dissolved fluorescent dye. For example, solvent polarity, pH, dissolved oxygen content, presence of ions, and ability of hydrogen bonding can strongly affect τ_{Decay}. This is the reason why decay time of a fluorophore should be specified together with the relevant environment.

Consider the case of Fluorescein, a widely used fluorophore. A single-exponential decay with a lifetime of 4.1 ns can be observed from Fluorescein only when dissolved in aqueous solution with pH larger than 10. Under these conditions, Fluorescein, a dicarboxylic acid, is completely dissociated, and it is in di-anionic form. When this dye is dissolved in slightly acidic buffer, the observed decay curve is an undefined mixture of emissions from three different forms of Fluorescein. Although the sensitivity of decay rate to the environment sometimes complicates analyses, it also enables probing of a fluorescent molecule's microscopic environment in a biological sample.

The Jabłoński diagram is also useful for understanding the concept of fluorescence quantum yield, Q_F. The definition of Q_F is intuitive: it is the ratio of the number of emitted photons to the number of absorbed excitation photons. If fluorescence were the only process depopulating the excited S_1 state, then all excitation photons would be converted to fluorescence and Q_F would equal 1. Owing to the competing non-radiative decay processes mentioned above, a fraction of the excited state population returns to ground state via alternative pathways, without photon emission. Observed quantum yields are often much less than 1. Using the corresponding rate constants, Q_F can be expressed as:

$$Q_F = k_F/(k_F + k_{\mathrm{NR}} + k_Q) = \tau_{\mathrm{Decay}}/\tau_F$$

The significance of the last equation becomes obvious after rearranging in the following way:

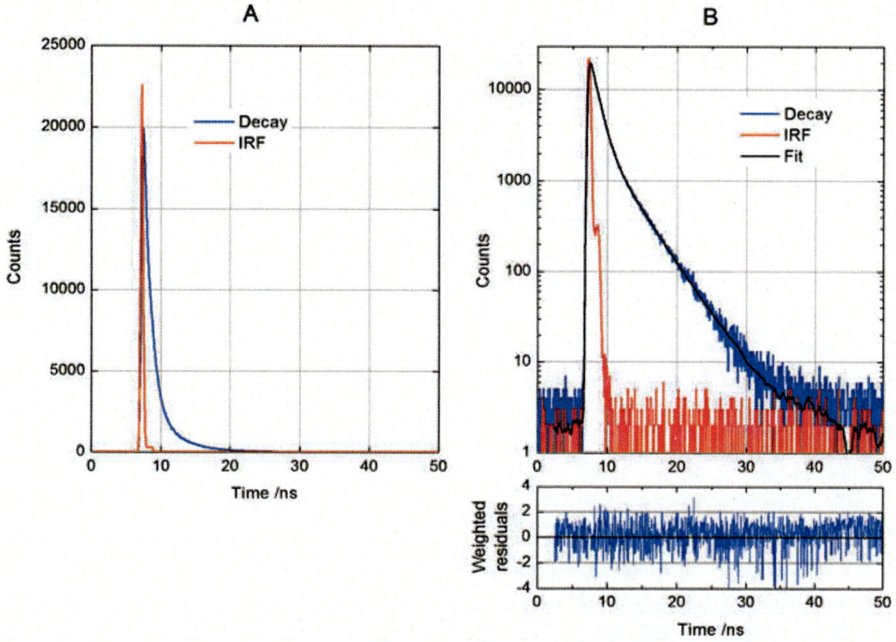

Fig. 3 An example decay and instrument response function (IRF) obtained by time-correlated single-photon counting (TCSPC) method. (**a**) Plotted on a linear intensity scale. (**b**) Plotted on a logarithmic intensity scale, also showing the fitted theoretical model decay function. The plot of weighted residuals at the bottom of the graph shows the deviation of experimental data from the fitted model decay

$$1/\tau_F = k_F = Q_F/\tau_{\text{Decay}}$$

and realizing that Q_F and τ_{Decay} are experimental observables. One must measure both quantities in order to calculate the fundamental fluorescence rate constant of the dye, k_F.

So far we assumed the simplest possible case, a mono-exponential decay that is initiated with an infinitely short light pulse. However, the technical reality of time-domain lifetime measurements is different. Laser pulses have a finite duration. This means the time zero, "the moment of excitation," is not defined precisely. Proper data analysis has to take into account the duration and time profile of the excitation as well, since it is modifying the decay curve, especially at the very beginning of the decay. An instrument's electronic components also have finite response times, and their overall effect has to be accounted for.

Figure 3 shows a real example: a measurement of fluorescence decay from a mixture of Oxazin1 and Oxazin4 fluorophores (two red emitting laser dyes) dissolved in ethanol, collected using a time-correlated single-photon counting (TCSPC) instrument. The excitation source was a laser diode emitting 635 nm (red) pulses with approximately 80 ps duration. Fluorescence with $\lambda > 650$ nm has been collected. In Fig. 3a, the data are plotted using a linear intensity scale. At a first sight, the decay curve (plotted with blue color) indeed resembles a standard exponential decay function. The red curve is the so-called instrument response function (IRF hereafter), which was obtained by measuring scattered laser photons using the same instrument shortly after the fluorescence decay measurement. The IRF indeed looks like a sharp, short pulse.

Figure 3b shows the same data, now plotted using a logarithmic intensity scale. This is the proper way to present and analyze TCSPC decay measurements. The logarithmic plot reveals many useful details. The decay curve is obviously not single exponential. If it were, then the fluorescence plot would be a straight line, whereas here we see that the blue curve bends. This is consistent with the expected double-exponential decay behavior. Oxazin1 and Oxazin4 dissolved in ethanol have lifetimes of 1 and 3.5 ns, respectively. We also clearly see a shoulder in the IRF (red curve). The excitation pulse shape is not symmetric. Both the IRF and decay curve have a small background (offset) due to unavoidable noise, so-called detector dark counts, collected during data acquisition. Although the laser pulses had approximately 80 ps optical duration, the collected IRF is considerably broader than that. It has more than 300 ps width at half of the peak value. This broadening of the acquired result is caused by the detector and instrument electronics. The fluorescence decay curve is subject to the same broadening. This is the reason we need both the fluorescence and IRF curves for the data analysis to evaluate the decay parameters of the fitted model. The measured fluorescence decay curve is analyzed by assuming it is a convolution of a theoretical decay function (e.g., single or double exponential) with the shape of the IRF. Our goal is to obtain the parameters of the theoretical decay function. The most common data analysis method for TCSPC data is called iterative re-convolution. This means generating decay curves as a convolution of the IRF with a selected model of the fluorescence decay. The result is then compared to the experimentally obtained decay data. The adjustable model parameters are iterated until the calculated decay matches the experimental one.

Figure 3b shows such a re-convoluted, fitted decay model, plotted in black. The bottom panel of Fig. 3b is the plot of weighted residuals. Its purpose is to show the deviation of experimental data points from the fitted theoretical decay curve. "Good fit" results in a plot of residuals that are randomly distributed around zero, with no obvious trends in the deviations.

But how do we know which decay model is appropriate to fit to the data? For an unknown sample, the appropriate decay model is also unknown. In the above example of a binary mixture of dyes, where each features a single-exponential decay, the choice of a double-exponential function was easy and logical. Observing anything else would be a surprise, demanding an explanation.

It is common to start with the simplest possible model, the mono-exponential decay. If the model does not fit, add – step by step – one or more exponential terms. However, the model has to be kept as simple as possible. Practically any decay curve can be very well described by using just five-exponential terms, but this does not necessarily mean that this a correct decay model. A decay curve can be non-exponential, for example, and there are also cases where there is a certain continuous distribution of decays times instead of several discrete lifetimes.

What can cause a multi-exponential, or *even more* complicated decay curve of a single fluorophore (probe or label)? One trivial reason is a presence of some fluorescent impurity in the sample. Another reason – very common in biology – is the heterogeneity of the emitter's environment. (As already mentioned above, the environment usually has a strong effect on the observed lifetime.)

Yet another possible reason is energy transfer, i.e., FRET (Förster resonant energy transfer), a process utilized in biology to measure molecular distances. Because of the popularity of this method, let us briefly explain the basic effects of FRET on the involved decay curves. FRET is the non-radiative transfer of excited state energy from an excited state molecule (the donor) to a second molecule (the acceptor). When the donor's fluorescence is detected, FRET acts as a quenching process and shortens the lifetime of the donor. Two things must be kept in mind: (1) Not all donor molecules are sufficiently close to an acceptor to undergo FRET; thus, a fraction of donors will emit with their original, unquenched lifetime. (2) The rate and efficiency of FRET is strongly distance dependent, so a distribution of donor–acceptor distances produces a distribution of observable donor lifetimes.

When acceptor fluorescence is monitored after selective excitation of the donor, FRET can be regarded as a relayed excitation of acceptor. The result is a rising component in the acceptor's fluorescence decay curve, which requires an additional exponential term with negative amplitude.

Fluorescence kinetics is a vast topic and in-depth introduction is beyond the scope of this chapter. The reader is referred to the classic textbooks of J. R. Lakowicz [11] J. N. Demas [12], and B. Valeur [13], A shorter introduction can be found, e.g., in [14]. For an overview of photon counting data analysis, see [15–17].

1.5 Time-Resolved Fluorescence Anisotropy

Together with wavelength and intensity, polarization of the excitation and emitted light are potentially informative parameters of a fluorescence measurement – see Sect. 1.3. When the excitation beam's polarization is controlled and specified, analysis of polarization properties of the emitted fluorescence provides useful characterization of the fluorophore molecule and about its mobility and interactions.

Every electronic transition (i.e., photon absorption and emission) has an associated transition moment vector. (This is a simplified, but usually entirely sufficient, approximation.) The different electronic absorption bands in a molecule have transition moment vectors of different magnitudes and orientations. The orientation of these vectors relative to the structure of the fluorophore is fixed and characteristic for each fluorescent molecule.

Let us have a look at what is special when linearly polarized light is used for excitation. For simplicity, consider a homogenous fluid solution at room temperature. Due to rotational Brownian motion, the orientation of molecules is random and

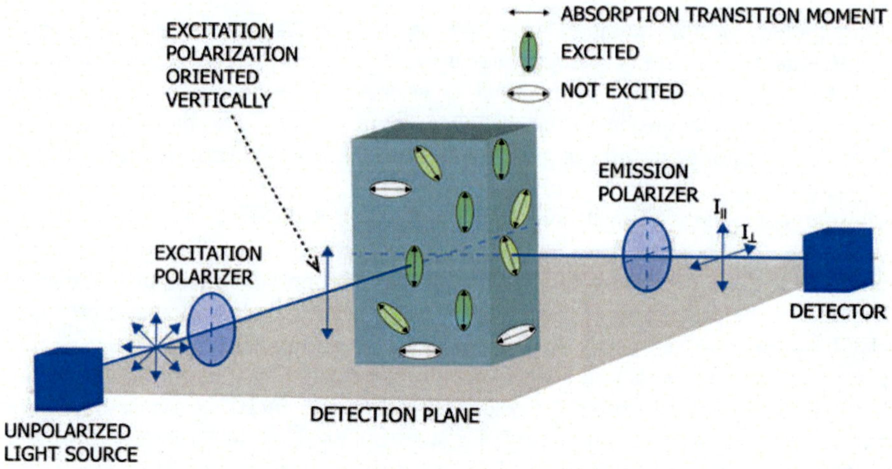

Fig. 4 Schematic representation of photo-selection and fluorescence anisotropy in spectrometer format (so called L-geometry)

fluctuating. At the instant the exciting light passes through solution, only those molecules whose absorption transition moment is close to parallel with the exciting light's polarization orientation (i.e., the light's electric field direction) can absorb the polarized exciting light. Linear polarized excitation does an orientation selection of molecules in the solution. This is called photo-selection. As a result, the initial population of excited state molecules has a well-defined set of alignments (Fig. 4).

This initial ensemble of excited, aligned fluorophores will start to decay to their electronic ground state. Simultaneously, the excited molecules undergo random rotations due to Brownian motion, thus randomizing their initial alignment. The length of time between an excited state's formation and its emission of a photon is random. (The decay kinetics discussed in previous Sect. 1.4 describes the statistical behavior of an ensemble, not of individual excited states.) The polarization of an emitted photon depends on the orientation of the fluorophore's emission transition moment at the instant of photon emission. The orientation of molecules that decay immediately after excitation is well-defined by the light absorption process (the molecule had no time to rotate). As more time elapses between excitation and photon emission, molecular rotation has more opportunity to randomize the orientations of the remaining excited molecules. Their increasingly random orientation is reported by the polarization plane of the emitted light. In other words, the fraction of various polarization planes of detected emission will change, and increasingly randomize, during a decay. Simply put, at the beginning of a decay, the emission will be strongly polarized. Toward the end of the decay of the whole ensemble, the detected emission usually will be almost depolarized. Measurement of the speed and kinetics of this change, the depolarization, is the subject of time-resolved fluorescence anisotropy. By analyzing the intensity of fluorescence in orthogonal polarization planes as the decay proceeds, we can get important new information on the structure and dynamics of the fluorophore molecule, as well as about its environment.

Note that standard fluorescence spectrometers use the so-called L-geometry observation mode: the emission is detected in a direction perpendicular to the excitation. The direction of these two beams (excitation and emission) thus defines a plane: the detection plane. During standard anisotropy measurements, the excitation beam polarization is oriented vertically relative to the detection plane, as illustrated in Fig. 4. This vertical polarization plane defines 0°. Thus, the perpendicular (90°) polarization orientation is actually oriented horizontally, which is parallel to the detection plane. In fluorescence spectrometer software and manuals, $I_{||}(t)$ and $I_{\perp}(t)$ are often written as $I^{VV}(t)$ and $I^{VH}(t)$, respectively. In the following text, we shall use the notation using $I_{||}(t)$ and $I_{\perp}(t)$.

Time-resolved fluorescence anisotropy is defined as [11, 18, 19]:

$$r(t) = \frac{I_{||}(t) - I_{\perp}(t)}{I_{||}(t) + 2 \cdot I_{\perp}(t)}$$

$I_{||}(t)$ and $I_{\perp}(t)$ are experimental observables: fluorescence intensity decay curves recorded with the emission polarizer set parallel to and perpendicular to the exciting light's polarization orientation, respectively. Parallel (0°) and perpendicular (90°) orientation is meant relative to the polarization plane orientation of excitation that defines 0°.

The parallel and perpendicular polarized decay curves must be recorded under identical conditions (other than the detection polarizer orientation) because our major interest is the tiny difference represented by the numerator. Note that this difference term is divided by a special sum of decays. It can be shown that this sum in the denominator is equivalent to the *total* fluorescence intensity decay, $I_T(t)$, which is independent of molecular motion and polarization effects. (Such a polarization independent decay curve, directly proportional to $I_T(t)$, is the primary interest during standard lifetime measurements. When using a time-resolved fluorimeter, the necessary conditions for standard lifetime measurements are usually accomplished with an emission polarizer set to the so-called magic angle, 54.7°, relative to the polarization plane of excitation.)

Since both the numerator and denominator have a physical dimension of intensity, $r(t)$ is a dimensionless quantity. Owing to the intensity normalization (denominator), anisotropy is independent of total emission intensity, which is a great advantage. Figure 5 shows an example of input data ($I_{||}(t)$ and $I_{\perp}(t)$ decay curves), as well as an anisotropy decay function $r(t)$ calculated directly, according to the formula presented above.

Figure 5 illustrates a fluorescent system where the anisotropy decay is clearly visible within the $I_{||}(t)$ and $I_{\perp}(t)$ decay curves. The high viscosity of ethylene glycol solvent at room temperature greatly slows the rotation of Coumarin6 dye molecules. The different shapes of $I_{||}(t)$ and $I_{\perp}(t)$ in Fig. 5a are apparent during the first few nanoseconds, but at later times, the two decays curves are identical, entirely

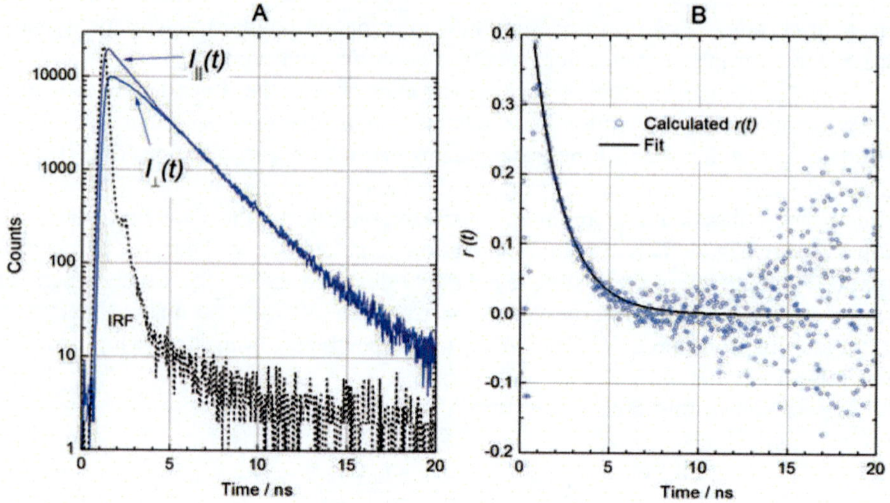

Fig. 5 Polarized intensity decay curves and calculated anisotropy decay of Coumarin6 dissolved in ethylene glycol at room temperature. (**a**) Parallel and perpendicular polarized intensity decay curves, as primary observables. The IRF is also shown for comparison, plotted with dotted line. (**b**) Anisotropy decay curve calculated point by point. A single-exponential decay model was fitted to an appropriate portion of *r*(*t*)

dominated by the spontaneous fluorescence decay of Coumarin6 with a $\tau_{\text{Decay}} =$ 2.3 ns in this sample. When those decay curves are substituted into the definition formula of *r*(*t*), we obtain the anisotropy decay depicted in Fig. 5b. The anisotropy decay displays the kinetics of the diminishing difference between $I_{||}(t)$ and $I_{\perp}(t)$. It is remarkable that a simple single-exponential decay curve with an apparent decay time of 1.6 ns can be fitted to *r*(*t*):

$$r(t) = r_0 \cdot e^{-t/\theta}$$

where the initial value of the decay, r_0, is called the fundamental anisotropy (of Coumarin6 molecule in this particular case) and θ is the molecule's rotational correlation time.

More viscous solvents or more rigid molecular environments (e.g., membrane or cellular environment) produce longer rotational correlation times and slower decay of fluorescence anisotropy. In the above example, *r*(*t*) approaches zero as time proceeds. This is expected, because rotation of dye molecules is not restricted in a fluid solvent, leading to complete randomization of molecular orientation. In biological samples we sometimes encounter restricted molecular motion leading to hindered rotation. An example is a rod-like probe molecule embedded in a lipid bilayer, or a fluorescent label covalently attached to a protein. In this case, *r*(*t*) may not reach 0 at long enough *t*. A non-zero residual anisotropy value, r_{∞}, is a sign of hindered rotation. In this case:

$$r(t) = (r_0 - r_\infty) \cdot e^{-t/\theta} + r_\infty$$

In theory, only molecules classified as spherical rotors exhibit single exponential $r(t)$. However, for small polar molecules dissolved in polar solvents, this model is a remarkably good approximation. A general $r(t)$ decay function can have up to five-exponential terms, but resolving (observing) more than two exponentials places extraordinary demands on the quality of input data $I_{||}(t)$ and $I_{\perp}(t)$. A typical $r(t)$ of a covalently labeled protein features a fast but hindered decay component corresponding to the "wagging" motion of the label and another, slower one due to local mobility of the labeled protein segment. The "rigid" rotational motion of an entire large protein molecule is typically too slow to create a measurable effect on $I_{||}(t)$ and $I_{\perp}(t)$.

The fundamental anisotropy, r_0, has a characteristic value for each fluorophore: r_0 is determined by the angle, β, between the molecule's absorption transition moment vector at the exciting wavelength and its emission transition moment vector at the detection wavelength. The fundamental anisotropy is independent of environment.

$$r_0 = \frac{2}{5}\left(\frac{3\cos^2\beta - 1}{2}\right)$$

This formula is valid for single-photon excitation, the most common case. It follows that the limiting values of r_0 are 0.4 and -0.2, for $\beta = 0°$ (the molecule's transition vectors are parallel) and for $\beta = 90°$ (the molecule's transition vectors are perpendicular), respectively. If the solution were frozen, blocking the rotation ($\theta \to \infty$), then there would be no observable kinetics of $r(t)$. Only a constant $r(t) = r_0$ value would be observed. In the example depicted in Fig. 5b, the estimated r_0 value is very close to 0.4, indicating that the two transition vectors are close to parallel in a Coumarin6 molecule (λ_{Exc} = 420–470 nm, λ_{Em} = 500–550 nm). It makes this dye a very good probe for anisotropy dynamics studies. An example of a fluorophore with $\beta = 90°$ is pyrene when excited into the S_2 excited state (structured absorption band between 300 and 340 nm) and its normal $S_1 \to S_0$ fluorescence ($\lambda > 370$ nm) is observed. $r(t)$ of pyrene in fluid solvent would show a rise (note: not a "decay") from -0.2 to zero. A hypothetical fluorophore that had an angle between orientation of absorption and emission transition vectors exactly of "magic angle" ($\beta = 54.7°$) would show zero anisotropy and depolarized emission, regardless of the nature of its environment.

The steady-state anisotropy value at a given observation wavelength, r, of a sample of interest in biology is usually easily measurable using standard fluorescence spectrometers equipped with polarizers. r and parameters θ, r_0, and τ_{Decay} accessible from time-resolved fluorescence measurements are related to each other. Thorough derivation and explanation of the underlying principles is beyond the scope of this introductory text. Nevertheless, let us discuss two useful relations:

$$r = \frac{r_0}{1 + \tau_{\mathrm{Decay}}/\theta}$$

and

$$\theta = \frac{\eta V}{RT}$$

where η, V, R, and T are viscosity of the environment, volume of the rotating particle, universal gas constant, and thermodynamic temperature, respectively.

The first equation is just an approximation that can be derived assuming single exponential $I_T(t)$ and $r(t)$ decays. It is one form of the Perrin equation that links together several observables. When combined with the second formula, it explains why the value measured for the steady-state anisotropy r increases and approaches r_0 when:

(a) The sample is cooled down. The increase of θ is caused not only by decreased T in the denominator but also by increased viscosity η at lower T.
(b) The sample is made more viscous, for example, by adding sucrose to the solution.
(c) The size of the rotating fluorophore (or assembly) is increased, for example by complexation to proteins.
(d) The fluorophore used as a reporter has very short τ_{Decay}. Such a dye can report only about the early stages of rotational diffusion upon excitation, when the anisotropy is still high.

It should be obvious that time-resolved anisotropy measurements provide a much more detailed picture of molecular motion than steady-state anisotropy measurements. However, an $r(t)$ function often cannot be obtained in the neat form depicted in Fig. 5b. Anisotropy can be evaluated only where there is sufficient fluorescence intensity. As $I_{\parallel}(t)$ and $I_{\perp}(t)$ intensities are vanishing, the directly calculated $r(t)$ function becomes noisy. Another problem is at the beginning of $r(t)$. Due to finite duration of the IRF, the beginning of $I_{\parallel}(t)$ and $I_{\perp}(t)$ decays are strongly affected by convolution effects, as explained previously in Sect. 1.4. These convolution effects blur the region of $t = 0$ and, thus, the apparent value of r_0 and the decay shape of any fast component of $r(t)$. The rotational correlation time of Coumarin6 in ethylene glycol was found to be $\theta = 1.6$ ns, which is much longer than the IRF of the instrument. It allowed us to neglect the convolution effects in $r(t)$. However, Coumarin6 in less viscous solvents, like ethanol or acetone at room temperature, rotates much more rapidly: θ is then in the order of 100 s of picoseconds. As a rule of thumb, when the rotational correlation time is shorter than the width of the IRF, a simple direct analysis of $r(t)$ (as depicted in Fig. 5b) becomes impossible. Note that deconvolution of IRF influence or re-convolution fitting of a model to a directly calculated $r(t)$ is not applicable, because $r(t)$ is a ratio function.

The state-of-the-art solution is to perform global (simultaneous) fitting of $I_{\parallel}(t)$ and $I_{\perp}(t)$ decays, instead of directly calculating $r(t)$ [20]. The shapes of $I_{\parallel}(t)$ and $I_{\perp}(t)$

decay curves, experimental observables illustrated in Fig. 5a, are not entirely independent of each other. They are linked together with the same $r(t)$ function (illustrated in Fig. 5b) whose parameters we seek to evaluate. They are further constrained by the pure, polarization independent, fluorescence intensity decay, $I_T(t)$ of the probe. It can be shown that:

$$I_{\|}(t) = [1 + 2 \cdot r(t)] \cdot I_T(t) \quad \text{and} \quad I_{\perp}(t) = [1 - r(t)] \cdot I_T(t)$$

regardless of particular mathematical models used to describe $r(t)$ and $I_T(t)$ [21, 22].

Global analysis in this context means the following: we have two input data sets, $I_{\|}(t)$ and $I_{\perp}(t)$. Based on prior knowledge about the fluorophore and expected behavior of rotation, we choose appropriate model functions for the anisotropy decay $r(t)$ and the fluorescence intensity decay $I_T(t)$. As an example, let us consider the simplest case of spherically symmetric rotor (molecule) with single-exponential fluorescence decay. The adjustable parameters of $r(t)$ are r_0 and θ. The adjustable parameters of $I_T(t)$ are τ_{Decay} and $I(0)$. These four parameters are the same in both $I_{\|}(t)$ and $I_{\perp}(t)$ decay curves. Global analysis means simultaneous fitting of two fairly constrained decay models to $I_{\|}(t)$ and $I_{\perp}(t)$ functions, respectively.

Substituting the simplest possible functional forms of $r(t)$ and $I_T(t)$ (single-exponential functions) into the formulas above, one obtains the resulting $I_{\|}(t)$ and $I_{\perp}(t)$ that are both double-exponential decays with identical component lifetimes. In both decay curves, one exponential component has a lifetime of τ_{Decay}. The other exponential term has an apparent lifetime of $\tau = (\tau_{\text{Decay}} \cdot \theta)/(\tau_{\text{Decay}} + \theta)$. The relative amplitudes of these exponentials are linked by the value of r_0.

This global analysis strategy is very robust and the formulas are relatively easy to extend for more complicated models of $r(t)$ and $I_T(t)$. The main advantage is that the analysis involves iterative re-convolution with an IRF, therefore short decay or correlation times, as well as r_0 can be resolved. However, the complexity of the fitted model rapidly increases. A double exponential $r(t)$ combined with a single exponential $I_T(t)$ leads to a triple-exponential model, while a double exponential $r(t)$ combined with a double exponential $I_T(t)$ produces a highly complex four-exponential model.

Technical remarks:

For the sake of simplicity, the discussion above assumed equal detection sensitivity for emissions with various polarizations. However, in a typical fluorescence spectrometer, the efficiency of light detection may vary as the detection polarization plane changes. This has to be taken into account: the so-called G-factor is a correction factor that expresses the ratio of detection sensitivities of parallel and perpendicular polarized signal components. Incorporation of this correction into the definition formula of anisotropy is straightforward: it involves a simple multiplication of $I_{\perp}(t)$ with the experimentally determined G-factor:

$$r(t) = \frac{I_{\|}(t) - G \cdot I_{\perp}(t)}{I_{\|}(t) + 2 \cdot G \cdot I_{\perp}(t)}$$

where G is the detection sensitivity for $I_{\|}(t)$ divided by sensitivity for $I_{\perp}(t)$.

In spectrometers using L-geometry observation, determination of this sensitivity ratio is relatively easy. It is enough to rotate the excitation polarization plane to 90°, that is, to horizontal. This leads to a photo-selection which is isotropic when viewed from the detector's side. In other words, there is no anisotropy in the emitted fluorescence. Under these conditions, any change of detected signal intensity when the emission polarizer is rotated from position corresponding to $I_{\|}$ to the angle corresponding to $I_{\perp}$ directly reflects the sensitivity of detection for these two principal polarization planes.

Performing quantitative anisotropy measurements using microscopes is considerably more complicated. The detection geometry is co-linear: the microscope objective delivers the excitation light as well as collects the emission. This makes it impossible to obtain isotropic photo-selection – necessary for G-factor determination – by rotating the polarization plane of excitation. Yet, another problem is that the excitation beam is entering to (and the emission is collected from) the focus area as a light cone under considerable angles. In effect, this creates and mixes various polarization planes. The problem is more serious for high magnification objectives with higher numerical apertures [23–25].

2 Time-Resolved Spectroscopy and Microscopy

2.1 *Time-Resolved Techniques and Their Development*

Time-resolved fluorescence spectra and fluorescence kinetics parameters provide important detailed information on a fluorescing molecule and its interactions with its environment. A variety of technical challenges arise in fluorescence measurements with high (ns–ps–fs) time resolution. For many years, achievable time resolution was mostly limited by the properties of photo-detectors. Enormous development of pulsed lasers in the last quarter of the twentieth century made short pulsed excitation available before there were sufficiently fast photo-detectors. Currently, there exists a pulsed laser and/or LED for any fluorescence spectroscopy needs, while detection with a high time resolution, dynamic range, and speed of measurement has been the main topic of improvements. The following list briefly summarizes the main development stages of the time-resolved fluorescence measurement instrumentation.

Direct Measurement of fluorescence decay with a photomultiplier (PMT) and oscilloscope. The time resolution, depending on instrumentation used, was typically in ns region. Such direct measurement was simple but had a low dynamic range. Cost of the instrumentation was mostly determined by the price of the oscilloscope.

Currently, fast digitizer boards, 10 GHz oscilloscopes, and fast photo-detectors together with data analysis software provide 10 ps resolution. So, this technique is used in research laboratories where the cost of the fast oscilloscope and complexity of the data analysis are not a problem.

Indirect Measurement with a Box-Car Integrator Better (faster) electronic switching circuits allowed us to cut-off or switch fluorescence decay signal at any chosen time. By integrating the rest of the decay signal after switching, and repeating this step by step through the whole fluorescence decay period, it is possible to reconstruct the decay profile. Compact instrument working on this principle had ns resolution and slightly better dynamic range than previous direct measurement. This technique is not really in use now, except for special purposes.

Time-Correlated Single-Photon Counting (TCSPC) Under conditions that detect just one photon from the whole fluorescence response following pulsed excitation, and using time-to-amplitude conversion, one can reconstruct an excitation–emission time difference histogram that corresponds to the fluorescence decay. Switching electronics and time-to-amplitude conversion units available in the 1970s allowed us to materialize such measurement system with ns resolution, which proved to have many advantages – among them high dynamic range and high accuracy. So, the implementation of TSCPC detection represented a breakthrough in time-resolved fluorescence spectroscopy. Currently after many improvements in switching and time-tagging electronics and introduction of ultrafast solid-state detectors, modern versions of TCSPC with ps resolution are the top-class techniques for exact fluorescence decay measurement. For that reason, Sect. 2.2.1 Time-Resolved Photon Counting is solely devoted to the description of TCSPC and some new variants of the photon counting-based detection.

Phase and Amplitude Modulation Technique This technique uses exciting light that is amplitude-modulated at various frequencies, and it measures a phase shift of the resulting modulated fluorescence emission intensity. Time resolution down to ps can be achieved, but this technique does not provide information on the character (exponential/non-exponential) of the fluorescence decay. Phase/amplitude modulation methods are now commercially available mainly in set-ups for FLIM microscopy (see below), or they are used in wide-field camera detection format. The key advantage of frequency-domain FLIM is its fast lifetime image acquisition, making it suitable for dynamic applications such as live cell research. The entire field of view is excited semi-continuously – using relatively broad excitation pulses – and read out simultaneously. Hence, frequency-domain lifetime imaging can be near instantaneous.

Streak-Camera Detection The most advanced commercial streak cameras have fs time resolution and can function in a single-shot regime! This type of streak camera is extremely expensive and provides only limited dynamic range. Currently, streak cameras are used in femto- and atto-second plasma research, XUV, and Rtg imaging studies.

Description of the techniques and instrumentation used in time-resolved fluorescence/spectroscopic studies can be found in many fluorescence-related textbooks, e.g., [11–13]. To follow a development of the top-class techniques in each historical stage, see, e.g., [26] and follow, e.g., a relevant chapter in each of the 19 volumes of *Springer Chemical Physics Ultrafast Phenomena* series published between 1978 and 2014.

2.2 *State-of-the-Art Techniques for ps–fs Time Resolution*

2.2.1 Time-Resolved Photon Counting

Arguably, this is the most widely used technique today. A fundamental advantage of the method is that it records the true shape of the decay curve (time evolution of emission intensity) without any assumption about the particular decay model. Another advantage is its inherent sensitivity. Counting and timing photons, one by one, means working at the quantum limit of light detection.

The basic idea is as follows: a fluorescence decay initiated with a short excitation pulse is a very fast process, typically completed in a few dozens of nanoseconds. Moreover, such a single decay is very weak in terms of signal intensity, because there are practical limits for the maximum excitation pulse energy. (The limitation today is typically not the available power of suitable sources, but comes from the sample: bleaching, photo-toxicity, local heating, etc., all to be avoided.) The solution is to excite the sample repetitively, with low energy pulses, many times. Instead of attempting to obtain a decay curve (shape) from a single excitation pulse, only one fluorescence photon is detected after each excitation pulse. High-performance timing electronics determines the time elapsed between excitation of the sample and detection of an emitted photon. An approximation of the true decay curve is obtained by constructing a histogram of these elapsed times (measured on ps–ns or longer time ranges) from many excitation–emission cycles. If excitation pulse rates are high enough (kHz in the past, but tens of MHz today), a reasonable approximation of the decay curve (i.e., histogram) with sufficient signal to background ratio can be achieved relatively quickly.

The general term "time-resolved photon counting" includes the well-known methods of time-correlated single-photon counting (TCSPC) and multi-channel scaling (MCS). The former was the breakthrough method in the past century to reach the time resolution necessary for fluorescence studies. MCS is another traditional pulse counting method used typically on microsecond and longer time scales. Both are popular and still widely used today, thus deserving a brief description here.

One can imagine the simplest, generic TCSPC data acquisition as a repeated stopwatch measurement. An excitation pulse starts the watch and the *first* detected photon stops it. The watch has a finite maximum time range divided into, say, thousands of time bins, very short discrete time intervals. Thus, the digitized readout of a time-delay measurement is an integer value from the interval $<$ 0..number of time bins>. Let us have a linear array of memory cells corresponding to these possible integer readout values. These cells are all empty at the beginning. When the stopwatch returns a digital value, the content of corresponding memory cell is incremented by one. Repeating this process many times, one finally obtains a histogram: the distribution of various time intervals between the excitation pulse and a subsequent photon arrival. As explained in the classic book [27] this distribution is nothing else than the shape of the decay curve we are after, provided the photons we have detected are really "*rare events*." So rare that in the above-described stopwatch measurement we can encounter only a single photon within the whole fluorescence response to one excitation pulse, and there are certainly no more "later" photons arriving in the same excitation–emission cycle. This requirement is easy to understand if one considers the unavoidable time period needed to process such a precise timing. That is called dead time of the electronics, lasting dozens of nanoseconds up to microseconds in the past. If there were more photons arriving after the start of the stopwatch, the first one, the earliest one, would be always timed, counted, and histogrammed, but the later ones would be always lost. Such a bias toward the first detected photon would introduce a systematic distortion of the recorded histogram. Early photons (actually, the first ones) arriving to the detector would be over-represented. This is traditionally called "pile-up effect" [27–29] and can (must) be avoided by limiting the photon counting rate. Considering Poisson statistics relevant for detection of rare events it can be shown [27] that if the photon detection probability (i.e., average photon counting rate) is just a few percent of the excitation pulse rate, then encountering more than one photon in a single excitation–emission (start–stop) cycle is negligible. In simple terms, the fluorescence signal must be so weak that – on average – one has to send hundreds of excitation pulses to the sample in order to obtain at least one (single) photon as a response to one excitation pulse. This may sound as a very inefficient approach, but in practice it is not a real limitation. Usually the low quantum yield of the sample, low overall detection efficiency, or the limited available (or allowable) excitation intensity ensures these conditions.

The above-described TCSPC process was the standard in the past, approximately up to the 80s of the past century, when common excitation sources had relatively low pulse repetition rates. It is important to realize that starting a stopwatch with every excitation pulse, knowing that most pulses will *not* lead to photon detection, is not really economical. The logical solution is to abandon this nicely causal measurement sequence: to reverse the meaning of start and stop signals. Let us start the stopwatch with the inherently rare detected single photon and stop it with the subsequent laser pulse, which is guaranteed to always be there! The recorded (histogrammed) time intervals will convey the same information; only the histogram will appear reversed in time. Reversing the time axis back to the natural one is trivial arithmetic, assuming the time interval between excitation pulses is constant, which is normally the case. In contemporary TCSPC electronics, the start pulses are those from photon detectors and stop pulses are derived from the pulsed excitation source (called SYNC signal). The common term for this modern scenario is therefore "reversed start–stop mode."

In comparison to the above-described classic TCSPC, multi-channel scaling, MCS, by design, detects more photons per excitation cycle. Consequently, more than one photon can be sorted into the same time bin in the same measurement cycle. The simple stopwatch analogy as described above is not valid here anymore. In MCS, an excitation pulse starts a so-called sweep, when detected photons are counted and sorted into subsequent time bins. Imagine it as a fast counting process in a fast-moving time window, which lasts just for a moment, as short as the time width of the bin. The number of photons counted during that time window is stored in the corresponding memory cell. The counter is reset to zero when the time window moves and continues counting for the next time bin, and so on, until the end of the sweep. Timing resolution of MCS cannot compete with TCSPC, but MCS is more efficient in terms of photons collected per excitation pulse.

As TCSPC instrumentation evolved, the electrical time resolution was increasing (down to a few ps), and the size of the necessary electronics was shrinking. Contemporary instruments are often computer boards, while they were as large as a cabinet around 1980, when TCSPC became widespread in photophysics. There is a growing need for multi-channel devices, that is, for timing electronics that can process signals from multiple detectors simultaneously. Multiplying the number of TCSPC devices was (and still is) very expensive, and their mutual synchronization is not trivial at all. A viable solution was the use of a so-called router unit that accepted inputs from several detectors. Those pulses were routed to a single TCSPC timing unit that processed them as they came, one by one. Using the routing information, the result was sorted into several histograms, one for each detector. The basic principle of TCSPC limits the maximum signal counting rate anyway, and this kind of statistical time-sharing of the TCSPC processing unit works very efficiently as far as the coincident detector signals (pulses) are still rare [30, 31] (Fig. 6).

However, the demand for truly parallel processing of multi-channel detection was increasing. Capturing coincident photons (or lack of their coincidence) is a basic requirement in order to study phenomena like photon antibunching, bunching, and cross-correlation in fluorescence correlation spectroscopy. Using a router unit and single-channel TCSPC electronics does not allow detection of photon coincidences: such events have to be discarded, because they cannot be reliably routed. Another disadvantage of the very convoluted timing process involving the cascade of TAC, ADC, and MCA units is the relatively long dead-time of this kind of instrument. Even in fairly modern instruments, dead-time is in the order of hundreds of nanoseconds, seriously limiting the maximum total photon detection rate.

The latest generation of TCSPC electronics uses time-to-digital converters. The job of TAC (time-to-amplitude converter) and ADC (analog-amplitude-to-digital converter) is performed in a single, integrated circuit, TDC (time to digital converter). TDCs are not only much faster; they cost significantly less than high-performance TAC- and ADC-based solutions [32, 33]. The role of MCA, which is sorting the measured start–stop times into histograms, can be performed using very fast FPGA (field programmable gate arrays) today [34]. This new technology made it possible to develop devices capable of truly parallel, simultaneous timing on several input channels [35–37].

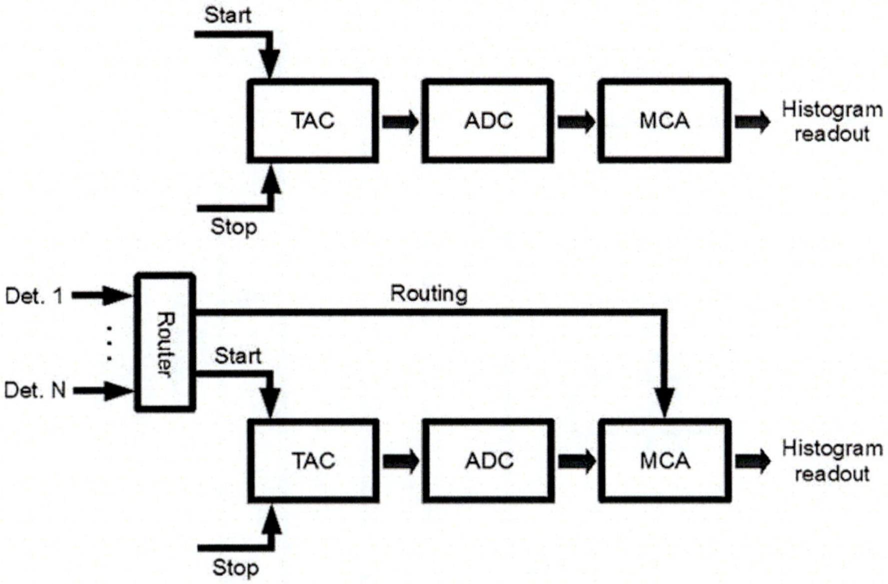

Fig. 6 Simplified block diagram of classical TCSPC electronics (top) and its extension with a router (bottom). Top: Start and stop signals are detector pulses and synchronization pulses from a pulsed light source, respectively. TAC is time-to-amplitude converter. It is based on a fast voltage ramp, started and stopped by corresponding input signals. Its output is a voltage value (amplitude). ADC is analog (amplitude) to digital converter, which converts the TAC output to a digital number. MCA, multi-channel analyzer, sorts these digital values into a histogram. Bottom: A router unit is able to collect signals from multiple detectors. Whenever an input pulse is received from any of the detectors, this signal is forwarded to the TAC as a start signal and the event is processed as usual. However, the router also generates routing information that is fed to the MCA. Based on this routing information, for each detector a separate histogram is collected

Yet another important development step was including a global clock into TCSPC electronics, measuring the absolute time of photon detection (measured from the start of data acquisition) on a macroscopic time scale [29, 32–34]. This feat is different from the fundamental process of picosecond precise (laser pulse to first photon) time-delay measurement, as performed in TCSPC. When the goal is to obtain a simple TCSPC histogram, i.e., the shape of a decay curve, only the various ps delays of photons measured from the laser pulses do matter. Histogramming is a kind of online data reduction scheme, when the order of photon arrival is irrelevant, and therefore that information is not preserved. However, there are many experimental scenarios (e.g., single-molecule detection, fluorescence correlation spectroscopy FCS, and as we shall see later, fluorescence lifetime imaging, FLIM) where the global detection time of a single photon is equally important as its ps delay.

Thus, modern TCSPC solutions became ultrafast time taggers [36, 37]. Instead of just histogramming delay times of special event pairs (laser pulse to photon detection), they are capable to attach time tags to each individual detected event. The full information content of the measurement is preserved (Fig. 7).

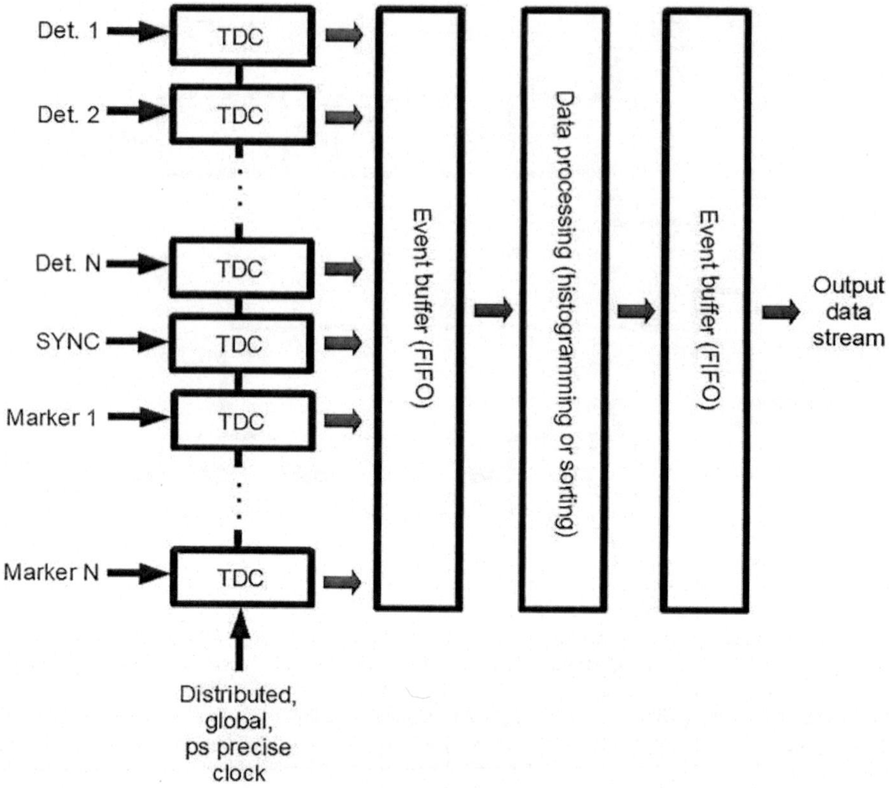

Fig. 7 Simplified block diagram of a modern multi-channel time tagger/TCSPC electronics

Detector (photon) pulses are basically treated in the same way as synchronization pulses from laser pulses, as well as any other timing relevant signal received from other devices. All these inputs generate "events" that are precisely time-tagged. The overall event rate may be very high; therefore, the time tags are entering a high-speed first-in-first-out (FIFO) event buffer, in order to be able to process large momentary event rates. The recorded events are pre-processed in high-speed programmable hardware. Finally, they are output in digital form, typically as an ordered list of events containing information about the kind of the signal (identification number of input channel) and their precise time tags.

The most generic data output is then a list of various kinds of events, each with its time-tag. Each entry of the list contains the identification number of the input channel where the event was detected. The pulsed laser's synchronization signal is treated as one of the possible events. Certain input channels can be dedicated to special events, e.g., frame, line, or pixel markers in the case of image scanning.

Classical TCSPC decay curve measurement can be then regarded as a special, very simple, case of processing of available event time tags: The time tag of a SYNC event (that marks the moment of excitation by a laser pulse) is subtracted from the time tag of a detected fluorescence photon. The result is equivalent to the delay time of photon measured, e.g., by old-style TAC-ADC-based electronics. These delays are histogrammed in post-processing of the original list of events. The advantage of obtaining a decay curve in this way is that the overall measurement time can be arbitrarily divided into shorter time periods, and a dedicated histogram can be calculated for each such a period. This is a very important feature with many applications.

Consider for instance a single molecule undergoing reversible conformational changes on millisecond time scale observed with a confocal microscope. If the conformation affects the lifetime of its fluorescence, the fluctuation of the lifetime can be recorded and studied in detail. Intensity traces, as well as lifetime traces, can be calculated; thus, the kinetics of those changes is revealed.

The same basic idea is involved in fluorescence lifetime imaging microscopy, FLIM. A FLIM image is always a composite image consisting of an intensity (grayscale) image overlaid with a false color layer, where the colors are mapped to different lifetime values. The most common way of getting a FLIM image using time-resolved photon counting involves scanning. (There are several scanning modalities: e.g., the sample or the microscope objective is moved with piezo-actuators, or the laser beam is deflected, as in laser scanning confocal microscopes.) The goal is to obtain a lifetime value (thus color) for every pixel of the image. This is not complicated at all, if we are able to map the motion of the scanner in time. Recording, time-tagging, and storing special marker events, namely line-start and line-stop markers in the list of "events," makes it possible to reconstruct the FLIM image in post-processing. It is enough to realize that the time tag of a photon identifies its spatial origin, since we know the moment of start and stop of every line during scanning. Assuming regular scanner motion, every photon in the event list can be unambiguously mapped to an image pixel, based on comparison of its time tag with that of the line markers. Once we sorted the photons into pixels, histogramming their delay times measured from SYNC pulses is straightforward. Analysis of these histograms, unique for each pixel, yields the lifetime value that is then mapped to a color.

The FLIM concept described above is the most common these days and is typically implemented using a confocal microscope. However, there are alternative devices that achieve a time-domain photon counting-based FLIM image without scanning, in wide-field observation mode. These are basically multi-pixel photon counting detectors with integrated timing electronics [38–41].

2.2.2 Fluorescence Up-Conversion (Decay Sampling)

Fluorescence decay sampling is a nonlinear optical technique. A schematic representation of how the up-conversion technique works is shown in Fig. 8a, taken from [42]. Fluorescence is excited by UV pump beam short pulse and focused to a nonlinear BBO crystal. Simultaneously, a fs 800 nm gate pulse, delayed by a chosen Δt after the pump pulse, overlaps with the fluorescence signal at the BBO crystal.

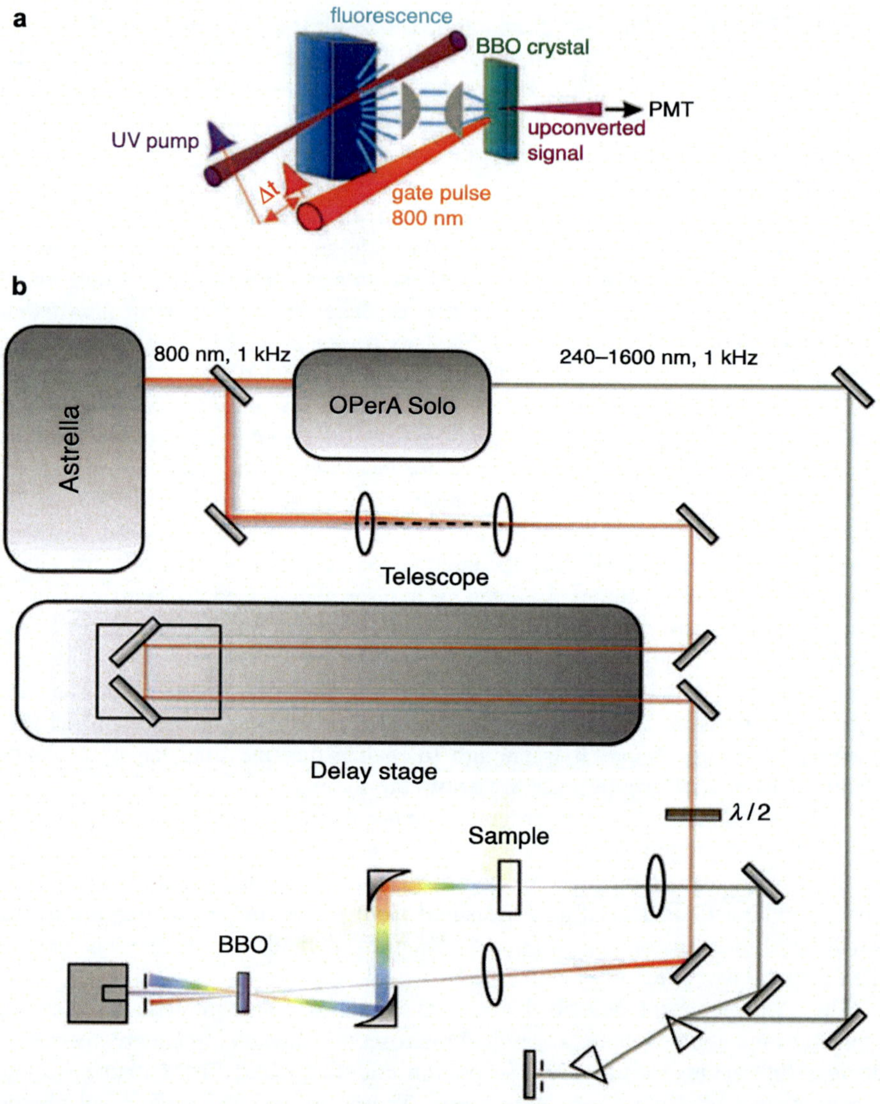

Fig. 8 (**a**) Scheme of fs-resolved fluorescence up-conversion and (**b**) the corresponding experimental setup. Reprinted from [42] published under an open access Creative Common CC BY license

Up-converted (sum) signal appears at the output of BBO only when both the fluorescence and probe signals are overlapping. This up-converted signal is detected by a PMT. A time delay between the pump pulse that excites fluorescence and the probe pulse is defined by the optical delay stage – see Fig. 8b. Thus, by changing the Δt delay while measuring intensity of the up-converted signal one can register a

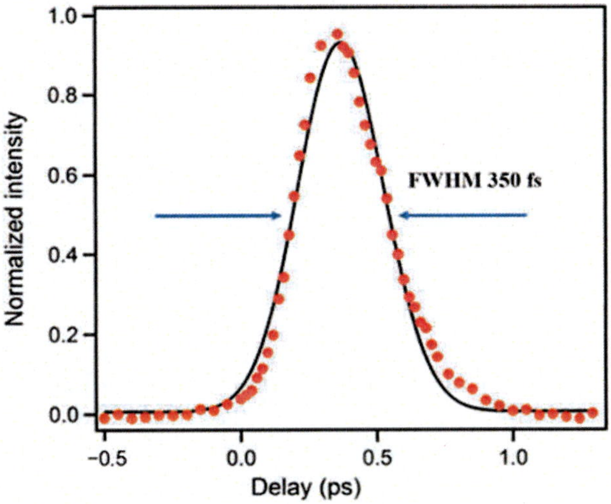

Fig. 9 A typical pulse response function of an up-conversion spectrofluorometer. Reprinted from [42] published under an open access Creative Common CC BY license

fluorescence decay profile. The time resolution of this technique is given by the profile (duration) of the probe pulse which – today – can be easily in tens of fs or shorter (but expensive laser set-up is needed for that). These days routine pulse response function of this technique is illustrated in Fig. 9 below.

Recently, fluorescence decay sampling technique was combined with other nonlinear optical spectroscopies. Such combination spectroscopy provides an extremely detailed information on molecular dynamics [43].

Recent progress in instrumentation and technology resulted in compact spectrofluorometers that combine femtosecond fluorescence up-conversion and TCSPC in one module, even with automated switching between those two techniques – e.g., HARPIA TF from Light Conversion [44].

3 Some Nonlinear Techniques and New Trends

3.1 Multi-photon Excitation Techniques

Multi-photon excitation is a widely used nonlinear technique for both fluorescence spectroscopy and microscopy.

When the excitation is done by a focused single beam, the quadratic dependence of the two-photon absorption on excitation light intensity leads to much smaller excited volume within the sample. In case of excitation by two laser beams, the probability of two-photon absorption depends on both spatial and temporal overlap of the incident beams. Two-photon and even three-photon excited FLIM is relatively common now [45, 46].

Using a quantum light – entangled two photons is the simplest case – as a powerful spectroscopic tool to reveal novel information about complex molecules is a new emerging field [47].

Great attention has been already devoted to absorption/excitation with entangled photon pairs (E2PA) at low photon flux – with the expectation that using entangled photon pairs could bring (large?) quantum advantages. A comprehensive comparison of so far experimental reports, together with a discussion of critical experimental features that influence each measurement of the E2PA cross-section, is in [48, 49].

Regardless of current ambiguities about expected quantum advantages, and regardless of both the experimental complexity and the advanced theory needed for correct interpretation of such spectroscopic results, there is no doubt that using a quantum light in nonlinear optical spectroscopies will bring a qualitatively new information, in particular about bio-macromolecules and bio-functional molecular aggregates.

Surprisingly, recent results show that time-energy entanglement of near-IR photons preserves through a thick biological media even at room temperature! [49] That brings about hope for entanglement-enhanced fluorescence imaging.

3.2 *Emerging New Techniques: Excitation by Structured Photons*

Interesting light patterns can be achieved by cleverly combining different electromagnetic modes. Such modal superposition can occur in the spatial and/or temporal domain. Resulting light patterns – called structured light or structured photons – provide new possibilities for many fields, including spectroscopy and microscopy. A comprehensive overview of both light shaping techniques and some application of structured light in microscopy is presented, e.g., in "Roadmap on multimode light shaping" [50].

3.3 *Combined Nonlinear Optical and Fluorescence Spectroscopies and Microscopy*

Since fs nonlinear 2D optical spectroscopies – comprising, e.g., four- and six-wave mixing and stimulated transitions – became kind of "advanced research laboratory standard," some such research setups are using fluorescence emission as "reporter" on the complex molecular excited states that were generated by the previous nonlinear processes. Currently, this approach demands complex and extensive instrumentation in combination with an advanced theory to interpret results correctly. But its potential to provide very detailed information on the excited states of complex molecular systems is obvious and fast advances in instrumentation are promising.

A recent example of this approach is the study of one-exciton and bi-exciton properties of molecular systems by using coherently and fluorescence-detected two-dimensional electronic spectra of increasing order of nonlinearity, for example, four-wave and six-wave mixing [51]. On top of comparison of fluorescence- and coherent-detection properties [52], they suggest a general relation between nonlinear fluorescence-detected signal and the order of nonlinear wave-mixing scheme.

The increasing number of studies that use (non-linear) fluorescence signal as a reporter on electronic and vibrational transient states properties of complex- and biomolecular systems indicates a new field for further fluorescence spectroscopy development.

Related nonlinear vibrational, electronic, or vibrational-electronic spectroscopy can now be carried out with multiple mode-locked lasers with highly stabilized repetition frequencies. Such arrangement allows extremely fast automatic scanning – i.e., asynchronous optical sampling – to investigate molecular relaxation processes in one sweep. An advanced overview of techniques of coherent nonlinear spectroscopy with multiple mode-locked lasers is in [53].

A very promising new approach that combines fluorescence and interferometric microscopy is described in [54].

They demonstrate that fluorescence microscopy combined with interferometric scattering (iSCAT) microscopy – a high-speed and time-unlimited imaging technique – can uncover, e.g., the real-time dynamics of nanoscopic nascent adhesions in living cells. See their instrumental setup in Fig. 10.

This study shows that the combination of fluorescence and interferometric scattering microscopy – high-speed and time-unlimited imaging technique – can uncover the real-time dynamics of nanoscopic processes such as cell adhesions onto the extracellular matrix.

4 Conclusions

After a brief introduction providing a historical view of fluorescence and its measurement, we have shown in this chapter that currently available methods and corresponding instrumentation of time-resolved fluorescence spectroscopy cover the whole possible time range of biological processes' dynamics – from steady-state to any real-time molecular dynamics, up to femtosecond region.

Currently, the most universal techniques for the time-resolved fluorescence studies are based on time-correlated single-photon counting approach with great improvements and expansions by using extremally fast single-photon timing and tagging, together with tunable short pulse excitation. But, regardless of the high quality of the data provided by such instrumentation (time resolution, dynamic range, reproducibility, etc.), misinterpretation of the data can occur if the technique

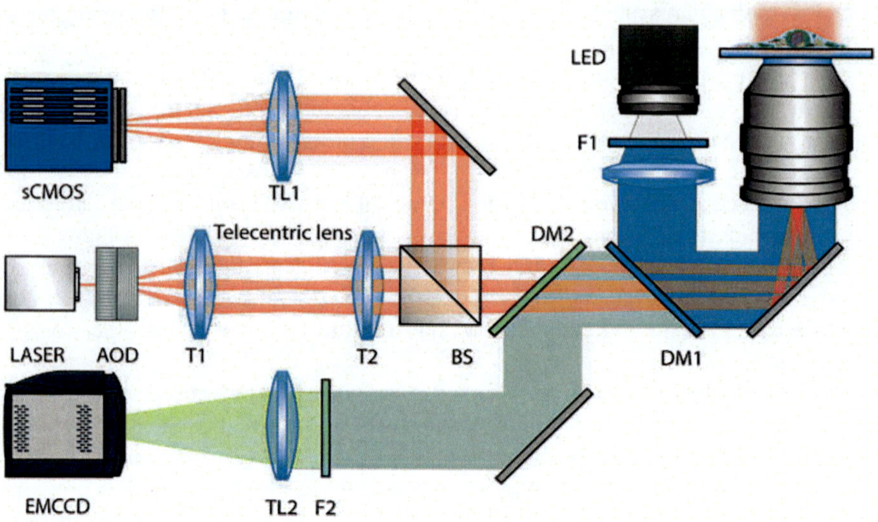

Fig. 10 Schematic representation of the combined setup of iSCAT and epifluorescence microscopy. The laser beam (red) steered by the two-axis AOD impinges on the sample surface through the 100× objective lens. The scanning beam is collimated by a set of telecentric lenses (T1, T2). The reflected and scattered lights are collected back through the objective lens, split with a 50R:50 T beam splitter (BS), and imaged onto the sCMOS camera via a tube lens (TL1). The optical bandpass filter (F1) only transmits light in the blue range out of the broad spectrum of the white LED to excite GFP. The emission filter (F2) only transmits light in the emission spectrum of GFP. The fluorescent signal is projected onto the EMCCD camera via a tube lens (TL2). Reprinted (adapted) with permission from Park J-S, Lee I-B, Moon H-M, Ryu J-S, Kong S-Y, Hong S-C, et al. Fluorescence-Combined Interferometric Scattering Imaging Reveals Nanoscale Dynamic Events of Single Nascent Adhesions in Living Cells. J Phys Chem Lett 2020 11(23):10233–41. Available from: https://pubs.acs.org/doi/10.1021/acs.jpclett.0c02103. Copyright 2020 American Chemical Society

is applied without reasonable understanding to both the technique's limitations and critical aspects and complexity of photophysics of the sample under investigation.

Thanks to the quality and a broad scope of fluorescence parameters that can be reliably measured with current instrumentation, fluorescent molecules – native or purposely incorporated into a functional biomolecular complex – have become efficient reporters about the structure, intra- and inter-molecular interactions, and/or biological functions of the molecular complex studied. It is possible to study dynamics of biomolecular functions in real time or study just one molecule at a time.

Tremendous instrumental and technology development of recent years allowed new combinations of fluorescence with other optical nonlinear spectroscopic techniques with extremely high time resolution (in fs region) or combinations of time-resolved nonlinear optical spectroscopies with microscopy up to a single-molecule resolution. Just a small illustration of such combined techniques (far from any systematic overview!) is in the last paragraph of this chapter. In this direction, we can certainly expect an explosive development in years to come.

Acknowledgments The authors are truly grateful to Matthew Zimmt for his constructive comments and feedback. Many thanks go to Min Cho and David Birch for valuable discussions and for providing us with some important recent results, and to Barbora Chmelova for help with illustrations and references.

References

1. Bernard CI, Newton HE (1958) A history of luminescence from the earliest times until 1900. Am Hist Rev [Internet] 63(4):937. https://www.jstor.org/stable/10.2307/1848952?origin=crossref
2. NobelPrize.org (1967) Manfred Eigen – biographical [Internet]. Nobel Prize Outreach AB 2022. https://www.nobelprize.org/prizes/chemistry/1967/eigen/biographical/
3. NobelPrize.org (1967) Ronald G.W. Norrish – biographical [Internet]. Nobel Prize Outreach AB 2022. https://www.nobelprize.org/prizes/chemistry/1967/norrish/biographical/
4. NobelPrize.org (1967) George Porter – biographical [Internet]. Nobel Prize Outreach AB 2022. https://www.nobelprize.org/prizes/chemistry/1967/porter/biographical/
5. Porter G, Reid ES, Tredwell CJ (1974) Time resolved fluorescence in the picosecond region. Chem Phys Lett [Internet] 29(3):469–472. https://linkinghub.elsevier.com/retrieve/pii/000926147485147X
6. NobelPrize.org (1999) Press release [Internet]. Nobel Prize Outreach AB 2022. https://www.nobelprize.org/prizes/chemistry/1999/press-release/
7. NobelPrize.org (2008) Press release [Internet]. Nobel Prize Outreach AB 2022. https://www.nobelprize.org/prizes/chemistry/2008/press-release/
8. NobelPrize.org (2014) Press release [Internet]. Nobel Prize Outreach AB 2022. https://www.nobelprize.org/prizes/chemistry/2014/press-release/
9. Horng ML, Gardecki JA, Papazyan A, Maroncelli M (1995) Subpicosecond measurements of polar solvation dynamics: Coumarin 153 revisited. J Phys Chem [Internet] 99(48):17311–17337. https://pubs.acs.org/doi/abs/10.1021/j100048a004
10. Sýkora J, Kapusta P, Fidler V, Hof M (2002) On what time scale does solvent relaxation in phospholipid bilayers happen? Langmuir [Internet] 18(3):571–574. https://pubs.acs.org/doi/10.1021/la011337x
11. Lakowicz JR (2006) Principles of fluorescence spectroscopy [Internet].3rd edn. Springer, Boston. http://link.springer.com/10.1007/978-0-387-46312-4
12. Demas JN (1983) Excited state lifetime measurements.1st edn. Academic Press, New York
13. Valeur B, Berberan-Santos MN (2013) Molecular fluorescence. Principles and applications.2nd edn. Wiley
14. Birch DJS, Chen Y, Rolinski OJ (2015) Fluorescence. In: Photonics, biomedical photonics, spectroscopy, and microscopy IV [Internet]. Wiley, pp 1–58. https://onlinelibrary.wiley.com/doi/10.1002/9781119011804.ch1
15. Patting M (2008) Evaluation of time-resolved fluorescence data: typical methods and problems. In: Standardization and quality assurance in fluorescence measurements I [Internet]. Springer, Berlin, pp 233–258. http://link.springer.com/10.1007/4243_2008_020
16. Bevington PR (1992) Data reduction and error analysis for the physical sciences. McGraw-Hill
17. O'Connor DV, Ware WR, Andre JC (1979) Deconvolution of fluorescence decay curves. A critical comparison of techniques. J Phys Chem [Internet] 83(10):1333–1343. https://pubs.acs.org/doi/abs/10.1021/j100473a019

18. Birch DJS, Imhof RE (1991) Time-domain fluorescence spectroscopy using time-correlated single-photon counting. In: Lakowicz JR (ed) Topics in fluorescence spectroscopy, vol 1. Springer, pp 64–71
19. Steiner RF (2002) Fluorescence anisotropy, theory and applications. In: Lakowicz JR (ed) Topics in fluorescence spectroscopy, vol 2, pp 1–51
20. Crutzen M, Ameloot M, Boens N, Negri RM, De Schryver FC (1993) Global analysis of unmatched polarized fluorescence decay curves. J Phys Chem [Internet] 97(31):8133–8145. https://pubs.acs.org/doi/abs/10.1021/j100133a005
21. Kapusta P, Erdmann R, Ortmann U, Wahl M (2003) Time-resolved fluorescence anisotropy measurements made simple. J Fluoresc 13:179–183
22. Trevor S, Kenneth GP (2015) A review of the analysis of complex time-resolved fluorescence anisotropy data. Methods Appl Fluoresc [Internet] 3(2):022001. https://iopscience.iop.org/article/10.1088/2050-6120/3/2/022001
23. Koshioka M, Sasaki K, Masuhara H (1995) Time-dependent fluorescence depolarization analysis in three-dimensional microspectroscopy. Appl Spectrosc [Internet] 49(2):224–228. http://journals.sagepub.com/doi/10.1366/0003702953963652
24. Schaffer J, Volkmer A, Eggeling C, Subramaniam V, Striker G, Seidel CAM (1999) Identification of single molecules in aqueous solution by time-resolved fluorescence anisotropy. J Phys Chem A [Internet] 103(3):331–336. https://pubs.acs.org/doi/10.1021/jp9833597
25. Devauges V, Marquer C, Lécart S, Cossec J-C, Potier M-C, Fort E et al (2012) Homodimerization of amyloid precursor protein at the plasma membrane: a homoFRET study by time-resolved fluorescence anisotropy imaging. PLoS One [Internet] 7(9):e44434. https://dx.plos.org/10.1371/journal.pone.0044434
26. Fleming GR (1986) Chemical applications of ultrafast spectroscopy. Oxford University Press
27. Desmond V, O'Connor DP (1984) Time-correlated single photon counting [Internet]. Academic Press. https://linkinghub.elsevier.com/retrieve/pii/B9780125241403X50011
28. Patting M, Wahl M, Kapusta P, Erdmann R (2007) Dead-time effects in TCSPC data analysis. In: Prochazka I, Migdall AL, Pauchard A, Dusek M, Hillery MS, Schleich WP (eds) Photon counting applications, quantum optics, and quantum cryptography, vol 6583, pp 72–81. http://proceedings.spiedigitallibrary.org/proceeding.aspx?doi=10.1117/12.722804
29. Becker W (2005) Castleman AW, Toennies JP, Zinth W (eds) Advanced time-correlated single photon counting techniques [Internet]. Springer series in chemical physics, vol 81. Springer, Berlin. http://link.springer.com/10.1007/3-540-28882-1
30. Birch DJS, McLoskey D, Sanderson A, Suhling K, Holmes AS (1994) Multiplexed time-correlated single-photon counting. J Fluoresc [Internet] 4(1):91–102. http://link.springer.com/10.1007/BF01876663
31. Birch DJS, Hungerford G, McLoskey D, Sagoo K, Yip P (2019) Instrumentation for fluorescence lifetime measurement using photon counting. Springer, pp 103–133. http://link.springer.com/10.1007/4243_2018_2
32. Wahl M, Erdmann R, Lauritsen K, Rahn H-J (1998) Cohn GE (ed) Hardware solution for continuous time-resolved burst detection of single molecules in flow, pp 173–178. http://proceedings.spiedigitallibrary.org/proceeding.aspx?articleid=933939
33. Wahl M, Ortmann U, Lauritsen K, Erdmann R (2002) Chin AK, Dutta NK, Herrick RW, Linden KJ, DJ MG (eds) Application of sub-ns pulsed LEDs in fluorescence lifetime spectroscopy, pp 171–178. http://proceedings.spiedigitallibrary.org/proceeding.aspx?articleid=874883
34. Wahl M, Rahn H-J, Gregor I, Erdmann R, Enderlein J (2007) Dead-time optimized time-correlated photon counting instrument with synchronized, independent timing channels. Rev Sci Instrum [Internet] 78(3):033106. http://aip.scitation.org/doi/10.1063/1.2715948
35. Wahl M (2014) Modern TCSPC electronics: principles and acquisition modes. In: Methods and applications [Internet]. Springer, pp 1–21. http://link.springer.com/10.1007/4243_2014_62
36. Wahl M, Rahn H-J, Röhlicke T, Kell G, Nettels D, Hillger F et al (2008) Scalable time-correlated photon counting system with multiple independent input channels. Rev Sci Instrum [Internet] 79(12):123113. http://aip.scitation.org/doi/10.1063/1.3055912

37. Wahl M, Röhlicke T, Kulisch S, Rohilla S, Krämer B, Hocke AC (2020) Photon arrival time tagging with many channels, sub-nanosecond deadtime, very high throughput, and fiber optic remote synchronization. Rev Sci Instrum [Internet] 91(1):013108. http://aip.scitation.org/doi/10.1063/1.5121412
38. Henderson RK, Johnston N, Mattioli Della Rocca F, Chen H, Day-Uei Li D, Hungerford G et al (2019) A 192x128 time correlated SPAD image sensor in 40-nm CMOS technology. IEEE J Solid-State Circuits [Internet] 54(7):1907–1916. https://ieeexplore.ieee.org/document/8681087/
39. Zickus V, Wu M-L, Morimoto K, Kapitany V, Fatima A, Turpin A et al (2020) Fluorescence lifetime imaging with a megapixel SPAD camera and neural network lifetime estimation. Sci Rep [Internet] 10(1):20986. http://www.nature.com/articles/s41598-020-77737-0
40. Hirvonen LM, Suhling K (2017) Wide-field TCSPC: methods and applications. Meas Sci Technol [Internet] 28(1):012003. https://iopscience.iop.org/article/10.1088/1361-6501/28/1/012003
41. Suhling K, Hirvonen LM, Becker W, Smietana S, Netz H, Milnes J et al (2019) Wide-field time-correlated single photon counting-based fluorescence lifetime imaging microscopy. Nucl Instruments Methods Phys Res Sect A Accel Spectrometers, Detect Assoc Equip [Internet] 942: 162365. https://linkinghub.elsevier.com/retrieve/pii/S0168900219309519
42. Cao S, Li H, Zhao Z, Zhang S, Chen J, Xu J et al (2021) Ultrafast fluorescence spectroscopy via upconversion and its applications in biophysics. Molecules [Internet] 26(1):211. https://www.mdpi.com/1420-3049/26/1/211
43. Xu J, Knutson JR (2008) Chapter 8: Ultrafast fluorescence spectroscopy via upconversion. In: Applications to biophysics, methods in enzymology [Internet], pp 159–183. https://linkinghub.elsevier.com/retrieve/pii/S0076687908034083
44. HARPIA-TF (2022) Femtosecond fluorescence upconversion and TCSPC module [Internet]. Light Conversion. https://lightcon.com/product/harpia-fluorescence-module/
45. Herman P, Lakowicz JR (2014) Lifetime-based imaging. In: Biomedical photonics handbook [Internet]2nd edn. CRC Press, pp 376–419. https://www.taylorfrancis.com/books/9781420085136/chapters/10.1201/b17290-16
46. So PTC, Hosseini P, Dong CY, Berland KM (2014) Two-photon excitation fluorescence microscopy. In: Biomedical photonics handbook [Internet]. CRC Press, pp 440–457. https://www.taylorfrancis.com/books/9781420085136/chapters/10.1201/b17290-18
47. Schlawin F, Dorfman KE, Mukamel S (2018) Entangled two-photon absorption spectroscopy. Acc Chem Res [Internet] 51(9):2207–2214. https://pubs.acs.org/doi/10.1021/acs.accounts.8b00173
48. Parzuchowski KM, Mikhaylov A, Mazurek MD, Wilson RN, Lum DJ, Gerrits T et al (2021) Setting bounds on entangled two-photon absorption cross sections in common fluorophores. Phys Rev Appl [Internet] 15(4):044012. https://link.aps.org/doi/10.1103/PhysRevApplied.15.044012
49. Lum DJ, Mazurek MD, Mikhaylov A, Parzuchowski KM, Wilson RN, Jimenez R et al (2021) Witnessing the survival of time-energy entanglement through biological tissue and scattering media. Biomed Opt Express [Internet] 12(6):3658. https://opg.optica.org/abstract.cfm?URI=boe-12-6-3658
50. Piccardo M, Ginis V, Forbes A, Mahler S, Friesem AA, Davidson N et al (2022) Roadmap on multimode light shaping. J Opt [Internet] 24(1):013001. https://iopscience.iop.org/article/10.1088/2040-8986/ac3a9d

51. Malý P, Mueller S, Lüttig J, Lambert C, Brixner T (2020) Signatures of exciton dynamics and interaction in coherently and fluorescence-detected four- and six-wave-mixing two-dimensional electronic spectroscopy. J Chem Phys [Internet] 153(14):144204. http://aip.scitation.org/doi/10.1063/5.0022743
52. Malý P, Lüttig J, Mueller S, Schreck MH, Lambert C, Brixner T (2020) Coherently and fluorescence-detected two-dimensional electronic spectroscopy: direct comparison on squaraine dimers. Phys Chem Chem Phys [Internet] 22(37):21222–21237. http://xlink.rsc.org/?DOI=D0CP03218B
53. Cho M (2021) Coherent nonlinear spectroscopy with multiple mode-locked lasers. J Phys Chem Lett [Internet] 12(42):10284–10294. https://pubs.acs.org/doi/10.1021/acs.jpclett.1c01492
54. Park J-S, Lee I-B, Moon H-M, Ryu J-S, Kong S-Y, Hong S-C et al (2020) Fluorescence-combined interferometric scattering imaging reveals nanoscale dynamic events of single nascent adhesions in living cells. J Phys Chem Lett [Internet] 11(23):10233–10241. https://pubs.acs.org/doi/10.1021/acs.jpclett.0c02103

A Quantitative Approach to Applications of Electronic Energy Transfer (EET)

Lennart B.-Å. Johansson

Contents

Abstract In brief, the Förster theory (FT) of electronic energy transfer (EET) between/among fluorescent and non-fluorescent molecules is discussed.

Here the EFT concerns applications for quantitative studies of biomolecular systems, in particular lipid membranes and proteins. The EET may take place within pairs of chromophores, as well as within random or regularly ordered distributions of chromophores. Described and exemplified are also relations between experimental data and theories.

Discussed finally are suggestions that aim at further extending the applicability of the FT.

Keywords EET in regular aggregates · Electronic energy transfer (EET) within pairs · Förster's theory · Two-particle approximation

Abbreviations

DAET Donor to acceptor energy transfer, also referred to as *hetero* FRET
DDEM Donor–donor energy migration, also referred to as *homo* fluorescence resonance energy transfer (FRET)

L. B.-Å.Johansson (✉)
Department of Chemistry, Umeå University, Umeå, Sweden
e-mail: Lennart.B.Johansson@umu.se

R. Šachl, M. Amaro (eds.), *Fluorescence Spectroscopy and Microscopy in Biology*,
Springer Ser Fluoresc (2023) 20: 87–98, https://doi.org/10.1007/4243_2022_28,
Published online: 14 August 2022

EET	Electronic energy transfer
FT	Förster's theory
MC	Monte Carlo

1 Background to Förster's Theory

Early in the twentieth century [1–3] experiments revealed that electronic energy might be transferred without transfer of matter of any kind, e.g. as in sensitised fluorescence in vapours. A detailed physical explanation was at first derived from quantum theory by Theodor Förster [4]. The theoretical treatment also covers similar phenomena observed among chromophoric/fluorescent molecules. Förster's explanation assumes *weak coupling* between electric transition dipole moments of electronically excited ($\widehat{\mu}_D$) and non-excited ($\widehat{\mu}_A$) species (cf. Fig. 1). Within this approach and by applying quantum mechanical time-dependent perturbation theory, Förster derived the rate constant of electronic transfer (ω) between an electronically excited molecule {i.e. the donor group (D)} and another compound being in its electronic ground state, named the acceptor (A). In fact, the process defines a specific fluorescence quenching process of an excited D-molecule, but unlike most other fluorescence quenching processes the rate of EET is rather simply related to experimental properties. The rate of transfer reads:

$$\omega = \frac{3}{2}\frac{1}{\tau_D}\kappa^2\left[\frac{R}{R_0}\right]^6 \tag{1}$$

Here τ_D stands for the fluorescence lifetime of the donor in absence of acceptors. The distance between D and A is R, whereas the Förster radius (R_0) can be calculated

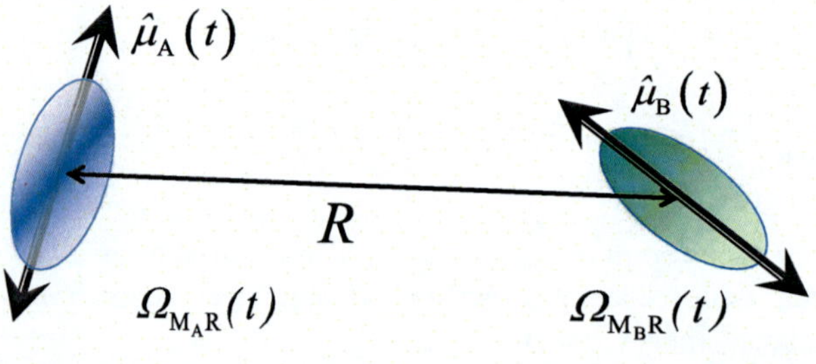

Fig. 1 Schematic indicating electric dipole–dipole coupling between two chromophoric molecules A and B. Here A and B denote the acceptor and B the donor, or both being donors, i.e. if A and B are of the same kind. The unit transition dipoles ($\widehat{\mu}_i$, i = A or B) are separated at a distance R with orientations described relative a common molecular frame of $\widehat{R}$, with mutual orientations indicated by Eulerian angles Ω_{M_iR} (i = A or B). The time-dependence accounts for molecular reorientations that may change on the timescales of EET and fluorescence

from the overlapping between the corrected normalised D-fluorescence $\{f_D(\lambda)\}$ and molar absorptivity $\{\varepsilon_A(\lambda)\}$ spectra;

$$R_0 = \left\{ \frac{9000(\ln 10)(2/3)\Phi J}{125\pi^5 n^5 N_A} \right\}^{1/6} \tag{2a}$$

$$J = \int_{\text{spectra}} \varepsilon_A(\lambda) f_D(\lambda) \lambda^4 d\lambda \tag{2b}$$

Φ, n and N_A denote the fluorescence quantum yield of the donor, the refractive index of the solvent and the Avogadro constant, respectively. Conveniently, κ^2 is taken to be the isotropic average, i.e. $\langle\kappa^2\rangle = 2/3$ for rotational motions being much faster than the transfer rates.

The κ^2-factor accounts for the mutual orientation of the coupled D and A electronic transition dipoles;

$$\kappa^2 = \left[\hat{\mu}_D \cdot \hat{\mu}_A - 3\left(\hat{\mu}_D \cdot \hat{R}\right)\left(\hat{R} \cdot \hat{\mu}_A\right) \right]^2 \tag{2c}$$

Here, $\hat{\mu}_A$, $\hat{\mu}_D$ and $\hat{R}$ denote unit vectors.

Notice that the above mechanism also explains energy transport between chromophoric molecules of the *same* kind, since their fluorescence and absorption spectra are usually overlapping. Thus, EET is possible from an excited D-molecule to another D molecule in its electronic ground state. We therefore have to distinguish between EET in two extreme cases, namely the reversible DD transfer and the irreversible DA transfer. The former process enables migration of the initially absorbed excitation energy among several donor molecules, i.e. between two or more donors and it is here referred to as a donor–donor energy migration (DDEM) in space.

From Förster's theory (i.e. Eqs. 1 and 2) it follows that spatial as well as orientational distributions can substantially influence the rate of EET.

2 Observing EET

Electronic energy transfer from an excited donor to an acceptor must shorten the D-fluorescence lifetime. It is therefore detectable by fluorescence lifetime measurements, or by the observed fluorescence spectral intensities, in presence and absence of acceptors. On the other hand, the reversible DD energy migration process implies that the number of excited donor molecules remains invariant, whereby the fluorescence relaxation time of donors remains equal to that in absence of coupling to ground-state donors. In most D-systems studied, however, the *mutual orientations* of a primary (or *initially*) and a secondarily excited D-molecule differ. Consequently the polarisation of the emitted fluorescent photon differs from that created by the initially excited donors. The DDEM process obviously acts similar as to a reorientation of the primarily excited D. Fluorescence depolarisation experiments provide information about *changed* orientations of excited molecules at the times of excitation and fluorescence emission. Thus, depolarisation experiments simultaneously report on the rates of reorientations, as well as that of energy migrations among donors. Therefore, a fundamental question reads; how to distinguish between the two processes? A further elaboration on this is presented in subsections below.

2.1 Analyses of EET Data Within Pairs of Chromophores

Frequently applications of EET concern the determination of inter-chromophoric distances, e.g. as related to macromolecular structure. Numerous studies have been reported in the literature (see, e.g., the textbook of Lakowicz [5] and papers cited therein). Structural intramolecular distances within, e.g., a protein can be estimated in EET experiments by means of using DA or DD pairs. This, is because typical Förster radii range between ca. 10–100 Å, i.e. intramolecular distances in, as well as dimensions of proteins. A reasonable objection to the use of EET in this context would be that structural information is also available from NMR and X-ray studies, and these techniques often provide a higher resolution. However, this is *not* generally true, because these methods exhibit limitations for reasons of inherent restrictions, as well as difficulties in preparing samples. For instance, it has been shown that structural changes upon adequate perturbations/interactions are available from EET experiments on suitably labelled proteins [6, 7].

Pairs of chromophores could be naturally present (= commonly referred to as intrinsic probes), or artificially introduced (extrinsic probes) into a macromolecule. Well-defined extrinsic labelling of proteins is possible by means of *site-specific mutagenesis* for introducing reactive amino acids, which can be specifically labelled with reactive chromophores. For this purpose, Cys-residues are useful in the labelling of proteins with sulfhydryl specific chromophores. If no suitable intrinsic probes are available in a protein, it is most convenient to introduce two donor groups [7]. Hereby one avoids different mixtures of non-specific labelling of two reactive positions with possible combinations and fractions of D-A, A-D, D-D and A-A pairs.

The excitation probability of the *initially* excited D group within a DD or a DA pair at a time t later $\{\chi(t)\}$ is according to Förster's theory given by:

$$\chi(t) = \frac{1}{2}[1 + \exp(-\gamma\langle\omega\rangle t)]F_{\mathrm{D}}(t), \tag{3}$$

where $\gamma = 1$ or 2 for DA or DD transfer, respectively. The bracket $\langle\omega\rangle$ indicates an orientational average of the interacting electronic transition dipoles. Furthermore, $F_{\mathrm{D}}(t)$ denotes the fluorescence relaxation in the absence of EET.

As pointed out above only fluorescence depolarisation experiments reveal DD electronic energy transfer, but also depend on the reorienting and transfer rates, as well as orientational *restrictions* of the donors. Hence, the rate of EET is influenced, since κ^2 depends on time if the transfer reorientations occur on similar time scales. To account for this an extended Förster theory was derived [8]. In this approach the excitation probability of the initially excited D within DD pairs reads;

$$\chi(t) = \frac{1}{2}\left\{\left\langle\left[1 + \exp\left(-2\int_0^t \omega(t')dt'\right)\right]\right\rangle\right\} F_D(t)t) \tag{4}$$

Equation 4 is the solution to the *stochastic* master equation of energy migration within a DD pair, which was derived from the stochastic Liouville equation [9]. Notice in this context that $\{\langle\ldots\rangle\}$ denotes a *stochastic average*. At times $t \to 0$ Eq. 4 becomes equal to Eq. 3.

In general, mutual reorientations within a pair of donors usually change due to reorienting motions at each D-binding site, e.g. in a protein.

The fluorescence anisotropy $\{r(t)\}$ constructed from data recorded by depolarisation experiments is then composed of different contributions, as given below [10]:

$$r(t) = \frac{r(0)}{2}\left[\rho_{AA}(t) + \rho_{BB}(t) + \rho_{AB}(t) + \rho_{BA}(t)\right]exp\left(-t/\phi_{c,glob}\right) \tag{5a}$$

As it stands, Eq. 5a also accounts for the overall global tumbling of the macromolecule ($\phi_{c,\ glob}$), which often can be neglected [10]. The initial excitation probability of D_A and D_B is equal (= 1/2). The fluorescence anisotropy of each donor in the absence of EET is described by $\rho_{ii}(t)$ (i = A or B). The contribution due to energy transfer from D_i to D_j is described by $\rho_{ij}(t)$. Taken together the ρ-terms are:

$$\rho_{ii}(t) = \left\langle P_2[\widehat{\mu}_i(0) \cdot \widehat{\mu}_i(t)]\ \frac{\chi(t)}{F_{D(t)}}\right\rangle\ i = A, B \tag{5b}$$

$$\rho_{ij}(t) = \left\langle P_2\left[\widehat{\mu}_i(0) \cdot \widehat{\mu}_j(t)\right]\ \frac{1-\chi(t)}{F_{D(t)}}\right\rangle\ i,j = A, B \text{ where } A \neq B \tag{5c}$$

$P_2\left[\widehat{\mu}_i(0) \cdot \widehat{\mu}_j(t)\right]$ are second rank Legendre polynomials which correlate the mutual orientations of the electronic transition dipole moments at times of excitation ($t = 0$) and emission ($t = t$). For more details in applying the extended Förster theory the reader is kindly recommended to consider references [9] and [10], as well as papers cited therein.

2.2 *Analyses of EET Data of Donors in Spatial Distributions*

For systems composed of many randomly distributed donors undergoing energy migration one faces additional questions concerning the interpretation of fluorescence depolarisation data. Firstly, how to account for the description of energy migration among several donors. This question involves the handling of many-donor interactions, but secondly also how to account for the influence of donor orientation rates and restrictions. The influence of translational motions in condensed media is usually negligible, since the fluorescence and energy transfer processes are much faster rates. However, by means of Monte Carlo (MC) simulations the description of these complicating circumstances can be overcome. For three-dimensional isotropic distributions of donors MC simulations have been compared with an analytical model previously derived [11]. It turns out that the agreement is very good at *reduced concentrations* of <2 [12]. The concept reduced concentration refers to the average number of interacting molecules (here donors) within a sphere defined by the Förster radius, R_0. In two-dimensional systems, the corresponding property refers to the average number of donors within a circle defined by R_0. For energy migration in two-dimensional *random* distributions, e.g. a model membrane or lipid bilayer, MC simulations [13] support a previously suggested model derived within the so-called two-particle approximation [14]. The two-particle approximation allows for deriving analytical equations for the time-dependent excitation probability of the *initially* excited donor $\{G^s(t)\}$. $G^s(t)$ depends on reduced concentrations (usually known) and certain mutual extreme orientations. For a lipid bilayer one needs to distinguish between inter- and intralayer donor energy migration. Within the Baumann-Fayer model [14] these contribute to the excitation probability of the initially excited donors yield as the joint probability:

$$G^s(t) = G^s_{\text{intra}}(t)G^s_{\text{inter}}(t) \tag{6}$$

A reasonable approach to modelling depolarisation experiments and to the fluorescence anisotropy reads:

$$r_{\text{DDEM}}(t) = G^s(t)r(t) \tag{7}$$

Here, $r(t)$ denotes the anisotropy obtained in the *absence* of energy migration, i.e. at low reduced concentrations.

In membranes composed of chemically different lipid species, the so-called rafts [15] may form. The rafts refer to regions comprising different lipid compositions. These regions mean non-uniform overall lipid distributions/compositions. From the studies of fluorophore-labelled lipids, MC analyses were adopted and indeed reveal heterogeneity in lipid membrane mixtures [16]. In a separate section of this book these aspects are further elaborated on.

2.3 *Analyses of EET Data Among Donors Within* Periodic *Spatial Distributions*

Some proteins form polymerised regular structures, i.e. arrangements with periodic spatial and orientation distributions of monomeric protein molecules. One important and well-known example is the muscle protein actin, which forms a helical structure, named F-actin [17] from monomeric actin, i.e. G-actin. For mixtures of mono-D-labelled monomers and wild type monomers, it is reasonable to assume a resulting non-covalent polymer with a statistically random distribution of labelled monomers. For different fractions of D-labelled monomers, one can investigate the influence of energy migration by means of fluorescence depolarisation experiments. Figure 2 displays data obtained for fractions of D-labelled G-actin ranging between 0.1 and 49 mol% [18]. As could be expected, the fluorescence anisotropy data reveal an increasing rate of energy migration with increasing D concentration. At the lowest concentration, however, there is an initial rapid relaxation of $r(t)$, which reaches a plateau region. The plateau region is compatible with the absence of energy migration, as well as the influence of F-actin and G-protein reorientations in the polymer on the timescale of fluorescence. Consequently, the anisotropy at <0.1 mol% represents F-actin containing *non-interacting* donor groups, where the initial relaxation can be ascribed to *local* reorientations of the donor group at its binding site. For a comparison, the anisotropy decays of donor labelled G-actin obtained at 277 and 293 K are also displayed in Fig. 2b. After a rapid initial relaxation a slower relaxation is observed which is compatible with the global orientation diffusion of the G-protein molecule.

The energy migration in the helical structure occurs among donors arranged within a repeated structural pattern, as is illustrated in Fig. 3. Every G-actin monomer exhibits a well-defined localisation with respect to its nearest neighbours. Because of the regular arrangement, it is therefore possible to use a limited number of unknown parameters to describe the relative monomer positions. These are defined in Fig. 3 and correspond to the distance from the helical axis (i.e. the C_∞-axis) to the labelled donor and that between labelled nearest D-groups along the C_∞-axis, denoted T_{xy} and T_z, respectively. The third parameter is the angle of rotation (θ) between nearest donor neighbours about the C_∞-axis. By means of MC simulations, it is possible to investigate possible migration paths and find the solutions that best fit to experimental data [18]. Notice, in common for all experimental data obtained on F-actin are the structural parameters (T_{xy}, T_z and θ) which should be invariant to the different degrees of labelling.

Data analyses are therefore suitable for a global analysis, i.e. all data are simultaneously fitted to the *same* model. The analyses require an efficient methodological approach, which enables the search of best-fit parameters in a multidimensional space. By using the so-called genetic algorithms this is possible [18, 19]. For further details, kindly consider reference [18] and papers cited therein.

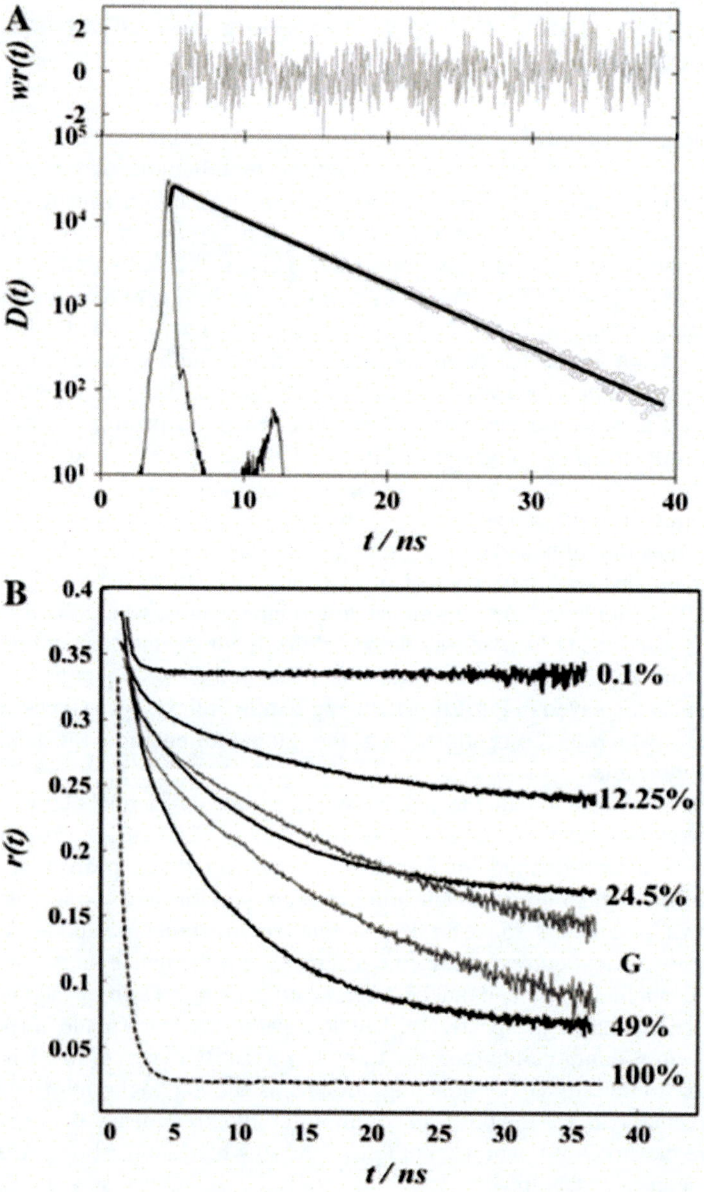

Fig. 2 Fluorescence depolarisation data obtained on fluorophore-labelled F-actin (BODIPY [18]). (**a**) The next most upper panel shows difference data $\{D(t)\}$ obtained for a 0.1 mol% BODIPY-labelled F-actin. The best-fit obtained, i.e. weighted residuals $\{wr(t)\}$ are displayed in the top panel. (**b**) A collection of fluorescence anisotropy $\{r(t)\}$ data calculated from depolarisation experiments with various labelling degree of F-actin samples (in *black*). The % refer to different mole fractions of BODIPY-labelled F-actin at 277 K. The dashed curve was obtained from simulations using the parameters extracted from global analyses and when assuming 100 mol% labelling of F-actin. The two decay anisotropy curves (in *blue*) are data obtained for G-actin at 277 K (upper) and 293 K (lower). Reprinted with kind permission of Biophysical Journal

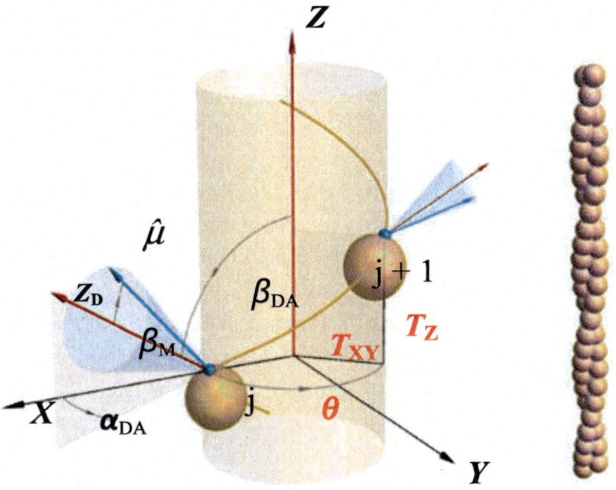

Fig. 3 Schematic showing the coordinate systems used to describe G-protein positions in regular structures forming helical, linear and ring-shaped aggregates. The Z_A axis coincides with the C_∞–axis of the aggregate, and T_{xy} denotes the distance from this axis to the position of a fluorescent group. The translational and rotational transformations between nearest protein neighbours are θ and T_z, respectively. The fluorophore undergoes local reorienting motions about an effective symmetry axis Z_D which transformed to the aggregate fixed frame by ($\propto_{DA}$, β_{DA})

3 Closing Comments

The aims of this chapter have been twofold; to provide a physical basis of electronic energy transfer, and to give a glimpse on the possibilities of using EET also as a *quantitative tool*, with focus on biomolecular applications on structure, function and dynamics. In this respect, I recall that few, if any luminescence quenching process, except for EET in the limit of weak coupling, can provide a quantitative connection between spectroscopic data and molecular properties.

Among several EET-applicants a crucial question concerns the orientation dependence of the coupling strength, i.e. the κ^2-factor, that can take values ranging between 0 (= i.e. forbidden EET) and 4. Another aspect deals with the influence of reorientation motions that can occur on the timescale of electronic coupling. In studies of systems with pairwise coupling, such as DA or DD labelled proteins, the handling of the κ^2-factor often needs to be carefully considered, preferably by means of independent experiments on donors and acceptors. The local motions and orientation restrictions of donors and acceptors can be determined from separate experiments on mono-labelled samples, which are then localised at adequate locations/positions in the structure investigated [10].

For DA systems with chromophores exhibiting *degenerate* transition dipoles in three dimensions, either of the D or the A group, the average κ^2-value will always be 2/3. Candidates of this kind are, for instance, complexes exhibiting charge-transfer transitions between metal ions and ligands of tetrahedral or higher symmetry.

Notice, it is sufficient if *one* of the components in a DA system fulfils this symmetry requirement. Hereby the rate of EET is solely dictated by the distance of separation (*R*).

A most powerful development for extending the applicability of EET on proteins would be to use "on-purpose designed" artificial fluorophoric amino acids. These would enable specific insertion at the level of expression of a protein, i.e. *site-specific mutagenesis*. The approach involves the development of a unique code for an *artificial* amino acid that corresponds to a specific codon, which enables insertion into the template sequence of the accessible primary sequence. In principle, any position in the protein structure is thereby replaceable by an artificial fluorophore. This approach would also solve the problem with specificity when labelling a protein with one donor and one acceptor. Furthermore, the methodology also makes it possible to label interior positions of a protein, which is tricky to perform.

Within the validity of Förster's mechanism EET experiments can be most informative for quantitative studies of structural *change,* e.g. in the presence of interacting components, such as substrates. This is a difficult task with NMR and X-ray methods, although these methods, no doubt, often can provide more detailed 3D structures. It should be emphasised that these established methods still are associated with severe limitations, not least concerning resolution and crystal quality. Thus, further developed/refined EET spectroscopy is a complementary tool, especially when insufficient structural information is unavailable by means of X-ray and NMR techniques.

References

1. Cario G (1922) Z Physik 10:185
2. Weiss J (1938) Nature 141:248
3. Weiss J (1939) Trans Faraday Soc 35:48
4. Förster T (1948) Ann Phys Berlin 2:55
5. Lakowicz JR (1999) Principles of fluorescence spectroscopy. Kluwer/Plenum, New York
6. Mikaelsson T, Ådén J, Wittung-Stafshede P, Johansson LB-Å (2014) Biophys J 107:401
7. Bergström F, Hägglöf P, Karolin J, Ny T, Johansson LB-Å (1999) PNAS 96:12477
8. Kubo R (1969) Stochastic processes in chemical physics. Advances in chemical physics, vol XVI, Shuler KE (ed). Wiley, New York
9. Johansson LB-Å, Edman P, Westlund P-O (1992) J Chem Phys 105:10896
10. Isaksson M, Hägglöf P, Håkansson P, Ny T, Johansson LB-Å (2007) PCCP 9:3914
11. Gouchanour CR, Andersen HC, Fayer MD (1979) J Chem Phys 70:4254
12. Engström S, Lindberg M, Johansson LB-Å (1988) J Chem Phys 89:204
13. Johansson LB-Å, Engström S, Lindberg M (1992) J Chem Phys 96:3844
14. Baumann J, Fayer MD (1986) J Chem Phys 85:4087
15. Hammond AT, Herbele FA, Baumgart T, Holowka D, Baird B, Feigenson GW (2005) PNAS USA 102:6320

16. Sachl R, Johansson LB-Å, Hof M (2012) Int J Mol Sci 13:16141
17. Lorentz M, Popp D, Holmes KC (1993) J Mol Biol 234:826
18. Marushchak D, Grenklo S, Johansson T, Karlsson R, Johansson LB-Å (2007) Biophys J 93:3291
19. Charbonneau P (1995) Astrophys J Suppl Ser 101:309

Single-Molecule FRET: Principles and Analysis

B. Israels, L. M. Lund, and V. Birkedal

Contents

Abstract Förster resonance energy transfer (FRET) is a powerful spectroscopic technique to study conformational changes of nucleic acids and proteins and their molecular interactions. When combined with a single-molecule approach, FRET has the distinct advantage that it can monitor the conformational heterogeneity and dynamics of individual molecules and enable the observation of short-lived molecular intermediates usually hidden in ensemble experiments. This in turn makes single-molecule FRET an interesting tool for dynamic structural biology.

This chapter presents the principles of single-molecule FRET spectroscopy and the added information it gives compared to ensemble FRET spectroscopy. We describe different experimental implementations, primarily focusing on intensity-based approaches. Fluorescence from single molecules requires careful experimental procedures to maximize the inherently low signal intensity, and meticulous data analysis, which is introduced in this chapter, to quantify FRET detection. We

B. Israels, L. M. Lund, and V. Birkedal (✉)
Department of Chemistry and Interdisciplinary Nanoscience Center (iNANO), Aarhus University, Aarhus, Denmark
e-mail: bisraels@inano.au.dk; linelund95@inano.au.dk; vicb@inano.au.dk

R. Šachl, M. Amaro (eds.), *Fluorescence Spectroscopy and Microscopy in Biology*, Springer Ser Fluoresc (2023) 20: 99–118, https://doi.org/10.1007/4243_2022_32,
Published online: 3 January 2023

comment on advantages and limitations of the technique, and its strength is illustrated by two application examples.

Keywords Conformational dynamics · Fluorescence microscopy · FRET spectroscopy · Quantitative FRET · Single-molecule FRET (smFRET)

1 Introduction

Life is dynamic and understanding its inner workings is a central scientific challenge. The cellular machinery is driven by interactions between DNA, RNA, proteins, and their environment. Molecular structure, binding affinity, and dynamics are each important regulators of these biochemical processes. A full view of this molecular dance, which allows resolving different molecular conformations, how they interconvert, and how molecules interact with each other, would greatly improve our mechanistic understanding of biology. It is unfortunately not currently possible to capture biological processes with atomistic spatial resolution and sub-millisecond time resolution. However, several techniques, alone or in combination, are striving to obtain high spatial and temporal resolutions toward dynamic structural biology [1]. They include, but are not limited to, X-ray crystallography, nuclear magnetic resonance (NMR), electron microscopy, molecular dynamics simulations, and Förster resonance energy transfer (FRET) approaches. The FRET mechanism describes non-radiative energy transfer through dipole–dipole coupling between a donor (D) and an acceptor (A) molecule [2, 3]. The energy transfer therefore strongly depends on the distance between the particular donor and acceptor pair typically in the range of 2–10 nm (Fig. 1). This approach is powerful in the context of dynamic structural biology as it allows distance measurements well below the optical diffraction limit, which can be used to identify and/or refine biomolecular structures. Furthermore, the coupling of FRET with single-molecule microscopy enables an exquisite quantification of dynamics.

In this chapter, we present an overview of smFRET principles and outputs and compare them to ensemble FRET experiments. We discuss experimental implementation basics focusing on intensity-based approaches and data analysis procedures so that the reader is introduced to the advantages and limitations of smFRET. Finally, we illustrate smFRET results by highlighting case studies in nucleic acid folding and protein–nucleic acid interactions.

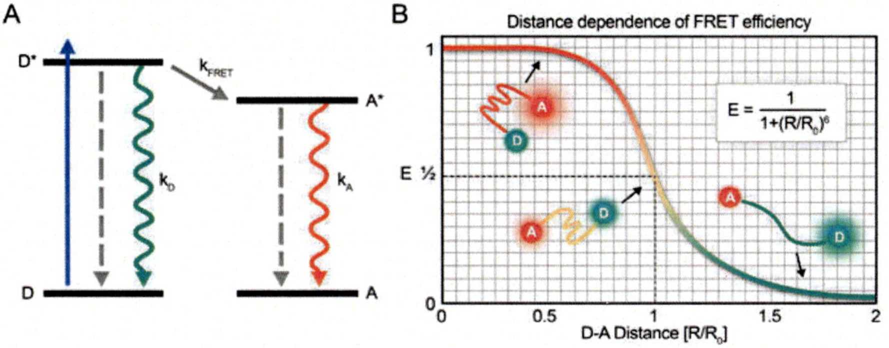

Fig. 1 (**a**) Illustration of the FRET mechanism. The donor (*D*) molecule is excited (blue arrow) and can relax from the excited state D^* by several channels: radiationless relaxation (dashed gray arrow), fluorescence (photon emission with a radiative rate k_D) (green wavy arrow), or energy transfer to the acceptor (*A*) molecule at rate k_{FRET}. The acceptor molecule can return to the ground state by emitting a photon with a radiative rate k_A (red wavy arrow) and/or by radiationless relaxation (dashed gray arrow). The FRET efficiency (*E*) is given by the rate of energy transfer between the donor and the acceptor, k_{FRET}, divided by the sum of all radiative and non-radiative relaxation rates of the donor. Often both the donor and the acceptor are fluorescent molecules; however, only the donor needs to be fluorescent. (**b**) Distance dependence of FRET efficiency (*E*). *R* is the distance between the donor (colored green) and the acceptor (colored red). R_0, the distance where the transfer efficiency is 50%, is the Förster radius specific for the donor–acceptor pair. The three different molecular conformations shown in the plot are illustrative of short, intermediate, and long donor–acceptor distances, respectively. The size of the glow around each *D*/*A* molecule is illustrative of the relative contribution of the green donor and the red acceptor for those conformations

2 Principles of Ensemble FRET

Most biomolecules are not intrinsically fluorescent and are typically modified by the addition of organic dyes to serve as FRET reporters on conformations. The dyes in a FRET pair are characterized by their chemical properties (molecular structure, size, charge distribution, hydrophobicity) and their electronic properties (spectral overlap of the emission spectrum of the donor and the absorption spectrum of the acceptor, extinction coefficients, quantum yields, and fluorescence lifetimes). The selection of suitable fluorophores for FRET depends on the nature of the biomolecule and experiment.

FRET spectroscopy at the ensemble level has been successfully used for many years to assess conformational changes, molecular interactions, and the average distance between fluorophores [3, 4]. A typical experimental approach based on measuring fluorescence intensities is illustrated in Fig. 2. The FRET efficiency, *E*, can be obtained by a number of experimental strategies, including quantifying the amount of donor fluorescence in the presence and absence of the acceptor, or by measuring the relative fluorescence of both the donor and the acceptor [5].

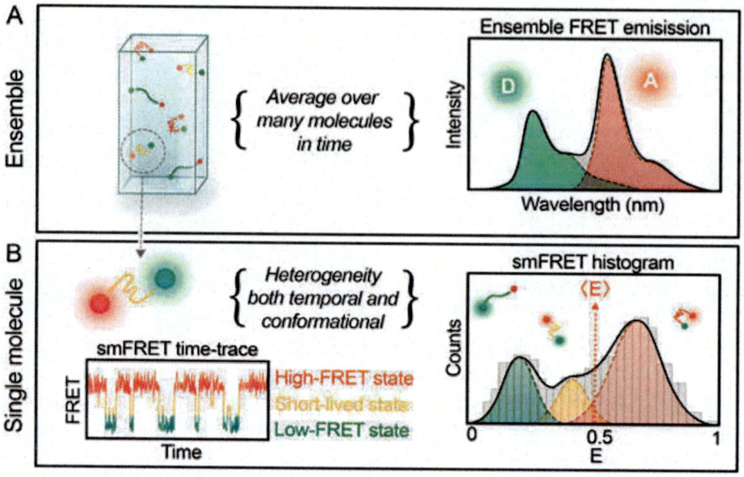

Fig. 2 Illustration of the contrast between ensemble measurements and single-molecule FRET. (**a**) An illustration of the ensemble example where many molecules are averaged over both conformations and time. This approach gives information about donor and acceptor chromophore emission using fluorescence spectroscopy yielding an average FRET value for the entire population of molecules. (**b**) An illustration showing how temporal and conformational heterogeneity in a sample can be resolved using smFRET. FRET time traces (called single-molecule trajectories) yield information about the distribution of molecules in different conformational states as well as the kinetics of state-to-state transitions. Conformational distribution information for a population of molecules is illustrated in the smFRET efficiency histogram revealing three subpopulations. The black and dashed lines illustrate Gaussian fits to the underlying conformational distribution smFRET data. The dotted red line marks the average FRET efficiency $\langle E \rangle$ from the histogram, which can be compared to the value derived from ensemble experiments depicted in panel **a**

Ensemble FRET experiments can yield information on the rate of conformational changes by making use of fast-mixing techniques before equilibrium is established [6]. FRET efficiencies can also be determined by using lifetime measurements, and in certain cases, this can enable obtaining the distribution of molecular distances in solution [7, 8].

Ensemble measurements are inherently limited to averaging molecular properties because an ensemble measurement reports on many molecules at once. This limitation is overcome with single-molecule microscopy.

3 Principles of smFRET: Beyond the Ensemble Average

Since most biological reactions happen via the action of single enzymes, DNA molecules, or RNA molecules, the biomolecular dynamics are inherently stochastic and unsynchronized. The ability to directly follow single-molecule fluorescence enables the observation of biological reactions in a manner that is typically not accessible through the use of conventional ensemble techniques [9]. While an ensemble FRET measurement often yields an average FRET value for the entire population of molecules, smFRET probes each molecule individually over time, resulting in heterogeneous FRET distributions in both time and conformational space (Fig. 2b). There are a number of ways to determine FRET efficiencies at the single-molecule level [10, 11]; it is often advantageous to use both donor and acceptor fluorescence intensities. From the first demonstration on dehydrated samples in 1996 [12], smFRET was developed, to allow solution-based in vitro studies of biomolecular interactions and conformational dynamics of nucleic acids and proteins [13–17]. More recently, smFRET investigations in cells have become possible [18, 19].

3.1 Biomolecular Conformations and Dynamics

Biomolecular dynamics cover a wide timescale (Fig. 3). smFRET offers the ability to track conformational changes of an individual molecule or biomolecular complex as a function of time (Fig. 2b) covering a sub-microsecond to hours-long timescale. This approach enables the investigation of chain dynamics and domain motion taking place in the sub-microseconds and microseconds range, nucleic acid and protein folding in the microseconds to minutes range, and processes that can take hours or longer, such as protein aggregation [24–26]. Assigning conformational

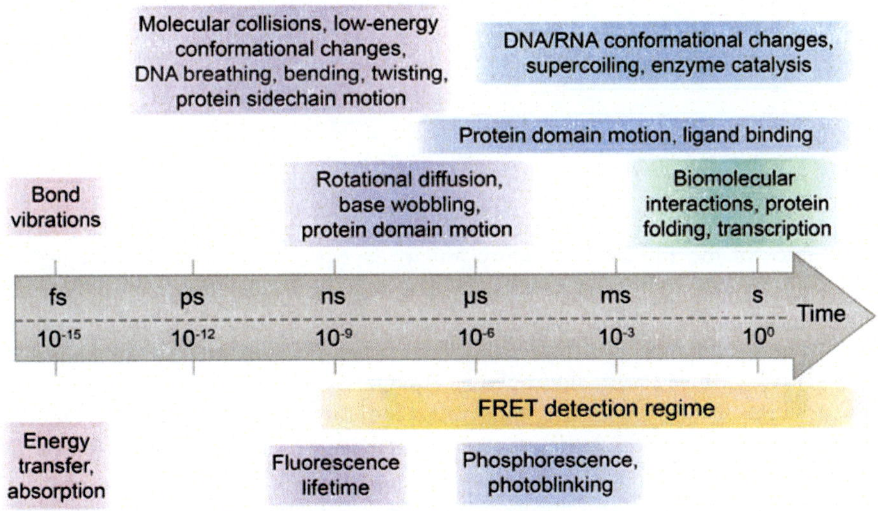

Fig. 3 Range of biomolecular dynamics. Dynamic processes [14, 20–22] (top) and photophysical processes [23] (bottom) correlated with their respective time intervals. smFRET can report on processes occurring over timescales from sub-microseconds to several hours resolving a broad spectrum of dynamics from nucleic acid conformational changes to protein folding and molecular interactions

states to each of the experimentally identified FRET efficiency states enables the identification of important short-lived transient states, unraveling reaction and dynamic exchange pathways [20], and coupling dynamic and structural information with biochemical functions [27, 28]. Intermediate structures and the allowed pathway dynamics are crucial for understanding the overall mechanism of molecular folding and biomolecular interactions.

3.2 *Quantitative FRET*

The concept of using FRET quantitatively to determine intermolecular distance was first described in 1967 by Stryer and Haugland as the "spectroscopic ruler" and applied at the ensemble level [3]. For many applications, smFRET is often used to report on conformational changes and their dynamics without determining absolute FRET efficiencies. Quantitative smFRET distances are essential for FRET-based structural studies. They yield quantitative information on the conformations of large, heterogeneous, and dynamic biomolecules and their complexes [29].

Converting smFRET measurements to absolute distances requires a number of critical steps that include the determination of absolute FRET efficiencies that are not instrument-dependent (see Sect. 4.3) and modeling of fluorophore dynamics [11, 30–32]. A study involving 20 laboratories around the world investigated the precision and accuracy of smFRET measurements [11]. The quantitative assessment of smFRET intensities matched well with the expected structural distances using several double-stranded DNA constructs. Thus, smFRET is an interesting tool for structure determination and validation of flexible and dynamic large structures, which are difficult to solve with other approaches.

4 Experimental Considerations for smFRET

The choice of the smFRET modality, microscope, and experimental schemes, as well as data analysis, depends on several factors including the type of molecule being studied, the nature and scale of the intermolecular interaction, the timescale of the dynamics, and the available equipment. smFRET requires the molecules of interest to be labeled with fluorophores that are typically separated by distances in the range of 2–10 nm and thus relies on robust and site-specific labeling [33–36]. FRET pairs with high extinction coefficients and high quantum yields are a suitable choice for smFRET experiments. An understanding of the intrinsic photophysical properties of fluorophores is also important as they can affect experimental results [37]. If the interaction involves only two molecules, a single-FRET pair may be sufficient. If the interaction has three or more simultaneous interactions, a three- or four-color FRET scheme may be more appropriate [38].

4.1 Microscope Configurations: Confocal and Widefield Microscopy

There are two general types of single-molecule FRET experimental modalities: confocal microscopy and wide-field total internal reflection fluorescence (TIRF) microscopy (Fig. 4a, c). Confocal geometries allow studying molecules freely diffusing in solution, one at a time (Fig. 4a, b). They are fundamentally limited to an observation time window corresponding to the time it takes for molecules to diffuse through the femtoliter volume excited by a laser, generally in the order of 10 ms or less [39].

If a chemical process or mechanism occurs over longer timescales, the molecules can be immobilized on a surface allowing for prolonged observation of many hours. Some commonly used methods to attach molecules to a microscope slide involve a biotin–streptavidin linkage [40]. If desired, the slide surface can be passivated to make it resistant to non-specific binding by molecules in the solution, for example, by the use of polyethylene glycol (PEG) [41, 42]. To image these surface-immobilized molecules, scanning confocal microscopy or wide-field microscopy can be used, the latter having the advantage to measure many molecules simultaneously. To selectively excite the molecules on the surface and not others in solution, and thus decrease background fluorescence, the phenomenon of total internal reflection (TIR) is utilized (Fig. 4c, d). TIRF can be implemented in a prism-based geometry and a microscope objective-based geometry.

4.2 Excitation Schemes and Multiplexed Detection

There are various ways to implement smFRET experiments that yield different information on conformational changes and dynamics. The techniques vary both in how the molecules are excited and in what signals are detected. The most common and simplest method for smFRET is single-pair (sp) FRET [12], which is compatible both with confocal and wide-field microscopy. spFRET uses single-color excitation of the donor molecule and simultaneous detection of the emission intensity of both donor and acceptor molecules. The absolute FRET efficiency, E, is given by Eq. 1.

$$E = \frac{I_{\mathrm{FRET}}}{I_{\mathrm{FRET}} + \gamma I_{D_{\mathrm{em}}|D_{\mathrm{ex}}}} \tag{1}$$

where I_{FRET} is the background-corrected fluorescence intensity of the acceptor dye resulting from energy transfer from the donor (see Sect. 4.3 for details of I_{FRET} determination). $I_{D_{\mathrm{em}}|D_{\mathrm{ex}}}$ and $I_{A_{\mathrm{em}}|D_{\mathrm{ex}}}$ are detected to determine FRET efficiencies (Fig. 5a) and denote the intensity of the donor or acceptor emission, respectively, upon donor excitation. γ is a normalization constant that accounts for differences in chromophore quantum-yield (Φ) and detector-efficiency (η) and is given by Eq. 2.

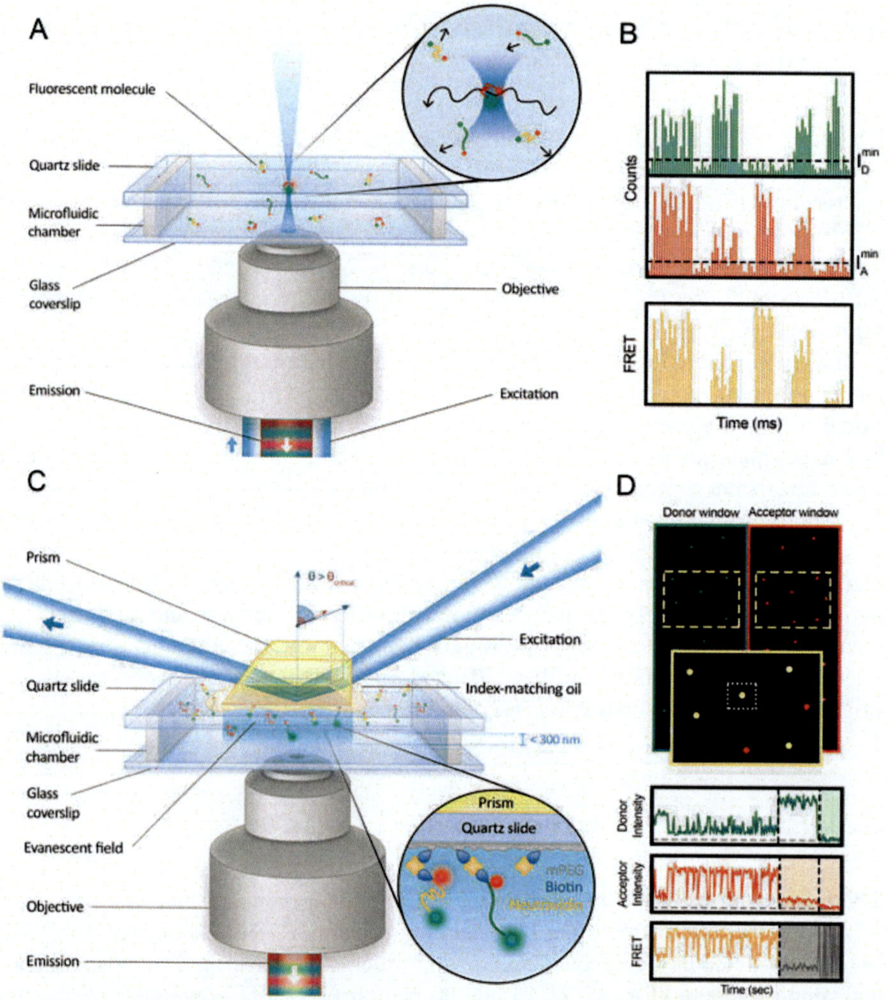

Fig. 4 Single-molecule FRET experimental schematics. (**a**) Example of a confocal microscope. The laser enters the back-end of the objective and is focused to a femtoliter volume inside the sample. Freely diffusing molecules are excited as they pass through the laser focal spot. Fluorescence from the molecules is collected through the same objective, focused through a pinhole to reduce out-of-plane background, and sent to point detectors. (**b**) Sample data from confocal FRET microscopy. Bursts of photons are collected by the donor detector (colored green) and the acceptor detector (colored red). All events with fluorescence intensity above a user-defined threshold (dashed black lines, $I_D^{\min}$ and $I_A^{\min}$) are interpreted as real events; from left to right: a high FRET, low FRET, acceptor-only, intermediate FRET, and donor-only signal. The yellow lines represent FRET values that make it into the final analysis after removing the low-intensity donor-only and acceptor-only subpopulations. (**c**) Example of a TIRF wide-field microscope: to perform prism-based TIRF, the incident light from a laser passes from the prism through index matching oil, into the quartz slide, and ultimately enters the sample at a super-critical angle. Alternatively, wide-field TIRF can be accomplished by exciting molecules bound to the bottom surface of a sample chamber in an objective-based geometry. The fluorescence from the sample is collected with an inverted

$$\gamma = \frac{\eta_A \Phi_A}{\eta_D \Phi_D} \tag{2}$$

where η_A and η_D are the detector efficiencies for the acceptor and the donor, and Φ_A and Φ_D are the quantum yields for the acceptor and the donor [11, 43–45].

spFRET is often used to report on conformational changes through relative FRET efficiencies. Using two lasers to excite both the donor and acceptor molecules yields additional information. One of these approaches is alternating-laser excitation (ALEX) [46, 47], which can be implemented both with confocal and wide-field microscopy. This technique allows superior detection of the low FRET efficiency subpopulation, which can be difficult to distinguish from the donor-only population that is often a dominant contribution around E~0 using spFRET. Donor-only single molecules are created due to acceptor photobleaching. Furthermore, ALEX enables directly identifying eventual acceptor photophysical effects and separating those from conformational changes. With the direct excitation of the acceptor molecule made possible, one can measure two additional quantities compared to spFRET: the intensity of the donor after acceptor excitation, $I_{D_{\mathrm{em}}|A_{\mathrm{ex}}}$, and the intensity of the acceptor after direct acceptor excitation, $I_{A_{\mathrm{em}}|A_{\mathrm{ex}}}$ (Fig. 5a). These quantities allow the determination of the stoichiometry ratio, S, which quantifies the ratio of donor to acceptor molecules and is given by Eq. 3.

$$S = \frac{I_{\mathrm{FRET}} + \gamma I_{D_{\mathrm{em}}|D_{\mathrm{ex}}}}{I_{\mathrm{FRET}} + \gamma I_{D_{\mathrm{em}}|D_{\mathrm{ex}}} + I_{A_{\mathrm{em}}|A_{\mathrm{ex}}}} \tag{3}$$

The stoichiometry ratio allows the set of single-molecule data to be categorized into subpopulations based on the amount of actively fluorescent donor and acceptor molecules present in a single-molecule measurement (Fig. 5b). A value of $S = 1$ indicates that the smFRET measurement has only a donor present and/or active, $S = 0.5$ means both donor and acceptor are present/active, and $S = 0$ means that only the acceptor is present/active. The subpopulation at $S = 0.5$ is the relevant distribution of FRET values in a sample.

Another way to implement excitation of both the acceptor and the donor is by using pulsed interleaved excitation (PIE), which is essentially a much faster version of ALEX using pulsed lasers. In PIE, laser pulses of different wavelengths are

Fig. 4 (continued) microscope and sent to a sensitive camera. The microscope slide is often attached to a nanometer-precision XY-stage that can move the slide to new positions to collect more data as immobilized molecules photobleach. (**d**) Sample data from TIRF FRET microscopy. A split-screen image of immobilized donor and acceptor molecules and an example single-molecule trajectory. The green-colored spots are the molecules visible only in the donor channel, and the red-colored spots are the molecules visible only in the acceptor channel. The yellow spots are molecules where the signal is observed in both channels; these are used to determine the FRET efficiency. Donor and acceptor photobleaching is indicated in the single-molecule trajectories by a non-white background

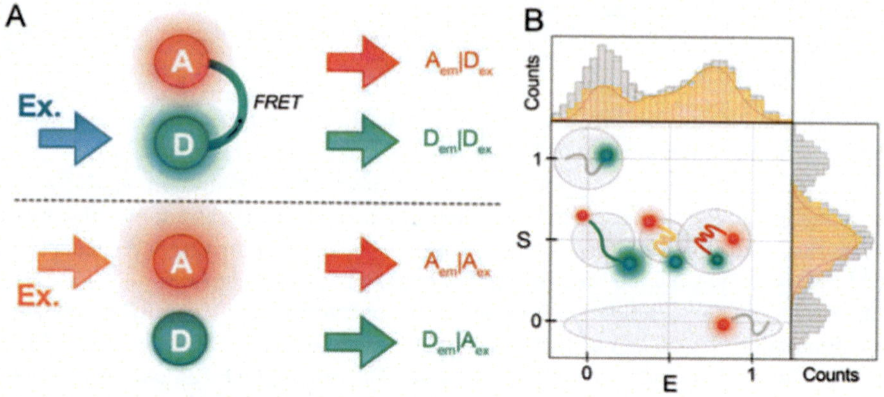

Fig. 5 (**a**) The four types of measurements in an ALEX experiment. Sequential donor and acceptor excitations give rise to the four detectable fluorescence signals, from top to bottom: $A_{em} \mid D_{ex}$ (acceptor emission upon donor excitation), $D_{em} \mid D_{ex}$ (donor emission upon donor excitation), $A_{em} \mid A_{ex}$ (acceptor emission upon acceptor excitation), and $D_{em} \mid A_{ex}$ (donor emission upon acceptor excitation), (**b**) *E-S* histogram. *E* (FRET efficiency) versus *S* (stoichiometry) histogram illustrating the division of measurements into three different categories: D-only ($S = 1$), distinct FRET populations ($S = 0.5$), and A-only ($S = 0$)

interleaved at a MHz rate such that the fluorescence emission caused by one wavelength laser is complete before the next pulse arrives [48]. This approach is used in combination with confocal microscopy and enables the mapping of each emitted photon to the laser which caused the excitation. PIE naturally enables obtaining fluorescence lifetimes of single molecules as the difference in time between the short laser pulse and the detected photon is measured. The simultaneous collection of fluorescence intensities and lifetimes enables increasing the information content of single-molecule experiments and even more parameters can be collected in an experiment including polarization anisotropy and fluorescence quantum yields [49, 50].

An additional way to increase the information content is to use additional fluorophores. smFRET with one donor and acceptor pair reports on one distance. Multi-color FRET is a way to monitor several distances simultaneously by incorporating more than two spectrally distinct fluorophores capable of transferring energy between them (Fig. 6). Three-color and four-color FRET is especially interesting for monitoring correlated changes in complex molecular dynamics or multi-component binding interactions. Single-molecule three-color FRET enabled the independent monitoring of the dynamics of two arms of a single DNA Holliday junction molecule [51], the conformation of a ligand and its binding to a target [52], and more. Four-color single molecule FRET has been demontrated [53]; at the time of this writing, there are only a few examples. Although information rich, the technical difficulty of executing a three- or four-color smFRET experiment is much higher than a two-color experiment. Multi-color FRET experiments typically make use of ALEX and require the spectral separation of the fluorescence from multiple chromophores.

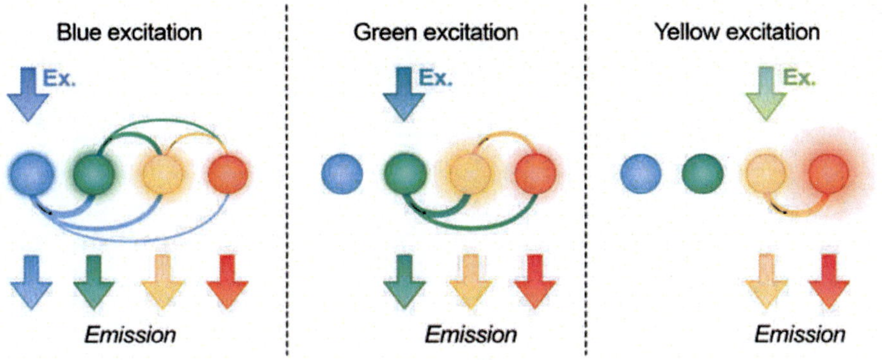

Fig. 6 Illustration of four-, three-, and two-color FRET. The four different colored spheres represent four spectrally distinct fluorophores with different excitation and emission spectra. The donor fluorophore is excited and can transfer energy to all other fluorophores either directly or via a chain of FRET events (colored lines in between spheres). The thickness of the line is proportional to the spectral overlap between fluorophores). Three- and four-color FRET experiments utilize multiple lasers consecutively to directly probe all the fluorophores and check for photobleaching

4.3 Data Analysis of Conformational Distributions and Dynamics

smFRET experiments yield data allowing the experimenter to follow conformational changes in a time-resolved manner. A qualitative view of conformational heterogeneity is obtained by determining the proximity ratio, directly using the measured signals $I_{D_{em}|I_{D_{ex}}}$ and $I_{A_{em}|D_{ex}}$ in Eq. 1, where I_{FRET} is given by $I_{A_{em}|D_{ex}}$ and $\gamma = 1$ (Fig. 7a).

Corrected smFRET data is necessary to obtain structural information about the different FRET populations. In order to transform the raw fluorescence-detected data into apparent and quantitative FRET measurements, various factors need to be accounted for (Fig. 7b, c). The fluorescence intensity of the FRET signal is given by:

$$I_{FRET} = I_{A_{exm}|D_{ex}} - \alpha I_{D_{em}|D_{ex}} - \delta I_{A_{em}|A_{ex}} \tag{4}$$

as the data is corrected for crosstalk, i.e., leakage (spectral bleed-through) of the donor into the acceptor channel, with the parameter α, and contributions from direct acceptor excitation with the δ-factor [45]. The α- and δ-factors can be experimentally extracted and shift the donor-only population to $E = 0$ and $S = 1$ and the acceptor-only populations to $E = 1$ and $S = 0$ yielding apparent FRET efficiencies (Fig. 7b) [11]. Quantitative FRET efficiencies can be obtained, using Eq. 1, after determining the γ-factor. In multi-color experiments an additional correction factor, the β-factor, corrects the S ratio [11].

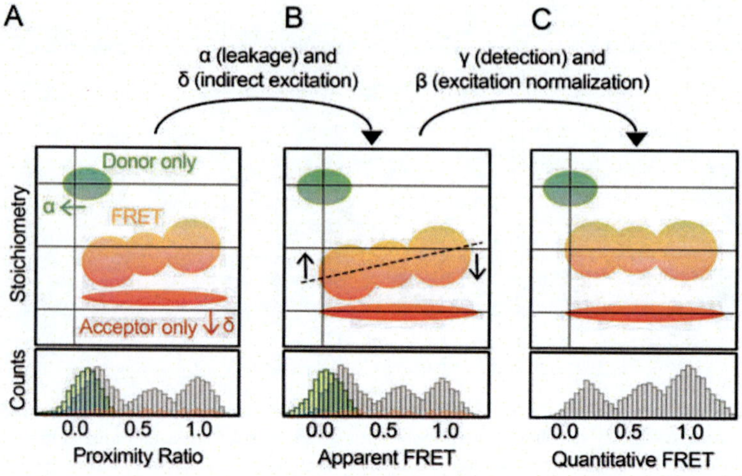

Fig. 7 Correction of single-molecule FRET data. Each panel shows the various correction parameters applied to single-molecule data sets following the procedure outlined in [11]. (**a**) First, background subtraction is performed yielding a proximity ratio distribution. (**b**) Next, donor leakage corrections (α) and indirect acceptor excitation corrections (δ) are performed yielding an apparent FRET distribution. (**c**) Lastly, detection normalization correction (γ) and excitation normalization correction (β) are applied to yield quantitative FRET distributions that can be turned into a distance distribution

If these corrections are implemented, the FRET efficiency goes from describing qualitative information about relative distances to reporting on quantifiable distances between fluorophores. Quantitative FRET can be used to obtain novel structural information that can be difficult to obtain with other techniques.

Knowledge of the conformational dynamics of a molecular system and the network of state-to-state transition rates inform on the molecule's free energy landscape. The outcome of such an analysis of smFRET data is a detailed model of relevant conformational transitions that a molecule can undergo in various conditions. The appropriate technique to analyze single-molecule conformational dwell times depends on the timescale of the measurements.

Fast timescale measurements performed on the sub-millisecond level are typically analyzed by calculating time-correlation functions of the signal. The rate and manner the time-correlation function decays from its initial value are related to molecular parameters such as the diffusion constant, the rate of conformational transitions, and the number of conformational macrostates [54, 55]. A maximum likelihood method used to analyze the photon-by-photon arrival times of smFRET experiments [56] has been shown to provide unique insights into conformational transitions of biomolecules, such as elucidating the time of protein folding transitions [24]. Additionally, different methods allow evaluating FRET distributions for the presence of dynamic heterogeneity and determining dynamics. They include the burst variance analysis (BVA) and photon distribution analysis (PDA) methods

[57, 58]. For fast dynamics, it is important to consider photophysical phenomena such as photoblinking [10, 59] and fluorescence enhancement due to steric hindrances induced by other molecules, such as protein-induced fluorescence enhancement (PIFE) [60, 61].

Hidden Markov modeling (HMM) is typically used for relatively slower measurements occurring on the millisecond to second timescale regime [62, 63]. HMM analysis ultimately will give an estimate of the number of distinct FRET states in a sample, the connectivity of states, and the transition rate between the states. For example, in the case of an intermolecular binding event, HMM analysis will yield on/off rates and can be used to calculate a binding constant. In more complex systems, dwell-time distributions (a histogram of the length of time spent in a macrostate) constructed from the HMM are fit with single-exponential decays to indicate the approximate rate constant of a state-to-state transition [64, 65].

The appropriate type of smFRET analysis ultimately depends on the nature and timescale of the experiment.

5 smFRET Application Examples

smFRET is a powerful approach for studying the function of biomolecules in heterogeneous mixtures and distinguishing subpopulations of conformers and the dynamics of their inter-conversion. Its application to nucleic acids and proteins has given important mechanistic insights to a number of areas of biological research recently reviewed in [13, 15, 66] and also within nanotechnology [67]. Here, we give an example from nucleic acid folding and DNA–protein interaction to illustrate how smFRET can be used to obtain molecular information in heterogeneous mixtures on reaction pathways and free energy landscapes.

DNA can form a range of secondary structures [68, 69]. A study by Aznauryan et al. focused on the complex folding pathway of the G-rich DNA sequences found at the telomeres [70], which can form specific types of non-canonical DNA secondary structures called G-quadruplexes. These structures can be very stable and are believed to participate in a number of biological processes [71]. By studying the DNA's conformational dynamics using TIRF-based smFRET, Aznauryan et al. identified the co-existence of several conformations with closely related stabilities and relatively slow conformational changes (Fig. 8a, b) and compiled a view of the folding energy landscape (Fig. 8c). The obtained thermodynamic and kinetic description of the folding process revealed parallel folding pathways through kinetic partitioning and trapping of the most stable conformation. Studies of this folding process using other biophysical techniques complement this view of the folding free energy landscape [72, 73]. DNA conformations, which are detectable by smFRET, can also be very transient [74].

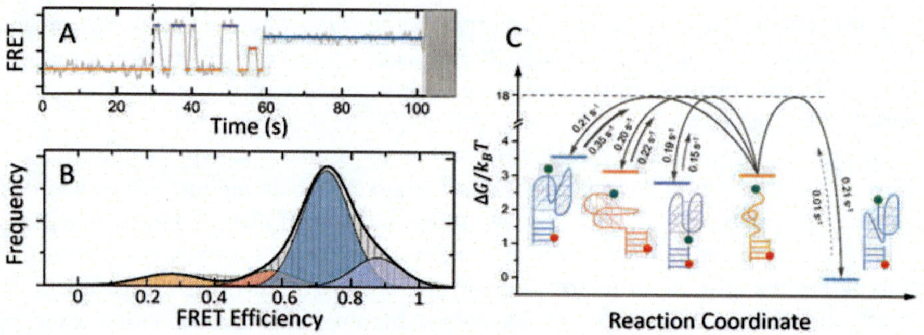

Fig. 8 DNA conformation dynamics with smFRET. (**a**) A smFRET trajectory illustrating conformational changes within a G-quadruplex-forming ssDNA strand. The trajectories were fit with a hidden Markov model using the variational Bayesian inference technique which selects the most likely N-state model. (**b**) A FRET efficiency histogram constructed from the smFRET trajectories of many molecules. The colored regions describe a four-Gaussian fit to the data, and the black line is the sum of the underlying fits. (**c**) One-dimensional free energy surface of G-quadruplex folding pathway constructed from smFRET data. The rate constants were extracted from single-exponential fits to the dwell-time distributions. Each structure represents a different G-quadraplex conformation. Adapted with permission from [70]

Proteins are capable of recognizing both specific nucleic acid sequences and structural conformations through direct and indirect readout mechanisms [75]. smFRET has had an enormous impact on the study of these interactions by uncovering key intermediates and important subpopulations not visible by ensemble measurements. A study by Hohlbein et al. used confocal single-molecule FRET to investigate protein/nucleic acid interactions involved with DNA replication in *Escherichia coli* [76]. The work showed that DNA polymerase I, the enzyme responsible for adding new nucleotides during DNA replication, has three distinct conformations: an open and closed conformation (corresponding to distinct FRET species observed before and after the correct insertion of a complementary nucleotide) and an intermediate conformation observed upon the incorporation of an incorrect nucleotide. These distinct conformations are important for the fidelity of DNA replication (Fig. 9). Later work by the same group used smFRET to obtain accurate distances for refining the structure of a DNA polymerase complex and in this way unveil a novel solution structure exhibiting a sharply bent DNA substrate [27]. The use of smFRET, and its combination with all-atomistic molecular dynamics simulations, is a very promising technique for quantifying structural dynamics. The bent DNA structure was also observed in live bacteria and a mechanism for substrate recognition was proposed [27]. *In cellulo* smFRET is complicated by the higher background from the crowded cellular environment and labeling difficulties, but is emerging as an exciting approach for studying molecular processes in cells [18, 19].

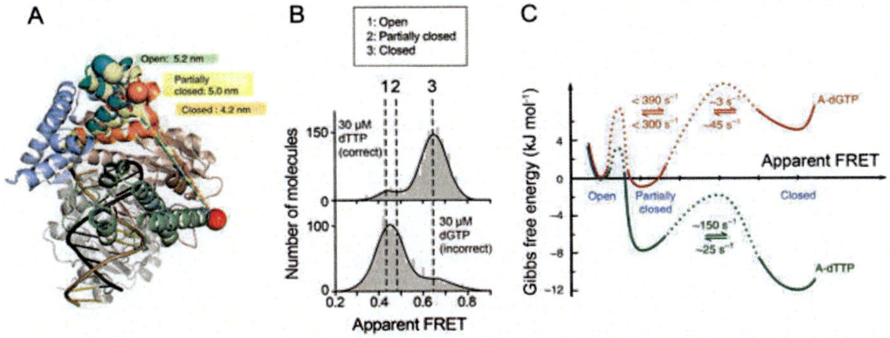

Fig. 9 Polymerase conformations and interactions with DNA studied with smFRET. (**a**) The structure of the polymerase-I enzyme around double-stranded DNA. The open and closed conformations are observed before and after the correct addition of a nucleotide. The proposed structure of the partially closed conformation is adopted when an incorrect nucleotide is added to the nascent DNA strand. (**b**) The FRET efficiency histograms corresponding to the correct base pair and incorrect base pair. The dotted lines represent the open and closed states observed when the correct base pair is added, and the dashed line is the partially closed state only observed upon the addition of the incorrect nucleotide. (**c**) Proposed free energy landscapes which show that binding of a nucleotide stabilizes the polymerase complex with the partially closed conformation, either weakly (incorrect base pair A-dGTP) or strongly (correct base pair A-dTTP), and that equilibration to the fully closed conformation is favored when the correct nucleotide is incorporated. Panels **a** and **c** are adapted with permission, and panel **b** is redrawn using data from [76]

6 Concluding Remarks

Single-molecule fluorescence enables investigating molecular processes and biological events while minimizing both temporal and conformational averaging. Over two decades of smFRET applications have given rise to unprecedented views into interaction mechanisms and conformational heterogeneities and dynamics.

Combining smFRET with other techniques can be used to increase the information content of experiments. For example, smFRET can be used in tandem with PIFE and stacking-induced fluorescence enhancement (SIFI) to increase sensitivity at close distances [61, 77]. Molecular dynamics simulations can be used in conjunction with smFRET for novel structure determination. smFRET can also be combined with force manipulation techniques, such as optical and magnetic tweezers, to monitor local conformational rearrangements amid global molecular manipulation [78].

On its own or in combination with other techniques, smFRET is a very powerful approach. We have given here an introduction to this method for investigating biomolecular structural dynamics.

Acknowledgments We thank the reviewer for his/her feedback. We would like to acknowledge support from the Novo Nordisk Foundation (grant NNF20OC0061417) and from the Danish Council for Independent Research (grant DFF-9040-00323B).

References

1. van den Bedem H, Fraser J (2015) Integrative, dynamic structural biology at atomic resolution–it's about time. Nat Methods 12:307–318. https://doi.org/10.1038/NMETH.3324
2. Förster T (1948) Zwischenmolekulare Energiewanderung und Fluoreszenz. Ann Phys 437:55–75. https://doi.org/10.1002/andp.19484370105
3. Stryer L, Hauglandt RP (1967) Energy transfer: a spectroscopic ruler. Proc Natl Acad Sci U S A 58:719. https://doi.org/10.1073/pnas.58.2.719
4. Latt SA, Cheung T, Blout ER (1964) Energy transfer. A system with relatively fixed donor-acceptor separation. J Am Chem Soc 87:995–1003. https://doi.org/10.1021/ja01083a011
5. Clegg RM (1992) Fluorescence resonance energy transfer and nucleic acids. Methods Enzymol 211:353–388. https://doi.org/10.1016/0076-6879(92)11020-j
6. Haas E (2005) The study of protein folding and dynamics by determination of intramolecular distance distributions and their fluctuations using ensemble and single-molecule FRET measurements. ChemPhysChem 6:858–870. https://doi.org/10.1002/cphc.200400617
7. Haas E, Wilchek M, Katchalski-Katzir E, Steinberg IZ (1975) Distribution of end-to-end distances of oligopeptides in solution as estimated by energy transfer. Proc Natl Acad Sci U S A 72:1807–1811. https://doi.org/10.1073/pnas.72.5.1807
8. Grinvald A, Haas E, Steinberg IZ (1972) Evaluation of the distribution of distances between energy donors and acceptors by fluorescence decay. Proc Natl Acad Sci 69:2273–2277. https://doi.org/10.1073/pnas.69.8.2273
9. Tinoco I, Gonzalez RL (2011) Biological mechanisms, one molecule at a time. Genes Dev 25: 1205–1231. https://doi.org/10.1101/GAD.2050011
10. Gopich IV, Szabo A (2012) Theory of the energy transfer efficiency and fluorescence lifetime distribution in single-molecule FRET. Proc Natl Acad Sci U S A 109:7747–7752. https://doi.org/10.1073/pnas.1205120109
11. Hellenkamp B, Schmid S, Doroshenko O et al (2018) Precision and accuracy of single-molecule FRET measurements – a multi-laboratory benchmark study. Nat Methods 15:669–676. https://doi.org/10.1038/s41592-018-0085-0
12. Ha T, Enderle T, Ogletree DF et al (1996) Probing the interaction between two single molecules: fluorescence resonance energy transfer between a single donor and a single acceptor. Proc Natl Acad Sci U S A 93:6264–6268. https://doi.org/10.1073/pnas.93.13.6264
13. Bacic L, Sabantsev A, Deindl S (2020) Recent advances in single-molecule fluorescence microscopy render structural biology dynamic. Curr Opin Struct Biol 65:61–68. https://doi.org/10.1016/J.SBI.2020.05.006
14. Lerner E, Barth A, Hendrix J et al (2021) FRET-based dynamic structural biology: challenges, perspectives and an appeal for open-science practices. elife 10:1–69. https://doi.org/10.7554/eLife.60416
15. Lerner E, Cordes T, Ingargiola A et al (2018) Toward dynamic structural biology: two decades of single-molecule Förster resonance energy transfer. Science 359(6373):eaan1133. https://doi.org/10.1126/science.aan1133
16. Sanders JC, Holmstrom ED (2021) Integrating single-molecule FRET and biomolecular simulations to study diverse interactions between nucleic acids and proteins. Essays Biochem 65: 37–49. https://doi.org/10.1042/EBC20200022
17. Walter NG, Huang CY, Manzo AJ, Sobhy MA (2008) Do-it-yourself guide: how to use the modern single-molecule toolkit. Nat Methods 5:475–489. https://doi.org/10.1038/nmeth.1215
18. Sustarsic M, Kapanidis AN (2015) Taking the ruler to the jungle: single-molecule FRET for understanding biomolecular structure and dynamics in live cells. Curr Opin Struct Biol 34:52–59. https://doi.org/10.1016/j.sbi.2015.07.001
19. Asher WB, Geggier P, Holsey MD et al (2021) Single-molecule FRET imaging of GPCR dimers in living cells. Nat Methods 18:397–405. https://doi.org/10.1038/s41592-021-01081-y

20. Fenwick RB, Esteban-Martín S, Salvatella X (2011) Understanding biomolecular motion, recognition, and allostery by use of conformational ensembles. Eur Biophys J 40(12): 1339–1355. https://doi.org/10.1007/S00249-011-0754-8
21. Fisette O, Lagüe P, Gagné S, Morin S (2012) Synergistic applications of MD and NMR for the study of biological systems. J Biomed Biotechnol. https://doi.org/10.1155/2012/254208
22. Zhong D (2007) Ultrafast catalytic processes in enzymes. Curr Opin Chem Biol 11:174–181. https://doi.org/10.1016/J.CBPA.2007.02.034
23. Jaffé HH, Miller AL (1966) The fates of electronic excitation energy. J Chem Educ 43:469–473. https://doi.org/10.1021/ED043P469
24. Chung HS, Eaton WA (2018) Direct protein folding transition path times from single molecule FRET. Curr Opin Struct Biol 48:30–39. https://doi.org/10.1016/j.sbi.2017.10.007
25. Kilic S, Felekyan S, Doroshenko O et al (2018) Single-molecule FRET reveals multiscale chromatin dynamics modulated by HP1α. Nat Commun 9. https://doi.org/10.1038/s41467-017-02619-5
26. Yang O, Ha T (2018) Single-molecule studies of ssDNA-binding proteins exchange. Methods Enzymol 600:463–477. https://doi.org/10.1016/bs.mie.2017.11.017
27. Craggs TD, Sustarsic M, Plochowietz A et al (2019) Substrate conformational dynamics facilitate structure-specific recognition of gapped DNA by DNA polymerase. Nucleic Acids Res 47:10788–10800. https://doi.org/10.1093/nar/gkz797
28. De Boer M, Gouridis G, Vietrov R et al (2019) Conformational and dynamic plasticity in substrate-binding proteins underlies selective transport in ABC importers. elife 8. https://doi.org/10.7554/ELIFE.44652
29. Kapanidis AN, Laurence TA, Nam KL et al (2005) Alternating-laser excitation of single molecules. Acc Chem Res 38:523–533. https://doi.org/10.1021/ar0401348
30. Beckers M, Drechsler F, Eilert T et al (2015) Quantitative structural information from single-molecule FRET. Faraday Discuss 184:117–129. https://doi.org/10.1039/C5FD00110B
31. Sindbert S, Kalinin S, Nguyen H et al (2011) Accurate distance determination of nucleic acids via Förster resonance energy transfer: implications of dye linker length and rigidity. J Am Chem Soc 133:2463–2480. https://doi.org/10.1021/ja105725e
32. Kalinin S, Peulen T, Sindbert S et al (2012) A toolkit and benchmark study for FRET-restrained high-precision structural modeling. Nat Methods 9:1218–1225. https://doi.org/10.1038/nmeth.2222
33. Kapanidis AN, Weiss S (2002) Fluorescent probes and bioconjugation chemistries for single-molecule fluorescence analysis of biomolecules. J Chem Phys 117:10953–10964. https://doi.org/10.1063/1.1521158
34. Zosel F, Holla A, Schuler B (2022) Labeling of proteins for single-molecule fluorescence spectroscopy. Methods Mol Biol 2376:207–233. https://doi.org/10.1007/978-1-0716-1716-8_12
35. Hanspach G, Trucks S, Hengesbach M (2019) Strategic labelling approaches for RNA single-molecule spectroscopy. RNA Biol 16:1119–1132. https://doi.org/10.1080/15476286.2019.1593093
36. Gust A, Zander A, Gietl A et al (2014) A starting point for fluorescence-based single-molecule measurements in biomolecular research. Molecules 19:15824–15865. https://doi.org/10.3390/molecules191015824
37. Ha T, Tinnefeld P (2012) Photophysics of fluorescent probes for single-molecule biophysics and super-resolution imaging. Annu Rev-Physchem 63:595–617. https://doi.org/10.1146/ANNUREV-PHYSCHEM-032210-103340
38. Hohng S, Lee S, Lee J, Jo MH (2014) Maximizing information content of single-molecule FRET experiments: multi-color FRET and FRET combined with force or torque. Chem Soc Rev 43:1007–1013. https://doi.org/10.1039/c3cs60184f
39. Kim JY, Kim C, Lee NK (2015) Real-time submillisecond single-molecule FRET dynamics of freely diffusing molecules with liposome tethering. Nat Commun 6:1–9. https://doi.org/10.1038/ncomms7992

40. Roy R, Hohng S, Ha T (2008) A practical guide to single-molecule FRET. Nat Methods 5:507–516. https://doi.org/10.1038/nmeth.1208
41. Chandradoss SD, Haagsma AC, Lee YK et al (2014) Surface passivation for single-molecule protein studies. J Vis Exp:4–11. https://doi.org/10.3791/50549
42. Paul T, Ha T, Myong S (2021) Regeneration of PEG slide for multiple rounds of single-molecule measurements. Biophys J:1–12. https://doi.org/10.1016/j.bpj.2021.02.031
43. McCann JJ, Choi UB, Zheng L et al (2010) Optimizing methods to recover absolute FRET efficiency from immobilized single molecules. Biophys J 99:961–970. https://doi.org/10.1016/j.bpj.2010.04.063
44. Hildebrandt LL, Preus S, Birkedal V (2015) Quantitative single molecule FRET efficiencies using TIRF microscopy. Faraday Discuss 184:131–142. https://doi.org/10.1039/c5fd00100e
45. Lee NK, Kapanidis AN, Wang Y et al (2005) Accurate FRET measurements within single diffusing biomolecules using alternating-laser excitation. Biophys J 88:2939–2953. https://doi.org/10.1529/biophysj.104.054114
46. Kapanidis AN, Lee NK, Laurence TA et al (2004) Fluorescence-aided molecule sorting: analysis of structure and interactions by alternating-laser excitation of single molecules. Proc Natl Acad Sci U S A 101:8936–8941. https://doi.org/10.1073/pnas.0401690101
47. Hohlbein J, Craggs TD, Cordes T (2014) Alternating-laser excitation: single-molecule FRET and beyond. Chem Soc Rev 43:1156–1171. https://doi.org/10.1039/c3cs60233h
48. Müller BK, Zaychikov E, Bräuchle C, Lamb DC (2005) Pulsed interleaved excitation. Biophys J 89:3508–3522. https://doi.org/10.1529/biophysj.105.064766
49. Eggeling C, Berger S, Brand L et al (2001) Data registration and selective single-molecule analysis using multi-parameter fluorescence detection. J Biotechnol 86:163–180. https://doi.org/10.1016/S0168-1656(00)00412-0
50. Rothwell PJ, Berger S, Kensch O et al (2003) Multiparameter single-molecule fluorescence spectroscopy reveals heterogeneity of HIV-1 reverse transcriptase:primer/template complexes. Proc Natl Acad Sci U S A 100:1655–1660. https://doi.org/10.1073/pnas.0434003100
51. Hohng S, Joo C, Ha T (2004) Single-molecule three-color FRET. Biophys J 87:1328–1337. https://doi.org/10.1529/biophysj.104.043935
52. Kim E, Lee S, Jeon A et al (2013) A single-molecule dissection of ligand binding to a protein with intrinsic dynamics. Nat Chem Biol 9:313–318. https://doi.org/10.1038/nchembio.1213
53. Lee J, Lee S, Ragunathan K et al (2010) Single-molecule four-color FRET. Angew Chem Int Ed 49:9922–9925. https://doi.org/10.1002/ANIE.201005402
54. Phelps C, Israels B, Marsh MC et al (2016) Using multiorder time-correlation functions (TCFs) to elucidate biomolecular reaction pathways from microsecond single-molecule fluorescence experiments. J Phys Chem B 120:13003–13016. https://doi.org/10.1021/acs.jpcb.6b08449
55. Santoso Y, Joyce CM, Potapova O et al (2010) Conformational transitions in DNA polymerase I revealed by single-molecule FRET. Proc Natl Acad Sci U S A 107:715–720. https://doi.org/10.1073/pnas.0910909107
56. Chung HS, Gopich IV (2014) Fast single-molecule FRET spectroscopy: theory and experiment. Phys Chem Chem Phys 16:18644–18657. https://doi.org/10.1039/c4cp02489c
57. Torella JP, Holden SJ, Santoso Y et al (2011) Identifying molecular dynamics in single-molecule fret experiments with burst variance analysis. Biophys J 100:1568–1577. https://doi.org/10.1016/j.bpj.2011.01.066
58. Kalinin S, Valeri A, Antonik M et al (2010) Detection of structural dynamics by FRET: a photon distribution and fluorescence lifetime analysis of systems with multiple states. J Phys Chem B 114:7983–7995. https://doi.org/10.1021/jp102156t
59. Gopich I, Szabo A (2005) Theory of photon statistics in single-molecule Förster resonance energy transfer. J Chem Phys 122:14707. https://doi.org/10.1063/1.1812746
60. Lerner E, Ploetz E, Hohlbein J et al (2016) A quantitative theoretical framework for protein-induced fluorescence enhancement-Förster-type resonance energy transfer (PIFE-FRET). J Phys Chem B 120:6401–6410. https://doi.org/10.1021/acs.jpcb.6b03692

61. Hwang H, Myong S (2014) Protein induced fluorescence enhancement (PIFE) for probing protein-nucleic acid interactions. Chem Soc Rev 43:1221–1229. https://doi.org/10.1039/c3cs60201j
62. Bronson JE, Fei J, Hofman JM et al (2009) Learning rates and states from biophysical time series: a Bayesian approach to model selection and single-molecule FRET data. Biophys J 97: 3196–3205. https://doi.org/10.1016/j.bpj.2009.09.031
63. McKinney SA, Joo C, Ha T (2006) Analysis of single-molecule FRET trajectories using hidden Markov modeling. Biophys J 91:1941–1951. https://doi.org/10.1529/biophysj.106.082487
64. Szoszkiewicz R, Ainavarapu SRK, Wiita AP et al (2008) Dwell time analysis of a single-molecule mechanochemical reaction. Langmuir 24:1356–1364. https://doi.org/10.1021/la702368b
65. Kinz-Thompson CD, Bailey NA, Gonzalez RL (2016) Precisely and accurately inferring single-molecule rate constants. Methods Enzymol 581:187–225. https://doi.org/10.1016/bs.mie.2016.08.021
66. Widom JR, Dhakal S, Heinicke LA, Walter NG (2014) Single-molecule tools for enzymology, structural biology, systems biology and nanotechnology: an update. Arch Toxicol 88:1965–1985. https://doi.org/10.1007/s00204-014-1357-9
67. Tsukanov R, Tomov TE, Liber M et al (2014) Developing DNA nanotechnology using single-molecule fluorescence. Acc Chem Res 47:1789–1798. https://doi.org/10.1021/ar500027d
68. Bandyopadhyay D, Mishra PP (2021) Decoding the structural dynamics and conformational alternations of DNA secondary structures by single-molecule FRET microspectroscopy. Front Mol Biosci 8:781. https://doi.org/10.3389/FMOLB.2021.725541/BIBTEX
69. Bochman ML, Paeschke K, Zakian VA (2012) DNA secondary structures: stability and function of G-quadruplex structures. Nat Rev Genet 13:770–780. https://doi.org/10.1038/nrg3296
70. Aznauryan M, Søndergaard S, Noer SL et al (2016) A direct view of the complex multi-pathway folding of telomeric G-quadruplexes. Nucleic Acids Res 44:11024–11032. https://doi.org/10.1093/nar/gkw1010
71. Varshney D, Spiegel J, Zyner K et al (2020) The regulation and functions of DNA and RNA G-quadruplexes. Nat Rev Mol Cell Biol 218(21):459–474. https://doi.org/10.1038/s41580-020-0236-x
72. Grün JT, Schwalbe H (2021) Folding dynamics of polymorphic G-quadruplex structures. Biopolymers:e23477. https://doi.org/10.1002/BIP.23477
73. Šponer J, Bussi G, Stadlbauer P et al (2017) Folding of guanine quadruplex molecules–funnel-like mechanism or kinetic partitioning? An overview from MD simulation studies. Biochim Biophys Acta, Gen Subj 1861:1246–1263. https://doi.org/10.1016/j.bbagen.2016.12.008
74. Israels B, Albrecht CS, Dang A et al (2021) Submillisecond conformational transitions of short single-stranded DNA lattices by photon correlation single-molecule Förster resonance energy transfer. J Phys Chem B. https://doi.org/10.1021/acs.jpcb.1c04119
75. Von Hippel PH, Berg OG (1986) On the specificity of DNA-protein interactions. Proc Natl Acad Sci USA 83:1608–1612. https://doi.org/10.1073/pnas.83.6.1608
76. Hohlbein J, Aigrain L, Craggs TD et al (2013) Conformational landscapes of DNA polymerase i and mutator derivatives establish fidelity checkpoints for nucleotide insertion. Nat Commun 4: 1–11. https://doi.org/10.1038/ncomms3131
77. Morten MJ, Lopez SG, Emilie Steinmark I et al (2018) Stacking-induced fluorescence increase reveals allosteric interactions through DNA. Nucleic Acids Res 46:11618–11626. https://doi.org/10.1093/nar/gky887
78. Hohng S, Zhou R, Nahas MK et al (2007) Fluorescence-force spectroscopy maps two-dimensional reaction landscape of the Holliday junction. Science 318:279–283. https://doi.org/10.1126/science.1146113

Principles of Fluorescence Correlation and Dual-Color Cross-Correlation Spectroscopy

Jan Ebenhan and Kirsten Bacia

Contents

Abstract Fluorescence Correlation Spectroscopy (FCS) is a non-invasive, highly sensitive technique for measuring the diffusive and photophysical properties of fluorescent species as well as their interactions. All of this information needs to be reliably extracted from the fluctuating fluorescence signal and interpreted in a theoretical framework. In this chapter, we describe the derivation of the basic equations governing FCS correlation curves. By pointing out their limitations and the underlying approximations and assumptions we hope to facilitate applications and the development of more elaborate models for more complex systems. Two detection channels are included to accommodate dual-color Fluorescence Cross-Correlation Spectroscopy. Moreover, we provide a generalized description for the separation of spatial movement and intramolecular change, taking translational diffusion (changes in position), rotational diffusion (changes in orientation) and fluorescence blinking (changes in the quantum mechanical state) into account. Since, experimentally, particles are often labeled with multiple fluorophores, besides multiple dynamics and multiple species of particles, multiple fluorophores per particle are also part of the description.

J. Ebenhan and K. Bacia (✉)
Institute of Chemistry and Charles-Tanford-Protein-Center, Martin-Luther-Universität Halle-Wittenberg, Halle, Germany
e-mail: jan.ebenhan@chemie.uni-halle.de; kirsten.bacia@chemie.uni-halle.de

R. Šachl, M. Amaro (eds.), *Fluorescence Spectroscopy and Microscopy in Biology*, Springer Ser Fluoresc (2023) 20: 119–138, https://doi.org/10.1007/4243_2022_33, Published online: 31 January 2023

Keywords Binding analysis · Confocal microscopy · Diffusion · Fluorescence correlation spectroscopy · Fluorescence cross-correlation spectroscopy · Fluorescence spectroscopy

Fluorescence correlation spectroscopy (FCS), like dynamic light scattering (DLS), is a non-invasive and single-molecule sensitive technique based on the temporal correlation of brightness fluctuations, with which primarily the size, concentration and photophysical properties of fluorescent particles can be investigated in small liquid volumes and at thermodynamic equilibrium. It was first described in 1972 by Elliot L. Elson, Douglas Magde, and Watt W. Webb [1].

FCS is an example of a concentration correlation spectroscopy method [2]. This type of spectroscopy is based on the measurement of fluctuations of the particle concentration in small subvolumes of a sample. In principle, FCS can employ any kind of measurand that depends on these concentrations, such as density, refractive index, extinction, etc. Notably, in DLS, the signal is the amplitude of the scattered light and thus depends on the phase distribution of the electric field and the spatial variation of the permittivity in the sample. In contrast, in FCS, the intensity of the radiation constitutes the signal. Hence, the fluorescence is separated from the irradiation and the detected fluorescence photons with all their measurable properties (number, arrival time, energy, polarization, momentum direction) constitute the signal.

By virtue of the high specificity of fluorescence labeling, FCS can probe one or more defined fluorescent species even in complex mixtures and living cells and measure physical and chemical characteristics of each of them. Static and dynamic phenomena thus become accessible, provided that a detectable change in the emitted photons occurs.

For an insight into the theoretical foundations of FCS, this chapter focuses on the derivation of the underlying equations and highlights the approximations and assumptions necessary therein (Sect. 1). In many FCS applications, the translational movement of the fluorescent particles, be it diffusive or directed, can be separated from their internal changes, e.g., photophysical blinking, rotation or chemical reactions. A generalized description reflecting this property will be given to disentangle contributions from both types of processes (see Sects. 2 and 3). Furthermore, the influence of multiple species of particles (Sect. 1) as well as multiple fluorophores per particle (Sect. 2) is considered.

Multiple dynamics, multiple species and multiple fluorophores per particle are often encountered experimentally. We, therefore, aim to describe the theoretical framework in a general, adaptable way, also pointing out the limitations. Examples of calibration and control experiments, which can be conducted to check applicability, can be found in more specialized protocols, see, e.g., [3–6].

1 Fluorescence Intensity

A certain intensity of the emitted radiation of the fluorophores (in short fluorescence intensity) causes a stream of photons, which in turn is converted into a sequence of counting events in the detector. For the correlation functions considered in FCS, the stochastic properties of these events primarily produce artifacts at small correlation times, where shot noise and detector afterpulses manifest as potentially strong positive correlations. In particular, the zero value is contaminated by the variance of the Poisson distributed counting events and is not useful as a measure of the true amplitude. The considerations in this chapter therefore refer to the correlations of the fluorescence intensity itself, which are caused by molecular processes in the sample, and also no longer draw a distinction between the fluorescence intensity before the optical system and the count rate of the detector. The quantum yield of the detection Φ_D is included in the molecular brightness Q and the signal is generally denoted by F.

The molecular brightness Q of a fluorophore is generally not a constant but may depend on a number of variables, including its location $\vec{r}$, its orientation $\vec{R}$ and its (quantum mechanical) state Z.

The total detected intensity can now be calculated using the so-called molecular detection function (MDF) W either as a spatial integral (with the differential volume element $d^3\vec{r} = dx\ dy\ dz$) over the particle density C of the fluorescent particles

$$F(t) = \int W\left(\vec{r}, t\right) Q\left(\vec{r}, \vec{R}, Z, t\right) C\left(\vec{r}, t\right) d^3\vec{r} \quad (1)$$

or as a sum

$$F(t) = \sum_i Q\left(\vec{r}_i, \vec{R}_i, Z_i, t\right) W\left(\vec{r}_i, t\right) \quad (2)$$

over the discrete contributions of individual particles i in the sample, where $\vec{r}_i$ denotes a particle's location, $\vec{R}_i$ its orientation and Z_i its (quantum mechanical) state.

Here, we use the discrete summation approach rather than the integral, because in this way the contributions of individual fluorophores are treated more intuitively and the probability distributions with respect to their motions and interactions are used in a more natural way. Conversely, the first approach does not require taking limits of infinite sums and it facilitates the description of spatial correlations in particles that are no longer diffusing independently as well as that of chemical reactions where the numbers and identities of particles are introduced as another random variable.

The expected value and the (cross-)correlation of the signal can now be defined starting from

$$\langle F(t)\rangle = \sum_i \left\langle Q\left(\vec{r}_i, \vec{R}_i, Z_i, t\right) W\left(\vec{r}_i, t\right)\right\rangle \tag{3a}$$

$$= \sum_i \iiint Q\left(\vec{r}_i, \vec{R}_i, Z_i, t\right) W\left(\vec{r}_i, t\right) P\left(\vec{r}_i, \vec{R}_i, Z_i, t\right) d^3\vec{r}_i d^3\vec{R}_i dZ_i \tag{3b}$$

$$\begin{aligned}\langle F_a(t)\cdot F_b(t+\tau)\rangle = \sum_{i,j} \int\int\int\int\int\int \\ [Q_a\left(\vec{r}_i, \vec{R}_i, Z_i, t\right) Q_b\left(\vec{r}_j, \vec{R}_j, Z_j, t+\tau\right) \\ P\left(\left\{\vec{r}_i, \vec{R}_i, Z_i, t\right\} \cap \left\{\vec{r}_j, \vec{R}_j, Z_j, t+\tau\right\}\right) \\ W_a\left(\vec{r}_i, t\right) W_b\left(\vec{r}_j, t+\tau\right) d^3\vec{r}_i d^3\vec{r}_j d^3\vec{R}_i d^3\vec{R}_j dZ_i dZ_j]\end{aligned} \tag{4}$$

where the indices a and b denote different channels, each with its own molecular brightness and MDF.

The angular brackets denote an ensemble average over all possible states of the system. In the FCS experiments, this average is realized by time averaging. Throughout this chapter, ergodicity is assumed to hold. Therefore, in the limit of infinite duration of the measurement, the time average approaches the ensemble average.

The autocorrelation and cross-correlation functions are defined by $a = b$ and $a \neq b$, respectively. In the technique of dual-color fluorescence cross-correlation spectroscopy (dcFCCS) described in this chapter, the channels represent different colors, i.e. different excited and detected regions of the spectrum. However, many other possibilities are conceivable or have already been realized, e.g., spatially shifted MDFs (dual-focus FCCS [7] and multi-focus FCS [8]) or detection of different polarizations (e.g., for measuring rotational diffusion, see [9] for a detailed derivation and [10] for an application to protein oligomerization). Also related to dual-focus correlation are image correlation techniques, such as ICS, RICS, and STICS (see chapter "Theoretical Insight into the Luminescence of Dyes and Pigments" of this book), which provide spatially resolved data in, e.g., biological samples. DcFCCS has also been combined with fast camera-based imaging to generate interaction maps (see [5]).

The joint distribution $P\left(\left\{\vec{r}_i, \vec{R}_i, Z_i, t\right\} \cap \left\{\vec{r}_j, \vec{R}_j, Z_j, t+\tau\right\}\right)$ indicates the probability density of finding at time t particle i at location $\vec{r}_i$ with orientation $\vec{R}_i$ in the (quantum mechanical) state Z_i and additionally at time $t + \tau$ particle j at location $\vec{r}_j$ with orientation $\vec{R}_j$ in the state Z_j. The molecular rotation contributes to the temporally correlated changes in the brightness of the fluorophore, because the orientations of the transition dipole moments of absorption and emission change relative to the electric field vector of the incident laser light and relative to the transmission direction of the polarization filter in front of the detector. The fluctuations arising from changes in the molecular orientation can be used to determine the rotational diffusion coefficient and thus the molecular size and shape. The molecular brightness is explicitly expressed here as a function of time and location, since photophysical phenomena can lead to a variable brightness, which itself exhibits a correlation and moreover depends on location due to the spatially varying laser intensity. In most cases, Q is proportional to the product of the occupation of the excited S_1 state and the rate constant k_f of the radiative transition to the ground state. Since the degree of occupation depends on the one hand on the intensity of the excitation, but on the other hand, is also limited to unity, the brightness must always be location-dependent and must approach a limiting value of saturation for increasing excitation. The exact functional dependence on intensity can be very complicated (see [11]), but in the simplest case (laser intensity constant in time; intersystem crossing (ISC) much slower than fluorescence emission; negligible antibunching) the result is [12]

$$Q \propto \frac{k_{ex}}{1 + \frac{k_{ex}}{k_{sat}}} \tag{5}$$

where

$$k_{sat} = \frac{k_{10}}{1 + \frac{k_{ISC}}{k_{phos}}} \tag{6}$$

Here k_{ex} is the rate of excitation for the fluorophore (proportional to the exciting laser intensity I), k_{sat} is the excitation rate for which half of the maximum fluorescence brightness is obtained, k_{ISC} and k_{phos} are the rate constants of ISC and phosphorescence, respectively, and k_{10} is the rate constant for the transition to the ground state (sum of radiative and non-radiative decays).

This treatment does not account for the possibility of photobleaching and presupposes knowledge of the (often quite complex) photophysics of the fluorophores in question. For practical applications, it is therefore customary to obtain FCS measurements under a range of excitation intensities to determine the maximum usable laser power, so as to neither introduce artifacts nor unnecessarily lower the signal-to-noise ratio [13].

If several species of different brightness exist in the sample, $F(t)$ is calculated as the sum of the individual intensities according to

$$F(t) = \sum_j F_j(t) = \sum_{i,j} Q_j\left(\vec{r}_{i,j}, \vec{R}_{i,j}, Z_{i,j}, t\right) W\left(\vec{r}_{i,j}, t\right). \tag{7}$$

Here, $\vec{r}_{i,j}$ denotes the position of the i-th particle of the j-th species. $\vec{R}_{i,j}$ and $Z_{i,j}$ denote the corresponding orientations and states.

2 Fluorescence Correlation

The fundamental equation of FCS is now the normalized correlation function

$$G_{a,b}(\tau) = \frac{\langle F_a(t) \cdot F_b(t+\tau)\rangle}{\langle F_a(t)\rangle\langle F_b(t)\rangle} - 1 = \frac{\langle \delta F_a(t) \cdot \delta F_b(t+\tau)\rangle}{\langle F_a(t)\rangle\langle F_b(t)\rangle} \tag{8}$$

where $\delta F(t) = F(t) - \langle F(t)\rangle$ is the deviation of the fluorescence intensity from the expected value. Here, only positive correlation times τ are considered, because for stationary processes $G_{a,\ b}(-\tau) = G_{b,\ a}(\tau)$ applies and differences between $G_{a,\ b}(\tau)$ and $G_{b,\ a}(\tau)$ only occur in the case of a non-stationary process.

In principle, the detected radiation field depends on the exact state and motion of all molecules in the sample. Since this problem is practically unsolvable, the following assumptions and simplifications are made, unless noted otherwise:

- The MDF is time-independent: $W\left(\vec{r}_i, t\right) \mapsto W\left(\vec{r}_i\right)$.
- The polarization is uniform. This assumption is more accurate the lower the numerical aperture of the objective.
- The process can be described as weakly stationary $\langle F_a(t) \cdot F_b(t+\tau)\rangle \mapsto \langle F_a(0) \cdot F_b(\tau)\rangle$.
- The sample is stationary and isotrop ($P\left(\vec{r}_i, \vec{R}_i, Z_i, t\right) \mapsto \frac{1}{4\pi} P\left(\vec{r}_i, Z_i\right)$ with normalization $\int d^3\vec{R}_i = 4\pi$).
- The particle density is homogeneous, implying that $P\left(\vec{r}_i\right) = \frac{1}{V}$ is the constant residence probability of the particles in the sample volume V.
- Individual fluorophores emit independently of one another, meaning there is no coherence of the fluorescence radiation and no energy transfer between the fluorophores as in FRET or fluorescence quenching.
- The particles diffuse, rotate and change their states independently of one another
$$(P\left(\left\{\vec{r}_i,\ \vec{R}_i,\ Z_i,\ t\right\} \cap \left\{\vec{r}_j,\ \vec{R}_j,\ Z_j,\ t+\tau\right\}\right)$$
$$= P\left(\vec{r}_i, \vec{R}_i, Z_i, t\right) \cdot P\left(\vec{r}_j, \vec{R}_j, Z_j, t+\tau\right)$$
for $i \neq j$).
- Rotation and diffusion do not depend on each other (absence of roto-translational coupling) and are not influenced by the state of the particle, implying that the particle does not dynamically change its shape ($P\left(\vec{r}_i, \vec{R}_i, Z_i, t\right) = P\left(\vec{r}_i, t\right) \cdot P\left(\vec{R}_i, t\right) \cdot P(Z_i, t)$).
- The translational and rotational diffusion coefficients do not depend on position or time, implying an absence of thermophoretic effects.

- Photophysical effects do not depend on spatial position ($Q\left(\vec{r}_i, \vec{R}_i, Z_i, t\right) \mapsto Q\left(\vec{R}_i, Z_i, t\right)$).
- The solvent contributes to the fluorescence only as background noise.

Thus Eq. 3b now simplifies to

$$\langle F(t)\rangle = \sum_i \iiint Q\left(\vec{R}_i, Z_i, t\right) W\left(\vec{r}_i\right) \frac{1}{4\pi V} d^3\vec{r}_i d^3\vec{R}_i dZ_i \tag{9a}$$

$$= \frac{\bar{Q}}{V}\sum_i \int W\left(\vec{r}_i\right) d^3\vec{r}_i = \frac{\bar{Q}}{V}\sum_i \Omega_1 = \frac{N}{V}\bar{Q}\Omega_1 \tag{9b}$$

$$\lim_{V\to\infty} \langle F(t)\rangle = C\bar{Q}\Omega_1 = cN_A\bar{Q}\Omega_1 \tag{10}$$

Here, the brightness averaged over all orientations and possible photophysical processes is defined as

$$\bar{Q} = \iint Q\left(\vec{R}, Z, t\right) \frac{1}{4\pi} d^3\vec{R}\, dZ \tag{11}$$

and the space integrals of the MDF are defined as

$$\Omega_{a,n} = \int_V \left[W_a\left(\vec{r}\right)\right]^n d^3\vec{r} \tag{12}$$

$$\Omega_{n,a;m,b} = \int_V \left[W_a\left(\vec{r}\right)\right]^n \left[W_b\left(\vec{r}\right)\right]^m d^3\vec{r}. \tag{13}$$

The channel index a in $\Omega_{a,\,n}$ can be omitted if only one channel is used.

All identical particles contribute equally. Their contribution is proportional to their number N in the volume V. For practical purposes, the sample size can be considered as infinitely large. Under these conditions, the contribution is proportional to the particle density C or molar concentration c (where $C = N_A\, c$ with the Avogadro constant N_A). Finally, in the presence of several species, the total average fluorescence becomes

$$\langle F(t)\rangle = N_A\Omega_1 \sum_j c_j \bar{Q}_j \tag{14}$$

For the evaluation of Eq. 4 the double sum over the particles of the same species is split into the correlations between identical particles and the correlations between different particles. For the former, the conditional probability $P(B|A) = \frac{P(A\cap B)}{P(A)}$ is introduced. For better distinguishability of the different positions of the same particle, the time dependence of position, orientation, and state is given in the form $x = x(0)$ and $x' = x(\tau)$:

$$\langle F_a(0) \cdot F_b(\tau)\rangle$$

$$= \sum_i \left\langle Q_a\left(\vec{R}_i, Z_i, 0\right) Q_b\left(\vec{R}'_i, Z'_i, \tau\right) W_a\left(\vec{r}_i\right) W_b\left(\vec{r}'_i\right)\right\rangle$$

$$+ \sum_{i \neq j} \left\langle Q_a\left(\vec{R}_i, Z_i, 0\right) Q_b\left(\vec{R}_j, Z_j, \tau\right) W_a\left(\vec{r}_i\right) W_b\left(\vec{r}_j\right)\right\rangle \tag{15a}$$

$$= \sum_i \left\langle Q_a\left(\vec{R}_i, Z_i, 0\right) Q_b\left(\vec{R}'_i, Z'_i, \tau\right)\right\rangle \cdot \left\langle W_a\left(\vec{r}_i\right) W_b\left(\vec{r}'_i\right)\right\rangle$$

$$+ \sum_{i \neq j} \left\langle Q_a\left(\vec{R}_i, Z_i, 0\right) W_a\left(\vec{r}_i\right)\right\rangle \cdot \left\langle Q_b\left(\vec{R}_j, Z_j, \tau\right) W_b\left(\vec{r}_j\right)\right\rangle \tag{15b}$$

$$= N\Big[\int\int\int\int Q_a\left(\vec{R}, Z, 0\right) Q_b\left(\vec{R}', Z', \tau\right) P\left(\vec{R}, Z, 0\right) \cdot$$

$$P\left(\left\{\vec{R}', Z', \tau\right\} | \left\{\vec{R}, Z, 0\right\}\right) d^3\vec{R}\, d^3\vec{R}'\, dZ dZ'\Big] \cdot \iint W_a\left(\vec{r}\right) W_b\left(\vec{r}'\right)$$

$$\frac{1}{V} P\left(\left\{\vec{r}', \tau\right\} | \left\{\vec{r}, 0\right\}\right) d^3\vec{r}\, d^3\vec{r}' + \frac{N(N-1)}{V^2} \overline{Q}_a \overline{Q}_b \Omega_{1,a} \Omega_{1,b} \tag{15c}$$

$$= \overline{Q}_a \overline{Q}_b \left(\Omega_{1,a;\,1,b} \frac{N}{V} \Theta(\tau) \Xi(\tau) + \Omega_{1,a} \Omega_{1,b} \frac{N(N-1)}{V^2}\right) \tag{15d}$$

In Eq. 15b the averages are simplified by exploiting the presumed independence of the position and the brightness correlations (first term) and by exploiting the independence of individual particles from each other (second term). In the first term of Eq. 15c the averages are made explicit as integrals over their respective variables and the summation over identical particles is replaced by their total number N. Since one self-correlation for each particle is already contained in the first term, the sum in the second term of Eq. 15b runs over a total of $N^2 - N = N(N - 1)$ identical contributions. In Eq. 15c the second term is replaced by a term that is proportional to $N(N - 1)$ and to the product of the average intensities in both channels (see Eq. 9b). In Eq. 15d, the brightness correlation function $\Theta(\tau)$ and the position correlation function $\Xi(\tau)$ are introduced. $\Theta(\tau)$ describes the dimensionless normalized correlations of the molecular brightness $Q_{(a,b)}$ due to changes of orientation and state. $\Xi(\tau)$ describes the dimensionless normalized correlations of the apparent brightness changes due to diffusion through the focus:

$$\Theta(\tau) = \frac{\langle Q_a(t) \cdot Q_b(t+\tau) \rangle}{\bar{Q}_a \bar{Q}_b}$$
$$= \frac{1}{\bar{Q}_a \bar{Q}_b} \int \int \int \int Q_a\left(\vec{R}, Z, 0\right) Q_b\left(\vec{R}', Z', \tau\right) P\left(\vec{R}, Z, 0\right)$$
$$\cdot P\left(\left\{\vec{R}', Z', \tau\right\} | \left\{\vec{R}, Z, 0\right\}\right) d^3 \vec{R} d^3 \vec{R}' dZ dZ' \quad (16)$$

$$\Xi(\tau) = \frac{1}{\Omega_{1,a;1,b}} \iint W_a\left(\vec{r}\right) W_b\left(\vec{r}'\right) P\left(\left\{\vec{r}', \tau\right\} | \left\{\vec{r}, 0\right\}\right) d^3 \vec{r} d^3 \vec{r}' \quad (17)$$

Since for infinite correlation times a complete decoupling of all fluctuations can be assumed, the conditional probability again attains the equilibrium value according to $P\left(\left\{\vec{R}', Z', \tau\right\} | \left\{\vec{R}, Z, 0\right\}\right) = P\left(\vec{R}', Z', 0\right) = \frac{1}{4\pi}$ and $P\left(\left\{\vec{r}', \tau\right\} | \left\{\vec{r}, 0\right\}\right) = P\left(\vec{r}', 0\right) = \frac{1}{V}$. The former, via Eq. 11, results in

$$\lim_{\tau \to \infty} \Theta(\tau) = 1 \quad (18)$$

and the latter in

$$\lim_{\tau \to \infty} \Xi(\tau) \propto \frac{1}{V}, \quad (19)$$

whereby an infinite sample thus leads to $\lim_{V \to \infty} \Xi(\tau) = 0$ and

$$\lim_{\tau \to \infty} \langle F_a(0) \cdot F_b(\tau) \rangle = C^2 \bar{Q}_a \bar{Q}_b \Omega_{1,a} \Omega_{1,b} \quad (20)$$

Neglecting the one particle that is correlated with itself in the second summand in Eq. 15d by approximating $N(N - 1) \approx N^2$ will only be noticeable in very small compartments, for example when measuring in bacteria or lipid vesicles. After substituting these results into Eq. 8, we finally obtain

$$G_{a,b}(\tau) = \frac{\bar{Q}_a \bar{Q}_b \left(\Omega_{1,a;1,b} \frac{N}{V} \Theta(\tau) \Xi(\tau) + \Omega_{1,a} \Omega_{1,b} \frac{N(N-1)}{V^2} \right)}{\left(\frac{N}{V} \bar{Q}_a \Omega_{1,a} \right) \left(\frac{N}{V} \bar{Q}_b \Omega_{1,b} \right)} - 1 \quad (21a)$$

$$\approx \frac{\bar{Q}_a \bar{Q}_b \left(\Omega_{1,a;1,b} C \Theta(\tau) \Xi(\tau) + \Omega_{1,a} \Omega_{1,b} C^2 \right)}{\left(C \bar{Q}_a \Omega_{1,a} \right) \left(C \bar{Q}_b \Omega_{1,b} \right)} - 1 \quad (21b)$$

$$= \frac{\bar{Q}_a \bar{Q}_b \Omega_{1,a;\,1,b} C \Theta(\tau) \Xi(\tau)}{(C\bar{Q}_a \Omega_{1,a})\,(C\bar{Q}_b \Omega_{1,b})} \tag{21c}$$

$$= \frac{1}{C} \frac{\Omega_{1,a;\,1b}}{\Omega_{1,a}\Omega_{1,b}} \Theta(\tau)\Xi(\tau) \tag{21d}$$

Here, the proportionality between variance and expected value of the Poisson distributed particle number in a given volume [14] becomes apparent: The numerator of the correlation function in Eq. 21c scales linearly with the concentration and quadratically with the molecular brightness. By normalizing with the square of the mean fluorescence intensity, which is proportional to the concentration squared and the molecular brightness squared, the latter no longer enters the equation, but the function as a whole increases inversely proportional with the concentration.

For practical reasons, the term $\frac{\Omega_{1,a;\,1,b}}{\Omega_{1,a}\Omega_{1,b}}$ is combined with the particle density and the so-called effective focal volume V_{eff} is defined as a ratio of space integrals over the MDF:

$$V_{\text{eff,x}} = \frac{\Omega_{1,a}\Omega_{1,b}}{\Omega_{1,a;\,1,b}} \tag{22}$$

$$V_{\text{eff},(a,\,b)} = \frac{\Omega^2_{1,(a,\,b)}}{\Omega_{2,(a,\,b)}} \tag{23}$$

Here $V_{\text{eff},\,(a,\,b)}$ denotes the effective focal volume for the autocorrelation of one of the two channels and $V_{\text{eff},\,\text{x}}$ that for their cross-correlation. By introducing the effective particle number

$$N_{\text{eff}} = C\,V_{\text{eff}} \tag{24}$$

the general form of the F(C)CS correlation function is given by

$$G_{a,b}(\tau) = \frac{1}{N_{\text{eff}}}\,\Theta(\tau)\,\Xi(\tau) \tag{25}$$

In the case of autocorrelation, V_{eff} represents $V_{\text{eff},\,(a,\,b)}$, in the case of cross-correlation V_{eff} represents $V_{\text{eff},\,\text{x}}$.

The diffusion behavior of the particles has no impact on the zero value of the position correlation function $\Xi(0)$, resulting in $\lim_{\tau\to 0} \Xi(\tau) \approx 1$. Provided that the brightness is constant or correlated on significantly shorter time scales, $\lim_{\tau\to 0} \Theta(\tau) \approx 1$ also holds. Therefore, the F(C)CS amplitude is given by

$$\mathfrak{A} = \lim_{\tau \to 0} G_{a,b}(\tau) = \frac{1}{N_{\text{eff}}} \tag{26}$$

Moreover, as no motion occurs for $\tau \to 0$,

$$\lim_{\tau \to 0} P\left(\left\{\vec{r}', \tau\right\} | \left\{\vec{r}, 0\right\}\right) = \delta\left(\vec{r}' - \vec{r}\right) \tag{27}$$

The double space integral thus reduces to a simple one in Eq. 17 and the origin of the normalization factor becomes apparent:

$$\lim_{\tau \to 0} \Xi(\tau) = \frac{1}{\Omega_{1,a;\,1,b}} \iint W_a\left(\vec{r}\right) W_b\left(\vec{r}'\right) \delta\left(\vec{r}' - \vec{r}\right) d^3\vec{r}\, d^3\vec{r}' \tag{28a}$$

$$= \frac{1}{\Omega_{1,a;\,1,b}} \int W_a\left(\vec{r}\right) W_b\left(\vec{r}\right) d^3\vec{r} = 1 \tag{28b}$$

Commonly, the mean fluorescence intensity from the experiment, $\langle F(t) \rangle$, is divided by the effective particle number, N_{eff}, which is readily obtained from the experimental autocorrelation curve, to yield a measure for the molecular brightness (counts per molecule, CPM), which is valid in the case of a single species. Using Eq. 10 for $\langle F(t) \rangle$ and Eqs. 22 and 24 for N_{eff}, we obtain for the counts per molecule

$$\eta = \frac{\langle F(t) \rangle}{N_{\text{eff}}} = \bar{Q} \frac{\Omega_2}{\Omega_1} \tag{29}$$

If several species of fluorophores i with their individual fluorescence intensities $F_{(a,b),\,i}$ and corresponding correlation functions $\Theta_i(\tau)$ and $\Xi_i(\tau)$ are present and if these fluorophores behave as independent particles in the same way as individual particles of a single species do and if no chemical reactions occur (or chemical reactions are slow compared to the time scale of the diffusion), then the following applies

$$\left\langle \delta F_{a,i}(t) \cdot \delta F_{b,j}(t+\tau) \right\rangle = \delta_{ij} \langle \delta F_{a,i}(t) \cdot \delta F_{b,i}(t+\tau) \rangle. \tag{30}$$

Equation 30 states that the fluorescence signals from different species i and j are uncorrelated. Signals are correlated if they originate from the same species ($i = j$). Since the bilinearity of the covariance leads to the additivity of the contributions of the individual species to the numerator of Eq. 8, we obtain

$$\left\langle F_{(a,\,b)}(t) \right\rangle = \sum_i \left\langle F_{(a,\,b),i}(t) \right\rangle \tag{31}$$

$$\delta F_{(a,\ b)}(t) = \sum_i \delta F_{(a,\ b),i}(t) \tag{32}$$

$$\langle \delta F_a(t) \cdot \delta F_b(t+\tau) \rangle = \sum_{i,j} \langle \delta F_{a,i}(t) \cdot \delta F_{b,j}(t+\tau) \rangle = \sum_i \langle \delta F_{a,i}(t) \cdot \delta F_{b,i}(t+\tau) \rangle \tag{33a}$$

$$= \sum_i G_{a,b,i}(\tau) \langle F_{a,i}(t) \rangle \langle F_{b,i}(t) \rangle = \sum_i \frac{1}{V_{\text{eff}} C_i} \, \Theta_i(\tau) \Xi_i(\tau) \left(\Omega_{1,a} C_i \bar{Q}_{a,i} \right) \left(\Omega_{1,b} C_i \bar{Q}_{b,i} \right) \tag{33b}$$

where Θ_i and Ξ_i are the brightness correlation and position correlation function for species i, respectively. In the last line Eqs. 8, 10, and 21d were used. Upon dividing by $\langle F_a(t) \rangle \langle F_b(t) \rangle$ and simplifying, the contributions of the individual species to the correlation functions can be expressed as follows, where $\mathfrak{A}_i$ are the partial amplitudes and φ_i the amplitude fractions:

$$G_{a,b}(\tau) = \frac{\sum_i \langle \delta F_{a,i}(t) \cdot \delta F_{b,i}(t+\tau) \rangle}{\sum_i \langle F_{a,i}(t) \rangle \cdot \sum_j \langle F_{b,j}(t) \rangle} = \mathfrak{A} \sum_i \varphi_i \Theta_i(\tau) \Xi_i(\tau) \tag{34}$$

$$\mathfrak{A} = \sum_i \mathfrak{A}_i = \sum_i \frac{1}{V_{\text{eff}}} \frac{C_i \bar{Q}_{a,i} \bar{Q}_{b,i}}{\sum_j C_j \bar{Q}_{a,j} \cdot \sum_k C_k \bar{Q}_{b,k}} \tag{35}$$

$$\varphi_i = \frac{\mathfrak{A}_i}{\mathfrak{A}} = \frac{C_i \bar{Q}_{a,i} \bar{Q}_{b,i}}{\sum_j C_j \bar{Q}_{a,j} \bar{Q}_{b,j}} \tag{36}$$

The aim of the last term in Eq. 34 is to separate the different correlation functions $\Theta_i(\tau)\Xi_i(\tau)$ associated with each species from the overall amplitude $\mathfrak{A}$, which is oftentimes the easiest parameter to measure and interpret. The partial amplitudes $\mathfrak{A}_i$ thus represent the (absolute) part of the amplitude belonging to species i, while the amplitude fractions φ_i represent the same value relative to the overall amplitude.

These equations are needed to analyze FCS measurements of mixtures and derive the concentrations from fitted amplitudes. It can be seen from the numerator of Eq. 35 that the amplitude fraction depends quadratically on the brightness, but linearly on the concentration of a species. This weighting causes a considerable bias toward the brighter species. If the brightnesses are not known beforehand, only apparent fractions or concentrations can be obtained and these will always underestimate the amount of dim particles in the sample.

The analysis of binding experiments relies on correlation amplitudes (see for example in [15, 16]), which can be evaluated using Eq. 35. However, to reliably determine these amplitudes, the time-dependent correlation should be analyzed by fitting Eq. 34 to the measured correlation curves. Therefore, a more detailed examination of the brightness correlation function $\Theta(\tau)$ and the position correlation function $\Xi(\tau)$ is necessary.

3 Brightness Correlation Function

The main contributions to the time-dependent brightness fluctuations of fluorophores are given by their rotation, interaction with other particles (e.g., FRET, fluorescence quenching), chemical and photochemical reactions (e.g., cis-trans isomerization, photobleaching) and photophysical processes (e.g., ISC, antibunching). All these phenomena can be used to obtain information about molecular parameters and interactions of particles with each other (shape and size from rotation, distances from FRET, binding energies from isomerization rates, oligomerization from antibunching, etc.).

If the probability density can be completely factorized and the molecular brightness depends multiplicatively on orientation and state, then $\Theta(\tau)$ can be decomposed into a product of rotation correlation and state correlation functions. The relevant processes in FCS applications are typically transitions between states of different brightnesses of a single fluorophore, which are often described as first-order unimolecular reactions (ISC, photobleaching, antibunching). If several fluorophores (number n_{F}) are attached to a particle, states with several excited fluorophores exist at the same time. If the fluorophores behave independently of each other, apart from their common diffusion, the sum of their contributions can be written in analogy to Eq. 30 as:

$$Q_{(a,\,b)}(t) = \sum_{i=1}^{n_{\mathrm{F}}} Q_{(a,\,b),i}(t) \tag{37}$$

$$\langle Q_a(t) \cdot Q_b(t+\tau) \rangle = \sum_{i,j}^{n_{\mathrm{F}}} \langle Q_{a,i}(t) \cdot Q_{b,j}(t+\tau) \rangle \tag{38a}$$

$$= \sum_{i=1}^{n_{\mathrm{F}}} \langle Q_{a,i}(t) \cdot Q_{b,i}(t+\tau) \rangle + \sum_{i \neq j}^{n_{\mathrm{F}}} \langle Q_{a,i}(t) \rangle \cdot \langle Q_{b,j}(t+\tau) \rangle \tag{38b}$$

$$= \sum_{i=1}^{n_{\mathrm{F}}} \bar{Q}_{a,i} \bar{Q}_{b,i} \Theta_i(\tau) + \sum_{i \neq j}^{n_{\mathrm{F}}} \bar{Q}_{a,i} \bar{Q}_{b,j} \tag{38c}$$

In Eq. 38c the definitions from Eqs. 11 and 16 were used. Inserting these equations back into Eq. 16 and writing

$$\bar{Q}_a\bar{Q}_b = \left(\sum_{i=1}^{n_F} \bar{Q}_{a,i}\right)\left(\sum_{j=1}^{n_F} \bar{Q}_{b,j}\right) = \sum_{i=1}^{n_F} \bar{Q}_{a,i}\bar{Q}_{b,i} + \sum_{i\neq j}^{n_F} \bar{Q}_{a,i}\bar{Q}_{b,j}, \tag{39}$$

we obtain for the total brightness correlation function

$$\Theta(\tau) = \frac{\sum_{i=1}^{n_F} \bar{Q}_{a,i}\bar{Q}_{b,i}\Theta_i(\tau) + \sum_{i\neq j}^{n_F} \bar{Q}_{a,i}\bar{Q}_{b,j}}{\sum_{i=1}^{n_F} \bar{Q}_{a,i}\bar{Q}_{b,i} + \sum_{i\neq j}^{n_F} \bar{Q}_{a,i}\bar{Q}_{b,j}}. \tag{40}$$

Subtracting one from each side and arranging yields

$$\Theta(\tau) - 1 = \sum_{i=1}^{n_F} \frac{\bar{Q}_{a,i}\bar{Q}_{b,i}}{\sum_{i=1}^{n_F} \bar{Q}_{a,i}\bar{Q}_{b,i} + \sum_{i\neq j}^{n_F} \bar{Q}_{a,i}\bar{Q}_{b,j}} [\Theta_i(\tau) - 1], \tag{41}$$

which for identical fluorophores can be simplified to

$$\Theta_{n_F}(\tau) = \frac{n_F\bar{Q}_a\bar{Q}_b\Theta_1(\tau) + n_F(n_F-1)\bar{Q}_a\bar{Q}_b}{n_F\bar{Q}_a\bar{Q}_b + n_F(n_F-1)\bar{Q}_a\bar{Q}_b} = \frac{\Theta_1(\tau)-1}{n_F} + 1. \tag{42}$$

It can be seen from n_F appearing in the denominator in this last expression that a larger number of individually acting fluorophores on one particle diminishes the influence of their brightness correlation function. Since the individual brightnesses do not fluctuate in a correlated manner, the total brightness of the particle represents an average value with decreased variance.

Photophysical processes usually depend on the intensity of the exciting radiation and thus on the position of the fluorophore in relation to the focus. Therefore, not all of the assumptions made above are fulfilled and diffusion and brightness are no longer decoupled. However, a simplification can be made for slow diffusion, in which case a local photophysical equilibrium is established at any position. Although Q still depends on $\vec{r}$, now the relationship $\left\langle Q_{(a,b)}\left(\vec{r}_i, \vec{R}_i, Z_i, t\right) W_a\left(\vec{r}_i\right)\right\rangle = \left\langle Q_{(a,b)}\left(\vec{r}_i, \vec{R}_i, Z_i, t\right)\right\rangle \cdot \left\langle W_a\left(\vec{r}_i\right)\right\rangle$ holds. Therefore, brightness correlations and positional correlations can still be separated.

A case of practical relevance is that of the three-level system (S_0, S_1 and T_1) with singlet-triplet transitions [17] and the rate constants k_{ex} for the excitation, k_{10} for the fluorescence and non-radiative decay, k_{ISC} for the intersystem crossing and k_{phos} for the phosphorescence.

The scheme in Fig. 1 shows the reactions which connect the populations of the three levels, [S_0], [S_1], [T]. These populations can be written as a vector, $\vec{\pi}(t)$. A corresponding system of deterministic linear differential equations with constant

Fig. 1 Scheme of the three-level system for photophysical processes used here. Arrows denote unimolecular reactions between the populations of the singlet and triplet states

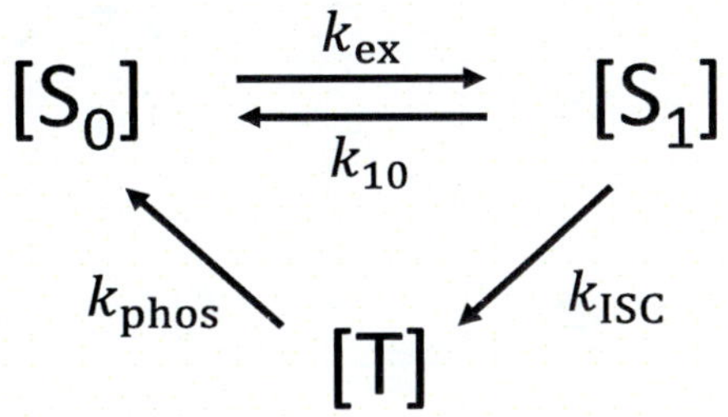

coefficients can be set up to describe the kinetics and can be formulated as a matrix equation:

$$\frac{d\vec{\pi}(t)}{dt} = \mathbf{Q}\,\vec{\pi}(t) \tag{43}$$

This results in a 3 × 3 matrix for **Q** with the rate constants given in the scheme.

$$\mathbf{Q} = \begin{pmatrix} -k_{ex} & k_{10} & k_{phos} \\ k_{ex} & -k_{10} - k_{ISC} & 0 \\ 0 & k_{ISC} & -k_{phos} \end{pmatrix}. \tag{44}$$

For the usual case $k_{10} \gg k_{ISC}$, k_{phos}, an approximate solution for the matrix equation is given by a sum of two exponential functions with characteristic time constants τ_{trip} and τ_{anti}, associated with triplet blinking and antibunching, respectively [11]:

$$\frac{1}{\tau_{anti}} = k_{ex} + k_{10} \tag{45}$$

$$\frac{1}{\tau_{trip}} = k_{phos} + \frac{k_{ex}\,k_{ISC}}{k_{ex} + k_{10}} \tag{46}$$

This leads to an expression for the individual correlation function $\Theta_1(\tau)$, which, by considering Eq. 42, results in [13]:

$$\Theta_{n_F}(\tau) = 1 + \frac{\bar{x}_T e^{-\tau/\tau_{trip}} - e^{-\tau/\tau_{anti}}}{n_F(1 - \bar{x}_T)} \tag{47}$$

Here, $\bar{x}_T$ represents the average fraction of particles in the triplet state.

From this general equation many different applications can be derived, for example, the calculation of protein oligomerization via antibunching (see [18, 19]), the detection of single-nucleotide differences in DNA via their influence on blinking times (see [20]) or the study of protonation dynamics in fluorescent proteins (see [21]).

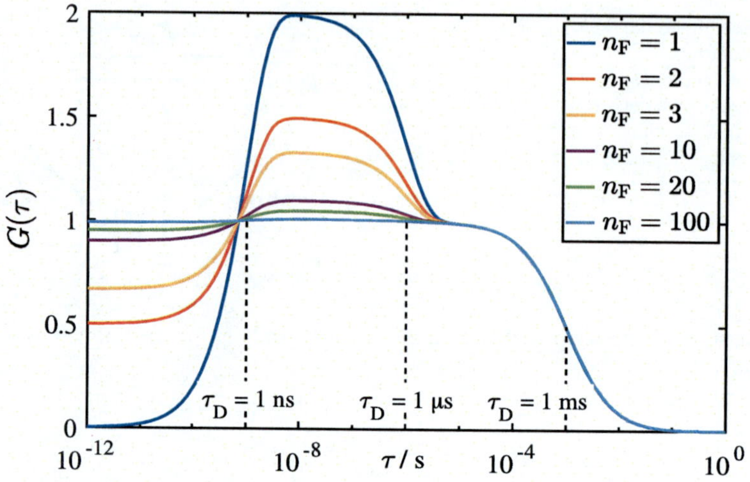

Fig. 2 Theoretical FCS correlation curves $G(\tau)$ for particles diffusing through a 3DG focus with structural parameter $S = 6$, diffusion time $\tau_D = 1$ ms and effective particle number $N_{eff} = 1$. The particles contain a varying number of fluorophores n_F given in the legend. The photophysics follows Eq. 47 with constants $\tau_{anti} = 1$ ns, $\tau_{trip} = 1$ µs and $\bar{x}_T = 0.5$

An important result that follows from Eq. 47 is the decreasing height of the blinking term in the correlation curves with increasing number of fluorophores n_F per particle (see Fig. 2). Therefore, the blinking becomes negligible in FCS measurements of lipid vesicles with a large number of lipidic fluorophores or nanoparticles containing a large number of dyes. In this case, the blinking term does not need to be included in the fitting procedure, whereby the uncertainty in the determination of the other parameters is reduced (Fig. 2).

4 Position Correlation Function

The basis of the calculation of the position correlation function is the knowledge of the time dependence of $P\left(\left\{\vec{r}', \tau\right\}|\left\{\vec{r}, 0\right\}\right)$, where the two endpoints, $\lim_{\tau \to 0} P\left(\left\{\vec{r}', \tau\right\}|\left\{\vec{r}, 0\right\}\right) = \delta\left(\vec{r}' - \vec{r}\right)$ and $\lim_{\tau \to \infty} P\left(\left\{\vec{r}', \tau\right\}|\left\{\vec{r}, 0\right\}\right) = \frac{1}{V}$ are already known. In addition, for homogeneous and isotropic diffusion which is not influenced by the laser focus,

$$P\left(\left\{\vec{r}', \tau\right\}|\left\{\vec{r}, 0\right\}\right) = P\left(|\vec{r}' - \vec{r}|, \tau\right) = P\left(|\Delta\vec{r}|, \tau\right) \tag{48}$$

can be assumed, so that a function of only two parameters remains to be determined. This probability density function describes the probability of finding a particle after a certain period of time τ at a certain distance $|\Delta \vec{r}|$ from its point of origin.

Generally, diffusion models can be classified by two criteria: Fickian diffusion with a mean square deviation (MSD) of the position that increases linearly in time, and Gaussian diffusion with a normal distribution of step sizes for the particle movement, whose variance is determined by the MSD. The simplest case of free diffusion of a particle in space, without any interfering interaction with other matter, fulfills both criteria. Non-Fickian diffusion with $\left\langle |\Delta \vec{r}(t)|^2 \right\rangle \propto t^\gamma$ is usually called anomalous diffusion [22]. Non-Gaussian yet Fickian diffusion takes place, for example, when diffusion processes with different diffusion coefficients are superimposed in complex structures [23]. Realistically, neither criterion will be fulfilled exactly, but the simple equations are used as approximations.

For isotropic Gaussian diffusion with independent motion in n spatial directions, the probability as a function of the vector $\Delta \vec{r} = (\Delta x_1, \Delta x_2, \ldots \Delta x_n)^\mathrm{T}$ is given by [24].

$$P\left(|\Delta \vec{r}|, \tau\right) = \prod_{i=1}^{n} \frac{e^{-\frac{|\Delta x_i(\tau)|^2}{2\left\langle |\Delta x_i(\tau)|^2 \right\rangle}}}{\sqrt{2\pi \left\langle |\Delta x_i(\tau)|^2 \right\rangle}} \tag{49a}$$

$$= \frac{e^{-\frac{|\Delta \vec{r}|^2}{2\left\langle |\Delta \vec{r}(\tau)|^2 \right\rangle / n}}}{\left(2\pi \left\langle |\Delta \vec{r}(\tau)|^2 \right\rangle / n\right)^{n/2}}. \tag{49b}$$

If a reference measurement for $\Xi(\tau)$ is available on a system containing particles that follow Fick's law of diffusion and whose diffusion coefficient is known, a one-to-one relation between mean square displacement $\left\langle |\Delta \vec{r}(\tau)|^2 \right\rangle$ and position correlation function $\Xi(\tau)$ exists, from which first $\Xi(\tau) = f\left(\left\langle |\Delta \vec{r}(\tau)|^2 \right\rangle\right)$ and by (numerical) inversion finally $\left\langle |\Delta \vec{r}(\tau)|^2 \right\rangle = f^{-1}(\Xi(\tau))$ can be obtained. Thus, anomalous diffusion can be investigated even without explicit knowledge of $W_{(a,b)}$, which can be used e.g. for the calculation of viscoelastic properties in passive microrheology [25]. By conducting FCS measurements on different spatial scales via multiple focus sizes (spot variation FCS), the nanoscopic material properties behind the anomalous diffusion can be investigated, such as lipid domains or protein and polymer meshworks, see for example [26–28].

Nonetheless, for practical applications of the correlation functions in FCS and FCCS experiments, it is convenient to devise a model for the MDF and calculate $\Xi(\tau)$ analytically. The simplest and most common model for the MDF in a confocal setup is a three-dimensional Gaussian function (3DG) with cylindrical symmetry, where we now use the vector $\vec{r} = (x, y, z)^\mathrm{T}$ and the center of focus is at $\vec{r} = \vec{0}$ and the ratio of the axial vs. lateral elongation of the focus is called the structural parameter (or structure parameter) $S = \omega_z/\omega_{xy}$ [29]:

$$W_{3DG}\left(\vec{r}\right) = e^{-\frac{2(x^2+y^2)}{\omega_{xy}^2}} \cdot e^{-\frac{2z^2}{\omega_z^2}} = e^{-\frac{2\left(x^2+y^2+\frac{z^2}{S^2}\right)}{\omega_{xy}^2}} \tag{50}$$

$$V_{\text{eff,x}} = \left(\frac{\pi}{2}\right)^{\frac{3}{2}}\left(\omega_{\text{xy},a}^2 + \omega_{\text{xy},b}^2\right)\sqrt{S_a^2\omega_{\text{xy},a}^2 + S_b^2\omega_{\text{xy},b}^2} \tag{51}$$

$$V_{\text{eff},(a,\ b)} = \pi^{\frac{3}{2}} S_{(a,\ b)}\omega_{\text{xy},(a,\ b)}^3. \tag{52}$$

Due to the Gaussian functions, the integration according to Eq. 17 can be solved analytically.

Introducing the diffusion time with the definition

$$\tau_{\text{D},(a,\ b)} = \frac{\omega_{\text{xy},(a,\ b)}^2}{4D} \tag{53}$$

as the characteristic time a particle takes to traverse the focus in lateral direction, one obtains for the position correlation function:

$$\Xi(\tau) = \left(1 + \frac{\tau}{(\tau_{\text{D},a} + \tau_{\text{D},b})/2}\right)^{-1}\left(1 + \frac{\tau}{(S_a^2\tau_{\text{D},a} + S_b^2\tau_{\text{D},b})/2}\right)^{-1/2}. \tag{54}$$

For the experimentally important case of species of different brightnesses with a single diffusion time diffusing in 3D the FCS correlation function (not including photophysical effects) is given by [29]

$$G_{(a,\ b)}(\tau) = \mathfrak{A}\frac{1}{1 + \frac{\tau}{\tau_{\text{D},(a,\ b)}}}\frac{1}{\sqrt{1 + \frac{\tau}{S_{(a,\ b)}^2\tau_{\text{D},(a,\ b)}}}} \tag{55}$$

If diffusion is restricted to a plane which is centered in the focus and perpendicular to the laser beam the MDF can be modeled as a two-dimensional Gaussian function (2DG). In this case, the FCS correlation function simplifies to

$$G_{(a,\ b)}(\tau) = \mathfrak{A}\frac{1}{1 + \frac{\tau}{\tau_{\text{D},(a,\ b)}}} \tag{56}$$

Figure 3 illustrates the typical shapes of FCS diffusion curves for the 2DG model and for the 3DG model, including different values of the structural parameter S.

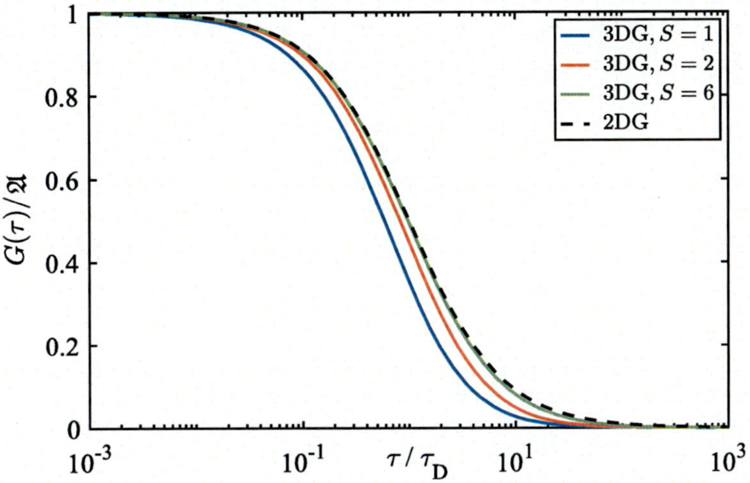

Fig. 3 Normalized theoretical FCS correlation curves, $G(\tau)/\mathfrak{A}$, for a single type of diffusing particles in the absence of photophysical effects. The MDF is described by a 3DG or 2DG function for diffusion in three or two dimensions, respectively. Correlation curves for the 3DG MDF with different structural parameters S and for the corresponding two-dimensional case (2DG) are shown. For illustration purposes, the values on the abscissa are given relative to the diffusion time τ_D. For increasing values of S, the curves for the 3DG case approach their two-dimensional counterpart

References

1. Magde D, Webb WW, Elson E (1972) Thermodynamic fluctuations in a reacting system – measurement by fluorescence correlation spectroscopy. Phys Rev Lett 29(11):705–708
2. Elson EL, Webb WW (1975) Concentration correlation spectroscopy – new biophysical probe based on occupation number fluctuations. Annu Rev Biophys Bioeng 4:311–334
3. Bacia K, Haustein E, Schwille P (2014) Fluorescence correlation spectroscopy: principles and applications. Cold Spring Harb Protoc 2014(7):709–725
4. Bacia K, Schwille P (2007) Practical guidelines for dual-color fluorescence cross-correlation spectroscopy. Nat Protoc 2(11):2842–2856
5. Krieger JW, Singh AP, Bag N, Garbe CS, Saunders TE, Langowski J, Wohland T (2015) Imaging fluorescence (cross-) correlation spectroscopy in live cells and organisms. Nat Protoc 10(12):1948–1974
6. Malchus N (2011) Fluorescence correlation spectroscopy: detecting and interpreting the mobility of transmembrane proteins in vivo. Curr Protoc Toxicol 48(1):2–19
7. Dertinger T, Pacheco V, von der Hocht I, Hartmann R, Gregor I, Enderlein J (2007) Two-focus fluorescence correlation spectroscopy: a new tool for accurate and absolute diffusion measurements. ChemPhysChem 8(3):433–443
8. Krmpot AJ, Nikolic SN, Oasa S, Papadopoulos DK, Vitali M, Oura M, Mikuni S, Thyberg P, Tisa S, Kinjo M, Nilsson L, Terenius L, Rigler R, Vukojevic V (2019) Functional fluorescence microscopy imaging: quantitative scanning-free confocal fluorescence microscopy for the characterization of fast dynamic processes in live cells. Anal Chem 91(17):11129–11137
9. Loman A, Gregor I, Stutz C, Mund M, Enderlein J (2010) Measuring rotational diffusion of macromolecules by fluorescence correlation spectroscopy. Photochem Photobiol Sci 9(5): 627–636

10. Oura M, Yamamoto J, Ishikawa H, Mikuni S, Fukushima R, Kinjo M (2016) Polarization-dependent fluorescence correlation spectroscopy for studying structural properties of proteins in living cell. Sci Rep 6:1–7
11. Widengren J, Mets U, Rigler R (1995) Fluorescence correlation spectroscopy of triplet-states in solution – a theoretical and experimental-study. J Phys Chem 99(36):13368–13379
12. Gregor I, Patra D, Enderlein J (2005) Optical saturation in fluorescence correlation spectroscopy under continuous-wave and pulsed excitation. ChemPhysChem 6(1):164–170
13. Petrov EP, Schwille P (2008) State of the art and novel trends in fluorescence correlation spectroscopy. In: Resch-Genger U (ed) Standardization and quality assurance in fluorescence measurements II: bioanalytical and biomedical applications. Springer, Heidelberg, pp 145–197
14. Landau LD, Lifshitz EM (1980) Statistical physics, Part 1. Pergamon Press, Oxford
15. Jakobowska I, Becker F, Minguzzi S, Hansen K, Henke B, Epalle NH, Beitz E, Hannus S (2021) Fluorescence cross-correlation spectroscopy yields true affinity and binding kinetics of plasmodium lactate transport inhibitors. Pharmaceuticals-Base 14(8):757
16. Krüger D, Ebenhan J, Werner S, Bacia K (2017) Measuring protein binding to lipid vesicles by fluorescence cross-correlation spectroscopy. Biophys J 113(6):1311–1320
17. Bernard J, Fleury L, Talon H, Orrit M (1993) Photon bunching in the fluorescence from single molecules – a probe for intersystem crossing. J Chem Phys 98(2):850–859
18. Sýkora J, Kaiser K, Gregor I, Bonigk W, Schmalzing G, Enderlein J (2007) Exploring fluorescence antibunching in solution to determine the stoichiometry of molecular complexes. Anal Chem 79(11):4040–4049
19. Temirov J, Werner J, Goodwin P, Bradbury A (2012) "Sizing" the oligomers of Azami Green fluorescent protein with FCS and antibunching. SPIE
20. Yeh HC, Puleo CM, Ho YP, Bailey VJ, Lim TC, Liu K, Wang TH (2008) Tunable blinking kinetics of Cy5 for precise DNA quantification and single-nucleotide difference detection. Biophys J 95(2):729–737
21. Mudalige K, Habuchi S, Goodwin PM, Pai RK, De Schryver F, Cotlet M (2010) Photophysics of the red chromophore of HcRed: evidence for cis-trans isomerization and protonation-state changes. J Phys Chem B 114(13):4678–4685
22. Hofling F, Franosch T (2013) Anomalous transport in the crowded world of biological cells. Rep Prog Phys 76(4):046602
23. Wang B, Kuo J, Bae SC, Granick S (2012) When Brownian diffusion is not Gaussian. Nat Mater 11(6):481–485
24. Berne B, Pecora R (1976) Dynamic light scattering. Wiley, New York
25. Rathgeber S, Beauvisage HJ, Chevreau H, Willenbacher N, Oelschlaeger C (2009) Microrheology with fluorescence correlation spectroscopy. Langmuir 25(11):6368–6376
26. Jin W, Simsek MF, Pralle A (2018) Quantifying spatial and temporal variations of the cell membrane ultra-structure by bimFCS. Methods 140:151–160
27. Šachl R, Bergstrand J, Widengren J, Hof M (2016) Fluorescence correlation spectroscopy diffusion laws in the presence of moving nanodomains. J Phys D Appl Phys 49(11):114002
28. Scipioni L, Lanzano L, Diaspro A, Gratton E (2018) Comprehensive correlation analysis for super-resolution dynamic fingerprinting of cellular compartments using the Zeiss Airyscan detector. Nat Commun 9:1–7
29. Krichevsky O, Bonnet G (2002) Fluorescence correlation spectroscopy: the technique and its applications. Rep Prog Phys 65(2):251–297

Part II
Fluorescence Microscopy: Basics and Advanced Approaches

Introduction to Fluorescence Microscopy

Radek Macháň

Contents

R. Macháň (✉)
Singapore Centre for Environmental Life Sciences Engineering (SCELSE), Nanyang Technological University (NTU), Singapore, Singapore
e-mail: radek.machan@ntu.edu.sg

R. Šachl, M. Amaro (eds.), *Fluorescence Spectroscopy and Microscopy in Biology*, Springer Ser Fluoresc (2023) 20: 141–200, https://doi.org/10.1007/4243_2022_34, Published online: 3 January 2023

Abstract Owing to its chemical specificity, high brightness, live-cell compatibility and other favourable characteristics, fluorescence is ubiquitously used as a source of contrast in biological microscopy. Fluorescence microscopes are also indispensable for many fluorescence spectroscopy techniques, such as single-molecule spectroscopy, fluorescence fluctuation spectroscopy and any form of fluorescence (micro)-spectroscopy performed in-situ in living cells and organisms. This makes fluorescence microscopy a key topic of this whole volume. This chapter aims at providing a general introduction to the topic, describing a generic fluorescence microscope and its essential components with more emphasis on objective lenses and their role in determining the resolution of microscopy images. The differences between widefield and confocal microscopes are explained and the later part of the chapter introduces briefly other variants of fluorescence microscopy and approaches to increase image resolution.

Keywords Confocal microscopy · Image resolution · Optical microscopy · Optical sectioning · Super-resolution imaging

1 Introduction

Fluorescence is one of the most commonly used sources of contrast in biological microscopy. Its uses range from the most routine assays, such as counting cells based on fluorescently labelled nuclei or counting live and dead bacteria using live/dead staining [1], to the most advanced forms of super-resolution or functional imaging [2, 3]. Fluorescence as a source of contrast in biological microscopy owes its popularity to several factors including:

1. *Chemical specificity*. Selective fluorescent labelling allows visualising the spatial distribution of specific structures or molecules of interest. Some fluorescent labels are specific to certain classes of molecules (for example DNA labelling dyes [4, 5]) or to certain types of environment (for example lipophilic dyes accumulating preferentially in non-polar environments such as lipid membranes or lipid droplets [6, 7]). Besides this type of low-level specificity there exist various labelling strategies to achieve more specific targeting. These include, for example the use of fluorescently labelled antibodies or DNA sequences, genetically encoding proteins of interest tagged either with fluorescent proteins or with specific anchor sequences for fluorescent label binding [8–10]. Last but not least, some biomolecules exhibit intrinsic fluorescence (for example

photosynthetic dyes or metabolic cofactors) and can be specifically imaged by fluorescence microscopy without the need for extrinsic labelling [11, 12]. Although indispensable in fluorescence microscopy studies, fluorophores and labelling strategies are beyond the scope of this chapter. More information on this topic can be found in chapter "Choosing Fluorescent Probes for Biology" of this volume.

2. *Multiplexing capability.* There exists a broad range of fluorophores with distinct excitation and/or emission spectra. Since signals from spectrally distinct fluorophores can be readily distinguished, it is possible to image simultaneously, or nearly simultaneously, several species of molecules or structures of interest selectively labelled by distinct fluorophores. Such images can be then used, for example, to evaluate colocalisation between different species of proteins or other biomolecules [13–15]. When linear spectral unmixing is used (see Sect. 2.2.1), even signals from fluorophores with highly overlapping spectra can be separated, allowing distinguishing of more than 100 different spectral signatures in fluorescence images [16].
3. *Environmental sensitivity.* Fluorescence brightness, spectra as well as fluorescence lifetimes of many fluorophores are sensitive to the fluorophore's environment. This allows fluorescence microscopy to image distributions of, for example, pH, polarity, viscosity, or the concentration of specific ions [17–19] inside of living cells or multicellular organisms.
4. *High brightness.* Fluorescence is typically several orders of magnitude brighter than Raman scattering, another source of chemically specific contrast in microscopy [20]. While fluorescence cannot compete with Raman scattering in the amount of chemical information it provides, its high brightness allows on one side direct observation of fluorescence in microscope eyepieces and, on the other side, it permits imaging and tracking of individual fluorophore molecules.
5. *Compatibility with living cells and organisms.* Like most other forms of optical microscopy, fluorescence microscopy is suitable for imaging living cells or multicellular organisms. Low excitation intensities are often sufficient to obtain enough fluorescence signal, resulting in low phototoxicity and making fluorescence microscopy suitable for live-cell and live-organism imaging over extended periods of time [21, 22].
6. *Technical simplicity and affordability.* In its simplest forms, fluorescence microscopy can be performed with very simple optical microscope setups, which are not fundamentally more complex than setups for brightfield optical microscopy, the simplest and oldest form of optical microscopy. Various low-cost designs of fluorescence microscopes or upgrades to brightfield microscopes have been proposed, including lens-free designs [23–32]. The technical simplicity of fluorescence microscope also ensures its straightforward combination with other optical microscopy modalities [33, 34].
7. *Super-resolution capabilities.* While in its basic forms, fluorescence microscopy offers the same resolution as other forms of optical microscopy, which means diffraction-limited at the best (see Sect. 3.1 for more details), there exist a number of techniques that allow fluorescence imaging with significantly better resolution. Most of these techniques rely on some features specific to fluorescence

microscopy such a switching of fluorophores between bright (fluorescent) and dark states or sparsity of features typical for fluorescent images. For more details, see Sect. 6 and chapter "STED and RESOLFT Fluorescence Nanoscopy".

So far, when talking about fluorescence as a source of contrast in microscopy, we have mainly considered fluorescence intensity as the source of contrast. Fluorescence intensity images inform us mainly about the spatial distribution of fluorescent probes and fluorescently labelled molecules of interest in the specimen and can be acquired even with the simplest forms of fluorescence microscopes. However, other characteristics of fluorescence can be also used to generate contrast in fluorescence microscopy and to provide different types of information about the system under study. As examples we can mention fluorescence lifetime used in fluorescence lifetime imaging microscopy (FLIM) (see chapter "Time-Resolved and Polarized Spectroscopy" for more details), fluorescence anisotropy [35, 36], fluorescence linear dichroism [37] or statistics of fluorescence intensity fluctuations employed in fluorescence correlation spectroscopy (FCS) imaging (see chapter "2D FCS Including Imaging FCS and Scanning FCS").

The term microscopy is usually associated with imaging of microstructures. However, fluorescence microscopes are also frequently used in non-imaging techniques as a device to generate small, well-defined detection volumes essential in FCS (see chapter "FCS and FCCS Principles and Applications") and other single-molecule fluorescence spectroscopies. While these techniques are often not generating images themselves, they frequently benefit from the imaging capabilities of fluorescence microscopes for positioning the detection volumes into specific regions of the specimen. This aspect has become more important as biomolecular studies move from in-vitro settings to measurements performed directly in living cells and organisms. Overall, these fluorescence spectroscopic or micro-spectroscopic techniques can be seen as a form of fluorescence microscopy in a broader sense, bringing us to the conclusion that this whole volume is dedicated largely to fluorescence microscopy. As many of more specialised modalities of fluorescence microscopy are, therefore, discussed in other chapters, this chapter limits itself only to general basics of fluorescence microscopy.

2 Epifluorescence Microscope

The most common architecture of a fluorescence microscope is an epifluorescence microscope, which employs episcopic, or in other words reflected light, mode of observation. In practice this means that the excitation light is delivered to the specimen by the same objective lens which is used for collecting the fluorescence light. A generic (widefield) epifluorescence microscope is schematically depicted in Fig. 1. An epifluorescence microscope can be built in an upright (the objective is above the specimen) as well as in an inverted configuration (the objective is below the specimen), whichever configuration is more suitable for the specimens of

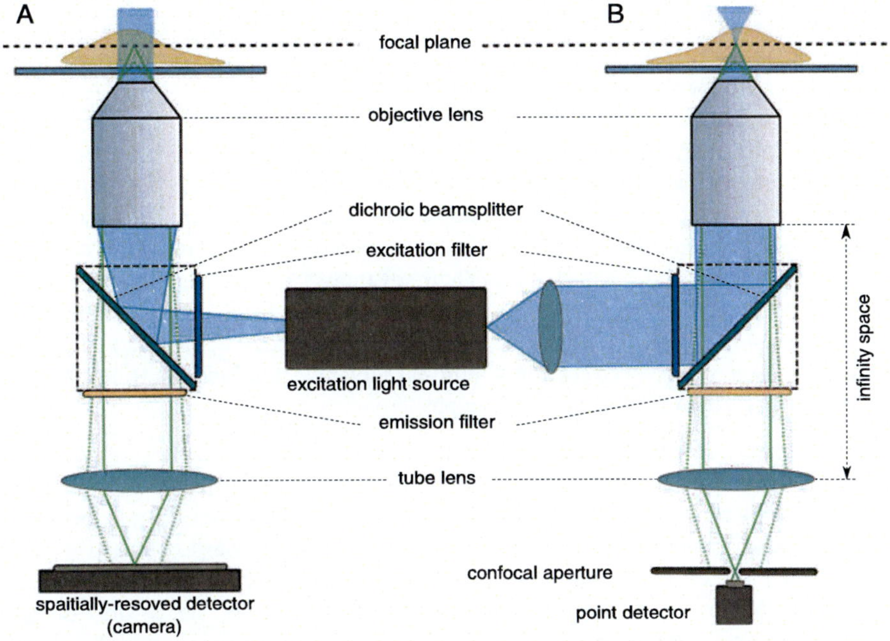

Fig. 1 A schematic drawing of the main components of an inverted widefield (**a**) and confocal (**b**) epifluorescence microscope. Refer to Sects. 2.1–2.4 for an explanation of the role of the individual components. Green lines schematically illustrate rays of fluorescence emitted from an in-focus fluorophore (solid lines) and an out-of-focus fluorophore (dotted lines). The rays from the in-focus fluorophore are focused in the primary image plane producing a sharp (diffraction limited) image of the fluorophore, while the unfocused rays from the out-of-focus fluorophore produce a blurry background. Confocal aperture in the confocal setup (**b**) stops most of the out-of-focus fluorescence from reaching the detector (see Sect. 4 for more details)

interest. The inverted configuration is preferred for imaging cells in culture dishes and similar systems, where small objects are found close to the bottom of a dish filled with a liquid medium. The upright configuration is more convenient for imaging larger non-transparent objects, for example intravital imaging in rodents. Microscope slides are easily imaged in either configuration. The schematic in Fig. 1 shows explicitly an inverted microscope; turning the figure by 180° gives a schematic of an upright microscope.

2.1 Light Sources

The excitation light in an epifluorescence microscope can be provided by lasers or incoherent light sources, often referred to as lamps. Lasers are used in confocal microscopy (see Sect. 4), while incoherent light sources are common in widefield fluorescence microscopy with the exception of some specialised widefield modalities such as total internal reflection fluorescence (TIRF) microscopy (see Sect. 5.3). Mercury or Xenon arc lamps producing white light covering a broad spectral range from ultraviolet (UV) to near infrared (NIR) used to be the most common light sources in widefield epifluorescence microscopy. Mercury metal halide lamps offered a similar spectrum with the advantages of a longer lifetime and easier operation. However, all these lamps have been largely rendered obsolete by solid-state light sources usually based on light-emitting diodes (LEDs). LEDs often emit in relatively narrow spectral ranges (although white-light LEDs are also available); therefore, LED-based light sources usually contain several LEDs which together cover the whole desired spectral range. Their advantages include very long lifetimes, high-intensity stability, high energy efficiency [38, 39] as well as fast modulation capabilities [40]. Beside the high efficiency of conversion of electric energy to light, solid-state light sources have three additional advantages which contribute to the overall energy efficiency of the microscope:

1. LEDs can be switched on and off nearly instantaneously. Therefore, they can be on only when excitation light is needed. In contrast, metal halide or arc lamps have quite long warm-up times and cannot be switched off, for example, between the acquisition of individual images; a mechanical shutter has to be used instead to prevent the excitation light from reaching the specimen when it is not needed.
2. The output power of LEDs can be regulated electronically, hence they do not need to emit at full power if only low excitation light intensity is required. Neutral density filters or aperture stops are used to control the excitation light intensity produced by metal halide or arc lamps, while the lamp itself emits at its full power.
3. Only a narrow spectral band of excitation light is usually used at a time to specifically excite certain fluorophores of interest. Since metal halide or arc lamps produce white light covering a broad spectrum, the desired spectral band is selected by an excitation filter (see Sect. 2.2.) which blocks the rest of the light from reaching the specimen. Most of the output of the lamp is, thus, typically wasted. On the other hand, since solid-state light sources usually contain multiple LEDs, each emitting only in a relatively narrow spectral band, only the LED emitting in the desired spectral band can be turned on.

The combination of nearly instantaneous on and off switching and narrow spectral bands (see points 1 and 3 above) makes LED-based light sources well suited for very fast multi-channel fluorescence imaging. Rapid switching between different excitation spectral bands, without the need for any mechanically moving parts, enables nearly simultaneous selective excitation of different fluorophore species in the specimen. To fully benefit from the high speed of excitation band switching, there must be no other mechanical movements involved in switching between different fluorophore channels; the microscope, therefore, needs to be equipped with a main beamsplitter and an emission filter that is suitable for all the fluorophores of interest (such as a polychroic mirror or a partially reflective mirror and a multi-band emission filter; see Sect. 2.2 for more details).

2.2 Beamsplitters and Filters

Light from an excitation light source (see Sect. 2.1) passes through an excitation filter to select the desired excitation spectral window. Excitation filters are not needed if the light source emits only in a narrow spectral band such as is the case with most lasers and some LEDs. In setups with multiple laser lines available for excitation, acousto-optical tuneable filters are sometimes employed instead of excitation filters to select a desired laser line or a combination of lines. Together with the ability of rapidly switching the excitation laser lines, they offer an additional capability of fast modulation of a laser line intensity [41].

The excitation light is then reflected by a beam splitter, the main or primary beam splitter, to the microscope objective. Fluorescence emitted by the specimen is collected by the same objective and passes through the primary beam splitter in order to reach the detection device, for example a digital camera or human eye. The primary beam splitter thus plays the key role of separating the excitation and emission light paths. It is usually a dichroic mirror which reflects light of wavelengths shorter than a cut-off wavelength and transmits light of longer wavelength. The separation is, therefore, based on the well-known phenomenon of Stokes shift, the red shift of a fluorophore's emission spectrum compared to its excitation spectrum. It is a very efficient way of separating the emission and excitation paths since the losses of emission as well as excitation light are minimal on a properly chosen dichroic mirror. The disadvantage of a dichroic mirror with a single cut-off wavelength is its limitation to a specific spectral window. When a specimen labelled by multiple, spectrally distinct fluorophores needs to be imaged, a different dichroic mirror has to be used for each fluorophore, which slows down the acquisition speed. This limitation is overcome by polychroic mirrors, which have several alternating reflection and transmission bands. They allow simultaneous or nearly simultaneous (for example with a light source allowing very fast switching of excitation spectral bands; see Sect. 2.1) excitation of distinct fluorophore species. Polychroic mirrors with 2 to 4 bands are commonly used in fluorescence microscopy and mirrors with higher numbers of bands are also available. With increasing number of spectral bands, the transmissive spectral windows have to be narrower, which reduces the amount of fluorescence light that can reach the detection device. The loss in fluorescence signal tends to be more prominent for fluorophores with broader spectra that are more likely to stretch further over the other reflective spectral bands.

Where full flexibility in the choice of excitation wavelengths and their combinations is important, an acousto-optical beam splitter can be used as the primary beam splitter [42]. The device acts as a polychroic mirror with arbitrarily tuneable reflection and transmission bands. The drawbacks are the increased cost and complexity of such a setup. Alternatively, a broad-band partially reflective mirror (for example with 10% reflectivity and 90% transmissivity across the whole visible spectrum) can be used as a low-cost alternative in such cases. Similarly to an acousto-optical beam splitters, it is suitable for any combination of excitation and emission bands. Its low efficiency in delivering excitation light to the objective can be compensated by increasing the excitation light source power. If the excitation source is powerful enough, even lower reflectivity to transmissibility ratios can be used to minimise the loss of fluorescence light on the beamsplitter. A limitation lies in its lack of spectral selectivity. Reflected or back-scattered excitation light passes the mirror equally well as fluorescence and all the burden of blocking the unwanted excitation light then falls on the element which typically follows after the primary beam splitter – an emission filter.

The role of the emission filter is to block unwanted reflected and scattered excitation light from reaching the detection device and to define the spectral band in which fluorescence is detected. It, therefore, increases the specificity of detecting the fluorescence signal of interest. Similarly to dichroic and polychroic mirrors, there exists a range of long-pass, single band-pass and multiple band-pass filters. Fluorescence spectra are usually quite broad, spanning ranges often exceeding 100 nm; therefore, compromises are usually needed between the requirements of specificity (enhanced by narrow-band filters) and maximising collected signal (increasing with broader transmissivity bands of the filters). When we want to specifically image individual fluorophore species in a specimen labelled by multiple fluorophore species, selective excitation of individual species by light of appropriately selected wavelengths applied one at a time is usually more efficient and preferred compared to attempting emission signal separation by narrowing down the emission filter transmission bands. When multiple fluorophore species are excited by the same wavelength, it is often impossible to detect the signal from an individual fluorophore species only, without contamination with signal from the other simultaneously excited species (so-called spectral crosstalk). Linear spectral unmixing introduced in Sect. 2.2.1 may be necessary in such a case to separate contributions from the individual fluorophore species.

Matched sets including an excitation filter, a dichroic or polychroic mirror and an emission filter, for a certain spectral range of fluorescence, are commonly mounted in filter cubes. The whole filter cubes are then exchanged when switching between fluorophores, either by rotating a filter cube turret or translating a filter cube slider. In some setups, excitation filters, beam splitters and emission filters are mounted in separate turrets and can be thus freely combined. This flexible arrangement is advantageous when two fluorophores (or a single environmental sensitive probe in two different environments) excited by the same wavelength but emitting at different spectral windows, which means exhibiting different Stokes shifts, are present in the specimen. A single excitation filter and dichroic mirror alongside with two different emission filters are sufficient in this case. An example is probing lipid membrane organisation with Laurdan, which changes its emission spectrum in response to the membrane phase [43].

In some cases, for example when the process under study is very fast, it is necessary to capture the signal from multiple fluorophores truly simultaneously. Polychroic mirrors as main beam splitters can help in such cases by enabling simultaneous excitation in different spectral bands; however, a single multi-band emission filter and a single-detection device are not sufficient in this case since they would not allow separation of the signals from individual fluorophores. Multiple detection devices are needed, each collecting light in a different spectral window. Secondary dichroic beam splitters and emission filters are then usually employed to spectrally separate the fluorescence signals. The individual detection devices can be, for example separate digital cameras or separate portions of the chip of a single camera. The latter approach, beside reducing the cost of the setup by decreasing the number of cameras needed, is advantageous in situations where the final experiment output is sensitive to camera noise characteristics, such as in fluorescence fluctuation spectroscopies [44]. Its obvious limitation is a reduction of the accessible field of view.

The choice of beam splitters and filters is an important step in a fluorescence microscopy experiment design. The choice is based on the excitation and emission spectra of fluorophores used to label the specimen, with the goals of maximising the portion of the emission spectrum detected and, in the cases when multiple fluorophore species are used, of minimising the spectral crosstalk between individual spectral channels. Fluorophore spectra databases are a useful tool for this task [45]. Note that matching the detection spectral window to the emission spectrum of a fluorophore is more important than matching the excitation wavelength to the fluorophore's excitation spectrum maximum. The reduced excitation efficiency, when using a wavelength outside of the excitation spectrum maximum, can be compensated by increased excitation light intensity. Light which is not absorbed by the specimen is not likely to cause it any harm. Therefore, if the excitation light is not strongly absorbed by other molecules in the specimen than the fluorophores of interest, the risk of photodamage and phototoxicity scales with the excitation rate of the fluorophores rather than the excitation light intensity itself. On the other hand, there are no easy ways to compensate for fluorescence light lost on sub-optimally chosen filters or beam splitters, other than, for example, extending image integration times (and thus the duration of the experiment), detection gain (and thus the noise in the image) or fluorophore excitation rate (and thus the risk of increased fluorophore bleaching and photodamage of the specimen).

Spectral crosstalk is minimised by selecting fluorophores with spectrally well-separated emission and/or excitation. However, the more different fluorescent labels are present in the specimen, the smaller the spectral separation between them can be, eventually leading to the presence of fluorophores with overlapping spectra. In such cases, dichroic mirrors and optical filters are not sufficient to unambiguously separate signals from such fluorescent labels. Linear spectral unmixing can be a solution to this problem.

2.2.1 Linear Spectral Unmixing

We assume that the fluorescence emission spectrum of a mixture of fluorophore species is a linear combination of emission spectra of the individual species weighted by the relative abundance of each fluorophore species. If we acquire a multi-channel image with C fluorescence channels, then in each pixel of the image, we get the following intensities in each channel:

$$I_c = \sum_{s=1}^{S} M_{cs}\, w_s$$

Index c denotes the fluorescence channel and goes from 1 to C, index s identifies the fluorophore species and goes from 1 to S, the number of the fluorophore species. w_s is the weight with which s-th fluorophore species contributes to the overall signal, or in other words, the relative abundance of the s-th fluorophore in the mixture. M_{cs} are the elements of the so-called mixing matrix which describes the contribution of each fluorophore species to the individual channels, that means their normalised spectral signatures. The mixing matrix can be found by acquiring images (with the same set of C channels) in samples containing each a single fluorophore species only.

By measuring the signal in all C channels, we obtain a set of C equations for S unknown weights w_s. This set of equations can be unambiguously solved only when C is larger than or equal to S. This means we need at least as many channels as we have fluorophore species. The unknown weights w_s are then obtained from the following equations:

$$w_s = \sum_{c=1}^{C} U_{sc}\, I_c$$

U_{sc} are the elements of the unmixing matrix, which can be found by matrix inversion (in the case when $C = S$) or pseudo-inversion (when $C > S$) of the mixing matrix. Finding the weights w_s for all pixels in the image gives us a new multi-channel image of S channels, each channel corresponding to the distribution of a single fluorophore species.

When spectra of two fluorophore species, i-th and j-th, are heavily overlapped, the corresponding mixing matrix elements M_{ci} and M_{cj} are likely very similar for all channels c, compromising the reliability of the unmixed weights w_i and w_j. Increasing the number of channels C may help in such cases by leveraging subtler difference in the fluorophore spectra. On the other hand, higher number of channels C implies narrower spectral bands for each channel, resulting in lower signal in each channel and, therefore, in compromised signal-to-noise ratio. Lower signal-to-noise ratio is another factor increasing the uncertainty in the unmixed weights.

Because we have assumed a linear mixing model, where the spectrum of a mixture of fluorophore species is a linear combination of spectra of the individual species, linear spectral unmixing fails where this assumption is not met. This is, for example the case of saturated pixels. If the pixel value in a channel exceeds the dynamic range of the image, the value is clipped at maximum of the dynamic range. This affects the ratios between intensities I_c in different channels and leads to incorrect weights w_s. The more a fluorophore species contributes to the saturated channel, the more will its weight be underestimated.

In some cases, the mixing matrix M_{cs} is not known. For example, when working with a sample exhibiting multiple spectral patterns of autofluorescence, it is not possible to measure spectra of each autofluorescence component separately. For such situations, various blind spectral unmixing algorithms have been proposed, which estimate spectral signatures of the individual fluorophore species based on the distribution of mixed spectra present in individual pixels of the image [46, 47].

2.3 Image Formation

So far, we have not touched on the image-forming optics, which form the core of any optical microscope. As mentioned earlier, in epifluorescence microscopy, the objective lens has the double role of delivering excitation light to the specimen and of imaging the emitted fluorescence. Modern optical microscopes use the so-called infinity-corrected optical system. In this system, the objective lens, called infinity-corrected objective lens, is not used alone to produce the magnified image. Instead, it collimates rays originating from the focal plane. The collimated, that means parallel, rays are then focused by another lens, called tube lens, to produce a magnified image. Optical filters and beamsplitters are typically located in the so-called infinity space between the objective lens and the tube lens. Since the fluorescence light is collimated in this space, plan parallel optical elements such as optical filters do not disturb its wavefront. The image formed by the tube lens, called the primary image, can be then viewed by the microscope eyepiece or captured by a spatially resolved detector such as a digital camera, with its sensor positioned in the primary image plane. In some setups, though, the image formed by the tube lens is relayed to the camera sensor by a telescope (a pair of lenses) to increase or decrease the magnification of the image or to gain space for other optical elements such as secondary beamsplitters and filters.

Objective lenses are critical in determining the quality and resolution of microscopy images. Because of its central importance, the whole Sect. 3 is dedicated to this topic.

2.4 (Widefield) Image Detection

The last component in the detection optical path is the detection device itself. As we said above, in widefield epifluorescence microscopy, it is usually a spatially resolved sensor, or in other words a digital camera. Such a sensor is a two-dimensional array of small photosensitive elements, which we can call the physical pixels of the sensor, or pixels for short in the context of this section which is dedicated to camera sensors. In other context the term pixel can also refer to image pixels, the unit building blocks of a digital image. While those two are clearly linked (typically the intensity values of image pixels are proportional to the signal detected by the camera physical pixels), there is a conceptual distinction between them.

For fluorescence microscopy, monochrome cameras are usually preferred because of their higher sensitivity. A typical colour camera sensor contains pixel-level colour filters that make different physical pixels sensitive to light of different colour. In the usual Bayer pattern, in a group of 4 (2 × 2) neighbouring physical pixels, one is sensitive to blue light, one to red light and two are sensitive to green light [48]. Since the spectral bands for fluorescence detection are typically defined by the emission filters, the spectral information provided by a colour camera is redundant and the pixel-level colour filters only lead to additional signal losses. Furthermore, red or blue fluorescence is detected only by a fourth of the physical pixels (1 pixel in each group of 4), resulting in a correspondingly reduced sensitivity in that part of the spectrum.

The first digital camera technology that become widespread in fluorescence microscopy were charge-coupled devices (CCDs). A CCD sensor is an array of metal-oxide-semiconductor (MOS) capacitors that are charged by photoelectric effect when exposed to light. These capacitors form the physical pixels of the sensor and their pitch in the array is the physical pixel size. The charges are then converted to voltage by a charge amplifier and the voltage is then digitised. A usual CCD uses a single charge amplifier for the whole chip, which means that the charges from all pixels need to be transferred to a single readout point. This is achieved by applying external voltage, which induces the charges to hop to a neighbouring pixel. Charges from whole lines of pixels are transferred in this way to the neighbouring lines. The line on the edge of the sensor area is transferred to a readout register in which the charges are moved towards the charge amplifier in a direction perpendicular to the line transfer direction. Once the charges from all pixels in the readout register have been read one by one, all lines are again shifted by one step, and a new line enters the readout register. This is repeated until each line has been read in the readout register. The process can be quite time-consuming for sensors with large pixel numbers and, unless a mechanical shutter is used, may result in image smearing caused by exposing the camera while the lines of pixels are being shifted towards the readout register. To avoid the need for a mechanical shutter, different CCD architectures with masked pixels have been developed. An interline CCD uses a sensor chip with every second row of pixels covered by an opaque metal mask. The charge accumulated in the active, non-masked, rows of pixels is transferred to the masked rows and then shifted to the readout register and read out while the next frame is being collected by the active rows. The disadvantage is the reduction of the sensor fill-factor to 50%. This can be partially compensated by the use of micro-lens arrays that focus the light to pixels in the active rows. In a frame transfer CCD, half of the sensor chip is covered by an opaque metal mask. The lines are quickly transferred to the masked area and then read out through the readout register while the next frame is being collected in the non-masked part of the sensor. This design ensures full fill-factor of the sensor at the cost of using a chip of twice the size. As such it is preferred in low-light applications, where high sensitivity is more important than low cost.

Among the advantages of CCDs are high quantum efficiency (QE, the fraction of incident photons contributing to charge generation), which can reach values above 90% and high uniformity of pixel noise characteristics ensured by using a single charge amplifier for all pixels [49, 50].

A modification of CCD sensor called electron-multiplying CCD (EM-CCD) was developed to increase the signal at very low-light levels. An amplification register, through which the charges pass on their way to readout, is added to the chip. High voltage is used to shift electrons in this register resulting in impact ionisation similarly to what happens in an avalanche photodiode. The higher the applied voltage in the amplification register, the larger the gain in electron number. Although the amplification is an additional source of noise, the large gain in signal it provides leads to significant improvements in signal-to-noise ratio at low-light levels. EM-CCDs thus became the sensors of choice for single-molecule fluorescence, calcium imaging and other low-light applications [51–54].

An alternative digital camera technology, which has become increasingly widespread in fluorescence microscopy, is complementary metal-oxide-semiconductor (CMOS) sensors. Their principal difference from CCDs lies in the fact that in a CMOS sensor, each pixel is equipped with its own charge amplifier. This makes charge readout from CMOS sensors faster than from CCDs; however, it also means that part of the sensor active area is lost to charge amplifier transistors. This problem can be partially solved by micro-lens arrays focusing light to the active area of each pixel. Another potential problem associated with pixel-wise readout is the non-uniformity in amplification and in noise caused by variations between the amplifiers at individual pixels. This used to limit the spread of CMOS sensors in scientific applications until the arrival of so-called scientific-grade CMOS (sCMOS) sensors, which feature low readout noise and high quantum efficiencies that do not lag behind CCDs. Another important development in CMOS sensor technology was the integration of field-programmable gate arrays which can be programmed to correct for sensor-specific pixel-to-pixel variations (such as so-called hot pixels or fixed-pattern noise) caused by non-uniform performance of the individual amplifiers [50, 55]. All these developments have made current CMOS sensors a viable alternative to CCDs and EM-CCDs. Being easier and cheaper to manufacture, especially at large formats, cameras with CMOS sensors have become the mainstream choice in fluorescence microscopy, pushing CCDs and EM-CCDs to the position of a niche technology reserved for highly specialised applications. sCMOS sensors are successfully used in single-molecule fluorescence and even low-cost industrial CMOS cameras or cell phone CMOS cameras have been shown to be capable of single-molecule detection [56–60].

3 Objective Lenses and Image Resolution

As mentioned in Sect. 2.3, the objective lens is a key component of any optical microscope and its performance determines the quality of the magnified image.

3.1 Resolution of a Fluorescence Microscope

The purpose of a microscope is to observe small details of the imaged objects. The resolution or resolving power of a microscope, that is the dimension of the smallest detail of the object that can be resolved in the image, is therefore a key indicator of the microscope performance. The higher the resolution, the smaller are the resolvable details, the better. Let us define for simplicity the resolution as the smallest distance between two parallel straight lines that can be resolved as separate lines. This is straightforward enough in theory, less so in an experiment at very high resolutions, where no resolution targets with parallel straight lines of well-defined spacing are readily available. Various approaches how to experimentally measure the resolution in such cases have been developed [61–70], but we will not enter into this topic here.

To characterise the resolution of a fluorescence microscope or in general any optical microscope, we will introduce the concept of the microscope's point spread function (PSF) defined as the image of a point-like object. A hypothetical perfect optical system images each point-like object as a single point, resulting in what we can call a geometrically perfect image. This is however not the case of real optical systems for reasons which we will discuss further.

This concept is particularly relevant in fluorescence microscopy, where each individual fluorophore in the specimen is an independent light source of negligible dimensions. A fluorophore can be approximated as a self-luminous point the image of which is the PSF. The image captured by the microscope is then the sum of images of all fluorophores within the field of view, that means a weighted (by brightness of individual fluorophores) sum of PSFs centred each at the geometrically perfect image of the respective fluorophore. This is mathematically described as a convolution of the geometrically perfect image with the PSF.

Such convolution results in blurring of the image; the broader the PSF, the more blurred becomes the image. Let us consider the hypothetical parallel straight lines; the image of each individual line will not be an infinitely thin geometrical line; it will have the width of the PSF (see Fig. 2). If we start from an infinitely thin line and keep broadening the PSF or decreasing the distance between the lines, the images of the lines will be increasingly overlapped as the PSF width increases with respect to the distance between the lines, until at some point it will not be possible to tell whether there are two closely spaced lines or just a single line.

The next question is what a microscope PSF looks like and by what factors is it determined. To answer it, we need to consider the nature of light as electromagnetic waves. When they propagate through apertures, waves exhibit phenomena referred to as diffraction. Since a microscope objective lens has a finite diameter, it represents an aperture in the optical system and the PSF can be described using the theory of diffraction on a circular (typical cross-section of objective lenses) aperture. The theory predicts that the PSF has the form of an Airy function, which consists of a main central peak and decaying concentric rings. Most of the photons contributing to the image are concentrated in the central peak of the Airy function, making it

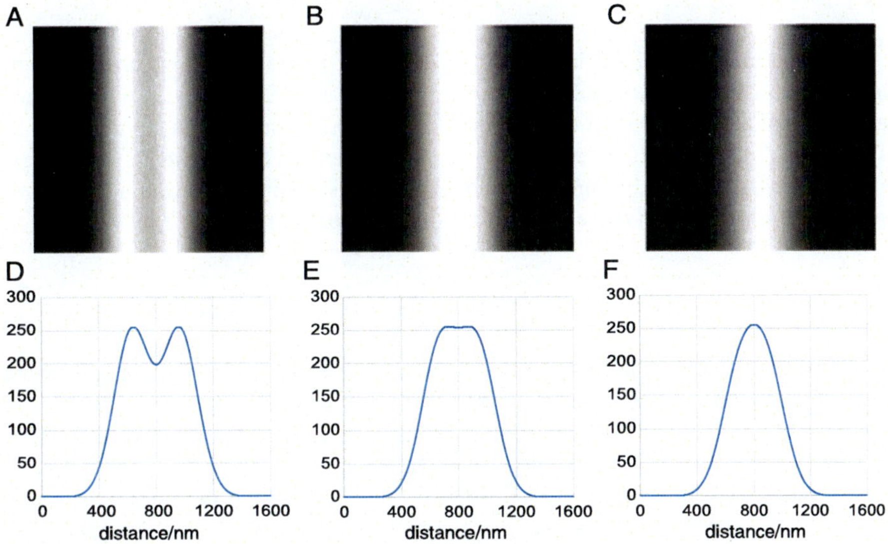

Fig. 2 Diffraction-limited resolution limit illustrated on simulated images of two thin parallel fluorescent lines (**a–c**) and the respective intensity profiles along the direction perpendicular to the fluorescent lines (**d–f**). The PSF width is $R_{PSF} = 339$ nm. The first case (**a** and **d**) corresponds to the Rayleigh criterion with distance between lines $d = 339$ nm, the second case (**b** and **e**) to Sparrow criterion ($d = 264$ nm) and the last case (**c** and **f**) to unresolved lines at $d = 200$ nm. The dimensions of the images are 1,600 × 1,600 nm. The PSF was for simplicity approximated by a Gaussian function. Intensities in the plots of intensity profiles are shown in relative units

possible to approximate the microscope PSF by a Gaussian function. The smaller the aperture is, the stronger is its effect on the propagation of the waves and the broader is the central peak of the PSF. From what we said earlier, this also means the lower is the resolution.

From the theory of diffraction of light on a circular aperture, it follows that the radius of the PSF central peak R_{PSF}, that is the radial distance between the maximum of the peak and the first minimum of the Airy function is:

$$R_{PSF} = 0.62\frac{\lambda}{NA}$$

Where λ is the wavelength of the detected light, the florescence wavelength in this case, and NA is the numerical aperture of the objective lens. Because of its influence on the microscope resolution, NA is a key parameter of microscope objective lens performance, and we will discuss it more in Sect. 3.2.

Considering the effect the PSF has on the image of the hypothetical parallel straight lines (see Fig. 2), we can deduce that the microscope resolution d, that is the smallest distance between two lines, at which they can be resolved as separate lines, is proportional to R_{PSF}:

$$d = q\, R_{PSF}$$

The commonly used Rayleigh resolution criterion uses $q = 1$, which means the lines are considered resolvable if the maximum of one PSF coincides with the first minimum of the other PSF [71, 72]. This may be unnecessarily strict, and we may be able to resolve the two lines as separate even if the centres of the respective PSFs are slightly closer. Ernst Abbe in his seminal work [73] derived a formula for microscope resolution, which corresponds to a slightly smaller q:

$$d = \frac{\lambda}{2\, NA}$$

Another commonly used criterion, Sparrow criterion, considers the line to be resolvable as long as the intensity profile along a direction perpendicular to the two fluorescent lines contains two separate maxima with a minimum in between, leading to q of approximately 0.78 (see Fig. 2b, e). In other words, there has to be a dip, no matter how small, in the intensity profile, in order to resolve the lines as separate [71, 72]. This can be considered as an ultimate limit to the resolution. Once the PSFs overlap to such an extent that the intensity profile contains just a single maximum, there is in principle no way how to tell that the image originates from two separate lines and not just from a single, possibly thicker, line (see Fig. 2c, f).

Considering the dip in the intensity profile brings us also to the influence of signal-to-noise ratio on the image resolution. As the distance between the fluorescent lines decreases, so does the difference between the maxima of the intensity profile and the local minimum in between them. When this difference, the depth of the dip, is smaller than the noise in the image, it may not be possible to resolve the dip in the intensity profile and hence to resolve the two fluorescent lines. Sparrow criterion may be therefore perceived as a limit which applies when the signal-to-noise ratio of the image approaches infinity.

Our discussion here dealt only with resolution of two lines which are both in the focal plane, or in other words with the lateral resolution of the microscope. Similar considerations apply also to the resolution along the optical axis. The axial resolution of a microscope objective lens is also determined by its *NA*. For practical purposes, we can consider the diffraction-limited resolution of a high *NA* objective to be approximately 0.2 μm in the lateral and approximately 0.6 μm in the axial direction [71].

3.2 *Resolution and Optical Aberrations*

The diffraction-limited resolution introduced in the preceding section can be achieved only with an objective lens well corrected for optical aberrations such as astigmatism, field curvature and spherical and chromatic aberrations. From a microscope user's perspective, the last two aberrations are the most important to be aware of.

Chromatic aberration stems from the dependence of refractive index on wavelength of light, known as the chromatic dispersion of a material. Chromatic dispersion of optical glass results in wavelength dependence of the magnification (lateral chromatic aberration) and focal length (axial chromatic aberration) of a lens. Designers of microscope objectives combine lenses made of glass with different chromatic dispersion to produce a lens assembly with minimal chromatic aberration. Increasing the number of lenses in the assembly allows them to match the magnification and focal length for more individual wavelengths and maintain sufficiently small chromatic aberration for any wavelength in between. Based on the number of wavelengths for which the aberration is corrected, microscope objective lenses are grouped into three categories: Achromats (2 wavelengths), Semi-apochromats or Fluars (3 wavelengths) and Apochromats (≥ 3 wavelengths with stricter limits on the residual aberration) [72, 74]. Besides the number of individual wavelengths for which the chromatic aberration is cancelled out, the different classes of objectives also differ in the spectral range over which the chromatic aberration correction extends. Objectives with a higher class of chromatic aberration correction may be needed when imaging multiple fluorescence channels spanning the whole visible spectrum or even exceeding the visible range, for example to near infrared.

Typically, the residual axial chromatic aberration (see Fig. 3) of corrected objective lenses is more pronounced than the lateral one. In widefield microscopes, the residual axial chromatic aberration can be easily compensated for by adjusting the objective to specimen distance for each channel. This is very straightforward with a motorised microscope, where this task can be automated and does not require the operator's intervention when switching fluorescence channels. It is of course essential to determine correctly the necessary focus shifts, for example by taking images of fluorescent beads labelled by multiple fluorophores which cover all the channels of interest (ideally in the same medium and sample carrier which are used in the actual experiments). This simple solution is, however, not always applicable to confocal microscopy, as we will mention in Sect. 4.4, making apochromatic objectives a preferred choice for confocal imaging.

Spherical aberration is inherent to all optical elements with spherical surfaces, which is the case of nearly all lenses. It is manifested by PSF broadening and therefore, by image blurring and loss of contrast. The general correction strategy relies on combining lenses with positive and negative spherical aberration to produce an assembly with minimal overall aberration. The aberration is wavelength dependent and similarly to chromatic aberration correction, in an apochromatic lens, the aberration is corrected for more individual wavelengths than in an achromatic lens [72, 74]. Beside the optical path within the objective lens, spherical aberration in the image is also affected by the optical path outside of the lens, particularly the part between the fluorophore and the objective lens. Here in general, the light propagates through the specimen, through a coverglass and through an immersion medium, which fills the space between the coverglass and the objective front lens. Of course, in some case there may be no coverglass in between the specimen and the objective lens, which case can be generalised as containing coverglass of zero thickness, or the "covereglass" may not be made of glass but, for example, of polystyrene. Objective lenses are designed to work with (generalised) coverglass of a particular thickness and refractive index and with immersion medium of a particular refractive index.

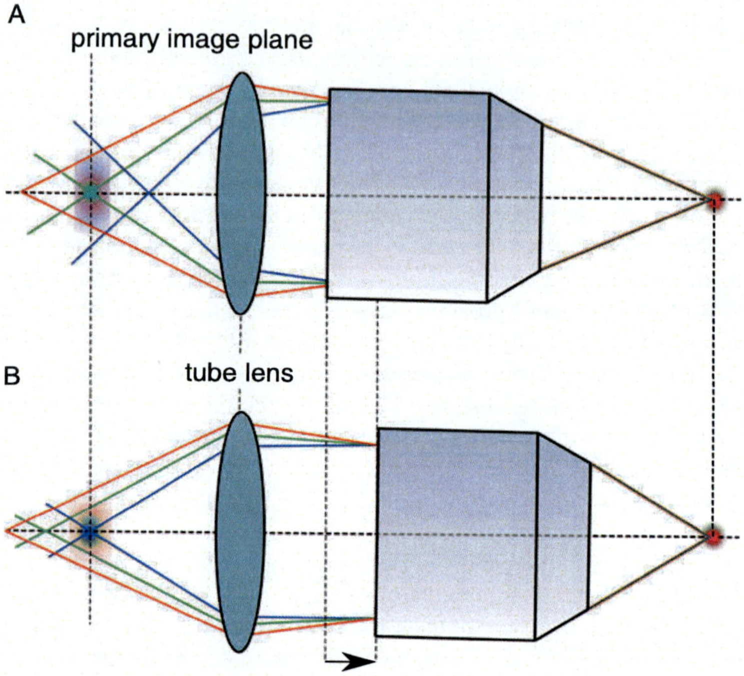

Fig. 3 A schematic illustration of axial chromatic aberration. Fluorophores located in the same plane in the specimen but emitting in different spectral regions are not imaged by the objective lens to the same plane. Thus, if fluorophores emitting in one spectral region, for example the green region (**a**), are sharply focused in the primary image (which also means that the rays of light emitted by them are collimated in the infinity space between the objective lens and the tube lens), fluorophores emitting in the other spectral regions are blurred. By refocusing the objective, fluorophores emitting in another spectral region, for example blue (**b**), can be brought into focus

Spherical aberration correction is compromised if either the coverglass or the immersion medium differ from the expectations. The effect increases with the objective *NA*. Many higher *NA* (> 0.5) lenses are therefore equipped with a correction collar that allows the user to tune the spherical aberration correction to work optimally with coverglass of different thickness and material and/or with different immersion media [72]. There are, for example, objectives which can be used with water, glycerol, or microscopic oil as immersion, as well as air immersion objectives which can accommodate a broad range of coverglass thickness from 0 to 1.2 or 1.5 mm. The quality of images produced by low *NA* objective lenses depends only marginally on the coverglass thickness; therefore, such lenses are usually not equipped with adjustable correction.

Other than the immersion medium and the coverglass, the specimen itself and its optical properties affect the spherical aberration correction [75]. In general, the correction performs the best if there is no refractive index difference between the specimen and the immersion medium. For this reason, specimens for high-resolution microscopy are often mounted in media with carefully tuned refractive index

[76]. The effects of the specimen refractive index increase with the length of the optical path within the specimen, or in other words, with the depth within the specimen at which we focus. When imaging, for example, a specimen in an aqueous medium, at shallow depths (units of μm), an oil immersion objective is likely to slightly outperform a water immersion one because of the former's higher *NA*; however, at larger depths (10s of μm and more) a water immersion objective is likely to largely outperform an oil immersion counterpart because of the increasing spherical aberration in the latter case.

Since the specimen-induced aberrations are specific to the specimen, more precisely to a specific location within the specimen, no generic correction scheme can be incorporated in the design of the objective lens. Adaptive optics offer a possible solution as they allow to tune to correction for each individual location in the specimen and different strategies have been developed for finding a good correction without adding too much to the time needed for image acquisition [77–80].

3.3 Objective Lens Numerical Aperture

NA of an objective lens is defined as the product of the immersion medium refractive index n and $\sin\theta$, where θ is the largest possible half angle of the cone of light that can enter the objective lens:

$$NA = n \; \sin\vartheta$$

The importance of the refractive index of the medium is illustrated in Fig. 4. It shows how rays of light are bent by refraction on interfaces between media of different refractive indices, as described by Snell's law. In this way, the oil immersion objective working with immersion medium of a higher refractive index effectively collects light emitted by the fluorophore into a larger solid angle. This means that the larger the *NA*, the higher is the objective's light collection efficiency and the brighter is the image formed by such a lens. This is a very important consideration in fluorescence microscopy where the photon budget is limited.

As we noted in Sect. 3.1, *NA* also determines the resolution of the objective lens. This, together with its impact on light collection efficiency, makes *NA* probably the single most important parameter describing objective lens performance. The influence of objective *NA* on microscope resolution follows also from an information-theory point of view. Higher resolution means higher information contents of the image; since photons are the information carriers in optical microscopy, collecting a larger portion of all photons emitted by the fluorophores in the microscope field of view does not increase only the brightness but also the resolution of the image.

From geometrical considerations, it follows that to maximise *NA*, it is necessary to minimise the distance between the fluorophore and the objective front lens, that is the working distance of the objective. This is not always possible because of steric constraints imposed by the specimen. In some situations, where long working

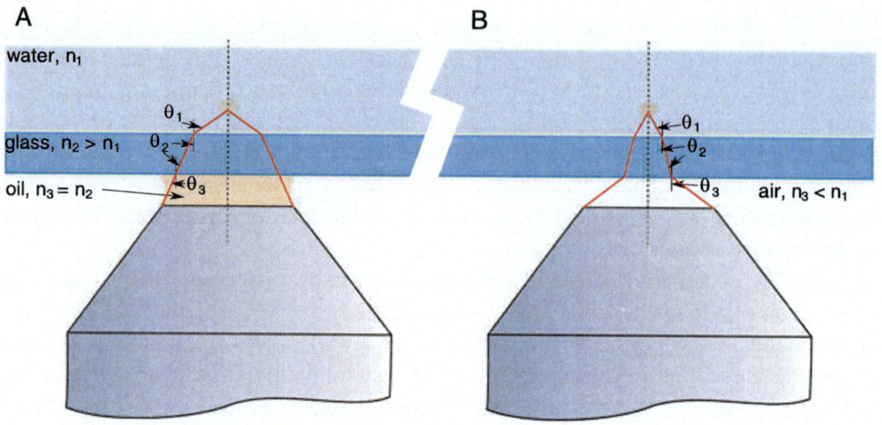

Fig. 4 A schematic illustration of the influence of the refractive index of the immersion medium on the light collection efficiency of an objective lens as expressed by the objective's *NA*. As the rays of light pass across interfaces between different media, the product $n \sin \theta$ remains constant as predicted by Snell's law. As a result, an oil immersion objective (**a**) can collect light emitted by the fluorophore into a larger solid angle than a similar air objective (**b**)

distance of the objective lens is necessary, *NA* has to be compromised and together with it the light collection efficiency and the resolution of the objective.

From the definition of *NA*, it is clear that since $\sin\theta < 1$, $NA < n$. This is true for a single-objective lens. However, by using multiple objective lenses, each with $NA < n$, it is possible to construct a microscope with overall higher *NA*, equal to the sum of the *NA*s of the individual objectives. In this way light emitted by the fluorophores into a larger solid angle can be collected and the resolving power of the microscope enhanced. This approach is referred to as 4Pi-microscopy in allusion to the full solid angle of 4 π, an ultimate goal that is in practice not achievable. High *NA* objective lenses tend to be quite bulky limiting practically the number of objectives that can be thus combined to two in most cases [71, 81, 82].

3.4 Resolution and Image Pixel Size

So far, we have only considered the details resolvable in the image generated by the microscope optics in the primary image plane. There is, however, one more step involved in capturing the image, which can also affect the resolution of the final digital image. In a widefield microscope, the primary image is captured by a camera sensor consisting of discrete physical pixels. Features in the primary image that are detected by a single physical pixel are not resolvable in the digital image. To discuss this topic, we will introduce the projected pixel dimension p_S, which is the physical pixel pitch p_C divided by M, the microscope's total magnification (the product of the magnifications of the objective lens and any additional magnifying or demagnifying

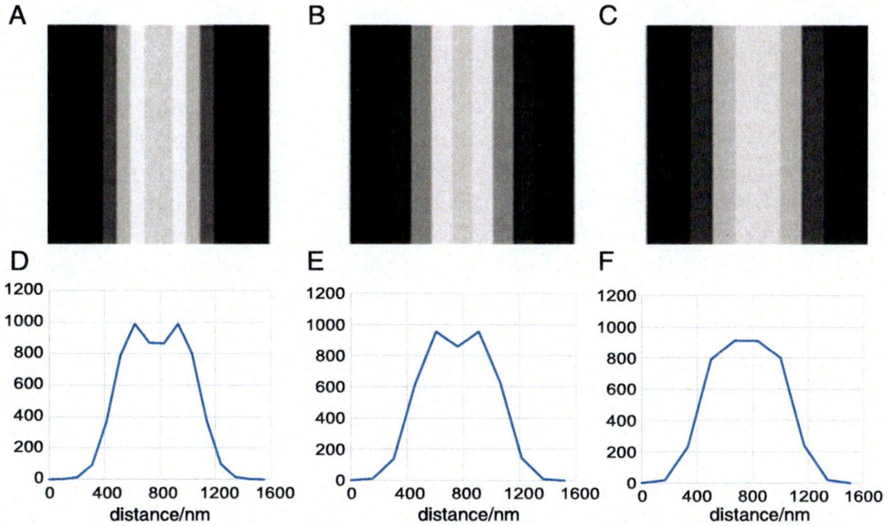

Fig. 5 The influence of the projected pixel size on image resolution illustrated on simulated images of two thin parallel fluorescent lines (**a–c**) and the respective intensity profiles along the direction perpendicular to the fluorescent lines (**d–f**). The simulated situation is identical to that shown in Fig. 2 **a** and **d** (the PSF width is $R_{PSF} = 339$ nm and the distance between the lines $d = 339$ nm); however, this time the effect of different projected pixel size p_s is simulated. Images and intensity profiles for $p_s = 103$ nm (**a** and **d**), 152 nm (**b** and **e**) and 167 nm (**c** and **f**) illustrate that the two lines can be resolved only if p_s is smaller than half of the distance between the lines. It should be noted also that in the case of pixel-size limited resolution, it is the distance of features in the primary image, rather than the distance of the actual structures in the specimen that matters. The overlap of the PSFs results in the intensity maxima being closer to each other (approximately 318 nm) than the distance of the underlying thin lines. The same effect can also lead to overestimation of resolution if it is measured directly as the shortest distance between resolvable features in an image. The dimensions of the images are 1,600 × 1,600 nm. The PSF was for simplicity approximated by a Gaussian function. Intensities in the plots of intensity profiles are shown in relative units

optics between the objective lens and the camera). It is, therefore, the dimension of the area in the sample, which is captured by a single camera pixel. Let us again consider the case of the two parallel lines, this time separated by a distance larger than the diffraction-limited resolution. As seen in Fig. 5c and f, when p_S is larger than half of the distance between the lines, the lines appear as a single line in the image. This is a demonstration of the Nyquist-Shannon sampling theorem. A diffraction-limited image is therefore obtained only if p_S is smaller than half of the resolution limit d, or in other words:

$$d > \frac{p_C}{2M}$$

4 Confocal Microscopy

One of the weaknesses of widefield epifluorescence microscopy, which we have discussed so far, is its susceptibility to high background originating from out-of-focus planes, especially when imaging thicker specimens. The cause of this problem is illustrated in Fig. 1a. Fluorescence is excited in the whole volume of the specimen that is illuminated by excitation light, which means that fluorophores in planes below and above the focal plane emit as bright fluorescence as fluorophores in the focal plane. Rays of light emanating from a fluorophore in the focal plane are collimated by an infinity corrected objective lens and then focused by a tube lens to the plane of a camera sensor, where they generate, in the ideal case, a diffraction-limited spot, the PSF of the microscope. Fluorophores located at different positions in the focal plane are imaged as PSFs at different positions in the plane of the camera sensor and together all these PSFs add up to a sharp, meaning diffraction limited, image. However, rays coming from fluorophore in out-of-focus planes are not collimated by the objective lens, they are either divergent or convergent in the infinity space. As a result, the signal from such fluorophores is spread over larger areas in the plane of the camera sensor, contributing blurry background to the image.

A commonly used solution to this problem is based on blocking the out-of-focus light by a spatial mask with an aperture positioned in such a way to selectively transmit the signal originating from the focal plane. This is the principle of confocal microscopy as illustrated in Fig. 1b. In this case a mask with a small aperture is placed in the primary image plane instead of a camera sensor. Rays originating from a fluorophore located in a corresponding spot in the focal plane are tightly focused in the plane of the mask and, therefore, pass through the aperture to the detector. On the other hand, light emanating from fluorophores in out-of-focus planes is defocused in the plane of the mask and only a small fraction of it can pass through the aperture.

We can notice that with a single aperture, only light coming from a single position within the focal plane can make it through the aperture. Rays from other locations within the focal plane, though focused in the plane of the mask, would not coincide with the aperture and would, thus, be completely blocked. This means that in such a configuration, we are not able to directly capture an image; instead, we are measuring the fluorescence intensity originating from a single small detection volume in the specimen. While this is clearly useful for various forms of fluorescence micro spectroscopy, such as single-point FCS, it is clearly not a complete solution to the problem of blurry background in widefield images. To get a confocal fluorescence image, we need to scan the specimen with the confocal detection volume and measure the fluorescence intensity for each position, that is for each pixel (or each voxel in the case of 3-dimensional imaging). Before we proceed to scanning the specimen, let us make a few additional observations:

1. Since fluorescence is collected only from a small effective detection volume, it would be wasteful to excite the whole volume of the specimen and subject it unnecessarily to photobleaching of fluorophores and photo-toxic effects. Instead, fluorescence is excited by a focused (typically laser) beam. The highest excitation

efficiency is achieved in the focus of the beam which is aligned to correspond to the effective detection volume defined by the aperture; the excitation beam focus and the aperture are therefore said to be confocal with respect to each other. The focused excitation further confines the effective excitation volume.

2. Since we are capturing the intensity from a single-detection volume at a time, there is no need for a spatially resolved light sensor. A single-point detector of light is enough in this case. Photomultiplier tubes (PMTs) are commonly used in point-scanning confocal microscopy (and other point-scanning microscopy modalities, see Sect. 5.1). PMTs with GaAsP cathodes have become very common, replacing, the previously most widespread, multi-alkali PMTs. GaAsP PMTs offer higher QE compared to multi-alkali PMTs (over 40% across most of the visible spectrum). Even higher QE is offered by single-photon avalanche photodiodes (SPADs), which however suffer from lower dynamic range and have been preferred predominantly for low-light photon counting applications [83]. Hybrid detectors combining technologies (and to a large extent also the advantages) of SPADs and PMTs are replacing SPADs in many microscopy applications. Another perspective technology of point-detectors are silicon photomultipliers (large arrays of SPADs), attractive for their larger active areas, higher dynamic range and low cost [84–86].
3. The size of the confocal aperture affects the efficiency of rejection of out-of-focus light as well as the light collection efficiency. As said earlier, the image of an in-focus fluorophore is the microscope PSF. Most of the light forming the image is concentrated in the PSF central peak. Setting the aperture size such that it admits the whole central peak of the PSF but not more, therefore, ensures that nearly all signal from the in-focus fluorophores is collected without admitting more out-of-focus light than necessary. Such aperture size is referred to as 1 Airy unit (1 AU), a relative unit deriving its name from the Airy function describing the PSF. Increasing the aperture size above 1 AU increases only marginally the collection efficiency of light from in-focus fluorophores, while having a larger impact on the out-of-focus light collection efficiency. In this way the rejection of out-of-focus light and the optical sectioning capability (see Sect. 4.5) of the confocal microscope are compromised with little gain in in-focus signal. On the other hand, apertures smaller than 1 AU effectively reduce the detection volume, thus, improving the resolution of the image in both the lateral and the axial direction [87–89]. This resolution improvement, however, comes at the cost of reduced signal since, even for a fluorophore located ideally in the objective lens focus, only a portion of the signal contained in the PSF central peak can be detected. Since the photon budget in biological fluorescence microscopy is usually quite limited, apertures of approximately 1 AU are usually chosen as a good compromise between requirements on resolution and sectioning on one side and on signal on the other side.
4. A detector with a small active area or an optical fibre can also act as a confocal aperture [90]. The disadvantage of the former solution is its limitation to a single detector. When using a mask with an aperture or an optical fibre, the signal can be spectrally divided by secondary dichroic mirrors or by a dispersive element (a prism or a grating) onto multiple detectors for simultaneous detection of distinct fluorophore species.

5. The pixel-wise manner of image acquisition makes point-scanning microscopy modalities the usual choice for implementing any type of micro spectroscopy imaging. Any type of spectroscopy can be readily applied on the signal that passes through the confocal aperture, which may not be easily doable when the whole image is captured at once by a camera. For these reasons, for example FLIM (see chapter "Time-Resolved and Polarized Spectroscopy") is typically performed in a point-scanning manner, although cameras capable of time-resolved fluorescence detection exist and have become commercially available and are likely to become more widespread [91, 92].

One possible approach to confocal scanning uses mechanical movement of the specimen with respect to the microscope objective, usually using a piezo positioner. Either the sample stage or the objective lens can be the moving part. Because of its slower speed, this form of scanning is rarely used in biological confocal microscopy. It is convenient, for example in correlative confocal fluorescence and atomic force microscopy (AFM), because the physical movement of the specimen with respect to the cantilever is required in AFM [93].

4.1 Laser Scanning Microscopy

Faster scanning is achieved by moving the excitation beam and leaving the objective lens and the specimen stationary. Microscopes employing this form of scanning are called laser scanning microscopes and are routinely used in biological imaging [94]. The scanning is realised by deflecting the laser beam to different points in the focal plane with a pair of mirrors rapidly moved by a galvanometer drive. When driven at their resonant frequencies, such mirrors can scan several thousand lines per second. Fluorescence emitted from the laser focus then follows the reverse path of the excitation laser beam and is deflected by the scanning mirrors (so-called de-scanned) to the confocal aperture. The confocality of the laser focus and the aperture is, therefore, maintained for an arbitrary position within the scanned area.

As with any other point-scanning microscopy technique, the speed of scanning determines the time needed to scan an image as well as the pixel exposure time. While fast scanning with resonant scanners gives short image acquisition time (or in other words high frame rates) which compare favourably with non-scanning imaging modalities, the pixel exposure times are necessarily extremely short and, considering typical fluorescence photon fluxes, repeated scanning with averaging is typically needed to collect sufficient signal. At a first glance such fast scanning rates may not, therefore, look particularly useful if they require repeated scanning, which reduces the effective frame rate. Yet, it has been shown that there are advantages to very fast, even if repeated, scanning. Most fluorophores exhibit some form of dark or dim states, to which they enter through the excited state. Under continuous excitation, there exists an equilibrium between the fluorophore populations in the bright and the dark states. Fluorophores in the dark state do not contribute to the fluorescence signal and their fraction, thus, reduces the effective fluorophore brightness. Since they enter the dark state through the excited state, in the absence of excitation, there is no dark state population (all fluorophores are in the ground state ready for excitation). This means that in a short time after the excitation has been turned on and before the equilibrium dark state population has been established, the effective fluorophore brightness is transiently higher. Going back to confocal scanning, the higher the scanning speed, the more is this transient phase represented in the pixel exposure time, giving typically a higher effective fluorophore brightness [95].

4.2 *Image Sampling and Magnification in Point-Scanning Microscopy Modalities*

In Sect. 3.4 we have discussed the influence of the projected pixel size p_S on the resolution of the captured digital image and concluded that in order not to compromise the optically achievable resolution d, p_S needs to be equal to or smaller than half of d. This is a general conclusion which holds for point-scanning microscopy modalities as well. However, unlike in widefield microscopy, where p_S is determined by the interplay between the microscope magnification and camera pixel pitch (and microscope operator thus can do little about it), in point-scanning microscopy modalities the microscope operator has a full control over p_S, which is flexibly determined by the scanning settings. This gives the microscope operator the power to intentionally (or unintentionally) over- or under-sample the image. Undersampling, using too large p_S, compromises the image resolution; oversampling, on the other hand, increases unnecessarily image acquisition time by scanning more lines and typically also compromises signal-to-noise ratio by reducing pixel integration time (unless scanning speed is reduced correspondingly, further increasing the acquisition time) and the effective area from which the signal in each pixel is collected. The optimal p_S is therefore only slightly smaller than $d/2$.

Another observation, worth making at this point, is the relative unimportance of objective lens magnification in point-scanning microscopy modalities. The actual magnification of the acquired image is defined by the sweep of the scanner, which can be nearly arbitrarily reduced to increase magnification. The objective lens magnification determines only the maximum dimensions of the scanned area, that is the minimum magnification possible. Since resolution and light collection efficiency is determined solely by the objective's *NA*, lower magnification objectives with high *NA* are the most flexible solution for point-scanning microscopy modalities and advantageous over higher magnification objectives of the same *NA*.

4.3 *Spinning Disk Confocal Microscopy*

The speed of point-scanning confocal microscopy is necessarily limited by the pixel-wise image acquisition, which, especially for larger fields of view, can significantly limit the achievable temporal resolution. An obvious way to overcome this limitation and increase the acquisition speed lies in multiplexing the confocal image acquisition, that means in acquiring signal at multiple points simultaneously. A variety of multiplexed confocal designs have been proposed including line-scanning confocals [96–100], which illuminate a whole line in the focal plane and replace the confocal aperture with a confocal slit; however, the most widespread design is a spinning disk confocal microscope [101, 102]. The design has quite a long history starting in the 1960s [103, 104], although it was only much later that is became widespread.

A typical spinning disk confocal microscope uses a disk with pinholes arranged in a spiral pattern rotating at a high speed. The disk has the dual purpose of defining the positions of excitation foci and rejecting out-of-focus fluorescence. As the disk rotates, the excitation foci move over the focal plane, until every point within the field of view has been exposed to excitation. The in-focus fluorescence, which passes through the disk, is focused on a camera sensor. As the camera is exposed throughout the rotation of the disk, a complete image of fluorescence from the focal plane builds up within the camera exposure time. Some designs feature a second disk with micro lenses, which focuses the excitation laser beam on the pinholes in the main disk. This ensures that the excitation light is not wasted and alleviates the requirements on laser power.

4.4 *Chromatic Aberration in Confocal Microscopy*

As we mentioned in Sect. 3.2, simple refocusing may not be a sufficient solution to compensate for residual axial chromatic aberration of the objective lens in confocal microscopy. Let us get back to that point and explain why it is so and at what situation it is particularly problematic. In a widefield microscope, fluorescence is excited uniformly in the entire depth of the specimen (provided the specimen is sufficiently transparent). Therefore, chromatic axial aberration is relevant only for fluorescence emission to make sure fluorescence in all channels is collected from approximately the same focal plane within the specimen. In a confocal microscope, on the other hand, the excitation beam is focused into the focal plane, making excitation most efficient in that plane. It is therefore desirable that axial chromatic aberration is corrected also for the excitation wavelength to ensure confocality of the excitation laser beam and the aperture for all channels.

This may can be problematic when working, for example with fluorophores excited in the UV part of the spectrum, such as many popular DNA labelling dyes emitting in the blue to green range of the visible spectrum. At confocal microscopes, such dyes are usually excited with a 405 nm laser line, which lies outside of the corrected range of Achromats, Semi-apochromats, and even older Apochromats. The axial chromatic aberration can increase quite steeply towards shorter wavelength outside of the corrected range resulting in axial distances of several μm between the foci of the 405 nm laser line and of those laser lines which lie within the corrected range (see Fig. 6a for a schematic illustration).

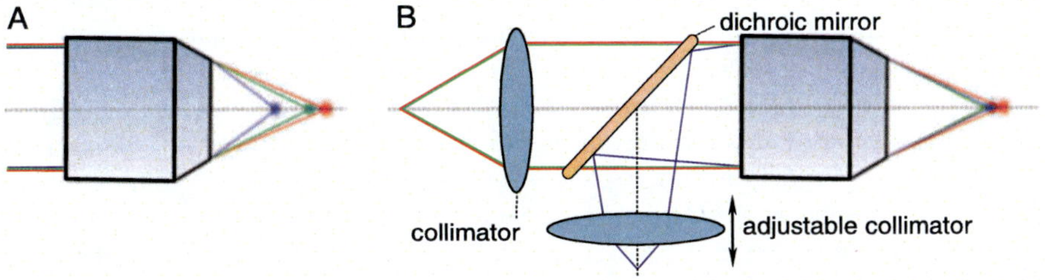

Fig. 6 A schematic illustration of axial chromatic aberration in confocal microscopy (**a**) and an approach to reducing it by using separate collimators for different laser lines. Axial chromatic aberration causes the foci of different laser lines to lie at different axial positions (**a**). This effect can be very significant for laser lines outside of the corrected spectral range of the objective lens. 405 nm laser line (represented in purple in the figure) is outside of the corrected range of many objective lenses; therefore, in many confocal setups, it is delivered to the objective following a separate optical path with a separate collimator (**b**). The beam entering the objective can be, thus, made divergent or convergent, resulting is moving the focus along the optical axis (the more divergent the beam is, the further is the focus from the objective lens and vice versa). By tuning the collimator for the 405 nm laser line, its focus can be overlapped with the foci of the remaining laser lines. A dichroic mirror is used to merge the optical path of the 405 nm laser line with the path of the other laser lines. Note that the depicted dichroic mirror is not the main dichroic mirror, which separates the excitation and emission paths and which was omitted for simplicity in the figure. The main dichroic mirror would be located between the depicted excitation dichroic and the objective lens

Alternatively, if the 405 nm line follows a different optical path with a different collimator than the rest of the laser lines, its focus can be tuned to overlap with the focus of the other laser lines by adjusting the dedicated collimator (see Fig. 6b). This solution has been adopted by many commercially available laser scanning microscopes.

4.5 *Optical Sectioning and Volumetric Confocal Imaging*

By rejecting out-of-focus fluorescence, a confocal image effectively contains the signal only from a thin section of the specimen (approximately 0.5 μm with high *NA* objective lenses) [105]. This property of confocal microscope is usually referred to as optical sectioning in contrast to actual physical sectioning (cutting the specimen into thin slices) often used to produce background-free images of thick specimens. By focusing the microscope into different planes within the specimen, the whole volume of the specimen can be imaged and a 3-dimensional image rendered from the optical sections. Focusing the microscope into different planes is usually realised by moving either the objective or the specimen stage along the optical axis. Alternatively, tuneable or variable-focus lenses can be employed to modify the focal length of the imaging optical system [106, 107]. This approach eliminates mechanical moving parts and allows for very fast switching between different focal planes.

However, by tuning the focal length of the imaging optics, the magnification, NA and aberration corrections are also affected, restricting the usable range of this approach.

Acquiring a 3-dimensional image of a relatively large specimen can be a time-consuming task as it requires capturing many optical sections, each of which may take a non-negligible time to capture, especially in the case of point-scanning confocal microscopes. Beside the time needed, photobleaching is another concern. Although the excitation in a confocal microscope is the most efficient in the focal plane, the excitation efficiency does not drop sharply away from that plane and fluorophores are excited, hence potentially bleached, across the thickness of several optical sections above and below the focal plane. Fluorophores in the sections imaged later in the course of the acquisition may be thus considerably bleached out by imaging of the preceding sections. In response to these concerns, several designs of multifocal confocal microscopes have been proposed, neither having yet become widely used [108–110]. Another technique addressing, among other, these concerns is lightsheet microscopy, also known as selective plane illumination microscopy (SPIM). SPIM has become a well-established alternative to confocal microscopy for 3-dimensional imaging and we will return to it in Sect. 5.2.

Another issue faced not only by confocal microscopy when performing 3-dimensional imaging, is the limited penetration of light into specimens that are not transparent. As the optical path through the specimen increases, larger fractions of both the excitation light and the fluorescence are scattered and/or absorbed by the specimen leading to decaying signal-to-noise ratio of the image. This is especially prominent in the case of optically dense samples such as tissue. A confocal aperture can help to some extent by screening out scattered photons, which would otherwise increase the background. However, when photons emitted by an in-focus fluorophore are scattered, their paths are altered and they are likely not to reach the confocal aperture and not to contribute to the detected signal. In this way, the signal is gradually lost in confocal microscopy with increasing depth in the specimen.

One way to overcome this limitation is based on making the specimen optically transparent by tissue clearing [111]. This approach is however not compatible with live samples. Other approaches utilise the fact that light scattering efficiency decreases with increasing wavelength. Infrared light thus penetrates considerably deeper into tissue or other non-transparent specimens than visible light. Near-infrared confocal microscopy is therefore better suited for imaging such specimens than confocal microscopy in the visible range [112–115]. The longer the excitation and emission wavelengths are, the deeper is the penetration depth. Practical limitations stem from the restricted choice of suitable fluorophores rather than from the microscopy hardware side. A popular alternative is multi-photon fluorescence microscopy which allows excitation of standard visible fluorophores by infrared light as we describe in Sect. 5.

Adaptive optics can also help to compensate for the effects of light scattering by the specimen [77–79, 116].

5 Other Modalities with Optical Sectioning Capabilities

5.1 Multi-Photon Fluorescence Microscopy

A fluorophore can be excited by simultaneously absorbing multiple (n) photons. Assuming all the n photons having the same wavelength and therefore the same energy (which is in general not necessary), each photon contributes $1/n$ of the excitation energy. Since photon energy is inversely proportional to the wavelength, the wavelength of each of the n photons can be n times longer than the wavelength of a single photon that would be needed to excite the molecule by single-photon absorption. Therefore, Stokes shift is reversed in the case of multi-photon fluorescence excitation and the fluorescence has shorter wavelength than the excitation. Visible fluorescence can be thus excited by infrared light. Since simultaneous absorption of n photons is needed, multi-photon excitation requires much higher excitation light intensities than single-photon excitation. The excitation efficiency in multi-photon fluorescence scales with the nth power of the excitation intensity, making it an example of non-linear optical phenomena (that means optical phenomena where the response magnitude, such as the fluorescence intensity, depends on the excitation intensity in a non-linear way). Obviously, the excitation intensity needed grows with the the photon number n. Practical considerations regarding the necessary excitation light sources limit the photon numbers n used in multi-photon fluorescence microscopy to 3 at most, with 2-photon microscopy being by far the most common case of multi-photon microscopy [117, 118].

To achieve excitation intensities high enough for multi-photon excitation to occur, the excitation photon flux is typically concentrated both in time and space. The former is realised by using pulsed lasers with very short pulses (approximately 100 fs) and high peak power. Besides tuneable Ti-sapphire lasers, which are the most common example of such excitation sources for multi-photon microscopy, single-wavelength fibre lasers have emerged as a cost-efficient alternative [119]. The latter is achieved by tightly focusing the excitation laser beam to a diffraction-limited spot, similarly to laser scanning confocal microscopy. Images are then acquired by point-by-point scanning as discussed in Sect. 4.1. Although most multi-photon microscopes rely on point-by-point scanning, 2-photon microscopy has been implemented also in widefield [120–125] or line-scanning [126, 127] configurations.

As mentioned in Sect. 4.5, multi-photon microscopy is commonly used for 3-dimensional imaging of highly scattering specimens, because of the increased penetration depth of infrared light. This may not seem at first a sufficient solution of the scattering issue since the fluorescence in multi-photon microscopy is in the visible range, hence equally susceptible to scattering as fluorescence in single-photon confocal microscopy. To fully understand the benefits of multi-photon microscopy, we need to realise that because of the non-linear nature of the excitation process, excitation efficiency decays sharply with increasing distance from the focal plane. This effectively ensures optical sectioning without the need for a confocal aperture. Removing the confocal aperture enables the collection of a larger fraction of photons emitted by in-focus fluorophores. To maximise the fraction of collected scattered fluorescence photons, the detectors in multi-photon microscopy are often placed as close as possible to the objective lens. Since the collected fluorescence light does not need to reach the confocal aperture, it does not have to pass through the scanner to be de-scanned as in laser scanning confocal microscopy. Instead, all the fluorescence collected by the objective lens is focused onto the detector active area. This configuration known as non-descanned detection is common in multi-photon microscopy [128, 129].

Another application that benefits from multi-photon microscopy is imaging of UV-excitable fluorophores, which cannot be efficiently excited at a confocal microscope. This is the case of many intrinsic fluorophores, making multi-photon microscopy a good solution for label-free fluorescence studies of cells and tissues [130–133].

5.2 *Lightsheet Microscopy*

Unlike the other fluorescence microscopy modalities introduced in this chapter, lightsheet microscopy or selective-plane illumination microscopy (SPIM) does not use episcopic geometry. Instead, the excitation and the emission light paths are oriented perpendicularly to each other as show schematically in Fig. 7. The concept appeared for the first time in 1903 [134], after which it has been practically forgotten until further developments about a century later [135, 136]. Since then, lightsheet microscopy has become a well-established technique for rapid 3-diemnsional fluorescence imaging, which had a large impact in many biological fields, in particular in developmental biology.The broad range of specimens and topics studied by lightsheet microscopy led to the developments of various microscope designs suited optimally for certain ranges of applications. Reviewing all the modalities of lightsheet microscopy is beyond the scope of this section; more can be found, for example in recent reviews [137, 138]. In essence, in a lightsheet microscope, the light from an excitation source (a laser in the modern implementations) is focused by illumination optics to generate a thin sheet of light (hence the name of the technique) which excites fluorescence in a layer of the specimen (an optical section) and the fluorescence is imaged by an objective lens placed perpendicular to the lightsheet and detected by a camera. In its simplest form, the illumination optics consist of a single cylindrical lens which focuses a collimated beam of light into a sheet of light (as opposed to the more common spherical lenses which focus a collimated beam into a single spot). Other optical elements can be added to produce thinner lightsheets or increase the extent of its usable area [139, 140]. By the usable area, we understand here the portion of the lightsheet around its focus where the thickness of the lightsheet can be considered reasonably uniform (see Fig. 7). Illumination is often provided from two or more opposing sides of the specimen to avoid shades resulting from limited penetration of the lightsheet into larger specimens [141, 142]. Various designs of single-objective lightsheet microscopes in which a

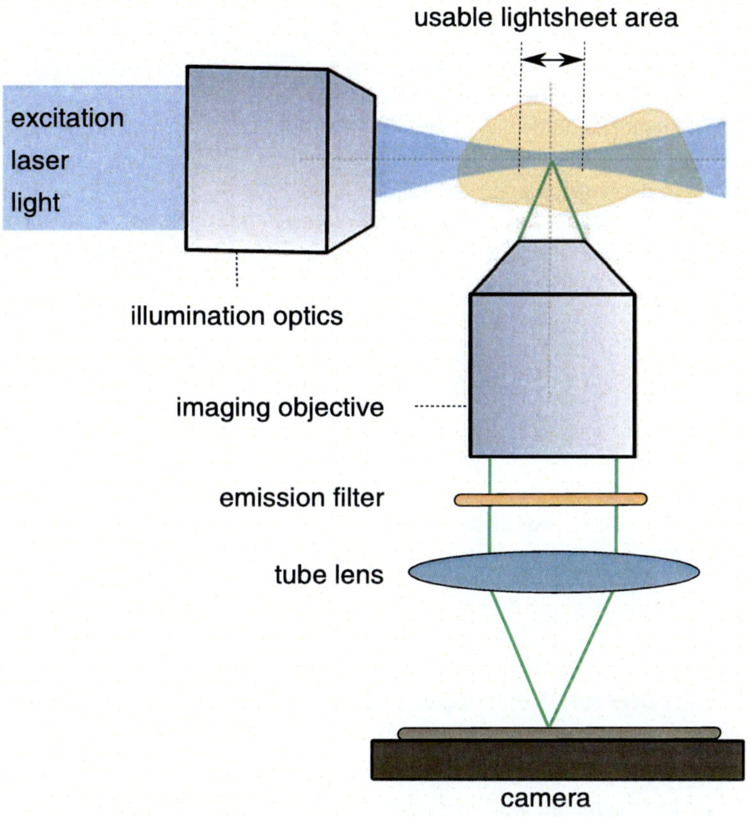

Fig. 7 A schematic illustration of a generic lightsheet microscope. The illumination optics produce a sheet of excitation light which excites fluorescence in a section of the specimen. The fluorescence is imaged by an objective lens placed perpendicularly to the lightsheet. See Sect. 5.2 for more details

single-objective lens provides excitation and detection similarly to a standard epifluorescence microscope, are a notable exception to this generic scheme [143–146]. Different variants of lightsheet microscopes differ in various aspects of their designs such as:

1. Orientation of the lightsheet (vertical, horizontal, or slanted) chosen according to the way the respective specimens are typically mounted.
2. Thickness and the extent of the usable area of the lightsheet. In general, the higher is the *NA* of the illumination optics that create it, the thinner is the lightsheet and at the same time the smaller is its usable area. For large specimens such as whole cleared organs, thicker lightsheets (several μm) with larger usable area are preferable; such setups are often referred to for historical reasons as ultramicroscopes [147]. Imaging of small specimens such as live cells, benefits from thinner, sub-μm optical sections [148]. To increase the illumination *NA*, thus

reducing the lightsheet thickness, objective lenses are usually added to the illumination path as the last optical element before the specimen. Thin lightsheets with increased usable areas can be produced from beams other than Gaussian beams, which are the most common beams produced by lasers. Notable examples are Bessel beam lightsheets [149] and lattice lightsheets produced by interference of multiple Bessel beams [148]. Such lightsheets improve the sectioning capability without compromising the field of view.

3. The mode in which the lightsheet is generated. The original lightsheet microscopes used a static lightsheet which illuminates simultaneously the whole optical section. Later an alternative scheme has been introduced in which the lightsheet is created by rapidly scanning a beam along a line [126, 149, 150]. This approach prevents striping artefacts which can occur when different parts of a static lightsheet interfere with each other after being perturbed by the specimen.

3-dimensional imaging with a lightsheet microscope is usually realised by mechanically moving either the specimen or the lightsheet with respect to the other. Different optical sections are therefore acquired sequentially like in a confocal microscope; however, non-illuminated sections are not subject to bleaching. Methods to increase the throughput of lightsheet imaging by parallel multi-plane imaging have been proposed [151–153].

5.3 *Total Internal Reflection Fluorescence Microscopy*

This technique uses the phenomenon of total internal reflection which can occurs at the interface of two media with different refractive indices. When light propagating from the medium of higher refractive index hits such an interface under a sufficiently large angle, the angle of refraction predicted by Snell's law is larger than a right angle and, in that case, there is no refracted beam propagating into the medium of lower refractive index. The beams are only reflected into the medium of higher refractive index, hence total internal reflection. However, wave optics predict that some energy of the impingent wave enters the medium of lower refractive index in the form of an exponentially decaying wave, called the evanescent wave (see Fig. 8a). The penetration depth can be tuned by the angle of incidence to be a fraction of the incident light wavelength. Since evanescent wave can excite fluorophores present only within a very thin layer of the lower refractive index medium close to the interface, total internal reflection fluorescence (TIRF) can be used for selective imaging of such fluorophores.

In practice, the higher refractive index medium is usually a microscope coverslip and the lower refractive index medium is an aqueous medium of the specimen. High *NA* oil immersion objectives are used to support illumination under sufficiently large angles. TIRF clearly outperforms confocal microscopy or lightsheet microscopy in terms of thinness of the optical section; however, the said section is limited to the vicinity of the coverslip surface. This makes TIRF the technique of choice for single-

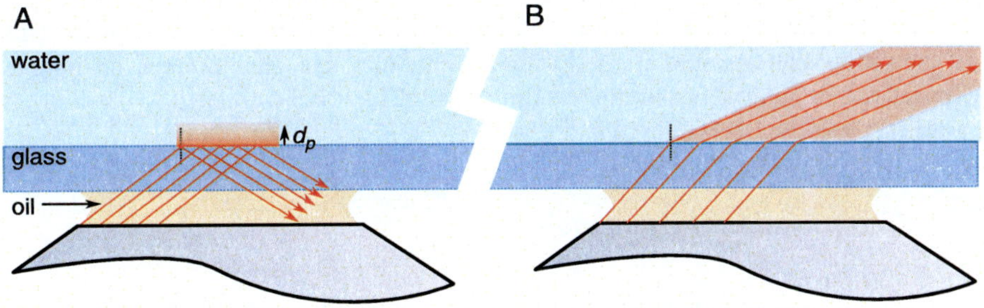

Fig. 8 A schematic illustration of TIRF (**a**) and HILO (**b**) microscopy. See the text of Sect. 5.3 for more details. d_p denotes the penetration depth of the evanescent field (**a**)

molecule imaging and cell membrane studies [44, 154–159]. When the incidence angle of the excitation beam is slightly larger than the critical angle needed for total internal reflection, refracted beams propagate into the specimen under a very large angle, that means in a direction nearly parallel to the interface. In this way, fluorescence can be excited in a section reaching deeper than the evanescent field penetration (see Fig. 8b). This approach is referred to as highly inclined and laminated optical sheet (HILO) and is useful when the structures of interest lie slightly further from the coverslip surface [160, 161]. A related approach called super-critical angle fluorescence (SAF) uses the fact that a part of the light emitted by fluorophores very close to optical interfaces propagates at directions nearly parallel with the interface. By collecting only light propagating under such angles, fluorophores located in a thin layer close to the interface are selectively imaged [162–165].

5.4 *Structured Illumination*

Structured illumination represents a widefield solution facilitating the discrimination of in-focus fluorescence from the out-of-focus background. It uses spatially modulated excitation illumination, with the spatially modulated pattern being focused in the objective focal plane. Hence the in-focus fluorescence is also spatially modulated and the modulation decreases with the distance from the focal plane. Since the fluorescence image is spatially modulated, multiple images with different positions of the modulated pattern need to be acquired to reconstruct a non-modulated image of the in-focus fluorescence. The sparser is the illumination pattern, the more efficient is the background removal, but at the same time, a higher number of modulated images is needed to reconstruct a non-modulated image. In the case of a pattern consisting of alternating bright and dark stripes of equal width, the minimal number of modulated images needed is three [166]. A related approach called aperture correlation microscopy uses a similar image reconstruction strategy and a spinning disk with slit apertures to produce the modulated excitation [167].

Note that structured illumination is also used to increase the resolution of microscopy images, as we will discuss in Sect. 6.3. In that case the requirements on excitation pattern spatial modulation frequency and the number of images acquired are in general higher than if only optical section is aimed at.

5.5 Computational Approaches

The easiest, though of course not necessarily the most powerful, way to remove out-of-focus background from a widefield fluorescence image is by simply subtracting an estimate of the background from the image [168, 169]. A common way to do this known as unsharpen mask estimates the background as a blurred version of the image itself. The image can be blurred, for example by replacing each pixel value with the median or a (weighted) average of pixels in its neighbourhood; the larger the neighbourhood considered, the more smoothed-out is the resulting image. The blurred image is then subtracted from the original one; diffuse background that has been little affected by the blurring is thus removed and only sharper features are preserved. An improved version of this approach, called nearest neighbours, utilises additional information present in a 3-dimensional image (a stack of images acquired for different axial positions of the focal plane) by subtracting from each image in the stack the blurred versions of its nearest neighbours, the images acquired at the adjacent planes below and above the image in question [168, 170]. These approaches and other ways of background subtraction such as the rolling ball algorithm, are very simple computationally and can deliver impressive-looking results; however, it should be remembered that they reduce the information content of the images rather than adding to it, potentially producing serious artefacts, and cannot be therefore considered in any way equivalent to hardware sectioning approaches [168].

Image deconvolution represents a more sophisticated, albeit more computationally intensive, approach to (not only) out-of-focus background removal from widefield images [168, 171, 172]. As the name suggests, deconvolution attempts to estimate what the image would look like if it has not been subject to convolution with the microscope PSF (see Sect. 3.1). Because of the axial extent of the PSF (which is considerably larger than its lateral extent) fluorophores in out-of-focus planes contribute to the image, producing blurry background. Using the knowledge of the PSF (which can be either predicted theoretically or characterised experimentally by imaging, for example small florescent beads), deconvolution can remove the contribution of the out-of-focus fluorophores. Similarly, deconvolution can also improve the contrast of small details in the image by reducing the blurring effects of the PSF in the lateral direction. As an example of such features, we can consider the two closely spaced lines in Fig. 2b, e which suffer from poor contrast because of the large overlap of the PSFs. In this way, deconvolution can effectively improve the image resolution and in this capacity, it is often applied even on images acquired with confocal microscopy and other microscopy modalities with optical sectioning capabilities [76, 171].

6 Resolution Beyond the Diffraction Limit

The resolution limit introduced in Sect. 3.1 applies to standard widefield optical microscopy without any a-priory assumptions about the specimen and, as Ernst Abbe himself noted in his seminal work on the topic, the limit does not necessarily apply under different experimental conditions [71, 73]. Since Abbe's times, various approaches to resolving structures finer than the diffraction limit have been explored. New technologies that became available in the second half of the twentieth century, particularly lasers and computers, enabled practical applications of such approaches, generally termed super-resolution microscopy, towards the end of the twentieth century [71]. This led to several ground-breaking papers published around the turn of the century, which greatly boosted the practical interest in super-resolution microscopy [173–181]. Since then, numerous new approaches have been proposed and many super-resolution techniques have become widely used and available in standard commercial microscopes. The field has become so broad, that this section can give only a very limited view of it; more information can be found in numerous reviews [182–187].

Super-resolution microscopy techniques circumvent the diffraction resolution limit by using usually one (or more) of these ingredients, which were not present in Abbe's theory: non-linear effects such as stimulated emission [173] or fluorescence saturation [188–190], assumptions of sparse fluorophore distribution [191–195], near-field optics [196–199] and quantum statistics of photons [199–203]. The above-mentioned fluorescence saturation is in general a problematic phenomenon in fluorescence microscopy. Because of the finite time spent by a fluorophore in the excited state before emitting a photon, the photon emission rate from a fluorophore is limited, regardless of the excitation intensity. At low excitation intensities, the fluorescence intensity grows approximately proportionally with increasing excitation intensity; however, at higher excitation intensities, the fluorescence intensity increase rate is lower than that of the excitation intensity, until it eventually completely saturates. Usually, excitation intensities in fluorescence microscopy are chosen to be safely below the saturation threshold since higher excitation intensities increase the risk of photodamage to the specimen without adequate increase in fluorescence signal. The use of fluorescence saturation for resolution enhancement demonstrates how otherwise problematic phenomena can be harnessed to enhance the capabilities of fluorescence microscopy.

The subsequent Sects. 6.1–6.5 introduce briefly the main branches of super-resolution microscopy with one notable exception, stimulated emission depletion microscopy (STED) and related techniques [173, 205–208]. Omission of this wide-spread and powerful super-resolution approach is explained by it being the topic of chapter "STED and RESOLFT Fluorescence Nanoscopy". Just very briefly, STED is a point-scanning super-resolution approach that can be implemented as a modification of a laser scanning confocal microscope. Excitation is provided by a diffraction-limited laser focus just like in a regular LSM; the difference lies in applying a second laser beam of longer wavelength, which induces stimulate emission of the fluorophores. The second beam is specially shaped to have zero intensity in its centre (so-called doughnut beam profile); hence, fluorophores in the centre of the excitation volume do not undergo stimulated emission and remain in the excited state. The collected fluorescence, therefore, originates from a reduced, sub-diffraction, volume. Note that saturation of the stimulated emission is essential for reaching sub-diffraction resolution.

6.1 *Near-Field Fluorescence Microscopy*

The diffraction limit of resolution holds for far-field optical microscopy, assuming the aperture of the objective lens as well as the distance between the objective lens and the focal plane are many times larger than the wavelength of light. This assumption held for all the microscopy modalities described so far in this chapter. However, it is not the case for near-field fluorescence microscopy, using a sub-wavelength aperture to deliver excitation light to and/or collect the fluorescence of fluorophores at sub-wavelength distances from the aperture. The resolution is limited only by the aperture size in such a scenario. The aperture can be experimentally realised, for example by the end of a tapered optical fibre. Because of the requirement of the sub-diffraction distance between the aperture and the fluorophores, the technique is limited to imaging surface structures, such as cell membranes [196–198]. The image is then captured by scanning the aperture along the surface of interest, collecting fluorescence from a tiny, sub-wavelength surface area at a time. This approach, called near-field scanning optical microscopy (NSOM) can be, therefore, regarded as a form of scanning-probe microscopy, together with, for example AFM.

Evanescent fields employed in TIRF and SAF microscopies are another example of near-field optical phenomena employed in fluorescence microscopy and providing sub-wavelength optical sectioning, which can be seen as a form of axial super-resolution.

6.2 *Single Molecule Localisation Microscopy (SMLM) and Fluorescence Fluctuation-Based Techniques*

SMLM is a conceptually remarkably simple, yet very powerful, super-resolution strategy that does not require any specialised microscopy setup. A widefield microscope with a sensitive camera capable of single-molecule detection is all that is needed [60, 209, 210]; confocal or lightsheet implementations are also possible [211–213]. To explain its principle, let us first consider an image of a single fluorophore, which is, as we said in Sect. 3.1, the PSF centred at the geometrically perfect image of the fluorophore. By finding the centre of the respective PSF, we can pinpoint the location of the fluorophore with, in principle, unlimited precision. In practice, the localisation precision is limited only by the signal-to-noise ratio of the image, not by the PSF width, and can reach nm order [214, 215], or even better if performed at cryogenic temperatures [216]. In this way, sparsely distributed fluorophores, separated ideally by distances much larger than the PSF width, can be imaged with deeply sub-diffraction resolution. However, such sparse fluorophore distributions are in general not sufficient for imaging arbitrary continuous structures. Nyquist-Shannon sampling theorem dictates that to resolve details as small as d, the structure needs to be labelled by fluorophores spaced not more than $d/2$ apart, unless of course a-priory assumptions about the structures can supplement the missing information. At such high fluorophore densities, it is, however, impossible to resolve PSFs corresponding to individual fluorophores and find their centres. The trick employed in SMLM to overcome this problem is illustrated in Fig. 9. If only a sparse subset of fluorophores is fluorescent at a time, positions of those fluorophores can be precisely localised. By acquiring a stack of many such images (frames), each providing a different random subset of fluorophore positions, we eventually accumulate enough fluorophore positions to resolve arbitrary structures with sub-diffraction resolution limited by the localisation precision of individual fluorophores.

This generic principle has various practical implementations differing in the ways how to ensure that only a sparse subset of fluorophores contributes to each frame. There exist two generic strategies, each with many variants varying in experimental details:

1. Employing transitions of fluorophores between a dark (non-fluorescent) and a bright (fluorescent) state. The fluorophores in this case are static, attached to the structure of interest being imaged. The original implementation of photoactivation localisation microscopy (PALM) used, for example, photoactivatable fluorescent protein molecules that reside initially mostly in a dark state from which they can be activated to a fluorescent state by UV light. With properly tuned intensity of the UV illumination, only a parse subset of fluorescent protein molecules is activated; the activated molecules are then imaged and, while doing so, irreversibly photobleached, allowing another sparse subset to be activated and imaged [176, 217]. Later, many new photoactivable fluorescent proteins have been introduced as well as photo-convertible fluorescent proteins (which change colour of their fluorescent emission upon, usually UV, illumination) and photo-switchable proteins, which can undergo repeated cycles of switching between bright and dark states [218–220]. Stochastic optical reconstruction microscopy (STORM), another example of this generic strategy, uses stochastic transitions between bright and dark states (referred to as blinking), exhibited by many organic fluorophores [175, 178, 221]. High excitation intensities and specific chemical environments are usually needed to induce suitable blinking dynamics [222, 223].
2. Employing transient binding of fluorophores to the structures of interest. The structure of interest is in this case not labelled by static fluorophores; rather, majority of fluorophores diffuse in the surrounding medium and only a small

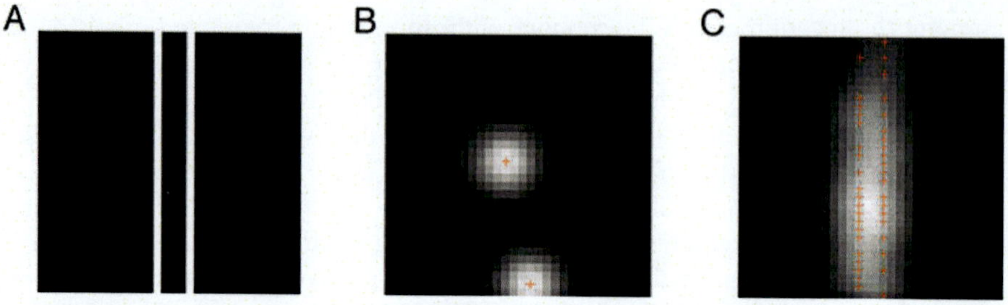

Fig. 9 An illustration of the principle of single molecule localisation microscopy (SMLM). The underlying structure consists of 2 thin fluorescently labelled parallel lines separated by the distance of 150 nm (**a**). When only a sparse subset of fluorophores emits light, the positions of those fluorophores can be localised with sub-diffraction accuracy as the centres of the respective PSFs (**b**). The last panel (**c**) shows an average of 100 such images, each containing the contribution from a random sparse subset of fluorophores. The average image has diffraction-limited resolution ($R_{PSF} = 339$ nm) and the two lines are unresolved; however, the overlayed fluorophore localisations from all of the 100 individual images (displayed as red crosses) clearly resolve the two lines. More localisations would be needed to unambiguously identify the structure as two parallel continuous lines unless the information contained in the image is supplemented by a-priori assumptions such as an expectation that the underlying structure consists of parallel continuous lines. The dimensions of the images are 1,600 × 1,600 nm. The PSF was for simplicity approximated by a Gaussian function

fraction of them binds to the structure of interest at a time. Non-bound fluorophores in the medium typically diffuse over long enough distances within the camera exposure time to make their images practically completely smeared out, thus preventing them from being localised. In each frame, a sparse pattern of fluorophores is visible, marking the positions of a sparse subset of possible binding sites. By accumulating the frames, increased number of binding sites is revealed, until the whole structure can be resolved [181, 224–226]. This approach is usually referred to as Points accumulation for imaging in nanoscale topography (PAINT). If the fluorescent molecules perform diffusive movement along the structure of interest, tracking their movement reveals simultaneously their dynamics as well as super-resolved details of the underlying structure (provided enough trajectories have been accumulated to cover the whole structure with sufficient density). This is the principle of a technique called single-particle tracking PALM (sptPALM) [227–230].

Note that the results of SMLM are not images, but rather sets of fluorophore coordinates (and other characteristics such as the amplitude and width of the respective PSF from which the fluorophore was localised). An image is generated from such data usually by convolution of the fluorophore localisations with a suitable narrow PSF, the width of which can be chosen to match the expected localisation precision [191, 231].

Figure 9 explicitly illustrates SMLM principle for localising fluorophores within the focal plane, thus achieving lateral super-resolution. However, SMLM can be extended to 3 dimensions by including axial localisation of the fluorophores with sub-wavelength precision. Fluorophores in the focal plane can be recognised according to the width of the corresponding PSFs being the narrowest (diffraction limited). The further a fluorophore is from the focal plane, the broader is the respective PSF, until it eventually becomes smeared out to such an extent that localisation of the fluorophore becomes impossible [232]. However, for a well-corrected objective lens, the broadening of the PSF is symmetrical both below and above the focal plane; thus, analysing the PSF width identifies only the absolute axial distance, but not the direction of the fluorophore's position from the focal plane. This limitation can be overcome by modifying the PSF to have distinct shapes for fluorophores located above and below the focal plane, respectively. This is practically realised by introducing, for example astigmatic lenses or phase masks into the emission light path before the camera [233–236]. For astigmatic lenses, the accessible axial range is below 1 μm; for some phase masks, it can exceed 10 μm [233]. Other schemes for 3-deimensional SMLM involve simultaneous acquisition of images in multiple focal planes or interferometric approaches [237–240].

To achieve high degrees of sparsity, favourable for precise localisation of individual fluorophores, fluorophores either need to spend most of the time in the dark states or most of the binding sites need to be vacant at all times (in the cases when strategy 2 is used). This leads to the need of collecting high numbers of frames (often on the order of thousands) to accumulate enough localisations, translating to long acquisition times. Movement of any structures of interest within the prolonged acquisition time compromises the resolution, potentially negating the resolution enhancement, beside possibly producing a distorted image. This can be particularly problematic in live-cell SMLM [241]. Localisation algorithms able to process overlapping PSFs relax the requirements on fluorophore sparsity, and thus the time needed for SMLM data acquisition [242–244].

A further reduction in the image acquisition time, often by more than an order of magnitude, is achieved by related super-resolution techniques employing stochastic fluorophore blinking to enhance image resolution without depending on localising individual fluorophores [60, 194, 245–251]. These techniques are similar to SMLM in many aspects, such as the experimental implementation and structure of the raw image data, which also consist of time series of frames, each containing contributions from a subset of fluorophores. However, in contrast to SMLM, these techniques yield directly a super-resolved image rather than a list of fluorophore positions. For example, super-resolution optical fluctuation imaging (SOFI) uses temporal cumulants of fluorescence fluctuations to generate the super-resolved image. While these techniques still require fluorophore intensities to fluctuate, this does not have to be on/off blinking as required in SMLM; transitions between a brighter and a darker state are sufficient. Since individual fluorophores are not localised, the techniques can deal with higher fluorophore densities; however, SMLM in general reaches higher resolution enhancement.

6.3 Structured Illumination Microscopy (SIM)

SIM represents a very generic framework for super-resolution imaging. To understand its principle, it is very helpful to introduce the concept of spatial frequency and Fourier transform of images. Spatial-domain representation of images is what we usually imagine under the term "microscopy image", a representation of light intensity distribution as a function of spatial coordinates in 2 or 3 dimensions. The spatial-domain representation can be translated to frequency domain by Fourier transform. Frequency domain representation of an image is easily understood for an image of periodically repeating structures, which are described in frequency domain by the number of repetitions per a unit of distance. It can be shown mathematically that any arbitrary, non-periodic, structure can be described as a series of harmonic functions of different spatial frequencies. Frequencies approaching infinity are needed to describe sharp edges; hence the image is blurred by placing an upper limit on the spatial frequency bandwidth. The performance of an imaging system is described in frequency domain by its optical transfer function (OTF), which is a Fourier transform of its PSF. In a simplified approximation, the objective lens aperture can be described as a low-pass filter with cut-off frequency of $f_d = 1/d$, where d is the spatial-domain resolution limit. Increasing image resolution is thus equivalent to increasing its high-frequency content.

In the traditional implementation of SIM [174], the specimen is excited by a periodic pattern of bright and dark stripes (generated for example by placing a diffraction grating into the excitation light path) at spatial frequency f_x, resulting in spatial modulation of fluorophore excitation efficiency. The fluorescence image is then determined by the distribution of fluorophores in the specimen as well as by the periodic excitation light modulation. This can be described in the frequency domain as interference of the spatial frequency f_x of the periodic excitation pattern with spatial frequencies describing fluorescent structures in the specimen, analogously to describing interference of electromagnetic or acoustic waves. Interference of two waves produces waves with frequencies equal to the sum and the difference between the frequencies of the individual waves. Let us consider a spatial frequency f_s associated with sub-diffraction structural details of the specimen. Such frequency is not present in a diffraction-limited image as it is outside of the bandwidth supported by the objective lens ($f_s > f_d$). However, if the differential frequency $f_s - f_x$ lies within the supported bandwidth, such structural details can be reconstructed from an image acquired with spatially modulated excitation. Since the excitation in epifluorescence microscopy is delivered by the same objective lens, the modulation frequency of the excitation light needs to be within the objective's bandwidth ($f_x \leq f_d$), leading to the following observation about f_s:

$$f_s f_x \leq f_d$$

$$f_s \leq 2 f_d$$

In other words, the maximum resolution enhancement of the traditional SIM implementation does not exceed factor of 2.

Multiple images for different positions and orientations of the excitation intensity are needed to generate a single super-resolved SIM image. The pattern needs to be shifted in space to ensure that every point in the field of view has been exposed to (sufficiently high) excitation intensity and rotated to ensure resolution is enhanced for arbitrarily oriented structures. The traditional SIM implementation uses at least 9 images (3 shifts for 3 rotations) to produce a super-resolved image; SIM reconstruction algorithms that can work with as few as 7 or 6 images have been also proposed [252, 253]. The modest numbers of images required compared to SMLM make SIM better suited for observing dynamical processes in living specimens. For the best SIM results, it is advisable to acquire images at multiple focal planes and use the 3-dimensional information in SIM reconstruction. Multifocal and lightsheet implementations of SIM for 3-dimensional imaging have been also described [254–256].

The traditional SIM reconstruction algorithm assumes the knowledge of the excitation patterns. This makes it sensitive to specimen-induced optical aberrations that tend to distort the excitation patterns. This issue is alleviated by blind reconstruction algorithms for SIM, which do not require any knowledge of the excitation patterns, making them applicable to the case of random excitation patterns as well [257–263].

Resolution enhancement exceeding factor of 2 can be achieved in SIM by harnessing non-linear optical effects [264, 265] such as fluorescence saturation [190]. Although the illumination pattern does not contain spatial frequencies exceeding f_d, the non-linear relationship between excitation and emission intensities results in higher frequency components being modulated on the fluorescence emission distribution. However, the high excitation intensities needed to produce significant florescence saturation can be harmful to living specimens.

The 2-times limit on resolution enhancement can be also circumvented if the illumination pattern does not have to be transmitted by the microscope objective [266–268]. Strategies employing photonic chips, surface plasmons or hyperbolic metamaterials to generate patterned illumination increase considerably the resolution enhancement without the need for a highly specialised microscopy setup [269–274]; their main limitation is the shallow penetration depth resulting from the use of near-field optical phenomena in generating the patterned excitation intensity distribution.

6.4 Confocal Microscopy and Related Approaches

As mentioned in Sect. 4, reducing the size of the confocal aperture below 1 AU improves the resolution of confocal microscopy achieving sub-diffraction resolution. The resolution enhancement, however, comes at the cost of reduced signal. Consequently, this approach is not widely used in biological fluorescence microscopy which often struggles against limited photon budget. Resolution improvement without the sacrifice of signal can be achieved in laser scanning microscopy by replacing the confocal aperture and a point detector with a spatially resolved detector, such as a camera or a detector array [87, 89, 202, 275–281]. The whole camera or detector array captures all signal available, while a single camera pixel or a single detector in the array acts as a small pinhole (smaller than 1 AU). The distribution of the signal across individual pixels or individual detectors, thus, provides information allowing computational reconstruction of a super-resolved image, for example by an algorithm termed pixel reassignment. The resolution enhancement is similar to standard SIM, that means approaching factor of 2. The principle of pixel reassignment can be also realised in a hardware version, termed photon reassignment then, employing an additional beam scanning step that sweeps the emission beam across the camera with a twice as large sweep (to match the resolution doubling) as the sweep of the scanner used to scan the laser focus across the specimen [88, 282–284]. This approach is usually termed re-scan confocal microscopy. In multifocal or spinning disk versions, an additional 2x magnifying microlens array is added to the emission path [285–287].

While approaches derived from confocal microscopy evolved from a different starting point than SIM, they are closely related. Both use sparse distributions of excitation intensity and generate a single super-resolved image from signal captured with series of different excitation intensity distributions (which together make sure that every point within the field of view has been exposed to excitation). Considering this similarity, all these techniques can be viewed as variants of a generalised SIM approach [71, 187, 192, 286–291].

6.5 Computational Approaches

Computational super-resolution methods, that can be applied as a post-processing step to an image acquired with any standard microscopy modality, are very attractive for their flexibility, absence of special experimental requirements and, consequently, affordability. Since a super-resolved image has higher information contents than a diffraction-limited image, the additional information needs to be derived from a-priori knowledge or assumptions. For example, image deconvolution uses the knowledge of microscope PSF (either determined experimentally or predicted theoretically based on the parameters of the microscope) to increase the image resolution by factors potentially exceeding 2, although in practice, the resolution enhancement is usually much more modest because of limited signal to noise ratio [205, 288, 289, 292]. Deconvolution delivers the best results if applied to a 3-dimensional image stack, in which case it can benefit from the knowledge of the PSF in all three dimensions. Therefore, 3-dimensional stacks are usually acquired for deconvolution processing even if just a single plane is of interest; the additional planes can be viewed as an additional source of information used in generating a super-resolved image of the plane of interest.

Techniques based on deep learning represent an attractive alternative to traditional deconvolution algorithms, which they can outperform in terms of computational speed as well as resolution enhancement [293]. They can be trained on sets of image pairs consisting each of a synthetic ground-truth image and the same image degraded by simulated microscope imaging (considering PSF and noise). The training process thus supplies a-priori knowledge about the microscope PSF as well as about the types of structures expected in the specimen (through the types of structures present in the synthetic training images). Although assumptions about PSF are used in the training process, the technique is much less sensitive to the actual PSF than traditional deconvolution algorithms, making it applicable even in cases where the details of the microscope PSF are not known [293].

Some computational techniques perform optimally only on images with a sufficient degree of sparsity in fluorescence intensity distribution (which is, however, hugely lower than the high degrees of sparsity required by SMLM) [193, 195]. This is a reasonable assumption in fluorescence microscopy where most images exhibit such levels of sparsity (featuring, for example small fluorescent puncta or filamentous fluorescent structures). Examples of such techniques include super-resolution radial fluctuations (SRRF) employing local gradient fields convergence (termed radiality) [294, 295] or mean-shift super-resolution (MSSR) employing mean-shift theory [296] to enhance image resolution. While both techniques are able to increase the resolution of individual images, their resolution enhancement is significantly increased if time series of images are available. The fluorescence intensity fluctuations captured in the time series provide additional information that can be converted to resolution enhancement. This links these powerful techniques to fluctuation-based approaches such as SOFI introduced in Sect. 6.2; however, their reduced dependence on intensity fluctuations makes them significantly more flexible and usable under a broader range of conditions [193, 294].

Sparsity in fluorescence images, beneficial for the performance of many computational super-resolution approaches, can be enhanced by a sparse distribution of excitation light [192, 288]. If sparse excitation distributions are used instead of homogeneous excitation, multiple images with different excitation distributions need to be captured sequentially to ensure that every point within the field of view has been exposed to excitation. Super-resolution techniques employing such a mode of excitation can be viewed as examples of generalised SIM (together with confocal approaches, see Sects. 6.3 and 6.4).

7 Concluding Remarks and Outlook

Because of its ubiquitous use in, not only, biological and biomedical sciences, fluorescence microscopy has developed into such a broad field that this chapter could provide only a very basic introduction which necessarily missed many important aspects of the topic and many noteworthy developments. Many of the more specialised topics can be found in other chapters of this volume, to which this chapter can serve as an introduction.

A very important topic that has been mostly avoided in this chapter is processing and analysis of microscopy images. An introduction to this topic can be found, for example in references [169, 297, 298]. Segmentation, identification of the objects of interest, represents a critical step in many image analysis workflows such as object counting, object tracking, object morphometry or object-based colocalisation analysis. This step is often relatively simple in fluorescence microscopy compared to other techniques such as brightfield or scanning electron microscopy. Since the objects of interest typically appear bright against dark background, a single global threshold is often enough to distinguish pixels belonging to the objects of interest from background pixels and simple global thresholding algorithm can be applied successfully in such cases [299, 300]. More challenging scenarios involving uneven background and/or uneven intensity distribution in the objects of interest require more complex segmentation approaches such as locally determined thresholds or increasingly popular deep-learning approaches [301, 302].

Deep learning is also very powerful in enhancing image quality, outperforming traditional deconvolution algorithms in terms of speed as well as image improvement [293, 303]. This is an example of a trend where the advances in computational power and image processing algorithms relax the requirements on the microscope hardware, sample preparation and on acquisition time, allowing obtaining images of comparable quality with simpler microscope setups and/or in a shorter time. The lower cost of increasing computational capacity compared to upgrading highly specialised microscopy hardware is another aspect of this trend, which has the potential of making high-performance fluorescence microscopy more affordable and widespread. All of this is particularly pronounced in the case of lens-free microscopy, which uses remarkably simple and cost-effective setups combined with computational image reconstruction. While lens-free microscopy has been mostly limited to transmitted light imaging, there have been also lens-free implementations of fluorescence microscopy [27–32].

Another obvious trend in fluorescence microscopy has been the increasing throughput of, especially 3-dimensional, imaging, driven by advances in lightsheet microscopy and digital camera technology. Increasingly larger specimen volumes can be imaged in shorter time and at higher resolution. At the same time, fluorescence microscopy tends to be gentler and less invasive to allow observation of living specimens for longer periods and under less artificial conditions as demonstrated, for example, by fluorescence micro-endoscopy or miniature head-mounted microscopes for brain imaging in freely moving rodents [304, 305]. The increasing size of the 3-dimensional datasets and the growing rate at which they are produced place high demands not only on the computational power needed for their real-time processing but also on data storage capacity required to keep them. To tackle the latter challenge, solutions have been proposed to reduce the size of the dataset while keeping all relevant information contents [306–308]. Such solutions may prove critical to sustaining the current trends in fluorescence microscopy in the long run [309].

References

1. Robertson J, McGoverin C, Vanholsbeeck F, Swift S (2019) Optimisation of the protocol for the LIVE/DEAD® BacLightTM bacterial viability kit for rapid determination of bacterial load. Front Microbiol 10:801. https://doi.org/10.3389/fmicb.2019.00801
2. Baddeley D, Bewersdorf J (2018) Biological insight from super-resolution microscopy: what we can learn from localization-based images. Annu Rev Biochem 87(1):965. https://doi.org/10.1146/annurev-biochem-060815-014801
3. Ishikawa-Ankerhold HC, Ankerhold R, Drummen GPC (2012) Advanced fluorescence microscopy techniques--FRAP, FLIP, FLAP, FRET and FLIM. Molecules 17(4): 4047–4132. https://doi.org/10.3390/molecules17044047
4. Wang Y, Schellenberg H, Walhorn V, Toensing K, Anselmetti D (2017) Binding mechanism of fluorescent dyes to DNA characterized by magnetic tweezers. Mater Today Proc 4:S218–S225. https://doi.org/10.1016/j.matpr.2017.09.190
5. Otto F, Tsou KC (1985) A comparative study of DAPI, DIPI, and Hoechst 33258 and 33342 as chromosomal DNA stains. Stain Technol 60(1):7–11. https://doi.org/10.3109/10520298509113885
6. Pagano RE, Ozato K, Ruysschaert J-M (1977) Intracellular distribution of lipophilic fluorescent probes in mammalian cells. Biochim Biophys Acta Biomembr 465(3):661–666. https://doi.org/10.1016/0005-2736(77)90282-6
7. Greenspan P, Mayer EP, Fowler SD (1985) Nile red: a selective fluorescent stain for intracellular lipid droplets. J Cell Biol 100(3):965–973. https://doi.org/10.1083/jcb.100.3.965
8. Horisawa K (2014) Specific and quantitative labeling of biomolecules using click chemistry. Front Physiol 5:457. https://doi.org/10.3389/fphys.2014.00457
9. Yan Q, Bruchez MP (2015) Advances in chemical labeling of proteins in living cells. Cell Tissue Res 360(1):179–194. https://doi.org/10.1007/s00441-015-2145-4
10. Thorn K (2017) Genetically encoded fluorescent tags. Mol Biol Cell 28(7):848–857. https://doi.org/10.1091/mbc.e16-07-0504
11. Schaefer PM, Kalinina S, Rueck A, von Arnim CAF, von Einem B (2019) NADH autofluorescence – a marker on its way to boost bioenergetic research. Cytom Part A 95(1): 34–46. https://doi.org/10.1002/cyto.a.23597
12. Oxborough K (2004) Imaging of chlorophyll a fluorescence: theoretical and practical aspects of an emerging technique for the monitoring of photosynthetic performance. J Exp Bot 55(400):1195–1205. https://doi.org/10.1093/jxb/erh145
13. Bolte S, Cordelières FP (2006) A guided tour into subcellular colocalization analysis in light microscopy. J Microsc 224(3):213–232. https://doi.org/10.1111/j.1365-2818.2006.01706.x
14. Adler J, Parmryd I (2010) Quantifying colocalization by correlation: the Pearson correlation coefficient is superior to the Mander's overlap coefficient. Cytom Part A 77(8):733–742. https://doi.org/10.1002/cyto.a.20896
15. Dunn KW, Kamocka MM, McDonald JH (2011) A practical guide to evaluating colocalization in biological microscopy. Am J Physiol Physiol 300(4):C723–C742. https://doi.org/10.1152/ajpcell.00462.2010
16. Valm AM, Oldenbourg R, Borisy GG (2016) Multiplexed spectral imaging of 120 different fluorescent labels. PLoS One 11(7):e0158495
17. Klymchenko AS (2017) Solvatochromic and fluorogenic dyes as environment-sensitive probes: design and biological applications. Acc Chem Res 50(2):366–375. https://doi.org/10.1021/acs.accounts.6b00517
18. Han J, Burgess K (2010) Fluorescent indicators for intracellular pH. Chem Rev 110(5): 2709–2728. https://doi.org/10.1021/cr900249z
19. Zhang X-X, Wang Z, Yue X, Ma Y, Kiesewetter DO, Chen X (2013) pH-sensitive fluorescent dyes: are they really pH-sensitive in cells? Mol Pharm 10(5):1910–1917. https://doi.org/10.1021/mp3006903

20. Palonpon AF et al (2013) Raman and SERS microscopy for molecular imaging of live cells. Nat Protoc 8(4):677–692. https://doi.org/10.1038/nprot.2013.030
21. Gritti N, Kienle S, Filina O, van Zon JS (2016) Long-term time-lapse microscopy of C. elegans post-embryonic development. Nat Commun 7(1):12500. https://doi.org/10.1038/ncomms12500
22. Kamei M, Weinstein BM (2005) Long-term time-lapse fluorescence imaging of developing zebrafish. Zebrafish 2(2):113–123. https://doi.org/10.1089/zeb.2005.2.113
23. Collins JT et al (2020) Robotic microscopy for everyone: the OpenFlexure microscope. Biomed Opt Express 11(5):2447. https://doi.org/10.1364/boe.385729
24. Merces GOT, Kennedy C, Lenoci B, Reynaud EG, Burke N, Pickering M (2021) The incubot: a 3D printer-based microscope for long-term live cell imaging within a tissue culture incubator. HardwareX 9. https://doi.org/10.1016/j.ohx.2021.e00189
25. Ishmukhametov RR, Russell AN, Wheeler RJ, Nord AL, Berry RM (2016) A simple low-cost device enables four epi-illumination techniques on standard light microscopes. Sci Rep 6 (February):1–13. https://doi.org/10.1038/srep20729
26. Stewart C, Giannini J (2016) Inexpensive, open source epifluorescence microscopes. J Chem Educ 93(7):1310–1315. https://doi.org/10.1021/acs.jchemed.5b00984
27. Coskun AF, Sencan I, Su T-W, Ozcan A (2010) Lensless wide-field fluorescent imaging on a chip using compressive decoding of sparse objects. Opt Express 18(10):10510. https://doi.org/10.1364/oe.18.010510
28. Pang S, Han C, Kato M, Sternberg PW, Yang C (2012) Wide and scalable field-of-view Talbot-grid-based fluorescence microscopy. Opt Lett 37(23):5018. https://doi.org/10.1364/ol.37.005018
29. Banik S et al (2021) Recent trends in smartphone-based detection for biomedical applications: a review. Anal Bioanal Chem:2389–2406
30. Kuo G, Linda Liu F, Grossrubatscher I, Ng R, Waller L (2020) On-chip fluorescence microscopy with a random microlens diffuser. Opt Express 28(6):8384. https://doi.org/10.1364/oe.382055
31. Sasagawa K, Kimura A, Haruta M, Noda T, Tokuda T, Ohta J (2018) Highly sensitive lens-free fluorescence imaging device enabled by a complementary combination of interference and absorption filters. Biomed Opt Express 9(9):4329. https://doi.org/10.1364/boe.9.004329
32. Shanmugam A, Salthouse CD (2014) Lensless fluorescence imaging with height calculation. J Biomed Opt 19(01):1. https://doi.org/10.1117/1.jbo.19.1.016002
33. Sandoz PA, Tremblay C, van der Goot FG, Frechin M (2019) Image-based analysis of living mammalian cells using label-free 3D refractive index maps reveals new organelle dynamics and dry mass flux. PLoS Biol 17(12):e3000553
34. Elsayad K et al (2016) Mapping the subcellular mechanical properties of live cells in tissues with fluorescence emission-Brillouin imaging. Sci Signal 9(435):rs5. https://doi.org/10.1126/scisignal.aaf6326
35. Clayton AHA, Hanley QS, Arndt-Jovin DJ, Subramaniam V, Jovin TM (2002) Dynamic fluorescence anisotropy imaging microscopy in the frequency domain (rFLIM). Biophys J 83(3):1631–1649. https://doi.org/10.1016/S0006-3495(02)73932-5
36. Suhling K, Levitt J, Chung P-H (2014) Time-resolved fluorescence anisotropy imaging. Methods Mol Biol 1076:503–519. https://doi.org/10.1007/978-1-62703-649-8_22
37. Timr Š et al (2014) Accurate determination of the orientational distribution of a fluorescent molecule in a phospholipid membrane. J Phys Chem B 118(4):855–863. https://doi.org/10.1021/jp4067026
38. Ladouceur A-M, Brown CM (2021) Fluorescence microscopy light source review. Curr Protoc 1(9):e243. https://doi.org/10.1002/cpz1.243
39. Mubaid F et al (2019) Fluorescence microscope light source stability. Histochem Cell Biol 151(4):357–366. https://doi.org/10.1007/s00418-019-01776-6
40. Heřman P, Maliwal B, Lin H, Lakowicz J (2001) Frequency domain microscopy with the LED as a light source. J Microsc 203:176–181. https://doi.org/10.1046/j.1365-2818.2001.00943.x

41. Bei L, Dennis GI, Miller HM, Spaine TW, Carnahan JW (2004) Acousto-optic tunable filters: fundamentals and applications as applied to chemical analysis techniques. Prog Quantum Electron 28(2):67–87. https://doi.org/10.1016/S0079-6727(03)00083-1
42. Birk H, Engelhardt J, Storz R, Hartmann N, Bradl J, Ulrich H (2002) Programmable beam-splitter for confocal laser scanning microscopy. Proc SPIE 4621
43. Parasassi T, De Stasio G, d'Ubaldo A, Gratton E (1990) Phase fluctuation in phospholipid membranes revealed by Laurdan fluorescence. Biophys J 57(6):1179–1186. https://doi.org/10.1016/S0006-3495(90)82637-0
44. Yavas S, Macháň R, Wohland T (2016) The epidermal growth factor receptor forms location-dependent complexes in resting cells. Biophys J 111(10). https://doi.org/10.1016/j.bpj.2016.09.049
45. Lambert TJ (2019) FPbase: a community-editable fluorescent protein database. Nat Methods 16(4):277–278. https://doi.org/10.1038/s41592-019-0352-8
46. Jiménez-Sánchez D, Ariz M, Morgado JM, Cortés-Domínguez I, Ortiz-de-Solórzano C (2020) NMF-RI: blind spectral unmixing of highly mixed multispectral flow and image cytometry data. Bioinformatics 36(5):1590–1598. https://doi.org/10.1093/bioinformatics/btz751
47. McRae TD, Oleksyn D, Miller J, Gao Y-R (2019) Robust blind spectral unmixing for fluorescence microscopy using unsupervised learning. PLoS One 14(12):e0225410
48. Bayer B (1976) Patent US3971065A, color imaging array, US3971065A
49. Janesick JR (2001) Scientific charge-coupled devices. SPIE Press, Bellingham
50. Stuurman N, Vale RD (2016) Impact of new camera technologies on discoveries in cell biology. Biol Bull 231(1):5–13. https://doi.org/10.1086/689587
51. Okten Z, Churchman LS, Rock RS, Spudich JA (2004) Myosin VI walks hand-over-hand along actin. Nat Struct Mol Biol 11(9):884–887. https://doi.org/10.1038/nsmb815
52. Namiki S, Ikegaya Y (2009) Current application and technology of functional multineuron calcium imaging. Biol Pharm Bull 32(1):1–9. https://doi.org/10.1248/bpb.32.1
53. Unruh JR, Gratton E (2008) Analysis of molecular concentration and brightness from fluorescence fluctuation data with an electron multiplied CCD camera. Biophys J 95(11): 5385–5398. https://doi.org/10.1529/biophysj.108.130310
54. Bag N, Sankaran J, Paul A, Kraut RS, Wohland T (2012) Calibration and limits of camera-based fluorescence correlation spectroscopy: a supported lipid bilayer study. ChemPhysChem 13(11):2784–2794. https://doi.org/10.1002/cphc.201200032
55. Fossum ER (1993) Active pixel sensors: are CCDs dinosaurs? Proc SPIE 1900. https://doi.org/10.1117/12.148585
56. Beier HT, Ibey BL (2014) Experimental comparison of the high-speed imaging performance of an EM-CCD and sCMOS camera in a dynamic live-cell imaging test case. PLoS One 9(1). https://doi.org/10.1371/journal.pone.0084614
57. Singh AP et al (2013) The performance of 2D array detectors for light sheet based fluorescence correlation spectroscopy. Opt Express 21(7):8652–8668. https://doi.org/10.1364/OE.21.008652
58. Diekmann R, Till K, Müller M, Simonis M, Schüttpelz M, Huser T (2017) Characterization of an industry-grade CMOS camera well suited for single molecule localization microscopy – high performance super-resolution at low cost. Sci Rep 7(1):1–10. https://doi.org/10.1038/s41598-017-14762-6
59. Mandracchia B, Hua X, Guo C, Son J, Urner T, Jia S (2020) Fast and accurate sCMOS noise correction for fluorescence microscopy. Nat Commun 11(1):1–12. https://doi.org/10.1038/s41467-019-13841-8
60. Diederich B et al (2020) Nanoscopy on the Chea(i)p. bioRxiv. https://doi.org/10.1101/2020.09.04.283085
61. Chao J, Ram S, Ward ES, Ober RJ (2009) A 3D resolution measure for optical microscopy. Proc IEEE Int Symp Biomed Imaging 5193252:1115. https://doi.org/10.1109/ISBI.2009.5193252

62. Nieuwenhuizen RPJ et al (2013) Measuring image resolution in optical nanoscopy. Nat Methods 10(6):557–562. https://doi.org/10.1038/nmeth.2448
63. Banterle N, Bui KH, Lemke EA, Beck M (2013) Fourier ring correlation as a resolution criterion for super-resolution microscopy. J Struct Biol 183(3):363–367. https://doi.org/10.1016/j.jsb.2013.05.004
64. Barentine AES, Schroeder LK, Graff M, Baddeley D, Bewersdorf J (2018) Simultaneously measuring image features and resolution in live-cell STED images. Biophys J 115(6):951–956. https://doi.org/10.1016/j.bpj.2018.07.028
65. Tortarolo G, Castello M, Diaspro A, Koho S, Vicidomini G (2018) Evaluating image resolution in stimulated emission depletion microscopy. Optica 5(1):32. https://doi.org/10.1364/optica.5.000032
66. Raab M et al (2018) Using DNA origami nanorulers as traceable distance measurement standards and nanoscopic benchmark structures. Sci Rep 8(1):5–13. https://doi.org/10.1038/s41598-018-19905-x
67. Lin R, Clowsley AH, Lutz T, Baddeley D, Soeller C (2019) 3D super-resolution microscopy performance and quantitative analysis assessment using DNA-PAINT and DNA origami test samples. Methods 174:56–71. https://doi.org/10.1016/j.ymeth.2019.05.018
68. Thevathasan JV et al (2019) Nuclear pores as versatile reference standards for quantitative superresolution microscopy. Nat Methods 16(10):1045–1053. https://doi.org/10.1038/s41592-019-0574-9
69. Descloux A et al (2018) Combined multi-plane phase retrieval and super-resolution optical fluctuation imaging for 4D cell microscopy. Nat Photonics 12(3). https://doi.org/10.1038/s41566-018-0109-4
70. Pilger C et al (2021) Super-resolution fluorescence microscopy by line-scanning with an unmodified two-photon microscope. Philos Trans R Soc A Math Phys Eng Sci 379(2199). https://doi.org/10.1098/rsta.2020.0300
71. Cremer C, Masters BR (2013) Resolution enhancement techniques in microscopy. Eur Phys J H 38(3):281–344. https://doi.org/10.1140/epjh/e2012-20060-1
72. Kubitscheck U (2013) Principles of light microscopy. In: Kubitscheck U (ed) Fluorescence microscopy: from principles to biological applications1st edn. Wiley-VCH, Weinheim
73. Abbe E (1873) Beiträge zur Theorie des Mikroskops und der mikroskopischen Wahrnehmung. Arch für Mikroskopische Anat 9(1):413–468. https://doi.org/10.1007/BF02956173
74. Zhang Y, Gross H (2019) Systematic design of microscope objectives. Part I: system review and analysis. Adv Opt Technol 8(5):313–347
75. Diel EE, Lichtman JW, Richardson DS (2020) Tutorial: avoiding and correcting sample-induced spherical aberration artifacts in 3D fluorescence microscopy. Nat Protoc 15(9):2773–2784. https://doi.org/10.1038/s41596-020-0360-2
76. Lam F, Cladière D, Guillaume C, Wassmann K, Bolte S (2019) Super-resolution for everybody: an image processing workflow to obtain high-resolution images with a standard confocal microscope. Methods 115:17–27. https://doi.org/10.1016/j.ymeth.2016.11.003
77. Ji N (2017) Adaptive optical fluorescence microscopy. Nat Methods 14(4):374–380. https://doi.org/10.1038/nmeth.4218
78. Rodríguez C, Ji N (2018) Adaptive optical microscopy for neurobiology. Curr Opin Neurobiol 50:83–91. https://doi.org/10.1016/j.conb.2018.01.011
79. Sahu P, Mazumder N (2020) Advances in adaptive optics–based two-photon fluorescence microscopy for brain imaging. Lasers Med Sci 35(2):317–328. https://doi.org/10.1007/s10103-019-02908-z
80. Gallagher J, Delon A, Moreau P, Wang I (2017) Optimizing the metric in sensorless adaptive optical microscopy with fluorescence fluctuations. Opt Express 25(13):15558. https://doi.org/10.1364/oe.25.015558
81. Hell SW, Stelzer EH, Lindek S, Cremer C (1994) Confocal microscopy with an increased detection aperture: type-B 4Pi confocal microscopy. Opt Lett 19(3):222. https://doi.org/10.1364/ol.19.000222

82. Hell S, Stelzer EHK (1992) Properties of a 4Pi confocal fluorescence microscope. J Opt Soc Am A 9(12):2159–2166. https://doi.org/10.1364/JOSAA.9.002159
83. Antolovic IM, Bruschini C, Charbon E (2018) Dynamic range extension for photon counting arrays. Opt Express 26(17):22234–22248. https://doi.org/10.1364/OE.26.022234
84. Modi MN, Daie K, Turner GC, Podgorski K (2019) Two-photon imaging with silicon photomultipliers. Opt Express 27(24):35830–35841. https://doi.org/10.1364/OE.27.035830
85. Ching-Roa VD, Olson EM, Ibrahim SF, Torres R, Giacomelli MG (2021) Ultrahigh-speed point scanning two-photon microscopy using high dynamic range silicon photomultipliers. Sci Rep 11(1):1–12. https://doi.org/10.1038/s41598-021-84522-0
86. Caccia M, Nardo L, Santoro R, Schaffhauser D (2018) Silicon photomultipliers and SPAD imagers in biophotonics: advances and perspectives. Nucl Instruments Methods Phys Res Sect A Accel Spectrometers Detect Assoc Equip 926:101–117. https://doi.org/10.1016/j.nima.2018.10.204
87. Huff J (2015) The Airyscan detector from ZEISS: confocal imaging with improved signal-to-noise ratio and super-resolution. Nat Methods 12(12):i–ii. https://doi.org/10.1038/nmeth.f.388
88. De Luca GMR et al (2013) Re-scan confocal microscopy: scanning twice for better resolution. Biomed Opt Express 4(11):2644. https://doi.org/10.1364/boe.4.002644
89. Sheppard CJR et al (2020) Pixel reassignment in image scanning microscopy: a re-evaluation. J Opt Soc Am A 37(1):154–162. https://doi.org/10.1364/JOSAA.37.000154
90. Ilev IK, Waynant RW (2000) A simple submicron confocal microscope with a fiberoptic output. Rev Sci Instrum 71(11):4161–4164. https://doi.org/10.1063/1.1322589
91. Becker W et al (2016) A wide-field TCSPC FLIM system based on an MCP PMT with a delay-line anode. Rev Sci Instrum 87(9). https://doi.org/10.1063/1.4962864
92. Webb SED et al (2002) A wide-field time-domain fluorescence lifetime imaging microscope with optical sectioning. Rev Sci Instrum 73(4):1898–1907. https://doi.org/10.1063/1.1458061
93. Tisler J et al (2009) Fluorescence and spin properties of defects in single digit nanodiamonds. ACS Nano 3(7):1959–1965. https://doi.org/10.1021/nn9003617
94. Amos WB, White JG (2003) How the confocal laser scanning microscope entered biological research. Biol Cell 95(6):335–342. https://doi.org/10.1016/S0248-4900(03)00078-9
95. Borlinghaus RT (2006) MRT letter: high speed scanning has the potential to increase fluorescence yield and to reduce photobleaching. Microsc Res Tech 69(9):689–692. https://doi.org/10.1002/jemt.20363
96. Sheppard CJR, Mao XQ (1988) Confocal microscopes with slit apertures. J Mod Opt 35(7): 1169–1185. https://doi.org/10.1080/09500348814551251
97. Kim S et al (2015) Spectrally encoded slit confocal microscopy using a wavelength-swept laser. J Biomed Opt 20(3):1–5. https://doi.org/10.1117/1.JBO.20.3.036016
98. Lee J, Miyanaga Y, Ueda M, Hohng S (2012) Video-rate confocal microscopy for single-molecule imaging in live cells and superresolution fluorescence imaging. Biophys J 103(8): 1691–1697. https://doi.org/10.1016/j.bpj.2012.09.014
99. Amos WB, White JG (1995) Pawley JB (ed) Direct view confocal imaging systems using a slit aperture BT – handbook of biological confocal microscopy. Springer, Boston, pp 403–415
100. Chong SP, Pant S, Chen N (2011) Line-scan focal modulation microscopy for rapid imaging of thick biological specimens. Opt Sens Biophotonics:83111C. https://doi.org/10.1364/ACP.2011.83111C
101. Gräf R, Rietdorf J, Zimmermann T (2005) Live cell spinning disk microscopy. In: Rietdorf J (ed) Microscopy techniques. Springer, Berlin, pp 57–75
102. Halpern AR, Lee MY, Howard MD, Woodworth MA, Nicovich PR, Vaughan JC (2022) Versatile, do-it-yourself, low-cost spinning disk confocal microscope. Biomed Opt Express 13(2):1102–1120. https://doi.org/10.1364/BOE.442087
103. Petráň M, Hadravský M, Egger MD, Galambos R (1968) Tandem-scanning reflected-light microscope. J Opt Soc Am 58(5):661–664. https://doi.org/10.1364/JOSA.58.000661
104. Egger MD, Petráň M (1967) New reflected-light microscope for viewing unstained brain and ganglion cells. Science 157(3786):305–307. https://doi.org/10.1126/science.157.3786.305

105. Borlinghaus R, Gröbler B (1998) Basic principles and applications of confocal laser scanning microscopy. In: Isenberg G (ed) Modern optics, electronics and high precision techniques in cell biology. Springer, Berlin, pp 33–53
106. Kang SY, Duocastella M, Arnold CB (2020) Variable optical elements for fast focus control. Nat Photonics 14(9):533–542. https://doi.org/10.1038/s41566-020-0684-z
107. Deguchi T, Bianchini P, Palazzolo G, Oneto M, Diaspro A, Duocastella M (2020) Volumetric Lissajous confocal microscopy with tunable spatiotemporal resolution. Biomed Opt Express 11(11):6293. https://doi.org/10.1364/boe.400777
108. Lee D-R, Kim Y-D, Gweon D-G, Yoo H (2013) Dual-detection confocal fluorescence microscopy: fluorescence axial imaging without axial scanning. Opt Express 21(15):17839. https://doi.org/10.1364/oe.21.017839
109. Badon A et al (2019) Video-rate large-scale imaging with multi-Z confocal microscopy. Optica 6(4):389. https://doi.org/10.1364/optica.6.000389
110. He K et al (2018) Computational multifocal microscopy. Biomed Opt Express 9(12):6477. https://doi.org/10.1364/boe.9.006477
111. Tuchin EA, Zhu VV, Genina D (2022) Handbook of tissue optical clearing: new prospects in optical imaging.1st edn. CRC Press, Boca Raton
112. Zubkovs V et al (2018) Spinning-disc confocal microscopy in the second near-infrared window (NIR-II). Sci Rep 8(1):13770. https://doi.org/10.1038/s41598-018-31928-y
113. Sun C, Wang Y, Zhang H, Qian J (2017) Near-infrared laser scanning confocal microscopy and its application in bioimaging. Opt Quant Electron 50(1):35. https://doi.org/10.1007/s11082-017-1309-8
114. Xia F et al (2021) Short-wave infrared confocal fluorescence imaging of deep mouse brain with a superconducting nanowire single-photon detector. ACS Photonics 8(9):2800–2810. https://doi.org/10.1021/acsphotonics.1c01018
115. Wang M, Chen N (2019) Three-dimensional cellular imaging in thick biological tissue with confocal detection of one-photon fluorescence in the near-infrared II window. J Biophotonics 12(7):e201800459. https://doi.org/10.1002/jbio.201800459
116. Zhang Y et al (2021) Instant FLIM enables 4D in vivo lifetime imaging of intact and injured zebrafish and mouse brains. Optica 8(6):885–897. https://doi.org/10.1364/OPTICA.426870
117. Diaspro A, Chirico G, Collini M (2005) Two-photon fluorescence excitation and related techniques in biological microscopy. Q Rev Biophys 38(2):97–166. https://doi.org/10.1017/S0033583505004129
118. Wang T, Xu C (2020) Three-photon neuronal imaging in deep mouse brain. Optica 7(8):947. https://doi.org/10.1364/optica.395825
119. Xu C, Wise FW (2013) Recent advances in fibre lasers for nonlinear microscopy. Nat Photonics 7(11):875–882. https://doi.org/10.1038/nphoton.2013.284
120. Yew EYS, Choi H, Kim D, So PTC (2011) Wide-field two-photon microscopy with temporal focusing and HiLo background rejection. Multiphot Microsc Biomed Sci XI 7903: 79031O. https://doi.org/10.1117/12.876068
121. Hwang JY et al (2011) Multimodal wide-field two-photon excitation imaging: characterization of the technique for in vivo applications. Biomed Opt Express 2(2):356. https://doi.org/10.1364/boe.2.000356
122. Escobet-Montalbán A et al (2018) Wide-field multiphoton imaging through scattering media without correction. Sci Adv 4(10):eaau1338. https://doi.org/10.1126/sciadv.aau1338
123. Macias-Romero C, Zubkovs V, Wang S, Roke S (2016) Wide-field medium-repetition-rate multiphoton microscopy reduces photodamage of living cells. Biomed Opt Express 7(4):1458. https://doi.org/10.1364/boe.7.001458
124. Palero J, Santos SICO, Artigas D, Loza-Alvarez P (2010) A simple scanless two-photon fluorescence microscope using selective plane illumination. Opt Express 18(8):8491. https://doi.org/10.1364/oe.18.008491

125. Lavagnino Z, Cella Zanacchi F, Ronzitti E, Diaspro A (2013) Two-photon excitation selective plane illumination microscopy (2PE-SPIM) of highly scattering samples: characterization and application. Opt Express 21(5):5998. https://doi.org/10.1364/oe.21.005998
126. Truong TV, Supatto W, Koos DS, Choi JM, Fraser SE (2011) Deep and fast live imaging with two-photon scanned light-sheet microscopy. Nat Methods 8(9):757–760. https://doi.org/10.1038/nmeth.1652
127. Zhang Y, Zhou T, Hu X, Xie H, Dai Q, Kong L (2019) Enhance imaging depth in wide-field two-photon microscopy by extended detection and computational reconstruction. Opt InfoBase Conf Pap Part F134(15):20117–20132. https://doi.org/10.1364/nlo.2019.ntu1b.1
128. Rosenegger DG, Tran CHT, LeDue J, Zhou N, Gordon GR (2014) A high performance, cost-effective, open-source microscope for scanning two-photon microscopy that is modular and readily adaptable. PLoS One 9(10). https://doi.org/10.1371/journal.pone.0110475
129. Bumstead JR (2018) Designing a large field-of-view two-photon microscope using optical invariant analysis. Neurophotonics 5(02):1. https://doi.org/10.1117/1.nph.5.2.025001
130. Cao R, Wallrabe HK, Periasamy A (2020) Multiphoton FLIM imaging of NAD(P)H and FAD with one excitation wavelength. J Biomed Opt 25(01):1. https://doi.org/10.1117/1.jbo.25.1.014510
131. Bower AJ, Li J, Chaney EJ, Marjanovic M, Spillman DR, Boppart SA (2018) High-speed imaging of transient metabolic dynamics using two-photon fluorescence lifetime imaging microscopy. Optica 5(10):1290. https://doi.org/10.1364/optica.5.001290
132. Jyothikumar V, Sun Y, Periasamy A (2013) Investigation of tryptophan–NADH interactions in live human cells using three-photon fluorescence lifetime imaging and Förster resonance energy transfer microscopy. J Biomed Opt 18(6):060501. https://doi.org/10.1117/1.jbo.18.6.060501
133. Shen B et al (2020) Label-free whole-colony imaging and metabolic analysis of metastatic pancreatic cancer by an autoregulating flexible optical system. Theranostics 10(4):1849–1860. https://doi.org/10.7150/thno.40869
134. Masters BR (2020) Masters BR (ed) Richard Zsigmondy and Henry Siedentopf's Ultramicroscope BT – superresolution optical microscopy: the quest for enhanced resolution and contrast. Springer, Cham, pp 165–172
135. Huisken J, Swoger J, Del Bene F, Wittbrodt J, Stelzer EHK (2004) Optical sectioning deep inside live embryos by selective plane illumination microscopy. Science 305(5686): 1007–1009. https://doi.org/10.1126/science.1100035
136. Stelzer EHK (2015) Light-sheet fluorescence microscopy for quantitative biology. Nat Methods 12(1):23–26. https://doi.org/10.1038/nmeth.3219
137. Wan Y, McDole K, Keller PJ (2019) Light-sheet microscopy and its potential for understanding developmental processes. Annu Rev Cell Dev Biol 35:655–681. https://doi.org/10.1146/annurev-cellbio-100818-125311
138. Stelzer EHK et al (2021) Light sheet fluorescence microscopy. Nat Rev Methods Prim 1(1). https://doi.org/10.1038/s43586-021-00069-4
139. Saghafi S, Becker K, Hahn C, Dodt HU (2014) 3D-ultramicroscopy utilizing aspheric optics. J Biophotonics 7(1–2):117–125. https://doi.org/10.1002/jbio.201300048
140. Li H, Wu Z, Yang Z, Zhanghao K, Xi P, Jin D (2021) Axially overlapped multi-focus light sheet with enlarged field of view. Appl Phys Lett 118(22). https://doi.org/10.1063/5.0049013
141. Krzic U, Gunther S, Saunders TE, Streichan SJ, Hufnagel L (2012) Multiview light-sheet microscope for rapid in toto imaging. Nat Methods 9(7):730–733. https://doi.org/10.1038/nmeth.2064
142. Aakhte M, Müller H-AJ (2021) Multiview tiling light sheet microscopy for 3D high-resolution live imaging. Development 148(18). https://doi.org/10.1242/dev.199725
143. Meddens MBM, Liu S, Finnegan PS, Edwards TL, James CD, Lidke KA (2016) Single objective light-sheet microscopy for high-speed whole-cell 3D super-resolution. Biomed Opt Express 7(6):2219. https://doi.org/10.1364/boe.7.002219

144. Ponjavic A, Ye Y, Laue E, Lee SF, Klenerman D (2018) Sensitive light-sheet microscopy in multiwell plates using an AFM cantilever. Biomed Opt Express 9(12):5863. https://doi.org/10.1364/boe.9.005863
145. Kumar M, Kishore S, McLean DL, Kozorovitskiy Y (2021) Crossbill: an open access single objective light-sheet microscopy platform. bioRxiv. https://doi.org/10.1101/2021.04.30.442190
146. Galland R, Grenci G, Aravind A, Viasnoff V, Studer V, Sibarita J-B (2015) 3D high- and super-resolution imaging using single-objective SPIM. Nat Methods 12(7):641–644. https://doi.org/10.1038/nmeth.3402
147. Dodt H-U et al (2015) Ultramicroscopy: development and outlook. Neurophotonics 2(4): 041407. https://doi.org/10.1117/1.nph.2.4.041407
148. Chen BC et al (2014) Lattice light-sheet microscopy: imaging molecules to embryos at high spatiotemporal resolution. Science 346(6208). https://doi.org/10.1126/science.1257998
149. Planchon TA et al (2011) Rapid three-dimensional isotropic imaging of living cells using Bessel beam plane illumination. Nat Methods 8(5):417–423. https://doi.org/10.1038/nmeth.1586
150. Keller PJ, Schmidt AD, Wittbrodt J, Stelzer EHK (2008) Reconstruction of zebrafish early embryonic development by scanned light sheet microscopy. Science 322(5904):1065–1069. https://doi.org/10.1126/science.1162493
151. Ren YX et al (2020) Parallelized volumetric fluorescence microscopy with a reconfigurable coded incoherent light-sheet array. Light Sci Appl 9(1). https://doi.org/10.1038/s41377-020-0245-8
152. Hofmann M, Ghebjagh SG, Lemke K, Sinzinger S (2021) Multi-sheet excitation and imaging of flow driven samples in an LSFM with a modified multi-focal diffractive lens. Opt InfoBase Conf Pap 2021:7–8. https://doi.org/10.1364/cosi.2021.cm1a.7
153. Zunino A et al (2021) Multi-plane encoded light-sheet microscopy with acousto-optics. Proc SPIE 11654
154. Axelrod D (1981) Cell-substrate contacts illuminated by total internal reflection fluorescence. J Cell Biol 89(1):141–145. https://doi.org/10.1083/jcb.89.1.141
155. Schneckenburger H (2005) Total internal reflection fluorescence microscopy: technical innovations and novel applications. Curr Opin Biotechnol 16(1):13–18. https://doi.org/10.1016/j.copbio.2004.12.004
156. Puttrich V, Rohlena J, Braun M, Lansky Z (2022) In vitro reconstitution of molecular motor-driven mitochondrial transport. Methods Mol Biol 2431:533–546. https://doi.org/10.1007/978-1-0716-1990-2_28
157. Woodward X, Stimpson EE, Kelly CV (2018) Single-lipid tracking on nanoscale membrane buds: the effects of curvature on lipid diffusion and sorting. Biochim Biophys Acta Biomembr 1860(10):2064–2075. https://doi.org/10.1016/j.bbamem.2018.05.009
158. Tannert A, Voigt P, Burgold S, Tannert S, Schaefer M (2008) Signal amplification between Gβγ release and PI3Kγ-mediated PI(3,4,5)P3 formation monitored by a fluorescent Gβγ biosensor protein and repetitive two component total internal reflection/fluorescence redistribution after photobleaching analysis. Biochemistry 47(43):11239–11250. https://doi.org/10.1021/bi800596b
159. Ruckstuhl T, Seeger S (2004) Attoliter detection volumes by confocal total-internal-reflection fluorescence microscopy. Opt Lett 29(6):569–571. https://doi.org/10.1364/OL.29.000569
160. Tokunaga M, Imamoto N, Sakata-Sogawa K (2008) Highly inclined thin illumination enables clear single-molecule imaging in cells. Nat Methods 5(2):159–161. https://doi.org/10.1038/nmeth1171
161. van't Hoff M, de Sars V, Oheim M (2008) A programmable light engine for quantitative single molecule TIRF and HILO imaging. Opt Express 16(22):18495–18504. https://doi.org/10.1364/OE.16.018495
162. Enderlein J, Ruckstuhl T, Seeger S (1999) Highly efficient optical detection of surface-generated fluorescence. Appl Optics 38(4):724–732. https://doi.org/10.1364/AO.38.000724

163. Ruckstuhl T, Verdes D, Winterflood CM, Seeger S (2011) Simultaneous near-field and far-field fluorescence microscopy of single molecules. Opt Express 19(7):6836. https://doi.org/10.1364/oe.19.006836
164. Winterflood CM, Ruckstuhl T, Reynolds NP, Seeger S (2012) Tackling sample-related artifacts in membrane FCS using parallel SAF and UAF detection. ChemPhysChem 13(16): 3655–3660. https://doi.org/10.1002/cphc.201200395
165. Ries J, Ruckstuhl T, Verdes D, Schwille P (2008) Supercritical angle fluorescence correlation spectroscopy. Biophys J 94(1):221–229. https://doi.org/10.1529/biophysj.107.115998
166. Langhorst MF, Schaffer J, Goetze B (2009) Structure brings clarity: structured illumination microscopy in cell biology. Biotechnol J 4(6):858–865. https://doi.org/10.1002/biot.200900025
167. Wilson T, Juškaitis R, Neil MAA, Kozubek M (1996) Confocal microscopy by aperture correlation. Opt Lett 21(23):1879–1881. https://doi.org/10.1364/OL.21.001879
168. Sticker M, Elsässer R, Neumann M, Wolff H (2020) How to get better fluorescence images with your widefield microscope: a methodology review. Micros Today 28(6):36–43. https://doi.org/10.1017/s155192952000156x
169. Russ JC, Neal FB (2018) The image processing handbook. CRC Press
170. Agard DA (1984) Optical sectioning microscopy: cellular architecture in three dimensions. Annu Rev Biophys Bioeng 13(1):191–219. https://doi.org/10.1146/annurev.bb.13.060184.001203
171. Cox G, Sheppard C (1993) Effects of image deconvolution on optical sectioning in conventional and confocal microscopes. Bioimaging 1(2):82–95. https://doi.org/10.1002/1361-6374(199306)1:2<82::AID-BIO3>3.3.CO;2-K
172. Manz W, Arp G, Schumann-Kindel G, Szewzyk U, Reitner J (2000) Widefield deconvolution epifluorescence microscopy combined with fluorescence in situ hybridization reveals the spatial arrangement of bacteria in sponge tissue. J Microbiol Methods 40(2):125–134. https://doi.org/10.1016/S0167-7012(99)00103-7
173. Hell SW, Wichmann J (1994) Breaking the diffraction resolution limit by stimulated emission: stimulated-emission-depletion fluorescence microscopy. Opt Lett 19(11):780–782. https://doi.org/10.1364/OL.19.000780
174. Gustafsson MGL (2000) Surpassing the lateral resolution limit by a factor of two using structured illumination microscopy. J Microsc 198(2):82–87. https://doi.org/10.1046/j.1365-2818.2000.00710.x
175. Rust MJ, Bates M, Zhuang X (2006) Sub-diffraction-limit imaging by stochastic optical reconstruction microscopy (STORM). Nat Methods 3(10):793–795. https://doi.org/10.1038/nmeth929
176. Betzig E et al (2006) Imaging intracellular fluorescent proteins at nanometer resolution. Science 313(5793):1642–1645. https://doi.org/10.1126/science.1127344
177. Heintzmann R, Cremer CG (1999) Laterally modulated excitation microscopy: improvement of resolution by using a diffraction grating:185–196. https://doi.org/10.1117/12.336833
178. Heilemann M, Van De Linde S, Mukherjee A, Sauer M (2009) Super-resolution imaging with small organic fluorophores. Angew Chem Int Ed 48(37):6903–6908. https://doi.org/10.1002/anie.200902073
179. Heilemann M et al (2002) High-resolution colocalization of single dye molecules by fluorescence lifetime imaging microscopy. Anal Chem 74(14):3511–3517. https://doi.org/10.1021/ac025576g
180. Lidke K, Rieger B, Jovin T, Heintzmann R (2005) Superresolution by localization of quantum dots using blinking statistics. Opt Express 13(18):7052–7062. https://doi.org/10.1364/OPEX.13.007052
181. Sharonov A, Hochstrasser RM (2006) Wide-field subdiffraction imaging by accumulated binding of diffusing probes. Proc Natl Acad Sci U S A 103(50):18911–18916. https://doi.org/10.1073/pnas.0609643104

182. Valli J, Garcia-Burgos A, Rooney LM, de Melo e Oliveira BV, Duncan RR, Rickman C (2021) Seeing beyond the limit: a guide to choosing the right super-resolution microscopy technique. J Biol Chem 297(1). https://doi.org/10.1016/j.jbc.2021.100791
183. Liu S, Hoess P, Ries J (2022) Super-resolution microscopy for structural cell biology. Annu Rev Biophys 51(1):301–326. https://doi.org/10.1146/annurev-biophys-102521-112912
184. Prakash K, Diederich B, Heintzmann R, Schermelleh L (2022) Super-resolution microscopy: a brief history and new avenues. Philos Trans R Soc A Math Phys Eng Sci 380(2220). https://doi.org/10.1098/rsta.2021.0110
185. Eggeling C, Willig KI, Sahl SJ, Hell SW (2015) Lens-based fluorescence nanoscopy. Q Rev Biophys 48(2):178–243. https://doi.org/10.1017/S0033583514000146
186. Jacquemet G, Carisey AF, Hamidi H, Henriques R, Leterrier C (2020) The cell biologist's guide to super-resolution microscopy. J Cell Sci 133(11):jcs240713. https://doi.org/10.1242/jcs.240713
187. Long BR, Robinson DC, Zhong H (2014) Subdiffractive microscopy: techniques, applications, and challenges background: diffraction limit in biological imaging. Wiley Interdiscip Rev Syst Biol Med 6(2):151–168. https://doi.org/10.1002/wsbm.1259.Subdiffractive
188. Humpolíčková J, Benda A, Macháň R, Enderlein J, Hof M (2010) Dynamic saturation optical microscopy: employing dark-state formation kinetics for resolution enhancement. Phys Chem Chem Phys 12(39). https://doi.org/10.1039/c0cp00059k
189. Zhang Y et al (2018) Super-resolution fluorescence microscopy by stepwise optical saturation. Biomed Opt Express 9(4):1613. https://doi.org/10.1364/boe.9.001613
190. Gustafsson MGL (2005) Nonlinear structured-illumination microscopy: wide-field fluorescence imaging with theoretically unlimited resolution. Proc Natl Acad Sci U S A 102(37): 13081–13086. https://doi.org/10.1073/pnas.0406877102
191. Lelek M et al (2021) Single-molecule localization microscopy. Nat Rev Methods Prim 1(1): 39. https://doi.org/10.1038/s43586-021-00038-x
192. Yu J-Y, Becker SR, Folberth J, Wallin BF, Chen S, Cogswell CJ (2018) Achieving superresolution with illumination-enhanced sparsity. Opt Express 26(8):9850. https://doi.org/10.1364/oe.26.009850
193. Opstad IS et al (2020) Fluorescence fluctuations-based super-resolution microscopy techniques: an experimental comparative study
194. Mukamel EA, Babcock H, Zhuang X (2012) Statistical deconvolution for superresolution fluorescence microscopy. Biophys J 102(10):2391–2400. https://doi.org/10.1016/j.bpj.2012.03.070
195. Manton JD, Xiao Y, Turner RD, Christie G, Rees EJ (2018) ELM: super-resolution analysis of wide-field images of fluorescent shell structures. Methods Appl Fluoresc 6(3). https://doi.org/10.1088/2050-6120/aac28e
196. Enderle T, Ha T, Chemla DS, Weiss S (1998) Near-field fluorescence microscopy of cells. Ultramicroscopy 71(1–4):303–309. https://doi.org/10.1016/s0304-3991(97)00075-2
197. Enderle T, Ha T, Ogletree DF, Chemla DS, Magowan C, Weiss S (1997) Membrane specific mapping and colocalization of malarial and host skeletal proteins in the plasmodium falciparum infected erythrocyte by dual-color near-field scanning optical microscopy. Proc Natl Acad Sci U S A 94(2):520–525. https://doi.org/10.1073/pnas.94.2.520
198. García-Parajó MF et al (2005) Near-field fluorescence microscopy. NanoBiotechnology 1(1): 113–120. https://doi.org/10.1385/NBT:1:1:113
199. Harootunian A, Betzig E, Isaacson M, Lewis A (1986) Super-resolution fluorescence near-field scanning optical microscopy. Appl Phys Lett 49(11):674–676. https://doi.org/10.1063/1.97565
200. Schwartz O, Oron D (2012) Improved resolution in fluorescence microscopy using quantum correlations. Phys Rev A 85(3):1–5. https://doi.org/10.1103/PhysRevA.85.033812
201. Schwartz O, Levitt JM, Tenne R, Itzhakov S, Deutsch Z, Oron D (2013) Superresolution microscopy with quantum emitters. Nano Lett 13(12):5832–5836. https://doi.org/10.1021/nl402552m

202. Tenne R et al (2019) Super-resolution enhancement by quantum image scanning microscopy. Nat Photonics 13(2):116–122. https://doi.org/10.1038/s41566-018-0324-z
203. Israel Y, Tenne R, Oron D, Silberberg Y (2017) Quantum correlation enhanced super-resolution localization microscopy enabled by a fibre bundle camera. Nat Commun 8:1–5. https://doi.org/10.1038/ncomms14786
204. Classen A, von Zanthier J, Scully MO, Agarwal GS (2017) Superresolution via structured illumination quantum correlation microscopy. Optica 4(6):580. https://doi.org/10.1364/optica.4.000580
205. Korobchevskaya K et al (2016) Intensity weighted subtraction microscopy approach for image contrast and resolution enhancement. Sci Rep 6(April). https://doi.org/10.1038/srep25816
206. Hell SW, Kroug M (1995) Ground-state-depletion fluorscence microscopy: a concept for breaking the diffraction resolution limit. Appl Phys B Lasers Opt 60(5):495–497. https://doi.org/10.1007/BF01081333
207. Eilers Y, Ta H, Gwosch KC, Balzarotti F, Hell SW (2018) MINFLUX monitors rapid molecular jumps with superior spatiotemporal resolution. Proc Natl Acad Sci 115(24). https://doi.org/10.1073/pnas.1801672115
208. Friedrich M, Gan Q, Ermolayev V, Harms GS (2011) STED-SPIM: stimulated emission depletion improves sheet illumination microscopy resolution. Biophys J 100(8):L43–L45. https://doi.org/10.1016/j.bpj.2010.12.3748
209. Prakash K (2021) Laser-free super-resolution microscopy. Philos Trans R Soc A Math Phys Eng Sci 379(2199):20200144. https://doi.org/10.1098/rsta.2020.0144
210. Mao H et al (2019) Cost-efficient nanoscopy reveals nanoscale architecture of liver cells and platelets. Nanophotonics. https://doi.org/10.1515/nanoph-2019-0066
211. Hosny NA, Song M, Connelly JT, Ameer-Beg S, Knight MM, Wheeler AP (2013) Super-resolution imaging strategies for cell biologists using a spinning disk microscope. PLoS One 8(10). https://doi.org/10.1371/journal.pone.0074604
212. Gustavsson A-K, Petrov PN, Lee MY, Shechtman Y, Moerner WE (2018) 3D single-molecule super-resolution microscopy with a tilted light sheet. Nat Commun 9(1). https://doi.org/10.1038/s41467-017-02563-4
213. Lu C-H et al (2019) Lightsheet localization microscopy enables fast, large-scale, and three-dimensional super-resolution imaging. Commun Biol 2(1):177. https://doi.org/10.1038/s42003-019-0403-9
214. Thompson RE, Larson DR, Webb WW (2002) Precise nanometer localization analysis for individual fluorescent probes. Biophys J 82(5):2775–2783. https://doi.org/10.1016/S0006-3495(02)75618-X
215. Yildiz A, Selvin PR (2005) Fluorescence imaging with one nanometer accuracy: application to molecular motors. Acc Chem Res 38(7):574–582. https://doi.org/10.1021/ar040136s
216. Weisenburger S et al (2017) Cryogenic optical localization provides 3D protein structure data with angstrom resolution. Nat Methods 14(2):141–144. https://doi.org/10.1038/nmeth.4141
217. Hess ST, Girirajan TPK, Mason MD (2006) Ultra-high resolution imaging by fluorescence photoactivation localization microscopy. Biophys J 91(11):4258–4272. https://doi.org/10.1529/biophysj.106.091116
218. Lippincott-Schwartz J, Patterson GH (2009) Photoactivatable fluorescent proteins for diffraction-limited and super-resolution imaging. Trends Cell Biol 19(11):555–565. https://doi.org/10.1016/j.tcb.2009.09.003
219. Shcherbakova DM, Sengupta P, Lippincott-Schwartz J, Verkhusha VV (2014) Photocontrollable fluorescent proteins for superresolution imaging. Annu Rev Biophys 43:303–329. https://doi.org/10.1146/annurev-biophys-051013-022836
220. Wang S, Moffitt JR, Dempsey GT, Xie XS, Zhuang X (2014) Characterization and development of photoactivatable fluorescent proteins for single-molecule-based superresolution imaging. Proc Natl Acad Sci U S A 111(23):8452–8457. https://doi.org/10.1073/pnas.1406593111
221. van de Linde S et al (2011) Direct stochastic optical reconstruction microscopy with standard fluorescent probes. Nat Protoc 6(7):991–1009. https://doi.org/10.1038/nprot.2011.336

222. Dempsey GT, Vaughan JC, Chen KH, Bates M, Zhuang X (2011) Evaluation of fluorophores for optimal performance in localization-based super-resolution imaging. Nat Methods 8(12): 1027–1036. https://doi.org/10.1038/nmeth.1768
223. Vogelsang J et al (2008) A reducing and oxidizing system minimizes photobleaching and blinking of fluorescent dyes. Angew Chem Int Ed 47(29):5465–5469. https://doi.org/10.1002/anie.200801518
224. Jungmann R, Avendaño MS, Woehrstein JB, Dai M, Shih WM, Yin P (2014) Multiplexed 3D cellular super-resolution imaging with DNA-PAINT and exchange-PAINT. Nat Methods 11(3):313–318. https://doi.org/10.1038/nmeth.2835
225. Schoen I, Ries J, Klotzsch E, Ewers H, Vogel V (2011) Binding-activated localization microscopy of DNA l. Nano Lett 11(9):4008–4011. https://doi.org/10.1021/nl2025954
226. Kuo C, Hochstrasser RM (2011) Super-resolution microscopy of lipid bilayer phases. J Am Chem Soc 133(13):4664–4667. https://doi.org/10.1021/ja1099193
227. Manley S et al (2008) High-density mapping of single-molecule trajectories with photoactivated localization microscopy. Nat Methods 5(2):155–157. https://doi.org/10.1038/nmeth.1176
228. Rossier O et al (2012) Integrins β1 and β3 exhibit distinct dynamic nanoscale organizations inside focal adhesions. Nat Cell Biol 14(10):1057–1067. https://doi.org/10.1038/ncb2588
229. Gao Y, Foo YH, Winardhi R, Tang Q, Yan J, Kenney L (2017) Charged residues in the H-NS linker drive DNA binding and gene silencing in single cells. Proc Natl Acad Sci 114: 201716721. https://doi.org/10.1073/pnas.1716721114
230. Kapanidis A, Uphoff S, Stracy M (2018) Understanding protein mobility in bacteria by tracking single molecules. J Mol Biol 430. https://doi.org/10.1016/j.jmb.2018.05.002
231. Ovesný M, Křížek P, Borkovec J, Švindrych Z, Hagen GM (2014) ThunderSTORM: a comprehensive ImageJ plug-in for PALM and STORM data analysis and super-resolution imaging. Bioinformatics 30(16). https://doi.org/10.1093/bioinformatics/btu202
232. Palayret M et al (2015) Virtual-'light-sheet' single-molecule localisation microscopy enables quantitative optical sectioning for super-resolution imaging. PLoS One 10(4):1–15. https://doi.org/10.1371/journal.pone.0125438
233. von Diezmann L, Shechtman Y, Moerner WE (2017) Three-dimensional localization of single molecules for super-resolution imaging and single-particle tracking. Chem Rev 117(11): 7244–7275. https://doi.org/10.1021/acs.chemrev.6b00629
234. Bon P et al (2018) Self-interference 3D super-resolution microscopy for deep tissue investigations. Nat Methods 15(6):449–454. https://doi.org/10.1038/s41592-018-0005-3
235. Rehman SA, Carr AR, Lenz MO, Lee SF, O'Holleran K (2018) Maximizing the field of view and accuracy in 3D single molecule localization microscopy. Opt Express 26(4). https://doi.org/10.1364/OE.26.004631
236. Wang W et al (2019) Generalized method to design phase masks for 3D super-resolution microscopy. Opt Express 27(3):3799. https://doi.org/10.1364/oe.27.003799
237. Hajj B et al (2014) Whole-cell, multicolor superresolution imaging using volumetric multifocus microscopy. Proc Natl Acad Sci U S A 111(49):17480–17485. https://doi.org/10.1073/pnas.1412396111
238. Oudjedi L et al (2016) Astigmatic multifocus microscopy enables deep 3D super-resolved imaging. Biomed Opt Express 7(6). https://doi.org/10.1364/BOE.7.002163
239. Sims RR et al (2020) Single molecule light field microscopy. Optica 7(9):1065. https://doi.org/10.1364/optica.397172
240. Shtengel G et al (2009) Interferometric fluorescent super-resolution microscopy resolves 3D cellular ultrastructure understanding molecular-scale architecture of cells requires determination of 3D locations of specific proteins with accuracy match-ing their nanometer-length sca. Proc Natl Acad Sci 106(9):3125–3130
241. Shroff H, Galbraith CG, Galbraith JA, Betzig E (2008) Live-cell photoactivated localization microscopy of nanoscale adhesion dynamics. Nat Methods 5(5):417–423. https://doi.org/10.1038/nmeth.1202

242. Holden SJ, Uphoff S, Kapanidis AN (2011) DAOSTORM: an algorithm for high-density super-resolution microscopy. Nat Methods 8(4):279–280. https://doi.org/10.1038/nmeth0411-279
243. Min J et al (2014) FALCON: fast and unbiased reconstruction of high-density super-resolution microscopy data. Sci Rep 4:4577. https://doi.org/10.1038/srep04577
244. Hugelier S et al (2016) Sparse deconvolution of high-density super-resolution images. Sci Rep 6. https://doi.org/10.1038/srep21413
245. Dertinger T, Colyer R, Iyer G, Weiss S, Enderlein J (2009) Fast, background-free, 3D super-resolution optical fluctuation imaging (SOFI). Proc Natl Acad Sci U S A 106(52): 22287–22292. https://doi.org/10.1073/pnas.0907866106
246. Dertinger T, Pallaoro A, Braun G, Ly S, Laurence TA, Weiss S (2013) Advances in superresolution optical fluctuation imaging (SOFI). Q Rev Biophys 46(2):210–221. https://doi.org/10.1017/S0033583513000036
247. Geissbuehler S, Bocchio NL, Dellagiacoma C, Berclaz C, Leutenegger M, Lasser T (2012) Mapping molecular statistics with balanced super-resolution optical fluctuation imaging (bSOFI). Opt Nanoscopy 1(1):4. https://doi.org/10.1186/2192-2853-1-4
248. Agarwal K, Macháň R (2016) Multiple signal classification algorithm for super-resolution fluorescence microscopy. Nat Commun 7:13752. https://doi.org/10.1038/ncomms13752
249. Deng Y, Sun M, Lin PH, Ma J, Shaevitz JW (2014) Spatial covariance reconstructive (SCORE) super-resolution fluorescence microscopy. PLoS One 9(4):1–9. https://doi.org/10.1371/journal.pone.0094807
250. Yahiatene I, Hennig S, Müller M, Huser T (2015) Entropy-based super-resolution imaging (ESI): from disorder to fine detail. ACS Photonics 2(8):1049–1056. https://doi.org/10.1021/acsphotonics.5b00307
251. Cox S et al (2011) Bayesian localization microscopy reveals nanoscale podosome dynamics. Nat Methods 9(2):195–200. https://doi.org/10.1038/nmeth.1812
252. Gong H, Guo W, Neil MAA (2021) GPU-accelerated real-time reconstruction in python of three-dimensional datasets from structured illumination microscopy with hexagonal patterns. Philos Trans R Soc A Math Phys Eng Sci 379(2199). https://doi.org/10.1098/rsta.2020.0162
253. Christensen CN, Ward EN, Lu M, Lio P, Kaminski CF (2021) ML-SIM: universal reconstruction of structured illumination microscopy images using transfer learning. Biomed Opt Express 12(5):2720–2733. https://doi.org/10.1364/BOE.414680
254. Abrahamsson S et al (2017) Multifocus structured illumination microscopy for fast volumetric super-resolution imaging. Biomed Opt Express 8(9):4135. https://doi.org/10.1364/boe.8.004135
255. Chang B-J, Perez Meza VD, Stelzer EHK (2017) csiLSFM combines light-sheet fluorescence microscopy and coherent structured illumination for a lateral resolution below 100 nm. Proc Natl Acad Sci 114(19). https://doi.org/10.1073/pnas.1609278114
256. Ströhl F, Kaminski CF (2019) A concept for single-shot volumetric fluorescence imaging via orthogonally polarized excitation lattices. Sci Rep 9(1):1–9. https://doi.org/10.1038/s41598-019-42743-4
257. Yeh L-H, Tian L, Waller L (2016) Structured illumination microscopy with unknown patterns and a statistical prior. https://doi.org/10.1364/BOE.8.000695
258. Christensen CN, Ward EN, Lio P, Kaminski CF (2020) ML-SIM: a deep neural network for reconstruction of structured illumination microscopy images, pp 1–9
259. Mudry E et al (2012) Structured illumination microscopy using unknown speckle patterns. Nat Photonics 6(5):312–315. https://doi.org/10.1038/nphoton.2012.83
260. Lukeš T et al (2014) Three-dimensional super-resolution structured illumination microscopy with maximum a posteriori probability image estimation. Opt Express 22(24):29805. https://doi.org/10.1364/oe.22.029805
261. Min J et al (2013) Fluorescent microscopy beyond diffraction limits using speckle illumination and joint support recovery. Sci Rep 3:1–6. https://doi.org/10.1038/srep02075

262. Jost A, Tolstik E, Feldmann P, Wicker K, Sentenac A, Heintzmann R (2015) Optical sectioning and high resolution in single-slice structured illumination microscopy by thick slice blind-SIM reconstruction. PLoS One 10(7). https://doi.org/10.1371/journal.pone.0132174
263. Chakrova N, Heintzmann R, Rieger B, Stallinga S (2015) Studying different illumination patterns for resolution improvement in fluorescence microscopy. Opt Express 23(24):31367. https://doi.org/10.1364/oe.23.031367
264. Lu-Walther HW et al (2016) Nonlinear structured illumination using a fluorescent protein activating at the readout wavelength. PLoS One 11(10):1–14. https://doi.org/10.1371/journal.pone.0165148
265. Zhang H, Zhao M, Peng L (2011) Nonlinear structured illumination microscopy by surface plasmon enhanced stimulated emission depletion. Opt Express 19(24):24783. https://doi.org/10.1364/oe.19.024783
266. Joseph J, Faiz KP, Lahrberg M, Tinguely JC, Ahluwalia BS (2020) Improving the space-bandwidth product of structured illumination microscopy using a transillumination configuration. J Phys D Appl Phys 53(4). https://doi.org/10.1088/1361-6463/ab4e68
267. Yilmaz H, van Putten EG, Bertolotti J, Lagendijk A, Vos WL, Mosk AP (2015) Speckle correlation resolution enhancement of wide-field fluorescence imaging. Optica 2(5):424. https://doi.org/10.1364/optica.2.000424
268. Cragg GE, So PTC (2000) Lateral resolution enhancement with standing evanescent waves. Opt Lett 25(1):46. https://doi.org/10.1364/ol.25.000046
269. Helle ØI, Dullo FT, Lahrberg M, Tinguely JC, Hellesø OG, Ahluwalia BS (2020) Structured illumination microscopy using a photonic chip. Nat Photonics 14(7):431–438. https://doi.org/10.1038/s41566-020-0620-2
270. Tang M et al (2022) High-refractive-index chip with periodically fine-tuning gratings for tunable virtual-Wavevector spatial frequency shift universal super-resolution imaging. Adv Sci 9(9):1–11. https://doi.org/10.1002/advs.202103835
271. Lee YU et al (2021) Metamaterial assisted illumination nanoscopy via random super-resolution speckles. Nat Commun 12(1). https://doi.org/10.1038/s41467-021-21835-8
272. Lee YU et al (2021) Organic hyperbolic material assisted illumination nanoscopy. Adv Sci 8(22):1–7. https://doi.org/10.1002/advs.202102230
273. Ponsetto JL et al (2017) Experimental demonstration of localized plasmonic structured illumination microscopy. ACS Nano 11(6):5344–5350. https://doi.org/10.1021/acsnano.7b01158
274. Wei F, Liu Z (2010) Plasmonic structured illumination microscopy. Nano Lett 10(7): 2531–2536. https://doi.org/10.1021/nl1011068
275. Bertero M, Boccacci P, Defrise M, De Mol C, Pike ER (1989) Super-resolution in confocal scanning microscopy: II. The incoherent case. Inverse Probl 5(4):441–461. https://doi.org/10.1088/0266-5611/5/4/003
276. Müller CB, Enderlein J (2010) Image scanning microscopy. Phys Rev Lett 104(19):1–4. https://doi.org/10.1103/PhysRevLett.104.198101
277. Sheppard CJR, Mehta SB, Heintzmann R (2013) Superresolution by image scanning microscopy using pixel reassignment. Opt Lett 38(15):2889. https://doi.org/10.1364/ol.38.002889
278. Schulz O et al (2013) Resolution doubling in fluorescence microscopy with confocal spinning-disk image scanning microscopy. Proc Natl Acad Sci U S A 110(52):21000–21005. https://doi.org/10.1073/pnas.1315858110
279. Sheppard C (1988) Super-resolution in confocal imaging. Opt Int J Light Electron Opt 80:53
280. Sheppard CJR, Castello M, Tortarolo G, Vicidomini G, Diaspro A (2017) Image formation in image scanning microscopy, including the case of two-photon excitation. J Opt Soc Am A 34(8):1339. https://doi.org/10.1364/josaa.34.001339
281. Korobchevskaya K, Lagerholm BC, Colin-York H, Fritzsche M (2017) Exploring the potential of Airyscan microscopy for live cell imaging. Photodermatol 4(3). https://doi.org/10.3390/photonics4030041

282. Gregor I, Spiecker M, Petrovsky R, Großhans J, Ros R, Enderlein J (2017) Rapid nonlinear image scanning microscopy. Nat Methods 14(11):1087–1089. https://doi.org/10.1038/nmeth.4467
283. Roth S, Sheppard CJR, Wicker K, Heintzmann R (2013) Optical photon reassignment microscopy (OPRA). Opt Nanoscopy 2(1):5. https://doi.org/10.1186/2192-2853-2-5
284. Castello M et al (2019) A robust and versatile platform for image scanning microscopy enabling super-resolution FLIM. Nat Methods 16(2):175–178. https://doi.org/10.1038/s41592-018-0291-9
285. Azuma T, Kei T (2015) Super-resolution spinning-disk confocal microscopy using optical photon reassignment. Opt Express 23(11):15003. https://doi.org/10.1364/oe.23.015003
286. York AG et al (2013) Instant super-resolution imaging in live cells and embryos via analog image processing. Nat Methods. https://doi.org/10.1038/nmeth.2687
287. Winter PW et al (2014) Two-photon instant structured illumination microscopy improves the depth penetration of super-resolution imaging in thick scattering samples. Optica 1(3):181. https://doi.org/10.1364/optica.1.000181
288. Yu J-Y, Narumanchi VV, Chen S, Xing J, Becker SR, Cogswell CJ (2020) Analyzing the super-resolution characteristics of focused-spot illumination approaches. J Biomed Opt 25(05):1. https://doi.org/10.1117/1.jbo.25.5.056501
289. Hayashi S (2016) Resolution doubling using confocal microscopy via analogy with structured illumination microscopy. Jpn J Appl Phys 55(8). https://doi.org/10.7567/JJAP.55.082501
290. Fallet C et al (2014) Conical diffraction as a versatile building block to implement new imaging modalities for superresolution in fluorescence microscopy. Nanoimaging Nanospectrosc II 9169:9–14. https://doi.org/10.1117/12.2061059
291. Fallet C et al (2015) A new method to achieve tens of nm axial super-localization based on conical diffraction PSF shaping. Single Mol Spectrosc Superresolution Imaging VIII 9331:65–75. https://doi.org/10.1117/12.2077712
292. Schrader M, Hell SW, van der Voort HTM (1996) Potential of confocal microscopes to resolve in the 50–100 nm range. Appl Phys Lett 69(24):3644–3646. https://doi.org/10.1063/1.117010
293. Weigert M et al (2018) Content-aware image restoration: pushing the limits of fluorescence microscopy. Nat Methods 15(12):1090–1097. https://doi.org/10.1038/s41592-018-0216-7
294. Culley S, Tosheva KL, Matos Pereira P, Henriques R (2018) SRRF: Universal live-cell super-resolution microscopy. Int J Biochem Cell Biol 101:74. https://doi.org/10.1016/j.biocel.2018.05.014
295. Gustafsson N, Culley S, Ashdown G, Owen DM, Pereira PM, Henriques R (2016) Fast live-cell conventional fluorophore nanoscopy with ImageJ through super-resolution radial fluctuations. Nat Commun 7:1–9. https://doi.org/10.1038/ncomms12471
296. García ET et al (2021) Nanoscopic resolution within a single imaging frame. bioRxiv:2021.10.17.464398
297. Cordelières F et al (2016) Bioimage data analysis. WILEY-VCH, Weinheim
298. Miura K, Sladoje N (2020) Bioimage data analysis workflows. Springer, Cham
299. Kurita T, Otsu N, Abdelmalek N (1992) Maximum likelihood thresholding based on population mixture models. Pattern Recogn 25(10):1231–1240. https://doi.org/10.1016/0031-3203(92)90024-D
300. Otsu N (1979) A threshold selection method from gray-level histograms. IEEE Trans Syst Man Cybern 9(1):62–66. https://doi.org/10.1109/TSMC.1979.4310076
301. Arzt M et al (2022) LABKIT: labeling and segmentation toolkit for big image data. Front Comp Sci 4. https://doi.org/10.3389/fcomp.2022.777728
302. Arganda-Carreras I et al (2017) Trainable Weka segmentation: a machine learning tool for microscopy pixel classification. Bioinformatics 33(15):2424–2426. https://doi.org/10.1093/bioinformatics/btx180
303. Krull A, Buchholz TO, Jug F (2019, 2019) Noise2void-Learning denoising from single noisy images. Proc IEEE Comput Soc Conf Comput Vis Pattern Recognit:2124–2132. https://doi.org/10.1109/CVPR.2019.00223

304. Mekhail SP, Arbuthnott G, Chormaic SN (2016) Advances in fibre microendoscopy for neuronal imaging. Opt Data Process Storage 2(1):30–42. https://doi.org/10.1515/odps-2016-0003
305. Supekar OD et al (2022) Miniature structured illumination microscope for in vivo 3D imaging of brain structures with optical sectioning. Biomed Opt Express 13(4):2530. https://doi.org/10.1364/boe.449533
306. Schmid B et al (2013) High-speed panoramic light-sheet microscopy reveals global endodermal cell dynamics. Nat Commun 4(1):2207. https://doi.org/10.1038/ncomms3207
307. Pomarico E et al (2020) Quantifying the effect of image compression on supervised learning applications in optical microscopy
308. Amat F, Höckendorf B, Wan Y, Lemon WC, McDole K, Keller PJ (2015) Efficient processing and analysis of large-scale light-sheet microscopy data. Nat Protoc 10(11):1679–1696. https://doi.org/10.1038/nprot.2015.111
309. Vopson MM (2020) The information catastrophe. AIP Adv 10(8):85014. https://doi.org/10.1063/5.0019941

STED and RESOLFT Fluorescent Nanoscopy

Andreas Bodén, Francesca Pennacchietti, and Ilaria Testa

Contents

Abstract Fluorescence microscopy is an invaluable tool in cell biology to investigate the functional, structural, and dynamical properties of biological specimens. For a long time, the resolution of fluorescence microscopes was thought to be fundamentally limited by diffraction. According to Abbe's law of diffraction, published in 1873, the smallest spatial details accessible with visible light are defined only by the optics of the microscope, i.e., numerical aperture and the wavelength of the light. However, the last 25 years of research have shown that it is possible to investigate even smaller structures using only visible light. This chapter covers the basic principles of coordinate-targeted switching techniques, a family of super-resolution microscopy methods. Furthermore, it provides an overview of the state-of-the-art strategies to push their ability toward faster and more efficient imaging of living cells and tissue.

A. Bodén, F. Pennacchietti, and I. Testa (✉)
Science for Life Laboratory, KTH Royal Institute of Technology, Stockholm, Sweden
e-mail: andreas.boden@scilifelab.se; francesca.pennacchietti@scilifelab.se; ilaria.testa@scilifelab.se

R. Šachl, M. Amaro (eds.), *Fluorescence Spectroscopy and Microscopy in Biology*, Springer Ser Fluoresc (2023) 20: 201–232, https://doi.org/10.1007/4243_2022_35, Published online: 3 January 2023

Keywords Fluorescence nanoscopy · Live-cell imaging · Photoswitching · Parallelization · Smart scanning

1 Introduction

Fluorescence microscopy is an invaluable tool for the investigation of spatial distribution and dynamics of cellular and tissue structures in biological samples. Fluorescence allows reaching high contrast and selective labelling of proteins of interest in a live-cell compatible manner. Two main strategies can be distinguished in fluorescence microscopy, according to the way the image is created. In the case of widefield microscopy, the whole sample is excited through an even and extended illumination, and the fluorescence is collected at once with your eye or on an imaging detector (Fig. 1a). Alternatively, in scanning approaches, the sample is probed point-by-point using a sharply confined spot of excitation light. To enhance the image quality of a scanning microscope, a pinhole is placed in the image plane of the detection path to spatially filter out and select primarily the light emitted from the focal plane of the sample (confocal microscope, Fig. 1b). Confocal microscopy introduces optical sectioning and increased spatial resolution compared to widefield approaches. Widefield systems can however image significantly faster than scanning microscopes on extended fields of view (FOV) and are simpler and cheaper to build and maintain. Both widefield and scanning microscopes are limited in spatial resolution by the wavelength of light, as first realized by Ernst Abbe in 1873 when he formulated the theory behind the diffraction limit [1]. The wave nature of light means that it cannot be focused to a spot smaller than roughly half its wavelength. This characteristic consequently sets a limit on the size and shape of the so-called point spread function (PSF) of a microscope, which describes the image of an isolated emitting point in the sample. If two adjacent emitting points are distanced significantly closer than the size of the PSF and imaged through an optical system, their PSFs will overlap and they will thus not be resolvable in the image.

Super-resolution fluorescence microscopy techniques emerged in the late 1990s, pushing the resolution beyond the diffraction limit imposed by the wave nature of light. The key to surpassing the diffraction limit lies in the possibility to switch a fluorescent molecule between an on and an off state [2–4] (Fig. 2a). This can be done either in a deterministic way, using the on/off switching to reduce the effective PSF, or in a stochastic way, by recording the emission of many individual molecules within a diffraction-limited spot one at a time (single molecule localization microscopy, SMLM) [5–7].

This chapter will focus on the deterministic methods, alternatively called coordinate-targeted switching approaches, since they rely on the point-by-point investigation of the sample typical of a scanning microscope. The optical suppression of the fluorescence can be achieved either by STimulated Emission Depletion (STED) [8, 9] or more generally by Reversible Saturable OpticaL Fluorescence

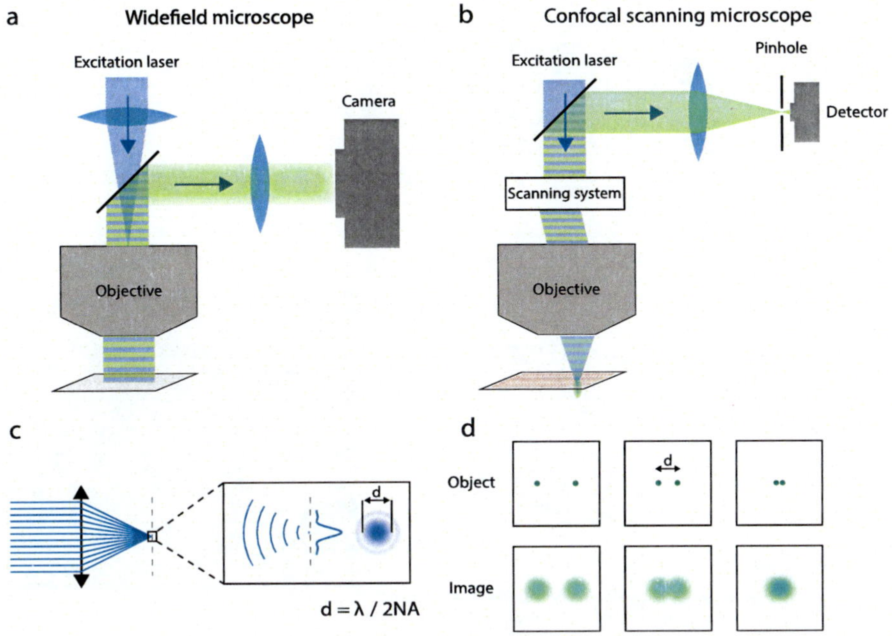

Fig. 1 Fluorescence microscopy. (**a**) A widefield fluorescence microscope illuminates the whole sample with excitation light, creating fluorescence emission that is imaged through the objective onto a camera. (**b**) An image can also be generated by sequentially scanning a focused probing beam over the sample, at each point measuring the amount of emitted fluorescence. In a confocal scanning microscope, the detected light is also passed through a pinhole to reject emission from out-of-focus planes. (**c**) The wave nature of light means that it cannot be focused onto an infinitesimally small spot. The smallest achievable focal point is around half the wavelength of the light that is focused. This limit is called the diffraction limit. (**d**) The image of an emitting point is called the point spread function (PSF) and its size is limited by the diffraction limit. Two adjacent emitting points can only be resolved if they reside further away from each other than about half the size of the PSF of the imaging system

Transitions (RESOLFT) [10–12]. STED is based on optically perturbing the electronic transition pathway that generates the fluorescence, and can therefore act on the ps-ns scale (Fig. 2d). In RESOLFT, the fluorescence properties of the molecule changes in response to a perturbation of its chemical configuration (Fig. 2e). Since the dark states involved in RESOLFT have a longer lifetime, the switching is notably slower (μs-ms scale) but requires magnitudes lower illumination intensities. Starting from a conventional confocal microscope, super-resolution can be achieved using STED or RESOLFT by superimposing a patterned illumination on top of the excitation illumination. The patterned illumination suppresses fluorescence emission from the outer regions of the diffraction-limited excitation spot and allows fluorescence emission only from the very central point where the suppressing light is zero. In this way, the effective excitation spot is shrunk, giving a finer probing of the sample structure (Fig. 2f). Expanding scanning systems to these super-resolution

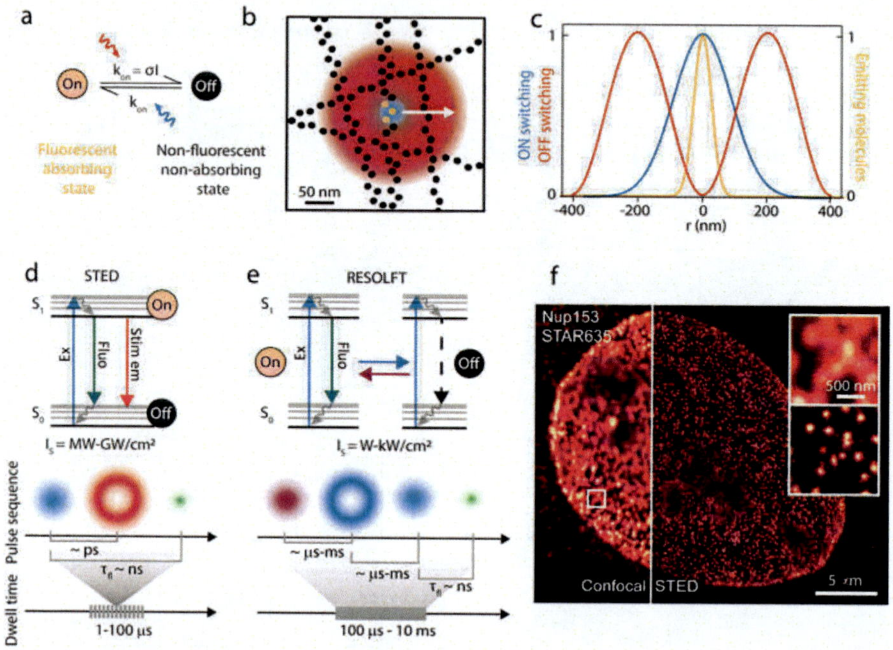

Fig. 2 Coordinate-targeted nanoscopy. (**a**) Illustration of the on/off switching between a fluorescent and a non-fluorescent generic molecular state. The on-to-off transition is induced by red light, while the off-to-on is induced by blue light. (**b**) Spatial distribution of the light and the state distribution of the molecules in the foci of a coordinate-targeted switching point-scanning microscope in 2D and (**c**) in 1D. To optimize on-state confinement (orange line), the minimum of the off-switching light (red line) should be exactly at the maximum of the on-switching (blue line). (**d**) Jablonski diagram illustrating electronic transitions of the standard pathway of fluorescence emission and how stimulating emission can inhibit the emission of fluorescence by forcing the molecule to emit the excess energy as stimulated emission photons. Below the state diagram is an illustration of the pulse sequence and dwell time for a STED recording. In the pulse sequence, the time delay between the different pulses and emission is highlighted. (**e**) Diagram illustrating a molecular transition where the molecule can instead be switched between two metastable molecular states, referred to as on and off state. The probes can be switched between the two states through illumination with different wavelengths of light. Below the state diagram, an example RESOLFT pulse scheme and dwell time is shown on a time scale. (**f**) Example of STED imaging. Confocal and STED images of the nucleus of a U2OS cell with inset showing zoomed-in area. The nuclear pore complex protein Nup153 has been labeled with Abberior STAR635P. Scale bars: 5 µm, and 500 nm, respectively

implementations however also adds significant complexity and comes with new challenges, both from an optical engineering perspective and in terms of sample preparation and label design.

STED and RESOLFT approaches were first introduced in a point-scanning implementation, where images are acquired by sampling over the FOV using a single probing spot. This adds a dependency between the temporal resolution and the imaged area (or volume) since the dwell time at each position is multiplied with the number of sampling points. Thus, point-scanning microscopy methods reach a high temporal resolution for small areas and volumes but imaging time quickly increases when scaling up the FOV. To overcome these limitations, multiple approaches to further increase the imaging speed have been proposed and demonstrated since the introduction of the methods, including minimizing the recording time and illumination doses by using smart scanning or by introducing light patterns to parallelize the acquisition. Smart scanning refers to systems where a single excitation spot is adaptively scanned over the sample, meaning that the scanning speed or illumination intensities are modulated in real time during the scan. Therefore, scanning of regions of the sample which do not contain structures of interest can either be completely avoided or illumination levels can be adapted to the structure. This increases the speed and/or lowers the illumination dose in an adaptable way for each FOV and its biological content. Smart or adaptive scanning systems have been demonstrated to improve performance both in STED and RESOLFT [13, 14]. Another approach to increase acquisition speed is to instead utilize the concept of parallelization where a large array of confined PSFs is generated in each illumination sequence. In these methods, the temporal resolution does not scale linearly with the FOV size, since the total sampling area is instead determined by the distance between two adjacent points in the array (which can be as close as 250 nm, limited only by diffraction). As a consequence, parallelized systems have a clear advantage when imaging larger areas.

In this chapter, we will describe the underlying concepts of STED and RESOLFT microscopy, and outline their strengths and limitations. In particular, we will review the approaches and ideas demonstrated to increase the imaging speed and decrease the illumination dose with the main goal of increasing their compatibility with live-cell imaging over extended periods of time.

2 Light-Induced Reversible State Transitions

Many state transitions can be used as reversible on/off switches. Their suitability for coordinate-targeted switching techniques depends on how efficiently molecules can be switched to a certain state at a specific time and be reversibly brought back to the original state with minimal loss. This translates into spatiotemporal control of the saturation of state populations within the fluorophore population which is necessary for successful imaging. In practice, the biological application will lend itself better to the use of one type of transition over another, as the application sets the requirements on illumination, light source, recording time, photobleaching, and imaging environment. Regardless of the nature of the transition, the effective excitation spot will be given by the equation:

$$d = \frac{\lambda}{2\text{NA}} \frac{1}{\sqrt{1 + I/I_S}} \tag{1}$$

where the intensity of saturation for the on-to-off transition is $I_S = k_{\text{on}}/\sigma_{\text{off}}$, with σ_{off} being the molecular cross-section for the transition and k_{on} accounting for the competing process that restores the initial fluorescent state [15]. The saturation intensity reports the intensity required to suppress the fluorescence to half its initial level and it is the key to obtaining the spatially sub-diffracted confinement of emitting molecules.

2.1 Reversible Electronic Transitions: Stimulated Emission

Following excitation from the ground state to the excited state, a fluorophore can release energy through spontaneous fluorescence emission or thermal relaxation. The return to the ground state can, alternatively, be induced by stimulated emission in an interaction with incoming photons. Illumination with light in which the photons have an energy equal to the energy gap between the excited and ground state will drive the electrons back to the ground state, resulting in the emission of photons of the same energy as the stimulating illumination, as illustrated in Fig. 2d. In STED, therefore, stimulated emission represents a competing process to fluorescence emission, effectively able to deplete the molecules from the excited state and thereby suppress the spontaneous fluorescence emission. Commonly used STED microscopes feature an illumination beam shaped like a doughnut to induce stimulated emission (depletion beam) overlapped onto a Gaussian-shaped beam of light used for fluorescence excitation. In the center, only the excitation light is present as the doughnut features a local "zero" here (Fig. 2c), and the molecules residing here will thus emit fluorescence. Instead, on the crests of the doughnut where the intensity of the STED beam is higher than the saturation level, I_s, the stimulated emission will prevail over fluorescence emission, effectively turning off the molecules in this peripheral area.

To efficiently drive the molecules to the off state, the stimulated emission transition has to outcompete spontaneous fluorescent emission, which has a lifetime of a few nanoseconds (1–5 ns). Combined with the low stimulated emission cross-section ($\sigma \approx 10^{-16}$ cm^2), this translates into a requirement of high photon flux ($I_s \sim$ MW–GW/cm^2) to achieve saturation of the on-to-off transition.

The relatively high intensity required by the STED beam sets specific demands on the fluorescent probes [16–18]. The main parameter for a probe to perform well in STED microscopy is photostability, meaning the ability of the probe to withstand several excitation-depletion cycles without significant photobleaching. The depletion wavelength depends on the emission spectrum of the fluorophore, and moving it toward the peak of emission gives a larger cross-section and therefore reduces the STED power needed [19, 20]. However, to avoid direct excitation by the depletion beam and to minimize interference with the fluorescence window, the depletion wavelength is generally moved to the red tail of the emission spectra. Even in such optimized depletion conditions, the STED image quality can be compromised by the interference of the STED beam with other photophysical pathways. This can lead to excited state absorption [21, 22] which, in turn, can result in a higher probability of bleaching.

Stimulated emission is a general process universal to all fluorophores. As such, STED has been demonstrated with many fluorophores of different nature, e.g., organic molecules [16, 23], fluorescent proteins [24–26], and inorganic nanocrystals [27–29] (e.g., nanodiamonds, quantum dots, and lanthanide upconversion nanoparticles). Besides matching the photophysical parameters optimal for STED imaging, the choice of probe is also linked to its biocompatibility. Cytotoxicity can arise from the intracellular oligomerization and aggregation of probes, and the high illumination power can induce photothermal effects around the probe as well as the formation of free radicals and singlet oxygen molecules. It is difficult to absolutely predict and directly detect the damage induced during imaging [30, 31]. Therefore, it is often estimated by assessing cell viability before and after the imaging experiment. Phototoxicity is highly dependent on the irradiation wavelength, therefore a common strategy is to shift the illumination to the red region of the spectra. This also has the additional advantage of being less susceptible to scattering and creating less autofluorescence, thus increasing the possible penetration depth. Labeling strategies used for live-cell and in vivo STED microscopy rely on genetically encoded markers like fluorescent proteins; self-labeling protein tags [32–35] (SNAP-, HALO- or CLIP-tags) or unnatural aminoacids [36] combined with cell-permeable dyes; or small cell-permeable molecules with high labelling specificity that can be for example drug-based or lipid-based (e.g., NileRed and SiR-Actin) [37].

The nature of the STED process and in particular the need to act within the fluorescent lifetime of the fluorophore sets general requirements on the hardware used in STED microscopes [22]. STED lasers need to deliver high illumination power to the sample during the lifetime of the excited state and thus be precisely timed after the excitation (delay in the order of ~ps to allow relaxation to the lowest excited state). Pulsed lasers with picosecond pulse widths match these power (MW–GW/cm^2) and time (~500 ps) requirements for optimal depletion. The fast time scale of stimulated emission depletion directly translates into fast imaging times. Pixel dwell times in STED microscopy can vary from 1–300 μs depending on the fluorophore brightness and density. This requires a fast and precise scanning system, often based on galvanometric mirrors, resonant scanners [38], or electro-optical scanners [39]. Additionally, the fast scanning of STED systems also requires fast detectors and electronics able to record and distinguish photons from adjacent scanning points. Common detector types for STED microscopes are, for example, photo-multiplying tubes (PMTs), avalanche photodiodes (APDs), or single-photon avalanche diodes (SPAD).

2.2 Reversible Molecular Transitions

RESOLFT encompasses all the coordinate-targeted super-resolution approaches that rely on metastable or long-lived fluorescent and non-fluorescent states, which can be reversibly and efficiently populated upon illumination with specific wavelengths. The use of molecular transitions between metastable states with long lifetimes, such as isomers, decreases the intensity required since the transitions and population saturation do not have to be induced on the nanosecond time scale. Compared to using electronic transitions, the intensities used while imaging can be decreased by several orders of magnitude. For example, in states with lifetimes on the microsecond timescale, the intensity required for saturation is on the order of a few W/cm^2.

RESOLFT techniques set strong requirements on the fluorescent label, especially in terms of switching kinetics and photobleaching resistance [40]. In STED microscopy, the use of electronic transitions on the nanosecond timescale makes the process quasi-instantaneous. In RESOLFT, the kinetics between the on and the off state reside within the microsecond–millisecond timescale and therefore directly influence the image recording time negatively (Fig. 1e). To be suitable for RESOLFT imaging, the conformational changes that underpin photoswitching needs to withstand many cycles. For example, to be able to acquire an image with an expected resolution of 50 nm in the focal plane, a step size of 25 nm is needed (according to Nyquist-Shannon sampling theory). Assuming that proteins within a 200×200 nm^2 area are switched on in every illumination cycle, the proteins will need to cycle at least $(200/25)^2 \approx 60$–70 times without a significant loss in fluorescence to allow for an image to be acquired. The ability of a probe to be photoswitched many times without significant loss of fluorescence is referred to as fatigue resistance.

Reversible on-off switching mechanisms can be found in reversibly photoswitchable fluorescent proteins (RSFPs) and organic photochromic fluorophores (Fig. 3). In GFP-like proteins, reversible photoswitching is generally a result of a photoinduced isomerization coupled with a protonation/deprotonation of the chromophore [41]. Three types of photoswitching can be identified according to how light controls the switch: positive, negative, and decoupled. The three types and their response to illumination schemes are illustrated in Fig. 3a–c.

In negative switchers, the excitation wavelength also induces off-switching (Figs. 1e and 3a). The coupling of the excitation and off-switching wavelength imposes a sequential recording scheme where, at first, an ensemble of molecules is switched to their emitting states. Subsequently, a spatially patterned light featuring a minimum in the center switches off all the molecules in the periphery and leaves only the ones in the center able to emit. At last, those are excited and their fluorescence is induced upon illumination with a third beam of the same wavelength (Fig. 3a). Negative RSFPs have varying off-switching kinetics affecting the dwell time used for imaging, from tens of μs to ms. For fast switchers like rsEGFP2 or Dronpa2, the dwell time is generally 300 μs in a point-scanning recording and 3 ms in a parallelized system where lower intensities are generally used. The most

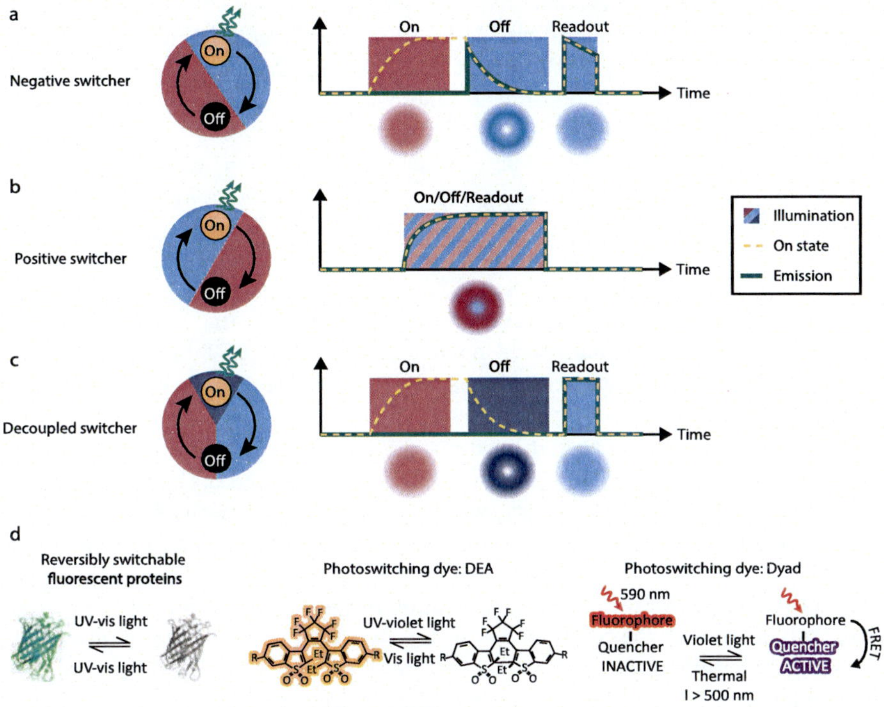

Fig. 3 Reversibly switchable probes. Example of photoswitching behavior and related effects on the pulse scheme used for RESOLFT. (**a**) Negative switchers are fluorescently excited by the same wavelength as that which causes them to switch off. (**b**) Positive switchers are fluorescently excited by the same wavelength as that which switches them on. (**c**) Decoupled switched are switched on and off and are excited by three separate wavelengths. (**d**) Example of used switching mechanisms for RESOLFT. In reversibly switchable fluorescent proteins, the switch is linked to a *cis/trans* isomerization combined with a protonation/deprotonation step or by a rearrangement of the hydrogen network. Among the photoswitching dyes, some of the molecular mechanisms used are either directly switching the dye between an "open" and a "closed" form or indirectly using this switching to enable or impair the fluorescence of a covalently-linked dye. The molecular structure and schemes are readapted from the original publications [59, 60]

common types of negative switchers have excitation maxima in the blue (~488 nm) and on-switching in the UV–violet. The ability to completely cycle a population within hundreds of microseconds together with their ability to withstand thousands of switching cycles made rsEGFP2 [12] and Dronpa2 [42] fundamental for the development of the RESOLFT concept and the possibility to achieve the best trade-off between temporal and spatial resolution. Additional variants such as rsGreen [43] and rsFolder [44] were also developed to enhance the expression levels in mammalian and bacterial cellular systems. The need for new and brighter labels led to the further development of the GMars family [45, 46], whose biphasic photobleaching trend can be used for prolonged timelapse imaging. The development of the rsGamillus family [47], instead, extended the imaging to biological processes in more acidic pH compared to the physiological one. Furthermore, the use of photoswitching mechanisms outside the GFP superfamily and into the bacterial photoreceptors, such as the LOV domain of YtvA from *Bacillus subtilis*, adds to the list a smaller oxygen-independent alternative compatible with anaerobic environments [48].

For the positive switchers, the excitation wavelength is coupled to the on-switching path while the off-switching is driven with violet light (Fig. 3b). All the reported positive switchers, such as Padron [49] and Kohinoor [50], are in the green spectral range, with on-switching and excitation around 488–500 nm and off-switching in the UV–violet range. With the faster switching speed and the different dependency on the wavelengths exhibited by the newly developed Padron2, Konen et al. demonstrated point-scanning RESOLFT imaging with illumination patterns superimposed both spatially and temporally and with pixel dwell times of ~500 μs [51].

For the decoupled switchers, the switching is fully decoupled from the excitation wavelength (Fig. 3c). The RSFP Dreiklang is switched on and off by illumination at ~365 nm and ~405 nm, respectively, while the excitation wavelength is at ~515 nm [52, 53]. In contrast to the other groups, the switching is the result of a reversible hydration/dehydration reaction that modifies the chromophore. In terms of imaging, these switching properties allow maximization of the photon budget of the protein, potentially increasing the SNR. However, the switching speed of current decoupled switchers is significantly slower and has so far required pixel dwell times of ~20–50 ms, thus liming their use for dynamic live-cell imaging.

The reported switching mechanisms are predominantly in the green region of the spectra and require the use of UV–violet light to a varying extent. To reduce any potentially phototoxic effects of the shorter wavelengths and thereby enhance the live-cell compatibility of RESOLFT, there is a continuous push to develop probes in the red region of the spectra. Examples of variants successfully used in RESOLFT are asFP595 [10] and rsCherryRev1.4 [54]. They are negative switchers, with excitation maxima in the orange (~590 nm) and a broad on-switching range between UV–violet and green (400–510 nm). The newly reported red fluorescent protein variants of rsFusionRed show a sevenfold decrease in off-switching time, allowing imaging speeds compatible with living cells and enough photostability to follow their dynamics over multiple frames [55].

Different photoswitching strategies have been identified also in synthetic dyes (Fig. 2d). The covalently-linked dye pair of Cy3 and Alexa647 can act as an organic photoswitcher, using the long-lived non-fluorescent state of the dye in a thiol buffer solution and the possibility to revert it by a different wavelength of light [56]. Fluorescence bis-sulfone diarylethenes have also been used for RESOLFT [57–59]. These compounds present an "open" and "closed" chemical form of the chromophore and upon illumination with UV–visible light (350–405 nm), they switch to the fluorescent isomer. Alternatively, the fluorescence of a dye can be modulated by a covalently-linked photoswitchable quencher [60]. This strategy employs a different class of photoswitchers called spironaphthoxazines, which can switch from a yellow spiro form to a deep violet merocyanine form upon 405 nm illumination, enabling Förster energy transfer and consequently quenching of the second fluorophore. Even if brighter than fluorescent proteins, the overall performance of synthetic dyes in RESOLFT microscopy is limited. The major challenges are still photobleaching, reduced number of cycles, and often the need for UV light or a specific environment for optimized switching. Nevertheless, the possibility to open up the technique to new labeling approaches for living cells (like SNAP-Tag or HaloTag technologies) is an area of interest in the field.

From the hardware point of view, the nature of the transitions used in RESOLFT eases the requirement of power and time compared to STED microscopes, and thus the requirements of the lasers. The much lower peak intensities (kW–W/cm^2) required for molecular switching are compatible with continuous wave (CW) lasers. A key technical aspect is also the possibility to digitally modulate the illumination on the μs-ms timescale, achievable both through commercial CW lasers or acuosto-optic modulators. The different pulse sequences that have been implemented in RESOLFT are summarized in Fig. 3.

3 Adaptive and Smart Scanning of the Illumination

In a point-scanning system, an image is acquired by scanning a probing beam over the FOV to be imaged, point-by-point and line-by-line (Fig. 4a). The residence time of the beam in each pixel is called the dwell time, t_s, while the step size from one point to the other (corresponding to the final pixel dimension) is called the scanning step, Δx. In the context of a single-point-scanning system, the imaging speed is determined by the number of scanning points, n_s, required to construct the image and the dwell time required to measure at each scanning position. The number of scanning points in turn depends on the scanning step size and the size of the imaged field of view (l_{fov}), while the dwell time required to collect sufficient signal depends on the illumination power, switching kinetics, brightness, and labelling density. In most traditional systems, the aforementioned imaging parameters, n_s, t_s, and l_{fov}, are set to fixed values before image acquisition begins. This means that the same parameters will be used for every scanning point, regardless of the sample properties at that point. Consider, for example, a large sample area in which labeled structures of interest are sparsely dispersed. Large regions of the area will not contain structure and thus not generate any signal, but will still be imaged using the same parameters as the regions containing structure. This results in unnecessarily long acquisition times and large total illumination dose also to samples containing small or sparse structures (Fig. 4a). Smart scanning or adaptive illumination systems feature a more flexible and sample-dependent scanning scheme, where the dwell times or illumination intensities are adjusted depending on the properties of the sample in the currently scanned region (Fig. 4b–c).

Adaptive illumination has been used to decrease the illumination dose of STED microscopy (Fig. 4b). In adaptive illumination STED microscopy, the power of the STED laser is modulated during the scan in response to the structure of the sample.

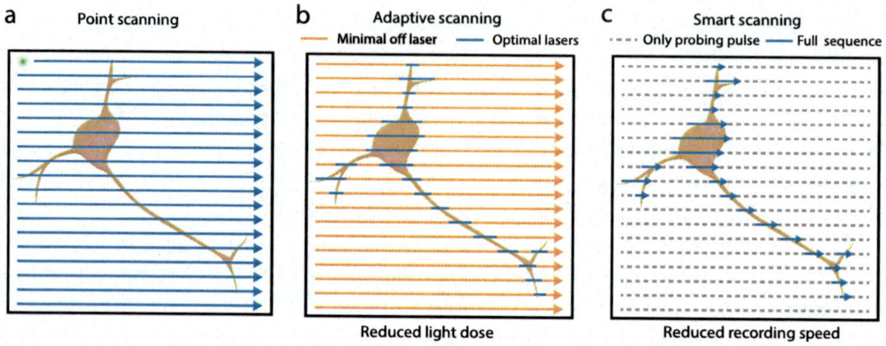

Fig. 4 Point-scanning approaches. (**a**) Conventional point scanning. By scanning a probing focal spot over the sample in a raster-scanning geometry and detecting the amount of fluorescence generated at each point, a point-scanning microscope builds up an image of the sample pixel-by-pixel. According to the Nyquist-Shannon criteria, the scan step should be no more than half the expected resolution. Therefore, if a 10× resolution increase is expected over a diffraction-limited image, the scanning steps required are 100× more in a 2D image. This leads to a significantly slower recording time and increased light dose on the sample. The size of the probing focal spot determines the level of detail i.e. the resolution of the image. (**b, c**) Point-scanning microscopy can be made more efficient by implementing adaptive or smart scanning approaches. (**b**) Adaptive scanning refers to the adaptation of illumination intensities to the sample structure. A weak intensity is used as an initial probe. When sample structure is detected, the illumination powers are increased to optimize image quality at regions containing structure of interest. (**c**) Smart scanning approaches also adapt the temporal pulse sequence at each pixel to the sample, essentially skipping over pixels that do not generate significant signal when probed with a probing pulse. Adaptive scanning minimizes the illumination dose on the sample and smart scanning additionally decreases the required scanning time

The sample is scanned with a diffraction-limited excitation spot with constant intensity. As long as the excitation spot does not generate fluorescence, no STED illumination is applied. When fluorescence exceeds a certain threshold the STED beam is activated (RESCue-STED) [61]. In DyMIN (Dynamic Intensity Minimum), the STED beam is instead switched on at increasing power the higher the fluorescence detected, and therefore the structure of interest more centered in the recorded area. This strategy minimizes the STED illumination, using the STED beam only when and at the intensity needed to maximize information instead of always at the maximum level independently from the characteristics of the sample. DyMin has been tested on periodic actin structures as well as the gephyrin clusters at inhibitory synapses of rat hippocampal neurons [14], with a reduction of total light dose on the sample up to 45-fold. Moreover, the more efficient illumination scheme allows the use of longer pixel dwell times without bleaching, enabling imaging at higher signal-to-noise ratio compared to conventional scanning.

When the regions of interest are within the scale of the diffraction limit, the concept of sample-dependent scanning can be adapted accordingly. In MINFIELD [62], a low-resolution image of a sparse sample is used to localize small substructures and define small regions of interest (~100 × 100 nm^2) that are sequentially imaged with the STED beam active. Since the region of interest containing the structure is smaller than the size of the STED beam (central zero FWHM ~ 280 nm), MINFIELD minimizes the exposure of fluorophores to STED illumination as they will not see the maximum intensity of the doughnut crest.

These approaches are efficient in decreasing the light dose to the sample. A further improvement toward live-cell compatibility is to speed up the acquisition with smart scanning approaches. Smart scanning implementations in RESOLFT involve using a short initial probing pulse to detect the presence of any fluorescent structure [13] in a specific scanning step. If the probing pulse generates a low signal, the region is ignored and the scan rapidly moves on to the next scanning position. If the probing pulse generates sufficient photons, the region is considered to be of interest, and the full RESOLFT pulse scheme is applied at the current scanning point (Fig. 4c). This smart scanning approach saves time in the acquisition by not dwelling on pixels that do not add information to the image, and the speed-up depends on the type of structure being imaged. The method reaches up to 6 times faster recordings and is an especially efficient approach when imaging, for example, neuronal protrusions or sparse point-like structures, where only a small fraction of the pixels (15%) in the final image will contain relevant information. Using smart scanning RESOLFT, Dreier et al. show peroxisome dynamics in small fields of view at imaging speeds of 2–5 Hz and mitochondria membranes at imaging speeds of 27 Hz, essential to remove motion artifacts. Furthermore, it was shown that the illumination dose could be decreased by 70–90% [13] since regions of the sample not containing labeled structures are not exposed to the RESOLFT imaging scheme but only to the probing illumination.

The use of fast scanning devices, like electro-optic deflectors (EODs), introduces the possibility of pixel hopping also in STED. FastRESCue STED [63] is a direct evolution of the RESCue-STED [61] technique. If no signal is detected, not only is the STED beam inactive but also the scan skips to the next pixel. This strategy increases the imaging speed by up to 4–5 times compared to RESCue-STED, with a decreased light dose of ~20% [63] in a biological sample compared to conventional STED.

4 Optimized Patterns and Adaptive Detection

The methods described above optimize the scanning strategies to improve image quality and speed while lowering illumination dose. Other methods instead focus on the illumination or detection side.

In the illumination, different spatial distributions of the STED beam have been explored. In particular, the use of one-dimensional depletion patterns instead of a two-dimensional doughnut shape allows for a more efficient depletion (due to steeper minima and consequently lower saturation light). A two-dimensional resolution improvement is then obtained by recording one-dimensional high-resolution images at multiple angles of the depletion pattern. For this reason, the strategy is called tomographic STED [64].

In the detection, by measuring the arrival time of photons at the detection site, information regarding their origin across the PSF can be inferred. Gated-STED is an implementation mainly applied to STED using CW depletion lasers [65]. The lower peak powers of CW lasers decreases the efficiency of stimulated emission and thus the STED beam has not had time to induce sufficient suppression of fluorescence at early times after the excitation pulse. With time-gated detection as in gated-STED, only the photons arriving in the later parts of the lifetime window and thus increasingly from the central parts of the PSF are used to construct the image, increasing the resolution. A similar concept is used in the SPLIT-STED implementation, where a phasor analysis is used to separate "high-resolution photons" (mainly from the central parts of the PSF) from "low-resolution photons" (mainly from the outer parts of the PSF) based on photon arrival times [66–68]. Optimized detection and photon assignment schemes like these enable high-resolution imaging to be performed with lower and less perturbing illumination intensities. Alternatively, the strategies can be used to improve the resolution without the need of increasing the illumination dose.

One additional strategy to compensate for photobleaching, therefore allowing long recording time, focuses on the probe and more specifically the labelling strategy. Combining the PAINT (Point Accumulation for Imaging in Nanoscale Topography) [69] and the STED approach, fluorogenic labels that transiently bind to their target structure can be used to obtain a constant exchange of labels between the solution and the target structure [70, 71]. As a consequence, the photobleached population of fluorophores is constantly exchanged with a new set of bright fluorophores, allowing prolonged STED recordings. The potential of this approach was demonstrated by extended timelapse and volumetric recording without degradation of the image quality. Alternatively, a photobleaching-immune approach has been developed for the study of the extracellular space of the brain with 3D STED microscopy. The entire extracellular fluid is labeled with a fluorescent marker to enable super-resolution shadow imaging (SUSHI [72]) of the extracellular space in living organotypic brain slices, and thus negative imaging of the complete cellular outlines. The large volume and continuous refreshing of fluorescent labels makes the technique unaffected by photobleaching and protected from phototoxicity since phototoxic molecules are produced outside of the cells.

5 Parallelization

Another way of increasing the imaging speed, especially for large FOV images, is through parallelization of the illumination and read-out (Fig. 5). In coordinate-targeted switching techniques, the parallelized acquisition means that the illumination confines and probes the fluorescence of the sample in multiple points at the same time. Using different types of light patterns, arrays of sharply confined emitting

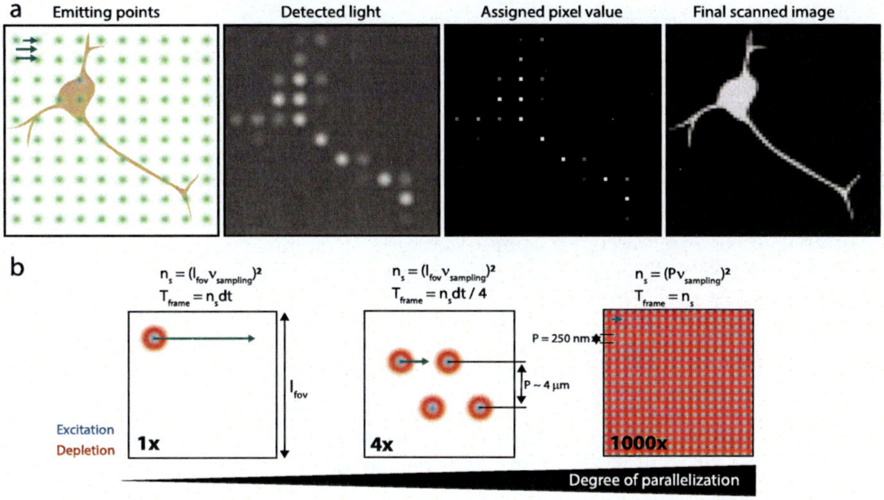

Fig. 5 From point scanning to parallelization. (**a**) Representation of a parallelized workflow. Parallelization of coordinate-targeted switching techniques relies on creating multiple emitting points in the sample that are imaged onto an extended sensor. If the emitting points are sufficiently far apart, the emission from the different points can be independently quantified and assigned to the corresponding pixel in the final image. Thanks to the parallelization, the number of scanning steps can be greatly reduced. (**b**) From a single scanning point, the degree of parallelization can be gradually increased by multiplexing the foci and doughnut patterns or pushed to the limit using large patterns generating tightly packed emitting spots separated only by a distance just above the diffraction limit

points are generated, each probing a spatially separate part of the sample. For arrays of emitting points covering the whole FOV, the scanning time is then decoupled from the FOV and instead defined by the distance between the individual illumination points. The array of emitting points needs to be read out at the same time but in spatially distinct pixels achievable either with multiple point detectors or using large camera sensors. The emission from each point is quantified by measuring the recorded fluorescence signal, which will be assigned to the corresponding pixel in the final image. Parallelized scanning has been proposed and implemented both in STED and RESOLFT systems.

The different characteristics of STED and RESOLFT introduce different challenges and limitations regarding the extent of parallelization. In STED, the scale of achievable parallelization is limited by the high intensity required for efficient depletion, while in RESOLFT the lower intensity allows for extended parallelization of the illumination. In both STED and RESOLFT, the temporal resolution of parallelized systems is to a certain extent limited by the current camera technology. Read-out times of state-of-the-art cameras are currently limited to a few milliseconds for the often required megapixel frames. In STED microscopy, the required pixel dwell time is usually only a few tens of microseconds. In point-scanning systems with practically instantaneous read-out, the pixel dwell time is thus usually also a few tens of microseconds. In a highly parallelized STED system, though, every scan step requires a camera frame to be read out, resulting instead in pixel dwell times on the millisecond time scale. Much of the gain from the reduced number of scan steps is therefore lost with the increased pixel dwell time. On the other hand, the switching cycle in RESOLFT systems requires several hundreds of microseconds or even milliseconds to achieve the on-state confinement in each scan step. The camera read-out time, therefore, increases the pixel dwell time by a relatively low amount and parallelization of RESOLFT systems can thus give a vast improvement in imaging speed compared to point-scanning systems. Nonetheless, even a parallelized RESOLFT system only outperforms a point-scanning STED system in terms of imaging speed if the side of a squared imaged region surpasses ~5–10 μm.

The use of cameras as detectors adds some additional complexity, but also offers more possibilities to the processing of the detected signal. In contrast to a single-point detector, which inevitably integrates all photons on the photosensitive area, a camera can capture images of the intensity distribution detected. This allows for different types of digital pinholing or more advanced statistical analysis of the detected signal.

5.1 Parallelized STED

STED has been demonstrated with different approaches and degrees of parallelization [73–75]. With four illumination spots at a distance of 5.7 μm, the first implementation could achieve only a moderate degree of parallelization. The moderate degree of parallelization however allowed for the continued use of point detectors, leading to an increased recording speed by four times compared to a single-point-scanning approach [74] (Fig. 5a). This however comes at the expense of system complexity and scalability of the approach. This together gives a four times faster system able to image structures down to ~35 nm (Fig. 5b, center).

A significantly higher degree of parallelization was achieved by introducing a new STED beam pattern exhibiting zero-intensity points over a large region and at a periodicity of 290 nm [73, 75] (Fig. 5b, right). Each zero-intensity point allows for highly confined emission creating a large array of emitting points. To detect the highly parallelized emission, point detectors are substituted by a camera sensor. As discussed, the slower reading time of a camera compared to a point detector in a STED microscope limits the pixel dwell time. For this reason, the 1,000-fold parallelization achieved in the implementations based on large interference patterns for depletion can increase the recording speed to seconds, but not to the theoretical recording time of 100 frames per second (potentially achievable considering the pixel dwell times of STED). Furthermore, to achieve such speeds, the extension of the parallelization requires high laser intensities to induce stimulated emission, demanding powerful laser sources or a smaller number of foci. The use of a spatial light modulator to create the optical lattice for the depletion could reach ~70 nm resolution over 3 × 3 μm^2 FOV [73]. Improvement on the laser source (low repetition rate and high pulse energy) could extend the FOV to 20 μm in diameter [75]. The continued development of laser and detector technology may further extend the current limit of parallelized STED.

5.2 *Parallelized RESOLFT*

In contrast to STED microscopy, the molecular switching required for RESOLFT microscopy leads to pixel dwell times on the order of hundreds of microseconds to milliseconds. Since this is on par with the speed of fast exposure times using state-of-the-art sCMOS cameras, significant gains can be made in terms of imaging speed by parallelizing the raster-scanned acquisition of RESOLFT microscopes. Many different implementations of parallelized RESOLFT exist, all aiming at achieving high degrees of parallelization by using illumination patterns covering large areas of the sample (up to 100 × 100 μm^2) [76–79].

5.2.1 Widefield-RESOLFT

The implementation of parallelized RESOLFT that gives the highest degree of parallelization uses homogeneous widefield illumination for the on-switching and read-out step of the pulse scheme and a superposition of orthogonal sinusoidal patterns in the off-switching step to achieve the confined emission. This implementation is termed Widefield-RESOLFT (wf-RESOLFT) [76] and is illustrated in Fig. 6. The distribution of confined emitting points in wf-RESOLFT is solely

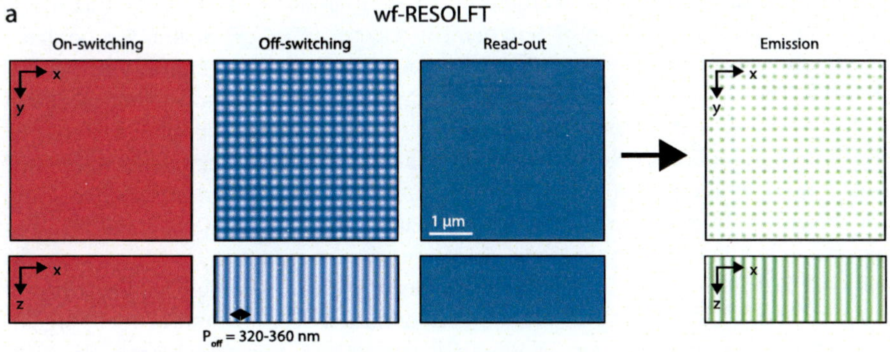

Fig. 6 wf-RESOLFT patterns. A high degree of parallelization can be achieved by switching on the whole FOV with a widefield illumination. A patterned illumination exhibiting an array of zeros switches off everything except the fluorophores residing in these zero-intensity points. A final widefield illumination excites fluorescence from the remaining on-state fluorophores. Scale bar: 1 μm

determined by the geometry of the off-switching pattern. To achieve sharply confined on-state points in a short time, it is beneficial to use an off-switching pattern with a steep intensity gradient surrounding the intensity zeroes. The shorter the periodicity of the pattern, the lower the intensity at its peaks is needed to achieve the same confinement [75, 80]. As an additional benefit, the contraction of the periodicity translates into an increased speed of recording, although it comes with the drawback of cross-talk between neighboring emitting spots, whose PSFs will start to overlap in the detected image. Overlapping PSFs on a parallelized detector can be separated using computational unmixing or deconvolution algorithms [76, 77], although the presence of noise will quickly deteriorate the image quality. In practice, this limits the accessibility of the technique to bright and flat samples, while axially extended samples where out-of-focus emission has a larger contribution or signal-to-noise is low are difficult to image. Balancing all these aspects, wf-RESOLFT has been implemented using off-switching patterns with a periodicity, P, of 320–360 nm [46, 76, 77], and with a massive improvement in throughput covering a FOV of 100 μm in diameter. Overall, the time of the RESOLFT pulse scheme multiplied by n_s (now defined as $(P/d_s)^2$ instead of $(l_{fov}/d_s)^2$ as in point scanning) leads to a frame time of around 3 s for a $l_{fov} = 100 \times 100\ \mu m^2$. The low illumination powers required for RESOLFT imaging makes wf-RESOLFT a suitable method for acquiring timelapse recordings of large and flat samples.

5.2.2 MoNaLISA or Multifoci Parallelized RESOLFT

To extend the parallelized approach to thicker 3D samples, the influence of out-of-focus emitting regions and the cross-talk between in-focus emitters needs to be minimized or eliminated. This has been achieved by independently modulating the periodicity of the on-switching and read-out illumination, named P_{on} and P_{ro}, from the one of the off-switching illumination, P_{off}, in the RESOLFT pulse scheme (Fig. 7). In practice, this can be done by creating an array of intensity maxima in the focal plane with the use of microlenses and relay optics. Spatially modulated on-switching light switches on primarily the fluorophores residing at these points (Fig. 7). The co-alignment of the intensity maxima of the on-switching multifoci with every minimum of the off-switching array generates sharply confined nanosized regions of fluorophores (Fig. 7). The additional multifoci read-out pattern co-aligned with the on-switching pattern is used to excite fluorescence from these spatially confined on-state molecules. The multifoci pattern in the on-switching and read-out step not only allows for better signal detection from in-focus emitters but also gives the system improved optical sectioning. Indeed, the non-overlapping PSFs of the emitting points allow for digital pinholing to discard most of the out-of-focus signal. Additionally, the combination of illumination patterns creates emission primarily from the focal plane of the system (Fig. 8a), both due to the two-step process of

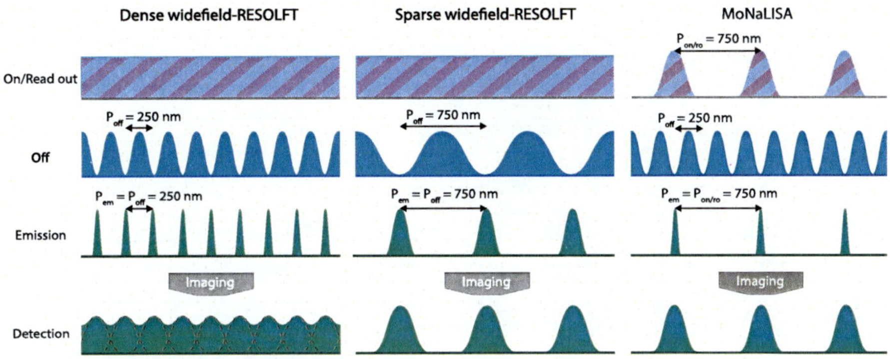

Fig. 7 wf-RESOLFT and MoNaLISA. The wf-RESOLFT configuration forces a trade-off between the ability of the off-switching pattern to confine the emission and the distance between adjacent emitting points. An off-switching pattern with high modulation will create sharply confined emitting spots, but at such a short distance from each other that they will be hard to distinguish in the diffraction-limited detection. Relaxing the modulation of the off-switching pattern will bring the emitting points further apart but will also compromise the confinement of the emitting spots. The MoNaLISA scheme alleviates this problem by decoupling the modulation of the off-switching pattern from the distance between adjacent emitting spots

on-switching and read-out but also since much of the out of focus on-switched molecules will be switched off by the off-switching pattern.

Having decoupled periodicities ($P_{on} = P_{ro} \neq P_{off}$) allows shrinking P_{off} down to the diffraction limit ~250 nm, without affecting the cross-talk of the emitting spots, which is defined by P_{on}. The MoNaLISA (Molecular Nanoscale Live Imaging with Sectioning Ability [78]) system delivers fast high-resolution imaging of living cells by balancing the trade-off between imaging speed and image quality and by providing optical sectioning through axially confined emission and digital pinholing. Fields of view of 50 × 50 μm^2 are recorded at around 1 Hz inside the three-dimensional structures of the cell. The enhanced contrast gained by the full integration of the emitting volume enables the use of fast and resistant RSFPs that due to lower brightness are generally difficult to image in other parallelized systems (Fig. 8b). Dim signal, like the endogenous level of expression of vimentin tagged with rsEGFP2 protein, can be visualized thanks to the improved contrast of MoNaLISA. In Fig. 8c, the contrast and resolution of MoNaLISA images are compared with widefield and enhanced confocal (using multifoci arrays without the off-switching pattern) images of the same structures.

5.2.3 3D Parallelized RESOLFT

Although the multifoci arrays used in the MoNaLISA microscope provide optical sectioning abilities, the axial resolution is still diffraction limited. The off-switching pattern used in these implementations does not provide any additional confinement

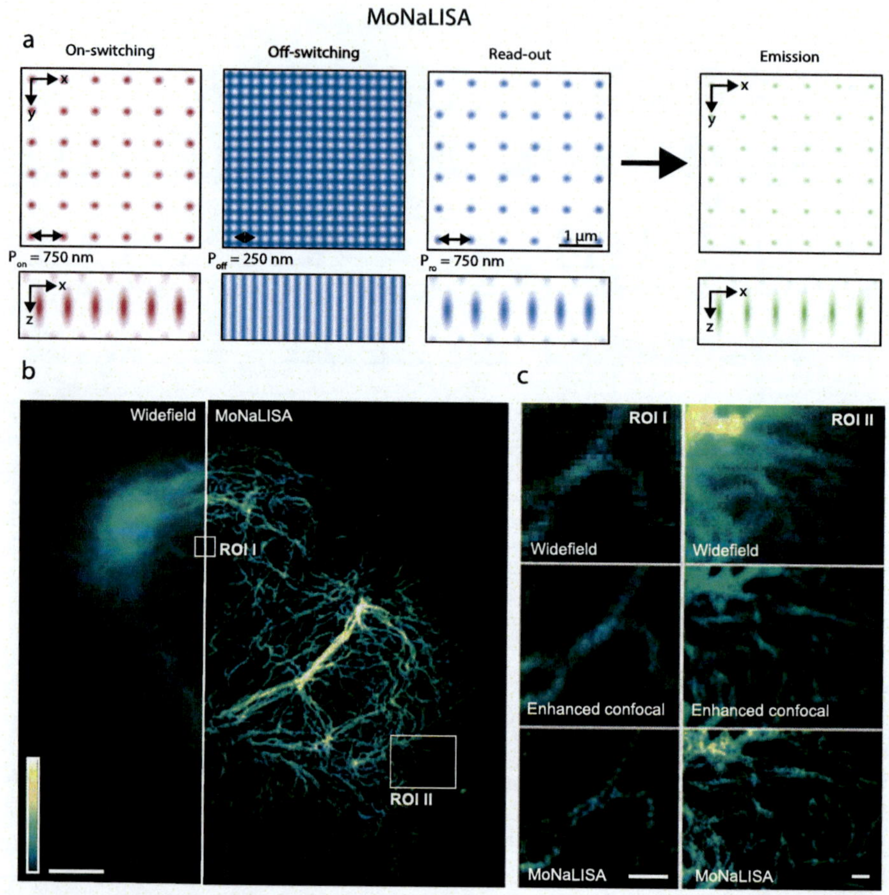

Fig. 8 MoNaLISA patterns and data. (**a**) In the MoNaLISA scheme, additional patterns are added into the pulse scheme causing the on-switching and read-out light to be confined to a sparser subset of zero-points in the off-switching pattern and also inducing an axially confined emission. (**b**) Comparison of a traditional widefield image with MoNaLISA image of U2OS cell endogenously expressing Vimentin-rsEGFP2. Scale bar: 5 µm. (**c**) Zoom-ins of ROI I and II showing a comparison of the resolution and contrast achieved with widefield imaging, enhanced confocal (using multifoci arrays without the off-switching pattern), and the full MoNaLISA scheme. Scale bars: 500 nm. Image adapted from Masullo et al. [78]

along the axial direction. Three-dimensional super-resolution has however been demonstrated using a more complex off-switching pattern that exhibits three-dimensionally confined zero-intensity points [79]. An example of a 3D-confining pattern is shown in Fig. 9a. This pattern, in combination with the multifoci array used for on-switching and read-out, creates emitting volumes that are confined in all three spatial dimensions. Probing the spatial distribution of fluorophores in 3D

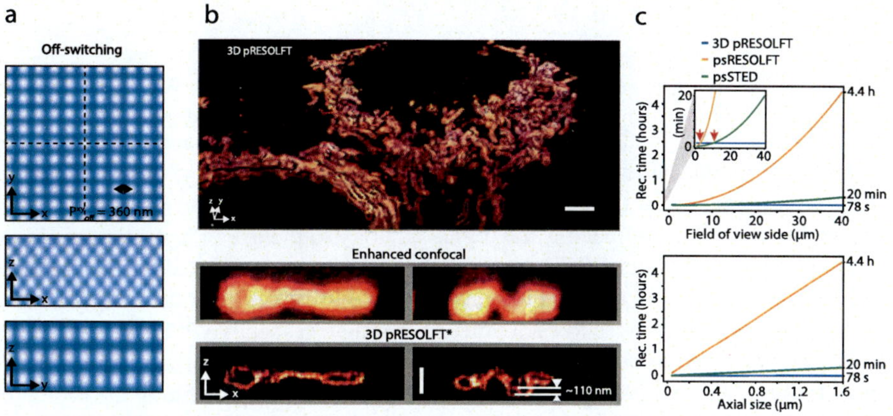

Fig. 9 3D pRESOLFT pattern and data. (**a**) Extension to 3D super-resolution can be achieved by creating an off-switching pattern that exhibits zero-volumes confined in all three dimensions, creating a 2D lattice structure in both the x–y, x–z and y–z planes. (**b**) Full volumetric recording of a mitochondrial network, together with an x–z slice of a single mitochondria compared with an enhanced confocal recording of the same structure showing the drastically improved resolution in all dimensions. Cells are U2OS expressing rsEGFP2-Omp25. Star (*) denotes deconvolved data. Scale bars: 2 μm (top) and 500 nm (bottom). (**c**) Graphs showing approximate imaging time taken for volumetric imaging using psRESOLFT, psSTED, and 3D pRESOLFT. The top graph assumes a fixed axial size of 1.6 μm with increasing lateral FOV. The bottom graph assumes a fixed 40 × 40 μm^2 lateral FOV with increasing axial size. Calculations assume standard pixel dwell times of the different techniques (psSTED: 30 μs, psRESOLFT: 400 μs, 3D pRESOLFT: 6 ms) and a scan step size of 50 nm

requires the sample to be raster scanned along the x-, y- and z-dimension with a scanning step satisfying the Nyquist-Shannon sampling criterion.

Examples of improved axial resolution on a point-scanning architecture have been proved by utilizing interference between two opposite objectives (isoSTED [81, 82], 4Pi-RESOLFT [83]) or using axial depletion patterns in a single objective configuration (3DSTED [9, 84]), shrinking the PSF down to 30–50 nm. This 3D parallelized RESOLFT (3D pRESOLFT) implementation has been shown to achieve a spatial resolution below 80 nm in all three dimensions. As such, it is a powerful tool for measuring the lateral and axial spatial distribution of proteins in micrometer thick sections of axially extended samples as shown in volumetric images of mitochondria of U2OS cells presented by Bodén et al. [79] and shown in Fig. 9b. Parallelization on a volumetric recording translates into a 200-fold improvement in the speed of recording, comparing the same RESOLFT dwell time (from 4.4 h for point-scanning RESOLFT to 78 s in 3D pRESOLFT, Fig. 9c).

5.3 *Digital Pinholing of Camera Data*

When multiple emitting points are imaged onto a camera, a consecutive step is needed to quantify the emissions from the emitting points and assign them to the correct pixels in the final image. The most straightforward approach is the use of a binary mask on the camera frame, aligned to the a priori knowledge of the geometry of the emitting points. If the emitting points are sufficiently far apart compared to the diffraction limit, this approach is the digital analogy to the use of a physical pinhole in a point-scanning system. This approach has been used in wf-RESOLFT systems with a Gaussian pinhole function instead of a binary mask [76]. If the images of the emitting points are too close to each other so that their respective PSFs overlap, very small pinholes need to be used in order to minimize cross-talk (around one-fifth of the Airy disc). Alternatively, the cross-talk can be unmixed by defining the problem as a system of linear equations to be solved through least-squares minimization or maximum likelihood estimations [75, 77]. High cross-talk between emitting regions reduces the total signal extractable from each spot. Alternatively, when the emitting points are non-overlapping, as in MoNaLISA and 3D pRESOLFT systems, they can be considered independent from each other. Therefore, each camera region can then be processed individually and the signal can be fully extracted and separated from the background contribution.

6 STED and RESOLFT for Live-Cell Imaging

STED and RESOLFT microscopy are versatile imaging tools that can be used for a plethora of different imaging applications in material and biological sciences including plants, bacterial, and mammalian systems. Their contribution to these disciplines is extensively covered in many recent reviews [85–87]. Here we limited our attention to the new strategies introduced in coordinate-targeted nanoscopy to push their ability toward live-cell imaging at the nanoscale. Within this goal, three aspects assume a central role: temporal resolution, phototoxicity, and cell accessibility. We summarized the techniques described in the previous paragraphs according to these metrics to give a comparative view of them (Table 1).

Since biological processes take place on many time scales, from milliseconds to hours, following sub-second dynamics without sacrificing spatial resolution or the overall cellular context represents an important challenge in biological imaging. STED microscopy, providing instantaneous nanoscale confinement, reaches extremely short pixel dwell times ($<$100 μs). Though parallelized approaches have been demonstrated, the most common implementation of STED still uses a single scanning point due to the requirements on power density and limitations imposed by current camera technology. This intrinsically links the temporal resolution of the technique to the size of the FOV recorded. For smaller fields of view, STED systems can reach tens of milliseconds temporal resolution enabling the study of fast

Table 1 Comparison of the super-resolution deterministic strategies for increased live-cell imaging compatibility

	Category	Technique name	Live-cell compatibility	Pixel dwell time	Frame time (n_s = 20 nm)	Light dose	Lateral resolution	Axial resolution	Specialized fluorophores	Key features	Ref
POINT SCANNING	Conventional	**psSTED**	Low	10–100 μs	20 μm in 10–100 s	50–900 kJ/cm^2 (775 nm)	~20–50 nm	~ 40–100 nm	No		
		psRESOLFT	Medium	100–500 μs	20 μm in 1–10 min	0.2 kJ/cm^2 (4, 405 nm)	~20–50 nm	~ 40–100 nm	Yes		
	Adaptive scanning	**RESCue**	Medium	X	20 μm in 10–100 s*	8 fold lower	50 nm	170 nm	No	Lower light dose	61
		Fast-RESCue	High	✓	20 μm in 2–20 s*	5–6 fold lower	50 nm	Not shown	No	Lower light dose, faster scanning	63
		MINFIELD	Medium	X	n/a	10–500 fold lower	20 nm	90 nm	No	Lower bleaching in small FOV	90
		DyMIN	Medium	X	20 μm in 10–100 s	~ 20 fold lower	29 nm	70 nm	No	Lower light dose	14
		SmartRESOLFT	High	✓	20 μm in 50–130 s	5 fold lower	40 nm	Not shown	Yes	Lower light dose, faster scanning	13
	Adaptive detection	**SPLIT-STED**	Medium	X	20 μm in 10–100 s	2 fold lower	60 nm	Not shown	No	Lower light dose	66,67
	Optimized pattern	**TomoSTED**	Medium	X	20 μm in 10–100 s	~ 2.5 fold lower	65 nm	Not shown	No	Lower light dose	64
	Refreshable probe/labeling	**SUSHI**	Medium	X	20 μm in 10–100 s	Same	50 m	200 nm	Yes	Prolonged imaging	72
		Exchangeable dyes STED	Medium	X	20 μm in 10–100 s	Same	60 nm	240 nm	Yes	Prolonged imaging	70
	Isotropic PSF	**isoSTED**	Low	X	20 μm in 10–100 s (1–8 min)	Same	50–60 nm	50–60 nm	No	Isotropic 3D resolution	81,82
		4Pi-RESOLFT	Medium	X	20 μm in 1–10 min (1–8 h)	Same	30–50 nm	30–50 nm	Yes	Isotropic 3D resolution	83
PARALLELIZED	STED-type	**4-points STED**	Low	✓	2.5–25 s	Same	35 nm	400–500 nm	No	Faster scanning	74
		1000-points STED	Low	X	20 μm in ~10 s	15 fold lower	30 nm	Not shown	No	Faster imaging on extended FOV	75
		SLM STED	Medium	X	2.8 μm in ~180 ms	Lower**	70 nm	Not shown	No	Faster imaging	73
	RESOLFT-type	**wfRESOLFT**	High	X	100 μm in ~ 6 s	Lower**	80 nm	Not shown	Yes	Faster imaging on extended FOV	76
		MoNaLISA	High	X	50 μm in ~ 5 s	Lower**	55 nm	320 nm	Yes	Faster volumetric imaging	78
		3D pRESOLFT	High	X	40 μm in ~ 8 s (7 min)	Lower**	70 nm	80 nm	Yes	Isotropic and faster imaging on extended FOV	79

(continued)

Table 1 (continued)

Table. Comparison of the super-resolution deterministic strategies for increased live-cell imaging compatibility.

The different techniques covered in the chapter are compared on the basis of several parameters. **Live-cell compatibility**: For live-cell compatibility we chose a scale from low to high, in which we considered both speed and light dose. In particular, fast and low-illumination techniques are considered as highly live-cell compatible (high), while slower and light-demanding techniques are identified as minimally live compatible (low). If only one of the two parameters is addressed by the technological development introduced, we define them as medium live-cell compatible. **Pixel dwell time:** Some of the reported techniques are based on shortening the average dwell time either based on decisions performed independently on each scanning step or multiplexing the reading in each scan step (✓), others maintain the dwell time fixed (✗). As a reference the typical dwell time for STED and RESOLFT are reported. **Frame time**: The time required to scan an image depends on the area to be scanned, the scan step, and the dwell time. For point-scanning approaches the area to be scanned corresponds to the FOV of interest, optical aberrations generally limit it to around $50 \times 50\ \mu m^2$, with reported optimized scanning up to $100 \times 100\ \mu m^2$ for STED microscopy[91]. The speed for the reported point-scanning techniques is assessed on a $20 \times 20\ \mu m^2$ FOV considering a scan step of 20 nm and a pixel dwell time of 10 – 100 μs for STED and 100–500 μs for RESOLFT. For the highly parallelized approaches, where the area scanned is decoupled from the FOV and only dependent on the periodicity of the pattern, the frame time and the achieved maximum FOV (throughput of the parallelization) is reported. For the techniques that have isotropic resolution (isoSTED, piRESOLFT, 3D pRESOLFT) the time needed to record a volume of $20 \times 20 \times 1\ \mu m^3$ with a scan step of 20 nm in all directions is reported. **Light dose:** Light doses are presented as comparative values, where RESOLFT implementations are compared to psRESOFLT and STED implementations to psSTED. The relative decrease reported is the one stated in the original publications. **Lateral and axial resolution:** The estimated resolution is reported in all dimensions if stated in the original publication. **Specialized fluorophores:** Some of the techniques rely on the use of specific labels or specialized labeling strategies. In particular, for RESOLFT, reversibly switchable fluorophores are required, while a transient binding is essential for STED techniques with exchangeable dyes. For SUSHI, a negative volumetric staining is used, instead of a direct link to a structure. **Key features:** Quick summary of the aspects that define the technique. **Ref:** Reference where the technique is reported and characterized.

*Frame time for (Fast-)RESCue STED is calculated considering a dwell time of 100 μs, while in the original publication a dwell time of 400 μs is generally considered, that would result in a frame time of 10 min and 2 min for RESCue and Fast-RESCue, respectively.

**The geometrical consideration on the repetitive off pattern vs single doughnut shape reported in Bergermann et al[75] translate to a varying degree to the other types of highly parallelized techniques, although the different implementations use different light patterns and spatial distributions for the different wavelengths in use. This is reflected in a complex change of the light dose to the sample, for an in-depth analysis on phototoxicity and the influence of the different wavelengths we refer to the original publications.

processes in small localized regions. RESOLFT on the contrary has intrinsically slower dwell time (>100 μs) but offers more modularity in recording strategy. The ability to work with lower intensities allows moving from single-point scanning to extended parallelized recordings.

The nanoscale resolution for both techniques also increases the need for sensitivity. Higher resolution ultimately means a decrease in the emitting volume and the number of fluorophores contributing to the signal. The universal nature of stimulated emission provides a great variety of fluorophores to choose from. Especially useful are the bright and photostable dyes that can push the resolving ability toward 20–30 nm. For RESOLFT, instead, the primary type of labels to date are the RSFPs that often offer a limited photon budget highly dependent on the illumination condition being used. When the system under study is dynamic, it is not only crucial to achieve high resolution on a snapshot of the sample, but also to describe its time evolution. The need for stable fluorophores to meet this demand is a common requirement among super-resolution techniques, as that photostability ultimately limits the sampling and the time window available.

Considerations on light damage in live-cell experiments are of great importance and a complex topic where many factors come into play. High irradiance on cells generally increases the risk of phototoxicity but how the light is distributed in space and time on the sample is also crucial to the effect. Slow scanning (i.e., long dwell time) induces more pronounced stress on the cells compared to fast scanning (i.e., short dwell time, compensated by line or frame averaging to reach a similar contrast [88]). The level of resolution achievable in the deterministic approach is tuned by the intensity of the off switching beam, which allows tuning of the power used according to the desired resolution. The smart scanning approaches push this concept of adaptability to the sample further by actively investigating the space and introducing a decision in the recording. In each step, a decision is taken whether to use the STED beam or not, or whether to stay or to move. Another parameter that can influence the level of phototoxicity in the sample is the wavelength of light in use, with red-shifted light being less toxic to the cells. Moving the STED beam toward the red has been crucial to increasing its live-cell compatibility. For RESOLFT almost all reported switching mechanisms involve the use of UV–violet light at different doses depending on the specific photophysics in use. The complex photophysics of some red RSFPs allows moving away from this requirement and completing the whole switching cycle with light above 500 nm. Active research into new switching mechanisms can further help to move away from this requirement. Especially for negatively switching RSFPs, the UV light dose is minimal and does not preclude imaging at different scales, from adherent cells to full organisms like *Drosophila* [89] or *C. Elegans* [13]. In general, RESOLFT microscopy with its significantly lower illumination intensities is a technique aimed primarily at live-cell imaging, allowing prolonged recordings of living cells and tissues. However, the performance of the technique is highly dependent on the use of a few select fluorescent markers. A broader adaptation to a wider range of biological questions is dependent on the continued development of new probes and illumination schemes.

7 Challenges and Outlooks

Coordinate-targeted switching approaches to super-resolution microscopy have in the last 20 years moved from being a niche technique, available only to highly specialized labs and for selected biological applications to being a technique widely used in life science institutes around the world. The psSTED microscope, for example, is today a workhorse in many imaging facilities. This is largely due to the versatility of the technique, easily extendable to multicolor and able to achieve high resolution in fixed, thin and thick samples. It also benefits from the possibility to use conventional dyes and to be added as an extra module onto a standard confocal microscope. However, for applications requiring long-term and minimally invasive monitoring of living cells, point-scanning systems often fall short due to the time taken to scan the image with a single point and the energy applied by focused light. Novel smart and adaptive scanning systems partially overcome this issue by optimizing the scanning trajectory and illumination sequence. Development in this direction is already available in some commercial microscopes and open-source initiative. RESOLFT microscopes are also commercialized, to some extent also with the parallelized recording (i.e., wfRESOLFT), even if these are less present in imaging facilities. The lack of commercial RESOLFT systems is likely due to the need for specific probes and that the technique is mainly targeted at live-cell imaging. Multicolor RESOLFT, though feasible, is also more challenging due to the need for multiple laser wavelengths for each RSFP requiring multiple lasers for multicolor imaging. Finally, the use of RESOLFT for live-cell imaging is still fairly recent, with the first successful results emerging only about 10 years ago.

Going from a single-point-scanning microscope to a parallelized system adds significant optical complexity to the setup but can provide orders of magnitude improvements in imaging speed and decrease of illumination dose. The added optical complexity of parallelized schemes arises particularly when different patterned illuminations are used in multiple steps of the illumination scheme to accommodate the need of specialized photoswitchable probes. In MoNaLISA and 3D pRESOLFT systems, three different illumination patterns need to be crafted and co-aligned to each other in order to generate high-quality images. Additionally, as in single-point-scanning approaches, the quality of the zeros in the off-pattern needs to be close to perfect in order to achieve sharp on-state confinement. Refractive index variations or other optical imperfections in the sample may cause aberrations to the illuminating beams distorting the geometry or degrading the quality of the zero-intensity points. The issues become more severe when imaging deeper into the sample. Imaging tens of micrometers inside tissue is thus challenging even with the improved optical sectioning provided by these systems. Future implementation of adaptive optics may provide solutions to restore the quality of the wave front inside the sample enabling high-resolution imaging further inside tissues. Creating patterns with controllable shape and maintained quality throughout thicker samples may also allow for even faster volumetric imaging with parallelization also along the axial dimensions.

Highly parallelized imaging schemes often rely on cameras for detection. Although they provide the properties that make this type of imaging possible, they also limit the systems in some aspects. One such limitation comes from the relatively long time required to read out the data from a camera frame. Even state-of-the-art sCMOS cameras require several milliseconds to read out a full camera frame (~4 MP). Since every scan step in a parallelized scanning sequence requires a frame to be read out, the camera read-out time is often a major bottleneck in the data acquisition scheme. In parallelized RESOLFT schemes, the camera read-out time usually accounts for about 20–50% of the pixel dwell time [55, 78, 79]. In parallelized STED systems, this relative time may be significantly higher due to the fast timescale of stimulated emission allowing for microsecond integration times. Cameras also introduce other challenges stemming from their noise properties. Not only do they exhibit non-zero amounts of read-out noise (not present in APDs), but the use of cameras can also introduce additional pixel-dependent noise. Considerations need to be taken to handles all these aspects and future camera and detector development will likely contribute greatly to further push these techniques in terms of image quality and speed.

Another challenge likely to be undertaken is to extend these super-resolution techniques to multi-channel or multicolor imaging. STED microscopy is commonly implemented in multicolor imaging of up to four colors [92]. RESOLFT, however, is still limited to the library of existing RSFPs. Finding multiple spectrally separated RSFPs with switching kinetics that allow for fast imaging remains a challenge. Additionally, if the different RSFPs are switched by different wavelengths, the illumination patterns need to be created and aligned in multiple beam paths adding again to the complexity of the optical system. Other ways to generate multicolor RESOLFT images have been proposed as being preferable, for example, by separating different RSFP labels by their switching kinetics [93, 94]. This has the benefit of requiring only one spectral illumination and detection channel. However, it still requires RSFPs that differ not in spectral but in photoswitching behavior, more specifically fast and slow off-switching kinetics. As a consequence, the imaging speed is limited by the switching speed of the slower switcher. Utilizing both spectral and dynamic parameters to separate channels may allow future implementations with three or more separable channels. More in general, the development of new switchable proteins in different regions of the spectra, with different photophysical behavior and with an added sensing module (e.g., calcium [95, 96] or pH [97]) will likely pave the way for new technical innovations and applications.

In this chapter, we covered the strategies used in deterministic super-resolution techniques to reach higher compatibility toward fast live-cell imaging. The deterministic coordinate-targeted switching approaches compose a subset of super-resolution fluorescence microscopy techniques. The distinction between coordinate-targeted approaches and stochastic localization techniques has however narrowed as techniques are emerging combining stochastic switching with scanning systems and illumination patterns. The concept was first introduced with the MINFLUX technique [98–100] and has been further adapted in other forms and to other applications [101–104]. The successful utilization of well-designed illumination strategies together with accurate knowledge of photophysical behavior will likely be the key to further refinement of optical systems for the interrogation of biological structures and dynamics. The continued improvement of such systems is essential for the advancement of biological research in the life sciences.

References

1. Abbe E (1873) Beiträge zur Theorie des Mikroskops und der mikroskopischen Wahrnehmung. Arch Mikrosk Anat 9(1):413–468
2. Hell SW, Kroug M (1995) Ground-state-depletion fluorscence microscopy: a concept for breaking the diffraction resolution limit. Appl Phys B Lasers Opt 60:495–497
3. Hell SW, Wichmann J (1994) Breaking the diffraction resolution limit by stimulated emission: stimulated-emission-depletion fluorescence microscopy. Opt Lett 19:780–782
4. Hell SW (2007) Far-field optical nanoscopy. Science 316:1153–1158
5. Hess ST, Girirajan TPK, Mason MD (2006) Ultra-high resolution imaging by fluorescence photoactivation localization microscopy. Biophys J 91:4258–4272
6. Rust MJ, Bates M, Zhuang XW (2006) Sub-diffraction-limit imaging by stochastic optical reconstruction microscopy (STORM). Nat Methods 3:793–795
7. Betzig E et al (2006) Imaging intracellular fluorescent proteins at nanometer resolution. Science 313:1642–1645
8. Hell SW, Wichman J (1994) Breaking the diffraction resolution limit by stimulated emission: stimulated-emission-depletion fluorescence microscopy. Opt Lett 19:780–782
9. Klar TA, Jakobs S, Dyba M, Egner A, Hell SW (2000) Fluorescence microscopy with diffraction resolution barrier broken by stimulated emission. Proc Natl Acad Sci 97:8206–8210
10. Hofmann M, Eggeling C, Jakobs S, Hell SW (2005) Breaking the diffraction barrier in fluorescence microscopy at low light intensities by using reversibly photoswitchable proteins. Proc Natl Acad Sci 102:17565–17569
11. Grotjohann T et al (2011) Diffraction-unlimited all-optical imaging and writing with a photochromic GFP. Nature 478:204–208
12. Grotjohann T et al (2012) rsEGFP2 enables fast RESOLFT nanoscopy of living cells. eLife 2012:1–14
13. Dreier J et al (2019) Smart scanning for low-illumination and fast RESOLFT nanoscopy in vivo. Nat Commun 10:556
14. Heine J et al (2017) Adaptive-illumination STED nanoscopy. Proc Natl Acad Sci 114:9797–9802
15. Sahl SJ, Schönle A, Hell SW (2019) Fluorescence microscopy with nanometer resolution: nanoscale resolution in far-field fluorescence microscopy. In: Hawkes PW, Spence JCH (eds) Springer handbook of microscopy. Springer, pp 1089–1143. https://doi.org/10.1007/978-3-030-00069-1_22
16. Wang L, Frei MS, Salim A, Johnsson K (2019) Small-molecule fluorescent probes for live-cell super-resolution microscopy. J Am Chem Soc 141:2770–2781
17. Stockhammer A, Bottanelli F (2021) Appreciating the small things in life: STED microscopy in living cells. J Phys D Appl Phys 54:033001
18. Jeong S, Widengren J, Lee J-C (2021) Fluorescent probes for STED optical nanoscopy. Nano 12:21
19. Bordenave MD, Balzarotti F, Stefani FD, Hell SW (2016) STED nanoscopy with wavelengths at the emission maximum. J Phys D Appl Phys 49:365102
20. Vicidomini G, Moneron G, Eggeling C, Rittweger E, Hell SW (2012) STED with wavelengths closer to the emission maximum. Opt Express 20:5225

21. Hotta J et al (2010) Spectroscopic rationale for efficient stimulated-emission depletion microscopy fluorophores. J Am Chem Soc 132:5021–5023
22. Vicidomini G, Bianchini P, Diaspro A (2018) STED super-resolved microscopy. Nat Methods 15:173–182
23. Zheng Q, Lavis LD (2017) Development of photostable fluorophores for molecular imaging. Curr Opin Chem Biol 39:32–38
24. Tønnesen J, Nadrigny F, Willig KI, Wedlich-Söldner R, Nägerl UV (2011) Two-color STED microscopy of living synapses using a single laser-beam pair. Biophys J 101:2545–2552
25. Matlashov ME et al (2020) A set of monomeric near-infrared fluorescent proteins for multicolor imaging across scales. Nat Commun 11:1–12
26. Willig KI, Stiel AC, Brakemann T, Jakobs S, Hell SW (2011) Dual-label STED nanoscopy of living cells using photochromism. Nano Lett 11:3970–3973
27. Han KY et al (2009) Three-dimensional stimulated emission depletion microscopy of nitrogen-vacancy centers in diamond using continuous-wave light. Nano Lett 9:3323–3329
28. Lesoine MD et al (2013) Subdiffraction, luminescence-depletion imaging of isolated, giant, CdSe/CdS nanocrystal quantum dots. J Phys Chem C 117:3662–3667
29. Liu Y et al (2017) Amplified stimulated emission in upconversion nanoparticles for super-resolution nanoscopy. Nature 543:229–233
30. Tosheva KL, Yuan Y, Matos Pereira P, Culley S, Henriques R (2020) Between life and death: strategies to reduce phototoxicity in super-resolution microscopy. J Phys D Appl Phys 53: 163001
31. Wäldchen S, Lehmann J, Klein T, van de Linde S, Sauer M (2015) Light-induced cell damage in live-cell super-resolution microscopy. Sci Rep 5:15348
32. Erdmann RS et al (2019) Labeling strategies matter for super-resolution microscopy: a comparison between HaloTags and SNAP-tags. Cell Chem Biol 26:584–592.e6
33. Keppler A et al (2003) A general method for the covalent labeling of fusion proteins with small molecules in vivo. Nat Biotechnol 21:86–89
34. Los GV et al (2008) HaloTag: a novel protein labeling technology for cell imaging and protein analysis. ACS Chem Biol 3:373–382
35. Gautier A et al (2008) An engineered protein tag for multiprotein labeling in living cells. Chem Biol 15:128–136
36. Arsić A, Hagemann C, Stajković N, Schubert T, Nikić-Spiegel I (2022) Minimal genetically encoded tags for fluorescent protein labeling in living neurons. Nat Commun 13:314
37. Bottanelli F et al (2016) Two-colour live-cell nanoscale imaging of intracellular targets. Nat Commun 7:1–5
38. Moneron G et al (2010) Fast STED microscopy with continuous wave fiber lasers. Opt Express 18:1302
39. Schneider J et al (2015) Ultrafast, temporally stochastic STED nanoscopy of millisecond dynamics. Nat Methods 12:827–830
40. Shcherbakova DM, Sengupta P, Lippincott-Schwartz J, Verkhusha VV (2014) Photocontrollable fluorescent proteins for superresolution imaging. Annu Rev Biophys 43: 303–329
41. Nienhaus K, Nienhaus GU (2016) Chromophore photophysics and dynamics in fluorescent proteins of the GFP family. J Phys Condens Matter 28:443001
42. Stiel AC et al (2007) 1.8 Å bright-state structure of the reversibly switchable fluorescent protein Dronpa guides the generation of fast switching variants. Biochem J 402:35–42
43. Duwé S et al (2015) Expression-enhanced fluorescent proteins based on enhanced green fluorescent protein for super-resolution microscopy. ACS Nano 9:9528–9541
44. El Khatib M, Martins A, Bourgeois D, Colletier JP, Adam V (2016) Rational design of ultrastable and reversibly photoswitchable fluorescent proteins for super-resolution imaging of the bacterial periplasm. Sci Rep 6:1–12
45. Wang S et al (2018) GMars-T enabling multimodal subdiffraction structural and functional fluorescence imaging in live cells. Anal Chem 90:6626–6634

46. Wang S et al (2016) GMars-Q enables long-term live-cell parallelized reversible saturable optical fluorescence transitions nanoscopy. ACS Nano 10:9136–9144
47. Shinoda H et al (2018) Acid-tolerant monomeric GFP from Olindias formosa. Cell Chem Biol 25:330–338.e7
48. Gregor C et al (2018) Novel reversibly switchable fluorescent proteins for RESOLFT and STED nanoscopy engineered from the bacterial photoreceptor YtvA. Sci Rep 8:2724
49. Andresen M et al (2008) Photoswitchable fluorescent proteins enable monochromatic multilabel imaging and dual color fluorescence nanoscopy. Nat Biotechnol 26:1035–1040
50. Tiwari DK et al (2015) A fast- and positively photoswitchable fluorescent protein for ultralow-laser-power RESOLFT nanoscopy. Nat Methods 12:515–518
51. Konen T et al (2021) The positive switching fluorescent protein Padron2 enables live-cell reversible saturable optical linear fluorescence transitions (RESOLFT) nanoscopy without sequential illumination steps. ACS Nano 15:9509–9521
52. Jensen NA et al (2014) Coordinate-targeted and coordinate-stochastic super-resolution microscopy with the reversibly switchable fluorescent protein dreiklang. ChemPhysChem 15:756–762
53. Brakemann T et al (2011) A reversibly photoswitchable GFP-like protein with fluorescence excitation decoupled from switching. Nat Biotechnol 29:942–950
54. Lavoie-Cardinal F et al (2014) Two-color RESOLFT nanoscopy with green and red fluorescent photochromic proteins. ChemPhysChem 15:655–663
55. Pennacchietti F et al (2018) Fast reversibly photoswitching red fluorescent proteins for live-cell RESOLFT nanoscopy. Nat Methods 15:601–604
56. Kwon J et al (2015) RESOLFT nanoscopy with photoswitchable organic fluorophores. Sci Rep 5:1–8
57. Nevskyi O et al (2018) Fluorescent diarylethene photoswitches—a universal tool for super-resolution microscopy in nanostructured materials. Small 14:1–12
58. Uno K et al (2021) Turn-on mode diarylethenes for bioconjugation and fluorescence microscopy of cellular structures. Proc Natl Acad Sci 118:e2100165118
59. Roubinet B et al (2016) Carboxylated photoswitchable diarylethenes for biolabeling and super-resolution RESOLFT microscopy. Angew Chem Int Ed 55:15429–15433
60. Frawley AT et al (2020) Super-resolution RESOLFT microscopy of lipid bilayers using a fluorophore-switch dyad. Chem Sci 11:8955–8960
61. Staudt T et al (2011) Far-field optical nanoscopy with reduced number of state transition cycles. Opt Express 19:5644
62. Göttfert F et al (2017) Strong signal increase in STED fluorescence microscopy by imaging regions of subdiffraction extent. Proc Natl Acad Sci 114:2125–2130
63. Vinçon B, Geisler C, Egner A (2020) Pixel hopping enables fast STED nanoscopy at low light dose. Opt Express 28:4516
64. Krüger J-R, Keller-Findeisen J, Geisler C, Egner A (2020) Tomographic STED microscopy. Biomed Opt Express 11:3139
65. Vicidomini G et al (2011) Sharper low-power STED nanoscopy by time gating. Nat Methods 8:571–573
66. Lanzanò L et al (2015) Encoding and decoding spatio-temporal information for super-resolution microscopy. Nat Commun 6:6701
67. Tortarolo G et al (2019) Photon-separation to enhance the spatial resolution of pulsed STED microscopy. Nanoscale 11:1754–1761
68. Wang L et al (2018) Resolution improvement in STED super-resolution microscopy at low power using a phasor plot approach. Nanoscale 10:16252–16260
69. Sharonov A, Hochstrasser RM (2006) Wide-field subdiffraction imaging by accumulated binding of diffusing probes. Proc Natl Acad Sci 103:18911–18916
70. Spahn C, Grimm JB, Lavis LD, Lampe M, Heilemann M (2019) Whole-cell, 3D, and multicolor STED imaging with exchangeable fluorophores. Nano Lett 19:500–505

71. Urban NT, Willig KI, Hell SW, Nägerl UV (2011) STED nanoscopy of actin dynamics in synapses deep inside living brain slices. Biophys J 101:1277–1284
72. Tønnesen J, Inavalli VVGK, Nägerl UV (2018) Super-resolution imaging of the extracellular space in living brain tissue. Cell 172:1108–1121.e15
73. Yang B, Przybilla F, Mestre M, Trebbia J-B, Lounis B (2014) Large parallelization of STED nanoscopy using optical lattices. Opt Express 22:5581
74. Bingen P, Reuss M, Engelhardt J, Hell SW (2011) Parallelized STED fluorescence nanoscopy. Opt Express 19:23716–23716
75. Bergermann F, Alber L, Sahl SJ, Engelhardt J, Hell SW (2015) 2000-fold parallelized dual-color STED fluorescence nanoscopy. Opt Express 23:211–211
76. Chmyrov A et al (2013) Nanoscopy with more than 100,000 'doughnuts'. Nat Methods 10: 737–740
77. Chmyrov A et al (2017) Achromatic light patterning and improved image reconstruction for parallelized RESOLFT nanoscopy. Sci Rep 7:44619
78. Masullo LA et al (2018) Enhanced photon collection enables four dimensional fluorescence nanoscopy of living systems. Nat Commun 9:1–9
79. Bodén A et al (2021) Volumetric live cell imaging with three-dimensional parallelized RESOLFT microscopy. Nat Biotechnol 39(5):609–618. https://doi.org/10.1038/s41587-020-00779-2
80. Rego EH et al (2012) Nonlinear structured-illumination microscopy with a photoswitchable protein reveals cellular structures at 50-nm resolution. Proc Natl Acad Sci U S A 109:1–9
81. Curdt F et al (2015) isoSTED nanoscopy with intrinsic beam alignment. Opt Express 23:30891
82. Siegmund R, Werner F, Jakobs S, Geisler C, Egner A (2021) isoSTED microscopy with water-immersion lenses and background reduction. Biophys J 120:3303–3314
83. Böhm U, Hell SW, Schmidt R (2016) 4Pi-RESOLFT nanoscopy. Nat Commun 7:10504
84. Harke B, Ullal CK, Keller J, Hell SW (2008) Three-dimensional nanoscopy of colloidal crystals. Nano Lett 8:1309–1313
85. Bond C, Santiago-Ruiz AN, Tang Q, Lakadamyali M (2022) Technological advances in super-resolution microscopy to study cellular processes. Mol Cell 82:315–332
86. Sahl SJ, Hell SW, Jakobs S (2017) Fluorescence nanoscopy in cell biology. Nat Rev Mol Cell Biol 18:685–701
87. Maglione M, Sigrist SJ (2013) Seeing the forest tree by tree: super-resolution light microscopy meets the neurosciences. Nat Neurosci 16:790–797
88. Kilian N et al (2018) Assessing photodamage in live-cell STED microscopy. Nat Methods 15: 755–756
89. Schnorrenberg S et al (2016) In vivo super-resolution RESOLFT microscopy of Drosophila melanogaster. elife 5:e15567
90. Göttfert F et al (2017) Strong signal increase in STED fluorescence microscopy by imaging regions of subdiffraction extent. Proc Natl Acad Sci 114:2125–2130
91. Alvelid J, Testa I (2020) Stable stimulated emission depletion imaging of extended sample regions. J Phys D Appl Phys 53:24001
92. Willig KI, Wegner W, Müller A, Clavet-Fournier V, Steffens H (2021) Multi-label in vivo STED microscopy by parallelized switching of reversibly switchable fluorescent proteins. Cell Rep 35:109192
93. Testa I, D'Este E, Urban NT, Balzarotti F, Hell SW (2015) Dual channel RESOLFT nanoscopy by using fluorescent state kinetics. Nano Lett 15:103–106
94. Valenta H et al (2021) Separation of spectrally overlapping fluorophores using intra-exposure excitation modulation. Biophys Rep 1:100026
95. Mishra K et al (2022) Genetically encoded photo-switchable molecular sensors for optoacoustic and super-resolution imaging. Nat Biotechnol 40(4):598–605. https://doi.org/10.1038/s41587-021-01100-5

96. Gielen V et al (2020) Absolute measurement of cellular activities using photochromic single-fluorophore biosensors. http://biorxiv.org/lookup/doi/10.1101/2020.10.29.360214. https://doi.org/10.1101/2020.10.29.360214
97. Seidel ZP, Wang JCK, Riegler J, York AG, Ingaramo M (2021) Relaxation sensors. Zenodo. https://doi.org/10.5281/zenodo.5810930
98. Balzarotti F et al (2017) Nanometer resolution imaging and tracking of fluorescent molecules with minimal photon fluxes. Science 355(6325):606–612
99. Eilers Y, Ta H, Gwosch KC, Balzarotti F, Hell SW (2018) MINFLUX monitors rapid molecular jumps with superior spatiotemporal resolution. Proc Natl Acad Sci 115:6117–6122
100. Gwosch KC et al (2020) MINFLUX nanoscopy delivers multicolor nanometer 3D-resolution in (living) cells. Nat Methods 17(2):217–224. https://doi.org/10.1101/734251
101. Cnossen J et al (2020) Localization microscopy at doubled precision with patterned illumination. Nat Methods 17(1):59–63. https://doi.org/10.1101/554337
102. Reymond L et al (2019) SIMPLE: structured illumination based point localization estimator with enhanced precision. Opt Express 27:24578
103. Jouchet P et al (2021) Nanometric axial localization of single fluorescent molecules with modulated excitation. Nat Photonics 15:297–304
104. Gu L et al (2019) Molecular resolution imaging by repetitive optical selective exposure. Nat Methods 16:1114–1118

Fluorescence Correlation Spectroscopy in Space and Time

Daniel Y. K. Aik and Thorsten Wohland

Contents

Abstract Fluorescence Correlation Spectroscopy (FCS) is a widely used technique to determine molecular dynamics and interactions. It uses observation volumes on the order of a femtolitre in size to distinguish the signal from single molecules against the background. As it is difficult to illuminate and specifically detect signals from such a small observation volume, FCS was originally conceived as a

D. Y. K. Aik
Department of Chemistry, National University of Singapore, Singapore, Singapore

Center for BioImaging Sciences, National University of Singapore, Singapore, Singapore

T. Wohland (✉)
Department of Chemistry, National University of Singapore, Singapore, Singapore

Center for BioImaging Sciences, National University of Singapore, Singapore, Singapore

Department of Biological Sciences, National University of Singapore, Singapore, Singapore
e-mail: twohland@nus.edu.sg

R. Šachl, M. Amaro (eds.), *Fluorescence Spectroscopy and Microscopy in Biology*, Springer Ser Fluoresc (2023) 20: 233–274, https://doi.org/10.1007/4243_2022_36,
Published online: 27 November 2022

single-spot measurement that measures mainly temporal information. Multiplexing was then achieved by sequential scanning and detecting different spots in a sample and thus also providing spatial information. With advances in technology, the introduction of different illumination and detection methods, and the emergence of super-resolution and light-sheet microscopy, new opportunities opened up to collect thousands of contiguous spots in a sample and thus provide high-resolution spatio-temporal information over a whole cross-section of a sample. This chapter describes the different 2D FCS modalities, their advantages and disadvantages, and some of their applications.

Keyword Fluorescence correlation spectroscopy · Image correlation spectroscopy · Imaging FCS · Scanning FCS

1 Introduction

Fluorescence Correlation Spectroscopy (FCS) combines various principles and concepts to provide a powerful technique to gain insights into molecular processes [1, 2]. First, as the name indicates, FCS is based on fluorescence, a technique that provides multiple advantages. A fluorophore excited in a particular wavelength range, determined by its excitation spectrum, will generally emit at a longer wavelength, defined by its emission spectrum. This shift in wavelength, the so-called Stokes shift, allows easy distinction of the fluorescence from the excitation wavelength. It is this property, in combination with its high absorption cross-section and quantum yield, that results in a signal-to-noise ratio in fluorescence measurements that allows detection of a single molecule against the scattering background of other processes. Furthermore, fluorescence is linear over a very wide concentration range, up to mM, covering most physiological relevant concentrations of biomolecules, making quantification simple. And by using fluorescent labels to tag molecules of interest, it is specific and selective in its observations.

The second concept is the concept of fluctuations, i.e. the deviation of the signal from its mean value. Fluctuations of system-characteristic molecular processes in an equilibrium system are driven by thermal energy. These molecular processes, in turn, can be analysed in detail by the properties of the fluctuations. The importance of this concept is sometimes reflected in the overarching name of Fluorescence Fluctuation Spectroscopy (FFS), which denotes all fluorescence modalities that are based on the measurement and analysis of fluorescence fluctuations, including FCS and many derived modalities.

Lastly, correlations are a statistical analysis tool that is used to extract information from the measured fluorescence fluctuations. Suppose one has an understanding of the fluctuation-creating process and its parameters. In that case, this can be combined with the mode of fluorescence excitation and detection to create theoretical models for the expected correlation functions. These theoretical models can be fitted to experimental data to extract values for the characteristic parameters. Although any process that can influence the fluctuations can be investigated by FCS, including chemical reactions, photophysical processes, rotation, and more, it mainly measures translational diffusion. Translational diffusion creates fluctuations by particles moving in and out of an observation volume. The correlation analysis can extract the diffusion coefficient and the average number of particles in the observation volume.

FCS was originally conceived as a single-point time trace measurement but was soon afterwards extended to spatial and spatiotemporal correlations. This extension was pursued along three interrelated paths. First, the concept of FCS, which measures temporal correlations at a single point, was transferred to spatial correlations within fluorescence images in Imaging Correlation Spectroscopy (ICS) [3]. ICS has no intrinsic time resolution and reports on spatial patterns that allow counting particles and determining particle cluster's sizes. Time resolution in ICS can be added by acquiring multiple images, with the time resolution being limited by the available image acquisition time. This was first achieved in temporal ICS, or TICS [4]. TICS is a compelling approach and was the first to be extended to spatiotemporal correlations for whole images in the form of Spatiotemporal ICS (STICS) [5] and other related techniques, as will be discussed in later sections.

The second approach was based on laser scanning. Flow influences the correlation function as it changes the way and the duration of particles moving through the observation volume [6]. As flow is in principle indistinguishable from a steady laser movement, the same correlation function applied to scanning FCS in which the laser beam is scanned through the sample at a constant speed [3, 7–10].

In scanning approaches, the confocal volume was moved at a particular speed and path through the sample. Although this required the use of more complex correlation functions, it provided several advantages. First, FCS is inherently only able to measure a process if fluctuations are created. In the classical approach, slowly moving or immobile particles cannot be observed when diffusion is measured, leaving crucial molecules in a sample inaccessible. By scanning the confocal volume, even stationary particles will contribute to the correlation function when the confocal volume is scanned over them, leading to characteristic fluctuations related to the scan speed instead of the diffusion coefficient [6]. Mobile particles still contribute to the correlation function with their processes now dependent on diffusion coefficient and scan speed. Second, scanning breaks the symmetry of the correlation function as the scanning process defines a direction within the sample. This is useful in the case that the process under study itself has a privileged direction. The classic example, in this case, is flow or active transport of particles.

In the case of flow, the autocorrelation function (ACF) derived from a stationary confocal volume will show the changing dynamics but cannot determine the flow direction. In contrast, scanning FCS will lead to different correlation functions depending on whether one scans along, against, or at an angle to the flow direction [11, 12]. It can thus determine flow velocities and profiles. Finally, the inherent time resolution in scanning, specified by the time the confocal detection dwells on a single pixel, is much higher than can be reached in the acquisition of whole images, as done in other spatiotemporal correlation techniques and thus can detect even very fast events.

To access the spatial information, one needs to know the position of the laser beam at each time point, i.e. the scan path needs to be known precisely. Scanning allows calculating correlations for each point of the laser trajectory. Koppel et al. used a confocal microscope to determine molecular mobility in solution and bilayers in so-called Scanning Concentration Correlation Spectroscopy (SCCS) [13]. In Position-Sensitive Scanning FCS (PSFCS), a circular trajectory was used to determine flow directions [14]. Using the inherent time structure in confocal images, Raster Image Correlation Spectroscopy (RICS) finally allowed to access molecular mobility in a whole image [15].

The third path was based on FCS multiplexing by collecting multiple temporal correlation functions simultaneously. Brinkmeier et al. used two-beam cross-correlation analysis to measure flow in microstructures [16, 17], which was subsequently extended to two-photon excitation [18]. A similar idea was implemented by detecting different parts of one single confocal volume, either using two pinholes [19] or using one pinhole but shifting the detectors with respect to the pinhole instead [20]. This was further optimized with modern instruments that use multi-element or multipoint detectors [21, 22].

Here also the advantage of combining temporal and spatial correlation functions was immediately recognized. A single-point measurement was able to determine flow speeds but could not determine the direction of flow, necessary for the determination of flow velocities and flow profiles. In the case of two points, however, the forward and backward correlation functions provide a means to determine flow directions and thus to measure flow profiles. Another advantage of multipoint FCS was that the distance between the points could be precisely controlled and thus was known. Since cross-correlations between the points provided information on how long particles needed to move from one point to the other and the distance between the points was known, the measurements required no calibration.

The measurements could be made so precise that even sub-nanometre differences in hydrodynamic radius could be measured by the so-called dual-focus FCS [23–25]. The extension to more points was first achieved only with a small number of points by either using multiple single-point detectors [26, 27] or using small multi-pixel devices that could be read out sufficiently fast to measure molecular dynamics in biological samples [26, 28].

This situation changed when it was shown that electron-multiplying charged-coupled device (EMCCD) and scientific complementary metal-oxide semiconductor (sCMOS) cameras could be read out sufficiently fast, at least for regions of interest on the order of 100–1,000 pixels [29–33]. Since then, the so-called Imaging FCS approaches have allowed the temporal correlation function for each pixel or the spatiotemporal correlation functions between any pixels or group of pixels to be calculated. Imaging FCS has been developed and applied to a wide range of samples, from solution measurements to cells and organisms [34, 35].

The three approaches thus resulted in methods to measure molecular spatiotemporal dynamics over whole images with different advantages and disadvantages as will be discussed in this chapter.

2 Image-Based Correlation Spectroscopy

Image Correlation Spectroscopy (ICS) is the spatial analogue to FCS. While FCS analyses patterns in time, which are interpreted as the dynamics of molecular processes, ICS analyses patterns in space in images or series of images (Fig. 1a) to

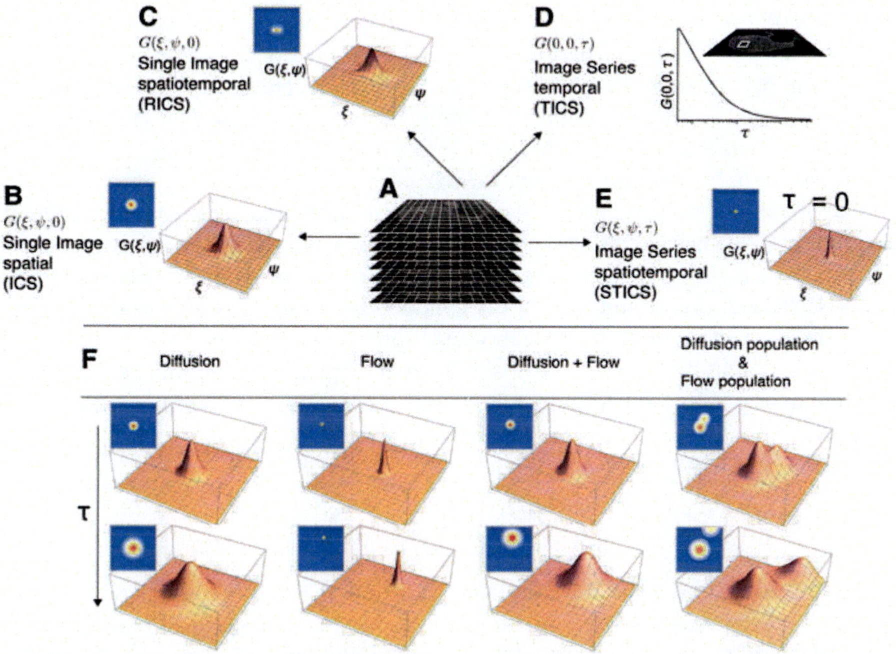

Fig. 1 Image-based correlation spectroscopy. (**a**) A single image or a time stack of images is recorded. (**b**) Single images can provide spatial resolution in ICS, providing data on the number of particles and cluster sizes. (**c**) If the image was scanned and contained inherent time resolution, it could be used in RICS to extract the number of particles at each ROI and spatiotemporal information, including diffusion coefficients, flow parameters, and binding interactions. (**d**) In a stack of frames, the temporal information can be extracted for each ROI in an image. (**e**) And in STICS the spatial and temporal information of the image stack is recovered. (**f**) Examples of processes in STICS: diffusion broadens the central peak of the correlation function; flow moves the peak with time; diffusion and flow of a single component shows a combination of correlation peak movement and broadening; and if there are two components, one diffusing and the other undergoing flow, then two correlation peaks will develop a broadening central peak and a second peak that broadens and moves away from the origin

yield information on numbers and sizes of structures. The combination of time and spatial correlation analysis leads to the general equation for the correlation function.

$$G_{ab}(\xi, \psi, \tau) = \frac{\left\langle \langle \delta I_a(x, \ y, \ t)\delta I_b(x+\xi, \ y+\psi, \ t+\tau)\rangle_{xy}\right\rangle_t}{\langle I_a\rangle_t \langle I_b\rangle_{t+\tau}} \tag{1}$$

Here $\delta I_a(x, y, t)$ is the intensity fluctuation in wavelength channel a at a particular position (x, y) at time t. $\delta I_b(x + \xi, y + \psi, t + \tau)$ is the intensity fluctuation in wavelength channel b at a particular position and time shifted by (ξ, ψ, τ) compared to $\delta I_a(x, y, t)$. The values (ξ, ψ, τ) are often called the spatial and temporal lags of the correlation. The angular brackets $\langle \ldots \rangle_p$ indicate averaging with respect to the parameter p. The indices a and b stand for different wavelength channels when performing dual colour-cross-correlation spectroscopy, for auto correlation analysis they can be omitted. This equation will be adapted in multiple ways, leading to various 2D FCS methods that emphasize the correlations' spatial, temporal, or spatiotemporal aspects. In principle, any imaging technique can provide data for these methods, including confocal scanning laser microscopy (CLSM), spinning disk confocal microscopy (SDCM), total internal reflection fluorescence microscopy (TIRFM), or light-sheet microscopy (LSM).

All these microscopy techniques contain, in principle, spatial and temporal information as images can either be collected as a time series, or in the case of CLSM, have an inherent time structure as pixels in the image are collected sequentially. However, the exploitation of the temporal information is only possible if the acquisition times are faster than the molecular processes to be observed. This section reviews different versions of ICS and how it expanded over time to arrive at spatiotemporal ICS (STICS), which fully analyses spatiotemporal information.

2.1 *Image Correlation Spectroscopy: ICS*

In the following, we write the correlation function again in the most general form with the indices a and b presenting images taken of the same sample in two different wavelength ranges. For $a = b$, we have an autocorrelation for ICS proper (Fig. 1b). If $a \neq b$, we speak of Image Cross-Correlation Spectroscopy (ICCS). ICS/ICCS takes input images and analyses them only in the spatial domain. Thus, the general function has no time component and takes the form

$$G_{ab}(\xi, \psi, 0) = \frac{\langle \delta I_a(x, \ y, \ t)\delta I_b(x+\xi, \ y+\psi, \ t)\rangle}{\langle I_a(t)\rangle \langle I_b(t)\rangle} \tag{2}$$

In this case, the correlation function can be calculated over the 2D Fourier Transforms (FT) of the image(s)

$$G_{ab}(\xi, \psi, 0) = FT^{-1}\{FT[Image_a] \times FT[Image_b]\} \tag{3}$$

In contrast to FCS, its imaging analogue ICS relies on the fluctuation of the intensity across an image, i.e. it is defined by the difference between pixel intensities at any point in space to the average intensity of an image. As in FCS, ICS functions are calculated for all possible pixel shifts (ξ, ψ) to provide a full spatial correlation function. However, it should be noted that in FCS, one calculates typically over many thousands if not millions of time points, while in ICS, the size of the sample ultimately limits its statistics.

The correlation in fluctuations arises from the non-random structures that are found in the image. As in microscopy, the smallest size of a structure in an image is given by the system's point spread function (PSF), whose radius is given by ω_0. It means that even a point source gives rise to spatial correlations of the size of the PSF. Therefore, pixels should be smaller than the PSF so that the correlations between pixels can be measured. In ICS the correlation function is, thus, fitted with a 2D Gaussian as an approximation for the PSF.

$$G_{ab}(\xi, \psi, 0) = G_{ab}(0, 0, 0)\ \exp\left\{-\frac{\xi^2 + \psi^2}{\omega_0^2}\right\} + G_{ab\infty} \tag{4}$$

Here $G_{ab}(0, 0, 0)$ is the amplitude of the spatial correlation function (SCF), which is inversely proportional to the number of particles in the observation area. $G_{ab\infty}$ is the convergence value for long distances. This is kept as a fit parameter to take account of possible incomplete decay of the SCF. The width of the SCF will be larger than the PSF if particles move during the acquisition of a single frame, something that is better treated in temporal measurements as discussed in later sections, and if there are non-random structures in the image of a size comparable or larger than the PSF.

ICS measures several important parameters. First, ICS determines the size of the PSF if one uses immobile particles that are much smaller than the size of the PSF itself [3]. Second, one can measure the number of particles in the observation area over the amplitude $G_{ab}(0, 0, 0)$. With the knowledge of the size of the observation area and the size of the image, one can count the number of particles [3, 36]. Note that the laser focus needs to be of comparable to the size of the fluorescent entity to capture the cluster density accurately [37].

In principle, several ICS measurements can be taken, and the temporal evolution of the system can be determined. However, the dynamic information is limited because long exposure times are used, or several images averaged to increase the signal-to-noise ratio. It has been shown that for reliable characterization, the temporal resolution must be at least 10 times shorter than the dynamics of interest. Therefore, the reverse is true; any diffusive processes with a time scale longer than the time resolution appear to be static and can be quantified by spatial ICS. In other words, immobile populations contribute to the SCF while fast diffusing populations add to the noise of the SCF. Like confocal FCS, ICS has a concentration limit above and below data fitting becomes difficult because of a decreasing signal-to-noise ratio. As a rule of thumb, densities of up to 100 particles per μm^2 in a 2D system can be quantified, and this holds for other related techniques [38, 39]. Further, spatial ICS is not a parametric free technique. The experimenter is required to adjust frame-time and the total number of images to compensate for photobleaching and signal-to-noise ratio. The counting capability of ICS combined with gradual photobleaching of fluorophores leads to the so-called pbICS [40] for determining the aggregate distribution of immobilized cluster as demonstrated in CHO cell transfected with green fluorescent protein (GFP)-tagged epidermal growth factor (EGF) receptor. In case of non-optimal acquisition settings, the experimenter can optimize for frame-time and signal-to-noise ratio during the data treatment stage.

A two-colour variant of spatial ICS, spatial image cross-correlation spectroscopy (spatial ICCS), has been introduced to measure colocalization of two different labelled species [41]. Two-colour variation has been employed to study the extent of interactions between two slow-moving species where different macromolecules labelled with two fluorophores of distinct emission wavelength are imaged simultaneously or alternatingly. In fact, ICS is employed behind the scenes in many algorithms. In camera-based microscopy (TIRFM and LSM), the principle of two-colour spatial ICCS doubles up as an alignment tool for the simultaneous two-colour experiment [42]. As it is a purely static method, it limits biology application to immobile and flat samples. An offline analysis tool is available with the JaCoP ImageJ Plugin [43]. Cerutti et al. have shown that ICS can be used to provide quality metrics from a single super-resolution image for evaluation. This method, called QuICS [44], quantitates image quality and can give useful hints on optimizing the imaging conditions.

Recently, the combination of super-resolution and ICCS has been explored. Oneto et al. combined ICCS with STED (STED-ICCS) [45] to estimate the nanoscale distance of nuclear sites with a spatial resolution down to ~50 nm by leveraging the shape of the cross-correlation function shift from the origin. This has been extended to Structured Illumination Microscopy and Image Cross-Correlation Spectroscopy (SIM-ICCS) [46]. Compared to single-molecule localization, the application of super-resolution and ICCS does not provide a complete statistical analysis of distances. Instead, the super-resolution implementation is helpful to analyse average distances between correlated particles in the region.

Up to now, we have discussed ICS mainly as it is performed in a confocal microscope. Nevertheless, it can also be performed in TIRFM with a camera as a detector and improved temporal resolution.

2.2 Temporal Image Correlation Spectroscopy: TICS

The temporal variant of image correlation spectroscopy (TICS) takes the same fluorescence microscopy image series as an input (Fig. 1d). Of all variations, it has the closest working principle to FCS. Unlike spatial ICS, TICS does not average several images. Instead, intensity fluctuations are recorded from frame to frame, correlated in time, and averaged over multiple pixels. This increases time resolution while keeping the signal-to-noise level high. The decay of the temporal correlation function (TCF) reflects the average time that particles require to move in and out of the observation volumes sampled across selected portions of an image; hence, the underlying molecular transport parameters (diffusion coefficient and flow speed) can be extracted [4, 47].

TICS captures the heterogeneity of molecular transport parameters and numbers across raster-scanned regions instead of single points in confocal FCS. However, the ability to sample temporal fluctuations in a larger space comes at the cost of reduced temporal resolution because of the limited confocal scanning speed, therefore imposing an upper limit on diffusion coefficients accessible to TICS. In other words, the fluorophore needs to be within the same beam focal area when the raster scan returns to the same position following the typical ~1 Hz frame rate.

The general correlation function for TICS is given by

$$G_{ab}(0,\ 0,\ \tau) = \frac{\left\langle \langle \delta I_a(x,\ \ y,\ \ t)\delta I_b(x,\ \ y,\ \ t+\tau)\rangle_{xy} \right\rangle_t}{\langle I_a\rangle_t \langle I_b\rangle_{t+\tau}} \tag{5}$$

TICS can also be performed as cross-correlation analysis [48, 49] if one records images in different wavelength channels a and b. For autocorrelation analysis in TICS we obtain the following function for diffusion [50].

$$G_{ab}^{\text{diff}}(0,\ 0,\ \tau) = G_{ab}(0,\ 0,\ 0)\left(1+\frac{4D\tau}{\omega_0^2}\right)^{-1} + G_{ab\infty} \tag{6}$$

where the parameters have the same definitions as discussed in the previous sections. For flow we obtain

$$G_{ab}^{\text{flow}}(0,\ 0,\ \tau) = G_{ab}(0,\ 0,\ 0)\ \exp\left\{-\frac{(v\tau)^2}{\omega_0^2}\right\} + G_{ab\infty} \tag{7}$$

where v represents the flow speed. For the case that a species of particles undergoes simultaneously flow and diffusion, we obtain a combination of the two previous equations.

$$G_{ab}^{\text{diff,flow}}(0,\ 0,\ \tau) = G_{ab}(0,\ 0,\ 0)\left(1+\frac{4D\tau}{\omega_0^2}\right)^{-1} exp\left\{-\frac{(v\tau)^2}{4D\tau+\omega_0^2}\right\} + G_{ab\infty} \tag{8}$$

The inclusion of temporal analysis in a stack of frames extends the information available from ICS and allows the measurements of diffusion at least of slow clusters [51]. Using two-photon temporal ICS, diffusion coefficient of as large as 1 μm/s in cell can be measured along with flow [47]. Interesting applications of TICS include the measurement of nanocarrier motion in live cells [49] or the determination of the mechanosensitive responses of integrin under different conditions over whole cells [52]. Moreover, as two-colour cross-correlation (TICCS) it characterized the recruitment of the proteins deleted in colorectal cancer (DCC) and UNC5B to the plasma membrane after netrin-1 activation [48]. The increased information content provided by TICS came at the cost of temporal and spatial resolution as the time resolution was limited by available acquisition times, and correlation functions were averaged over a user-defined region to improve the signal-to-noise ratio.

2.3 *Spatiotemporal Image Correlation Spectroscopy: STICS*

ICS captures effectively static information from the SCF of an image (Fig. 1b). In contrast, similar to FCS, TICS analyses the fluctuations of particles diffusing through an observation volume (Fig. 1d). As observation volumes are typically rotationally symmetric, no information on the direction of the movement of the particles is captured [53]. Measuring flow directions is possible by cross-correlating pixels intensity in space [51], and thus one needs to capture both spatial and temporal information. In addition, the characterization of flow in a sample requires the determination of multiple parameters. Flow in the sample might not be homogeneous, and a spatially resolved flow profile needs to be measured, comprising information about magnitude and direction, ideally at each point in the acquired image. Spatiotemporal ICS, or STICS, is the first 2D FCS method that captures both spatial and temporal correlations in a temporal series of images as indicated in Eq. 1 and thus can provide complete information of flow profiles [5, 54].

In principle, the spatiotemporal correlation function (STCF) is a function of three (two spatial and one temporal) if not four (three spatial and one temporal) variables. However, as this is difficult to picture, we restrict ourselves to two spatial and one temporal variable and describe the development of the two-dimensional spatial STCF as a function of the temporal variable. In the following discussion of the STCF, we will differentiate several cases. The first will be a static sample without any movement (Fig. 1e). The second will be a sample where particles undergo only diffusion (Fig. 1f(left)). The third sample will have particles under active transport (Fig. 1f (2nd from left)). Here we will assume that these particles are exclusively transported but do not diffuse. Lastly, we will look at particles undergoing flow, assuming that these particles are transported and diffuse (Fig. 1f (3rd from left)). The general equation for STICS is given by

$$G_{ab}(\xi, \psi, \tau) = G_{ab}(0,0,0)\left(1 + \frac{4D\tau}{\omega_0^2}\right)^{-1} \exp\left\{\frac{(\xi - v_x\tau)^2 + (\psi - v_y\tau)^2}{4D\tau + \omega_0^2}\right\} + G_{ab\infty} \tag{9}$$

which is the extension of the equation of TICS (Eq. 8) to include the spatial lags ξ and ψ for the x- and y-directions.

In the case of a static sample, STICS reduces to ICS with a time-invariant STCF that peaks at the origin of the spatial variables with the width of the PSF if the particles are much smaller than the PSF. If particles or structures imaged are on the same scale or larger than the PSF, width widens. In the presence of a diffusing particle population, the amplitude of the STCF will remain centred at the origin but with a wider peak representing the movement of the particles during the lag time. In this case, the STCF will increase in width with the increase in lag time, and the rate of the increase in width depends on the mean square displacement of the diffusing particles. In the case of actively transported particles, the STCF will not change shape but move with longer lag times further away from the origin as determined by the transport velocity. If flow is present, the STCF will displace from the origin with time as determined by the flow velocity, but it will also extend in width due to diffusion. Lastly, if multiple populations exhibit different dynamics, then multiple peaks will develop, with each peak representing the characteristics of the movement of one of the populations (Fig. 1f (right)). In practice, the resolution of these multiple populations is not always easy and removing the immobile fraction by Fourier or moving average filtering in the time domain is often necessary to reveal the dynamic population [50].

STICS has been used to analyse flow in microfluidic channels [55], transport of proteins in cells [56–58], kinetics of protein networks [59], and fast and confined diffusion in bacteria [60]. It has also been implemented with two-photon microscopy [61] to measure protein flow in developing *C. elegans* embryos. Furthermore, Pandzic et al. showed that STICS can be used with photo-activation to measure the diffusion of various membrane proteins and that it provides very similar data compared to single-particle tracking (SPT) in photo-activation localization microscopy [62].

As with other correlation methods, STICS can be conducted in a cross-correlation modality between different wavelength channels, leading to spatiotemporal image cross-correlation spectroscopy (STICCS) [63]. STICCS can determine protein interactions, and potentially even transient interactions [63].

In addition to STICS, spatiotemporal dynamics has also been addressed in some related techniques. ICS has been combined with single-particle tracking (SPT) in the so-called particle ICS (PICS), which circumvents the limitations of both SPT and ICS [64]. It can measure at high particle densities, unlike SPT, but has a high spatial resolution, unlike ICS. In another combination of SPT and correlation analysis, the so-called Tracking Image Correlation or TrIC, Dupont et al. analysed 3D tracking data with correlation analysis providing a tool for dynamic colocalization analysis in 3D [65]. Ashdown et al. demonstrated TIRF-SIM imaging can be combined with STICS to quantify molecular flow on subresolution length scales [66]. The high spatiotemporal resolution of these techniques makes them excellent members of the 2D and even 3D FCS toolset.

2.4 k-Space Image Correlation Spectroscopy: kICS

The image in a fluorescence microscope is the convolution of the microscope PSF and the actual distribution of the fluorescent-tagged molecules in the sample. We saw this in the fact that the PSF always limits the STCF in STICS. A common approach to disentangle deconvolutions is transforming the data into Fourier space where the convolution operation is simplified to a product making data treatment much simpler. It is used in k-space ICS (kICS), ICS in the spatial frequency domain, to separate various contributions to the correlation function. By performing the correlation in k-space, Kolin et al. have shown that they can determine the PSF and the diffusion coefficient independently [67].

The second problem in 2D FCS techniques is photobleaching as a whole cross-section of a sample is illuminated with elevated laser intensities – compared to simple imaging – to reach a sufficient signal-to-noise ratio when reading out data at high frame rates. Photobleaching causes a steady decrease in average fluorescence intensity with time. Bleaching can have multiple effects. If bleaching of a molecule is much slower than the time a particle requires to traverse a pixel observation volume, then there is only a general decrease in the overall intensity, which will lead to artefacts in the correlation functions with a time scale determined by the bleaching time. However, if bleaching is so high that fluorescent molecules have a non-negligible probability of being bleached during the transition of a pixel observation area, then the average transit time of molecules will be underestimated, and the diffusion coefficient will be overestimated. The latter form of bleaching is sometimes called cryptic photobleaching.

Over the years, there have been various ways to correct for photobleaching as none of the methods allows analysing uncorrected datasets without imposing bias and artefacts. One way to combat the observational bleaching effects is detrending by normalizing to the first recorded intensity value. Kolin et al. recover accurate diffusion parameters via TICS if the bleaching process is well-characterized as a function of time by a fit function [67]. It is important to note that detrending intensity trace is not the solution to correct cryptic bleaching. In principle, an optimum laser power exists which varies as a function of the diffusion coefficient and observation area. A quick and easy laser power calibration protocol would be a step towards automated fluorescence microscopy [68].

In the presence of photobleaching, kICS has a significant advantage. Under the spatiotemporal image correlation domain, STICS and kICS are similar in their working principle, with some key differences. In k-space, image correlation does not directly correlate fluctuations of image pixel intensities, and instead, it calculates the time-correlation function from a spatially Fourier transformed image. By doing so, kICS overcomes STICS limitations due to photobleaching and photophysics and can separate photophysics and photobleaching from dynamics [69]. Although initially developed for CLSM, kICS can also be applied to TIRFM. Furthermore, kICS can also work with scattering signals, as was shown in the characterization of the intracellular dynamics of gold nanoparticles [70].

Brandao et al. have shown that, in principle, kICS can extract ligand–receptor binding kinetics when certain conditions are met [71]. First, the photophysics, including bleaching and blinking, should be independent of binding and be measurable before the actual experiment. This is fulfilled for many fluorophores but needs to be verified. Second, only receptor-bound ligand should be visible. This second condition can be fulfilled when, for instance, measuring ligand binding to membrane receptors on a TIRFM with relatively slow acquisition rates. The diffusion coefficient for ligands in solution compared to the diffusion coefficient of membrane receptors differs by a factor of about 100. Thus, the fast-moving ligands will contribute a uniform background signal compared to the much slower-moving receptors. Under these conditions, when the photophysics is known, the primary source of unknown fluctuations are the binding and unbinding events of the ligand to the receptor, which can be determined with kICS. Receptor–to-receptor binding and interaction of membrane proteins with membrane domains can also be determined this way [72–74]. This, together with the fact that kICS can measure at a wide range of fluorophore densities, in contrast to other techniques like SPT, that can work only at low concentrations, makes kICS a very versatile technique to measure membrane events.

Originally, ICS variants were conducted at limited spatial and temporal resolution. Acquisition times at video rates allow only measurement of slow diffusing particles, e.g. membrane proteins. But they were also performed at limited spatial resolution as a minimum dimension is necessary to calculate spatial correlations. Nevertheless, this is not a fundamental limit as modern cameras allow fast read-out with time resolutions less than 100 μs per region of interest, containing hundreds to thousands of pixels. The pixels and frames can then be binned to optimize the data for different evaluation techniques, including 2D FCS modalities. With the new technology and advances in computation, these techniques will be able to also measure cytosolic or extracellular protein diffusion in cells and organisms and can be extended to 3D.

2.5 *Image Mean Squared Displacement: iMSD*

Biological systems are generally inhomogeneous, leading to anisotropy in transport and diffusion, as particles encounter different environments in different directions or are hindered by obstacles. In that case, directional analysis of particle mobility is required. This can be provided in scanning and imaging approaches, in which, e.g., flow profiles were measured, as discussed in the next section. But this can also be addressed in image-based approaches by analysing the spatial cross-correlations between different points in an image. For this purpose, a technique called image Mean Squared Displacement (iMSD) was introduced [75–77].

iMSD is a STICS modality in which the spatial correlations in all directions are analysed in time with respect to a particular point. This can be achieved in scanning systems as well as in multi-pixel detection systems. The calculations are performed on a polar grid, and values are calculated over a limited number of angles. This allows averaging more pixels the larger the distance from the central pixel, improving the signal-to-noise ratio. This kind of analysis allows to determine a direction-dependent measurement of transport and diffusion and provides a connectivity map of a sample, identifying obstacles to diffusion and transport [78].

In another application, iMSD was used to evaluate the spatial analogue of the FCS diffusion laws [79, 80] to analyse the diffusive modes of molecules in the cytoplasm [76]. If diffusion is length scale-dependent, e.g., when obstacles and trapping sites are present, then an analysis of the diffusion coefficient with length scale can identify diffusion mode. In analogy, different diffusive modes measured over the same length scale but at different time scales will show a similar dependency, which can be used in iMSD at different scan speeds to identify diffusive modes. In work combining most of the ICS techniques, including iMSD and RICS (see next section) discussed here, Hendrix et al. analysed the HIV-1 Gag polyprotein assembly in live cells using a customized confocal system [81]. This work shows the differences in applications of the various ICS techniques but also demonstrates that many of them can be applied to the same data.

Finally, iMSD was recently also implemented on a confocal microscope with an Airyscan [21], or multi-element single-photon avalanche diode [22] detector, allowing much faster acquisitions and widening the possibilities to measure dynamics with iMSD.

2.6 Pair Correlation Function: pCF

Pair correlation function (pCF) analysis is a special case of spatiotemporal correlations, in which two points in space are correlated in time. If the two points are sufficiently far apart so that their observation volumes do not overlap, the resulting pCF will possess a peak at the time it takes fluorescent probes to move from one location to the other [82]. In principle, this approach can be applied to any data that has some inherent spatial and temporal information, including line scanning, multi-focus, and imaging approaches. The probability that a particle that originated at (x_0, y_0, z_0) at time t is found at location (x_1, y_1, z_1) at a time $t + \tau$ is given by the diffusion propagator

$$P(r,\tau) = \frac{1}{(4\pi D\tau)^{3/2}} \exp\left\{-\frac{(x_1-x_0)^2+(y_1-y_0)^2+(z_1-z_0)^2}{4D\tau}\right\} \tag{10}$$

At a particular distance $r = \sqrt{(x_1-x_0)^2+(y_1-y_0)^2+(z_1-z_0)^2}$ between two points (x_0, y_0, z_0) and (x_1, y_1, z_1), this function will peak for a time τ that depends on the diffusion coefficient D and which can be calculated from the mean squared displacement, $\langle r^2 \rangle = 6Dt$. The corresponding pair correlation is

$$G(\tau) = \frac{\langle \delta I(x_0, y_0, z_0, t)\delta I(x_1, y_1, z_1, t+\tau)\rangle_t}{\langle I(x_0, y_0, z_0, t)\rangle_t \langle I(x_1, y_1, z_1, t)\rangle_{t+\tau}} \quad (11)$$

where the strength of the correlation depends on how many particles have diffused from the first to the second point at time τ as described by the diffusion propagator.

If there are obstacles between the two locations, the peak will occur later depending on how long a probe will need to bypass the obstacle. In the extreme case that a barrier exists that prevents probes from translocating from one point to the other, the pCF will be zero. This principle was used to measure molecular transport across nuclear pore complexes [83], and in the cellular nucleus at different chromatin densities, during different cell cycle stages, and during DNA repair [84–86]. This work nicely demonstrated the advantage of spatiotemporal correlations as single-point FCS only measures local mobility but not long-distance transport. The use of pCF allows the determination of intracellular as well as membrane organization [87, 88] and was used in plants to measure the movement of transcription factors [89]. The method was extended to 2D to measure anisotropic movement [78]. It should be noted that instead of calculating the temporal pCF for two positions, one can also calculate the spatial pCF at a particular time, describing the variation of molecular densities in space [90].

3 Scanning FCS

FCS was originally conceived as a single-point measurement, but it was early on realized how sample flow or laser scanning can provide extra information [6]. In the most simple case, flow or scanning increases sample throughput [91, 92]. But of interest in this chapter is the inclusion of spatial information into the originally temporal FCS measurements. We will therefore progress from simple scanning FCS approaches that provide 1D spatiotemporal information, to more complex scanning patterns that provide 2D spatiotemporal information, to finally RICS (Fig. 1c) that uses the inherent time structure in confocal images to obtain the same information but from readily commercially available instrumentation [15]. RICS thus made spatiotemporal correlations widely available. The different approaches have their own advantages and disadvantages and show different ways of monitoring fluctuations depending on the data collection method.

3.1 Scanning FCS in the Presence of Immobile Particles

FCS is insensitive to immobile or very slowly moving particles as these do not cause fluctuations and thus cannot be detected in the correlation functions. However, a fluctuating signal even from immobile particles can be obtained when one either

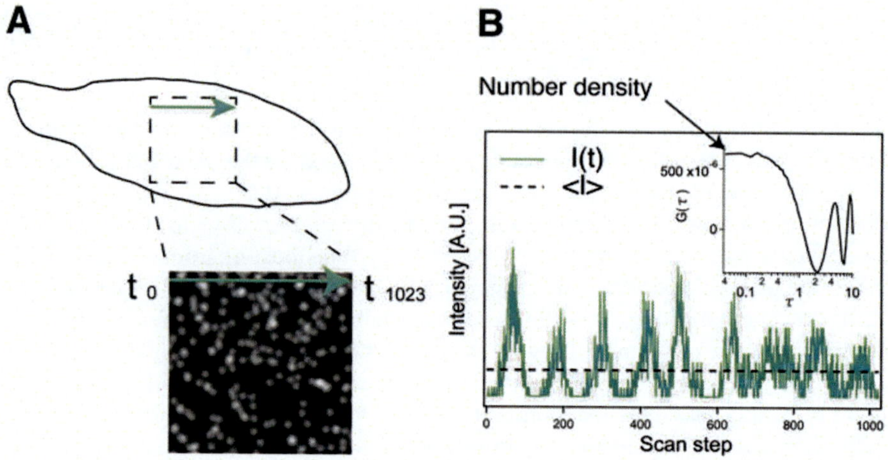

Fig. 2 Principles of scanning FCS for immobile particles. (**a**) An illumination spot with calibrated beam radius (w_0) is scanned across a sample either by confocal or stage scanning at a known speed. (**b**) The start of the line scan t_0 is synchronized with the data acquisition. The TCF decay shows the characteristic scan speed, while the amplitude is related to the particle density

moves the sample, e.g. by stage scanning or by driving some movement in the sample, or by scanning the beam. Weissman et al. developed a similar concept to measure molecular weights of macromolecule [93] although it differed in experimental configuration.

In 1986, the first implementation of Scanning Fluorescence Correlation Spectroscopy (SFCS) arose by translating samples linearly and horizontally by a translation stage while fixing the illumination beam (Fig. 2a). Petersen et al. successfully employed this configuration to extract aggregation/cluster size and the density of clusters from a line segment of a cell [7], aggregation of virus glycoproteins [8], ligand binding [9], and EGFR distribution on the cell plasma membranes with an average of 130 receptors per cluster and 7–8 clusters per μm^2 [10].

FCS statistical accuracy [94] depends on the signal-to-noise ratio and thus on the number of photons detected per detection interval and the square root of data points.

For SFCS, one can scan slowly and thus increase the counts per second. However, the scan range is limited by the physical dimension of the sample (~20 μm in the case of mammalian cells), while the characteristic fluctuation width is defined by the dimension of the focused laser beam, yielding at best some tens of fluctuations per scan. Therefore, multiple independent scans need to be averaged to improve the accuracy of immobile particle analysis.

To understand how this is done, let us take a look at the relevant correlation functions. The basic correlation function for confocal microscopes describing diffusion in 2D is

$$G_{2D\ \text{diffusion}}(\tau) = \frac{1}{N}\left(1 + \frac{4D\tau}{\omega_0^2}\right)^{-1} + G_\infty \tag{12}$$

where D is the diffusion coefficient and w_0 is the e^{-2} radius of the laser focus. In the case of sample flow or when the laser focus is scanned at a constant speed in the focal plane, and the absence of diffusion the equation changes to [6].

$$G_{2D\ \text{flow}}(\tau) = \frac{1}{N}\ \exp\left\{-\frac{\left(v_x^2 + v_y^2\right)\tau^2}{\omega_0^2}\right\} + G_\infty \tag{13}$$

where v_x and v_x describe the flow/scan speed in the x and y direction. The only unknown in this equation is the number of particles N, which measures the number of particles per observation volume observed on average and which can be obtained by fitting the experimental curve (Fig. 2b). If one has an estimate of the total number of monomers in the sample, then the aggregate size can be determined. However, one has to assume that the fluorescence is not altered in the aggregates compared to the monomeric state, that all molecules are fluorescent, and that photobleaching is negligible, as otherwise, the estimates will be biased to lower numbers.

Scanning FCS for the characterization of immobile particles provided estimates of the mean number of virus glycoprotein aggregates from Sindbis virus and vesicular stomatitis virus [8] and was later extended to measure cell surface receptor aggregation [9, 10]. The method was applied to 2D CLSM images [39] and extended to other microscopy technique which does not necessarily require scanning. For instance, Number and Brightness analysis [95] revealed the presence of a large bright immobile aggregate within a heterogeneous region of a sample.

3.2 *Scanning FCS in the Presence of Flow*

If flow is present in a sample in which the particles themselves also diffuse, then we obtain a correlation function which is a combination of the ones derived for flow and for diffusion:

$$G_{2D\ \text{diff, flow}}(\tau) = \frac{1}{N}\left(1 + \frac{4D\tau}{\omega_0^2}\right)^{-1} \exp\left\{-\frac{v^2\tau^2}{4D\tau + \omega_0^2}\right\} + G_\infty \tag{14}$$

However, as the confocal volume is rotationally symmetric, Eq. 14 does not allow us to determine the flow velocity $\vec{v}_{\text{flow}}$ and direction of flow (θ), as the correlation function is independent of the flow direction. The flow direction and thus the velocity can nevertheless be determined if one measures the correlation function with a stationary beam to determine the flow speed (Fig. 3a) and then in a second

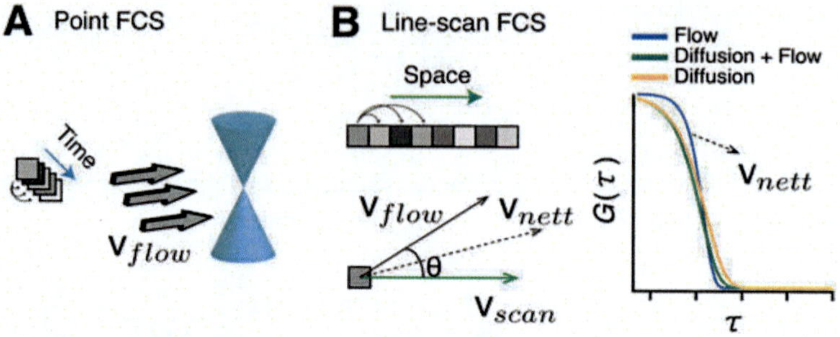

Fig. 3 Flow measurement. (**a**) Using a stationary beam, the flow speed can be determined as particles are driven through the focus and thus influence the shape and width of the autocorrelation function. (**b**) Scanning the beam in the sample leads to a characteristic change in the ACF as the flow (ν_{flow}) and scan (ν_{scan}) add up to a new effective speed (ν_{nett}) from which the angle between scan and flow can be determined

measurement (Fig. 3b) determines the correlation function in the presence of a known scan speed [11]. In that case, the correlation function changes to

$$G_{2D\ \text{diff,flow,scan}}(\tau) = \frac{1}{N}\left(1 + \frac{4D\tau}{\omega_0^2}\right)^{-1} \exp\left\{ - \frac{\left(\vec{v}_{\text{flow}} + \vec{v}_{\text{scan}}\right)^2 \tau^2}{4D\tau + \omega_0^2} \right\} + G_\infty \tag{15}$$

and one can determine the velocity according to the following equation:

$$\vec{v}^{\,2}_{\text{nett}} = \left(\vec{v}_{\text{flow}} + \vec{v}_{\text{scan}}\right)^2 = \vec{v}^{\,2}_{\text{flow}} + \vec{v}^{\,2}_{\text{scan}} + 2\vec{v}_{\text{flow}}\vec{v}_{\text{scan}} \cos\theta \tag{16}$$

The sequential strategy to determine flow and its directionality can be extended to 3D with three measurements, a stationary measurement to determine the flow speed and one in plane and on axial scan [12].

As a rule of thumb, the separation of diffusion and flow (Fig. 3c) is possible if the average time it takes a particle to traverse the focal volume by diffusion and by flow is within a factor 10 of each other. Otherwise, the faster process dominates to the extent that the slower one cannot be determined anymore. Line-scan FCS was used to characterize flow profiles in microfluidic channels and blood flow direction in living zebrafish with an accuracy down ±10° [11].

Cross-correlation has proven to be a valuable tool to elucidate vectorial flow information and, in the case of diffusion, yields absolute values without knowledge of the PSF of the system. The technique has been applied to a flow system with two spatially separated volumes using scattering of two wavelengths [96] or fluorescence with a single wavelength [17]. Since the maximum cross-correlation is proportional to the average time a molecule takes to move from a specific location to another, one can tune the time resolution of the technique by changing the distance between the two foci. As in line-scan FCS, directional information can be extracted by changing the direction of the placement of the point in the sample.

As line-scan FCS determines flow characteristics only in one direction, circular-scan FCS was developed. In circular-scan FCS, one can cross-correlate opposite points on the scan path to obtain directional information. Circular-scan FCS thus produces a map of molecular flows indicating if barriers are present [97, 98]. Since the laser beam is scanned repeatedly over the different locations, information is limited to the scan orbit with a resolution given by the size of the laser focus. The equation for circular-scan FCS is given by

$$G_{2D\ \text{diff,circular scan}}(\tau) = \frac{1}{N}\left(1+\frac{4D\tau}{\omega_0^2}\right)^{-1} exp\left\{-\frac{4\rho^2 \sin\left(\frac{\omega t}{2}\right)^2 \tau^2}{4D\tau+\omega_0^2}\right\} + G_\infty \quad (17)$$

In circular-scan FCS, the experimenter can choose radius (ρ) and angular scan frequency (ω) of the scan and produce intensity traces for each position on the scan and has the option to evaluate the data as auto- or cross-correlations as required. There are at least two ways to perform data analysis to study sample dynamics quantitatively.

Ruan et al., for instance, leveraged temporal and spatial information along the circular path independently [99]. With this approach, the confocal volume is scanned circularly at speed much faster than the particle needs to diffuse through a single location, resulting in a set of correlation functions from all points along the scanned path. This method recovers diffusion coefficients in giant unilamellar vesicle (GUV) membranes up to 20 $\mu m^2/s$ with the same precision as conventional FCS measurement. On the other hand, Petrasek et al. used the inherent spatiotemporal information in circular scanning to address the need for calibration of the focal volume and the problem of focusing on membranes. They suggested spatially cross-correlating points on the circular-scan path with a known diameter 2 ρ and a size smaller than 1 μm. This was successfully applied to determine diffusion coefficients ranging from dyes in solution [100] to slower diffusion in model membranes [101]. This approach to measuring diffusion is independent of the size of the focal volume and is not sensitive to the position of the membrane with respect to the focal plane but comes at the cost of more limited spatial information (Eq. 17).

While both scanning paths can be applied, circular scanning has some essential advantages. First, the statistics are no longer limited to the scanned path. Second, a broader range of dynamics can be observed [101] compared to earlier work [11]. Third, the autocorrelation can be easily computed with the existing algorithm without synchronizing data acquisition with the line scan.

3.3 *Scanning FCS as a Tool for Multiplexing and Avoidance of Artefacts*

Scanning FCS with line or circular patterns (Fig. 4a) can also provide FCS measurements at multiple points, i.e. along the scan path. The spatial realization first came by continuously scanning the laser beam across a sample in a line (Fig. 4c) or circle (Fig. 4d) while the emitted light from each spatial location was recorded in sequence, providing an intensity carpet. The intensity time-trace at each spatial location is then correlated in time with the temporal resolution given by the scanning frequency. At each location, one thus obtains information about diffusion.

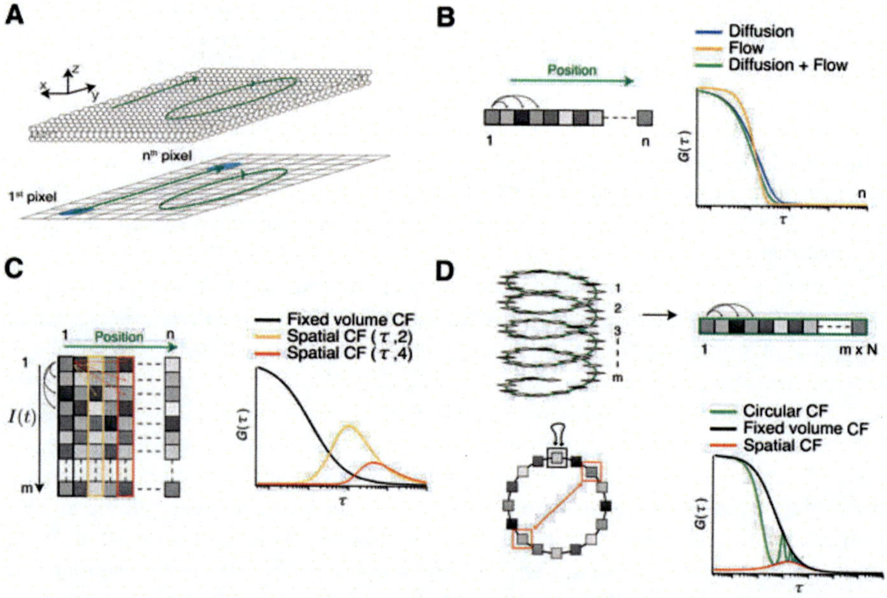

Fig. 4 **(a)** Laser scanning can be performed in various patterns, here shown as linear or circular scan. **(b)** A linear scan can be analysed continuously in time without explicitly using spatial information. In this case, a single autocorrelation function is calculated for all scan repeats. While this is experimentally simple, it provides only information on diffusion and flow speed. **(c)** Spatial information can be used by arranging multiple scans in an intensity carpet ordered according to the position in the sample along with the scan and different times they were recorded in the repeat scans. Auto- and spatial cross-correlation functions provide information about diffusion, flow, and connectivity in the sample. **(d)** In circular scanning, the same information can be obtained. Calculating the time autocorrelation function shows characteristic repeating peaks when the laser beam returns to the same point (*green line*). Two extremes envelop this function, first the autocorrelation function for a beam with the radius of the scan circle (*black line*), and second the spatial cross-correlation function between two points on opposite ends of the circle (*red line*). The two enveloping correlation functions are obtained by setting the *sin* function in Eq. 17 to either 0 or 1

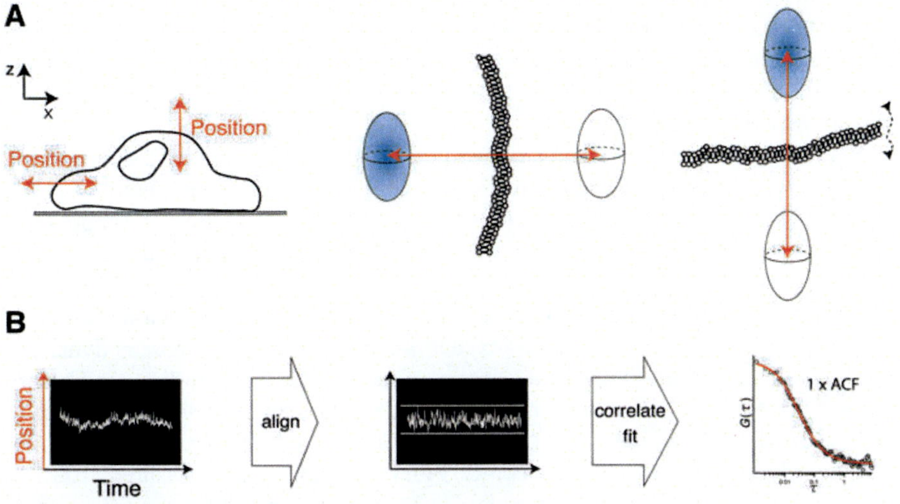

Fig. 5 Line scanning to reduce artefacts of sample movement. (**a**) The observation volume can be scanned laterally or axially across a fluctuating membrane. (**b**) Once per scan, the observation volume is on the membrane. These points are selected, aligned in time and then correlated to provide a correlation function. The time resolution, in this case, is limited to the time for one scan. However, the correlation function does not suffer from sample movement artefacts as only points on the membrane are correlated

This approach allows FCS multiplexing, but at the expense of temporal resolution since the observation volume returns to the same spot only once per scan. When 10–100 points are scanned, diffusion coefficients of about 0.05–1 $\mu m^2/s$ can be determined, enough to cover lipid membrane measurements and slow-moving complexes in solution. Software to process local ACFs from confocal scanning data is available as FoCuS-scan [102]. An alternative approach is represented by scanning the beam slowly across the sample. In this case, one can calculate the correlations within short time segments. The segmented FCS data can be combined to extract average diffusion values in specific compartments.

Di Bona et al. have used this approach to detect the small variations of diffusion of GFP in heterochromatin vs euchromatin [103]. Overall, SFCS is particularly useful for slower dynamics and is insensitive to photobleaching as the illumination beam spends very little time per observation volume.

Besides reducing sensitivity to photobleaching, scanning FCS can also be employed to remove movement artefacts. If the sample to be measured moves within a limited range, then scanning the laser beam over this range can ensure that the sample is captured at least once per scan (Fig. 5a). Cellular plasma membranes, for instance, have a tendency to undulate or move perpendicular to the membrane surface. In a point measurement, this will lead to large fluctuations in the signal due to the membrane movement, and thus artefacts in the correlation function [104]. By scanning the confocal volume perpendicular to the membrane, one can locate the position of the membrane in each scan. By aligning the data collected when the beam is on the membrane, their temporal correlation yields information about processes slower than the scan rate. In this modality, line-scan FCS sacrifices the multiplexing aspect to produce a single correlation function that is free from movement artefact (Fig. 4b).

Recent application of scanning a line in the z-direction, i.e. the optical axis of the microscope, has also been used to overcome sample movement and photobleaching artefacts [105] and was used to determine equilibrium dissociation constants (*Kd*) of ligand–receptor interactions in the Wnt signalling pathway at near-native amounts [106].

Scanning with two foci of different wavelengths, either in continuous-wave excitation or with pulsed interleaved excitation, can determine membrane dynamics and binding interactions with the advantage of reduced artefacts as discussed before [104]. This was used in combination with atomic force microscopy (AFM) to study rafts in model membranes [107]. Moreover, in live zebrafish embryos, in vivo binding affinities of Fgf8 to its receptors, Fgfr1 and Fgfr4 were measured [108]. In combination with STED, called axial ls-STED-FCS, time resolution on the μs-scale and spatial resolution of ~50 nm were achieved using tunable acoustic gradient index of refraction lenses [109]. Dual-colour dual-focus line scanning was applied in Wnt signalling pathway study to quantify ligand–receptor concentration, and diffusion coefficient without the need for a separate observation volume calibration [110].

3.4 Raster Image Correlation Spectroscopy: RICS

Scanning FCS utilizes the temporal and spatial information in a scan. In these applications, typically, the scan speed was uniform along a line or circular path. Nevertheless, even a confocal image contains an inherent time structure, with different time constants and the two perpendicular scan directions, that can calculate correlation functions. The fast scan direction is similar to the line-scan FCS with small delays τ_p between the acquisition of neighbouring pixels. However, when moving to the next line, the delay between the pixels of neighbouring lines, τ_l is much longer, depending on the time to scan one line. RICS uses this inherent time structure to calculate spatiotemporal correlation functions over small regions of interest (Fig. 6a).

To see how this works, let us look at the time structure of a confocal image. The time delay between any two pixels depends on their distance on the image grid. If we call the spatial lag, measured in number of pixels, in the fast scan direction ξ and the spatial lag in the slow scan direction (i.e. the lag in number of lines) ψ, then the time delay between these two pixels will be given by $\tau_p.\ \xi + \tau_l.\ \psi$, the RICS correlation function can thus be expressed as a correlation function of the spatial lags, which inherently defines the time dependence.

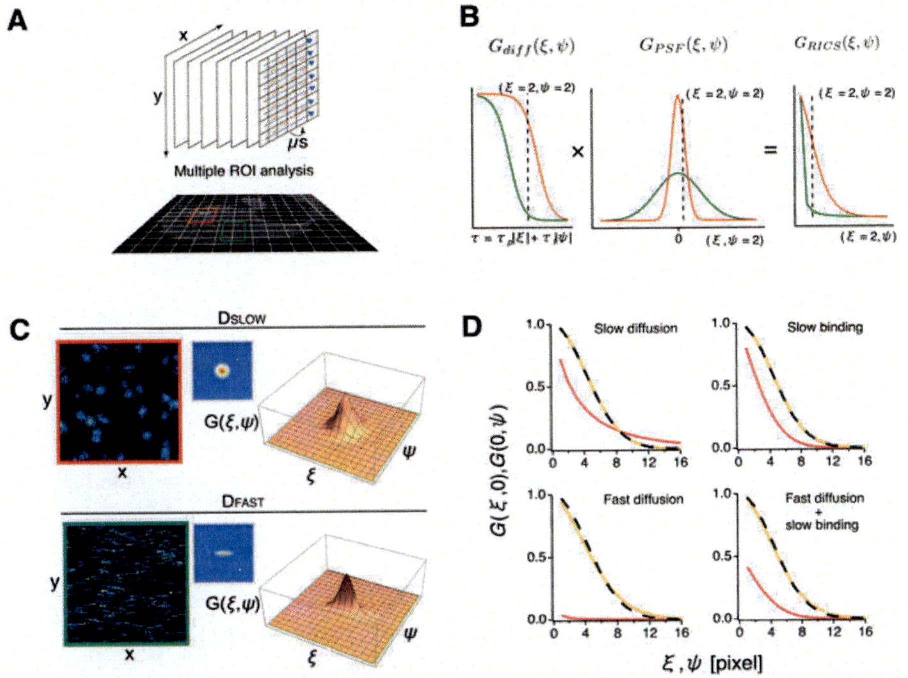

Fig. 6 Schematic illustration of a RICS experiment. (**a**) Rastered scanned image(s) can be analysed over multiple regions of interest (ROI) to create parameter maps. (**b**) The first two graphs show the different RICS correlation function components for diffusion and the PSF. The third graph shows the combination of the two. The red and green lines show the characteristic correlation function over the slow and fast regions. (**c**) RICS simulation and analysis for slow and fast-moving particles performed with simFCS3 [111]. The graph on the right shows the resulting 2D spatiotemporal correlation functions. (**d**) RICS allows binding to be discriminated against diffusion, provided binding/unbinding is much slower than diffusion. RICS cross-sections along and perpendicular to the scanned direction are plotted in orange and pink. The dotted curves stand for the PSF profile

$$G_{RICS}(\xi, \psi) = \frac{\langle \delta I(x, y)\delta I(x+\xi, y+\psi)\rangle}{\langle I\rangle^2} \tag{18}$$

The RICS correlation function (Fig. 6b (right)) contains now at least two contributions. First is the decay of the correlation function for diffusion (Fig. 6b (left)) due to the time delay between two pixels. Second, the broadening of the PSF (Fig. 6b (middle)) due to particle movement as we have seen before in the section on ICS.

In confocal FCS the correlation function for diffusion in 3D is given by (Magde 1974):

$$G_{FCS}(\tau) = \frac{\langle \delta I(t)\delta I(t+\tau)\rangle}{\langle I\rangle^2} = \frac{1}{N}\left(1+\frac{4D\tau}{\omega_0^2}\right)^{-1}\left(1+\frac{4D\tau}{\omega_z^2}\right)^{-\frac{1}{2}} + G_\infty \tag{19}$$

Here G_∞ is the value of the correlation function at long times, which is ideally 0 but is typically used as fitting parameter as the ideal value is only reached for infinitely long times. In RICS, due to its particular time structure, this component is

$$G_{\text{diff}}(\xi,\psi)=\frac{1}{N}\left(1+\frac{4D(\tau_p|\xi|+\tau_l|\psi|)}{\omega_0^2}\right)^{-1}\left(1+\frac{4D(\tau_p|\xi|+\tau_l|\psi|)}{\omega_z^2}\right)^{-\frac{1}{2}} \tag{20}$$

The second component describing the broadening of the PSF due to particle diffusion is given by

$$G_{PSF}(\xi,\psi)=exp\left\{-\frac{a^2(\xi^2+\psi^2)}{4D(\tau_p|\xi|+\tau_l|\psi|)+w_0^2}\right\} \tag{21}$$

where we denote the size of a pixel by a, note that if there is no diffusion, this equation will describe the PSF (Eq. 4). Putting this together, we obtain the correlation function for RICS

$$\begin{aligned}G_{\text{RICS}}(\xi,\psi)\\=\frac{1}{N}\left(1+\frac{4D(\tau_p|\xi|+\tau_l|\psi|)}{\omega_0^2}\right)^{-1}\left(1+\frac{4D(\tau_p|\xi|+\tau_l|\psi|)}{\omega_z^2}\right)^{-\frac{1}{2}}\\\times\exp\left\{-\frac{a^2(\xi^2+\psi^2)}{4D(\tau_p|\xi|+\tau_l|\psi|)+w_0^2}\right\}+G_\infty\end{aligned} \tag{22}$$

Other processes, e.g. blinking or binding, can be included in the correlation function as well [112]. This correlation function is very similar to the correlation functions for STICS (Eq. 9) and scanning FCS (Eq. 15). However, in RICS, we have explicitly taken the third axial dimension into account, as RICS can acquire sufficiently fast to measure diffusion in solution (Fig. 6c). An interesting point about RICS is that it covers different temporal ranges as the temporal delays in different image directions are different by 1–2 orders of magnitude. Moreover, these time ranges can be tuned by the scanning speed to acquire the data in a suitable range for the probe under investigation. Note that for RICS the pixel size is set smaller than the PSF, and multiple frames are averaged, especially in the case of small regions of interest. Other measures such as filtering immobile structures to detect better-diffusing particles were performed in the time domain by averaging images and subtracting the average image from each frame [15].

A two-colour variant called cc-RICS has been used to quantitatively evaluate the fraction of intact DNAs with both green and red fluorescence in living cells [81, 113, 114]. Compared to FCS/FCCS, such image-based RICS/cc-RICS has the advantage to allow monitoring of diffusion dynamics, reaction kinetics, and molecular interaction at multiple ROIs [115–117].

As RICS is implemented on a confocal microscope, it can be easily integrated with stimulated emission depletion (STED) microscopy providing simultaneous super-resolution and dynamics measurements [118, 119]. It is by now also possible to scan random areas [115]. For more details on its implementation and evaluation of RICS characteristics, we refer the interested reader to the literature [111, 120].

The fact that RICS can be used with commercially available confocal microscopes without any customization is a major advantage, and it consequently has been widely used. Among others, RICS has been used to detect binding (Fig. 6d) to adhesions and scaffolds [112, 121] and to measure lipid dynamics in cell membranes [122–125].

4 Multipoint and Imaging FCS

The idea to record multiple points for FCS simultaneously is an obvious extension of single-point confocal FCS, although not easily implemented as mentioned earlier. The availability of array detectors with a large number of detection elements and sufficiently fast read-out changed that situation. By using a spinning disk confocal microscope with a resting spinning disk to create multiple confocal areas, this was extended to ~1,000 detection elements [126]. This approach has recently culminated in the so-called massively parallel FCS (Fig. 7a), where a diffractive element creates a grid of confocal points and the signal is detected by recently developed single-photon avalanche diode (SPAD) arrays, where each pixel functions as its own pinhole and detector [127, 128]. This approach has the advantage of having large numbers of detection elements and having excellent time resolution sufficient even to measure fluorescence lifetimes, which other array detectors, especially cameras, cannot reach. Oasa et al. combined FLIM with massively parallel FCS (mpFCS) to characterize the diffusion across a cell of oligodendrocyte transcription factor 2 [129].

However, all these approaches use confocal illumination schemes. As the laser beams in confocal schemes are highly focused and traverse the whole sample as they converge before and diverge after the focal point, they create significant cross-talk for neighbouring pixels (Fig. 7a), a problem long known for spinning disk confocal microscopes [130]. For FCS this sets various lower limits on the minimum spacing between pixels required to avoid cross-talk between neighbouring pixels [29]. If pixels, and thus laser beams, are too close, the signal detected in one pinhole will stem from multiple laser beams, increasing the cross-talk and the effective observation volume, making the measurement of fluctuations necessary for FCS impossible. But even if ACFs can be recorded for each pinhole, the pixels need to be far enough apart to avoid an influence of the sample observed by one pixel to influence the shape of the ACF of the neighbouring pixel. One needs at least two photons from a particle for a correlation. Therefore, as long as a pinhole can detect two photons from a molecule in a neighbouring laser beam, it will lead to extra correlations in the ACF. However, more stringently, to avoid an influence of the cross-talk on the amplitude

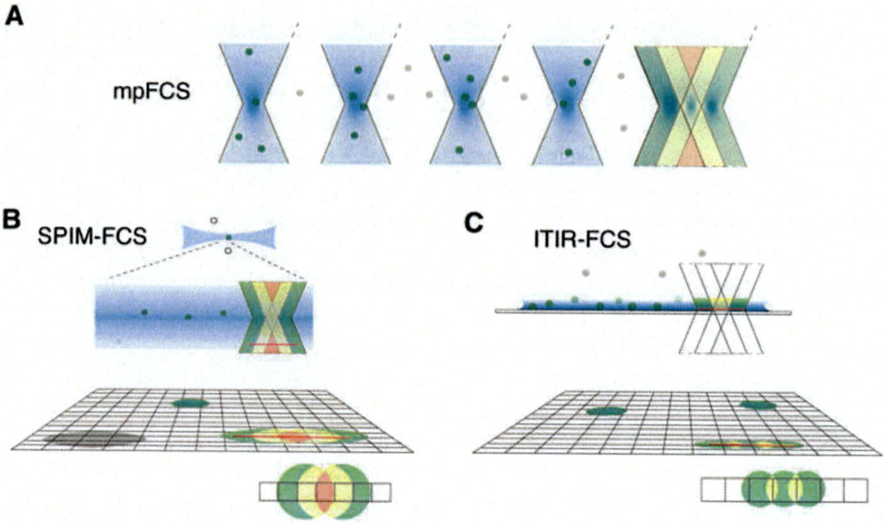

Fig. 7 Illumination scheme of various FCS multiplexing methods. (**a**) Simultaneous excitation and detection of fluorescent molecules in multiple confocal spots (mpFCS). The observation volumes have to be placed at a certain distance to avoid cross-talk. On the right, three observation volumes are placed contiguously in the sample. As indicated by the colours, some parts of the sample are illuminated and detected by all three observation volumes (*red*), some by two (*yellow*), and some by only one *green*. Note that because the laser beam in a confocal setup propagates through the whole sample, and thus the amount of cross-talk depends on the sample thickness, which is typically on the μm scale. (**b**) The use of light-sheet illumination and array detectors in SPIM-FCS. As the illuminating light sheet is only ~1 μm thick, cross-talk is limited, as indicated again by the colour-coded observation volumes, and contiguous pixels can be used. Nevertheless, the cross-talk will change the correlation function and needs to be accounted for. Cross-talk between pixels is indicated at the bottom of the graph. (**c**) In Imaging Total Internal Reflection FCS (ITIR-FCS), optical sectioning is based on the confinement of the evanescent wave generated at the glass/water interface. In ITIR-FCS the cross-talk is limited by both, the sample which is typically a membrane with ~5 nm thickness, or if measuring in solution by the exponentially decaying evanescent wave of ~100 nm thickness. Cross-talk between pixels is indicated at the bottom of the graph

(not the width or shape) of the ACFs, the pixels need to be even further apart. Although single photons are not enough for a correlation, uncorrelated single photons will contribute to the background and thus lower the ACF amplitude. These restrictions limit the number of pixels that can be used in confocal approaches.

A partial solution to the problem was found with new detection schemes that illuminate only a cross-section of limited thickness in a specimen. This can come in the form of TIRF (Fig. 7c (top)) with only ~100 nm above the cover slide illuminated [32, 131], or in light-sheet microscopy (Fig. 7b (top)) with light-sheet thicknesses of ~1 μm [132]. This minimizes cross-talk between pixels as only single sections of the sample are illuminated, and no signal from the bulk or other parts of the sample is excited and detected. This then allows the use of cameras with contiguous pixels as detectors [29, 30]. It should be noted that neighbouring pixels will still be correlated to some extent due to the finite size of the PSF, and the image of one molecule can

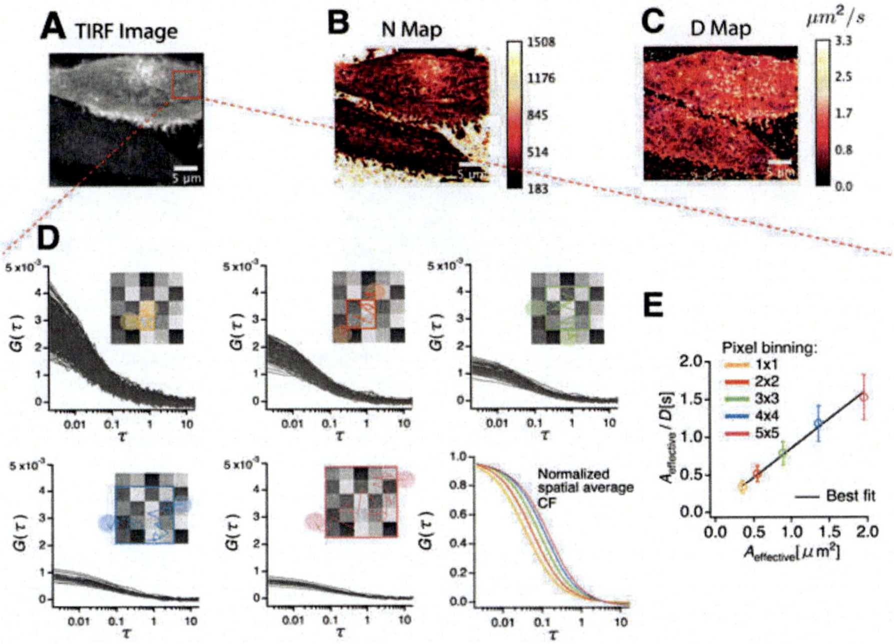

Fig. 8 Study of sample organization by Imaging FCS. (**a**) TIRFM image of PMT-mEGFP showing two CHO-K1 cells at different expression levels (TIRFM image displayed as average intensity z-projection for clarity). Each pixel was temporally correlated and fitted, and quantitative parameters are rendered as heat maps for (**b**) the number of particles *N* and (**c**) the diffusion coefficient *D*. (**d**) Imaging FCS allows pinhole size selection during data treatment by pixel binning. The autocorrelation function amplitude decreases *(left top) to (middle bottom)*. Furthermore, its width increases (*bottom right*) with the increase of the effective observation area. (**e**) Diffusion and its length-scale dependency to identify diffusive modes

always span at least neighbouring pixels (Fig. 7b, c (bottom)), and imaging FCS takes account of these effects in their data evaluation. But this effect is limited and is similar to the effect of the PSF in imaging. However, camera-based FCS, although allowing for the recording of contiguous pixels in a sample and thus creating full spatial maps (Fig. 8b, c), can do so only at a much lower time resolution. Most experiments in Imaging FCS have been performed at ~ms time resolution, although a resolution down to 0.02 ms has been reached [33, 133].

Imaging FCS was calibrated against a range of other techniques to ensure consistency of its results with the main techniques used to measure diffusion and diffusion modes [134–137] and was applied in a variety of contexts [35]. Cooper et al. measured the diffusion at liquid–solid interfaces and determined molecular interaction kinetics with surfaces [138]. Xu et al. measured diffusion within nanotubes and determined the influence of charge on molecular transport [139], Erstling et al. determined the photophysics of nanoparticles [140], and several applications for the characterization of lipid bilayers were reported [134, 141]. In

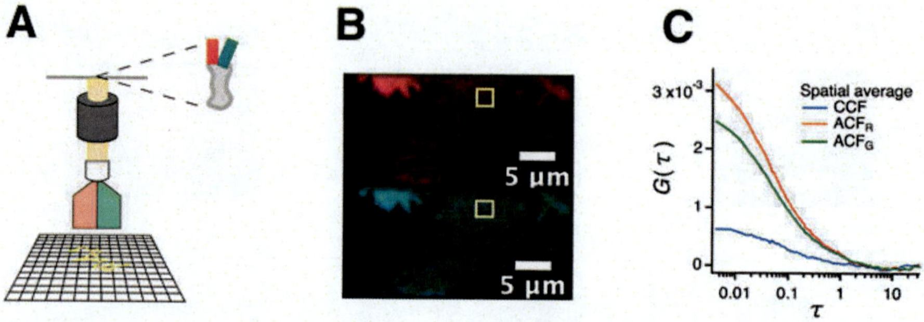

Fig. 9 DC-ITIR-FCCS modality [136]. (**a**) Schematic of TIRFM instrumentation coupled with an image splitter device to capture images of PMT-mEGFP-mApple into wavelength channels at 500 fps. (**b**) Image of double-labelled plasma membrane targeting PMT-mEGFP-mCherry displayed on two halves of the camera chip. (**c**) The double-labelled protein yields a positive cross-correlation in HEK cells; the auto- (*green* and *red*) and cross-correlation (*blue*) functions displayed are the spatial average across a selected ROI (in *yellow*)

the life sciences the interaction of peptides with lipid bilayers [142] and cell membranes [143] and the characterization of membrane receptor diffusion are typical applications of Imaging FCS [144–146]. By pooling several Imaging FCS measurements, Bag et al. obtained very precise diffusion coefficient measurements of lipids and membrane proteins in live cells [147, 148]. Imaging FCS was also used in developing small organisms to determine cytoplasmic and membrane diffusion [133, 149–151], and in biofilms to characterize diffusion in dependence of size and charge of diffusants [152]. In the form of Imaging FCCS (Fig. 9) it measured interactions of proteins in live cells [149, 153, 154].

This brings us to the question of what advantages or extra information these FCS modalities provide. The obvious answer is multiplexing. The approaches described in this section provide thousands of points routinely simultaneously, and even more than one million points in a single measurement have been achieved (Wohland unpublished data). The measurements can also be performed with two colours allowing to perform FCCS with a single array detector with excitation either in a continuous [149, 153–155] or interleaved manner (Wohland unpublished data).

A further advantage is that – similar to what was earlier mentioned in confocal microscopy with multipoint detectors for a single detection element – the data on array detectors can detect different parts of a sample and determine whether there is existing flow or transport [29] or characterize anisotropic diffusion [156]. Furthermore, the pixels on a camera can be binned to provide observation areas and volumes of different sizes [29]. The FCS diffusion law (Fig. 8e) can be used to determine the length-scale dependence of diffusion and thus determine the particular diffusion mode the particle is undergoing [79, 157, 158]. In 2D FCS and especially Imaging FCS this can be easily achieved as only one set of data is needed, and various binning sizes (Fig. 8d) can be applied to the existing data to create different observation areas [134, 141, 145, 159–161]. Thus, FCS diffusion law analysis can be derived from a single measurement and can be even performed on multiple areas of a cell [145]. Nevertheless, these approaches based on array detectors have wider applicability. By recording the maximal possible amounts of photons with as high spatial and temporal resolution as possible, one is not restricted to perform exclusively FCS but can evaluate the data in each or between any pixels or group of pixels with whatever analytical tools available. So it was already shown that FCS can be combined with super-resolution microscopy [162, 163] and simultaneously with other fluctuation or single-molecule fluorescence techniques [135, 162, 163].

Another approach that provides binding kinetics and that is at the origin of many of the techniques discussed here is the combination of TIRFM with FCS with single-point detection [164–167]. Although not strictly 2D, several studies have demonstrated the application of TIRF-FCS to measure biomolecular diffusion in membranes and ligand–receptor binding kinetics [164, 165, 168]. Extension to an imaging version could extend binding kinetics analyses to whole surfaces. A modality that integrates over a whole image to characterize reversible binding has already been established [169].

Finally, we need to discuss the differences between these schemes and how the data is evaluated. The 2D FCS modalities based on confocal schemes can use the already developed formulas for confocal FCS. This includes not only the parts for diffusion but as well for photophysical and other processes. Here, spatial information can be exploited by calculating spatial cross-correlations. In the case of two observation volumes with negligible cross-talk, the spatial cross-correlation function is given by [170]:

$$G(\tau, d) = \frac{1}{N}\left(1 + \frac{4D\tau}{w_0^2}\right)^{-1}\left(1 + \frac{4D\tau}{z_0^2}\right)^{-\frac{1}{2}} exp\left\{-\frac{d_x^2 + d_y^2}{4D\tau + w_0^2} - \frac{d_z^2}{4D\tau + z_0^2}\right\} \tag{23}$$

Here, d_x, d_y, and d_z are the distances between the two observation volumes in x, y, and z directions, and the other parameters are as previously defined.

As typical in confocal FCS, the observation volume is approximated here by a 3D Gaussian. However, in the case of 2fFCS, this approximation is not sufficient anymore, and a more sophisticated model needs to be used that requires numerical integration [23, 25]. 2fFCS is more robust against typical FCS artefacts, including optical aberrations, e.g., introduced by cover glass thickness variations or mismatch of refractive indices between mounting and immersion medium, and saturation effects stemming from the use of high laser powers for excitation [23, 171]. In the case of TIRF and SPIM-based illuminations with cameras as detectors, the excitation and detection change significantly compared to the confocal case. First, in the confocal case, the excitation profile is a Gaussian function and thus falls off as one moves away from the centre of the focus. In the TIRF and SPIM cases, the excitation intensity is constant in the focal plane and varies only in the axial direction. In addition, while in confocal FCS typically round pinholes are used, pixels on cameras are typically quadratic, which changes and simplifies the derivation of the FCS fit models [172]. Solutions for single-spot TIRF-FCS correlation functions taking into account the full 3D geometry and also binding interactions are available [166]. In the case of camera-based FCS, assuming measurements in a membrane, i.e., a strictly 2D layer in the focal plane, the correlation function has an analytic solution [173–175]. We provide here the basic correlation function for two square pixels assuming only diffusion and refer the reader to literature for special cases [156]. If we define the function p as

$$p(\tau, d) = \sqrt{\frac{4D\tau + w_0^2}{\pi}} \exp\left\{\frac{d^2}{4D\tau + w_0^2}\right\} + d \times erf\left[\frac{d}{\sqrt{4D\tau + w_0^2}}\right] \tag{24}$$

where d is the distance between two pixels in x- or y-direction, then we can write the function describing the shape of the correlations function separately for the x- and y-directions as

$$g_x(\tau) = (p(\tau,\ a + r_x) + p(\tau,\ a - r_x) - 2p(\tau,\ r_x)) \tag{25}$$

$$g_y(\tau) = (p(\tau,\ a + r_y) + p(\tau,\ a - r_y) - 2p(\tau,\ r_y)) \tag{26}$$

The function for the 2D case is then just the product of the correlations for the x- and y-directions

$$g(\tau) = g_x(\tau) g_y(\tau) \tag{27}$$

The ACF is obtained by setting the distances r_x and r_y to zero. Finally, we normalize the correlation function with the effective observation volume

$$G(\tau) = \frac{1}{N} \frac{g(\tau)}{g(0)} + G_\infty \tag{28}$$

Normalization ensures that the ACF amplitude is inversely proportional to the number of particles. For the CCF, the peak amplitude is proportional to the particles moving from one pixel to the other. A comparison with the ACFs of the two spatially correlated areas is necessary to deduce the number of cross-correlated particles, similar to dual-colour FCCS. If the two areas to be cross-correlated are not single pixels, then the spatial cross-correlations between two arbitrary binned pixel areas are just the sum of all possible CCFs between all pixels in the two areas [156]. Models for 3D measurements in SPIM-FCS [33, 149, 155] have also been derived, although they neglected cross-talk between pixels. Taking into account cross-talk between pixels, as can be necessary in 3D measurements, requires numerical integration as analytical solutions are no longer obtainable (in preparation).

If the characteristic length scale over which diffusion is measured is precisely known, the methods become essentially calibration-free. In confocal FCS, the precise dimensions of the confocal volume are typically unknown and are alignment dependent on which FCS is very sensitive. Thus, they need to be calibrated each measurement day. However, Imaging FCS modalities are much less alignment dependent as the illumination intensity is constant, the pixel size and their distance are exactly known, and the microscope PSF of an optical setup, which is an essential determinant of the observation volume of a pixel, typically does not change over time. Thus, Imaging FCS can directly determine the diffusion coefficient (Fig. 8c) without extra calibrations, and D is a fitting parameter in the model. In contrast, in confocal FCS the characteristic time a molecule needs to traverse the confocal observation volume is fitted and later recalculated into diffusion coefficients using a calibration measurement.

However, camera-based methods have also disadvantages. First, they are typically much slower with time resolutions in the millisecond range. This has been addressed by using the newly available SPAD arrays, which can measure faster [21, 22] and can even determine fluorescence lifetimes [129]. In addition, the observation volume in light-sheet microscopy for 3D measurements is generally larger than in the confocal case, limiting the concentration range and the signal-to-noise ratio accessible in SPIM-FCS. This results from the lower numerical aperture of the illumination objectives used in SPIM-FCS, mainly for practical reasons due to space constraints, making it impossible to align two high NA objectives. The lower limit in concentration has been addressed by using two objectives with equal but somewhat lower numerical aperture [176] and by SPIM using a single objective (soSPIM) for illumination, and detection [177]. Finally, the amount of data acquired in any of the 2D FCS methods is typically large as thousands of pixels are recorded over thousands of frames, and thus more storage and faster data evaluation are required. In the last years, several groups have therefore developed GPU based algorithms that speed up data evaluation by 1–2 orders of magnitude [128, 163, 178] and analyse data in real time [179].

5 Concluding Remarks

Correlation-based approaches are compelling tools to extract information from signals by analysing signal fluctuations around their mean value. By using fluorescence as the signal, the measurements become specific and provide single-molecule sensitivity. The resulting Fluorescence Correlation Spectroscopy measurements yield information on molecular transport, actions, and interactions. As FCS was a single-point measurement, it was multiplexed to measure multiple FCS curves simultaneously and was extended to include spatial correlations to obtain more information from a single measurement. We grouped these measurements here under the name of two-dimensional FCS, as they measure some or all points in a 2D sample cross-section and exploit spatial, temporal, or spatiotemporal correlations.

With new developments in microscope, detector, and computational technology and advances in data evaluation strategies, many of these techniques can be now performed on standard commercially available microscopes without extra modifications. Moreover, the publication of more open-source software tools [179] and reviews around the topic [34, 180] makes two-dimensional FCS easily accessible to a broad spectrum of users (for a list of software, see [175]). Finally, these methods are ideal to be combined with various microscopy and super-resolution techniques [66, 118, 122, 162, 163, 181–184] presenting the possibility to record the structure and dynamics of the system under investigation simultaneously and thus creating a much more complete picture of molecular events in live cells and organisms.

References

1. Magde D, Elson E, Webb WW (1972) Thermodynamic fluctuations in a reacting system – measurement by fluorescence correlation spectroscopy. Phys Rev Lett 29:705–708
2. Ehrenberg M, Rigler R (1974) Rotational Brownian motion and fluorescence intensify fluctuations. Chem Phys 4:390–401
3. Petersen NO, Höddelius PL, Wiseman PW, Seger O, Magnusson KE (1993) Quantitation of membrane receptor distributions by image correlation spectroscopy: concept and application. Biophys J 65:1135–1146
4. Wiseman PW, Brown CM, Webb DJ, Hebert B, Johnson NL, Squier JA, Ellisman MH, Horwitz AF (2004) Spatial mapping of integrin interactions and dynamics during cell migration by image correlation microscopy. J Cell Sci 117:5521–5534
5. Hebert B, Costantino S, Wiseman PW (2005) Spatiotemporal image correlation spectroscopy (STICS) theory, verification, and application to protein velocity mapping in living CHO cells. Biophys J 88:3601–3614
6. Magde D, Webb WW, Elson EL (1978) Fluorescence correlation spectroscopy. III. Uniform translation and laminar flow. Biopolymers 17:361–376
7. Petersen NO (1986) Scanning fluorescence correlation spectroscopy. I. Theory and simulation of aggregation measurements. Biophys J 49:809–815
8. Petersen NO, Johnson DC, Schlesinger MJ (1986) Scanning fluorescence correlation spectroscopy. II. Application to virus glycoprotein aggregation. Biophys J 49:817–820
9. St-Pierre PR, Petersen NO (1990) Relative ligand binding to small or large aggregates measured by scanning correlation spectroscopy. Biophys J 58:503–511
10. St-Pierre PR, Petersen NO (1992) Average density and size of microclusters of epidermal growth factor receptors on A431 cells. Biochemistry 31:2459–2463
11. Pan X, Yu H, Shi X, Korzh V, Wohland T (2007) Characterization of flow direction in microchannels and zebrafish blood vessels by scanning fluorescence correlation spectroscopy. J Biomed Opt 12:1–10
12. Pan X, Shi X, Korzh V, Yu H, Wohland T (2009) Line scan fluorescence correlation spectroscopy for three-dimensional microfluidic flow velocity measurements. J Biomed Opt 14:1–6
13. Koppel DE, Morgan F, Cowan AE, Carson JH (1994) Scanning concentration correlation spectroscopy using the confocal laser microscope. Biophys J 66:502–507
14. Skinner JP, Chen Y, Müller JD (2005) Position-sensitive scanning fluorescence correlation spectroscopy. Biophys J 89:1288–1301
15. Digman MA, Sengupta P, Wiseman PW, Brown CM, Horwitz AR, Gratton E (2005) Fluctuation correlation spectroscopy with a laser-scanning microscope: exploiting the hidden time structure. Biophys J 88:L33–L36
16. Brinkmeier M, Dörre K, Riebeseel K, Rigler R (1997) Confocal spectroscopy in microstructures. Biophys Chem 66:229–239

17. Brinkmeier M, Dörre K, Stephan J, Eigen M (1999) Two-beam cross-correlation: a method to characterize transport phenomena in micrometer-sized structures. Anal Chem 71:609–616
18. Dittrich PS, Schwille P (2002) Spatial two-photon fluorescence cross-correlation spectroscopy for controlling molecular transport in microfluidic structures. Anal Chem 74:4472–4479
19. Jaffiol R, Blancquaert Y, Delon A, Derouard J (2006) Spatial fluorescence cross-correlation spectroscopy. Appl Optics 45:1225–1235
20. Pan X, Foo W, Lim W, Fok MHY, Liu P, Yu H, Maruyama I, Wohland T (2007) Multifunctional fluorescence correlation microscope for intracellular and microfluidic measurements. Rev Sci Instrum 78:53711
21. Scipioni L, Lanzanó L, Diaspro A, Gratton E (2018) Comprehensive correlation analysis for super-resolution dynamic fingerprinting of cellular compartments using the Zeiss Airyscan detector. Nat Commun 9:5120
22. Slenders E, Castello M, Buttafava M, Villa F, Tosi A, Lanzanò L, Koho SV, Vicidomini G (2021) Confocal-based fluorescence fluctuation spectroscopy with a SPAD array detector. Light Sci Appl 10:31
23. Dertinger T, Pacheco V, von der Hocht I, Hartmann R, Gregor I, Enderlein J (2007) Two-focus fluorescence correlation spectroscopy: a new tool for accurate and absolute diffusion measurements. ChemPhysChem 8:433–443
24. Müller CB, Loman A, Richtering W, Enderlein J (2008) Dual-focus fluorescence correlation spectroscopy of colloidal solutions: influence of particle size. J Phys Chem B 112:8236–8240
25. Goossens K, Prior M, Pacheco V, Willbold D, Müllen K, Enderlein J, Hofkens J, Gregor I (2015) Accurate diffusion coefficients of organosoluble reference dyes in organic media measured by dual-focus fluorescence correlation spectroscopy. ACS Nano 9:7360–7373
26. Gösch M, Serov A, Anhut T, Lasser T, Rochas A, Besse PA, Popovic RS, Blom H, Rigler R (2004) Parallel single molecule detection with a fully integrated single-photon 2x2 CMOS detector array. J Biomed Opt 9:913–921
27. Yamamoto J, Mikuni S, Kinjo M (2018) Multipoint fluorescence correlation spectroscopy using spatial light modulator. Biomed Opt Express 9:5881–5890
28. Rochas A, Gosch M, Serov A, Besse PA, Popovic RS, Lasser T, Rigler R (2003) First fully integrated 2-D array of single-photon detectors in standard CMOS technology. IEEE Photon Technol Lett 15:963–965
29. Kannan B, Har JY, Liu P, Maruyama I, Ding JL, Wohland T (2006) Electron multiplying charge-coupled device camera based fluorescence correlation spectroscopy. Anal Chem 78: 3444–3451
30. Burkhardt M, Schwille P (2006) Electron multiplying CCD based detection for spatially resolved fluorescence correlation spectroscopy. Opt Express 14:5013–5020
31. Sisan DR, Arevalo R, Graves C, McAllister R, Urbach JS (2006) Spatially resolved fluorescence correlation spectroscopy using a spinning disk confocal microscope. Biophys J 91: 4241–4252
32. Kannan B, Guo L, Sudhaharan T, Ahmed S, Maruyama I, Wohland T (2007) Spatially resolved total internal reflection fluorescence correlation microscopy using an electron multiplying charge-coupled device camera. Anal Chem 79:4463–4470
33. Singh AP, Krieger JW, Buchholz J, Charbon E, Langowski J, Wohland T (2013) The performance of 2D array detectors for light sheet based fluorescence correlation spectroscopy. Opt Express 21:8652–8668
34. Bag N, Wohland T (2014) Imaging fluorescence fluctuation spectroscopy: new tools for quantitative bioimaging. Annu Rev Phys Chem 65:225–248
35. Singh AP, Wohland T (2014) Applications of imaging fluorescence correlation spectroscopy. Curr Opin Chem Biol 20:29–35
36. Srivastava M, Petersen NO (1996) Image cross-correlation spectroscopy: a new experimental biophysical approach to measurement of slow diffusion of fluorescent molecules. Methods Cell Sci 18:47–54

37. Wiseman PW, Petersen NO (1999) Image correlation spectroscopy. II. Optimization for ultrasensitive detection of preexisting platelet-derived growth factor-beta receptor oligomers on intact cells. Biophys J 76:963–977
38. Costantino S, Comeau JWD, Kolin DL, Wiseman PW (2005) Accuracy and dynamic range of spatial image correlation and cross-correlation spectroscopy. Biophys J 89:1251–1260
39. Rappaz B, Wiseman PW (2013) Image correlation spectroscopy for measurements of particle densities and colocalization. Curr Protoc Cell Biol 59:4.27.1–4.27.15
40. Lajevardipour A, Clayton A (2019) pbICS microscopy technique for determining oligomeric state
41. Comeau JW, Costantino S, Wiseman PW (2006) A guide to accurate fluorescence microscopy colocalization measurements. Biophys J 91:4611–4622
42. Crites TJ, Chen L, Varma R (2022) A TIRF microscopy technique for real-time, simultaneous imaging of the TCR and its associated signaling proteins. J Vis Exp 61:3892
43. Bolte S, Cordelières FP (2006) A guided tour into subcellular colocalization analysis in light microscopy. J Microsc 224:213–232
44. Cerutti E, D'Amico M, Cainero I, Dellino GI, Faretta M, Vicidomini G, Pelicci PG, Bianchini P, Diaspro A, Lanzanò L (2021) Evaluation of sted super-resolution image quality by image correlation spectroscopy (QuICS). Sci Rep 11:20782
45. Oneto M, Scipioni L, Sarmento MJ, Cainero I, Pelicci S, Furia L, Pelicci PG, Dellino GI, Bianchini P, Faretta M, Gratton E, Diaspro A, Lanzanò L (2019) Nanoscale distribution of nuclear sites by super-resolved image cross-correlation spectroscopy. Biophys J 117:2054–2065
46. Cainero I, Cerutti E, Faretta M, Dellino GI, Pelicci PG, Diaspro A, Lanzanò L (2021) Measuring nanoscale distances by structured illumination microscopy and image cross-correlation spectroscopy (SIM-ICCS). Sensors 21
47. Wiseman PW, Squier JA, Ellisman MH, Wilson KR (2000) Two-photon image correlation spectroscopy and image cross-correlation spectroscopy. J Microsc 200:14–25
48. Gopal AA, Rappaz B, Rouger V, Martyn IB, Dahlberg PD, Meland RJ, Beamish IV, Kennedy TE, Wiseman PW (2016) Netrin-1-regulated distribution of UNC5B and DCC in live cells revealed by TICCS. Biophys J 110:623–634
49. Penjweini R, Deville S, Maghsoudi OH, Notelaers K, Ethirajan A, Ameloot M (2019) Investigating the effect of poly-l-lactic acid nanoparticles carrying hypericin on the flow-biased diffusive motion of HeLa cell organelles. J Pharm Pharmacol 71:104–116
50. Kolin DL, Wiseman PW (2007) Advances in image correlation spectroscopy: measuring number densities, aggregation states, and dynamics of fluorescently labeled macromolecules in cells. Cell Biochem Biophys 49:141–164
51. Petersen NO, Brown C, Kaminski A, Rocheleau J, Srivastava M, Wiseman PW (1998) Analysis of membrane protein cluster densities and sizes in situ by image correlation spectroscopy. Faraday Discuss 111:289–243
52. Mets RD, Wang I, Balland M, Oddou C, Moreau P, Fourcade B, Albiges-Rizo C, Delon A, Destaing O (2019) Cellular tension encodes local Src-dependent differential β 1 and β 3 integrin mobility. Mol Biol Cell 30:181–190
53. Altmann SM, Lenne PF (2002) Forced unfolding of single proteins. Methods Cell Biol 68: 311–335
54. Nayal A, Webb DJ, Brown CM, Schaefer EM, Vicente-Manzanares M, Horwitz AR (2006) Paxillin phosphorylation at Ser273 localizes a GIT1–PIX–PAK complex and regulates adhesion and protrusion dynamics. J Cell Biol 173:587–589
55. Travagliati M, Girardo S, Pisignano D, Beltram F, Cecchini M (2013) Easy monitoring of velocity fields in microfluidic devices using spatiotemporal image correlation spectroscopy. Anal Chem 85:8080–8084
56. Coppola S, Pozzi D, Sanctis SCD, Digman MA, Gratton E, Caracciolo G (2013) Quantitative measurement of intracellular transport of nanocarriers by spatio-temporal image correlation spectroscopy. Methods Appl Fluoresc 1:15005

57. Potvin-Trottier L, Lord ND, Vinnicombe G, Paulsson J (2016) Synchronous long-term oscillations in a synthetic gene circuit. Nature 538:514–517
58. Harrison PL, Heath GR, Johnson BRG, Abdel-Rahman MA, Strong PN, Evans SD, Miller K (2016) Phospholipid dependent mechanism of smp24, an α-helical antimicrobial peptide from scorpion venom. Biochim Biophys Acta 1858:2737–2744
59. Nie J, Xu C, Jin J, Aka JA, Tempel W, Nguyen V, You L, Weist R, Min J, Pawson T, Yang XJ (2015) Ankyrin repeats of ANKRA2 recognize a PxLPxL motif on the 3M syndrome protein CCDC8. Structure 23:700–712
60. Rowland DJ, Tuson HH, Biteen JS (2016) Resolving fast, confined diffusion in bacteria with image correlation spectroscopy. Biophys J 110:2241–2251
61. Petrášek Z, Hoege C, Hyman AA, Schwille P (2008) Two-photon fluorescence imaging and correlation analysis applied to protein dynamics in C. elegans embryo
62. Pandžić E, Rossy J, Gaus K (2015) Tracking molecular dynamics without tracking: image correlation of photo-activation microscopy. Methods Appl Fluoresc 3:14006
63. Toplak T, Pandzic E, Chen L, Vicente-Manzanares M, Horwitz AR, Wiseman PW (2012) STICCS reveals matrix-dependent adhesion slipping and gripping in migrating cells. Biophys J 103:1672–1682
64. Semrau S, Schmidt T (2007) Particle image correlation spectroscopy (PICS): retrieving nanometer-scale correlations from high-density single-molecule position data. Biophys J 92: 613–621
65. Dupont A, Stirnnagel K, Lindemann D, Lamb DC (2013) Tracking image correlation: combining single-particle tracking and image correlation. Biophys J 104:2373–2382
66. Ashdown G, Cope A, Wiseman P, Owen D (2014) Molecular flow quantified beyond the diffraction limit by spatiotemporal image correlation of structured illumination microscopy data. Biophys J 107:L21–L23
67. Kolin DL, Ronis D, Wiseman PW (2006) k-Space image correlation spectroscopy: a method for accurate transport measurements independent of fluorophore photophysics. Biophys J 91: 3061–3075
68. Waithe D, Brown JM, Reglinski K, Diez-Sevilla I, Roberts D, Eggeling C (2020) Object detection networks and augmented reality for cellular detection in fluorescence microscopy. J Cell Biol 219:e201903166
69. Kolin DL, Costantino S, Wiseman PW (2006) Sampling effects, noise, and photobleaching in temporal image correlation spectroscopy. Biophys J 90:628–639
70. Bouzin M, Sironi L, Chirico G, D'Alfonso L, Inverso D, Pallavicini P, Collini M (2015) An intermittent model for intracellular motions of gold nanostars by k-space scattering image correlation. Biophys J 109:2246–2258
71. Brandão HB, Sangji H, Pandžić E, Bechstedt S, Brouhard GJ, Wiseman PW (2014) Measuring ligand–receptor binding kinetics and dynamics using k-space image correlation spectroscopy. Methods 66:273–282
72. Kure JL, Andersen CB, Mortensen KI, Wiseman PW, Arnspang EC (2020) Revealing plasma membrane nano-domains with diffusion analysis methods. Membranes 10:314
73. Arnspang EC, Schwartzentruber J, Clausen MP, Wiseman PW, Lagerholm BC (2013) Bridging the gap between single molecule and ensemble methods for measuring lateral dynamics in the plasma membrane. PLoS One 8:e78096
74. Arnspang EC, Login FH, Koffman JS, Sengupta P, Nejsum LN (2016) AQP2 plasma membrane diffusion is altered by the degree of AQP2-S256 phosphorylation. Int J Mol Sci 17:1804
75. Di Rienzo C, Gratton E, Beltram F, Cardarelli F (2013) Fast spatiotemporal correlation spectroscopy to determine protein lateral diffusion laws in live cell membranes. Proc Natl Acad Sci 110:12307
76. Di Rienzo C, Piazza V, Gratton E, Beltram F, Cardarelli F (2014) Probing short-range protein Brownian motion in the cytoplasm of living cells. Nat Commun 5:5891

77. Digiacomo L, Digman MA, Gratton E, Caracciolo G (2016) Development of an image mean square displacement (iMSD)-based method as a novel approach to study the intracellular trafficking of nanoparticles. Acta Biomater 42:189–198
78. Di Rienzo C, Cardarelli F, Di Luca M, Beltram F, Gratton E (2016) Diffusion tensor analysis by two-dimensional pair correlation of fluorescence fluctuations in cells. Biophys J 111:841–851
79. Wawrezinieck L, Rigneault H, Marguet D, Lenne P-F (2005) Fluorescence correlation spectroscopy diffusion laws to probe the submicron cell membrane organization. Biophys J 89:4029–4042
80. Masuda A, Ushida K, Okamoto T (2005) New fluorescence correlation spectroscopy enabling direct observation of spatiotemporal dependence of diffusion constants as an evidence of anomalous transport in extracellular matrices. Biophys J 88:3584–3591
81. Hendrix J, Baumgärtel V, Schrimpf W, Ivanchenko S, Digman MA, Gratton E, Kräusslich H-G, Müller B, Lamb DC (2015) Live-cell observation of cytosolic HIV-1 assembly onset reveals RNA-interacting Gag oligomers. J Cell Biol 210:629–646
82. Digman MA, Gratton E (2009) Imaging barriers to diffusion by pair correlation functions. Biophys J 97:665–673
83. Cardarelli F, Gratton E (2010) In vivo imaging of single-molecule translocation through nuclear pore complexes by pair correlation functions. PLoS One 5:e10475
84. Hinde E, Cardarelli F, Digman MA, Gratton E (2010) In vivo pair correlation analysis of EGFP intranuclear diffusion reveals DNA-dependent molecular flow. Proc Natl Acad Sci 107:16560
85. Hinde E, Cardarelli F, Digman MA, Kershner A, Kimble J, Gratton E (2011) The impact of mitotic versus interphase chromatin architecture on the molecular flow of EGFP by pair correlation analysis. Biophys J 100:1829–1836
86. Hinde E, Kong X, Yokomori K, Gratton E (2014) Chromatin dynamics during DNA repair revealed by pair correlation analysis of molecular flow in the nucleus. Biophys J 107:55–65
87. Baum M, Erdel F, Wachsmuth M, Rippe K (2014) Retrieving the intracellular topology from multi-scale protein mobility mapping in living cells. Nat Commun 5:4494
88. Niklas P, Elina S, Bagawath S, Sofia J, Enrico G (2019) Pair correlation analysis maps the dynamic two-dimensional organization of natural killer cell receptors at the synapse. ACS Nano 13:14274–14282
89. Clark NM, Hinde E, Winter CM, Fisher AP, Crosti G, Blilou I, Gratton E, Benfey PN, Sozzani R (2016) Tracking transcription factor mobility and interaction in arabidopsis roots with fluorescence correlation spectroscopy. Elife 5:e14770
90. Sengupta P, Jovanovic-Talisman T, Skoko D, Renz M, Veatch SL, Lippincott-Schwartz J (2011) Probing protein heterogeneity in the plasma membrane using PALM and pair correlation analysis. Nat Methods 8:969–975
91. Koltermann A, Kettling U, Bieschke J, Winkler T, Eigen M (1998) Rapid assay processing by integration of dual-color fluorescence cross-correlation spectroscopy: high throughput screening for enzyme activity. Proc Natl Acad Sci 95:1421
92. Bieschke J, Giese A, Schulz-Schaeffer W, Zerr I, Poser S, Eigen M, Kretzschmar H (2000) Ultrasensitive detection of pathological prion protein aggregates by dual-color scanning for intensely fluorescent targets. Proc Natl Acad Sci 97:5468
93. Weissman M, Schindler H, Feher G (1976) Determination of molecular weights by fluctuation spectroscopy: application to DNA. Proc Natl Acad Sci U S A 73:2776–2780
94. Koppel DE (1974) Statistical accuracy in fluorescence correlation spectroscopy. Phys Rev A 10:1938–1945
95. Unruh JR, Gratton E (2008) Analysis of molecular concentration and brightness from fluorescence fluctuation data with an electron multiplied CCD camera. Biophys J 95:5385–5398
96. Xia KQ, Xin YB, Tong P (1995) Dual-beam incoherent cross-correlation spectroscopy. J Opt Soc Am A 12:1571–1578

97. Cardarelli F, Lanzano L, Gratton E (2011) Fluorescence correlation spectroscopy of intact nuclear pore complexes. Biophys J 101:L27–L29
98. Cardarelli F, Lanzano L, Gratton E (2012) Capturing directed molecular motion in the nuclear pore complex of live cells. Proc Natl Acad Sci 109:9863
99. Ruan Q, Cheng MA, Levi M, Gratton E, Mantulin WW (2004) Spatial-temporal studies of membrane dynamics: scanning fluorescence correlation spectroscopy (SFCS). Biophys J 87: 1260–1267
100. Petrásek Z, Schwille P (2008) Precise measurement of diffusion coefficients using scanning fluorescence correlation spectroscopy. Biophys J 94:1437–1448
101. Petrášek Z, Derenko S, Schwille P (2011) Circular scanning fluorescence correlation spectroscopy on membranes. Opt Express 19:25006–25021
102. Waithe D, Schneider F, Chojnacki J, Clausen MP, Shrestha D, de la Serna JB, Eggeling C (2018) Optimized processing and analysis of conventional confocal microscopy generated scanning FCS data. Methods 140–141:62
103. Bona MD, Mancini MA, Mazza D, Vicidomini G, Diaspro A, Lanzanò L (2019) Measuring mobility in chromatin by intensity-sorted FCS. Biophys J 116:987–999
104. Ries J, Schwille P (2006) Studying slow membrane dynamics with continuous wave scanning fluorescence correlation spectroscopy. Biophys J 91:1915–1924
105. García-Sáez AJ, Carrer DC, Schwille P (2010) Fluorescence correlation spectroscopy for the study of membrane dynamics and organization in giant unilamellar vesicles. Methods Mol Biol 606:493–508
106. Eckert AF, Gao P, Wesslowski J, Wang X, Rath J, Nienhaus K, Davidson G, Nienhaus GU (2020) Measuring ligand-cell surface receptor affinities with axial line-scanning fluorescence correlation spectroscopy. Elife 9:e55286
107. Chiantia S, Ries J, Kahya N, Schwille P (2006) Cover picture: combined AFM and two-focus SFCS study of raft-exhibiting model membranes (ChemPhysChem 11/2006). ChemPhysChem 7:2229–2229
108. Ries J, Yu SR, Burkhardt M, Brand M, Schwille P (2009) Modular scanning FCS quantifies receptor-ligand interactions in living multicellular organisms. Nat Methods 6:643–645
109. Gao P, Nienhaus GU (2021) Axial line-scanning stimulated emission depletion fluorescence correlation spectroscopy. Opt Lett 46:2184–2187
110. Dörlich RM, Chen Q, Hedde PN, Schuster V, Hippler M, Wesslowski J, Davidson G, Nienhaus GU (2015) Dual-color dual-focus line-scanning FCS for quantitative analysis of receptor-ligand interactions in living specimens. Sci Rep 5:10149
111. Rossow MJ, Sasaki JM, Digman MA, Gratton E (2010) Raster image correlation spectroscopy in live cells. Nat Protoc 5:1761–1774
112. Digman MA, Gratton E (2009) Analysis of diffusion and binding in cells using the RICS approach. Microsc Res Tech 72:323–332
113. Digman MA, Wiseman PW, Horwitz AR, Gratton E (2009) Detecting protein complexes in living cells from laser scanning confocal image sequences by the cross correlation raster image spectroscopy method. Biophys J 96:707–716
114. Sasaki A, Yamamoto J, Jin T, Kinjo M (2015) Raster image cross-correlation analysis for spatiotemporal visualization of intracellular degradation activities against exogenous DNAs. Sci Rep 5:14428
115. Hendrix J, Dekens T, Schrimpf W, Lamb D (2016) Arbitrary-region raster image correlation spectroscopy. Biophys J 111:1785–1796
116. Mikuni S, Yamamoto J, Horio T, Kinjo M (2017) Negative correlation between the diffusion coefficient and transcriptional activity of the glucocorticoid receptor. Int J Mol Sci 18
117. Scipioni L, Di Bona M, Vicidomini G, Diaspro A, Lanzanò L (2018) Local raster image correlation spectroscopy generates high-resolution intracellular diffusion maps. Commun Biol 1:10
118. Eggeling C, Ringemann C, Medda R, Schwarzmann G, Sandhoff K, Polyakova S, Belov VN, Hein B, Middendorff CV, Schönle A, Hell SW (2009) Direct observation of the nanoscale dynamics of membrane lipids in a living cell. Nature 457:1159–1162

119. Hedde PN, Dörlich RM, Blomley R, Gradl D, Oppong E, Cato ACB, Nienhaus GU (2013) Stimulated emission depletion-based raster image correlation spectroscopy reveals biomolecular dynamics in live cells. Nat Commun 4:2093
120. Longfils M, Smisdom N, Ameloot M, Rudemo M, Lemmens V, Fernández GS, Röding M, Lorén N, Hendrix J, Särkkä A (2019) Raster image correlation spectroscopy performance evaluation. Biophys J 117:1900–1914
121. Norris SCP, Humpolíčková J, Amler E, Huranová M, Buzgo M, Macháň R, Lukáš D, Hof M (2011) Raster image correlation spectroscopy as a novel tool to study interactions of macromolecules with nanofiber scaffolds. Acta Biomater 7:4195–4203
122. Honigmann A, Mueller V, Ta H, Schoenle A, Sezgin E, Hell SW, Eggeling C (2014) Scanning STED-FCS reveals spatiotemporal heterogeneity of lipid interaction in the plasma membrane of living cells. Nat Commun 5:5412
123. Honigmann A, Sadeghi S, Keller J, Hell SW, Eggeling C, Vink R (2014) A lipid bound actin meshwork organizes liquid phase separation in model membranes. Elife 3:e01671
124. Benda A, Ma Y, Gaus K (2015) Self-calibrated line-scan STED-FCS to quantify lipid dynamics in model and cell membranes. Biophys J 108:596–609
125. Moens PDJ, Digman MA, Gratton E (2015) Modes of diffusion of cholera toxin bound to GM1 on live cell membrane by image mean square displacement analysis. Biophys J 108: 1448–1458
126. Needleman DJ, Xu Y, Mitchison TJ (2009) Pin-hole array correlation imaging: highly parallel fluorescence correlation spectroscopy. Biophys J 96:5050–5059
127. Vitali M, Bronzi D, Krmpot AJ, Nikolić SN, Schmitt FJ, Junghans C, Tisa S, Friedrich T, Vukojević V, Terenius L, Zappa F, Rigler R (2014) A single-photon avalanche camera for fluorescence lifetime imaging microscopy and correlation spectroscopy. IEEE J Sel Top Quantum Electron 20:344–353
128. Krmpot AJ, Nikolić SN, Oasa S, Papadopoulos DK, Vitali M, Oura M, Mikuni S, Thyberg P, Tisa S, Kinjo M, Nilsson L, Terenius L, Rigler R, Vukojević V (2019) Functional fluorescence microscopy imaging: quantitative scanning-free confocal fluorescence microscopy for the characterization of fast dynamic processes in live cells. Anal Chem 91:11129–11137
129. Oasa S, Krmpot AJ, Nikolić SN, Clayton AHA, Tsigelny IF, Changeux J-P, Terenius L, Rigler R, Vukojević V (2021) Dynamic cellular cartography: mapping the local determinants of oligodendrocyte transcription factor 2 (OLIG2) function in live cells using massively parallel fluorescence correlation spectroscopy integrated with fluorescence lifetime imaging microscopy (mpFCS/FLIM). Anal Chem 93:12011–12021
130. Toomre D, Pawley JB (2006) Disk-scanning confocal microscopy. Springer, Boston, pp 221–238
131. Ohsugi Y, Kinjo M (2009) Multipoint fluorescence correlation spectroscopy with total internal reflection fluorescence microscope. J Biomed Opt 14:14030
132. Wohland T, Shi X, Sankaran J, Stelzer EHK (2010) Single plane illumination fluorescence correlation spectroscopy (SPIM-FCS) probes inhomogeneous three-dimensional environments. Opt Express 18:10627–10641
133. Struntz P, Weiss M (2015) Multiplexed measurement of protein diffusion in Caenorhabditis elegans embryos with SPIM-FCS. J Phys D Appl Phys 49:44002
134. Bag N, Sankaran J, Paul A, Kraut RS, Wohland T (2012) Calibration and limits of camera-based fluorescence correlation spectroscopy: a supported lipid bilayer study. ChemPhysChem 13:2784–2794
135. Harwardt M-LIE, Dietz MS, Heilemann M, Wohland T (2018) SPT and imaging FCS provide complementary information on the dynamics of plasma membrane molecules. Biophys J 114: 2432–2443
136. Macháň R, Foo YH, Wohland T (2016) On the equivalence of FCS and FRAP: simultaneous lipid membrane measurements. Biophys J 111:152–161
137. Gupta A, Phang IY, Wohland T (2020) To hop or not to hop: exceptions in the FCS diffusion law. Biophys J 118:2434–2447

138. Cooper JT, Peterson EM, Harris JM (2013) Fluorescence imaging of single-molecule retention trajectories in reversed-phase chromatographic particles. Anal Chem 85:9363–9370
139. Xu H, Nagasaka S, Kameta N, Masuda M, Ito T, Higgins DA (2016) Imaging fluorescence correlation spectroscopy studies of dye diffusion in self-assembled organic nanotubes. Phys Chem Chem Phys 18:16766–16774
140. Erstling JA, Hinckley JA, Bag N, Hersh J, Feuer GB, Lee R, Malarkey HF, Yu F, Ma K, Baird BA, Wiesner UB (2021) Ultrasmall, bright, and photostable fluorescent core–shell aluminosilicate nanoparticles for live-cell optical super-resolution microscopy. Adv Mater 33:2006829
141. Bag N, Yap DHX, Wohland T (2014) Temperature dependence of diffusion in model and live cell membranes characterized by imaging fluorescence correlation spectroscopy. Biochim Biophys Acta 1838:802–813
142. Gupta A, Marzinek JK, Jefferies D, Bond PJ, Harryson P, Wohland T (2019) The disordered plant dehydrin Lti30 protects the membrane during water-related stress by cross-linking lipids. J Biol Chem 294:6468–6482
143. Bag N, Ali A, Chauhan VS, Wohland T, Mishra A (2013) Membrane destabilization by monomeric hIAPP observed by imaging fluorescence correlation spectroscopy. Chem Commun 49:9155–9157
144. Bag N, Huang S, Wohland T (2015) Plasma membrane organization of epidermal growth factor receptor in resting and ligand-bound states. Biophys J 109:1925–1936
145. Bag N, Ng XW, Sankaran J, Wohland T (2016) Spatiotemporal mapping of diffusion dynamics and organization in plasma membranes. Methods Appl Fluoresc 4:34003
146. Azbazdar Y, Ozalp O, Sezgin E, Veerapathiran S, Duncan AL, Sansom MSP, Eggeling C, Wohland T, Karaca E, Ozhan G (2019) More favorable palmitic acid over palmitoleic acid modification of Wnt3 ensures its localization and activity in plasma membrane domains. Front Cell Dev Biol 7:281
147. Bag N, Holowka DA, Baird BA (2020) Imaging FCS delineates subtle heterogeneity in plasma membranes of resting mast cells. Mol Biol Cell 31:709–723
148. Bag N, Wagenknecht-Wiesner A, Lee A, Shi SM, Holowka DA, Baird BA (2021) Lipid-based and protein-based interactions synergize transmembrane signaling stimulated by antigen clustering of IgE receptors. Proc Natl Acad Sci 118:e2026583118
149. Krieger JW, Singh AP, Bag N, Garbe CS, Saunders TE, Langowski J, Wohland T (2015) Imaging fluorescence (cross-) correlation spectroscopy in live cells and organisms. Nat Protoc 10:1948–1974
150. Ng X, Teh C, Korzh V, Wohland T (2016) The secreted signaling protein Wnt3 is associated with membrane domains in vivo: a SPIM-FCS study. Biophys J 111:418–429
151. Dhasmana D, Veerapathiran S, Azbazdar Y, Nelanuthala AVS, Teh C, Ozhan G, Wohland T (2021) Wnt3 is lipidated at conserved cysteine and serine residues in zebrafish neural tissue. Front Cell Dev Biol 9:671218
152. Sankaran J, Tan NJHJ, But KP, Cohen Y, Rice SA, Wohland T (2019) Single microcolony diffusion analysis in Pseudomonas aeruginosa biofilms. NPJ Biofilms Microbiomes 5:35
153. Pernuš A, Langowski J (2015) Imaging Fos-Jun transcription factor mobility and interaction in live cells by single plane illumination-fluorescence cross correlation spectroscopy. PLoS One 10:e0123070
154. Yavas S, Macháň R, Wohland T (2016) The epidermal growth factor receptor forms location-dependent complexes in resting cells. Biophys J 111:2241–2254
155. Krieger JW, Singh AP, Garbe CS, Wohland T, Langowski J (2014) Dual-color fluorescence cross-correlation spectroscopy on a single plane illumination microscope (SPIM-FCCS). Opt Express 22:2358–2375
156. Sankaran J, Manna M, Guo L, Kraut R, Wohland T (2009) Diffusion, transport, and cell membrane organization investigated by imaging fluorescence cross-correlation spectroscopy. Biophys J 97:2630–2639
157. Lenne P-F, Wawrezinieck L, Conchonaud F, Wurtz O, Boned A, Guo X-J, Rigneault H, He H-T, Marguet D (2006) Dynamic molecular confinement in the plasma membrane by microdomains and the cytoskeleton meshwork. EMBO J 25:3245–3256

158. Favard C, Wenger J, Lenne PF, Rigneault H (2011) FCS diffusion laws in two-phase lipid membranes: determination of domain mean size by experiments and Monte Carlo simulations. Biophys J 100:1242–1251
159. Huang H, Simsek MF, Jin W, Pralle A (2015) Effect of receptor dimerization on membrane lipid raft structure continuously quantified on single cells by camera based fluorescence correlation spectroscopy. PLoS One 10:e0121777
160. Ng XW, Bag N, Wohland T (2015) Characterization of lipid and cell membrane organization by the fluorescence correlation spectroscopy diffusion law. CHIMIA Int J Chem 69:112–119
161. Veerapathiran S, Wohland T (2018) The imaging FCS diffusion law in the presence of multiple diffusive modes. Methods 140:140–150
162. Kisley L, Brunetti R, Tauzin LJ, Shuang B, Yi X, Kirkeminde AW, Higgins DA, Weiss S, Landes CF (2015) Characterization of porous materials by fluorescence correlation spectroscopy super-resolution optical fluctuation imaging. ACS Nano 9:9158–9166
163. Sankaran J, Balasubramanian H, Tang WH, Ng XW, Röllin A, Wohland T (2021) Simultaneous spatiotemporal super-resolution and multi-parametric fluorescence microscopy. Nat Commun 12:1748
164. Thompson NL, Burghardt TP, Axelrod D (1981) Measuring surface dynamics of biomolecules by total internal reflection fluorescence with photobleaching recovery or correlation spectroscopy. Biophys J 33:435–454
165. Thompson NL, Axelrod D (1983) Immunoglobulin surface-binding kinetics studied by total internal reflection with fluorescence correlation spectroscopy. Biophys J 43:103–114
166. Hassler K, Anhut T, Rigler R, Gösch M, Lasser T (2005) High count rates with total internal reflection fluorescence correlation spectroscopy. Biophys J 88:L01–L03
167. Hassler K, Leutenegger M, Rigler P, Rao R, Rigler R, Gösch M, Lasser T (2005) Total internal reflection fluorescence correlation spectroscopy (TIR-FCS) with low background and high count-rate per molecule. Opt Express 13:7415–7423
168. Thompson NL, Steele BL (2007) Total internal reflection with fluorescence correlation spectroscopy. Nat Protoc 2:878–890
169. Mücksch J, Blumhardt P, Strauss MT, Petrov EP, Jungmann R, Schwille P (2018) Quantifying reversible surface binding via surface-integrated fluorescence correlation spectroscopy. Nano Lett 18:3185–3192
170. Weidemann T, Wachsmuth M, Tewes M, Rippe K, Langowski J (2002) Analysis of ligand binding by two-colour fluorescence cross-correlation spectroscopy. Single Mol 3:49–61
171. Müller CB, Loman A, Pacheco V, Koberling F, Willbold D, Richtering W, Enderlein J (2008) Precise measurement of diffusion by multi-color dual-focus fluorescence correlation spectroscopy. Europhys Lett 83:46001
172. Ries J, Petrov EP, Schwille P (2008) Total internal reflection fluorescence correlation spectroscopy: effects of lateral diffusion and surface-generated fluorescence. Biophys J 95: 390–399
173. Sankaran J, Bag N, Kraut RS, Wohland T (2013) Accuracy and precision in camera-based fluorescence correlation spectroscopy measurements. Anal Chem 85:3948–3954
174. Sankaran J, Shi X, Ho LY, Stelzer EH, Wohland T (2010) ImFCS: a software for imaging FCS data analysis and visualization. Opt Express 18:25468–25481
175. Wohland T, Maiti S, Macháň R (2020) An introduction to fluorescence correlation spectroscopy
176. Capoulade J, Wachsmuth M, Hufnagel L, Knop M (2011) Quantitative fluorescence imaging of protein diffusion and interaction in living cells. Nat Biotechnol 29:835–839
177. Singh AP, Galland R, Finch-Edmondson ML, Grenci G, Sibarita JB, Studer V, Viasnoff V, Saunders TE (2017) 3D protein dynamics in the cell nucleus. Biophys J 112:133–142
178. Yoshida S, Schmid W, Vo N, Calabrase W, Kisley L (2021) Computationally-efficient spatiotemporal correlation analysis super-resolves anomalous diffusion. Opt Express 29: 7616–7629

179. Aik DYK, Wohland T (2022) Microscope alignment using real-time imaging FCS. Biophys J 121:2663–2670
180. Yu L, Lei Y, Ma Y, Liu M, Zheng J, Dan D, Gao P (2021) A comprehensive review of fluorescence correlation spectroscopy. Front Physiol 9:644450
181. Leutenegger M, Ringemann C, Lasser T, Hell SW, Eggeling C (2012) Fluorescence correlation spectroscopy with a total internal reflection fluorescence STED microscope (TIRF-STED-FCS). Opt Express 20:5243–5263
182. Ashdown GW, Burn GL, Williamson DJ, Pandžić E, Peters R, Holden M, Ewers H, Shao L, Wiseman PW, Owen DM (2017) Live-cell super-resolution reveals F-actin and plasma membrane dynamics at the T cell synapse. Biophys J 112:1703–1713
183. Ashdown GW, Owen DM (2018) Spatio-temporal image correlation spectroscopy and super-resolution microscopy to quantify molecular dynamics in T cells. Methods 140–141:112–118
184. Schnitzbauer J, Wang Y, Zhao S, Bakalar M, Nuwal T, Chen B, Huang B (2018) Correlation analysis framework for localization-based superresolution microscopy. Proc Natl Acad Sci 115:3219

Part III
Applications of Fluorescence Spectroscopy and Microscopy to Biological Membranes

Determination of Biomolecular Oligomerization in the Live Cell Plasma Membrane via Single-Molecule Brightness and Co-localization Analysis

Clara Bodner and Mario Brameshuber

Contents

Abstract Oligomerization of biomolecules on the plasma membrane drives vital cellular functions. Despite the importance of detecting and characterizing these molecular interactions, there is an apparent lack of proper techniques capable of unraveling the exact composition of multimolecular complexes directly on the (live) cell membrane, particularly when present at high surface densities. In this chapter, commonly applied single-molecule fluorescence microscopy approaches are reviewed in terms of their suitability for detecting molecular aggregates down to

C. Bodner and M. Brameshuber (✉)
TU Wien, Institute of Applied Physics, Biophysics, Vienna, Austria
e-mail: bodner@iap.tuwien.ac.at; brameshuber@iap.tuwien.ac.at

R. Šachl, M. Amaro (eds.), *Fluorescence Spectroscopy and Microscopy in Biology*, Springer Ser Fluoresc (2023) 20: 277–304, https://doi.org/10.1007/4243_2022_25, Published online: 4 August 2022

the smallest biologically relevant unit: a molecular dimer. Amongst these techniques, Thinning Out Clusters while Conserving Stoichiometry of Labeling (TOCCSL) – a modality based on single-molecule brightness and co-localization analysis – is discussed in detail. The chapter lists basic principles of TOCCSL, its implementation on single-molecule fluorescence microscopes, analysis of brightness and co-localization-based microscopy data, and guidelines on how to choose parameters and labels to determine the number of labeled subunits within a multimolecular assembly. At the end, a tabular list of TOCCSL applications conducted so far is provided.

Keywords Brightness analysis · Co-localization analysis · Multimolecular complexes · Plasma membrane structure · Single-molecule microscopy · Thinning out clusters while conserving stoichiometry of labeling (TOCCSL)

1 Biomolecular Interactions on the Plasma Membrane

Life is all about interactions: from a simple handshake or a meaningful eye contact between humans, down to complex interactions between single cells and organelles, further down to the interactions of individual (bio)molecules, and even further down to nucleonic interactions between the smallest building blocks of life. On the intermediate length scale, molecular interactions drive the specific assembly of biomolecules into multimolecular structures. One of the major matrices for hosting these biomolecular interactions is the plasma membrane, as it represents the interface of a biological cell and the extracellular environment. Interactions occur between molecules of the same kind (homo-interactions), or between different species (hetero-interactions) and result in different aggregation states, ranging from the simplest form – a homo-dimer – to large multimolecular structures hosting a diverse set of biomolecules (see Fig. 1 for examples of biomolecular interactions on the plasma membrane of mammalian cells).

Some interactions between molecules are extremely stable and occur already before molecules reach the plasma membrane, as it is required for the correct function of various protein complexes. For example, only the complete assembly of the octameric T cell receptor CD3 complex (TCR/CD3) results in its sorting to the plasma membrane and facilitates highly sensitive and specific antigen-recognition during immune cell signaling [1].

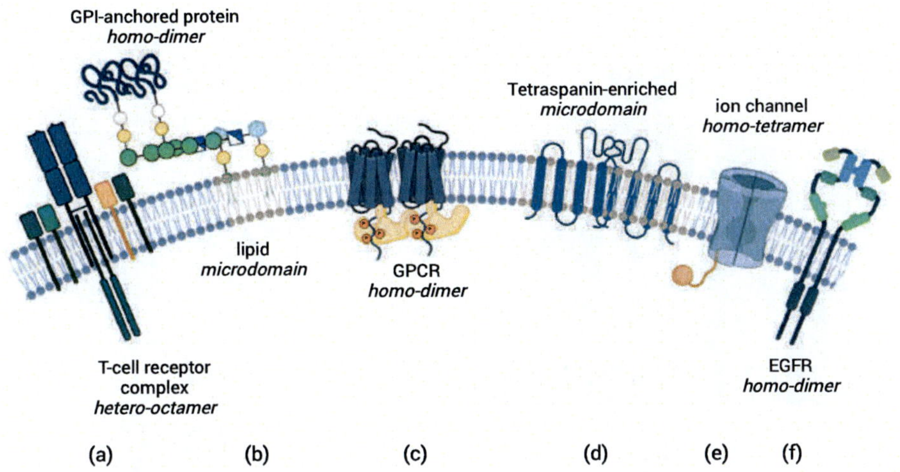

Fig. 1 Examples of biomolecular interactions and assemblies. From left to right: (**a**) The TCR/CD3 complex consists of three hetero-dimers with extracellular domains (αβ, δε, and γε subunits) and the ζ-chain homo-dimer with intracellular immunoreceptor tyrosine-based activation motifs (ITAMs). The αβ hetero-dimer is responsible for binding the antigen presented via MHC class I or II proteins on antigen-presenting cells (APC). (**b**) A class of proteins anchored via a GPI moiety is targeted to the extracellular side of the plasma membrane and the homo-association of such proteins is ascribed important cellular signaling functions. The lipid anchor is also responsible for allocating these proteins to lipid microdomains. (**c**) The minimal functional form of GPCR is represented by a homo- or hetero-dimer. (**d**) Another type of microdomain is reported to be enriched in proteins containing four transmembrane domains – so-called tetraspanins. Via homo-associations, these tetraspanins can form large networks on the cell membrane. (**e**) Most ion channels are homo- or hetero-tetramers and responsible for the controlled conductance of various ions through the plasma membrane. (**f**) The epidermal growth factor receptor (EGFR) is a transmembrane protein and described to form homo-dimers responsible for signaling. The receptor plays an important role in many cancer types, where mutations influence EGFR expression or activity. Created with BioRender.com

In this chapter, TCR/CD3 will serve as a model system representing a multimolecular complex. The α-β hetero-dimer, which is responsible for antigen binding, the δ-ε and γ-ε hetero-dimer, as well as the ζ-ζ homo-dimer containing intracellular domains transmitting the signal of antigen recognition (see Fig. 2) represent the eight subunits. Different techniques will be discussed in terms of their capability to detect the number of ε- and β-subunits directly at their place of action: the plasma membrane of a living T cell.

Stable homo- or hetero-association of four subunits represents the functional state of many ion channels, for example, the voltage-gated potassium channels [3, 4], with the composition being critical for correct function.

Interactions might also occur directly on the plasma membrane leading – in the simplest form – to a transient homo-dimer. For guanine nucleotide-binding (G-)

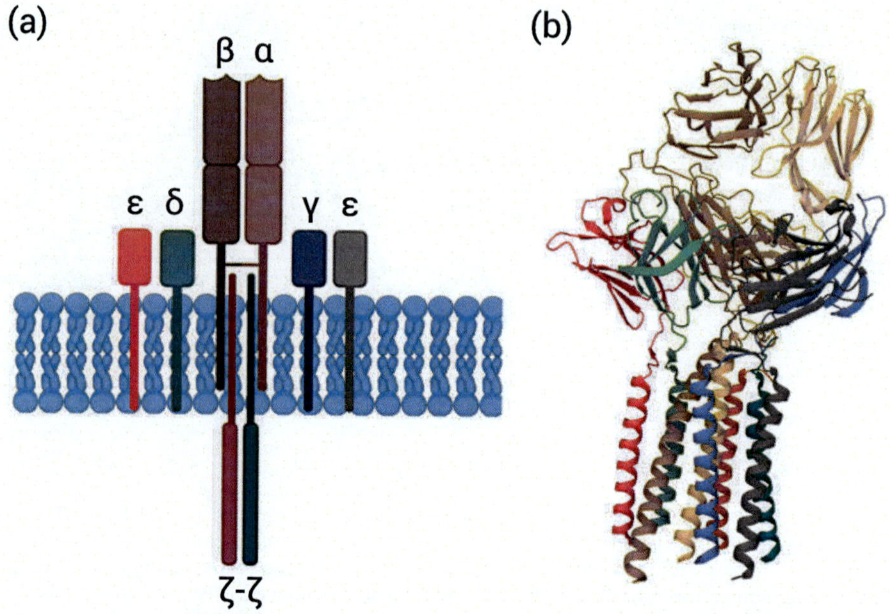

Fig. 2 Structure of the αβTCR/CD3 complex. (**a**) Sketch of the subunit assembly showing the α-β, δ-e, and γ-ε hetero-dimer, as well as the intracellular ζ-ζ homo-dimer. (**b**) Extracellular domains and transmembrane helices of the human TCR/CD3 complex as determined by cryo-electron microscopy [2]. Color coding same as in (**a**). PDB ID 6JXR. Created with BioRender.com

protein-coupled receptors (GPCRs) – a major target for pharmaceutical drugs – such a dimer was found to be the minimal functional structure, however, with considerable variations as the existence of hetero-dimers and the formation of larger complexes was also described to be required for signaling [5]. Another family of proteins, receptor tyrosine kinases, was described to form homo- and hetero-dimeric structures, as well as much larger complexes, which in turn activate the intracellular kinase domains to initiate signaling [6].

Much more transient interactions on the plasma membrane with lifetimes below 200 ms have been described for glycosyl-phosphatidylinositol (GPI) anchored proteins, including CD59, Thy1 and DAF [7, 8] and have been ascribed essential cellular functions. Larger biomolecular assemblies like tetraspanin-enriched microdomains [9] or lipid microdomains ("rafts") [10] were found to contain hundreds of different constituents and play important roles for the function of adhesion receptors and the compartmentalization of enzymatic activities, and are discussed as platforms for signaling, exocytosis, and endocytosis, respectively.

While the importance of these biomolecular assemblies for cellular functions is indisputable, there is an apparent lack of proper technologies capable of unraveling the exact composition of multimolecular complexes directly on the (live) cell membrane. Why is it so much harder to look at the interaction of two molecules than to take a picture of two close-by humans shaking hands? The main reason is, that human interactions happen on the same length (and time) scale for both, the observer and the observed, while the scales differ up to 9 orders of magnitude between single molecules and a human observer. One commonly used approach to tackle this discrepancy is to modify molecules of interest by binding a fluorescent molecule of similar size and use ultrasensitive fluorescence microscopy to directly visualize single molecules and multimolecular complexes.

2 Detecting Biomolecular Assemblies Directly on the Plasma Membrane

Equipped with powerful light sources and a sensitive detection camera, fluorescence microscopes are nowadays routinely used to observe single fluorescent molecules and should in principle be capable of detecting the smallest unit of a biomolecular assembly – a molecular dimer.

For the example of the TCR/CD3 complex, which contains two ε-subunits (the two outer subunits shown in Fig. 2), ideally all ε-subunits on the surface of a T cell can be fluorescently labeled. With an average expression level of 10^4 TCRs per cell and a T cell diameter of ~5 μm, the TCR density yields around 130 TCR complexes located within 1 μm^2 of the cell membrane. The average distance between TCRs is $1/\sqrt{130\ \mu m^{-2}} = 90$ nm, which is far below the optical resolution limit of a light microscope defined by the width of the point spread function (PSF) of the optical system. The typical size of the PSF, given by $0.61\frac{\lambda}{NA}$, with λ being the emission wavelength and NA the numerical aperture of the used objective, yields ~200 nm for visible light and the use of a high NA-objective. Avoiding overlapping PSFs hence limits molecule densities to values below one molecule per μm^2 for the unambiguous detection of individual fluorophores. Thus, the density of the TCR is about 1,000-fold too high for the direct imaging of ε-subunit dimers. The experimenter would observe a homogeneous fluorescence signal throughout the plasma membrane of non-stimulated T cells. Which strategies applicable in single-molecule fluorescence microscopy would now allow for the detection of these TCR ε-dimers at high surface densities?

2.1 *How About Decreasing the Label Density?*

Decreasing the label density on the cell membrane by a factor of ~1,000 for the observation of individual diffraction-limited signals can be realized by photobleaching [11], for example, by increasing the laser power and illumination time, by reducing the label concentration [12], or by photoactivating a subset of molecules [13]. Importantly, the label density will be reduced stochastically using these approaches, i.e., most TCR ε-dimers will become invisible due to the loss or lack of detectable labels or become apparent monomers with only one fluorescent label per ε-dimer visible. Only a minor fraction of detectable ε-dimers will remain with both ε-subunits carrying fluorescent labels. Hence, the likelihood for observing an ε-dimer will decrease substantially. Assuming 100% ($p_{\text{coloc}} = 1$) ε-dimers and a reduction of the surface density by a factor of 1,000 ($p_{\text{label}} = 10^{-3}$), an experiment would yield a probability for co-localization of the two ε-subunits of

$$p = \frac{p_{\text{label}} p_{\text{coloc}}}{1 + p_{\text{coloc}}(1 - p_{\text{label}})} \approx \frac{p_{\text{label}}}{2} = 5 \times 10^{-4}.$$

In words, only 5 out of 10,000 visible ε-subunits would be paired with another visible ε-subunit within the same TCR/CD3 complex. This small fraction cannot be straightforward detected with standard single-molecule fluorescence microscopy techniques. Detection of such low fractions demands for highly quantitative microscopy tools and enormous amounts of data to gain sufficient statistics for a reliable proof of the existence of a few ε-dimers.

2.2 *Would One of the Single-Molecule Localization Microscopy (SMLM) Modalities Work for Detecting Molecular Dimers?*

Photoactivated localization microscopy (PALM) [14, 15] and (direct) stochastic optical reconstruction microscopy ((d) STORM) [16, 17] are based on the fact, that a single, separated molecule can be localized with a much higher precision – even down to the sub-nanometer level [18] – than the diffraction limit of the optical system. These so-called SMLM or superresolution microscopy approaches (for a recent review see [19]) are nowadays routinely used to resolve cellular structures well below the diffraction limit (see Chap. 10 "Quantitative Photoactivated Localization Microscopy of Membrane Receptor Oligomers" for details on SMLM techniques). In contrast to the aforementioned strategies for decreasing the label density per image, in SMLM all molecules of interest are fluorescently labeled but are not visible at the same time in the detection channel. Only a few of these fluorescent molecules are stochastically activated by switching them into a bright or red-shifted fluorescent state and can be detected as diffraction-limited signals. Detected molecules are deactivated by either switching them back to a dark or blue-shifted state, or by terminal photobleaching. Tens of thousands of these activation-detection-deactivation cycles increase the probability of detecting a high fraction of – for the TCR as an example – ε-subunits. Localizing each individual signal in all recorded frames finally yields a reconstructed image with a theoretical resolution in the order of the localization precision. Optimizing photon yield and background noise could result in a superresolution image with a resolution one order of magnitude lower than the 90 nm average distance of TCR molecules. An ideal quantitative superresolution experiment would hence allow to detect individual ε-dimers. What was neglected so far?

1. Diffusional motion of the TCR/CD3 complex on the cell membrane heavily influences the reconstructed image. Due to the stochastic activation process, the individual ε-subunits of one TCR complex are imaged at different experimental time points – and hence different positions on the cell membrane – and the assignment of both subunits to the same complex is impossible. With a diffusion coefficient of ~0.04 $\mu m^2/s$ [20], the total recording time of the superresolution image would need to be shorter than 0.6 ms to virtually freeze the motion of the structure at the resolution of 10 nm and to observe ε-dimers. However, the photon budget needed for 10 nm resolution as well as technical limitations result in typical recording times of several seconds.
2. Chemical fixation is often performed to "freeze" the movement and to record a snapshot of the oligomerization state of certain proteins. This is also possible for the TCR/CD3 complex, and superresolution images can be analyzed in terms of clustering and oligomerization [21]. However, the repeated detection and subsequent localization of one and the same molecule within the localization precision limit, as well as the presence of a dark fraction of fluorophores renders quantitative superresolution imaging down to small nanoclusters or even individual dimers extremely challenging (see [22] for a review on the difficulties in detecting nanocluster). While there are several algorithms available to account for very well-defined blinking patterns of fluorescent molecules, the presence of only a few outliers can change the interpretation of results completely [21, 23]. Even with perfectly behaving fluorophores, i.e., clearly defined blinking patterns or no blinking at all, residual motion of proteins despite fixation needs to be considered [24]. One possible solution to overcome residual motion and potential chemical fixation artifacts is to conduct superresolution microscopy experiments at cryogenic temperatures [25–27]. Current technical developments and first applications look promising for using cryoSMLM for the routine detection and characterization of biomolecular assemblies at the plasma membrane in the near future.

2.3 *How about Using Stimulated Emission Depletion (STED) Imaging?*

STED microscopy [28, 29] (for reviews see, e.g., [30–32] has proved to be a meaningful approach to significantly improve the resolution of fluorescence images without the necessity of utilizing the stochastic blinking of fluorophores. Instead, fluorescence in the periphery of a focused excitation beam is efficiently quenched by a depletion laser (see Chap. 7 "STED and RESOLFT Fluorescent Nanoscopy" for more details on STED microscopy). Nowadays, a spatial resolution of ~60 nm can be routinely achieved in STED microscopy which, keeping the TCR as an example, enables the direct detection of individual TCR/CD3 complexes labeled via their ε-subunits on chemically fixed T cells [21]. Despite STED being a scanning technique, in which the excitation and depletion lasers are concomitantly moved over the sample, joining scan lines are recorded fast enough for neglecting residual motion of molecules on fixed cells. Frame rates achieved nowadays even allow STED imaging on live cells and the detection of biomolecular clustering on length scales larger than the resolution limit of STED microscopy. Determination of the composition of protein oligomers using STED by directly visualizing all labeled constituents is in most cases prevented by the limited resolution. However, unambiguously counting the exact number of dye molecules per molecular complex is possible via analysis of coinciding photon arrival times (photon antibunching, PA) [33–35] and enables the direct detection of ε-dimers within the TCR/CD3 complex in fluorescence correlation spectroscopy (FCS) experiments [20]. Combining STED imaging with PA seems ideal for the stoichiometric analysis of protein complexes at high surface densities, but requires costly and specialized equipment [36]. A recently introduced STED-related approach based on minimal photon fluxes (MINFLUX) might be an alternative method for the direct observation of molecular complexes at nanometer resolution [37, 38].

2.4 *Thinning Out Clusters While Conserving Stoichiometry of Labeling (TOCCSL)*

In contrast to the above-mentioned strategies i–iii, an easy-to-implement single-molecule fluorescence modality termed TOCCSL [39, 40] allows to determine the number of labeled subunits within a multimolecular assembly – also at high surface densities on the membrane of live cells. The approach is based on a FRAP (fluorescence recovery after photobleaching [41, 42]) protocol, with the main difference, that individual oligomers are detected at the onset of the very recovery process when the density of molecules is low enough to resolve single entities. In the following, the TOCCSL principle, experimental strategies, and considerations, as well as on overview of TOCCSL applications are discussed.

3 The TOCCSL Concept

3.1 *Principle*

In the plasma membrane, the physiological surface density of biomolecules is usually too high to resolve individual fluorescently labeled protein clusters, as their size and the average distance between two adjacent molecular aggregates are smaller than the resolution limit of light microscopy. The single-molecule fluorescence microscopy method TOCCSL [40, 43] is used to study the stoichiometry of cell surface proteins (or lipids) in life cells and model systems at physiologically high densities. The principle of TOCCSL is shown in Fig. 3.

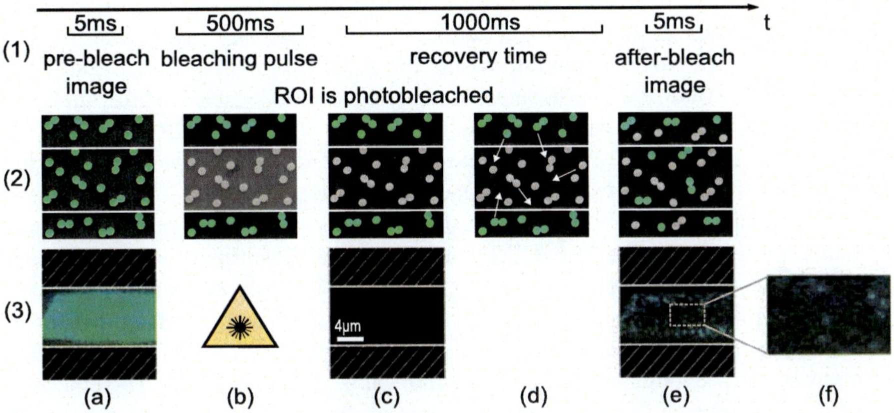

Fig. 3 A typical TOCCSL imaging protocol. (1) Illumination timing scheme. (2) Schematic illustration: (**a**) A control image (pre-bleach image) is taken at low laser intensity, which is used to determine the molecule density on the cell surface. (**b**) A short high-intensity laser pulse is applied which causes the region of interest (ROI), defined by an aperture, to be irreversibly photobleached. (**c**) A few milliseconds after the bleaching pulse, another control image can be taken to ensure that all molecules within the field of view are completely photobleached. (**d**) At the onset of the recovery process fluorescent molecules from the shielded area diffuse into the photobleached region and can be imaged as diffraction-limited spots (**e**). (**f**) Zoomed region from (**e**). (3) Example of a TOCCSL experiment. CHO cells expressing GPI-anchored SNAP were labeled with Alexa Fluor 488. Note that the contrast is identical for pre-bleach image, control image, and after-bleach image

One can consider the TCR/CD3 protein complex labeled via the two ε-subunits as an example (see Fig. 3. Note, that a more general example of a molecule existing as monomer and homo-dimer is sketched). Using TOCCSL, the surface density of active fluorophores in a small region of the cellular plasma membrane is transiently depleted by applying a high-intensity laser pulse (several kW/cm^2) to a defined area of the cell surface using an aperture. By doing so, all fluorophores within the illuminated region are irreversibly photobleached, while the stoichiometric composition of TCR/CD3 complexes remains unaltered: both labels of ε-subunits are either photobleached or are intact because they were not exposed to laser light. Due to Brownian motion, fluorescently labeled TCR/CD3 molecules from outside the aperture-defined area of the biomembrane can diffuse into the photobleached region during a short recovery time (several hundred ms up to several seconds). Using ultrasensitive fluorescence microscopy, single protein clusters can be imaged as diffraction-limited spots at the onset of this recovery process.

3.2 *Imaging Protocol*

Figure 3 illustrates a typical TOCCSL experiment with the timing protocol for sample illumination shown on the top. First, an image at low laser intensity with illumination time t_{ill} is taken in total internal reflection (TIR) configuration to report the initial situation of high surface density. A high-intensity laser pulse is then applied in TIR or non-TIR configuration for a bleaching time t_{bleach}. During a short recovery time t_{rec}, molecules from the shielded area diffuse into the photobleached region, where they are imaged directly after t_{rec} in TIR configuration. Both laser intensity and t_{bleach} are chosen such, that all active fluorophores within the aperture-defined area are fully photobleached as fast as possible, to avoid incompletely or partially bleached protein clusters, where one or more protein subunits remain fluorescent. For the TCR/CD3 complex partial bleaching of ε-dimers would yield a wrong composition of some TCR/CD3 complexes containing only one apparent-fluorescent ε-subunit. t_{rec} is chosen in a way that the surface density of single fluorescently labeled protein clusters in the inner region of the photobleached area remains low enough, such that single molecules can be resolved. The choice of t_{rec} and t_{ill} depends on the diffusion coefficient of the observed biomolecule and for t_{rec} it also depends on the surface density as well as the chosen aperture-size (for more details, see Sect. 5).

3.3 *Brightness Analysis*

Generally, the image of a point-like source of light is represented by its PSF, which exhibits an Airy disc in the center and decreasing intensities towards the radial direction. For this reason, a two-dimensional Gaussian function represents a good approximation for the PSF of a single point-like emitter, e.g., a single dye molecule or a fluorescent protein. The location of a fluorescent molecule can be determined, e.g., by identifying the center of the Gaussian fit. While the exact position and its precision are important parameters for SMLM, the integral of the Gaussian is used for the analysis of TOCCSL images and yields the overall number of emitted photons during t_{ill}, i.e., the brightness of the molecular complex. It is also possible to calculate the brightness directly from raw images via the sum of individual pixel values at the position of diffraction-limited signals corrected by the local background. Besides providing a more realistic photon estimation, this is particularly useful for long illumination times, where diffusion of molecular complexes yields asymmetric PSFs [44]. Note that often pixel values are not displayed in units of photons but rather counts and hence need to be corrected by considering the (electron multiplication) gain and electrons-per-count conversion factor.

In TOCCSL, the statistical distribution of fluorescently labeled subunits per protein cluster is determined via brightness analysis of the diffraction-limited single-molecule signals, which are imaged according to their PSF [40]. The photon emission of a single fluorescent molecule is stochastic. Thus, the number of photons F detected from a single fluorescent emitter cannot be determined precisely but is characterized by the probability density function (pdf) $\rho_1(F)$. $\rho_1(F)dF$ is the probability that the number of detected photons determined by a brightness measurement of a single-molecule signal lies within the interval $[F, F + dF]$. To determine the pdf

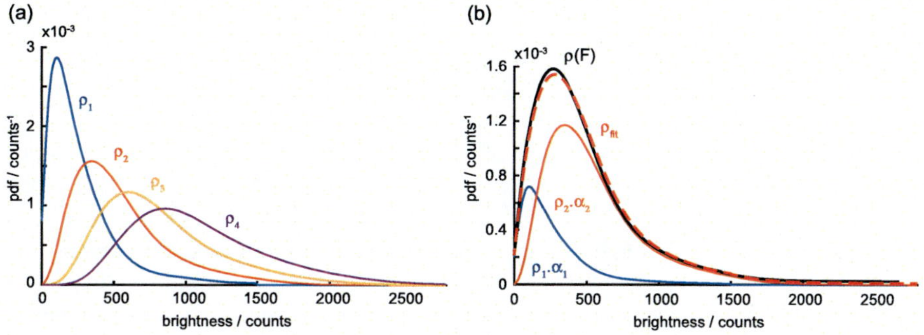

Fig. 4 Brightness-based analysis of a TOCCSL experiment. (**a**) By measuring the brightness distribution of single molecules, ρ_1, the corresponding distributions of 2, 3, or 4 molecules can be calculated via convolution integrals assuming independent emitters. Note that all distributions are normalized to 1. (**b**) The normalized pdf of measured fluorescence signals of protein clusters, $\rho(F)$, is fitted by a linear combination of various N-mer contributions, ρ_{fit}. Since all distributions are normalized, the area under contributing pdfs directly yields the fraction of N-mers. For the example shown, $\rho(F)$ is decomposed into a fraction of 25% monomers (α_1) and 75% dimers (α_2). Shown data is based on a simulation of 500 and 1,000 log-normal distributed brightness values [45] for ρ_1 and $\rho(F)$, respectively

of a single emitter, ρ_1, the fluorescence brightness values of individual fluorophores can be acquired experimentally by stepwise decreasing the emitted fluorescence of protein clusters through photobleaching of the sample, until most fluorophores are irreversibly photobleached and a few single-molecule signals exhibit their lowest brightness value (monomeric signal). Single molecules can also be tracked throughout the photobleaching process to ensure that the lowest brightness level within one photobleaching trace is used to determine ρ_1. If possible, under-labeling of molecular subunits might also yield the brightness distribution of single-molecule signals. In the case of the TCR/CD3 complex, decreasing the label density by a factor of ~1,000 will yield a good approximation of the ε-subunit monomer distribution with a neglectable fraction of ε-dimers contributing to this distribution (see Sect. 2.1).

Assuming higher order oligomers of order N, the brightness distribution of N co-localized independent emitters, $\rho_N(F)$, can be determined recursively from the single-dye brightness pdf ρ_1 as a series of convolution integrals $\rho_N(F) = \int \rho_1(F')\rho_{N-1}(F - F')dF'$ (see Fig. 4a).

When recording the fluorescence of a mixed monomer and higher order N-mer population, the resulting brightness distribution, $\rho(F)$, is based on a linear combination of the previously determined individual pdfs of each N-mer population, $\rho_N(F)$, weighted with the fraction of the respective oligomer of degree N, α_N:

$$\rho(F) = \sum_{N=1}^{N_{\max}} \alpha_N \times \rho_N(F) \tag{1}$$

with

$$\sum_{N=1}^{N_{\max}} \alpha_N = 1 \tag{2}$$

The pdf of the fluorescence signal of protein clusters, $\rho(F)$, is determined from a TOCCSL experiment. By non-linear least squares fitting of $\rho(F)$ with Eq. (1) the relative N-mer fractions α_N can be calculated (see Fig. 4b for an example).

For determining the stoichiometry of the TCR/CD3 complex, the β-subunit was fluorescently labeled and ρ_N was determined in a TOCCSL experiment [20]. ρ_I was obtained from the same experiment after prolonged and repeated photobleaching or, alternatively, from recording T cells with 150-fold substoichiometric TCR labeling. Analysis yielded a small fraction (3% ± 4%) of potential dimeric TCRβ dimers. The aforementioned labeling of the ε-subunits served as a positive-control: using TOCCSL, 74% ± 4% of TCR/CD3 complexes returning to the photobleached area contained two ε-subunits. Taken together, the stoichiometry of the TCR/CD3 complex with one β and two ε-subunits could be confirmed, however, with a small uncertainty regarding the potential presence of β-subunit dimers.

In a general TOCCSL assay, accidental coincidence of protein clusters can lead to the false-positive detection of higher order oligomers and thus, to an overestimation of protein clusters with higher subunit count, a phenomenon that is more pronounced at high protein surface densities. Additionally, diffraction of the laser at the aperture edges can result in incompletely photobleached protein clusters. Thus, when analyzing TOCCSL after-bleach images, only the innermost area of the photobleached region is selected, and only the signals within that area are used for further analysis. Recovered single molecules can additionally be tracked over consecutive frames to determine their diffusion coefficient as well as for distinguishing true-positive and false-positive higher order oligomers (see more details in Sect. 5).

3.4 Two-Color TOCCSL

The TOCCSL protocol can also be applied in two colors [46] using two spectrally different fluorophores and two lasers of different wavelength that match the excitation maxima of the two fluorophores. In comparison to the one-color implementation of TOCCSL, where oligomerization is determined by brightness analysis of recovered single-molecule signals, the two-color TOCCSL implementation allows for direct counting of protein subunits via co-localization and is used to quantify relative fractions of monomer and dimer populations. Due to the different emission wavelengths of the two spectrally distinct fluorophores, their signal can be separated by an image splitter into two detection channels on the same camera chip (see Fig. 5 for the principle and an example of two-color TOCCSL). After irreversibly photobleaching all fluorescent molecules within an aperture-defined region, fluorescently labeled protein clusters diffuse back into the field of view during a recovery time. The recovered molecules are then imaged in both color channels.

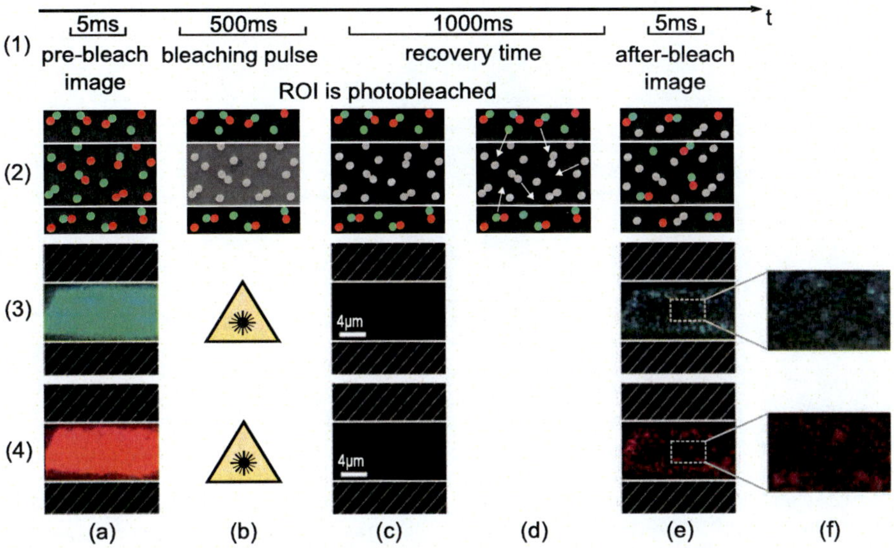

Fig. 5 A typical two-color TOCCSL imaging protocol. (1) Illumination timing Scheme. (2) Schematic illustration: (**a**) Control images are taken simultaneously in both color channels. (**b**), (**c**) The identical region of interest, defined by an aperture, is irreversibly photobleached in both color channels. (**d**) At the onset of the recovery process, fluorescent clusters carrying both spectrally distinct fluorescent molecules diffuse into the photobleached region. (**e**) The two spectrally different emission signals are imaged separately on the same camera chip. (**f**) Zoomed region from (**e**). (3) Example of a two-color TOCCSL experiment. CHO cells expressing GPI-anchored SNAP were labeled with Alexa Fluor 488 and Alexa Fluor 647 (green emission channel, i.e., Alexa Fluor 488 signals). (4) Two-color TOCCSL recordings of CHO SNAP-GPI cells labeled with Alexa Fluor 488 and Alexa Fluor 647 (red emission channel, i.e., Alexa Fluor 647 signals)

3.5 Co-localization Analysis

The two-color TOCCSL after-bleach images are used for co-localization analysis (see Fig. 6 for an example). To register both emission channels, a calibration is performed before each experiment. Multi-color beads, immobilized on a glass slide, are localized in the two spectrally separated detection channels. A correction matrix is determined, which specifies the necessary shift, stretch, and rotation of the second emission channel in respect to the first emission channel in order to register the position of the same bead imaged in both emission colors with accuracies of <20 nm.

For the co-localization analysis of two-color TOCCSL recordings, recovered single-molecule signals are localized in both detection channels and their positions are corrected using an affine transformation with the parameters determined by using the fluorescent beads. Signals found in both color channels are counted as co-localizations if their mutual distance is smaller than a predefined search-radius *R*. The algorithm can be tested by co-localizing single multi-color beads. The right choice of *R* depends on the inter-molecular distances of the observed protein

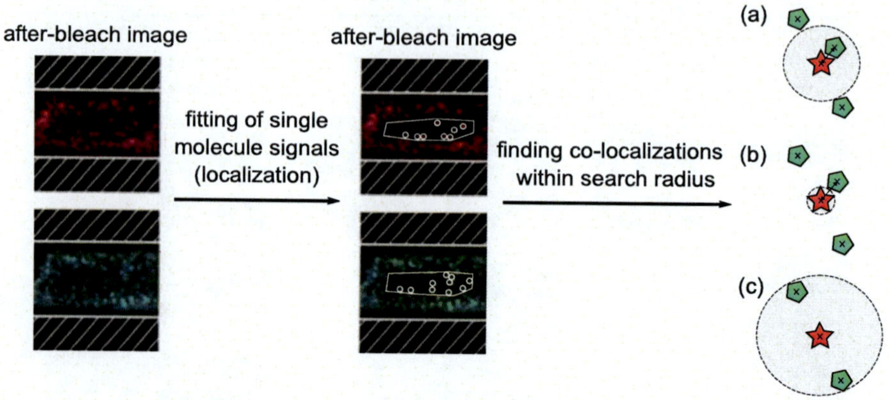

Fig. 6 After-bleach images are used for analysis: Recovered signals within selected analysis regions are localized, and the coordinates of both emission channels are corrected using multi-color fluorescent beads. (**a**) Pairs of two spectrally different fluorescence signals are counted as co-localizations if they are found within a predefined search-radius (dashed lines denote the borders of the search-region). (**b**) If the search-radius is chosen too small, dimer subunits do not co-localize and are falsely counted as monomers. (**c**) If the search-radius if chosen too large, true monomers co-localize and are counted as dimers (false-positive dimers). The rate of detected false-positive co-localizations strongly depends on the molecules density after recovery and increases towards the aperture edges. With the right choice of analysis region and recovery time, the number of false-positive dimers stays significantly below the number of true-positive dimers

complex, the emission wavelength of the fluorophore, the resolution of the microscopy setup, and the quality of fitting the positions. If R is chosen too small, protein dimers are not detected during co-localization analysis. In particular, if both color channels are not recorded simultaneously but rather temporally separated, diffusion of molecular complexes requires a larger search-radius. In contrast, if R is chosen too large, randomly encountered monomers co-localize and are detected as dimers, termed false-positive dimers (for more details, see Sect. 5).

To correct for false-positive dimers, coordinates of one color channel are mirrored alongside the x and y axis through the center of mass of all signals and the search algorithm for co-localizations is applied again, yielding the number of false-positive dimers. Dimer fractions are calculated considering all co-localized signals, subtracting all false-positive co-localized signals, and correcting for non-equimolar labeling as well as for unlabeled protein subunits. This correction is described in detail in [47]. Note, that in two-color TOCCSL experiments, only a fraction of true co-localized molecules can be detected directly, as depicted in Fig. 7.

The two-color implementation of TOCCSL was applied to exclude the presence of a minor fraction of TCRβ dimers. To this end, the β-subunit was labeled with a 1:1 mix of green and red fluorophores, and TCR oligomerization was quantified by co-localization analysis. For TCRβ, 0.4% of detected single-molecule events were visible in both color channels, i.e., were classified as true co-localized. In contrast,

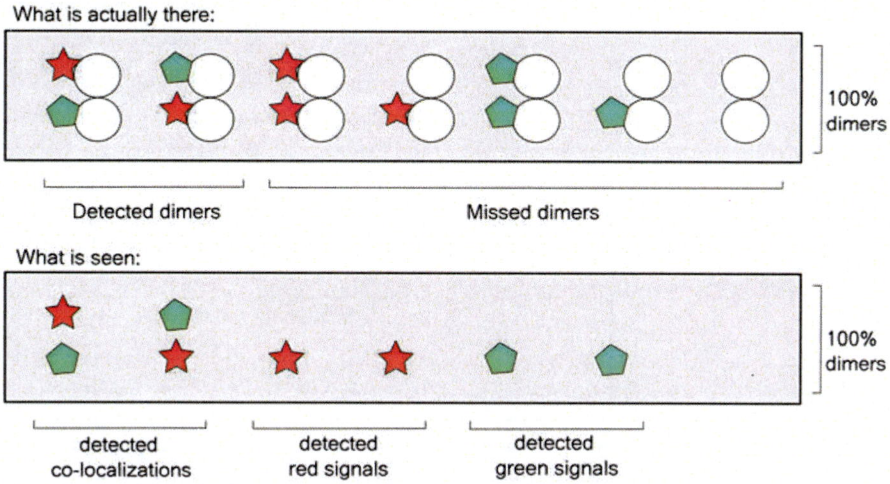

Fig. 7 Not all dimers are detected during co-localization analysis: Dimers carrying two spectrally identical fluorescent molecules as well as partially labeled dimers (labeling efficiency <1) are falsely counted as monomers as they appear only in one emission channel. Taking into account the number of detected co-localizations and the number of detected red and green emission signals, the dimer fraction determined from co-localization analysis can be corrected for non-equimolar labeling, for unlabeled protein subunits as well as for protein dimers carrying two identical fluorescent labels [47]

labeling of both of the ε-subunits or β-subunits via two different epitopes resulted in co-localization fractions two orders of magnitude higher [20].

The two-color TOCCSL assay can be combined with single-molecule FRET analysis for a more detailed characterization of molecular assemblies, e.g., by determination of their lateral extensions or inter-molecular distances. A combination with Single-Particle Tracking (SPT)/co-tracking of molecules in both emission channels achieves superior detection efficiencies, with the detection of only one co-localized trajectory out of hundred observations being statistically sufficient to rule out a false-positive observation [46].

4 Microscopy Setup

An exemplified microscopy system applicable for TOCCSL experiments is presented in Fig. 8.

The shown TOCCSL setup is based on a single-molecule fluorescence microscope with objective-based TIRF illumination. For precise and fast execution of the applied illumination timing protocol, directly modulated laser diodes are used in combination with a fast I/O module. Alternatively, the lasers can be modulated by acousto-optical modulators and mechanical shutters.

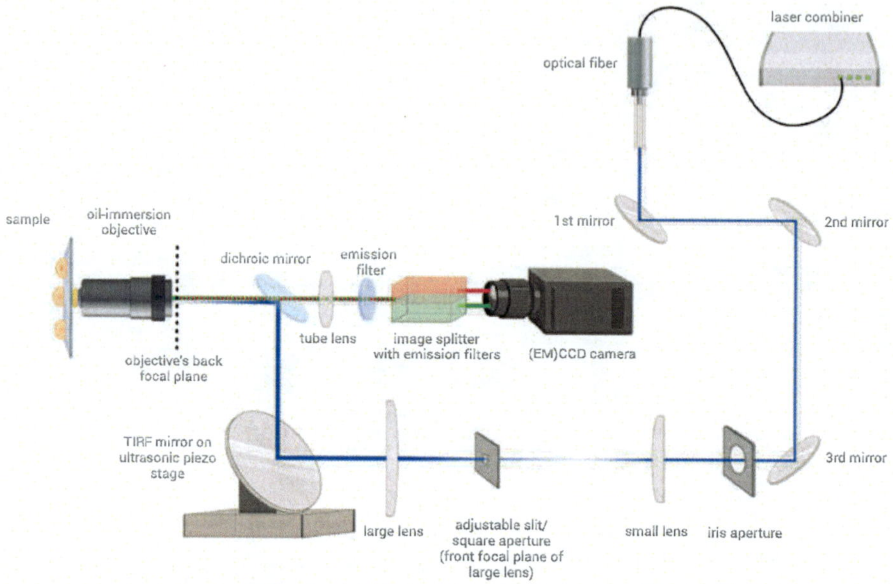

Fig. 8 Schematic illustration of a typical TOCCSL setup. Inside a multi-color laser combiner laser beams of differing wavelengths are overlaid with dichroic mirrors and coupled into an optical output fiber. The laser beams pass three mirrors and an iris aperture which is used to center the back-reflections originating from all following optical elements and to overlay them with the initial incoming beam. The beam is expanded and focused into the back focal plane of a high-NA oil immersion objective by using a two-lens system after being coupled into the microscope via a periscope. An adjustable slit/square aperture is placed at the front focal plane of the second lens (large lens), to confine the illuminated area at the sample plane. A mirror mounted to a piezo table at the bottom of the periscope allows for fast switching between Non-TIR and TIR modes. Excitation and emission light pass the same high-NA objective, which enables TIR excitation. A dichroic mirror separates the excitation beam from the emitted light. The emission light is further filtered by an emission band-pass filter suited to the used fluorophore and featuring an optical density of >6 to block totally reflected laser light. The emission signal is focused onto an (EM)CCD or sCMOS camera via a tube lens. For multi-color imaging, light is separated into two detection channels by an image splitter that is equipped with an additional dichroic mirror and optionally, two more emission filters

To obtain homogeneous illumination at the sample plane, the laser beam is focused into the back focal plane of the objective by a two-lens system. To ensure that the laser is entering the two-lens system parallel to the optical axis of the first lens, it is first guided via three adjustment-mirrors. The focal distance of the two-lens system must be calculated taking into account simple geometric optics (see Fig. 9).

Starting with the thin lens equations $\frac{1}{f_1} = \frac{1}{x_{o1}} + \frac{1}{x_{i1}}$ and $\frac{1}{f_2} = \frac{1}{x_{o2}} + \frac{1}{x_{i2}}$ for both lenses, they can be rewritten as $x_{i1} = \frac{x_{o1} \cdot f_1}{x_{o1} - f_1}$ and $x_{i2} = \frac{x_{o2} \cdot f_2}{x_{o2} - f_2}$.

With $d = x_{i1} + x_{o2} \Rightarrow x_{o2} = d - x_{i1}$ and x_{i1} inserted from above the front focal length of the two-lens system is given by

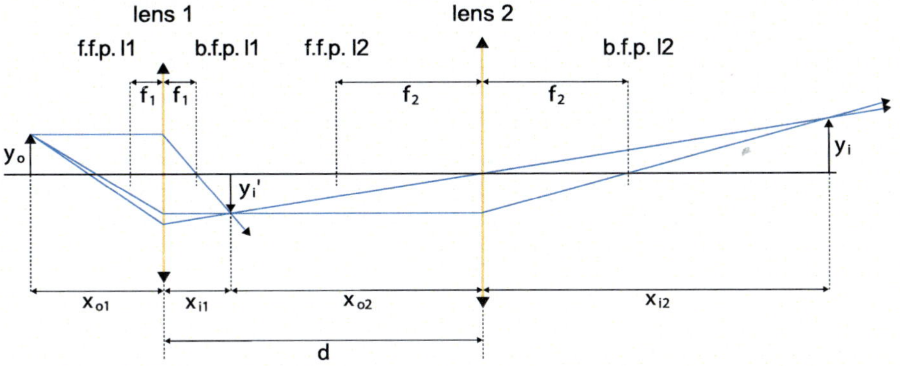

Fig. 9 Schematic illustration of a two-lens system. f.f.p. l1/2 and b.f.p. l1/2 denote the front focal plane and the back focal plane, respectively, and f_1/f_2 the focal length of the respective lens. The collimated (parallel) rays are focused at a distance $x_{o1} + x_{i1}$ from the laser source, where an intermediate image of height y_i' is formed. Subsequently, the beam is focused once more at a distance $x_{o2} + x_{i2}$ from the intermediate image, creating the final image of height y_i. x_{o1} and x_{o2} denote the distances from the object to the respective lens, x_{i1} and x_{i2} the distances from the respective lens to the image, and y_o the object height. The right choice of lenses and their respective distance d from each other ensures that the beam is focused into the back focal plane of the objective

$$x_{i2} = \frac{\left(d - \frac{x_{o1} \cdot f_1}{x_{o1} - f_1}\right) \cdot f_2}{d - \frac{x_{o1} \cdot f_1}{x_{o1} - f_1} - f_2} \tag{3}$$

Due to the collimated laser emitting a parallel beam, the object y_o can be considered to be infinitely far away and thus

$$f.f.l. = \lim_{x_{o1} \to \infty} x_{i2} = \frac{f_2 \cdot (d - f_1)}{d - (f_1 + f_2)} \tag{4}$$

The two lenses have to be chosen such, that the front focal length of the two-lens system is at the exact position of the objective's back focal plane. The laser is coupled into the microscope body via a periscope. Emission and excitation light both pass the same objective. To separate the excitation light from the Stokes-shifted emission light it, a dichroic mirror is used that reflects the excitation beam but transmits the spectrum of the emitted light. The emitted light is filtered from reflected excitation light and background by distinct emission filters.

For the implementation of TIRF, a mirror is used to shift the beam parallel to the optical axis. For this purpose, the mirror is mounted to a high-speed piezo table positioned behind the two-lens system, which allows for fast switching between TIR and non-TIR configuration (below 30 ms). Bleaching in non-TIR configuration and switching to TIR configuration when recording the recovery of single fluorescent molecules improves the signal-to-noise ratio, especially for cellular systems with high background signal due to intracellular fluorescence.

To restrict the excitation/illumination during imaging and photobleaching to a well-defined area of the cellular plasma membrane, a field stop, i.e., an adjustable slit/square aperture, is placed into the laser beam path at a sample-conjugated plane. Photobleaching of arbitrary shapes can be implemented by using a xy-laser scanning unit [48]. Images are recorded with an electron-multiplying charge-coupled device (EMCCD) or a scientific complementary metal-oxide-semiconductor (sCMOS) camera, that exhibit high read-out speed at low noise. For the two-color implementation of TOCCSL an image splitter is used to record the emitted fluorescence of the two spectrally different fluorophores on separate positions of the same camera chip.

5 Choice of Parameters

5.1 Choice of Recovery Time, Search-Radius and Analysis Region

The choice of t_{rec} depends on the mobility of the observed protein complexes, as well as its surface density and the size of the aperture. It should be chosen such, that multiple single-molecule signals per analysis region can be analyzed to obtain sufficient statistics [39]. However, the higher t_{rec}, the more molecules diffuse back into the field of view, which ultimately leads to an increase in surface density and thus to a higher fraction of false-positive co-localizations that contribute to the overestimation of higher order oligomers.

The fraction of false-positives scales with the co-localization search-radius as well as the protein surface density ρ. For a TOCCSL assay, assuming all protein clusters are randomly distributed over the surface, the probability for one of them to be located within a threshold distance R around another protein cluster is given by [46].

$$fp = 1 - \exp\left(-R^2 \cdot \pi \cdot \rho\right) \tag{5}$$

As the recovery proceeds from the aperture edges, the surface density after recovery increases in radial direction within the photobleached area. As fp depends on the surface density, it increases towards the aperture edges. Thus, molecules close to the aperture edges are excluded from the analysis. The higher t_{rec}, the more signals fill the photobleached spot, yielding an increase in fp. To keep fp at an approximately constant and low value, the area of the analysis region should be chosen smaller with increasing t_{rec}. The right choice of t_{rec} and an appropriate selection of the analysis area ensures that fp remains significantly below the number of detected true-positive dimers. For small fractions of true-positive dimers, this means that an extreme dilution of detected molecules and thus, the selection of a small analysis region is needed to ensure even smaller false-positive rates. This results in a reduced number of detectable events. Consequently, a high number of experiments is required.

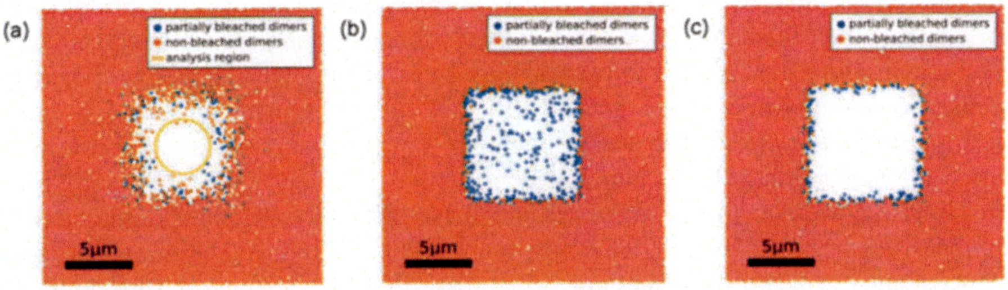

Fig. 10 Simulated after-bleach images for an initial protein surface density of $\rho = 100$ dimers/μm^2, a diffusion coefficient of $D = 0.5$ μm^2/s, a photobleaching intensity of $I = 5.32$ kW/cm^2 and each protein subunit carrying one GFP molecule (labeling efficiency = 1). (**a**) Simulation for a population of 100% dimers, $t_{\text{bleach}} = 400$ ms and $t_{\text{rec}} = 1$ s. The ideal choice of analysis region ensures that <20% false-positive dimers are detected. (**b**) If the chosen bleaching time is too short (e.g., $t_{\text{bleach}} = 100$ ms) partially bleached molecules remain within the photobleached region at $t_{\text{rec}} = 0$ s. (**c**) The edge zone of the aperture-defined region exhibits a high concentration of partially bleached dimers at $t_{\text{rec}} = 0$ s for a typical experimental bleaching time of 400 ms. Parts of this figure were taken from [49]

Figure 10a illustrated the right choice for the analysis region with <20% false-positive detections.

To further reduce *fp*, single-molecules can be tracked over consecutive images. Signals, that co-localize by coincidence move randomly and will separate over time, whereas true-positive oligomers exhibit a correlated movement and will remain associated (see also Sect. 3.5).

The surface density of protein clusters can be determined from the pre-bleach images of the TOCCSL sequence, by dividing the mean intensity per μm^2 by the corresponding single-molecule brightness obtained from the Gaussian fit. The protein surface densities in the after-bleach images can be determined by counting the number of analyzed diffraction-limited spots and dividing it by the area of the analysis region.

5.2 *Choice of Photobleaching Time*

Photobleaching is a stochastic process and depends on the illumination intensity at the sample plane as well as t_{bleach}. Due to diffraction at the aperture edges, the photobleaching probability is also a function of the molecules' coordinates within the photobleached area. The aperture-limited laser intensity profile, used in a TOCCSL experiment, typically exhibits an oscillating plateau bordered by an exponentially decreasing edge region, in which the photobleaching probability of fluorophores is reduced.

An appropriate choice of t_{bleach} and laser intensity is required, so that all fluorescently labeled molecules within the photobleached area of the sample are irreversibly switched off. The laser intensity and time needed for photobleaching strongly depend on the imaged fluorophore and its photobleaching probability and must be determined experimentally. For this purpose, a control image is recorded immediately after photobleaching. If the control image exhibits detectable single-molecule signals or high fluorescent background, the photobleaching intensity and t_{bleach} have to be increased (see Fig. 10b for an example of t_{bleach} being too short and how it affects the detection of molecular dimers).

However, protein clusters diffuse in and out of the aperture-defined region during photobleaching. Additionally, they experience an overall reduced laser intensity and consequently, a decreased bleaching probability at the aperture edges. Thus, increasing the photobleaching intensity is advantageous over increasing t_{bleach}, as the extended illumination at the aperture edges increases the number of incompletely photobleached protein clusters, with one or more protein subunits remaining fluorescently labeled (see Fig. 10c for an example of partial bleaching at the edges of the aperture). On the other hand, the laser intensity during imaging should be kept low to further avoid partial photobleaching of protein clusters. The ideal laser intensity during imaging depends on the observed fluorophore as well as the signal-to-noise ratio after photobleaching. To further reduce fluorescent background, especially in live cell experiments, photobleaching in non-TIR configuration can be performed. The contrast of recovered single-molecule signals at the bottom of the cell membrane is always increased in TIR configuration.

5.3 *Fluorescent Labels*

As for all fluorescence microscopy experiments, the fluorescent label and the conjugation method need to be chosen with care. In the following, a few considerations are listed on how to find the optimum labeling strategy for TOCCSL experiments:

5.3.1 No Unspecific Binding/Detection of Fluorophores

Unspecific but stable binding of labels to the cover-glass does not affect the outcome of a TOCCSL experiment because of the photobleaching step during the execution of the TOCCSL protocol. Only the initial apparent surface density is increased by these immobile events. More important is to avoid unspecific interactions of organic dyes with the bilayer or glycocalyx of the plasma membrane. Here, fluorophores can diffuse back into the analyzed region and alter the result. As long as the fluorophores are highly hydrophilic, i.e., soluble in the medium used in the experiment, these membrane interactions can be neglected. Examples of organic dyes showing strong or nearly no preferences to interact with synthetic membranes can be found in [50].

Using fluorescent proteins (FPs) often causes unspecific detection events and an increased background due to a potential high abundance in the cytosol and inner membranes of the ER and the Golgi apparatus. Photobleaching in non-TIR configuration efficiently decreases this background but cannot avoid the false-positive detection of FPs located in membrane-proximal vesicles. Single-particle tracking of recovered events allows to distinguish free diffusing molecules of interest from directed motion of vesicles via mean-square-displacement versus time-lag analysis [51].

5.3.2 Fast and Efficient Photobleaching While Keeping a Good Signal-to-Noise Ratio

The ideal fluorescent label for TOCCSL bleaches instantly upon high-power laser irradiation while concomitantly yielding a high number of photons in the after-bleach images recorded at low-power excitation. Practically, bleaching-stable dyes as well as fluorophores prone to blinking should be avoided. Most FPs compatible with single-molecule imaging, and non-blinking/non-bleaching-stable organic dyes are suited for one- or two-color TOCCSL applications. Rather than the countless number of different dyes nowadays offered, the availability of appropriate high-power and directly modulated laser sources is limited, rendering 488 nm or 640 nm laser diodes with ~200 mW output power optimal TOCCSL light sources.

5.3.3 Minimal-Invasive Conjugation to Biomolecule of Interest Possible

The type of fluorescent label determines the method used for conjugation to the biomolecule of interest. While FPs can be conjugated on a genetic level directly to proteins, organic dyes need to be conjugated via other molecules to label the target. In both cases, the conjugation must not affect the function of the biomolecule, e.g., the correct binding of ligands, enzymatic activity, or function of kinases. Another important aspect is the size of the fluorescent label, which should be as small as possible to avoid steric hindrance or size exclusion effects. While an antibody is about 180 kD in size, an antibody fragment (Fab) is only ~60 kD, and a single chain variable fragment (scFv) just about 28 kD (see Fig. 11 for a comparison of labels). The latter two are preferred, because they are monovalent binders. In contrast, an antibody is divalent and can lead to cross-linking of target molecules. Typical sizes of FPs or other genetic labels such as SNAP, CLIP, or Halo-tags (for a comparison of structure and kinetics, see [52]) are in the range of 20–30 kD. For the latter class, organic dyes with ~1 kD are enzymatically linked via the functional protein tags. Another category of small binders involves fluorescently labeled single domain nanobodies with a size of only ~15 kD. These binders are directed against FPs, as for the GFP-nanobody [53], or against peptides, as for the ALFA [54]- Spy [55]- or spot [56]-tag nanobody.

The close proximity of organic dyes to amphiphilic biomolecules also strongly influences their behavior; the influence can be minimized by, e.g., using neutral crosslinkers such as polyethylene glycol (PEG) [57].

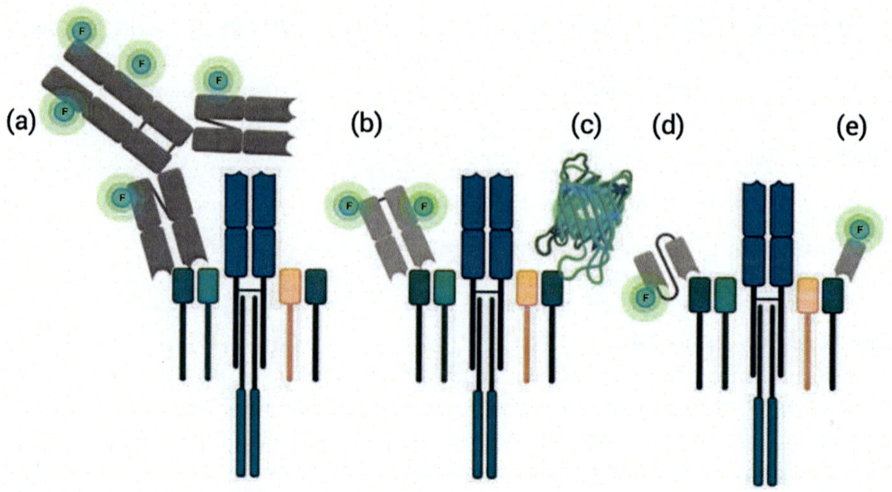

Fig. 11 Examples for labeling the TCR/CD3 complex. (**a**) Primary, NHS-labeled IgG antibody with a molecular weight (M_W) of ~150 kD, which features two epitope binding sites. (**b**) Via enzymatic digestion, Fabs with M_W ~ 50 kD can be generated and randomly labeled with NHS-functionalized organic dyes. (**c**) Genetically fused fluorescent proteins or tags (in the shown example GFP with a M_W ~ 28 kD) yield a 1:1 labeling stoichiometry, but can influence functionality, expression, and targeting of the protein complex. (**d**) Engineered and site-specifically labeled scFv represents a small (M_W ~ 27 kD), stoichiometric binder. (**e**) Nanobodies, only half the size of an scFv (M_W ~ 15 kD), are one of the smallest probes, which can be functionalized with exactly one fluorophore. For TOCCSL, small, monomeric, and site-specifically labeled probes (**c**–**e**) are preferred. Created with BioRender.com

In addition, also the conjugation of organic dyes to binder molecules like Fabs needs to be minimal invasive: if Fabs carry more than two organic dyes linked via NHS-labeling, the binding kinetics are strongly affected and in most cases only Fabs with no or just one dye are able to bind the protein of interest with high affinity [58].

5.3.4 Labeling Stoichiometry

Ideally, every molecule of interest carries exactly one fluorophore to allow straightforward brightness-based analysis of one-color TOCCSL experiments. This is less important for co-localization-based two-color TOCCSL experiments if the brightness in individual color channels is not considered. Genetically attaching a (monomeric)FP or protein/peptide tag to the biomolecule of interest causes in most cases a 1:1 labeling stoichiometry of biomolecule:FP/organic dye. However, for all protein tags, the maturation efficiency needs to be considered. Slow folding of proteins would deprive them from detection. In addition, labeling efficiency plays an important role for tags, where functionalized dyes need to be enzymatically linked.

1:1 labeling stoichiometries are also achieved by using nanobodies or scFvs. Via free cysteines, maleimide-functionalized dyes can be site-specifically attached. In principle, also Fabs labeled non-specifically at primary amines via succinimidylester-functionalized dyes can be used for brightness analysis. However, care must be taken when recording the monomer-brightness distribution. The distribution will include single- and double-labeled Fabs and will differ significantly from a distribution of a single dye. Using secondary antibody labeling or fluorescent primary antibodies should be avoided due to their multi- and divalent binding affinity, respectively. In addition, quenching is observed for many dyes, if more than two dyes are present per antibody.

For determining the stoichiometry of the TCR/CD3 complex, the organic dye molecules Alexa Fluor 488 and Alexa Fluor 647 conjugated to an scFv against the ε- or β-subunit were chosen as the best fitting label for the following reasons:

1. Both organic dyes are hydrophilic and show a very low preference for interacting with membranes [50].
2. Alexa Fluor 488/647 is bright and easy to bleach due to its high/medium quantum yield and medium/high extinction coefficient.
3. The combination of dyes allows for efficient detection of (single-molecule) FRET (Förster radius $R_0 = 5.6$ nm).
4. scFv can be site-specifically and efficiently conjugated with exactly one fluorophore (protein-to-dye ratio ranged between 0.95 and 1.0 [20]).
5. The chosen scFvs bind efficient and strong (half-life time is 44 min [20]) and are only 1/6 of the size of a regular antibody.

6 TOCCSL Applications

One- and two-color TOCCSL applications have been used to address the stoichiometry of various plasma membrane proteins. In the following, a tabular and chronologically overview of TOCCSL applications is provided, which should serve as a guide to select certain parameters and labeling strategies and represent a source for finding further literature on TOCCSL (Tables 1 and 2).

Table 1 One-color TOCCSL applications

Protein/system	Labeling	Study/finding	Source
Anti-DNP antibody anchored to a supported lipid bilayer via DNP-DPPE	FITC via NHS-labeling, on average 4.5 FITC/ antibody	TOCCSL proof-of-principle; labeling stoichiometry is conserved in TOCCSL	[40]
Lck-CFP-YFP on T24 cells	Yellow fluorescent protein (YFP)	Lck moves as small associates (monomer, dimers, and higher order multimers) in the membrane	[59]
GPI-anchored GFP (mGFP-GPI) on CHO cells Bodipy(FL)-GM1 on Jurkat T cells	Monomeric enhanced green fluorescent protein (mGFP), BodipyFL	mGFP-GPI forms cholesterol-dependent homo-dimers, Bodipy-GM1 clusters are present on Jurkat T cells	[43]
Bodipy-GM1 in a supported lipid bilayer and clustered by cholera toxinB (CTX-B)	BodipyFL	On average 1.4 Bodipy-GM1 molecules are bound per CTX-B	[46]
pMHC II (IEk/MCC(C))-Cy5 anchored to a supported lipid bilayer	Cy5 via maleimide chemistry	pMHC class II is monomeric at densities up to 500 molecules per μm^2	[60]
Orai1-mGFP	mGFP	Orai1 diffuses as homo-tetramer in the plasma membrane of T24 cells	[44]
Human serotonin transporter (hSERT)-mGFP on the plasma membrane of CHO cells	mGFP and fluorescent inhibitor JHC 1–64	Surface density independent and stable oligomerization of hSERT	[61]
mGFP-GPI on the plasma membrane of CHO cells	mGFP-GPI	mGFP-GPI homo-association is released by addition of oxidized phospholipids	[62]
Human serotonin transporter (hSERT)-mGFP in the ER & plasma membrane of CHO cells	mGFP	SERT oligomerization at the plasma membrane depends on PIP2 levels; SERT subunits rearrange in the ER	[63]
CD3ε and CD3β on murine primary T cells adhered to supported lipid bilayers	scFv-Alexa Fluor 647 coupled via maleimide chemistry	The TCR/CD3 complex consists of one CD3β and two CD3ε subunits; there exists no higher TCR oligomers on the membrane of resting T cells	[20]
Human dopamine transporter (hDAT) in the plasma membrane of CHO cells	mGFP	Monomers and dimers of hDAT coexist, no higher oligomers; dimers are stable over several minutes	[64]
mGFP-GPI on the plasma membrane of CHO cells	mGFP	Antimicrobial peptides influence the homo-association of mGFP-GPI	[65]

Table 2 Two-color TOCCSL applications

Protein/system	Labeling	Study/finding	Source
Bodipy-GM1 in a supported lipid bilayer and clustered by cholera toxin B (CTX-B)	BodipyFL-GM1 and CTX-B-Alexa Fluor 647	On average 1.4 Bodipy-GM1 molecules are bound per CTX-B	[46]
CD3ε and CD3β on murine primary T cells adhered to supported lipid bilayers	scFv-Alexa Fluor 488 and 647 coupled via maleimide chemistry	The TCR/CD3 complex consists of one CD3β and two CD3ε subunits; there exists no higher TCR oligomers on the membrane of resting T cells	[20]
ErbB2 and ErbB3 in CHO cells	Fab-Alexa Fluor 488 and 647 linked via NHS-ester	ErbB3 is activated via homodimerization and heterodimerization with ErbB2	[47]

References

1. Clevers H, Alarcon B, Wileman T, Terhorst C (1988) The T cell receptor/CD3 complex: a dynamic protein ensemble. Annu Rev Immunol 6:629–662
2. Dong D et al (2019) Structural basis of assembly of the human T cell receptor–CD3 complex. Nature 573:546–552
3. Ottschytsch N, Raes A, Van Hoorick D, Snyders DJ (2002) Obligatory heterotetramerization of three previously uncharacterized Kv channel α-subunits identified in the human genome. Proc Natl Acad Sci U S A 99:7986–7991
4. Yellen G (2002) The voltage-gated K+ channels and their relatives. Nature 419:35–42
5. Milligan G (2004) G protein-coupled receptor dimerization: function and ligand pharmacology. Mol Pharmacol 66:1–7
6. Lemmon MA, Schlessinger J (2010) Cell signaling by receptor tyrosine kinases. Cell 141: 1117–1134
7. Kusumi A, Tsunoyama TA, Hirosawa KM, Kasai RS, Fujiwara TK (2014) Tracking single molecules at work in living cells. Nat Chem Biol 10:524–532
8. Suzuki KGN et al (2012) Transient GPI-anchored protein homodimers are units for raft organization and function. Nat Chem Biol 8:774–783
9. Yáñez-Mó M, Barreiro O, Gordon-Alonso M, Sala-Valdés M, Sánchez-Madrid F (2009) Tetraspanin-enriched microdomains: a functional unit in cell plasma membranes. Trends Cell Biol 19:434–446
10. Simons K, Ikonen E (1997) Functional rafts in cell membranes. Nature 387:569–572
11. Douglass AD, Vale RD (2005) Single-molecule microscopy reveals plasma membrane microdomains created by protein-protein networks that exclude or trap signaling molecules in T cells. Cell 121:937–950
12. Vrljic M, Nishimura SY, Brasselet S, Moerner WE, McConnell HM (2002) Translational diffusion of individual class II MHC membrane proteins in cells. Biophys J 83:2681–2692
13. Manley S et al (2008) High-density mapping of single-molecule trajectories with photoactivated localization microscopy. Nat Methods 5:155–157
14. Betzig E et al (2006) Imaging intracellular fluorescent proteins at nanometer resolution. Science 313:1642–1645
15. Hess ST, Girirajan TP, Mason MD (2006) Ultra-high resolution imaging by fluorescence photoactivation localization microscopy. Biophys J 91:4258–4272
16. Rust MJ, Bates M, Zhuang X (2006) Sub-diffraction-limit imaging by stochastic optical reconstruction microscopy (STORM). Nat Methods 3:793–795
17. Heilemann M et al (2008) Subdiffraction-resolution fluorescence imaging with conventional fluorescent probes. Angew Chem Int Ed Engl 47:6172–6176

18. Pertsinidis A, Zhang Y, Chu S (2010) Subnanometre single-molecule localization, registration and distance measurements. Nature 466:647–651
19. Lelek M et al (2021) Single-molecule localization microscopy. Nat Rev Methods Prim 1:39
20. Brameshuber M et al (2018) Monomeric TCRs drive T cell antigen recognition. Nat Immunol 19:487–496
21. Rossboth B et al (2018) TCRs are randomly distributed on the plasma membrane of resting antigen-experienced T cells. Nat Immunol 19:821–827
22. Baumgart F et al (2019) What we talk about when we talk about nanoclusters. Methods Appl Fluoresc 7:13001
23. Platzer R et al (2020) Unscrambling fluorophore blinking for comprehensive cluster detection via photoactivated localization microscopy. Nat Commun 11:4993
24. Tanaka KAKK et al (2010) Membrane molecules mobile even after chemical fixation. Nat Methods 7:865–866
25. Hoffman DP et al (2020) Correlative three-dimensional super-resolution and block-face electron microscopy of whole vitreously frozen cells. Science 367(6475):eaaz5357
26. Weisenburger S et al (2014) Cryogenic colocalization microscopy for nanometer-distance measurements. ChemPhysChem 15:763–770
27. Kaufmann R et al (2014) Super-resolution microscopy using standard fluorescent proteins in intact cells under cryo-conditions. Nano Lett 14:4171–4175
28. Hell SW, Wichmann J (1994) Breaking the diffraction resolution limit by stimulated emission: stimulated-emission-depletion fluorescence microscopy. Opt Lett 19:780–782
29. Klar TA, Jakobs S, Dyba M, Egner A, Hell SW (2000) Fluorescence microscopy with diffraction resolution barrier broken by stimulated emission. Proc Natl Acad Sci U S A 97: 8206–8210
30. Hell SW (2007) Far-field optical nanoscopy. Science 316:1153–1158
31. Sahl SJ, Hell SW, Jakobs S (2017) Fluorescence nanoscopy in cell biology. Nat Rev Mol Cell Biol 18:685–701
32. Blom H, Widengren J (2017) Stimulated emission depletion microscopy. Chem Rev 117:7377–7427
33. Kimble HJ, Dagenais M, Mandel L (1977) Photon antibunching in resonance fluorescence. Phys Rev Lett 39:691–695
34. Sýkora J et al (2007) Exploring fluorescence antibunching in solution to determine the stoichiometry of molecular complexes. Anal Chem 79:4040–4049
35. Ta H, Kiel A, Wahl M, Herten DP (2010) Experimental approach to extend the range for counting fluorescent molecules based on photon-antibunching. Phys Chem Chem Phys 12: 10295–10300
36. Ta H et al (2015) Mapping molecules in scanning far-field fluorescence nanoscopy. Nat Commun 6:7977
37. Balzarotti F et al (2017) With minimal photon fluxes. Science 355:606–612
38. Schmidt R et al (2021) MINFLUX nanometer-scale 3D imaging and microsecond-range tracking on a common fluorescence microscope. Nat Commun 12:1478. https://doi.org/10.1038/s41467-021-21652-z
39. Brameshuber M, Schutz GJ (2012) Detection and quantification of biomolecular association in living cells using single-molecule microscopy. Methods Enzym 505:159–186
40. Moertelmaier M, Brameshuber M, Linimeier M, Schütz GJ, Stockinger H (2005) Thinning out clusters while conserving stoichiometry of labeling. Appl Phys Lett 87:1–3
41. Edidin M, Zagyansky Y, Lardner TJ (1976) Measurement of membrane protein lateral diffusion in single cells. Science 191:466–468
42. Axelrod D, Koppel DE, Schlessinger J, Elson E, Webb WW (1976) Mobility measurement by analysis of fluorescence photobleaching recovery kinetics. Biophys J 16:1055–1069
43. Brameshuber M et al (2010) Imaging of mobile long-lived nanoplatforms in the live cell plasma membrane. J Biol Chem 285:41765–41771

44. Madl J et al (2010) Resting state Orai1 diffuses as homotetramer in the plasma membrane of live mammalian cells. J Biol Chem 285:41135–41142
45. Mutch SA et al (2007) Deconvolving single-molecule intensity distributions for quantitative microscopy measurements. Biophys J 92:2926–2943
46. Ruprecht V, Brameshuber M, Schütz GJ (2010) Two-color single molecule tracking combined with photobleaching for the detection of rare molecular interactions in fluid biomembranes. Soft Matter 6:568–581
47. Váradi T et al (2019) Homo- and heteroassociations drive activation of ErbB3. Biophys J 117: 1935–1947
48. Belyy V et al (2017) PhotoGate microscopy to track single molecules in crowded environments. Nat Commun 8:13978
49. Kiesenhofer D (2017) Development of Monte Carlo simulations for characterization and optimization of TOCCSL experiments. TU Wien
50. Hughes LD, Rawle RJ, Boxer SG (2014) Choose your label wisely: water-soluble fluorophores often interact with lipid bilayers. PLoS One 9:e87649
51. Wieser S, Schutz GJ (2008) Tracking single molecules in the live cell plasma membrane-Do's and Don't's. Methods 46:131–140
52. Wilhelm J et al (2021) Kinetic and structural characterization of the self-labeling protein tags HaloTag7, SNAP-tag, and CLIP-tag. Biochemistry 60:2560–2575
53. Rothbauer U et al (2006) Targeting and tracing antigens in live cells with fluorescent nanobodies. Nat Methods 3:887–889
54. Götzke H et al (2019) The ALFA-tag is a highly versatile tool for nanobody-based bioscience applications. Nat Commun 10:1–12
55. Khairil Anuar INA et al (2019) Spy&Go purification of SpyTag-proteins using pseudo-SpyCatcher to access an oligomerization toolbox. Nat Commun 10:1–13
56. Virant D et al (2018) A peptide tag-specific nanobody enables high-quality labeling for dSTORM imaging. Nat Commun 9:1–14
57. Honigmann A et al (2014) Scanning STED-FcS reveals spatiotemporal heterogeneity of lipid interaction in the plasma membrane of living cells. Nat Commun 5:5412
58. Szabo A et al (2018) The effect of fluorophore conjugation on antibody affinity and the photophysical properties of dyes. Biophys J 114:688–700
59. Schwarzenbacher M et al (2008) Micropatterning for quantitative analysis of protein-protein interactions in living cells. Nat Methods 5:1053–1060
60. Huppa JB et al (2010) TCR-peptide-MHC interactions in situ show accelerated kinetics and increased affinity. Nature 463:963–967
61. Anderluh A et al (2014) Single molecule analysis reveals coexistence of stable serotonin transporter monomers and oligomers in the live cell plasma membrane. J Biol Chem 289: 4387–4394
62. Brameshuber M et al (2016) Oxidized phospholipids inhibit the formation of cholesterol-dependent plasma membrane nanoplatforms. Biophys J 110:205–213
63. Anderluh A et al (2017) Direct PIP 2 binding mediates stable oligomer formation of the serotonin transporter. Nat Commun 8:14089
64. Das AK et al (2019) Dopamine transporter forms stable dimers in the live cell plasma membrane in a phosphatidylinositol 4,5-bisphosphate independent manner. J Biol Chem 294:5632–5642
65. Schromm AB et al (2021) Cathelicidin and PMB neutralize endotoxins by multifactorial mechanisms including LPS interaction and targeting of host cell membranes. Proc Natl Acad Sci 118:e2101721118

Quantitative Photoactivated Localization Microscopy of Membrane Receptor Oligomers

Marina S. Dietz and Mike Heilemann

Contents

Abstract Optical super-resolution microscopy allows the visualization of cellular structures with a spatial resolution of a few tens of nanometers and has revolutionized our understanding in cell biology. However, the spatial resolution achieved with to-date super-resolution microscopy methods in cells is not sufficient to optically resolve proteins within densely packed protein clusters, which themselves represent relevant functional assemblies in cells. Single-molecule localization microscopy (SMLM) offers an opportunity to retrieve this information by analyzing the kinetics of on-off-switching ("blinking") observed in the fluorescence emission signatures of single fluorophores. We report the theoretical background of kinetics-based molecular quantification of SMLM data, discuss fluorescent probes and methods for protein labeling, and showcase applications in biology.

Keywords Membrane receptor oligomerization · Molecular quantification · Photoactivatable fluorescent proteins · Photoactivated localization microscopy · Single-molecule imaging · Single-molecule localization microscopy

---

M. S. Dietz and M. Heilemann (✉)
Institute of Physical and Theoretical Chemistry, Goethe-University Frankfurt, Frankfurt, Germany
e-mail: dietz@chemie.uni-frankfurt.de; heilemann@chemie.uni-frankfurt.de

R. Šachl, M. Amaro (eds.), *Fluorescence Spectroscopy and Microscopy in Biology*, Springer Ser Fluoresc (2023) 20: 305–322, https://doi.org/10.1007/4243_2022_37,
Published online: 3 January 2023

Abbreviations

DNA-PAINT	DNA-based point accumulation for imaging in nanoscale topography
*d*STORM	*Direct* stochastic optical reconstruction microscopy
FCS	Fluorescence correlation spectroscopy
FPALM	Fluorescence photoactivated localization microscopy
FRET	Förster resonance energy transfer
HGF	Hepatocyte growth factor
InlB	Internalin B
N&B	Number and brightness analysis
paFP	Photoactivatable fluorescent protein
PAINT	Point accumulation for imaging in nanoscale topography
PALM	Photoactivated localization microscopy
pcFP	Photoconvertible fluorescent protein
PC-PALM	Pair-correlation photoactivated localization microscopy
PLAD	Preligand assembly domain
PSF	Point spread function
QAFKA	Quantitative algorithm for fluorescence kinetics analysis
qPAINT	Quantitative point accumulation for imaging in nanoscale topography
qPALM	Quantitative photoactivated localization microscopy
qSMLM	Quantitative single-molecule localization microscopy
RTK	Receptor tyrosine kinase
SMLM	Single-molecule localization microscopy
STORM	Stochastic optical reconstruction microscopy
TIRF	Total internal reflection fluorescence
TLR4	Toll-like receptor 4
TNFR1	Tumor necrosis factor receptor 1
TNFα	Tumor necrosis factor alpha

1 Introduction

Single-molecule localization microscopy (SMLM) is a super-resolution technology that generates images by mapping the positions of single fluorophores [1, 2]. While diffraction-limited microscopy reads out the total fluorescence signal of a sample in one take, SMLM sequentially detects the fluorescence emission of single fluorophores. This requires thinning out the fluorescence signal of a sample by spatially and temporally isolating the emission light of fluorophores in a sample. This can be achieved by using photoswitchable fluorophores and stochastic activation of only a small subset of fluorophores or by using fluorophore labels that transiently bind to a target and that are supplied at sufficiently low concentration

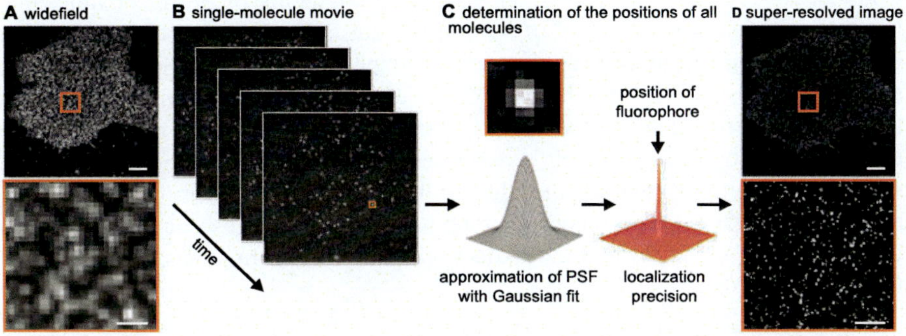

Fig. 1 Principle of single-molecule localization microscopy. (**a**) Diffraction-limited wide-field image of a cell labeled for a membrane receptor. (**b**) To circumvent the diffraction limit of light, a single-molecule movie is recorded by separating spatially close single-molecule emission events in time. (**c**) From the single-molecule movie, the positions of single fluorophores are determined. The point spread function (PSF) of each single emission event is approximated by a two-dimensional Gaussian function. The maximum of the Gaussian distribution reports the position of the fluorophore with a precision mainly depending on the inverse photon number. The localizations of all fluorophores of a single-molecule movie are collected and (**d**) enable the generation of a super-resolved image. Scale bars 5 μm, zoom-in scale bars 1 μm

in an imaging buffer. In order to determine the position of single fluorophores, the center of mass can be calculated by fitting a Gaussian function to the point spread function (PSF) of a single emitter ("localization"). A statistically sufficient number of single-molecule emission events provides a set of coordinates that, plotted in a 2D histogram, yields a reconstructed super-resolved image of the underlying structure (Fig. 1).

Various methods that build on this principle were developed in the last nearly two decades, often distinguished by the underlying experimental realization of separating fluorescence signal in time and space, e.g. (*direct*) stochastic optical reconstruction microscopy ((*d*)STORM) [3, 4], (fluorescence) photoactivated localization microscopy ((F)PALM) [5, 6], point accumulation for imaging in nanoscale topography (PAINT) [7], DNA-based PAINT (DNA-PAINT) [8], and others [9]. The spatial resolution of SMLM is, on the one hand, related to the precision of determining the center of mass, which scales inversely with the square root of the number of photons [10]. On the other hand, the structural resolution is determined by the labeling density; applying the Shannon-Nyquist theorem demands for a label density of at least twice as high as the highest spatial frequency that can be resolved in an image (the spatial frequency is the inverse of the spatial resolution) [11]. Practically, SMLM achieves a spatial resolution of a few tens of nanometers in cells.

SMLM has had a significant impact on cell biology research by enabling the visualization of cellular structures with nanoscale spatial resolution (we refer to recent reviews [2, 9, 12]). A limitation of SMLM is that the spatial resolution achieved in cells is not yet sufficient to visualize the spatial organization of proteins within dense assemblies. Such protein assemblies often constitute central functional hubs in a cell and it is desirable to understand their formation, function, and degradation. Examples are membrane-associated signaling platforms or molecular machines, which are composed of several proteins densely packed into homo- or often heteromeric nano-clusters. In order to assess the protein composition within such assemblies, SMLM can be extended towards quantitative SMLM (qSMLM) [13]. One experimental realization is the analysis of the kinetics of fluorescence emission of fluorophores, which is described by equations from chemical kinetics and is thus directly related to molecule numbers. This article reports the quantitative analysis of emission events recorded for photoactivatable fluorescent proteins (quantitative PALM, qPALM) and the extraction of molecular numbers for membrane protein complexes.

2 Theoretical Background of qPALM

2.1 Photophysics

A fundamental requirement of quantitative PALM is the mathematical description of fluorescence blinking kinetics. The fluorescence blinking kinetics of photoactivatable fluorescent proteins were among the first investigated in the context of SMLM [14–20]. One flavor of quantitative PALM builds on the approximation of the number of fluorescent blinking events with kinetic equations [17–20]. A blinking event is defined as a recurrent fluorescence emission event following photoactivation. Thus, for a photoactivatable or photoconvertible fluorescent protein, the number of blinking events is calculated from the total number of emission events of a single molecule minus one event (for a treatment of photoswitchable organic fluorophores, see Discussion). The photophysical behavior of fluorescent proteins can be summarized in a simple four-state model (Fig. 2a). Here, the number of blinking events corresponds to the number of transitions between a fluorescent and a dark state that can be reversibly populated several times before a molecule photobleaches irreversibly.

2.2 Kinetic Models

A histogram of blinking events N is generated by collecting single emission events that are extracted from a large number of clusters (Fig. 2b, c). Here, a cluster refers to "localization clouds" that are constituted by either one fluorophore-labeled protein (monomer), or multiple fluorophore-labeled proteins (oligomer), and yields a characteristic distribution (Fig. 2d). For clusters that are homogeneously constituted of one fluorophore-labeled proteins (i.e., a monomeric protein), this histogram is described by the geometric distribution, a probability distribution which has a single parameter p reporting the probability that the fluorophore does not photobleach (Eq. 1):

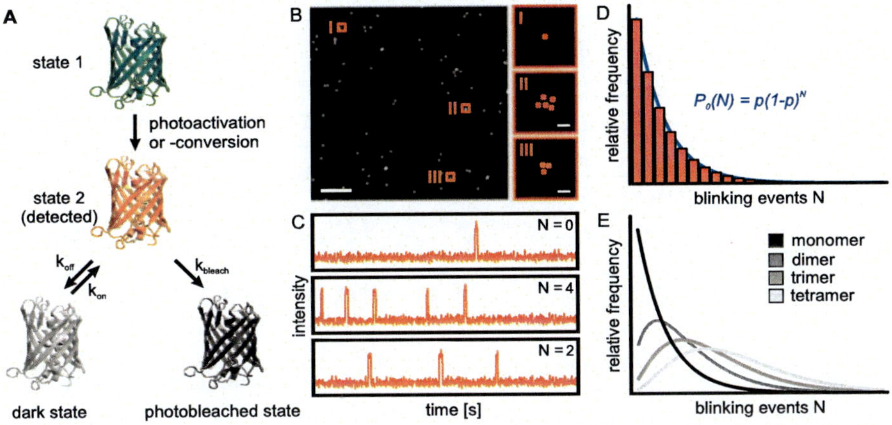

Fig. 2 Quantitative photoactivated localization microscopy. (**a**) Four-state model of a photoactivatable (paFP) or photoconvertible fluorescent protein (pcFP). At the beginning of a PALM experiment, the fluorescent protein is either in a non-activated state (paFP) or in a fluorescent state that is not detected (pcFP). Using violet light the fluorescent protein is transferred into a second fluorescent state which is detected during the measurement. Most fluorescent proteins can reversibly switch into a transient dark state introducing blinking into the time trace of a single fluorescent protein. Finally, the fluorescent protein irreversibly photobleaches. The presented crystal structure is mEOS4b (PDB 6YLS). (**b**) After recording a single-molecule movie, a super-resolved image is reconstructed from which localization clusters are selected and evaluated with regard to their localizations. Scale bar 1 μm. (**c**) Intensity-time traces allow counting the blinking events per single localization clusters. (**d, e**) Distribution of the number of blinking events generated from all localization clusters and fitted with a kinetic model function (see Sect. 2.2). Kinetic model functions are characteristic for a particular oligomeric state (**e**). The relative frequency (or empirical probability) is the absolute frequency (blinking events per localization cluster) normalized by the total number of events

$$P_0(N) = p(1-p)^N \tag{1}$$

The histogram of single-molecule emission events is constructed by collecting the number of emission events from a significantly large number of single fluorophores, typically a few hundreds. The parameter p is determined from approximating this histogram with Eq. 1. It was determined for several fluorescent proteins and, depending on the photophysical characteristics, yielded values between 0.1 and 0.8 [18, 21, 22] (see also Table 1).

The above model (Eq. 1) approximates the blinking histogram of a population of single fluorophores (which we will refer to as monomers). Extending this approach towards oligomers requires consideration of incomplete detection of multiple fluorophores in the same cluster. For this purpose, the kinetic model that describes the blinking histogram was extended by a binomial term with a second parameter q, with $(1-q)$ reporting the detectability of the fluorophore [18]. The detectability is the probability that a fluorophore is not detected, e.g. due to an immature chromophore inside a fluorescent protein, premature photobleaching, or incomplete labeling of the target protein. If we consider a true dimer, a detection efficiency of 50% for a given

Table 1 Blinking properties of different photoconvertible and photoactivatable fluorescent proteins determined with qPALM

Fluorescent protein	p	q	Reference
mEos2	0.30 ± 0.01		[17]
CD86-mEos2 (HeLa)	0.28 ± 0.01		[17]
CTLA4-mEos2 (HeLa)	0.28	0.30 ± 0.04	[18]
CD86-mEos2 (HEK293)	0.32		[22]
CTLA4-mEos2 (HEK293)	0.32	0.29	[22]
CD86-mEos2 (Cos-7)	0.29 ± 0.002		[23]
CTLA4-mEos2 (Cos-7)	0.29	0.26 ± 0.009	[23]
mEos3.2	0.30 ± 0.01, 0.32 ± 0.01		[21]
mEos3.2 + 100 mM MEA	0.17 ± 0.01		[21]
synthetic mEos3.2 dimer	0.32	0.36 ± 0.03	[21]
CD86-mEos3.2 (HeLa)	0.27 ± 0.01		[21]
CTLA4-mEos3.2 (HeLa)	0.27	0.39 ± 0.01	[21]
mEos4b	0.34 ± 0.01		[24]
CD86-mEos4b (HEK293T)	0.27 ± 0.01		[24]
CTLA4-mEos4b (HEK293T)	0.27	0.35	[24]
mMaple3	0.28 ± 0.01		[21]
mMaple3 + 100 mM MEA	0.56 ± 0.01		[21]
Dendra2	0.54 ± 0.01		[21]
PAmCherry2	0.77 ± 0.01		[21]

The p and q values for different fluorescent proteins determined in cell lines (brackets) or on single-molecule surfaces are listed (q values were determined using the p values of monomeric fluorescent proteins with values given). Errors are given as standard errors of the mean

fluorophore would yield a detection of 25% of clusters with two intact fluorophores, and the degree of dimerization would be underestimated without incorporating q into the kinetic model. This extension allows writing down a kinetic model for (m + 1)-mers (Eq. 2, Fig. 2e):

$$P_m(N) = \sum_{k=0}^{\min(m,\,N)} \binom{m}{k} \binom{N}{k} q^{m-k} (1-q)^k p^{k+1} (1-p)^{N-k} \tag{2}$$

The photobleaching probability p and the detectability q are determined experimentally by measuring reference samples as calibration standards. Suitable cellular reference samples are (membrane) proteins of known stoichiometry, such as monomeric CD86 and dimeric CTLA-4 fused to the respective fluorescent protein and expressed in the cell line of interest [17, 18]. Other options are synthetic or genetic dimers in which two fluorescent proteins are bridged by a DNA or peptide linker, respectively [21]. A third option is to make a reasonable estimate on the detectability of a fluorescent protein, e.g., by referring to published values reported for similar experiments; this might be useful if relative changes in the oligomeric state are sufficiently helpful in an experiment, rather than absolute numbers.

3 qPALM Experiment

3.1 Labeling

The design of a qPALM experiment begins with a suitable strategy to label a target protein with a photoactivatable or photoconvertible fluorescent protein. This can be achieved with transient transfection of a plasmid coding for the target protein fused to the fluorescent protein [17, 22]. As this might lead to heterogeneous expression levels and overexpression, an alternative route is to generate a cell line where the target protein coupled to the fluorescent protein is stably transfected and expressed at low concentration [25, 26] or where the protein of interest is directly labeled at the chromosomal locus, e.g., via CRISPR/Cas9 [27] or CRISPR/Cas12a [24]. Genomic labeling ensures an endogenous expression level of the target protein, yet is more time consuming.

Fluorescent proteins suitable for qPALM are largely compatible with those used for PALM imaging, including the fluorescent protein families of mEos [28–31], Dendra [32, 33], and mMaple [34, 35]. The blinking parameters p of some of these fluorescent proteins were previously characterized [21] and are summarized in Table 1.

3.2 Data Acquisition and Evaluation

The data acquisition in qPALM experiments is similar to conventional SMLM experiments. Single-molecule sensitive widefield microscopes with total internal reflection fluorescence (TIRF) illumination are the first choice. In qPALM, it is important to ideally detect all emission events of each fluorescent protein, which can be best realized by turning on laser excitation after starting the data acquisition and using a sufficiently long recording time. Adjusting the correct imaging plane is conducted in brightfield illumination or in a sample region that is not imaged afterwards. The intensity of the photoactivation laser (often 405 nm) is kept low enough so that the point spread functions of single emitters do not overlap. In the case of oligomers, the measurement is continued until no further emission events occur and all fluorescent proteins are bona fide detected.

In order to extract the oligomeric state of protein clusters, a multi-step analysis is conducted. First, single-molecule emission events are localized and a super-resolved image is reconstructed. Fluorescence emission signals recorded around the same position and in successive images likely belong to the same single emission event and are linked to a single localization (spatio-temporal linking, applying spatial filters for localization uncertainty). From the super-resolved image, localization clusters are selected by applying filters for size, roundness, and sufficient spatial distance to other clusters. The number of emission events detected in these localization clusters is extracted and histogrammed. The resulting distribution is next fitted with the general expression of the kinetic model (Eq. 2) (for a detailed protocol, we refer to Krüger et al. [36] and the supplementary material of Baldering et al. [21]). An automated qPALM analysis using a "quantitative algorithm for fluorescence kinetic analysis" (QAFKA) was recently introduced [37].

This procedure also allows the analysis of mixtures of different oligomers by linear combinations of the respective kinetic equations. For example, a mixture of monomers and dimers is described by Eq. 3 where f is the fraction of monomers.

$$P_{0,1}(N) = fp(1-p)^N + (1-f)p(1-p)^{N-1}(Np\,(1-q) + (1-p)q) \quad (3)$$

As a result, qPALM reports relative fractions of oligomers (obtained from a given data set) and cannot attribute a particular oligomeric state to an individual cluster.

We note that the proposed model for qPALM analysis assumes the detection of at least one fluorescent protein within a nano-cluster [18]. This approach neglects the existence of nano-clusters in which none of the fluorescent proteins is detected; a full statistical analysis including all scenarios of non-emitting fluorophores was developed to account for this [38] and increases the accuracy of the analysis, especially for higher oligomers.

The quantitative analysis can also be applied to *d*STORM data, in which organic fluorophores are operated as photoswitches and cycle between an "on" and an "off" state until photobleaching. A revised model for describing these blinking data was developed [39].

3.3 Reference Structures for Calibration

For the accurate determination of oligomeric states, the blinking parameters of the fluorescent proteins have to be determined in the relevant cell line. This includes the bleaching probability p and the detection efficiency q. These parameters can be determined from calibration standards, i.e., reference structures with known oligomeric state, e.g., monomeric and dimeric proteins (Fig. 3).

A simple method for determining the p value of a fluorescent protein is to measure single-molecule emission events on a surface (Fig. 3a). For this purpose, the fluorescent protein is applied at low concentration to a surface coated with poly-L-lysine. This straightforward procedure quickly reports on the photobleaching probability parameter p; values of several photoactivatable/-convertible fluorescent proteins are listed in Table 1.

The detection efficiency q is determined from a dimeric reference. For in vitro measurements, these dimers can be generated either synthetically or genetically (Fig. 3a) [21]. Synthetic dimers of two fluorescent proteins can be generated by introducing a linker, e.g., an oligonucleotide or a peptide, carrying two affinity tags (e.g., a trisNTA tag) that binds the fluorescent protein carrying a complementary tag

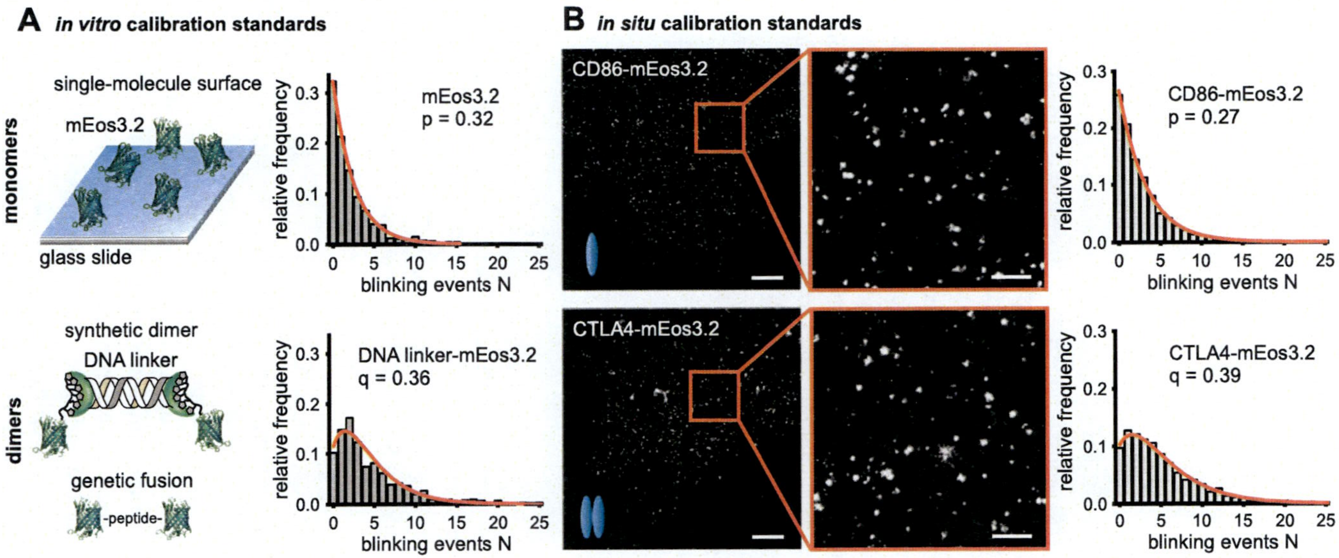

Fig. 3 Calibration standards for quantitative PALM experiments. (**a**) The photobleaching probability p and the detection efficiency q of photoactivatable or photoconvertible fluorescent proteins can be determined in vitro. For a monomeric reference, single fluorescent proteins are dispersed on a cover slide at single-molecule concentration (top left), and the number of blinking events is extracted, histogrammed, and fitted with Eq. (1), reporting the bleaching probability (top right). Dimeric reference structures can be constructed by coupling two fluorescent proteins synthetically to a DNA or peptide linker using affinity tags. Alternatively, a genetic fusion of two fluorescent proteins can be generated (bottom left). The detection probability q is obtained by fitting the blinking distribution of a dimeric reference structure with Eq. (2) using the previously determined parameter p (bottom right). (**b**) In situ qPALM measurements benefit from the determination of p and q directly in cells, which can be achieved using monomeric and dimeric reference proteins. Examples are the monomeric CD86 and the dimeric CTLA4 fused to the fluorescent protein (here mEos3.2). The blinking histogram is approximated with the appropriate fit functions. Scale bars 2 μm, zoom-in scale bars 500 nm. Adapted from Baldering et al. [21, 38] Copyright CC BY-NC-SA 3.0

(e.g., a His tag). Another possibility is to genetically design a fusion protein consisting of two fluorescent proteins connected by a short peptide linker.

We note that the blinking parameters of fluorescent proteins depend on their chemical environment. Reducing agents, such as β-mercaptoethylamine used in *d*STORM experiments, influence the photobleaching probability p of fluorescent proteins [21, 40]. This represents a method to tune the blinking of fluorescent proteins towards, e.g., a higher number of blinking events, which may be beneficial in a specific experiment.

p and q values obtained from single-molecule surfaces are a good estimate for cell measurements and are usually sufficient for quantification when relative changes of the oligomeric state of a target protein are needed. However, if more accurate numbers are desired, these parameters may be determined directly *in cellulo* from calibration measurements of membrane proteins with known stoichiometry, typically monomeric or dimeric proteins (Fig. 3b). The selected dimeric protein ideally shows no endogenous expression in the target cell system, in order to avoid heterodimers of labeled and unlabeled proteins that would affect the detection efficiency parameter q. Two proteins that were shown to be applicable to many widely used cell lines are monomeric CD86 and dimeric CTLA4. To determine the blinking parameters of the selected fluorescent protein, it is genetically fused to CD86 or CTLA4. The plasmid is then transiently transfected such that a low expression level is obtained in the target cell line. q values determined in previous studies are summarized in Table 1.

4 Quantification of Membrane Protein Oligomers with qPALM

qPALM is particularly suited to determine the oligomeric state of a protein in the plasma membrane and was applied to membrane receptors such as the toll-like receptor 4 (TLR4) [22], tumor necrosis factor receptor 1 (TNFR1) [25, 26], MET receptor [24], and the anion channel SLAH3 [23]. To highlight the capabilities of qPALM, we discuss how the oligomeric state of the two membrane receptors TNFR1 and MET is linked to their function.

One of the first biological targets addressed with qPALM was the membrane receptor TNFR1, which is involved in essential processes such as cell proliferation, inflammation, and cell death [41]. Its oligomeric state in the presence and absence of TNFα ligand is being discussed in the literature [42]. Using qPALM, the oligomeric state of TNFR1 was studied in its physiological environment in the plasma membrane of eukaryotic cells [25]. For this purpose, a stable cell line expressing TNFR1-mEos2 was generated. The receptor oligomerization was analyzed in unstimulated and ligand-stimulated cells (Fig. 4a). qPALM analysis reported monomeric and dimeric TNFR1 in untreated cells, supporting the model of an equilibrium between monomers and dimers [42]. Upon TNFα binding, most TNFR1 were found as

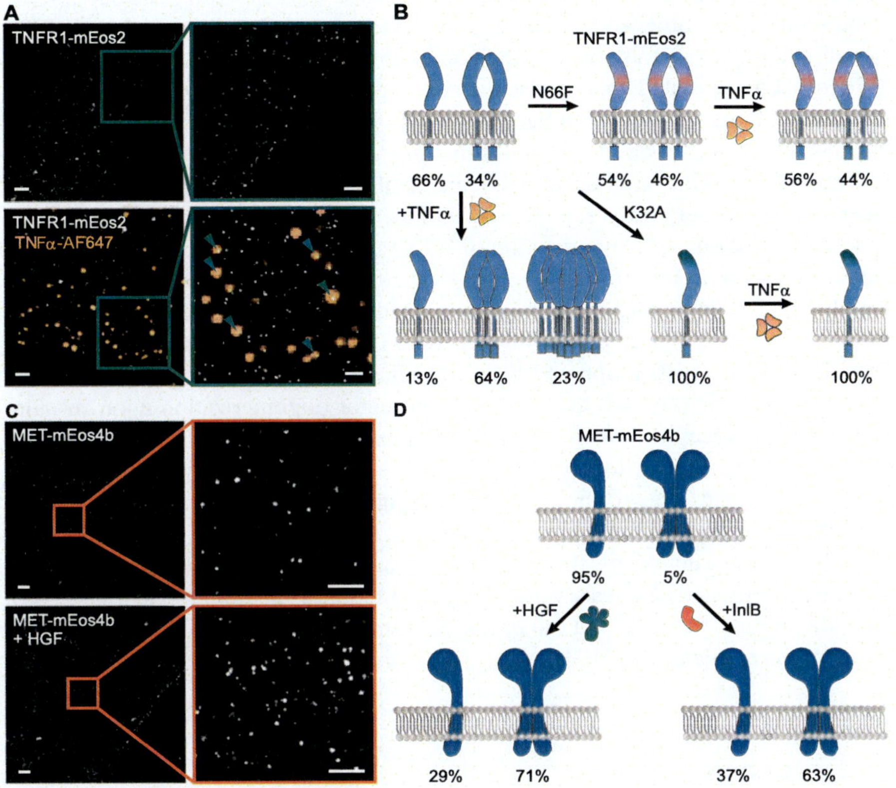

Fig. 4 qPALM analysis of the membrane receptors TNFR1 and MET. (**a**) PALM images of TNFR1-mEos2 in TNFR1/2 double knockout mouse embryonic fibroblasts in the resting state (top) and after TNFα-Alexa Fluor 647 treatment (bottom). In the zoom-in of the ligand-stimulated cell, colocalization of the ligand TNFα (orange) and TNFR1 (white) are highlighted by green arrow heads. (**b**) Mechanistic scheme of TNFR1 oligomerization in the resting state and upon ligand stimulation (TNFα). Mutation of the ligand-binding domain (N66F) and in the PLAD (K32A) strongly affect the oligomerization of TNFR1 and prevent ligand activation. (**c**) PALM images of MET-mEos4b in a CRISPR/Cas12a-generated HEK293T cell line either in the resting state (top) or in the HGF-stimulated state (bottom). (**d**) Mechanistic scheme of MET receptor oligomerization. Upon treatment with the ligands HGF and InlB, the dimer fraction of MET increases. Scale bars 2 μm, zoom-in scale bars 1 μm

trimers and higher order oligomers (Fig. 4b). Additional qPALM experiments investigated the role of two mutants of TNFR1: first, a mutation in the preligand assembly domain (PLAD) which drives weak dimerization between two TNFR1 yielded exclusively monomeric TNFR1. Second, a mutation in the ligand-binding site resulted in monomeric and dimeric TNFR1, of similar frequency than in unstimulated cells, while no trimers or higher order oligomers were found in ligand-treated cells (Fig. 4b). This study supported the model that in the absence of ligand, TNFR1 interacts via the PLAD promoting the formation of receptor dimers. The signaling-active species that form upon TNFα binding are supposedly trimers and higher order oligomers. Next to revealing a molecular model on TNFR1 signaling complexes, the effect of the competitive inhibitor zafirlukast was studied [26]. qPALM revealed that zafirlukast inhibits TNFR1 dimerization as well as formation of higher receptor oligomers upon TNFα stimulation, which provides information on the mode of action of this drug.

Receptor tyrosine kinases (RTKs) are another important class of membrane receptors responsible for essential cellular functions such as growth, proliferation, and differentiation [43]. Dimerization of RTKs upon ligand binding is the assumed model of activation for most RTKs [43]. However, also predimerization of RTKs in the absence of activating ligands was reported [44, 45]. The hepatocyte growth factor receptor MET is an RTK with important functions in vertebrate development as well as tissue regeneration and wound healing [46]. qPALM experiments under various conditions revealed the changes of receptor oligomerization upon stimulation. The MET receptor was stoichiometrically labeled with mEos4b using CRISPR/Cas12a [24, 47]. This fusion protein allowed the quantitative analysis of the endogenous oligomeric organization of MET (Fig. 4c). qPALM revealed that in unstimulated cells, MET largely is monomeric, while stimulation with the physiological ligand hepatocyte growth factor (HGF) or with the bacterial ligand internalin B (InlB) significantly increases the dimer population to a similar extent (Fig. 4d). This observation suggests that the bacterial protein InlB activates MET in a similar way like the native ligand HGF.

5 Discussion

qPALM builds on the analysis of blinking parameters of photoactivatable or photoconvertible fluorescent proteins. However, we note that fluorescent proteins may exhibit a more complex photophysical behavior than the here assumed four-state model, which is an active area of research (see, e.g., [15]). For the fluorescent protein mEos4b, a long-lived dark state in the photoconverted state was identified, and at the same time, a strategy to revert this dark state to the fluorescent state was reported [16]. An accurate inclusion of these phenomena may increase the accuracy of qPALM in future work. At the same time, it may also be interesting to explore alternative photoactivation pathways or site-specific mutations of fluorescent proteins [48], and in general the manipulation of blinking parameters by chemical reagents [40] and light [16].

qPALM analysis is relatively time-consuming, since the localization clusters are selected by several sequential filtering and selection steps that are performed manually. Therefore, automated analysis of qPALM data is desirable. Recently, a fully automated quantitative algorithm for fluorescence kinetics analysis (QAFKA) was developed that determines the positions of PSFs in a single-molecule movie, extracts characteristic features, and delivers the stoichiometry [37]. This automated analysis pipeline is a promising step in the direction of a time-efficient and more unbiased analysis.

The concept of the here presented approach of qSMLM can be extended to photoswitchable organic fluorophores. This requires accounting for the fact that organic fluorophores are typically in a bright state in the beginning of an experiment [39]. In addition, organic fluorophores are not shielded from the nanoenvironment like the fluorophores that are embedded within the barrel-like cage of fluorescent proteins. This renders the photophysical properties more susceptible to changes of the chemical environment in the vicinity of the fluorophore. Alternating blinking properties of organic dyes in the direct neighborhood of some amino acids such as tryptophan [49] or varying nucleobase environments in DNA origami [39] were reported. A possible solution are protein tags that bind and to some extent "embed" organic fluorophores inside the protein tag, such that a more homogenous nanoenvironment is generated. It was shown that the SNAP tag [50] is a promising approach to achieve this "shielding," demonstrated in qSMLM experiments that measured the oligomeric state of the μ-opioid G protein-coupled receptor [51].

The presented approach of qSMLM builds on the analysis of blinking parameters using a mathematical model to approximate the number of blinking events recorded for single protein clusters. Other variants of qSMLM were reported in the literature that use similar or slightly varied analysis procedures to extract molecule numbers from single-molecule blinking analysis of photoswitchable fluorophores [52], or alternatively from analyzing the binding kinetics of transiently binding fluorophore labels (quantitative PAINT, qPAINT) [53, 54]. qPAINT provides a higher number of emission events per single target through repetitive binding events of fluorophore labels to a target. This enables the extraction of molecular numbers from even a single cluster, which is not possible with most photoswitchable fluorescent proteins or organic fluorophores due to the few emission events. In addition, qPAINT can cover a larger dynamic range by tuning fluorophore-label concentrations in the buffer. On the other hand, qPAINT requires an additional labeling step, e.g., a DNA-labeled antibody targeting a protein, and a fluorophore-labeled, sequence complementary single-stranded DNA targeting the antibody. This may lead to unspecific detection events and/or non-stoichiometric labeling, which is largely avoided when using fluorescent proteins. A combination of both, direct genetic and stoichiometric labeling of a target and single cluster readout, would be desirable.

A variety of other fluorescence microscopy and spectroscopy methods is available to study protein oligomerization in cells. Similar to qPALM, pair-correlation analysis can be performed on super-resolved PALM data (PC-PALM) and yields information on cluster size, density, and protein numbers [55]. Unlike qPALM and qPAINT, which derive molecule numbers from the analysis of photophysical or binding kinetics, PC-PALM does not require well-separated localization clusters and therefore works well at high protein densities. In number and brightness (N&B) analysis, the average brightness per particle as well as the average number of particles per pixel in a fluorescence image is determined from the measurement of fluorescence fluctuations [56]. The oligomeric state can be determined from the brightness of a particle and its comparison with the brightness of a monomer. Similar to qPALM, the probability that a particle will not fluoresce is needed in the analysis. N&B analysis has been used in different studies to determine molecule numbers as well as protein oligomerization in live cells (reviewed in [57]). A similar quantitative analysis can be performed with fluorescence correlation spectroscopy (FCS). FCS reports on the concentration of fluorophores, as well as on the fluorescence intensity per molecule, so that the formation of larger oligomers can be tracked. This technique was also applied to living cells [58, 59] and can be combined with imaging [60]. Imaging FCS was recently extended into a multi-modal imaging tool combining quantitative analysis and super-resolution imaging [61]. When comparing FCS-based methods with qPALM, it is worth noting that qPALM is most useful for observing small oligomers (e.g., dimers), while FCS is well suited for detecting higher order clusters. Lastly, protein oligomerization is also accessible by measuring short-range spectroscopic interactions, such as in Förster resonance energy transfer (FRET). In order to assess homo-oligomers, FRET between identical fluorophores (homoFRET) can be applied, which builds on the measurement of fluorescence anisotropy and has been used to track receptor oligomerization [62].

Acknowledgments The authors gratefully acknowledge funding by the German Science Foundation (grants SFB 902, SFB 1177, HE 6166/17-1, GRK 2566 project number 414985841), LOEWE (Frankfurt Cancer Insititute), and the Volkswagen Foundation (grant 91067-9). We thank all past and present members of the Single-Molecule Biophysics group and our collaborators for their valuable contributions to this research.

References

1. Heilemann M (2010) Fluorescence microscopy beyond the diffraction limit. J Biotechnol 149(4):243–251
2. Lelek M, Gyparaki MT, Beliu G, Schueder F, Griffié J, Manley S, Jungmann R, Sauer M, Lakadamyali M, Zimmer C (2021) Single-molecule localization microscopy. Nat Rev Methods Primers. https://doi.org/10.1038/s43586-021-00038-x
3. Heilemann M, van de Linde S, Schüttpelz M, Kasper R, Seefeldt B, Mukherjee A, Tinnefeld P, Sauer M (2008) Subdiffraction-resolution fluorescence imaging with conventional fluorescent probes. Angew Chem 47(33):6172–6176
4. Rust MJ, Bates M, Zhuang X (2006) Sub-diffraction-limit imaging by stochastic optical reconstruction microscopy (STORM). Nat Methods 3(10):793–795
5. Betzig E, Patterson GH, Rachid Sougrat O, Lindwasser W, Olenych S, Bonifacino JS, Davidson MW, Lippincott-Schwartz J, Hess HF (2006) Imaging intracellular fluorescent proteins at nanometer resolution. Science 313(5793):1642–1645
6. Hess ST, Girirajan TPK, Mason MD (2006) Ultra-high resolution imaging by fluorescence photoactivation localization microscopy. Biophys J 91(11):4258–4272
7. Sharonov A, Hochstrasser RM (2006) Wide-field subdiffraction imaging by accumulated binding of diffusing probes. Proc Natl Acad Sci 103(50):18911–18916. https://doi.org/10.1073/pnas.0609643104
8. Jungmann R, Steinhauer C, Scheible M, Kuzyk A, Tinnefeld P, Simmel FC (2010) Single-molecule kinetics and super-resolution microscopy by fluorescence imaging of transient binding on DNA origami. Nano Lett 10(11):4756–4761

9. Sauer M, Heilemann M (2017) Single-molecule localization microscopy in eukaryotes. Chem Rev 117(11):7478–7509
10. Mortensen KI, Stirling Churchman L, Spudich JA, Flyvbjerg H (2010) Optimized localization analysis for single-molecule tracking and super-resolution microscopy. Nat Methods 7(5): 377–381
11. van de Sebastian L, Wolter S, Heilemann M, Sauer M (2010) The effect of photoswitching kinetics and labeling densities on super-resolution fluorescence imaging. J Biotechnol 149(4): 260–266
12. Magrassi R, Scalisi S, Cella Zanacchi F (2019) Single-molecule localization to study cytoskeletal structures, membrane complexes, and mechanosensors. Biophys Rev 11(5):745–756
13. Dietz MS, Heilemann M (2019) Optical super-resolution microscopy unravels the molecular composition of functional protein complexes. Nanoscale 11(39):17981–17991
14. Annibale P, Vanni S, Scarselli M, Rothlisberger U, Radenovic A (2011) Identification of clustering artifacts in photoactivated localization microscopy. Nat Methods 8(7):527–528
15. Avilov S, Berardozzi R, Gunewardene MS, Adam V, Hess ST, Bourgeois D (2014) In cellulo evaluation of phototransformation quantum yields in fluorescent proteins used as markers for single-molecule localization microscopy. PLoS One 9(6):e98362
16. Zitter D, Elke DT, Mönkemöller V, Hugelier S, Beaudouin J, Adam V, Byrdin M, Van Meervelt L, Dedecker P, Bourgeois D (2019) Mechanistic investigation of mEos4b reveals a strategy to reduce track interruptions in sptPALM. Nat Methods 16(8):707–710
17. Fricke F, Beaudouin J, Eils R, Heilemann M (2015) One, two or three? Probing the stoichiometry of membrane proteins by single-molecule localization microscopy. Sci Rep 5 (September):14072
18. Hummer G, Fricke F, Heilemann M (2016) Model-independent counting of molecules in single-molecule localization microscopy. Mol Biol Cell 27(22):3637–3644
19. Lee S-H, Shin JY, Lee A, Bustamante C (2012) Counting single photoactivatable fluorescent molecules by photoactivated localization microscopy (PALM). Proc Natl Acad Sci U S A 109(43):17436–17441
20. Rollins GC, Shin JY, Bustamante C, Pressé S (2015) Stochastic approach to the molecular counting problem in superresolution microscopy. Proc Natl Acad Sci U S A 112(2):E110–E118
21. Baldering TN, Dietz MS, Gatterdam K, Karathanasis C, Wieneke R, Tampé R, Heilemann M (2019) Synthetic and genetic dimers as quantification ruler for single-molecule counting with PALM. Mol Biol Cell 30(12):1369–1376
22. Krüger CL, Zeuner M-T, Cottrell GS, Widera D, Heilemann M (2017) Quantitative single-molecule imaging of TLR4 reveals ligand-specific receptor dimerization. Sci Signal 10(503). https://doi.org/10.1126/scisignal.aan1308
23. Lehmann J, Jørgensen ME, Fratz S, Müller HM, Kusch J, Scherzer S, Navarro-Retamal C et al (2021) Acidosis-induced activation of anion channel SLAH3 in the flooding-related stress response of Arabidopsis. Curr Biol 31(16):3575–85.e9
24. Baldering TN, Karathanasis C, Harwardt M-LIE, Freund P, Meurer M, Rahm JV, Knop M, Dietz MS, Heilemann M (2021) CRISPR/Cas12a-mediated labeling of MET receptor enables quantitative single-molecule imaging of endogenous protein organization and dynamics. iScience 24(1):101895
25. Karathanasis C, Medler J, Fricke F, Smith S, Malkusch S, Widera D, Fulda S et al (2020) Single-molecule imaging reveals the oligomeric state of functional TNFα-induced plasma membrane TNFR1 clusters in cells. Sci Signal 13(614). https://doi.org/10.1126/scisignal.aax5647
26. Weinelt N, Karathanasis C, Smith S, Medler J, Malkusch S, Fulda S, Wajant H, Heilemann M, van Wijk SJL (2021) Quantitative single-molecule imaging of TNFR1 reveals Zafirlukast as antagonist of TNFR1 clustering and TNFα-induced NF-κB signaling. J Leukoc Biol 109(2): 363–371
27. Khan AO, Simms VA, Pike JA, Thomas SG, Morgan NV (2017) CRISPR-Cas9 mediated labelling allows for single molecule imaging and resolution. Sci Rep 7(1):8450

28. McKinney SA, Murphy CS, Hazelwood KL, Davidson MW, Looger LL (2009) A bright and photostable photoconvertible fluorescent protein. Nat Methods 6(2):131–133
29. Paez-Segala MG, Sun MG, Shtengel G, Viswanathan S, Baird MA, Macklin JJ, Patel R et al (2015) Fixation-resistant photoactivatable fluorescent proteins for CLEM. Nat Methods 12(3): 215–218. 4 p following 218
30. Wiedenmann J, Ivanchenko S, Oswald F, Schmitt F, Röcker C, Salih A, Spindler K-D, Ulrich Nienhaus G (2004) EosFP, a fluorescent marker protein with UV-inducible green-to-red fluorescence conversion. Proc Natl Acad Sci U S A 101(45):15905–15910
31. Zhang M, Chang H, Zhang Y, Junwei Y, Lijie W, Ji W, Chen J et al (2012) Rational design of true monomeric and bright photoactivatable fluorescent proteins. Nat Methods 9(7):727–729. https://doi.org/10.1038/nmeth.2021
32. Chudakov DM, Lukyanov S, Lukyanov KA (2007) Using photoactivatable fluorescent protein Dendra2 to track protein movement. BioTechniques 42(5):553–563
33. Gurskaya NG, Verkhusha VV, Shcheglov AS, Staroverov DB, Chepurnykh TV, Fradkov AF, Lukyanov S, Lukyanov KA (2006) Engineering of a monomeric green-to-red photoactivatable fluorescent protein induced by blue light. Nat Biotechnol 24(4):461–465
34. McEvoy AL, Hoi H, Bates M, Platonova E, Cranfill PJ, Baird MA, Davidson MW, Ewers H, Liphardt J, Campbell RE (2012) mMaple: a photoconvertible fluorescent protein for use in multiple imaging modalities. PLoS One 7(12):e51314
35. Wang S, Moffitt JR, Dempsey GT, Sunney Xie X, Zhuang X (2014) Characterization and development of photoactivatable fluorescent proteins for single-molecule-based superresolution imaging. Proc Natl Acad Sci U S A 111(23):8452–8457
36. Krüger C, Fricke F, Karathanasis C, Sebastian DMM, Hummer G, Heilemann M (2017) Molecular counting of membrane receptor subunits with single-molecule localization microscopy. In: Single molecule spectroscopy and superresolution imaging X, vol 10071. SPIE
37. Saguy A, Baldering TN, Weiss LE, Nehme E, Karathanasis C, Dietz MS, Heilemann M, Shechtman Y (2021) Automated analysis of fluorescence kinetics in single-molecule localization microscopy data reveals protein stoichiometry. J Phys Chem B 125(22):5716–5721
38. Baldering TN, Bullerjahn JT, Hummer G, Heilemann M, Malkusch S (2019) Molecule counts in complex oligomers with single-molecule localization microscopy. J Phys D Appl Phys 52(47):474002. https://doi.org/10.1088/1361-6463/ab3b65
39. Karathanasis C, Fricke F, Hummer G, Heilemann M (2017) Molecule counts in localization microscopy with organic fluorophores. ChemPhysChem 18(8):942–948. https://doi.org/10.1002/cphc.201601425
40. Endesfelder U, Malkusch S, Flottmann B, Mondry J, Liguzinski P, Verveer PJ, Heilemann M (2011) Chemically induced photoswitching of fluorescent probes – a general concept for super-resolution microscopy. Molecules 16(4):3106–3118. https://doi.org/10.3390/molecules16043106
41. Ting AT, Bertrand MJM (2016) More to life than NF-κB in TNFR1 signaling. Trends Immunol 37(8):535–545
42. Kucka K, Wajant H (2020) Receptor oligomerization and its relevance for signaling by receptors of the tumor necrosis factor receptor superfamily. Front Cell Dev Biol 8:615141
43. Lemmon MA, Schlessinger J (2010) Cell signaling by receptor tyrosine kinases. Cell 141(7): 1117–1134
44. Dietz MS, Haße D, Ferraris DM, Göhler A, Niemann HH, Heilemann M (2013) Single-molecule photobleaching reveals increased MET receptor dimerization upon ligand binding in intact cells. BMC Biophys 6(1):6
45. Hofman EG, Bader AN, Voortman J, van den Heuvel DJ, Sigismund S, Verkleij AJ, Gerritsen HC, Henegouwen PMVB (2010) Ligand-induced EGF receptor oligomerization is kinase-dependent and enhances internalization. J Biol Chem 285(50):39481–39489
46. Birchmeier C, Birchmeier W, Gherardi E, Vande Woude GF (2003) Met, metastasis, motility and more. Nat Rev Mol Cell Biol 4(12):915–925

47. Fueller J, Herbst K, Meurer M, Gubicza K, Kurtulmus B, Knopf JD, Kirrmaier D et al (2020) CRISPR-Cas12a–assisted PCR tagging of mammalian genes. J Cell Biol 219(6). https://doi.org/10.1083/jcb.201910210
48. Turkowyd B, Balinovic A, Virant D, Gölz HG, Carnero FC, Endesfelder M, Bourgeois D, Endesfelder U (2017) A general mechanism of photoconversion of green-to-red fluorescent proteins based on blue and infrared light reduces phototoxicity in live-cell single-molecule imaging. Angew Chem 56(38):11634–11639
49. Nanguneri S, Flottmann B, Herrmannsdörfer F, Thomas K, Heilemann M (2014) Single-molecule super-resolution imaging by tryptophan-quenching-induced photoswitching of phalloidin-fluorophore conjugates. Microsc Res Tech 77(7):510–516. https://doi.org/10.1002/jemt.22349
50. Keppler A, Kindermann M, Gendreizig S, Pick H, Vogel H, Johnsson K (2004) Labeling of fusion proteins of O6-alkylguanine-DNA alkyltransferase with small molecules in vivo and in vitro. Methods. https://doi.org/10.1016/j.ymeth.2003.10.007
51. Möller J, Isbilir A, Sungkaworn T, Osberg B, Karathanasis C, Sunkara V, Grushevskyi EO et al (2020) Single-molecule analysis reveals agonist-specific dimer formation of μ-opioid receptors. Nat Chem Biol 16(9):946–954
52. Zanacchi FC, Manzo C, Alvarez AS, Derr ND, Garcia-Parajo MF, Lakadamyali M (2017) A DNA origami platform for quantifying protein copy number in super-resolution. Nat Methods 14(8):789–792
53. Baker MAB, Nieves DJ, Hilzenrat G, Berengut JF, Gaus K, Lee LK (2019) Stoichiometric quantification of spatially dense assemblies with qPAINT. Nanoscale 11(26):12460–12464
54. Jungmann R, Avendaño MS, Dai M, Woehrstein JB, Agasti SS, Feiger Z, Rodal A, Yin P (2016) Quantitative super-resolution imaging with qPAINT. Nat Methods 13(5):439–442
55. Sengupta P, Jovanovic-Talisman T, Skoko D, Renz M, Veatch SL, Lippincott-Schwartz J (2011) Probing protein heterogeneity in the plasma membrane using PALM and pair correlation analysis. Nat Methods 8(11):969–975
56. Digman MA, Dalal R, Horwitz AF, Gratton E (2008) Mapping the number of molecules and brightness in the laser scanning microscope. Biophys J 94(6):2320–2332
57. Papini C, Royer CA (2018) Scanning number and brightness yields absolute protein concentrations in live cells: a crucial parameter controlling functional bio-molecular interaction networks. Biophys Rev 10(1):87–96. https://doi.org/10.1007/s12551-017-0394-z
58. Kitamura A, Kubota H (2010) Amyloid oligomers: dynamics and toxicity in the cytosol and nucleus. FEBS J 277(6):1369–1379
59. Takahashi Y, Yuma Okamoto H, Popiel A, Fujikake N, Toda T, Kinjo M, Nagai Y (2007) Detection of polyglutamine protein oligomers in cells by fluorescence correlation spectroscopy. J Biol Chem 282(33):24039–24048
60. Krieger JW, Singh AP, Bag N, Garbe CS, Saunders TE, Langowski J, Wohland T (2015) Imaging fluorescence (cross-) correlation spectroscopy in live cells and organisms. Nat Protoc 10(12):1948–1974
61. Sankaran J, Balasubramanian H, Tang WH, Ng XW, Röllin A, Wohland T (2021) Simultaneous spatiotemporal super-resolution and multi-parametric fluorescence microscopy. Nat Commun 12(1):1748
62. Bader AN, Hofman EG, Voortman J, Henegouwen PMVB, Gerritsen HC (2009) Homo-FRET imaging enables quantification of protein cluster sizes with subcellular resolution. Biophys J 97(9):2613–2622

Diffusion Measurements at the Nanoscale with STED-FCS

Falk Schneider and Erdinc Sezgin

Contents

Abstract The cellular plasma membrane, a seemingly chaotic structure composed of thousands of lipids and proteins, empowers cellular life and represents a versatile hub for signalling. Transient interactions of membrane components integrate to large-scale membrane properties such as packing, tension, curvature, or viscosity and reorganise the membrane in response to physiological needs. How do nanoscale interactions and heterogeneities determine cellular plasma membrane properties? This question is not trivial to address with technologies limited by the diffraction of light. Here, we discuss Stimulated Emission Depletion (STED) nanoscopy and its combination with fluorescence correlation spectroscopy (FCS, STED-FCS) as a technique to shed light onto the relevant molecular scales. While imaging with STED allows for tremendous insights into the spatial domain of organisation, the combination of STED with FCS provides increased temporal resolution and yields insights into the nature of the molecular motion on the relevant scales, allowing us to study nanoscale diffusion modes. We provide a comprehensive overview of the

F. Schneider
Translational Imaging Center, University of Southern California, Los Angeles, CA, USA
e-mail: falkschn@usc.edu

E. Sezgin (✉)
Science for Life Laboratory, Department of Women's and Children's Health, Karolinska Institute, Solna, Sweden
e-mail: erdinc.sezgin@ki.se

R. Šachl, M. Amaro (eds.), *Fluorescence Spectroscopy and Microscopy in Biology*, Springer Ser Fluoresc (2023) 20: 323–336, https://doi.org/10.1007/4243_2022_27,
Published online: 4 August 2022

field, introduction to the technique and discuss common pitfalls with its implementation as well as alternative methodologies to measure the diffusion modes on the nanoscale.

Keywords Diffusion modes · Hindered diffusion · Membrane heterogeneity · Nanoscale diffusion · STED-FCS

1 Introduction

Every mammalian cell is enclosed by a plasma membrane (PM) which serves as a semi-permeable boundary to the surrounding medium and as a reaction platform composed of thousands of different types of lipids and proteins [1, 2]. The PM is known to be heterogeneously organised over a wide range of temporal and spatial scales ranging from nanometres to micrometres and nanoseconds to tens of seconds, respectively [3–5]. Fascinatingly, short-lived, nanoscale interactions and the heterogeneous membrane organisation contribute to its functions, to the point that the membrane should be pictured as much more than a disorganised fluid [6]. Deciphering the details of membrane organisation allows us to better understand signalling mechanisms, biophysical principles of cellular life, dysregulations of physiological processes and ultimately a variety of diseases. However, the dynamic nature of membrane organisation is challenging to access directly. Imaging methodologies can only provide snapshots or time-averaged insights into the principles of organisation [7]. Short-lived interactions integrating to the overall collective behaviour escape standard imaging technologies. Thus, in membrane research, the use of dynamic techniques such as fluorescence recovery after photobleaching (FRAP), single particle tracking (SPT), or fluorescence correlation spectroscopy (FCS) is very common [4]. While these methodologies provide insights into the behaviour of molecular motion and interactions, they struggle to probe interactions on the relevant scales: standard confocal, total internal reflection (TIR) or wide field setups cannot resolve two objects when they are closer than ~200 nm to each other [8]. Yet, lipids, proteins, and even protein clusters are <10 nm in size and their interactions may only last several milliseconds. [3] Even considering larger scales, for example, the formation of functional signalling platforms, lipid domains, or lipid rafts are well beyond what a standard microscope can elucidate. Consequently, the biology dictates that we need to employ technologies that allow us to probe interactions below the diffraction limit of light – on the nanoscale.

2 A Brief Introduction to Super-Resolution Microscopy

Several super-resolution methodologies have emerged over the past two decades and the development of photoactivated localisation microscopy (PALM), stochastic optical reconstruction microscopy (STORM) and stimulated emission depletion (STED) microscopy has been awarded with the Nobel Prize in Chemistry in 2014 [9, 10]. While it's often referred to as "breaking the diffraction limit", these techniques rather use smart tricks to circumvent it [8]. In single molecule localisation microscopy (SMLM) methods (such as PALM and STORM) the exact position of a fixed, isolated molecule in space can be approximated with sub-diffraction accuracy by fitting the emission profile of a single fluorophore to a model (for example, to a Gaussian in 2D). Of course, standard labelled samples contain more than thousands of single molecules very close to each other (closer than the diffraction limit and the microscope's resolution). Thus, the key is to switch them on and off to only allow isolated emitters at a time (i.e., per camera frame) which can be achieved chemically by tuning buffer conditions or photo-physically with a switching laser. A similar concept was later employed in the so-called points accumulation for imaging in nanoscale topography (PAINT) technologies, where emitters are only fluorescent when inserted into the membrane or in DNA-PAINT only transiently bound to a probe with a complimentary DNA-oligo [11, 12]. This acquisition strategies come with the necessity for long acquisition times and many frames (with sparse emitters) to reconstruct the position of all molecules within the sample which makes it challenging to combine with live-cell imaging and especially for studying PM dynamics. STED nanoscopy also employs the trick of switching molecules on and off but uses a spatial pattern (central zero) of a depletion beam to achieve this (Fig. 1a) [13, 14]. Like in confocal imaging this beam is then scanned to construct an image with resolution dictated by the depletion beam efficiency and laser power (Fig. 1b). By itself, also STED microscopy suffers from rather slow acquisition times per frame as it is based on laser scanning acquisitions (Fig. 1c). However, it can be straightforwardly combined with fluorescence correlation spectroscopy in a point measurement fashion which makes it an instant super-resolution technique with very high temporal resolution [15].

FCS only requires intensity data over time to report on the diffusion dynamics in the observation volume or in membranes in the observation spot (Fig. 2a). By parking the beam in the membrane and recording and analysing the intensity fluctuations, the underlying dynamics can be revealed. The STED beam constrains the observable fluorescence and the effective observation spot/volume to a sub-diffraction size (Figs. 1 and 2a). Consequently, the resulting intensity fluctuations used for the FCS analysis report directly on the nanoscale diffusion dynamics [16].

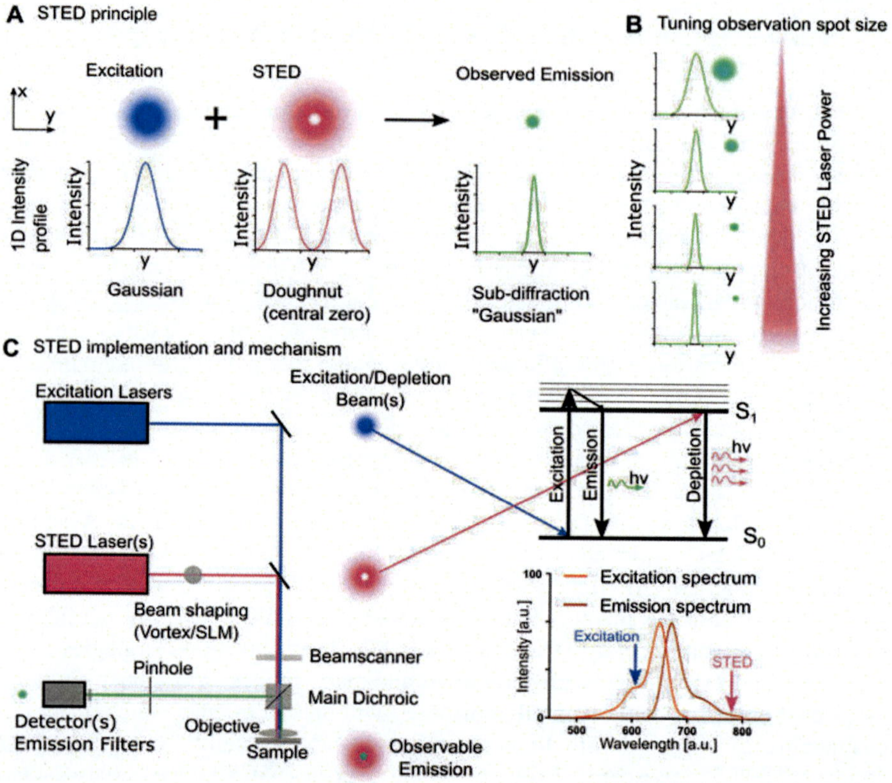

Fig. 1 STED microscopy. (**a**) Principle of STED microscopy. The excitation beam (blue) is superimposed with a doughnut-shaped STED beam (magenta). Fluorescence is depleted, cancelled, by the STED beam in the periphery of the excitation spot but the central zero of the STED beam allows to observe emission from the centre. The resulting, effective, emission or observation spot (green) is smaller than the diffraction limited excitation spot (blue). (**b**) The effective observation spot size, i.e., the resolution of the STED acquisition can be tuned by increasing the STED laser power. (**c**) Simplified implementation of a STED microscope (left) and mechanism of action (right). The excitation beam (blue) is combined with the doughnut-shaped STED beam (magenta) and scanned across the sample. Fluorescence is recorded by a point detector. Fluorescence excitation acts on the ground state S_0 which causes red-shifted emission after vibrational relaxations. The STED beam acts on the excited state S_1 and forces the molecule back to the ground state S_0

3 STED-FCS as a Window to Nanoscale Dynamics

STED-FCS allows for directly measuring the diffusion dynamics with a given resolution tuneable by the STED beam laser power which enables to measure diffusion dynamics on different length scales [7, 16]. This concept has been exploited before in spot-variation FCS (svFCS) where the back aperture of the objective is underfilled and the observation area enlarged [17, 18]. Thus, in svFCS the dynamics from confocal to longer length-scales (up to μm) can be revealed with

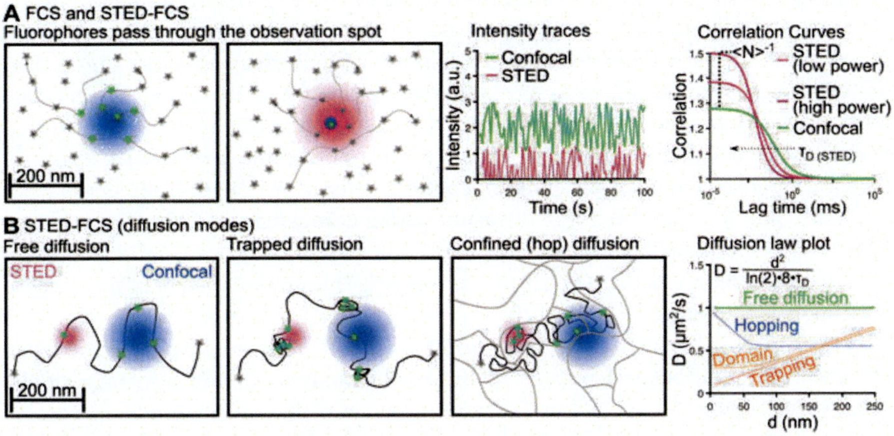

Fig. 2 Fluorescence correlation spectroscopy combined with STED offers to measure nanoscale diffusion modes. (**a**) Principle of FCS and STED-FCS. Fluorophores (grey stars) can be excited by the confocal excitation beam (blue, emitting molecules are depicted as green stars). When the confocal excitation beam is overlaid with a STED beam, the resulting observation spot is smaller. The recorded intensity profiles in confocal (green) and STED (magenta) illustrate lower counts for STED illumination as less molecules are excited than in the confocal case. Autocorrelation of the intensity traces results in the correlation curves which reveal the underlying dynamics. The transit time τ_D which is the average time a molecule needs to cross the observation spot is larger for the big confocal observation spot and shorter for the smaller STED observation spot (scaling with STED laser power). (**b**) STED-FCS diffusion modes. Molecules can undergo a variety of diffusion behaviours: They can simply diffuse freely (left), undergo trapped diffusion with transient halts in their diffusion path (binding to an interaction partner, middle), or their diffusion can be confined by a meshwork structure such as the cortical actin cytoskeleton (hop diffusion right). The diffusion law plot reveals this diffusion modes by plotting the apparent diffusion coefficient D against the observation spot diameter d (full width at half max of the observation spot). The size of the observation area can be tuned by the STED laser power

high temporal resolution. Plotting the diffusion time or diffusion coefficient against observation length-scale introduces the so-called diffusion law plot [18, 19]. The diffusion law gives insights how diffusion scales with space. It is very powerful in discriminating free diffusion from constrained diffusion or trapped diffusion and can help estimating sizes of domains or constraint length-scales. With this, the diffusion law plot is also similar to plotting the mean-squared-displacement (MSD) against time in SPT, which can also reveal hindrances in molecular diffusion [20].

STED-FCS takes the principles of the diffusion law to the nanoscale. By increasing the STED laser power, the observation spot sizes shrinks and the molecular transit time deceases (FCS diffusion time, τ_D, through the focus is shorter as the volume is smaller, Fig. 2a) [15, 16]. By calibration of the resolution, we can calculate the apparent diffusion coefficient *D*, $D = \frac{d^2}{8 \cdot \ln(2) \cdot \tau_D}$, and plot it against observation spot diameter *d* (defined as the full width at half max of the point spread function) [21]. Molecules undergoing free, Brownian, diffusion will show no change in diffusion coefficient with observation spot diameter *d*. However, any sort of hindrances will result in a deviation from this behaviour: Trapped diffusion, meaning nanoscale interactions with short halts in the diffusion path, will result in a reduction of apparent diffusion coefficient *D* with decreasing observation spot diameter as the trapping events become more pronounced in smaller observation areas (Fig. 2b). In contrast, hop diffusion, also termed compartmentalised diffusion, where molecules are constrained within a semi-permeable mesh, will cause an increase in *D* with smaller observation spot diameter as its length-scale approaches mesh size and starts probing free diffusion within compartments [22]. The presence of nanoscale domains appears as a decrease in apparent diffusion coefficient as the observation diameter decreases (similar to trapping) but stays constant or increases at very short length-scales [7] (Fig. 2b).

STED-FCS has extensively been used in the membrane field to investigate presence, distribution and impact of nanoscale hindrances, the influence of the actin cytoskeleton on membrane organisation, and ultimately cellular processes such as signalling [16, 23, 24]. The first experimental verification of the presence of nanoscale hindrances with STED-FCS were the measurements from Eggeling et al. in 2009 [16]. By comparing the diffusion mode of a fluorescently tagged (Atto647N) sphingomyelin (SM) and phospholipid (PE) completely different behaviours were observed and attributed to SM's ability to bind to other molecules in the PM [16]. The origin of this trapping behaviour was linked to the cortical actin cytoskeleton and cholesterol [16, 21, 23]. As a matter of fact, cortical actin cytoskeleton generally plays a crucial role in trapping and hopping of molecules; however, not every anomalous diffusion mode is dependent on it. For instance, ganglioside GM1 exhibits domain-like diffusion in cells, and this diffusion mode is preserved in cytoskeleton-free cell-derived vesicles (Fig. 3) [23]. Trapped diffusion has also been observed for Ras-kinases in tumour cells or for molecules in the context of HIV infection [24, 25]. To elucidate the origin of hindered diffusion, extensive studies on model membrane systems have been performed to, for instance, investigate the influence of phase separation, phase preference or presence of pore-forming proteins on the diffusion law [26–28].

In addition to obtaining the diffusion law in the membrane, STED-FCS can be used to reveal the diffusion in small entities (i.e., smaller than the diffraction limit), such as viruses, or virus budding sites [25, 29, 30]. Besides on 2D systems such as membranes, STED-FCS in solution is a rapidly developing field. With the use of new depletion patterns and adaptive optical elements such as spatial light modulators the diffusion law can also be measured in the three dimensional space with applications in measuring cytosolic trapping or interactions in liquid–liquid phase separated systems [31–34].

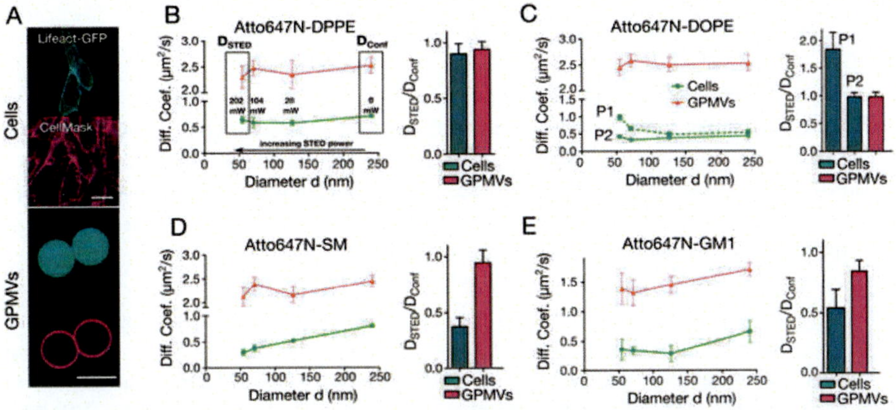

Fig. 3 Nanoscale diffusion modes of lipids in cells and cell-derived vesicles devoid of the cortical actin cytoskeleton. (**a**) PtK2 cells labelled with Cell Mask as PM stain and Lifeact-GFP (top) show cortical actin cytoskeleton underneath the PM while giant plasma membrane vesicles (GPMVs) derived from these cells (bottom) do not show the organized actin cytoskeleton. Diffusion mode of (**b**) Atto647N-labelled DPPE, (**c**) Atto647N-labelled DOPE, (**d**) Atto647N-labelled SM and (**e**) Atto647N-labelled GM1. Atto647N-DOPE shows both hop (pool 1, P1) and free diffusion (pool 2, P2) in live PtK2 cells, but only free diffusion in GPMVs. Atto647N-DPPE shows free diffusion both in cells and GPMVs. Atto647N-SM shows confined diffusion in cells and free diffusion in GPMVs. Atto647N-GM1 shows domain-like diffusion both in cells and in GPMVs. The diffusion behaviour can quickly be assessed by plotting the diffusion coefficient in STED (D_{STED}) divided by the diffusion coefficient in confocal (D_{Conf}). Scale bars 10 μm. Indicated STED laser powers in **b** were measured at the back focal plane. Figure adapted from ref. [23]

4 Practical Considerations and Implementations of STED-FCS

Performing STED-FCS experiments relies in the first instance only on having a STED microscope which is basically an extended laser scanning confocal microscope (Fig. 1b). However, many practical considerations have to be taken into account and we will briefly address the most important ones here and refer the reader for more details to [7].

1. Fluctuation data acquisition

   Fluctuation data can be acquired and processed in many different ways. The data can be acquired as simple point measurement over time (t-scan) and then post-processed [35]. Alternatively, the detector can be connected to a hardware correlator which outputs the auto-correlation curves ready to be fitted. While this is convenient, it does not allow for any photon filtering or post processing. It is also important to consider the data type saved and exported. Photon counting data allow for additional flexibility like lifetime gating in post-processing but also increase the size of the files as well the time to correlate and obtain the raw correlation curves [7, 36, 37]. It is useful to have a robust display of the correlation curves while acquiring the data.

2. Calibration

   For accurate and interpretable diffusion law measurements it is crucial to calibrate the observation spot size in confocal and at different STED laser powers. This can be achieved by imaging sub-diffraction fluorescent beads or by using a lipid bilayer as a freely diffusing system. Calibration should be performed daily and should not vary. Changes in apparent spot-sizes indicate drifts or misalignment of the optical system.

3. Experimental requirements for STED-FCS

   Most important requirement for a successful STED-FCS experiment is a good fluorescent label. The dye should have high molecular brightness and should be easily "depletable" (i.e., work well with STED). The molecular brightness is important as it determines the signal to noise ratio (SNR) in FCS [38, 39]. The brighter the dye, the better the signal, which is especially important when considering the short transit times under high STED laser beam intensity when only a few photons can be collected [39]. Ideally, the resolution should be calibrated using the same dye as in the experiment (see above for Calibration). In general, high-quality optics corrected for optical aberrations are required in STED microscopy to achieve low background levels. In STED-FCS higher background levels cause uncorrelated signal which dampens the autocorrelation function's amplitude complicating data analysis. Choice of depletion wavelength can have a great influence on the achievable resolution. Usually a wavelength far red-shifted from the emission maximum is chosen (Fig. 1c). The optimal wavelength is also a property of the fluorescent dye and needs to be determined empirically.

4. Data analysis and processing

   As mentioned above, the raw data type depends on the acquisition mode. If the data are acquired as photon streams they need to be correlated, if the data are acquired using a hardware correlator, they are ready to be analysed. The autocorrelation curves are fitted to theoretical models derived for various physical situations such as 2D or 3D diffusion, blinking, binding, etc. Fitting the data needs to be performed with care and requires some training and expertise. As a general rule: Always fit the data to the simplest model (fewest fitting parameters) possible. From the fits the transit time τ_D as well as the average number of molecules N and the molecular brightness can be extracted. The transit time as average time of molecules to transverse the focus relates inversely to the apparent diffusion coefficient D (Fig. 2b). The average number of molecules can be used to measure concentration of molecules, for example, as particle per μm^2 in membranes, yet, this is not reliable at higher STED powers since noise can dominate the amplitude. The molecular brightness can be used to determine degree of oligomerisation of the investigated fluorescent particle.

To see examples of good and bad curves as well as good and bad fitting, please see ref. [7]

5 Variations of STED-FCS

As the basis of STED-FCS is a STED laser scanning microscope, many other fluorescence fluctuation techniques can be combined with STED to shed light on the nanoscale which is a unique feature of the technology.

As already mentioned before, the fluctuation data can be acquired not only from a single point but also from multiple points in space: by scanning the confocal (or STED) focus quickly along one line or circle, fluctuation data for multiple points in space (pixels) are obtained. This approach is termed scanning STED-FCS and has helped to elucidate the spatial heterogeneity in nanoscale diffusion hindrances [40–42]. It allows mapping diffusion times or coefficient over space. By interleaving confocal and STED excitation, line-interleaved excitation scanning STED-FCS (LIESS-FCS) maps the diffusion modes across space [43]. Instead of only acquiring a line over time, a series of frames can be acquired and analysed by raster image correlation spectroscopy (RICS) in conjunction with STED [44, 45]. As a bigger area is integrated, the nanoscale dynamics are a result of the robust average. The combination with pair-correlation analysis allows the identification nanoscale obstacles in the diffusion path and reveals interconnectivity [46].

With the increased availability of commercial STED microscopes, we anticipate that the combination with various fluctuation analyses will become of more widespread use [7, 39]. In addition, the increased availability of functional probes, reporting on biophysical properties such as viscosity or lipid order has become more common and offers complementary insights using STED imaging as well as spectroscopy [37, 47–49].

6 Alternative Ways to Investigate Nanoscale

The investigation of nanoscale dynamics by STED microscopy and STED-FCS has proven as a powerful tool to elucidate membrane heterogeneities. While ease of use and ease of implementation as well as the extension of the fluctuation techniques are convenient, employing the high-power STED beam for long-term measurements can have negative effects on sensitive samples [50]. Additionally, STED microscopy relies on dyes that have the right photo-physical properties and requires specialised equipment for STED beam shaping [7].

The gold standard for nanoscale diffusion dynamics (first elucidated by STED-FCS) has been the free diffusion of phosphatidylethanolamine (PE) and the trapped diffusion of sphingomyelin (SM) in the PM of PtK2 cells [16, 21, 23, 43]. This very different nanoscale behaviour is not apparent by imaging (homogeneous PM stain in both cases) [7]. This PE/SM system has now been extensively used to demonstrate the power of alternative techniques to investigate nanoscale diffusion and hindrances.

To trick the diffraction limit, fluorescence in the periphery of the observation area needs to be rejected. In STED, this is achieved photo-physically by fluorescence depletion. In another clever attempt to do this physically, nano-apertures have been combined with FCS. In near-field scanning optical microscopy (NSOM) FCS (NSOM-FCS), the fluorescence is excited through a nanometric aperture [51]. By using different sizes of the aperture, FCS data on different scales can be obtained. Consistently, with the results from STED-FCS, very different behaviours were observed for PE and SM in the PM.

Building on the NSOM approach, nanoantennas have been employed to observe diffusion dynamics in sub-diffraction areas [52, 53]. By fabricating arrays of nanoantennas to confine illumination as well as to enhance fluorescence in the electromagnetic near-field, intensity fluctuations over time can be observed and the nanoscale diffusion dynamics probed [54]. Changing the size or distance of the antennas allows to probe dynamics at different length-scales. This method was applied to measure the diffusion of PE and SM in the PM and the results are in agreement with the previous measurements from NSOM-FCS and STED-FCS.

The above techniques are reporting on true nanoscale interactions by (photo-) physically restricting the area of collected light. NSOM-FCS and the nanoantenna based approaches, however, are near-field approaches necessitating the direct contact of sample and optics. As an alternative to STED-FCS and the near-field optics, confocal-based sFCS in combination with statistical analysis can be used to probe hindrances in diffusion [55]. Large sets of measurements can pick up on small fractions of different sub-diffusive behaviours and can be used to differentiate Brownian from trapped diffusion which has been experimentally shown using the PE/SM system. An additional alternative approach could be the image mean-squared-displacement (iMSD) methodology which accordingly takes the fluctuations from a larger area of pixels (and their statistics) into account [56, 57]. However, to the best of our knowledge, this has not been used systematically to study the PE/SM system in Ptk2 cells.

SPT methods could of course be used to study the nanoscale diffusion dynamics [20]. However, to obtain sufficient SNR and localisation precision, large structures such as fluorescent beads, quantum dots or gold particles need to be employed potentially disturbing the system under study [58]. Very fast frame-rates are necessary and great care needs to be taken during data analysis in terms of model selection and correction for camera blur [20, 59–61]. Yet, especially the combination of SPT with scattering techniques such as interferometric scattering (iSCAT) microscopy is attractive as the sample does not bleach and bleaching is arguably one of the biggest causes for artefacts in fluorescence and fluorescence fluctuation analysis [62].

7 Future Developments

The wealth of technology to study nanoscale dynamics is astonishing and developing further every day. The mapping of diffusion modes in LIESS-FCS (STED-FCS based) opened the door to spatially resolved diffusion maps [43, 63]. Pushing the spatial component further to image correlation methods will undoubtedly allow to resolve diffusion maps across entire cells at the nanoscale. Yet, one crucial component missing is the direct identification of interactions in a two-colour fashion. Due to the high spectral cross-talk and technical requirements for gating and pulse-interleaving schemes for excitation and depletion, STED fluorescence cross correlation spectroscopy (STED-FCCS) has not been realised yet. Though, we anticipate that such a tool will be invaluable and enable us to directly proof and characterise nanoscale interactions and heterogeneities.

Another very promising tool to study the nanoscale is MINFLUX which can be seen as a particle tracking approach with very high temporal and spatial resolution [64]. It avoids or minimises the typical issues of SPT such as photobleaching or the need for large labels and can easily achieve high frame-rates and large track length [65, 66]. Yet, many particles need to be sampled and the stability of the biological sample assured. We anticipate MINFLUX to evolve as a new gold standard for super-resolved tracking methodologies.

We are living in the exciting times of super-resolution microscopy enabling us to look beyond the curtain of diffraction. The discussed methods and their applications to nanoscale diffusion phenomena represent an extremely exciting field of study. However, all these technologies have been mostly applied on model membrane systems and immortalised cell lines. Future work will need to be undertaken to bring nanoscale diffusion measurements into tissues and whole organisms. Labelling the molecules in living cells and tissues still remains a bottleneck for STED and STED-FCS. STED-capable fluorescent proteins that can easily be implemented in living tissues will without doubt increase the applicability of STED and STED-FCS in the future. In addition, the combination of omics technologies with the biophysical characterisation of PM properties will bring us in a new age of research.

Acknowledgements We thank SciLifeLab Advanced Light Microscopy facility and National Microscopy Infrastructure (VR-RFI 2016-00968) for their support on imaging. ES is supported by grants from Swedish Research Council Starting Grant (2020-02682), from Karolinska Institutet and from the SciLifeLab National COVID-19 Research Program, financed by the Knut and Alice Wallenberg Foundation. FS acknowledges support by EMBO (ALTF 849-2020) and HFSP (LT000404/2021-L).

Author Contributions ES and FS wrote the manuscript and prepared the figures. Both authors contributed equally to the chapter.

References

1. Sezgin E, Levental I, Mayor S, Eggeling C (2017) The mystery of membrane organization: composition, regulation and roles of lipid rafts. Nat Rev Mol Cell Biol 18:361–374
2. Semrau S, Schmidt T (2009) Membrane heterogeneity – from lipid domains to curvature effects. Soft Matter 5:3174
3. Kadir SR et al (2021) Nanoscape, a data-driven 3D real-time interactive virtual cell environment. eLife 10:1–36
4. Jacobson K, Liu P, Lagerholm BC (2019) The lateral organization and mobility of plasma membrane components. Cell 177:806–819
5. Schneider F, Colin-York H, Fritzsche M (2021) Quantitative bio-imaging tools to dissect the interplay of membrane and cytoskeletal actin dynamics in immune cells. Front Immunol 11:1–13
6. Nicolson GL (2014) The fluid—mosaic model of membrane structure: still relevant to understanding the structure, function and dynamics of biological membranes after more than 40 years. Biochim Biophys Acta Biomembr 1838:1451–1466
7. Sezgin E et al (2019) Measuring nanoscale diffusion dynamics in cellular membranes with super-resolution STED–FCS. Nat Protoc 14:1054–1083
8. Sezgin E et al (2017) Super-resolution optical microscopy for studying membrane structure and dynamics. J Phys Condens Matter 273001:273001
9. Hell SW et al (2015) The 2015 super-resolution microscopy roadmap. J Phys D Appl Phys 48: 443001
10. Vangindertael J et al (2018) Methods and applications in fluorescence - an introduction to optical super-resolution microscopy for the adventurous biologist. Methods Appl Fluoresc 6:55
11. Sharonov A, Hochstrasser RM (2006) Wide-field subdiffraction imaging by accumulated binding of diffusing probes. Proc Natl Acad Sci 103:18911–18916
12. Jungmann R et al (2014) Multiplexed 3D cellular super-resolution imaging with DNA-PAINT and exchange-PAINT. Nat Methods 11:313–318
13. Hell SW, Wichmann J (1994) Breaking the diffraction resolution limit by stimulated emission: stimulated-emission-depletion fluorescence microscopy. Opt Lett 19:780
14. Klar TA, Hell SW (1999) Subdiffraction resolution in far-field fluorescence microscopy. Opt Lett 24:954–956
15. Kastrup L, Blom H, Eggeling C, Hell SW (2005) Fluorescence fluctuation spectroscopy in subdiffraction focal volumes. Phys Rev Lett 94:178104
16. Eggeling C et al (2009) Direct observation of the nanoscale dynamics of membrane lipids in a living cell. Nature 457:1159–1162
17. Lenne P-F et al (2006) Dynamic molecular confinement in the plasma membrane by microdomains and the cytoskeleton meshwork. EMBO J 25:3245–3256
18. Wawrezinieck L, Rigneault H, Marguet D, Lenne P-F (2005) Fluorescence correlation spectroscopy diffusion laws to probe the submicron cell membrane organization. Biophys J 89: 4029–4042
19. Šachl R, Bergstrand J, Widengren J, Hof M (2016) Fluorescence correlation spectroscopy diffusion laws in the presence of moving nanodomains. J Phys D Appl Phys 49:114002
20. Lagerholm BC, Andrade DM, Clausen MP, Eggeling C (2017) Convergence of lateral dynamic measurements in the plasma membrane of live cells from single particle tracking and STED-FCS. J Phys D Appl Phys 50:063001

21. Mueller V et al (2011) STED nanoscopy reveals molecular details of cholesterol- and cytoskeleton-modulated lipid interactions in living cells. Biophys J 101:1651–1660
22. Andrade DM et al (2015) Cortical actin networks induce spatio-temporal confinement of phospholipids in the plasma membrane – a minimally invasive investigation by STED-FCS. Sci Rep 5:11454
23. Schneider F et al (2017) Diffusion of lipids and GPI-anchored proteins in actin-free plasma membrane vesicles measured by STED-FCS. Mol Biol Cell 28:1507–1518
24. Guzmán C et al (2014) The efficacy of raf kinase recruitment to the GTPase H-ras depends on H-ras membrane conformer-specific nanoclustering. J Biol Chem 289:9519–9533
25. Chojnacki J et al (2017) Envelope glycoprotein mobility on HIV-1 particles depends on the virus maturation state. Nat Commun 8:545
26. Sarangi NK, Roobala C, Basu JK (2017) Unraveling complex nanoscale lipid dynamics in simple model biomembranes: insights from fluorescence correlation spectroscopy in super-resolution stimulated emission depletion mode. Methods 140–141:198–211
27. Honigmann A, Mueller V, Hell SW, Eggeling C (2013) STED microscopy detects and quantifies liquid phase separation in lipid membranes using a new far-red emitting fluorescent phosphoglycerolipid analogue. Faraday Discuss 161:77–89
28. Sarangi NK et al (2016) Super-resolution stimulated emission depletion-fluorescence correlation spectroscopy reveals nanoscale membrane reorganization induced by pore-forming proteins. Langmuir 32:9649–9657
29. Favard C et al (2019) HIV-1 Gag specifically restricts PI(4,5)P2 and cholesterol mobility in living cells creating a nanodomain platform for virus assembly. Sci Adv 5(10):eaaw8651
30. Urbančič I et al (2018) Lipid composition but not curvature is the determinant factor for the low molecular mobility observed on the membrane of virus-like vesicles. Viruses 10:415
31. Barbotin A, Galiani S, Urbančič I, Eggeling C, Booth MJ (2019) Adaptive optics allows STED-FCS measurements in the cytoplasm of living cells. Opt Express 27:23378
32. Barbotin A, Urbančič I, Galiani S, Eggeling C, Booth M (2020) Background reduction in STED-FCS using a bi-vortex phase mask. ACS Photon. https://doi.org/10.1021/acsphotonics.0c00388
33. Shakya A, King JT (2018) Non-Fickian molecular transport in protein-DNA droplets. ACS Macro Lett 7:1220–1225
34. Galiani S et al (2021) Diffusion and interaction dynamics of the cytosolic peroxisomal import receptor PEX5. Biophys Rep 2(2):100055. https://doi.org/10.1016/j.bpr.2022.100055
35. Waithe D et al (2017) Optimized processing and analysis of conventional confocal microscopy generated scanning FCS data. Methods 140–141:62–73
36. Clausen MP et al (2015) A straightforward approach for gated STED-FCS to investigate lipid membrane dynamics. Methods 88:67–75
37. Lanzanò L et al (2017) Measurement of nanoscale three-dimensional diffusion in the interior of living cells by STED-FCS. Nat Commun 8:65
38. Wohland T, Rigler R, Vogel H (2001) The standard deviation in fluorescence correlation spectroscopy. Biophys J 80:2987–2999
39. Schneider F et al (2020) High photon count rates improve the quality of super-resolution fluorescence fluctuation spectroscopy. J Phys D Appl Phys 53:164003
40. Honigmann A et al (2014) Scanning STED-FCS reveals spatiotemporal heterogeneity of lipid interaction in the plasma membrane of living cells. Nat Commun 5:5412
41. Benda A, Ma Y, Gaus K (2015) Self-calibrated line-scan STED-FCS to quantify lipid dynamics in model and cell membranes. Biophys J 108:596–609
42. Maraspini R, Beutel O, Honigmann A (2017) Circle scanning STED fluorescence correlation spectroscopy to quantify membrane dynamics and compartmentalization. Methods 140–141: 188–197
43. Schneider F et al (2018) Nanoscale spatiotemporal diffusion modes measured by simultaneous confocal and stimulated emission depletion nanoscopy imaging. Nano Lett 18:4233–4240

44. Hedde PN et al (2013) Stimulated emission depletion-based raster image correlation spectroscopy reveals biomolecular dynamics in live cells. Nat Commun 4:1–8
45. Garcia E, Bernardino de la Serna J (2018) Dissecting single-cell molecular spatiotemporal mobility and clustering at focal adhesions in polarised cells by fluorescence fluctuation spectroscopy methods. Methods 140–141:85–96
46. Bianchini P, Cardarelli F, Di Luca M, Diaspro A, Bizzarri R (2014) Nanoscale protein diffusion by STED-based pair correlation analysis. PLoS One 9:e99619
47. Sezgin E et al (2017) Polarity-sensitive probes for superresolution stimulated emission depletion microscopy. Biophys J 113:1321–1330
48. Carravilla P et al (2021) Long-term STED imaging of membrane packing and dynamics by exchangeable polarity-sensitive dyes. Biophys Rep 1:100023
49. Urbančič I et al (2021) Aggregation and mobility of membrane proteins interplay with local lipid order in the plasma membrane of T cells. FEBS Lett. https://doi.org/10.1002/1873-3468.14153
50. Kilian N et al (2018) Assessing photodamage in live-cell STED microscopy. Nat Methods 15:755–756
51. Manzo C, van Zanten TS, Garcia-Parajo MF (2011) Nanoscale fluorescence correlation spectroscopy on intact living cell membranes with NSOM probes. Biophys J 100:L8–L10
52. Winkler PM et al (2017) Transient nanoscopic phase separation in biological lipid membranes resolved by planar plasmonic antennas. ACS Nano. https://doi.org/10.1021/acsnano.7b03177
53. Regmi R et al (2017) Planar optical nanoantennas resolve cholesterol-dependent nanoscale heterogeneities in the plasma membrane of living cells. Nano Lett 17:6295–6302
54. Winkler PM, García-Parajo MF (2021) Correlative nanophotonic approaches to enlighten the nanoscale dynamics of living cell membranes. Biochem Soc Trans 0:1–13
55. Schneider F et al (2018) Statistical analysis of scanning fluorescence correlation spectroscopy data differentiates free from hindered diffusion. ACS Nano 12:8540–8546
56. Di Rienzo C, Gratton E, Beltram F, Cardarelli F (2013) Fast spatiotemporal correlation spectroscopy to determine protein lateral diffusion laws in live cell membranes. Proc Natl Acad Sci 110:12307–12312
57. Di Rienzo C, Gratton E, Beltram F, Cardarelli F (2016) Spatiotemporal fluctuation analysis: a powerful tool for the future nanoscopy of molecular processes. Biophys J 111:679–685
58. Clausen MP, Christoffer Lagerholm B (2011) The probe rules in single particle tracking. Curr Protein Pept Sci 12:699–713
59. Kusumi A et al (2005) Paradigm shift of the plasma membrane concept from the two-dimensional continuum fluid to the partitioned fluid: high-speed single-molecule tracking of membrane molecules. Annu Rev Biophys Biomol Struct 34:351–378
60. Reina F, Eggeling C, Lagerholm BC (2021) Lipid tracking at kilohertz sampling rates on live cell membranes through interferometric scattering microscopy. bioRxiv. https://doi.org/10.1101/2021.08.06.455401
61. Reina F et al (2021) TRAIT2D: a software for quantitative analysis of single particle diffusion data. bioRxiv 5390. https://doi.org/10.1101/2021.03.04.433888
62. Reina F (2018) Applications of interferometric scattering (iSCAT) microscopy to single particle tracking in model and cell membranes. Doctoral dissertation, University of Oxford
63. Bernabé-Rubio M, Bosch-Fortea M, Alonso MA, Bernardino de la Serna J (2021) Multidimensional and spatiotemporal correlative imaging at the plasma membrane of live cells to determine the continuum nano-to-micro scale lipid adaptation and collective motion. Methods 193:136–147
64. Balzarotti F et al (2017) Nanometer resolution imaging and tracking of fluorescent molecules with minimal photon fluxes. Science 355:606–612
65. Eilers Y, Ta H, Gwosch KC, Balzarotti F, Hell SW (2018) MINFLUX monitors rapid molecular jumps with superior spatiotemporal resolution. Proc Natl Acad Sci 115:6117–6122
66. Schmidt R et al (2021) MINFLUX nanometer-scale 3D imaging and microsecond-range tracking on a common fluorescence microscope. Nat Commun 12:1–12

Single-Molecule Microscopy Methods to Study Mitochondrial Processes

Timo Dellmann, Anna Kostina, and Ana J. Garcia Saéz

Contents

Abstract Mitochondria are essential organelles of eukaryotic cells with key functions in metabolism, apoptosis, and signaling. As a result, impaired mitochondrial function has been associated with numerous diseases. In order to understand mitochondrial processes, it is fundamental to gain knowledge about their structure and microcompartmentalization, including the function, organization, and dynamics of their protein, nucleic acid, and lipid components. A number of recent groundbreaking advances in fluorescence microscopy enable the study of mitochondrial biology with unprecedented detail. Among them, new methods based on single-molecule and super-resolution microscopy allow us to study mitochondrial structures, protein organizations, and dynamics. Here, we discuss the advantages and disadvantages of different single-molecule microscopy methods to study individual proteins in fixed and living cells in the background of mitochondrial processes, in situ.

Keywords Bax · DNA-PAINT · MINFLUX · Mitochondria · SIM · Single-molecule microscopy · Single-particle tracking · STORM · Super-resolution microscopy

Timo Dellmann and Anna Kostina contributed equally to this work.

T. Dellmann, A. Kostina, and A. J. Garcia Saéz (✉)
CECAD Research Center, Institute of Genetics, University of Cologne, Cologne, Germany
e-mail: tdellma1@uni-koeln.de; akostina@uni-koeln.de; ana.garcia@uni-koeln.de

R. Šachl, M. Amaro (eds.), *Fluorescence Spectroscopy and Microscopy in Biology*, Springer Ser Fluoresc (2023) 20: 337–354, https://doi.org/10.1007/4243_2022_23,
Published online: 10 August 2022

1 Introduction

Mitochondria are key organelles in eukaryotic cells that carry out essential functions in cellular metabolism, signaling, and the execution of cell death. The central role of mitochondrial function is demonstrated by the multiple disorders associated with dysfunction of this organelle. Mitochondria present a complex structure, composed by a double membrane system: the mitochondrial outer membrane (MOM) and the mitochondrial inner membrane (MIM), separated by the intermembrane space and with the MIM enclosing the mitochondrial matrix. This sophisticated organization allows for compartmentalization of the different mitochondrial functions. Furthermore, mitochondria adopt a tubular structure that organizes into a dynamic network in the cytosol, due to continuous fission and fusion of the organelle, as well as transport along the microtubule cytoskeleton. Importantly, mitochondrial structure, dynamics, and intracellular localization are important for the organelle function and their alteration has also been linked with disease.

During the last decades, fluorescence light microscopy has contributed to advance our understanding of the relationship between mitochondrial dynamic organization and function. More recently, the developments of single-molecule fluorescence methods have pushed even further the experimental approaches available to investigate mitochondrial biology by making accessible the analysis of the structural organization and dynamic behavior of mitochondrial components with unparalleled temporal and spatial resolution. Here we discuss the principles and applications of single-molecule microscopy to mitochondrial research.

In order to study mitochondrial processes at the single-molecule level using fluorescence microscopy, fluorescently labeled mitochondrial proteins have to be first localized. This implies the use of a microscopy technique that reveals the localization of individual proteins below the diffraction limit, given by Abbe's law:

$$d = \frac{\lambda}{2\mathrm{NA}} = \frac{\lambda}{2(n \times \sin \alpha)}$$

here λ is the wavelength of light (monochromatic excitation of the fluorophore), n is the refractive index of the imaging medium and NA is the numerical aperture of the used objective.

One such technique is total internal reflection fluorescence (TIRF) microscopy, which is based on the principle of total internal reflection (TIR). TIR only works when the illumination laser encounters a boundary to a medium in the sample with a lower refractive index. Here, the refractive behavior of the excitation laser beam is described by Snell's Law:

$$n_1 \sin \theta_1 = n_2 \sin \theta_2$$

Where, n_1 is the higher refractive index and n_2 the lower refractive index and θ is the angle of the excitation laser in medium 1 or 2. The angle θ of the excitation laser beam must be larger than the critical angle to be totally internal reflected. Total internal reflection of the excitation laser, in an inverted microscope produces an evanescent field/wave, exciting the fluorochromes in close proximity to the coverslip. The intensity *I*(*z*) of the evanescent field in the sample decays exponentially along the *x*-axis by the following equation:

$$I(z) = I(0)e^{-z/d}$$

Here, d is the distance of the fluorochrome to the coverslip. Furthermore, the penetration depth d_p of the evanescent field can be calculated as following:

$$d_p = \frac{\lambda}{4\pi}\left(n_1^2 \sin^2\theta - n_2^2\right)^{-1/2},$$

where λ is the wavelength of the excitation laser, the angle θ of the illumination and the refractive index of the coverslip/glass n_1 and the cultivation medium of the sample n_2. Thus, TIRF microscopy limits the excitation of fluorescent molecules to about 200 nm along the axial axis of the sample, above the coverslip and into the sample. This lowers the background fluorescence resulting from out of focus, because the excitation only takes place in the evanescent field. Thus, the signal-to-background ratio (SBR) is increased compared to e.g. the Epi-fluorescence, resulting in a better identification of individual fluorophores as long as the single fluorescent signals fulfill the Rayleigh criterion. If the illumination angle is equal or larger than the critical angle, the microscope setup is used in the highly inclined optical (HILO) sheet mode [1] and if the illumination light follows the axial axis of the objective, epi-fluorescence is used. It is important to note that most of the mitochondrial network in a cell is above the evanescent field. Thus, the HILO illumination (also called pseudo-TIRF) is the illumination of choice to provide optimal SBR for single-molecule studies of mitochondrial processes resulting in a bright fluorescent signal with a low background level.

Today, most TIRF microscopes are inverted microscopes, and the excitation laser beam must be focused to the back focal plane of the TIRF objective. Once the fluorophores in the sample are excited, they emit photons. The fluorescent light is collected by the TIRF objective and passes optical filters and is guided to a highly sensitive EMCCD or sCMOS camera installed in the emission light path of the objective. Important to note is that the pixel size of a camera used for single-molecule localization fulfills the Nyquist-Shannon theorem [2, 3]. If single fluorescent molecules are excited in the sample and if their fluorescence is bright enough to be distinguishable from the noise of the camera and the background of the sample, the light of the fluorescent molecules can be recorded as single signals by the camera.

Therefore, bright fluorochromes are the fluorophore of choice in single-molecule microscopy. The brightness of a fluorochrome can be calculated as follows:

$$\text{brightness} = \varepsilon \times \Phi,$$

where ε is the extinction coefficient of the fluorochrome and Φ is its quantum yield. Today many fluorescent molecules can be used in order to perform single-molecule microscopy at mitochondria. Nevertheless, fluorescent dyes are in most cases brighter than fluorescent proteins. Thus, choosing the best fluorescent label possible is fundamental for single-molecule localization microscopy (SMLM). In order to record single fluorescent signals of individual molecules, several different techniques can be used, here we will discuss the most common once: (1) fluorescent photoactivation localization microscopy fPALM/PALM [4–6], (2) stochastic optical reconstruction microscopy (STORM) [7–9], (3) DNA-Based Point Accumulation for Imaging in Nanoscale Topography (DNA-PAINT) microscopy [10] and (4) if SMLM is done in living cells, also single-particle tracking (SPT) [11, 12] can be performed. fPALM/PLAM uses photoactivatable or photoconvertable fluorescent proteins. STORM exploits the blinking of fluorescent dyes, which can be bound to the protein of interest (POI) via a tag or an anti-/nanobody. In contrast, DNA-PAINT uses the on- and off-state of binding and unbinding of a fluorescently labeled single-stranded DNA, often called the imager strand, to another single-stranded DNA, the handler strand, coupled to the POI. The handler strand in a DNA-PAINT experiment can also be coupled to the POI via a tag or an anti-/nanobody. Here, a very interesting approach is to use the ALFA-tag [13] or tags like SnapTag and HaloTag [14, 15]. Single-molecule localization microscopy is also the basis for SPT, and in this case highly photostable fluorescent tags with minimal dark states are preferred for the tracking of the signal. Important for all these techniques is that, in each frame of a movie, distinguishable individual signals have to be recorded, as already mentioned above. The minimal distance between two single fluorescent signals to be localized as two different singals is described by the Rayleigh criterion d_r:

$$d_r = \frac{1.22\lambda}{2\text{NA}} = \frac{0.61\lambda}{n \times \sin\alpha} = \frac{0.61\lambda}{\text{NA}}$$

Additionally, a low uncertainty in the localization of single fluorescent signals is improving the result in all mentioned techniques above. This uncertainty σ can be calculated as following,

$$\sigma = \sqrt{\frac{s^2}{N} + \frac{a^2/12}{N} + \frac{4\sqrt{\pi s^3 b^2}}{aN^2}}$$

where N is the number of gathered photons of the single fluorophore, s is the standard deviation of a Gaussian function for fitting the point spread function (PSF) of the fluorophore, a is the pixel size of the camera used, and b is the background of the image.

2 Single-Molecule Localization Microscopy

Single-molecule localization microscopy is the term used to encompass a number of microscopy methods, including e.g. STORM, PALM, and DNA-PAINT. The basic principle for these stochastic functional techniques is the imaging of scattered groups containing single molecules with the distance between them higher than the Rayleigh criterion. Given that biological samples are densely labeled with fluorescent proteins or dyes, precise localization of all the fluorophores in the sample is not directly accessible with standard illumination strategies due to the signal overlay and an ensemble readout. Single-molecule localization microscopy can help to solve this issue by controlling rate of fluorophors emitting a fluorescent signal or in DNA-PAINT the ON/OFF-state of the fluorophore. Currently, the minimum distance between two fluorophores that is possible to distinguish in the range of 5–10 nm. However, the highest resolution can be obtained only if fluorophores are separated in time and space. Thus, using the correct concentration of the fluorophores and the ON/OFF-switching of the fluorescent signals in the sample and the correct exposure time enables to reveal the biological structure with highest resolution possible.

2.1 PALM and STORM of Mitochondrial Proteins

Super-resolution imaging techniques like PALM and STORM are based on detection and time-resolved localization of randomly excited singe fluorophores [4–6, 9]. During a measurement, fluorophores can be photoswitched between a dark state and a fluorescent state [16]. Under certain laser illumination conditions, a subset of the fluorophores can be activated and their emission is recorded (and later used for localization), over many cycles of excitation/emission so that optimally all fluorophores of a biological structure are detected. The subsequent reconstruction of a super-resolution image is obtained in a similar manner with PALM, STORM and DNA-PAINT, by localizing the emitted signal with a 2D-Gaussion fit and reconstruction of the biological structure by combining the localizations of the individual fluorophores (Fig. 1).

To ensure that only a small subset of fluorophores is switched into the fluorescent state, both the fluorophore photophysical properties and the illumination protocol need to be optimized. Super-resolution microscopy with the accurate localization of single molecules is based on the separation of the signal of individual fluorescent molecules. The light intensity for illumination during the experiment is adjusted in a way that the sparsely activated fluorophores are not closer than the Rayleigh criterion, so that the position of each fluorophore can be determined with an accuracy of up to 1 nm. PALM is based on the use of e.g. photoactivatable GFP, which is

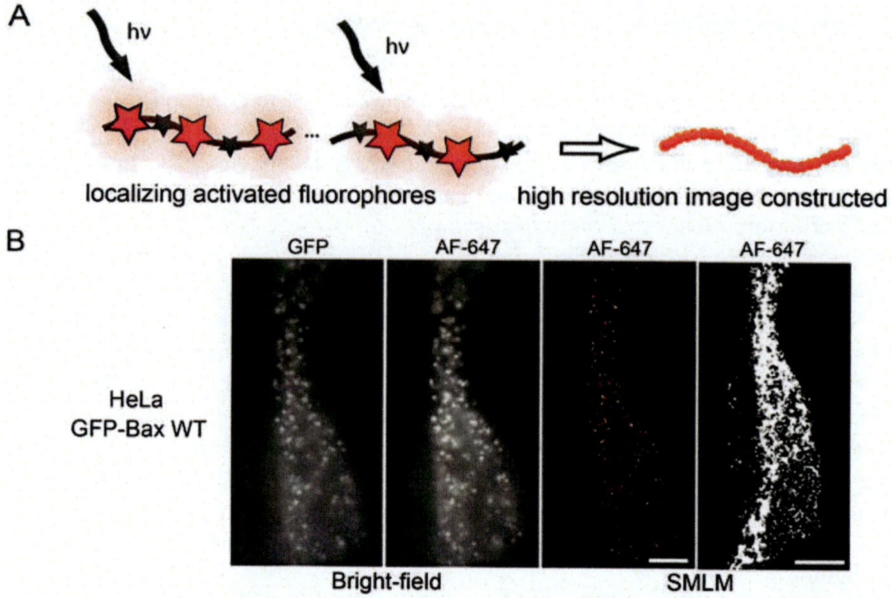

Fig. 1 Scheme of the STORM approach. (**a**) left – The illumination protocol excites only a subset of individual fluorophores in the population while others stay in the dark state and the fluorescence is collected in each frame. In the next cycle, another subset of fluorophores turn to the excited state for N number of frames collected. For image analysis, the individual fluorophores in each frame are localized with high accuracy and combined to build the super-resolution image on the right. (**b**) Representative reconstructed super-resolution image of GFP-Bax in HeLa cells stained with AF647-anti-GFP nanobodies. The image was acquired on fixed cells 3 h after induction of apoptosis. Scale bar, 5 μm. (Figure adapted from: [17])

normally stable in a dark state, and a fraction of the fluorophores can be switched into the fluorescent conformation by UV light. In STORM, fluorophores with strong blinking properties are used, like Atto647, so that they can be brought into a dark state from which only a small subset of them will be excited in a controlled manner and emit photons.

STORM has been used to reveal the organization of the proapoptotic protein Bax and its assembly into line, arc and ring structures during mitochondrial apoptosis, in situ [17] (Fig. 2). Furthermore, STORM can be upgraded to all three dimensions to perform 3D STORM, by installing a cylindrical lens into the emission light path of a TIRF microscope. 3D STORM of mitochondrial proteins allows the reconstruction of entire mitochondrial networks with the additional information of single mitochondrial proteins and their organization in all three dimensions [8].

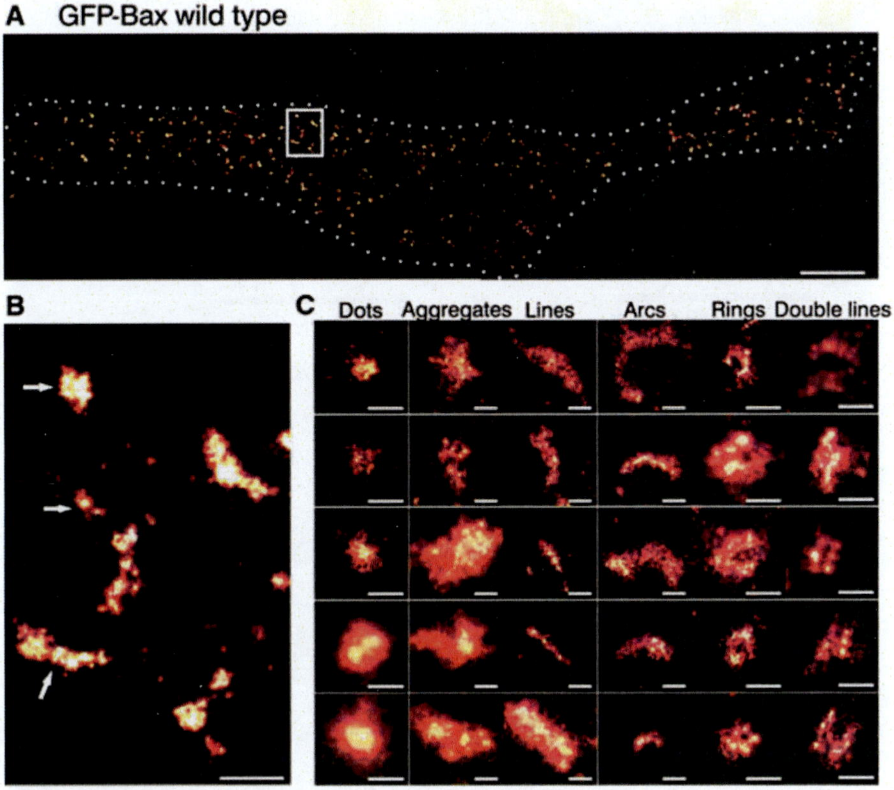

Fig. 2 STORM image revealing non-random structures of Bax in apoptotic mitochondria. (**a**) STORM image of Bax in a HeLa cell (dotted line) 3 h after apoptosis, induced by staurosporine. (**b**) Representative Bax structures on mitochondria during apoptosis, revealing supramolecular assemblies classified into dots, aggregates, lines, double lines, arcs and ring structures. Scale bars: (**a**) 5 μm, (**b**) 100 nm. (Figure adapted from: [17])

2.2 *DNA-Based Point Accumulation for Imaging in Nanoscale Topography (DNA-PAINT)*

A different approach for switching the fluorophores between an ON- or OFF-state on the observed target is DNA-PAINT. As with PALM and STORM, DNA-PAINT microscopy is also based on the fact that the centroid of the PSF of a fluorescent signal can be precisely determined. The methods above based on the fluorophore localization use the photophysical switching molecules between excited and relaxed states, negotiated by the chemical reactions or by the photon's excitation. The DNA-PAINT approach is different. The target in this case is labeled with a single-stranded DNA (ssDNA). The complementary ssDNA labeled with the fluorophore can then quickly hybridize to the docking DNA from the target, thereby leading to efficient labelling of the POI (Fig. 3).

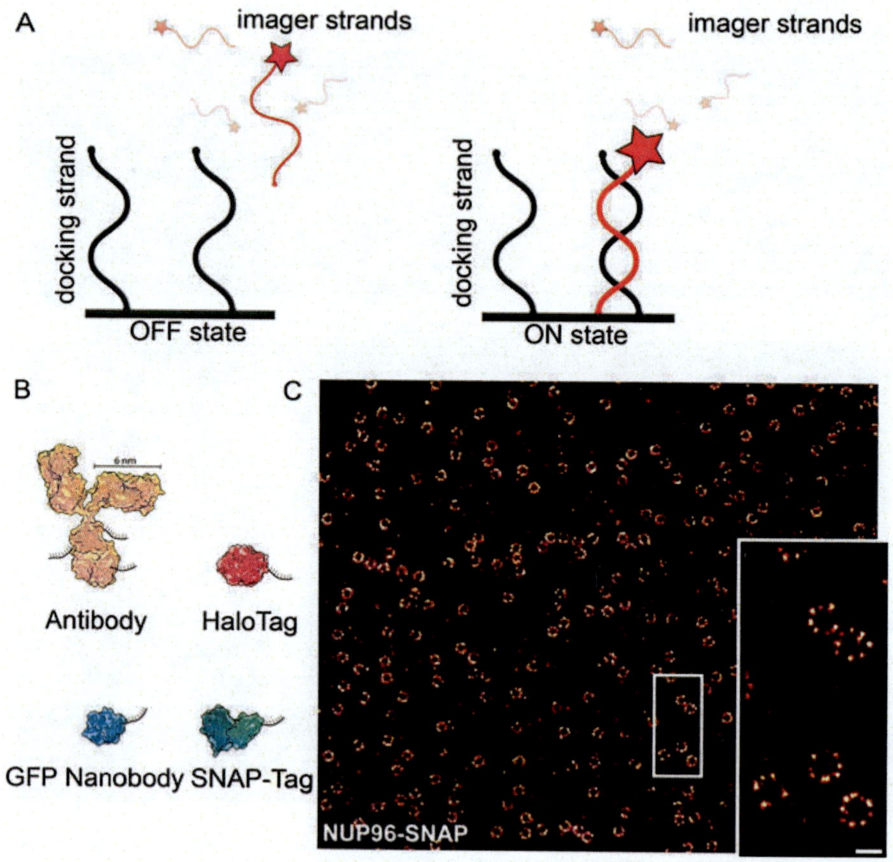

Fig. 3 Scheme of the DNA-PAINT approach. (**a**) left: The OFF-state, a signal is not detected, unbound imager strands carrying a fluorophore are defused away; right: ON state, the imager strand hybridized to the docking side, and the signal is detected. (**b**) comparison of different labelling probes: antibody, HaloTag, GFP Nanobody and SNAPTag. (**c**) DNA-PAINT overview image of NUP96-SNAP in U2OS cells. Scale bar: 5 μm. (Figure adapted from: Schlichthaerle et al. 2019)

The lengths of the imager strands and the sequence are selected to have a short lifetime in the bound state with the docking strands – a single-stranded DNA extension from the sample. During the binding event – hybridization of docking and imager strands – the fluorophore on the imager strand is exited during ON-time τ_b (bright) while the signal is recorded by a camera. The association rate k_{on} is calculated from the lifetime τ_d (dark) via $1/_{\tau_d} = k_{\mathrm{on}}c$, where c is the concentration of imager strand. The dissociation rate $k_{\mathrm{off}} = 1/_{\tau_b}$ depends on the duplex length. Adjusting the temperature, sequence, length and concentration/ratio of a DNA duplex allows optimization of the ON- and OFF-states of the fluorophores, so that they can be excited for a short enough time to avoid bleaching but long enough to record a signal with a high SBR. Since the fluorescent single-molecule signal can be detected and localized only when the fluorophore is located in the focal plane, the effect of blinking fluorescence similar to STORM is created. The unstable duplex allows multiple hybridisation events. Thus, one advantage of DNA-PAINT over STORM and PALM is that bleached fluorophores are exchanged within the sample by non-bleached imager strands, thereby extending the possible measurement time as well as the achieved spatial resolution. Furthermore, multicolour imaging of different proteins of interest can be achieved with dyes of the same wavelength by simply exchanging the imager strand, which are then measured in a sequential manner. Thus the same dye can be used for different proteins by using different DNA duplex constructs.

Anti- or nanobodies can be attached to ssDNA by modifying antibodies through the usage of click chemistry [18, 19]. This modification opens the possibility to use the DNA-PAINT approach for a variety of targets in the cell, similar to anti- or nanobodies in STORM but with the already mentioned advantages. With the given advantages mentioned above this opens the field for highly multiplexed imaging in biological samples with nanoscale resolution [20]. Important to note is that, immunolabeling for SMLM done with IgG antibodies cause large linkage errors (~10 nm, or ~20 nm for indirect immunolabeling) [21]. These errors are less when using nanobodies.

Recently, DNA-PAINT has been combined with SPT, which allows monitoring of single molecules for extended durations [22]. However, the use for mitochondrial processes has not yet been shown and will be a challenge for the future.

3 Single-Particle Tracking (SPT)

SPT is based on the single-molecule localization of individual fluorescent signals and in their reidentification in the following frames of a recorded movie, in order to track a fluorescent signal over time (Fig. 4a–c) [23–29]. In contrast to fPALM/PALM, STORM and DNA-PAINT, SPT works best with high frame rates, in order to gain a high time resolution, which allows accurate tracking of the POI and thereby reveals the accurate spatio-temporal behavior of the POI (Fig. 4d–f). Using a high frame rate results in less photons per frame of the individual signals, which is counterproductive to the localization of the fluorescent signal. Thus, especially in SPT, bright fluorophores with a high photostability are advantageous. Thereby photostable fluorophores result in long trajectories. This allows to study the spatiotemporal organization of an induvidual POI in different microcompartments. In order to gain single fluorescent signals, the POI can be genetically fused to a photoactivatable protein resulting in sptPALM [26, 30]. Thus, a subset of fluorophores can be activated and their fluorescent signal can be recorded. Another approach is to fuse the mitochondrial POI to a HaloTag or SNAPtag and label only a subset of the tagged proteins in order to gain single signals [23]. An important factor for this method is a posttranslational, substoichiometric labeling of the mitochondrial proteins of interest via a HaloTag or SnapTag with membrane-permeable dyes.

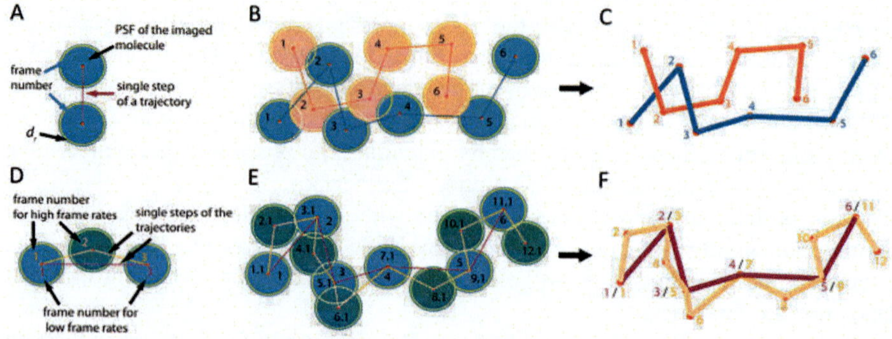

Fig. 4 Scheme of single-particle tracking. (**a**) The fluorescent signals are localized by determining the center of their PSF. (**b**) Reidentification of localizations of the same POI at different time points results in (**c**). Chronological series of single particles results in their trajectories (red and blue). (**d**) A Higher frame rate result in more localizations of a POI. (**e/f**) High frame rates improve the tracking result of the POI by providing more steps and thus a better resolved spatiotemporal information (purple versus orange)

Single-molecule tracking allows the study of mitochondrial proteins in isolated yeast mitochondria [31]. Furthermore, SPT of mitochondrial proteins can also be performed in situ, [23, 32] in correlation with the metabolic activity of mitochondria [33]. Additionally, 3D SPT is possible by engineering a double-helix point spread function (DH-PSF) [34, 35]. Multicolor SPT of different mitochondrial proteins at the same time is also possible [36, 37].

SPT of mitochondrial proteins via this method is done by using the HILO illumination method, which allows illumination of fluorophores above the evanescent field. Nevertheless, SPT of mitochondrial proteins performs best in the cell periphery. Together with the use of bright dyes, a high SBR can be achieved and results in a high localization precision. Thereby SPT allows the study of protein organization and micro-compartmentalization in the different mitochondrial locations [23, 38] (Fig. 5). Mitochondrial proteins of the MOM show a random movement along the MOM, while proteins of the oxidative phosphorylation system (OXPHOS) follow the MIM shape revealing the orientation of cristae. Furthermore, proteins that are part of the mitochondrial inner membrane organization system/mitochondrial contact site complex (MINOS/MICOS), such as Mic60, show a confined diffusion. The same is true for the translocase of inner membrane subunit 23 (Tim23). Especially in the case of proteins of the MIM, SPT of mitochondrial proteins reveals functional micro-compartments of continuous mitochondrial membranes [32, 39].

Besides single-molecule studies of mitochondrial proteins in living cells, SPT can also be done in supported lipid bilayers (SLBs), in vitro. This provides the advantage of studying individual protein–protein interactions directly in model membrane systems. Here, the assembly and regulation of proteins and their interaction partners can be studied [40]. This allows the analysis of the stoichiometry of Bax oligomers and the activation of Bax by cBid and the dissociation of Bax oligomers by Bcl-XL [40]. Furthermore, single lipid organization can be studied in a SLB [41].

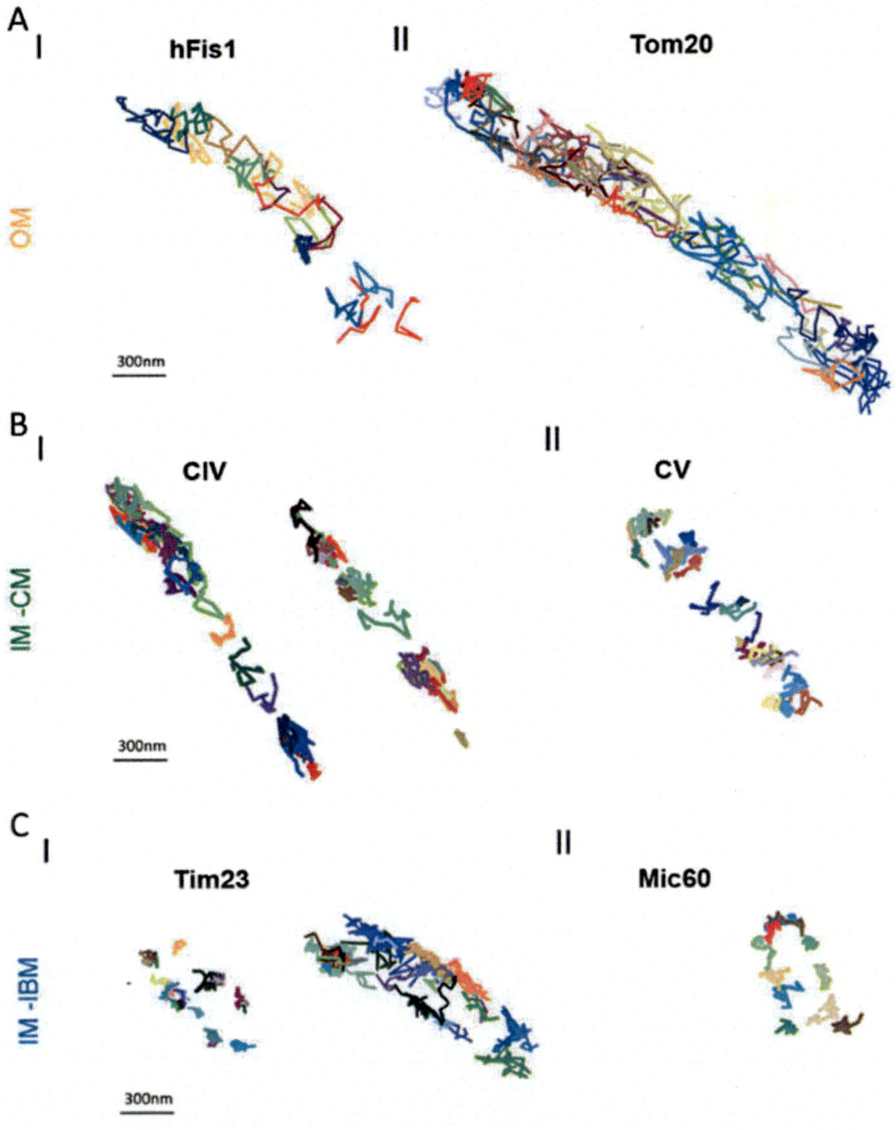

Fig. 5 Single-particle tracking of mitochondrial proteins. (**a**) Trajectories of the outer mitochondrial membrane proteins human fission factor 1 (hFis1) and translocase of outer membrane subunit 20 (Tom20). (**b**) Trajectory maps of proteins of the OXPHOS complex: the cytochrome c oxidase subunit CoxVIIIa (CIV) and F_1F_O ATP synthase subunit-γ (CV). (**c**) Trajectories of the MIM proteins, translocase of inner membrane subunit 23 (Tim23) and MINOS/MICOS complex protein 60 (Mic60). (Figure adapted from: [38])

4 Summary

Investigations of the mitochondrial proteins and their function to understand mitochondrial processes are a central aspect of cell biology. Recent technological advances in fluorescence microscopy approaches have opened a new range of methods for mitochondrial research. Nevertheless, visualization of protein behavior by single-molecule microscopy remains challenging. Here, a variety of super-resolution microscopy methods enable the protein localization with a precision of ~1 nm [21]. DNA-PAINT, STORM, and fPALM/PALM rely on the accurate localization and the detection of single fluorophores. The localization precision depends on the number of gathered photons, which responsively depends on the brightness of fluorophores and the SBR, as well as on the exposure time used [6, 10, 42]. This implies the use of bright fluorescent labels and long exposure times and a low background, but also an accurate determination of the PSF center. Interestingly, tagging the POIs to a SNAPtag or HaloTag as well as labeling via a nanobody allows the use of organic fluorescent dyes with best performance in STORM, DNA-PAINT, SPT, or even MINFLUX [43, 44]. Furthermore, in STORM and MINFLUX blinking behavior of the used dye can be adjusted and optimized. Comparably, in DNA-PAINT the ON and OFF rate of the imager strand can be changed by using different DNA-sequences. In addition, DNA-PAINT depends on the binding and unbinding of the imager strand and therefore allows long-term imaging of a sample, regardless of bleaching of single imager strands. Beside its possibility to perform highly multiplexed imaging, this is one of the main advantages of DNA-PAINT experiments as it results in a high resolution of a biological structure.

However, revealing the protein organization via STORM and DNA-PAINT is limited to fixed cells. In contrast, SPT is a powerful tool to visualize the spatio-temporal dynamics of mitochondrial processes and the involved proteins in living cells. SPT allows the study of the dynamics of individual proteins over time, while techniques like fluorescent recovery after photobleaching (FRAP) and fluorescent correlation spectroscopy (FCS) can only provide average information about the observed area of the cell and millisecond temporal resolution [45]. Nevertheless, a key aspect is that SPT has to be correlated with cellular processes such as apoptosis, metabolism, membrane dynamics, or signaling pathways, in order to answer biological questions of mitochondrial processes and proteins or lipids involved in it. An interesting option here is to use Dual-Color SPT in order to reveal co-diffusion of proteins as it was done with interferon receptors in the plasma membrane [46]. Thus, the interaction of mitochondrial proteins and their interplay with mitochondrial processes could possibly reveal in living cells.

The combination of different techniques, like STED-PAINT, opens new possibilities to increase the image resolution of fixed cells [47]. Another imaging approach combining two super-resolution techniques is called MINFLUX. This method provides three-dimensional resolution at the size of the fluorescent molecule. The "torus" or "doughnut" shaped excitation beam can be used in fixed or living cells. It creates a movable excitation beam featuring an intensity minimum. The fluorophores are then switched individually like in STORM or fPALM. It was demonstrated that fast three-dimension imaging and simultaneous two-color registration is possible, which is interesting for live cell applications [48].

Taken together, when planning the study of proteins involved in mitochondrial processes, it is important to determine what the specific imaging requirements are before choosing which microscopy method to use. To study protein organization with super-resolution at the single-molecule level, STORM and DNA-PAINT are still the methods of choice, since both techniques reveal protein complexes with high resolution. It is best to compare the performance of both methods in a protein complex of the nucleus – the nuclear pore complex (NPC) [49] consisting of around 30 different proteins called the nucleoporins (NUPs) [50]. Here, DNA-PAINT and STORM have successfully revealed the octameric structure of the single NUPs [14, 15, 51]. DNA-PAINT can reach tens of thousands of photons per blinking event, which is hard to achieve using STORM [14, 15]. In contrast, STORM is so far the only method that revealed the 3D architecture of the NPC [49, 51, 52]. Theoretically, this should also be possible with DNA-PAINT.

Interestingly, MINFLUX microscopy is also capable of revealing this octameric organization of NUPs [44]. MINFLUX imaging achieved a lateral localization precision below 3 nm and an axial localization precision below 2 nm. Thus, MINFLUX allows the highest localization precision in all three dimensions compared to other SMLM techniques; DNA-PAINT, STORM, and SPT. However, MINFLUX is still at its infancy and further work will be needed to bring this method to the cell biologist. Recently, it has been shown that dual-color 3D MINFLUX nanoscopy allows to reveal organellar substructures with a 3D localization precision of ~5 nm in mitochondria, and indicates that multiple Mic60 molecules arrange in a ring-like structure with a diameter of 40–50 nm, suggesting that Mic60 surrounds individual crista junctions [53]. Here, it was demonstrated to what extent MINFLUX will permit a closer look in the field of mitochondrial processes in respect of mitochondrial protein organization. MINFLUX allows to improve the temporal resolution of imaging and the 3D resolution. Thus, it is the most suitable technique to study protein assemblies and to reveal their formation due to cellular processes in all three dimensions via its nanometer-scale isotropic localization precision. Additionally, SPT with MINFLUX is possible [54, 55]. Here, MINFLUX allowed the study of the diffusion of the DPPE-ATTO 647 N lipid at RT in an SLB with a temporal acquisition with a mean of 117 μs per localization. This demonstrated what is technically possible with SPT by MINFLUX. One of the next challenges is to investigate the spatio-temporal dynamics of mitochondrial proteins or lipids by MINFLUX, in order to make use of the big advance in fluorescent microscopy provided by MINFLUX, namely, both the high localization precision and high temporal resolution of MINFLUX and thus to provide knowledge about e.g. the assembly of proteins into multiprotein complexes. Here, it will be of great interest to track multiple signals over time simultaneously.

In contrast to the mentioned techniques, another not yet mentioned technique here is structured illumination microscopy (SIM). SIM was used to reveal cristae dynamics in long-term experiments by using Hessian SIM [56]. Here, even the fusion events of two individual cristae into one was revealed. Hessian SIM allows to use less laser power compared to the conventional SIM resulting in less bleaching of the fluorescent label with high frame rates enabling to use a high spatio-temporal resolution of 88 nm and 188 Hz [56]. Thus, Hessian SIM is a promising microscopy method to investigate the organization of proteins and mitochondrial sub-compartments over time. This can allow to reveal mitochondrial processes with a higher temporal resolution even in longer time windows. The possibility to record mitochondrial processes in long time ranges with high temporal resolutions is important in biological processes in order to understand the full mechanism in living cells.

Finally, one of the next challenges in the coming future will be to determine the formation and dynamic of mitochondrial protein organizations of large macromolecular complexes like the MINOS/MICOS complex, the apoptotic pore, OXPHOS protein complexes and supercomplexes, or the TOM-TIM machinery. Here, the use of STORM, DNA-PAINT, MINFLUX, and SIM in fixed and living cells, in 2D and 3D, hold strong promises to shed new light on the nanoscale organization of mitochondrial proteins in relation to the mitochondrial processes in which they participate, in order to answer new biological relevant questions. Today the competition between the described methods results from their individual advantages in the triangle of spatial resolution, temporal resolution and sensitivity/phototoxicity to answer key questions in cell biology of high biological relevance. Thus, today a combination of different techniques enables to consider mitochondrial processes from all sides is still the best to fully understand the behavior of mitochondrial proteins and structures. Therefore, correlative light-electron microscopy is a very interesting method to reveal mitochondrial membrane structures and the localization of mitochondrial proteins within as well as their influence on mitochondrial sub-structures with nanometric resolution. Here, the next step can be to combine super-resolution microscopy and electron microscopy to enable a more detailed view on mitochondrial structures and mitochondrial sub-molecular protein organizations.

References

1. Tokunaga M, Imamoto N, Sakata-Sogawa K (2008) Highly inclined thin illumination enables clear single-molecule imaging in cells. Nat Methods 5:159–161
2. Nyquist H (1928) Certain topics in telegraph transmission theory. Trans Am Inst Electr Eng 47: 617–644
3. Shannon CE (1949) Communication in the presence of noise. Proc IRE 37:10–21
4. Hess ST, Gould TJ, Gunewardene M, Bewersdorf J, Mason MD (2009) Ultra-high resolution imaging of biomolecules by fluorescence photoactivation localization microscopy (FPALM). Methods Mol Biol 544:483–522
5. Betzig E et al (2006) Imaging intracellular fluorescent proteins at nanometer resolution. Science 313:1642–1645

6. Hess ST, Girirajan TPK, Mason MD (2006) Ultra-high resolution imaging by fluorescence photoactivation localization microscopy. Biophys J 91(11):4258–4272
7. Jungmann R, Steinhauer C, Scheible MB, Kuzyk A, Tinnefeld P, Simmel FC (2010) Single-molecule kinetics and super-resolution microscopy by fluorescence imaging of transient binding on DNA origami. Nano Lett 10(11):4756–4761
8. Huang B, Jones SA, Brandenburg B, Zhuang X (2008) Whole-cell 3D STORM reveals interactions between cellular structures with nanometer-scale resolution. Nat Methods 5: 1047–1052
9. Rust MJ, Bates M, Zhuang X (2006) Sub-diffraction-limit imaging by stochastic optical reconstruction microscopy (STORM). Nat Methods 3:793–796
10. Schnitzbauer J et al (2017) Super-resolution microscopy with DNA-PAINT. Nat Protoc 12(6): 1198–1228
11. Kusumi A, Sako Y, Yamamoto M (1993) Confined lateral diffusion of membrane receptors as studied by single particle tracking (nanovid microscopy). Effects of calcium-induced differentiation in cultured epithelial cells. Biophys J 65:2021–2040
12. Qian H, Sheetz MP, Elson EL (1991) Single particle tracking. Analysis of diffusion and flow in two-dimensional systems. Biophys J 60:910–921
13. Götzke H et al (2019) The ALFA-tag is a highly versatile tool for nanobody-based bioscience applications. Nat Commun 10:4403
14. Nieves DJ et al (2019) tagPAINT: covalent labelling of genetically encoded protein tags for DNA-PAINT imaging. R Soc Open Sci 6(12):191268
15. Schlichthaerle T et al (2019) Direct visualization of single nuclear pore complex proteins using genetically-encoded probes for DNA-PAINT. Angew Chem Int Ed 58:13004–13008
16. Heilemann M, Sauer M (2017) Single-molecule localization microscopy in eukaryotes. Chem Rev 117(11):7478–7509
17. Salvador-Gallego R et al (2016) Bax assembly into rings and arcs in apoptotic mitochondria is linked to membrane pores. EMBO J 35:389–401
18. Gierlich J et al (2006) Click chemistry as a reliable method for the high-density postsynthetic functionalization of alkyne-modified DNA. Org Lett 8(17):3639–3642
19. Thorek DLJ, Elias DR, Tsourkas A (2009) Comparative analysis of nanoparticle-antibody conjugations: carbodiimide versus click chemistry. Mol Imaging 8(4):221–229
20. Agasti SS, Wang Y, Schueder F, Sukumar A, Jungmann R, Yin P (2017) DNA-barcoded labelling probes for highly multiplexed Exchange-PAINT imaging. Chem Sci 8(4):3080–3091. https://doi.org/10.1039/C6SC05420J
21. Lelek M, Gyparaki MT, Beliu G et al (2021) Single-molecule localization microscopy. Nat Rev Methods Primers 1:39
22. Stehr F et al (2021) Tracking single particles for hours via continuous DNA-mediated fluorophore exchange. Nat Commun 12:4432
23. Appelhans T et al (2012) Nanoscale organization of mitochondrial microcompartments revealed by combining tracking and localization microscopy. Nano Lett 12:610–616
24. Deich J, Judd EM, McAdams HH, Moerner WE (2004) Visualization of the movement of single histidine kinase molecules in live Caulobacter cells. PNAS 101:15921–15926
25. Gahlmann A, Moerner WE (2014) Exploring bacterial cell biology with single-molecule tracking and super-resolution imaging. Nat Rev Microbiol 12:9–22
26. Manley S et al (2008) High-density mapping of single-molecule trajectories with photoactivated localization microscopy. Nat Methods 5:155–157
27. Schütz GJ, Schindler H, Schmidt T (1997) Single-molecule microscopy on model membranes reveals anomalous diffusion. Biophys J 73:1073–1080
28. Sergé A, Bertaux N, Rigneault H, Marguet D (2008) Dynamic multiple-target tracing to probe spatiotemporal cartography of cell membranes. Nat Methods 5:687–694
29. Shen H et al (2017) Single particle tracking: from theory to biophysical applications. Chem Rev 117:7331–7376

30. Subach FV, Patterson GH, Renz M, Lippincott-Schwartz J, Verkhusha VV (2010) Bright monomeric photoactivatable red fluorescent protein for two-color super-resolution sptPALM of live cells. J Am Chem Soc 132:6481–6491
31. Kuzmenko A et al (2011) Single molecule tracking fluorescence microscopy in mitochondria reveals highly dynamic but confined movement of Tom40. Sci Rep 1:195
32. Appelhans T, Busch K (2017) Single molecule tracking and localization of mitochondrial protein complexes in live cells. Methods Mol Biol 1567:273–291
33. Salewskij K et al (2020) The spatio-temporal organization of mitochondrial F1FO ATP synthase in cristae depends on its activity mode. Biochim Biophys Acta Bioenerg 1861:148091
34. Pavani SRP et al (2009) Three-dimensional, single-molecule fluorescence imaging beyond the diffraction limit by using a double-helix point spread function. PNAS 106:2995–2999
35. von Diezmann L, Shechtman Y, Moerner WE (2017) Three-dimensional localization of single molecules for super-resolution imaging and single-particle tracking. Chem Rev 117:7244–7275
36. Beinlich FRM, Drees C, Piehler J, Busch KB (2015) Shuttling of PINK1 between mitochondrial microcompartments resolved by triple-color superresolution microscopy. ACS Chem Biol 10: 1970–1976
37. Appelhans T, Beinlich FRM, Richter CP, Kurre R, Busch KB (2018) Multi-color localization microscopy of single membrane proteins in organelles of live mammalian cells. J Vis Exp 136:57690. https://doi.org/10.3791/57690
38. Appelhans T, Busch KB (2017) Dynamic imaging of mitochondrial membrane proteins in specific sub-organelle membrane locations. Biophys Rev 9:345–352
39. Kondadi AK, Anand R et al (2020) Cristae undergo continuous cycles of membrane remodelling in a MICOS-dependent manner. EMBO Rep 21:e49776
40. Subburaj Y et al (2015) Bax monomers form dimer units in the membrane that further self-assemble into multiple oligomeric species. Nat Commun 6:8042
41. Cheney PP, Weisgerber AW, Feuerbach AM, Knowles MK (2017) Single lipid molecule dynamics on supported lipid bilayers with membrane curvature. Membranes (Basel) 7:15
42. Jungmann R et al (2014) Multiplexed 3D cellular super-resolution imaging with DNA-PAINT and exchange-PAINT. Nat Methods 11(3):313–318
43. Balzarotti F et al (2017) Nanometer resolution imaging and tracking of fluorescent molecules with minimal photon fluxes. Science 355:606–612
44. Gwosch KC et al (2020) MINFLUX nanoscopy delivers 3D multicolor nanometer resolution in cells. Nat Methods 17:217–224
45. Landes CF, Kelly KF, Moringo NA, Tauzin LJ, Hoener BS, Shuang B, Wang W (2016) Super temporal-resolved microscopy (STReM). J Phys Chem Lett 7(22):4524–4529
46. Wilmes S et al (2015) Receptor dimerization dynamics as a regulatory valve for plasticity of type I interferon signaling. J Cell Biol 209:579–593
47. Spahn C et al (2018) Whole-cell, 3D, and multicolor STED imaging with exchangeable fluorophores. Nano Lett 19(1):500–505
48. Hell SW, Ries J, Ellenberg J, Hoess P, Balzarotti F, Pape JK, Gwosch KC (2020) MINFLUX nanoscopy delivers 3D multicolour nanometer resolution in cells. Nat Methods 17:217–224
49. Thevathasan JV, Kahnwald M, Cieśliński K, Hoess P, Peneti SK, Reitberger M, Heid D, Kasuba KC, Hoerner SJ, Li Y, Wu Y-L, Mund M, Matti U, Pereira PM, Henriques R, Nijmeijer B, Kueblbeck M, Sabinina VJ, Ellenberg J, Ries J (2019) Nuclear pores as versatile reference standards for quantitative superresolution microscopy. Nat Methods 16(10):1045–1053. https://doi.org/10.1038/s41592-019-0574-9
50. Ibarra A, Hetzer MW (2015) Nuclear pore proteins and the control of genome functions. Genes Dev 29:337–349
51. Szymborska A et al (2013) Nuclear pore scaffold structure analyzed by super-resolution microscopy and particle averaging. Science 341:655–658
52. Sabinina VJ et al (2021) Three-dimensional superresolution fluorescence microscopy maps the variable molecular architecture of the nuclear pore complex. MBoC 32:1523–1533

53. Pape JK et al (2020) Multicolor 3D MINFLUX nanoscopy of mitochondrial MICOS proteins. PNAS 117:20607–20614
54. Eilers Y, Ta H, Gwosch KC, Balzarotti F, Hell SW (2018) MINFLUX monitors rapid molecular jumps with superior spatiotemporal resolution. PNAS 115:6117–6122
55. Schmidt R et al (2021) MINFLUX nanometer-scale 3D imaging and microsecond-range tracking on a common fluorescence microscope. Nat Commun 12:1478
56. Huang X, Fan J, Li L et al (2018) Fast, long-term, super-resolution imaging with hessian structured illumination microscopy. Nat Biotechnol 36:451–459

Transient State (TRAST) Spectroscopy and Imaging: Exploiting the Rich Information Source of Fluorophore Dark State Transitions in Biomolecular and Cellular Studies

Jerker Widengren

Contents

Abstract Dark state transitions of fluorophores are central for all forms of fluorescence-based, single-molecule and super-resolution microscopy and spectroscopy. While these transitions typically are a limiting factor in single-molecule studies, they are a prerequisite for all super-resolution imaging techniques. In this chapter, an additional aspect of reversible dark state transitions is highlighted, namely that they can be used to sense a manifold of biomolecular environments, dynamics, and interactions. By Fluorescence Correlation Spectroscopy such transitions can be analyzed in a straightforward manner, but high time-resolution and single-molecule detection (SMD) conditions are required. To overcome these limitations and make monitoring of dark state transitions widely applicable for studies on biological samples, we have developed a technique called transient state (TRAST) imaging. In TRAST, fluorophore dark state transitions are monitored via the time-averaged fluorescence intensity, and how it varies with the modulation of the excitation light. Here, the concept of TRAST is described, and how it can be experimentally realized within different microscope modalities. Lastly, examples of biological applications are given, demonstrating how biologically relevant

J. Widengren
Experimental Biomolecular Physics, Department of Applied Physics, KTH, Royal Institute of Technology, Stockholm, Sweden
e-mail: jwideng@kth.se

R. Šachl, M. Amaro (eds.), *Fluorescence Spectroscopy and Microscopy in Biology*, Springer Ser Fluoresc (2023) 20: 355–374, https://doi.org/10.1007/4243_2022_26,
Published online: 4 August 2022

environmental and molecular interaction parameters can be monitored in solutions and in live cells, which are difficult, if possible at all, to follow via regular fluorescence readout parameters.

Keywords Biomolecular environments · Fluorescence · Fluorophore · Photophysics · Spectroscopy

1 Introduction

Fluorescence maintains the position as a dominating readout for cellular and biomolecular studies. This has been further strengthened by the remarkable development of fluorescence imaging and spectroscopy methods, offering single-molecule detection (SMD) sensitivity and super-resolution microscopy (SRM) with nanometer-scale imaging resolution (Nobel prize 2014 [1]). Fluorescence-based SRM, SMD, and SMD-based Fluorescence correlation spectroscopy (FCS) methods offer a unique combination of sensitivity, specificity and resolution in time and space, allowing intricate molecular interaction and localization patterns in cells to be resolved [2, 3]. Yet, the development of new fluorescence methods is not exhausted, but still very active (see, e.g., [4–8] for reviews). Interestingly, many of the major bottlenecks, and thereby also the basis for many of the advances, are to be found in the photophysical aspects, rather than on the optics/hardware side [4, 7].

What sensitivity, readout speed and resolution that can be reached in fluorescence-based spectroscopy and microscopy ultimately depend on the ability to detect a high number of fluorescence photons per molecule, in total and per time unit. This makes parameters such as fluorescence quantum yield, q_F, and photobleaching quantum yields, q_B, critical. Moreover, fluorophore blinking, caused by population dynamics of long-lived, non-fluorescent, triplet (T_1), photo-oxidized ($\dot{R}^+$), photo-reduced ($\dot{R}^-$) or photo-isomerized states, is also of central importance to all forms of fluorescence-based SMD and SRM. In SMD studies blinking is an important factor, which can influence the performance in several ways. Transitions to dark, non-fluorescent states can significantly reduce the molecular brightness and signal-to-background conditions and several of these dark states have also been found to be precursor states of permanent photobleaching. In addition, fluorophore dark state transitions can also obscure observations of single-molecule dynamic events of interest, if occurring on similar timescales. While the brightness of fluorescence molecular emitters is an important figure of merit in most SRM techniques, blinking, or the switching of fluorescence emitters on and off, is at the same time also an absolute prerequisite for all forms of SRM [4–7]. Thus, for several reasons, photophysical aspects is and will likely remain a central issue in SMD and SRM research. Also the spectacular, next-generation SRM techniques now entering the stage [9] critically rely on photophysics and properly controlled fluorophore blinking, and also here play an equally important role as the optics/instrumentation itself [4, 7].

Beyond their relevance for SMD and SRM, an additional, intriguing aspect of fluorophore blinking events is the longer lifetimes of the underlying dark states, ~10^{-6}–10^{-3} s compared to ~10^{-9} s for fluorescence lifetimes. With 10^3 to 10^6 times longer lifetimes and a correspondingly longer time to interact with the immediate environment, these dark states are often highly environment sensitive. Transitions to and from dark long-lived states are thus capable to reflect microenvironments as well as biomolecular dynamics and interactions in a sensitive way. Interestingly, while multi-parameter detection is an attractive strategy, used in a broad range of fluorescence applications in the biomedical field, multi-parameter readouts, including fluorescence blinking, and the whole set of readout parameters coupled to the underlying dark state transitions, have until recently only been considered to a very limited extent. Transitions into more long-lived dark states take place in several different categories of fluorescence emitters, including organic fluorophores, fluorescent proteins, and various kinds of nanoparticles. In this review, we will focus on transitions in organic fluorophore molecules (summarized in Fig. 1), describe how such transitions can be monitored in a broadly applicable manner, and how such monitoring can provide microenvironmental and molecular interaction parameters which are difficult, if at all possible, to retrieve from traditional fluorescence parameters (intensity, lifetime, emission wavelength and polarization/anisotropy).

2 Transient State Monitoring by FCS and Other Methods

Transient absorption spectroscopy (or flash photolysis, FP), invented more than 70 years ago [10], has for many years allowed detailed studies of photo-induced, transient states of fluorophores. Another alternative, restricted to studies of triplet state population dynamics, is to use the triplet state phosphorescence (room temperature phosphorescence, RTP) [11]. A third alternative is FCS. These three categories of methods to analyze transient state dynamics all have their strengths and weaknesses. In transient absorption spectroscopy (FP) the population kinetics of various states are monitored via their absorption by a separate probing beam, following an excitation pulse [12]. However, the absorption spectra of these transient states can often overlap with those of other photo-induced states, making it difficult to separate them from each other [13]. The technique is also relatively technically complicated, is mainly restricted to cuvette experiments, and lacks the sensitivity for measurements at low (<μM) concentrations. Its use for biomolecular studies is thus limited. The emission originating either directly (phosphorescence) or indirectly (delayed fluorescence) from the first excited triplet state offers some attractive features for biomolecular studies [11]. The long decay times make room temperature phosphorescence (RTP) well suited for monitoring of slow rotational motions, e.g., of membrane proteins by luminescence anisotropy [14], and to probe subtle changes in environmental conditions (viscosities, accessibilities of quenchers, polarities,

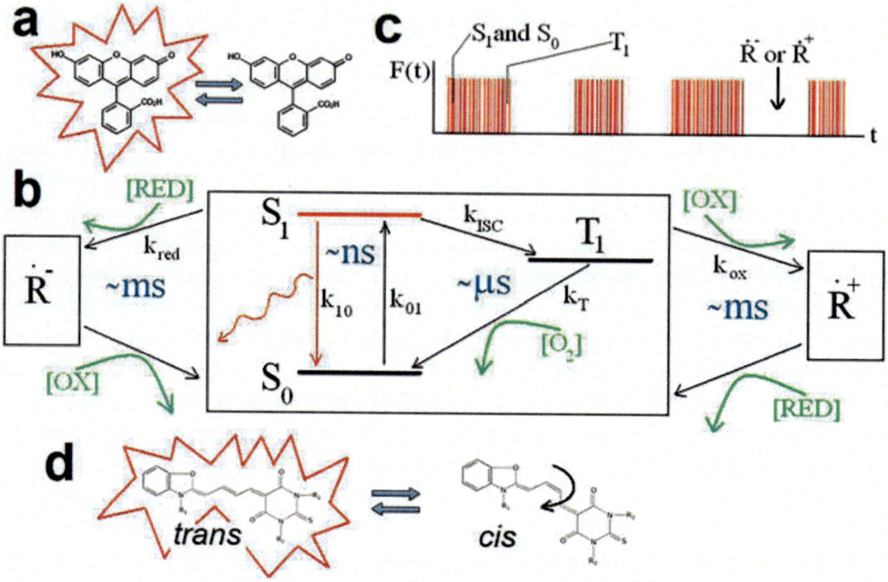

Fig. 1 Common reversible dark state transitions within organic fluorophore molecules. **(a)** Fluorophore blinking follows upon transitions to and from the dark states. **(b)** Representative scheme of underlying states and transitions, where S_0 and S_1 denote singlet ground and first excited states, T_1 the lowest triplet state, and $\dot{R}^+$ and $\dot{R}^-$ photo-oxidized and photo-reduced radical states, respectively. Rates: k_{01} and k_{10} are the excitation and de-excitation rates between S_0 and S_1, k_{ISC} denotes the intersystem crossing rate from S_1 to T_1, and k_T the triplet decay rate back to S_0. Typical timescales of the transitions refer to a typical organic fluorophore in an air-saturated aqueous environment and are indicated in blue. Oxidants [OX], reductants [RED], and molecular oxygen [O_2], as indicated in green, can strongly influence the lifetimes of $\dot{R}^-$, T_1, and $\dot{R}^+$. They can also promote the photo-oxidation and photo-reduction rates (k_{ox} and k_{red}), when present at mM concentrations and higher. With oxidants and reductants present in lower (µM) concentrations, the transitions to and from $\dot{R}^+$ and $\dot{R}^-$ are typically much slower than the transitions between S_0, S_1, and T_1 (within the black box), which then often can be considered as time-averaged on the timescale of the radical state transitions. Fluorescence emission only occurs via excited singlet state decay from S_1 to S_0 (marked red). **(c)** Schematic time trace of the fluorescence from a fluorophore exposed to constant excitation. Short (µs) interruptions are due to transient population of T_1, long ones (ms) to reversible radical state ($\dot{R}^+$ and $\dot{R}^-$) formation. **(d)** Reversible, photo-induced isomerization between a fluorescent, *trans* form and a non-fluorescent, *cis* form of a cyanine dye. The isomerization rates are highly viscosity dependent

etc.), revealing structural and dynamic information of biological macromolecules [15]. However, the RTP signal is weak, and specific RTP probes are scarce and cannot easily be loaded into cells [16]. Coupled to the long-lived RTP emission is also its susceptibility to dynamic quenching by oxygen and trace impurities. Therefore, RTP is largely restricted to deoxygenized, carefully prepared samples, limiting its applications for biomolecular monitoring.

By fluorescence fluctuation analysis with FCS, several of the limitations with FP and RTP readouts can be overcome. In FCS, fluorescence fluctuations are recorded from a sparse number of molecules at a time, passing through a small detection volume, typically comprising a focused laser excitation beam in a confocal arrangement [17] (Fig. 2a). From the recorded fluctuations, as individual fluorescent molecules diffuse into and out of the detection volume, translational diffusion coefficients and absolute concentrations of the fluorescent molecules can be extracted (with knowledge of the dimensions of the FCS detection volume). Similarly, transitions to and from dark states of the fluorescent molecules inside the FCS detection volume can also be determined, from the additional fluorescence fluctuations they generate (Fig. 2b). Taking a simple, standard FCS experiment as an example (Fig. 2a), performed on fluorescent molecules in a solution, which are excited by a focused laser and detected in a confocal arrangement, the total detected fluorescence from the sample volume can be expressed as:

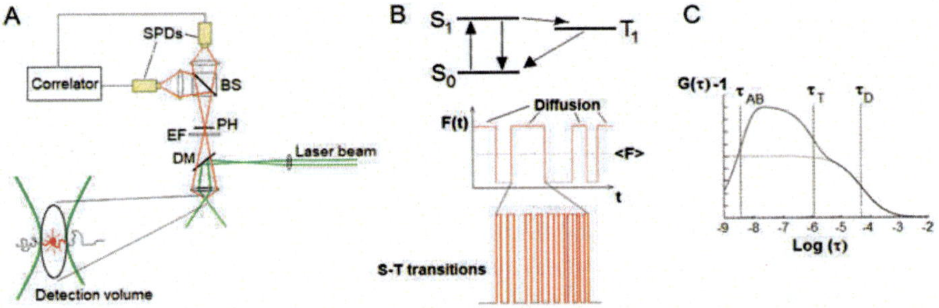

Fig. 2 Measurement of fluorophore dark state transitions by FCS. (**a**) Typical experimental setup for FCS. The beam of a continuous wave (CW) laser is fed into a confocal detection unit. The laser beam is deflected by a dichroic mirror (DM) and then focussed by the microscope objective down to the diffraction limit. The dimensions of the laser beam, together with the collection efficiency function of the microscope define the detection volume (lower left) from which fluorescence is collected. The fluorescence is transmitted through a dichroic mirror (DM), is spectrally and spatially filtered by an emission filter (EM) and a pinhole (PH), located in the image plane. A beam splitter divides the fluorescence into two paths, and fluorescence photons are eventually detected by single-photon counting photodiodes (SPDs). The photocurrents of the SPDs are fed into a PC-based correlator. (**b**) Schematic view of fluctuations in the fluorescence intensity time trace, F(t), as detected by the setup in A from a fluorophore solution, and with the fluorophores freely diffusing into and out of the open confocal detection volume, and undergoing singlet-triplet state transitions on a time scale much faster than the diffusion passage times through the detection volume. (**c**) Calculated FCS curve, given by Eq. (2), showing the time-dependent part of the ACF ($G(\tau)$-1) from a fluorophore undergoing three-dimensional diffusion through an observation volume (Fig. 2a) as well as transitions to and from a non-fluorescent triplet state (Fig. 2b). τ_{AB} and τ_T denote the anti-bunching and the triplet state relaxation times. τ_D refers to the average diffusion time of the fluorophores through the detection volume. The dotted line shows the diffusion-generated part of the FCS curve, denoted $G_D(\tau)$ in Eq. (2)

$$F(t) = q_D \int \mathrm{CEF}(\overline{r}) cF(\overline{r}, t) dV \tag{1}$$

Here, q_D is the detection quantum yield of the instrument, and $\mathrm{CEF}(\overline{r})$ is the collection efficiency function [18]. c is the concentration of fluorescent molecules. Dark state population kinetics of the fluorophores can then be extracted from the fluorescence intensity fluctuations and analyzed in terms of a normalized autocovariance function, $G(\tau)$, typically referred to as the autocorrelation function (ACF). For fluorophore molecules undergoing diffusion, and in addition also transitions between their ground and first excited singlet states (S_0, S_1) and a dark triplet state (T_1) (Fig. 2b) [19]:

$$\begin{aligned} G(\tau) &= \frac{< F(t)F(t+\tau) >}{< F(t)>^2} \\ &= \frac{1}{N(1-\overline{T})} G_D(\tau)\left[1 - \overline{T} + \overline{T}\exp\left(-\tau/\tau_T\right) - \exp\left(-\tau/\tau_{AB}\right)\right] + 1 \end{aligned} \tag{2}$$

Here, square brackets denote time average, N is the average number of fluorescent molecules in the detection volume, $G_D(\tau)$ represents the diffusion-dependent part of $G(\tau)$, and $\overline{T}$ is the average steady-state probability for the fluorescent molecules within the detection volume to be in their triplet states. τ_{AB} and τ_T denote the antibunching and the triplet state relaxation times, respectively.

Transient state monitoring by FCS offers a favorable combination of a high signal level (given by the readout of fluorescence photons) and an outstanding environmental sensitivity (given by the long lifetimes of the transient states). It can be realized with a comparatively simple setup, is applicable to a relatively wide range of samples, and can monitor a broad range of fluorescence blinking phenomena, including triplet [19], photo-ionized [20], and photo-isomerized [21, 22] states of fluorophores (Fig. 1). These transient state parameters can provide additional, sensitive information about local oxygen concentrations, redox environments, micro-viscosities, and molecular interactions, which are only very weakly, if at all, reflected in traditional fluorescence parameters (intensity, lifetime, polarization, and wavelength). However, FCS measurements rely on detection of spontaneous fluorescence fluctuations from individual molecules, i.e., SMD conditions are required. This puts high demands on the detection sensitivity and noise suppression. Since the fluctuations of the transient states typically take place in the μs time range, a high time resolution of the detection is also required. Moreover, since low fluorescent signals can typically not be compensated by higher concentrations of fluorescent molecules, FCS measurements rely on a high fluorescence brightness of the investigated molecules [23]. Taken together, FCS measurements put high demands on the detection sensitivity, time resolution, and noise suppression, which limits the applicability of FCS for transient state analyses in biological samples.

3 Transient State (TRAST) Spectroscopy/Imaging: Basic Concept

To overcome the limitations, as encountered in FP, RTP as well as in FCS measurements, our research group has devised a method – transient state (TRAST) monitoring, making transient state measurements more broadly applicable under biologically relevant conditions [24, 25]. The principle of TRAST is illustrated in Fig. 3: The time-averaged fluorescence intensity, $<F>$, from a sample subject to time-modulated excitation is recorded. By systematically varying the excitation modulation, and from how $<F>$ varies with the excitation modulation characteristics, kinetic information about photo-induced, non-fluorescent transient states can be obtained. Like FCS, TRAST combines the detection sensitivity of fluorescence with the environmental sensitivity of long-lived, non-fluorescent states. However, unlike FCS, TRAST is not dependent on the detection of stochastic fluorescence fluctuations from individual molecules. Therefore, TRAST measurements can move away from SMD requirements, and are not limited to the use of fluorophores with high photostability and brightness, nor to samples offering high signal-to-background conditions. The detection quantum yield, q_D, of the instrument is also not critical. Since TRAST also moves the time-resolution requirements from the detection to the excitation side, multi-pixel registration of time-averaged fluorescence can be performed with regular sCMOS and CCD cameras in a parallelized, spatially resolved manner.

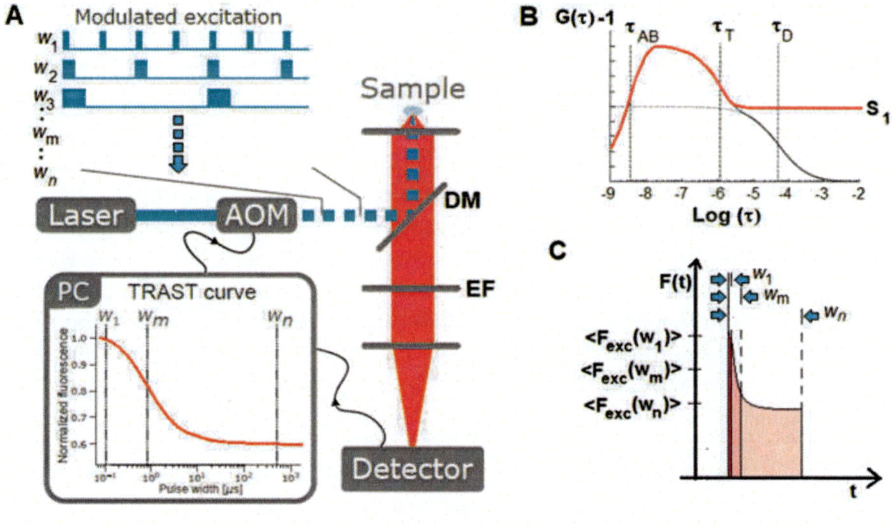

Fig. 3 **(a)** Simplified experimental setup and concept for TRAST. **(b)** Red line: Population probability of the S_1 state, following onset of excitation at time $t = 0$, for a fluorophore undergoing singlet-triplet transitions, as depicted in Fig. 2b. **(c)** Principal drawing of the fluorescence intensity measured in a TRAST experiment for different rectangular excitation pulse durations, w. See main text for further descriptions

For a transient state to be detectable with TRAST, it must be photo-induced and have a different luminescence brightness than the excited singlet state, S_1, within a selected spectral range. Such states are found in most fluorophores, and essentially all long-lived transient states can be considered non-luminescent. Also, the phosphorescence from triplet states is typically negligible, compared to the fluorescence from the S_1 state. Moreover, triplet state phosphorescence is easily quenched and is emitted in a different (longer) wavelength range than the fluorescence. Thus, in most cases, only fluorescence generated upon decay from S_1 to the ground singlet state, S_0 (as depicted in the electronic state model in Fig. 1) needs to be considered. The emitted fluorescence intensity, $F(t)$, will then be proportional to the S_1 population probability, $S_1(t)$:

$$F(t) = k_{10} q_f S_1(t) \tag{3}$$

Here, k_{10} denotes the decay rate from S_1 to S_0, and q_f the fluorescence quantum yield. For a fluorophore subject to constant excitation starting at $t = 0$, $S_1(t)$ can in a general form be described by:

$$S_1(t) = \frac{k_{01}}{k_{01} + k_{10}} \left[1 - e^{-\lambda_{ab} t} - \sum_{i=1}^{p} \left[A_i - A_i e^{-\lambda_i t}\right]\right] \tag{4}$$

Here, k_{01} is the excitation rate from S_0 to S_1, given by $\boldsymbol{k_{01} = \sigma \Phi_{exc} = \sigma I_{exc}/h\upsilon}$, with σ denoting the excitation cross-section, Φ_{exc} the excitation photon flux, I_{exc} the excitation intensity and $h\nu$ the excitation photon energy. p denotes the number of different photo-induced non-fluorescent states involved (states other than S_0 and S_1). λ_{ab} and λ_i are the eigenvalues, i.e., the rates of the relaxation modes of $S_1(t)$ upon onset of constant excitation, and A_i the related amplitudes, reflecting the population buildup of the different photo-induced non-fluorescent states. With a suitable initial condition, λ_i and A_i can be described analytically, as functions of the rate parameters in the electronic state model applicable to the fluorophores studied [24]. Assuming that the population of all photo-induced states is zero at $\boldsymbol{t = 0}$, the initial condition becomes $\boldsymbol{S_0(t = 0) = 1}$, i.e., $\boldsymbol{S_1(t = 0) = 0}$. Alternatively, if several states are populated before excitation, the sum of their populations must be 1 at $t = 0$. For most fluorescent molecules, equilibration between S_0 and S_1 after onset of excitation takes place within the time range of the fluorescence lifetime (ns), while the dark state relaxations ($1/\lambda_i$) typically occur on a μs-ms time scale. The S_0-S_1 equilibration time, referred to as the anti-bunching time [26–28] $\boldsymbol{\tau_{ab} = 1/\lambda_{ab}}$, is typically given by [29]:

$$\tau_{ab} = \frac{1}{k_{01} + k_{10}} \tag{5}$$

In TRAST experiments, the modulation of the excitation intensity can in principle be systematically varied in several ways. Typically, square-wave excitation pulse trains are applied, with low duty cycles (1:100 or lower) to avoid dark state population pile-up effects, and with the time-averaged fluorescence monitored as a function of the duration, w, of the excitation pulses (Fig. 3a). The evolution of $S_1(t)$, or $F(t)$, after onset of a square-wave excitation pulse at $t = 0$ is exemplified in Fig. 3b (red curve), for a fluorophore having three electronic states (S_0, S_1 and T_1), and described by the same state model as in Fig. 2b. Essentially, for a free fluorophore solution subject to a constant I_{exc} in time and space, this curve corresponds to the initial, time-dependent part of the ACF in Fig. 2c, to the part which is not generated by fluorophore diffusion. For a square-wave excitation in a TRAST experiment, and for the ACF recorded in a corresponding FCS measurement with the same constant I_{exc}, the same initial condition applies; Since the ACF reflects the probability to detect a second fluorescence photon from a fluorescent molecule in the sample, given that a first fluorescence photon is detected from it at $\tau = 0$, means that the fluorophore must be in S_0 at $\tau = 0$ (unless there are several independent emitters on the fluorescent molecule). Likewise, before onset of a square-wave excitation pulse in a TRAST experiment, the fluorophore can generally be considered to be in the S_0 state.

In a TRAST experiment, for a sample of fluorescent molecules studied in a confocal setup, subject to a square-wave excitation pulse train with n pulses of duration w and period T, the detected time-averaged fluorescence, $\langle F(w)\rangle$, can then be expressed as [24, 25, 30, 31]:

$$\langle F(w)\rangle = \frac{1}{n \cdot T} \iiint c q_D q_f \mathrm{CEF}(\overline{r}) k_{10} \sum_{i=1}^{n} \left(\int_0^w S_{1,i}(\overline{r}, t) dt \right) dV \tag{6}$$

Here $n{\cdot}T$ is the total duration of the excitation pulse train, c the concentration of the fluorescent species, q_D the overall detection quantum yield of the instrument, and CEF the collection efficiency function of the confocal setup. The term $S_{1,i}(\overline{r}, t)$ denotes the probability that a fluorescent molecule, located at $\overline{r}$ in the confocal detection volume, is in its excited singlet state at time t after onset of the i:th excitation pulse, and its time dependence can be described by Eq. (4). Dividing $\langle F(w)\rangle$ with the pulse train duty cycle ($\eta = w/T$) yields the average fluorescence intensity within an excitation pulse:

$$\langle F_{\mathrm{exc}}(w)\rangle = \langle F(w)\rangle / \eta \tag{7}$$

$\langle F_{exc}(w)\rangle$ normalized to 1 for pulse durations $|\lambda_{ab}|^{-1} << w << |\lambda_i|^{-1}$, denoted $\langle F_{exc}(w)\rangle_{norm}$, represents the averaged population of S_0 and S_1, within the pulse duration, and over the detection volume:

$$\langle F_{\mathrm{exc}}(w)\rangle_{\mathrm{norm}} = \langle S_0(w) + S_1(w)\rangle = \frac{\iiint \mathrm{CEF}(\overline{r}) \frac{k_{01}(\overline{r}) + k_{10}}{k_{01}(\overline{r})} \left(\frac{1}{w} \int_0^w S_{1,i}(\overline{r}, t) dt \right) dV}{\iiint \mathrm{CEF}(\overline{r}) dV} \tag{8}$$

From Eq. (8), and with knowledge of $I_{exc}(\overline{r}) = h\upsilon\Phi_{exc}(\overline{r})$ yielding $k_{01}(\overline{r}) = \sigma\Phi_{exc}(\overline{r})$, CEF($\overline{r}$), and the electronic state model, the transient state rate parameters can then be determined from the pulse duration dependence of $\langle F_{exc}(w)\rangle_{norm}$, making up a so-called TRAST-curve [24, 25, 30, 31] (Fig. 3a). Approximating the average excitation rate in the detection volume by:

$$\overline{k}_{01} = \frac{\int k_{01}(\overline{r})\overline{S}_1(\overline{r})\mathrm{CEF}(\overline{r})dV}{\int \overline{S}_1(\overline{r})\mathrm{CEF}(\overline{r})dV}, \tag{9}$$

where $\overline{S}_1(\overline{r}) = k_{01}(\overline{r})/(k_{10} + k_{01}(\overline{r}))$, i.e., the S_1 population at onset of excitation, after equilibration between the singlet states, but before dark state buildup. A simplified expression for $\langle F_{exc}(w)\rangle$ can be obtained, given by

$$\langle F_{exc}(w)\rangle_{norm} = \frac{1}{w}\frac{k_{10} + \overline{k}_{01}}{\overline{k}_{01}}\int_0^w S_1(t)dt, \tag{10}$$

with $S_1(t)$ given by Eq. (4), and according to the electronic state model applicable for the particular fluorophores studied.

4 TRAST: Some Experimental Realizations and Applications

Above, we have taken as an example the implementation of TRAST in a confocal setup, where the fluorescence of a sample is induced by a time-modulated stationary laser excitation beam, and the time-averaged fluorescence intensity is detected in a likewise stationary confocal detection volume. As discussed above, the experimental requirements for TRAST are much more relaxed than in FCS. Therefore, the TRAST approach can be made compatible with a range of other modalities for excitation time-modulation, combined with average fluorescence intensity detection. Such modalities include time-modulated wide-field excitation, moving arrays of laser foci or laser excitation fringe patterns, as well as various spatial confinement strategies of the excitation and/or the detection [24, 25, 31]. Apart from realizing TRAST in a stationary confocal microscope arrangement [24, 30, 32], as outlined above, and TRAST implemented with wide-field microscopy (Fig. 4a) [33–35], other experimental setups demonstrated for TRAST include total-internal-reflection microscopy with evanescent-field excitation [36], light-sheet excitation microscopy [37, 38], and two-photon-excitation microscopy [39]. Evidently, the time-modulated excitation experienced by a stationary sample can also be generated by translation of the sample with respect to the excitation, or vice versa. In this way, the excitation need not be idle, and in particular for low duty cycle excitation, a larger sample

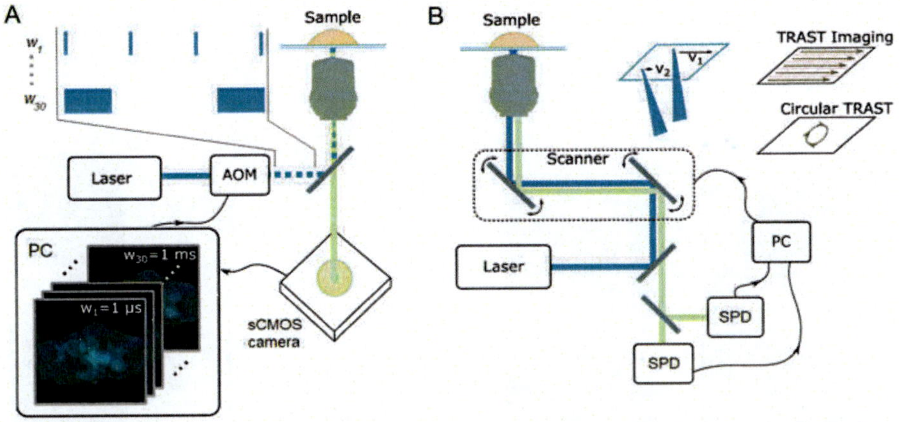

Fig. 4 TRAST imaging implemented with wide-field microscopy (**a**) and confocal laser scanning microscopy (CLSM) (**b**). Blue lines: Excitation laser beam paths, with different scanning speeds applied for CLSM (v_1, v_2, ...) as a means to vary the excitation pulse duration in the different locations in the sample. In the wide-field setup, the excitation light is stationary and modulated in time by an acousto-optical modulator (AOM). Green lines: Paths for the generated fluorescence light

volume can be interrogated within the same period of time. Based on this concept, TRAST imaging can also be well implemented in a confocal laser scanning microscope (CLSM) (Fig. 4b) [31].

Relaxed requirements on noise suppression in the samples, on the fluorescence brightness of the molecules studied, as well as on the detection quantum yield and time-resolution of the instrument, makes TRAST broadly applicable, and it has been demonstrated for studies in solution [24, 31, 36], live cells [33–35, 37] and in bacterial biofilms [38]. Moreover, since high fluorescence brightness is not required, also less bright, auto-fluorescent molecules can be studied [32, 39]. Figures 5 and 6 illustrate different applications of TRAST imaging of live cells, which allow additional microenvironmental information within the cells to be imaged, which are difficult, if possible at all to image via regular fluorescence parameters (intensity, emission wavelength, anisotropy/polarization, fluorescence lifetime). Figure 5a shows wide-field TRAST microscopy images of MCF7 (breast cancer) cells, displaying the triplet decay rates of Eosin fluorophores loaded into the cells [33]. Eosin and most other organic fluorophores undergo intersystem crossing into long-lived, dark triplet states. Triplet states are readily quenched by oxygen, and their lifetimes depend strongly on the local oxygen concentration (~μs in an air-saturated aqueous solutions, ~ms in the absence of oxygen). Triplet state lifetimes of fluorophores thus reflect local oxygen concentrations in a very sensitive manner. By wide-field TRAST imaging of singlet-triplet population kinetics of fluorophores in live cells, it is thus possible to image local oxygen concentrations, and oxygen metabolism, on a sub-cellular level. Figure 5a shows regular fluorescence (middle row) and triplet decay TRAST images (lower row) of MCF7 cells. In

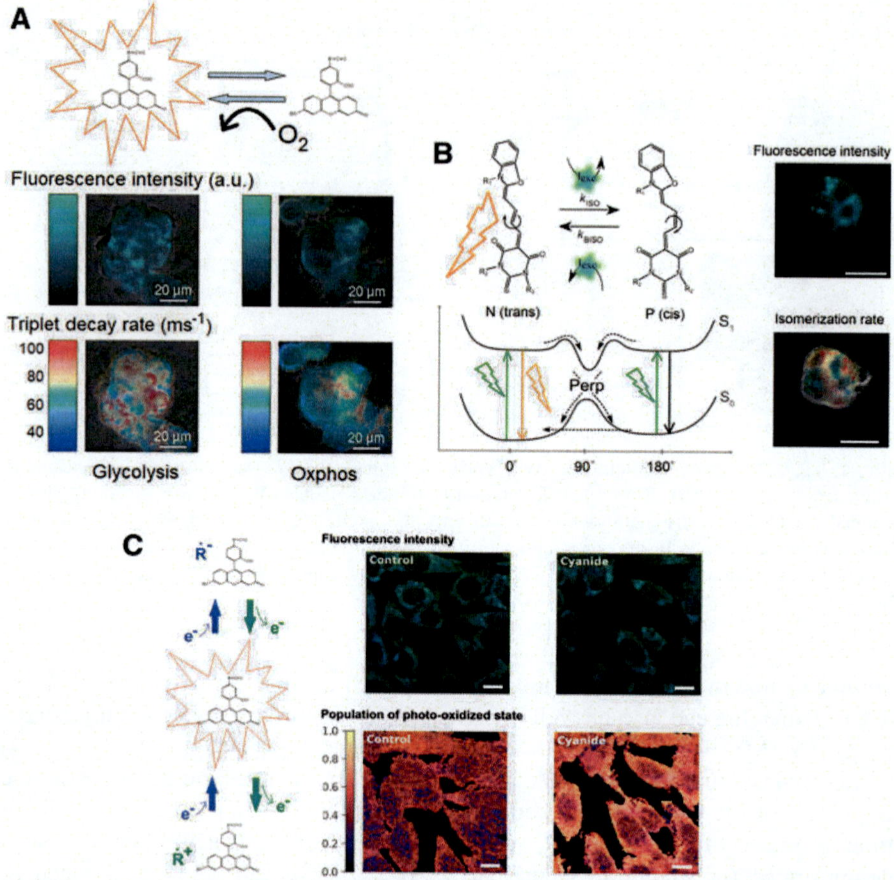

Fig. 5 TRAST imaging of metabolic, redox and microenvironmental conditions in live cells. Triplet state (**a**), photo-isomerization (**b**) and redox-state kinetics (**c**) of added fluorophores (**a** and **b**) or auto-fluorescent NADH (**c**) measured in live cells. (**a**) Regular fluorescence (middle row) and triplet decay TRAST images (lower row) of MCF7 cells, loaded with a high triplet yield fluorophore (Eosin). The triplet decay rate (k_T) of the fluorophore depends almost exclusively, and linearly, on the local oxygen concentration (as depicted in top row). Cells producing ATP primarily by glycolysis (left column), as is often the case for cancer cells, consume less oxygen. Thus, the local oxygen concentration can remain higher, and also the local k_T rates are higher. Cells with normal ATP production via oxidative phosphorylation (right column) consume more oxygen, local oxygen concentrations are lowered, with lower k_T rates as a result. See main text and [33] for further details. (**b**) Left: Excitation-driven isomerization (k_{ISO}) and back-isomerization (k_{BISO}) of a cyanine fluorophore (MC540), leading to transitions between a fluorescent, *trans*, conformation (N) and a non-fluorescent, *cis*, conformation (P). Right: Fluorescence (top) and TRAST (bottom) image of a live cell (MCF7), to which lipophilic MC540 was added and accumulated in the plasma membrane. The TRAST image shows the isomerization rate of MC540 on the cell, which reflects the local fluidity of the plasma membrane. See main text and [34] for further details. (**c**) Label-free TRAST imaging of mouse myoblast cells, using two-photon excitation of the auto-fluorescent co-enzyme NAD(P)H in the cells. In the imaging the excitation pulse duration was varied by the scanning speed of the excitation laser beam over the sample (as in Fig. 4b). Left: Schematic view of photo-induced transitions into transient radical states ($\dot{R}^+$ and $\dot{R}^-$) of a fluorescent molecule. For NAD(P)H, mainly transitions into $\dot{R}^+$ need to be considered. Right: Fluorescence (top row) and TRAST (lower row)

the experiments different growth media were used, so that oxidative phosphorylation (Oxphos) was either ongoing in the cells, or was instead funneled into glycolysis. This metabolic switch away from Oxphos is characteristic for cancer cells (the Warburg effect) and leads to a lower oxygen consumption [41–45]. While no distinction between the two categories of cells was possible to do from the fluorescence images, a higher triplet decay rate (reflecting a higher local oxygen concentration and a lower oxygen consumption) could be seen in the TRAST images taken from cells focused on glycolysis, i.e., TRAST triplet state imaging allows cells with a cancer-specific metabolism to be distinguished from cells undergoing regular metabolism. Figure 5b illustrates how excitation-driven *trans-cis* photo-isomerization of cyanine dyes can be imaged by wide-field TRAST, using the same procedure as in Fig. 5a and as described above (Fig. 4a). Trans-cis photo-isomerization can be well characterized by FCS (if SMD conditions are at hand) [21, 22] and the isomerization rates are sensitive to several environmental factors, including local viscosity and steric constraints experienced by the fluorophores. More recently, this environmental sensitivity attributed to isomerization of cyanine dyes has been exploited to monitor biomolecular conformations and interactions, where so-called protein-induced, or nucleic acid-induced fluorescence enhancement (PIFE/NAIFE) is detected via FCS, or indirectly via fluorescence intensity and lifetime measurements [46, 47]. Wide-field TRAST imaging allows the isomerization of cyanine dyes to be directly monitored in live cells, on a sub-cellular level. Figure 5b shows TRAST images of the isomerization rates of the lipophilic dye Merocyanine 540 (MC540), reflecting the local viscosity of the cellular membranes [34]. Figure 5c illustrates how TRAST imaging also can monitor photo-induced redox reactions (photo-oxidation or photo-reduction) of fluorescence emitters [39]. These reactions often take place on a somewhat slower timescale than singlet-triplet transitions ([48, 49], see also Fig. 1) and can provide information on the local redox environments of the emitters. The TRAST images in Fig. 5c were recorded by laser scanning microscopy (by the approach shown in Fig. 4b), where the fluorescence of the auto-fluorescent co-enzyme NAD(P)H was recorded upon two-photon excitation and at different scanning speeds of the excitation laser [39]. These images illustrate that TRAST can also be applied on less bright emitters, including auto-fluorescent compounds, and that label-free, live cell TRAST imaging is possible.

As a general remark to the examples in Fig. 5a, c, it can be noted that altered cellular metabolism and redox control are central features underlying development and progression of many diseases, including diabetes, infections, autoimmune diseases and cancer [41–44]. In cellular oncology, reprogramming of energy metabolism is considered a "hallmark of cancer" [50]. Yet, many details remain

Fig. 5 (continued) images, where the color scale of the TRAST images is proportional to the buildup of the $\dot{R}^{+}$ state of NAD(P)H in the cells. The buildup is increased in cells exposed to the mitochondrial blocker cyanide (right column). See main text and [39] for additional details un-resolved, and the mechanisms are far from understood. In this context, fluorescence blinking parameters, highly sensitive to local cellular oxygen concentrations and redox conditions, can add important information regarding the metabolic and redox status of cells, and can thereby possibly contribute to new diagnostics of different forms of cancer, and other diseases where cellular metabolism and redox control are central.

As a final example, Fig. 6 (right) shows how TRAST imaging can be applied to characterize low-frequency, intermittent interactions between molecules in cellular membranes, by monitoring the quenching of dark transient states of a membrane fluorophore by spin labels in the same membranes. Traditionally, dynamic quenching of fluorescence emission is a widely used approach for biomolecular interaction and conformation studies [51]. In such quenching studies, the fluorescence signal can typically be detected with excellent sensitivity. However, most fluorophores have excited state lifetimes in the range of nanoseconds, whereas bimolecular diffusion-mediated processes in biological membranes typically take place on three to six orders of magnitude longer time scales. For studies of intermittent, collisional interactions between biomolecules, particularly in membranes, the fluorescence lifetimes are thus often too short to allow diffusion-controlled

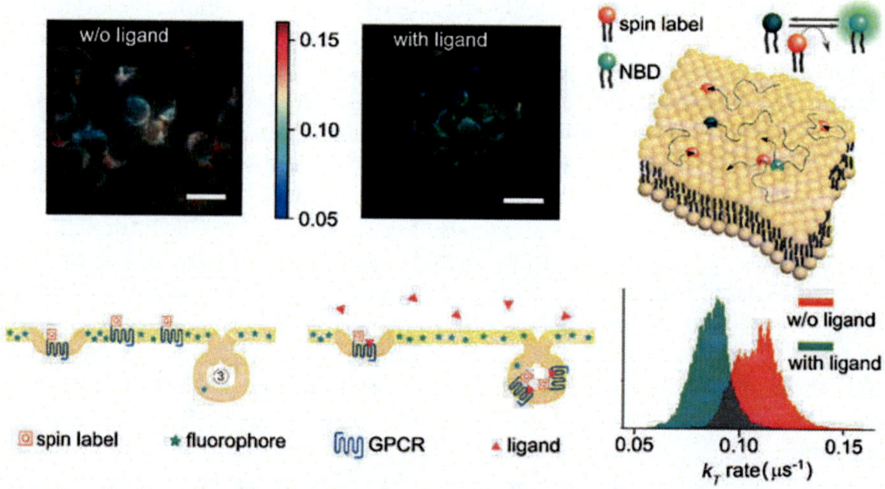

Fig. 6 TRAST imaging of low-frequency molecular encounters in membranes. Top right: Collisional encounters between fluorophore-labeled and spin-labeled molecules in a membrane can occur at time scales (µs to ms) which are much slower than the fluorescence lifetimes of most fluorophores (~ns). However, reversible singlet-triplet state transitions within the fluorophores (Fig. 1) can be readily affected. Left and middle: TRAST images of the triplet decay rate of NBD in the plasma membranes of 293T cells. The plasma membranes also contained spin marker (TEMPO) labeled GPCRs (NK1R). Upon binding of a ligand (SP), the GPCRs are internalized, and the spin labels are removed from the plasma membrane, which leads to a lower NBD triplet state decay rate, k_T, in the membrane. Bottom right: Cumulative histograms of k_T rates, as imaged by TRAST, with and without ligand added to the cells. See main text, and [40] for further details

collisional encounters with quenchers to significantly affect the fluorophore emission. Fluorescent probes with longer fluorescence lifetimes are rare, in particular if they both need to have a high fluorescence brightness and to be efficiently quenched by contact [52]. As a way to overcome these drawbacks, quenching of long-lived triplet states of fluorophores can be monitored via fluorescence, combining the sensitivity of fluorescence detection with the environmental sensitivity of the long-lived triplet and other dark transient states (Fig. 1). Based on this principle, we have shown in previous work how FCS measurements, monitoring fluorescence intensity fluctuations resulting from transitions between singlet and triplet states in membrane-bound fluorophores (Fig. 2), can be used to quantify quenching of the fluorophore triplet states by spin labels in the membranes [53]. Thereby, molecular encounters between the fluorophores and spin labels in the membranes can be monitored at ms^{-1} to μs^{-1} frequencies, at which practically no quenching of the fluorescence intensity occurs. More recently, based on the same principle, we introduced TRAST to follow the interaction between a spin-labeled G-protein coupled receptor (GPCR) (the neurokinin 1 receptor, NK1R) and lipids in the plasma membranes of live cells, labeled with the fluorophore NBD (7-nitrobenz-2-oxa-1,3-diazole-4-yl). NK1R is involved in several physiological and pathophysiological functions and is an important therapeutic target [54]. In its resting state, NK1R has been found to form clusters in the plasma membrane of 293T cells [55]. After activation with its natural ligand, substance P (SP), the receptor localization then changes significantly [56]. This reorganization is principally the result of the fast internalization of activated receptors. We found that the transient state kinetics of NBD labels in the plasma membranes differed significantly upon NK1R activation by its agonist SP. This reflects differences in the transient interactions between the GPCR and the labeled lipids, which could be attributed to modifications in the membrane environment upon activation. The observed changes could be monitored via the quenching rates, both on a whole-cell level, but also in a spatially resolved manner with sub-cellular resolution (Fig. 6, upper left and middle images).

From a more general perspective, GPCRs have a central role in transducing extracellular signals into intracellular responses and represent a majority of all drug targets. Lately, studies of GPCRs by, e.g., X-ray crystallography and cryo-EM have dramatically advanced the knowledge of structure-function relationships, but a complete understanding of GPCR function is impossible to attain without added information about GPCRs dynamics [2]. Here, GPCR–lipid interactions play an important role, but are often transient in character, and occur too intermittently to be readily observable by available technologies. By TRAST however, such protein–lipid interactions can be followed and imaged [40].

5 Concluding Remarks

Fluorophore dark state transitions are both a concern and a prerequisite for fluorescence-based SMD and SRM techniques, and considerable effort has been spent on investigating these phenomena. Yet, an additional aspect of fluorophore blinking has been largely overlooked, namely the strong environmental sensitivity and that fluorophore blinking properties thereby can provide additional fluorescence-based readout parameters. In this review, the principles of the TRAST technique are presented, and examples are given on how this technique can be applied to monitor and image different dark state transitions, reflecting microenvironmental conditions and molecular interactions not detectable via other fluorescence parameters. Alterations in such local cellular conditions, including local oxygen concentrations, redox conditions, and local fluidities, mobilities and molecular interactions, have been found to be both manifestations and driving forces behind several disease conditions, and thus of large biomedical relevance. It is therefore important to have adequate, widely applicable means to monitor such conditions, and here the TRAST approach can play an important role.

Acknowledgments The author likes to acknowledge the important contributions of present and past co-workers in the group, in the development and applications of the TRAST technique. This work was supported by funds from the Swedish Research Council (VR), the Knut and Alice Wallenberg Foundation (KAW), and the Swedish Foundation for Strategic Research (SSF).

References

1. Ehrenberg M (2014) Super-resolved fluorescence microscopy, scientific background on the Nobel prize in chemistry 2014. The Royal Swedish Academy of Sciences
2. Gusach A, Maslov I, Luginina A, Borshchevskiy V, Mishin A, Cherezov V (2020) Beyond structure: emerging approaches to study GPCR dynamics. Curr Opin Struct Biol 63:18–25
3. Jacobson K, Liu P, Lagerholm BC (2019) The lateral organization and mobility of plasma membrane components. Cell 177:806–819
4. Blom H, Widengren J (2017) Stimulated emission depletion microscopy. Chem Rev 117:7377–7427
5. Gupta A, Sankaran J, Wohland T (2019) Fluorescence correlation spectroscopy: the technique and its applications in soft matter. Phys Sci Rev 4. https://doi.org/10.1515/psr-2017-0104
6. Su D, Hou YM, Dong CQ, Ren JC (2019) Fluctuation correlation spectroscopy and its applications in homogeneous analysis. Anal Bioanal Chem 411:4523–4540
7. Schermelleh L, Ferrand A, Huser T, Eggeling C, Sauer M, Biehlmaier O, Drummen GPC (2019) Super-resolution microscopy demystified. Nat Cell Biol 21:72–84
8. Farka Z, Mickert MJ, Pastucha M, Mikusova Z, Skladal P, Gorris HH (2020) Advances in optical single-molecule detection: en route to supersensitive bioaffinity assays. Angew Chem Int Ed 59:10746–10773
9. Gwosch KC, Pape JK, Balzarotti F, Hoess P, Ellenberg J, Ries J, Hell SW (2020) MINFLUX nanoscopy delivers 3D multicolor nanometer resolution in cells. Nat Methods 17:217-+
10. Porter G (1950) Flash photolysis and spectroscopy. A new method for the study of free radical reactions. Proc R Soc A 200:284–300

11. Vanderkooi JM, Calhoun DB, Englander SW (1987) On the prevalence of room-temperatur protein phosphorescence. Science 236:568–569
12. Van Amerongen H, Van Grondelle R (1995) Transient absorption-spectroscopy in study of processes and dynamics in biology. Biochem Spectrosc 246:201–226
13. Lorenc M, Ziolek M, Naskrecki R, Karolczak J, Kubicki J, Maciejewski A (2002) Artifacts in femtosecond transient absorption spectroscopy. Appl Phys B Lasers Opt 74:19–27
14. Jovin TM, Vaz WLC (1989) Rotational and translational diffusion in membranes measured by fluorescence and phosphorescence methods. Methods Enzymol 172:471–513
15. Cioni P, Strambini GB (2002) Tryptophan phosphorescence and pressure effects on protein structure. BBA-Protein Struct M 1595:116–130
16. Yu Q, Huang TC, Li YP, Wei HJ, Liu SJ, Huang W, Du J, Zhao Q (2017) Rational design of a luminescent nanoprobe for hypoxia imaging in vivo via ratiometric and photoluminescence lifetime imaging microscopy. Chem Commun 53:4144–4147
17. Widengren J, Mets Ü (2001) Conceptual basis of fluorescence correlation spectroscopy and related techniques as tools in bioscience. In: Zander C, Enderlein J, Keller RA (eds) Single molecule detection in solution: methods and applications. Wiley-VCH, Heidelberg, pp 69–120
18. Rigler R, Mets U, Widengren J, Kask P (1993) Fluorescence correlation spectroscopy with high count rate and low-background - analysis of translational diffusion. Eur Biophys J Biophys Lett 22:169–175
19. Widengren J, Mets Ü, Rigler R (1995) Fluorescence correlation spectroscopy of triplet-states in solution - a theoretical and experimental-study. J Phys Chem 99:13368–13379
20. Widengren J, Dapprich J, Rigler R (1997) Fast interactions between Rh6G and dGTP in water studied by fluorescence correlation spectroscopy. Chem Phys 216:417–426
21. Widengren J, Schwille P (2000) Characterization of photoinduced isomerization and back-isomerization of the cyanine dye Cy5 by fluorescence correlation spectroscopy. J Phys Chem A 104:6416–6428
22. Widengren J, Seidel CAM (2000) Manipulation and characterization of photo-induced transient states of Merocyanine 540 by fluorescence correlation spectroscopy. Phys Chem Chem Phys 2: 3435–3441
23. Koppel DE (1974) Statistical accuracy in fluorescence correlation spectroscopy. Phys Rev A 10: 1938–1945
24. Sanden T, Persson G, Thyberg P, Blom H, Widengren J (2007) Monitoring kinetics of highly environment sensitive states of fluorescent molecules by modulated excitation and time-averaged fluorescence intensity recording. Anal Chem 79:3330–3341
25. Widengren J (2010) Fluorescence-based transient state monitoring for biomolecular spectroscopy and imaging. J R Soc Interface 7:1135–1144
26. Kimble HJ, Dagenais M, Mandel L (1977) Photon anti-bunching in resonance fluorescence. Phys Rev Lett 39:691–695
27. Ehrenberg M, Rigler R (1974) Rotational brownian-motion and fluorescence intensity fluctuations. Chem Phys 4:390–401
28. Kask P, Piksarv P, Mets U (1985) Fluorescence correlation spectroscopy in the nanosecond time range - photon antibunching in the dye fluorescence. Eur Biophys J Biophys Lett 12:163–166
29. Mets U, Widengren J, Rigler R (1997) Application of the antibunching in dye fluorescence: measuring the excitation rates in solution. Chem Phys 218:191–198
30. Hevekerl H, Tornmalm J, Widengren J (2016) Fluorescence-based characterization of non-fluorescent transient states of tryptophan - prospects for protein conformation and interaction studies. Sci Rep 6:35052
31. Sanden T, Persson G, Widengren J (2008) Transient state imaging for microenvironmental monitoring by laser scanning microscopy. Anal Chem 80:9589–9596
32. Tornmalm J, Widengren J (2018) Label-free monitoring of ambient oxygenation and redox conditions using the photodynamics of flavin compounds and transient state (TRAST) spectroscopy. Methods 140:178–187

33. Spielmann T, Xu L, Gad AKB, Johansson S, Widengren J (2014) Transient state microscopy probes patterns of altered oxygen consumption in cancer cells. FEBS J 281:1317–1332
34. Chmyrov V, Spielmann T, Hevekerl H, Widengren J (2015) Trans-Cis isomerization of lipophilic dyes probing membrane microviscosity in biological membranes and in live cells. Anal Chem 87:5690–5697
35. Geissbuehler M, Spielmann T, Formey A, Marki I, Leutenegger M, Hinz B, Johnsson K, Van De Ville D, Lasser T (2010) Triplet imaging of oxygen consumption during the contraction of a single smooth muscle cell (A7r5). Biophys J 98:339–349
36. Spielmann T, Blom H, Geissbuehler M, Lasser T, Widengren J (2010) Transient state monitoring by total internal reflection fluorescence microscopy. J Phys Chem B 114:4035–4046
37. Mucksch J, Spielmann T, Sisamakis E, Widengren J (2015) Transient state imaging of live cells using single plane illumination and arbitrary duty cycle excitation pulse trains. J Biophotonics 8:392–400
38. Karampatzakis A, Sankaran J, Kandaswamy K, Rice SA, Cohen Y, Wohland T (2017) Measurement of oxygen concentrations in bacterial biofilms using transient state monitoring by single plane illumination microscopy. Biomed Phys Eng Express 3:035020
39. Tornmalm J, Sandberg E, Rabasovic M, Widengren J (2019) Local redox conditions in cells imaged via non-fluorescent transient states of NAD(P)H. Sci Rep 9:15070
40. Tornmalm J, Piguet J, Chmyrov V, Widengren J (2019) Imaging of intermittent lipid-receptor interactions reflects changes in live cell membranes upon agonist-receptor binding. Sci Rep 9: 18133
41. Cairns RA, Harris IS, Mak TW (2011) Regulation of cancer cell metabolism. Nat Rev Cancer 11:85–95
42. Heiden MGV, Cantley LC, Thompson CB (2009) Understanding the Warburg effect: the metabolic requirements of cell proliferation. Science 324:1029–1033
43. Kroemer G, Pouyssegur J (2008) Tumor cell metabolism: Cancer's Achilles' heel. Cancer Cell 13:472–482
44. Seyfried TN, Shelton LM (2010) Cancer as a metabolic disease. Nutr Metab 7:7
45. Ward PS, Thompson CB (2012) Metabolic reprogramming: a cancer Hallmark even Warburg did not anticipate. Cancer Cell 21:297–308
46. Pace NA, Hennelly SP, Goodwin PM (2021) Immobilization of cyanines in DNA produces systematic increases in fluorescence intensity. J Phys Chem Lett 12:8963–8971
47. Rashid F, Raducanu VS, Zaher MS, Tehseen M, Habuchi S, Hamdan SM (2019) Initial state of DNA-dye complex sets the stage for protein induced fluorescence modulation. Nat Commun 10:2104
48. Chmyrov A, Sanden T, Widengren J (2010) Iodide as a fluorescence quencher and promoter-mechanisms and possible implications. J Phys Chem B 114:11282–11291
49. Widengren J, Chmyrov A, Eggeling C, Lofdahl PA, Seidel CAM (2007) Strategies to improve photostabilities in ultrasensitive fluorescence spectroscopy. J Phys Chem A 111:429–440
50. Hanahan D, Weinberg RA (2011) Hallmarks of cancer: the next generation. Cell 144:646–674
51. Lakowicz JR (2006) Principles of fluorescence spectroscopy. Springer, New York
52. Melo E, Martins J (2006) Kinetics of bimolecular reactions in model bilayers and biological membranes. A critical review. Biophys Chem 123:77–94
53. Stromqvist J, Chmyrov A, Johansson S, Andersson A, Maler L, Widengren J (2010) Quenching of triplet state fluorophores for studying diffusion-mediated reactions in lipid membranes. Biophys J 99:3821–3830
54. Garcia-Recio S, Gascon P (2015) Biological and pharmacological aspects of the NK1-receptor. Biomed Res Int 2015:495704
55. Meyer BH, Segura JM, Martinez KL, Hovius R, George N, Johnsson K, Vogel H (2006) FRET imaging reveals that functional neurokinin-1 receptors are monomeric and reside in membrane microdomains of live cells. Proc Natl Acad Sci U S A 103:2138–2143

56. Veya L, Piguet J, Vogel H (2015) Single molecule imaging deciphers the relation between mobility and signaling of a prototypical G protein-coupled receptor in living cells. J Biol Chem 290:27723–27735

The Analysis of In-Membrane Nanoscopic Aggregation of Lipids and Proteins by MC-FRET

Barbora Chmelová, Jana Humpolíčková, Kvido Stříšovský, and Radek Šachl

Contents

Abstract Many lipids and membrane proteins spontaneously co-cluster and oligomerize in cellular plasma membranes into larger (functional) units, whose detailed characterization requires high spatial resolution. In this contribution, we introduce a powerful spectroscopy/microscopy approach called MC-FRET

Barbora Chmelová and Jana Humpolíčková contributed equally to this work.

B. Chmelová
J. Heyrovský Institute of Physical Chemistry of the Academy of Sciences of the Czech Republic, Prague, Czech Republic

Faculty of Mathematics and Physics, Charles University, Prague, Czech Republic
e-mail: barbora.chmelova@jh-inst.cas.cz

J. Humpolíčková and K. Stříšovský
Institute of Organic Chemistry and Biochemistry of the Academy of Sciences of the Czech Republic, Prague, Czech Republic
e-mail: jana.humpolickova@uochb.cas.cz; kvido.strisovsky@uochb.cas.cz

R. Šachl (✉)
J. Heyrovský Institute of Physical Chemistry of the Academy of Sciences of the Czech Republic, Prague, Czech Republic
e-mail: radek.sachl@jh-inst.cas.cz

R. Šachl, M. Amaro (eds.), *Fluorescence Spectroscopy and Microscopy in Biology*, Springer Ser Fluoresc (2023) 20: 375–400, https://doi.org/10.1007/4243_2022_29,
Published online: 4 August 2022

developed for the analysis of in-membrane nanoscopic aggregation of lipids and proteins in biological membranes. The approach is based on Förster resonance energy transfer (FRET) occurring in membranes between fluorescently labelled lipids/proteins and subsequent analysis of data by Monte-Carlo simulations. The following applications of MC-FRET are presented here: (1) determination of lipid nanodomain sizes and their surface density; (2) characterization of inter-leaflet organization of lipid nanodomains and (3) the analysis of dimerization of in-membrane proteins.

Keywords Dimers · FRET · Gangliosides · GPMVs · Kappa squared · MC-FRET · Membrane protein–protein interactions · Nanodomains · Oligomerization

Abbreviations

A	Acceptor
ACF	Autocorrelation function
B.-F.	Bauman-Fayer model
bSM	Brain sphingomyelin
C	Surface concentration
CF-PEG-DSPE	Carboxyfluorescein-2-distearoyl-sn-glycero-3-phosphoethanolamine-*N*-[amino (Polyethylene glycol) 2000]
CTxB	Cholera toxin β
D	Donor
d	The distance between two bilayer leaflets
DGS-NTA(Ni)	1,2-Dioleoyl-sn-glycero-3-[(*N*-(5-amino-1-carboxypentyl) iminodiacetic acid)succinyl]
DHE	Dehydroergosterol
DHPE FITC	Fluorescein dihexadecanoylphosphatidylethanolamine
DHPE TR	Texas red dihexadecanoylphosphatidylethanolamine
DiI	1,1'-Dioctadecyl-3,3,3',3'-tetramethylindocarbocyanine perchlorate
DiOC18	Dioctadecyloxacarbocyanine perchlorate
DLPC	Dilauroylphosphatidylcholine
DOPC	1,2-Dioleoyl-sn-glycero-3-phosphocholine
DOPG	1,2-Dioleoyl-sn-glycero-3-phospho-glycerol
DPhPC	Diphytanoylphosphatidylcholine
DSPC	1,2-Distearoyl-sn-glycero-3-phos-phocholine
DSPC	Distearoylphosphatidylcholine
FCCS	Fluorescence cross-correlation spectroscopy
FP	Fused proteins
FRET	Förster resonance energy transfer
GCPII	Glutamate carboxypeptidase II
g-GM_1	Ganglioside GM_1 labelled in the headgroup with FL-BODIPY

$G_i(t)$	Survival probability function
GPMVs	Giant plasma membrane vesicles
GUVs	Giant unilamellar vesicles
GWALP23	Acetyl-GGALW^{5}LAL^{8}ALALALAL^{16}ALW^{19}LAGA-amide
iSCAT	Interferometric scattering microscopy
κ^2	Kappa squared
K_d	Dissociation constant
K_D	Distribution constant
L_0	The closest protein–protein distance
MC-FRET	FRET analysed by Monte-Carlo simulations
MSD	Mean square displacement
N_A	Numerical aperture
NBD-DLPE	*N*-(7-nitrobenz-2-oxa-1,3-diazol-4-yl)-dilauroylphosphatidylethanolamine
PALM	Photo-activated localization microscopy
PGPC	1-Palmitoyl-2-glutaryl-*sn*-glycero-3-phosphocholine
PIP2	Phosphatidylinositol 2
POPC	Palmitoyl-oleoyl-phosphocholine
POVPC	1-Palmitoyl-2-(5′-oxo-valeroyl)-*sn*-glycero-3-phosphocholine
PSM	Palmitoyl-sphingomyelin
Q_{DA}/Q_D	Quantum yield in the presence/absence of acceptors
R_0	Förster radius
r-GM_1	Ganglioside GM_1 labelled in the headgroup with 564/570-BODIPY
Rh-PEG-DSPE	Rhodamine101–2-distearoyl-*sn*-glycero-3-phosphoethanolamine-*N*-[amino (Polyethyleneglycol) 2000]
SM	Sphingomyelin
STED	Stimulated emission depletion microscopy
TCSPC	Time-correlated single photon counting
TOE	Tryptophan oleoyl ester
TRF	Time-resolved fluorescence
η	FRET efficiency

1 Introduction

Biophysical experiments performed in the last decades have shown that nanoscale distribution of lipids and proteins in cellular membranes is rather complicated and far from being completely homogeneous (see, for instance, the following reviews) [1, 2]. On the contrary, many lipids and membrane proteins spontaneously co-cluster and oligomerize into larger functional units, which may be essential for proper functioning of biological membranes [2]. To give a few examples for all, gangliosides self-assemble in plasma membranes into several tens of nanometre large lipid nanodomains, forming a platform for ligand to receptor interactions [3, 4]. Similarly, phosphatidylinositol-bisphosphates (PIP2), involved in cellular signalling, also co-cluster and form nanoscopic domains [5, 6]. Or, interestingly, trans-membrane proteins engaged in the respiratory chain cycle oligomerize into huge super-molecular complexes, which seem important for the proper function of the respiratory chain cycle [7].

These findings were achieved by the rapid development of advanced biophysical approaches that can detect and characterize nanoscopic objects in the membrane by offering high spatial and temporal resolution. These high-resolution approaches undoubtedly include fluorescence techniques such as Förster Resonance Energy Transfer (FRET) [8], Photo-Activated Localization Microscopy (PALM) [9], Stochastic Optical Reconstruction Microscopy [10] or Stimulated Emission Depletion Microscopy (STED) [11], but also label-free techniques such as Interferometric SCATtering microscopy (iSCAT) [12, 13], X-ray, neutron scattering, electron or atomic force microscopy and others [14, 15]. Whereas most of the above-mentioned fluorescence techniques are reviewed in other chapters of this book, this chapter focuses on the applications of FRET in the analysis of lipid nanodomain formation and in-membrane oligomerization of proteins.

The approach is based on the use of lipids and proteins that are labelled by either fluorescence donors or acceptors and the fact that their self-assembly will change the originally homogeneous donor and acceptor distributions into heterogeneous ones, bringing the donors closer to the acceptors, or taking them apart (Fig. 1). As a result, the efficiency of FRET will either increase or decrease, which can be quantified by either measuring the fluorescence intensity of the donors, or more precisely, their corresponding time-resolved fluorescence (TRF) decays. The analysis of TRF data by a novel analytical approach called MC-FRET [16, 17] (FRET analysed by Monte-Carlo simulations) enables then to measure lipid nanodomain sizes with nanometre precision, or, in case of protein oligomerization, to characterize the oligomeric state.

The structure of this chapter is as follows: we first review the classical analytical model that is used to analyse FRET between donors and acceptors that are homogeneously distributed in lipid bilayers (Sect. 2), introduce the MC-FRET approach developed for the analysis of FRET between heterogeneously distributed probes where analytical approaches fail (Sect. 3.1) and then describe the following applications of MC-FRET: (1) estimation of lipid nanodomain sizes and their surface density (Sect. 3.2), (2) discrimination between inter-leaflet coupled and inter-leaflet independent or anti-registered nanodomains (Sect. 3.3) and (3) quantification of protein dimerization in the native membrane (Sect. 3.4). Overall, a careful analysis of this approach reveals that MC-FRET can detect nanodomains as small as a few nanometres and determine whether the nanodomains are inter-leaflet coupled or independent [18]. Moreover, FRET allows for quantitative description of protein–protein interactions in planar lipid bilayers. The conclusions drawn in the following quantitative studies [19–21] specify in detail the facts that have to be considered when studying protein dimerization in non-planar membranes, such as Golgi, for example.

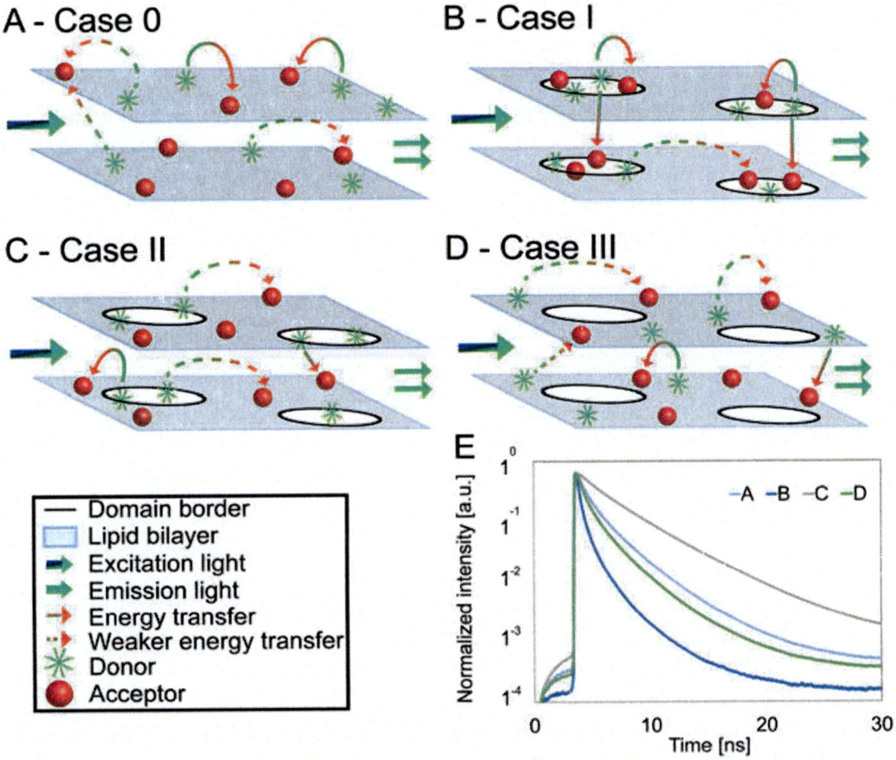

Fig. 1 (**a–d**) Possible arrangements of donors and acceptors in respect of the nanodomains and (**e**) corresponding time-resolved fluorescence decays of donors in the presence of acceptors. (**a**) A nanoscopically homogeneous bilayer with randomly distributed donors and acceptors; (**b–d**) A nanoscopically heterogeneous bilayer with (**b**) both donors and acceptors exhibiting increased affinity to the nanodomains; (**c**) donors localized in the nanodomains and acceptors excluded from them and (**d**) both donors and acceptors excluded from the nanodomains

2 FRET Between Homogeneously Distributed Donors and Acceptors

A cellular plasma membrane approximated by a lipid bilayer consists of two parallel sheets. Under the assumption that each leaflet contains homogeneously distributed donors and acceptors, FRET will occur both within one leaflet and between the two leaflets at the same time (Fig. 1). Whereas the first process is referred to as *intra*-FRET and depends only on the surface concentration of acceptors, the latter one is known as *inter*-FRET and depends in addition on the distance between the two bilayer leaflets d. In the absence of FRET, the donor fluorescence intensity $F_D(t)$ can be expressed as:

$$F_D(t) = \sum_i \alpha_i \exp\left(-\frac{t}{\tau_i}\right). \tag{1}$$

Here, α_i denotes the preexponential factors and τ_i the corresponding fluorescence lifetimes. To express mathematically the kinetics of fluorescence deexcitation of the donors in the presence of acceptors, it is convenient to introduce a so-called survival probability function *G*(*t*) describing the probability that a donor is still found in the excited state at the time *t* after the excitation. When several independent deexcitation processes occur at the same time, the total probability function $G_{TOT}(t)$ equals the product of $G_i(t)$ functions for individual deexcitation processes:

$$G_{TOT}(t) = \prod G_i(t). \tag{2}$$

By assuming dynamic limit conditions, where movements of dipoles are much faster than an energy transfer event, Bauman and Fayer showed that *G*(*t*) for *intra*-FRET, G_{intra} [22], equals

$$\ln G_{intra}(t) = -C_2 \Gamma\left(\frac{2}{3}\right)\left(\frac{t}{\langle\tau_D\rangle}\right)^{1/3}. \tag{3}$$

Here, Γ denotes the gamma function, $\langle\tau_D\rangle$ the intensity averaged fluorescence lifetime of donors in the absence of acceptors and C_2 the reduced surface concentration of acceptors in one leaflet. This concentration represents the average number of acceptors within the area of πR_0^2 and is related to the acceptor surface concentration *C*(A) by: $C_2 = \pi R_0^2 C(\mathrm{A})$. In the same work, the authors derived *G*(*t*) for *inter*-FRET, $G_{inter}(t)$ [22].

$$\ln G_{inter}(t) = -\frac{C_2}{3}\left(\frac{d}{R_0}\right)^2\left(\frac{2\mu}{3}\right)^{1/3}\int_0^{2/3\mu}(1-e^{-s})s^{-4/3}ds, \tag{4}$$

where θ_r is the angle between the bilayer normal and the vector connecting the locations of the donor and acceptor dipoles, $\mu = 3t\left(\frac{R_0}{d}\right)^6 \frac{1}{2\langle\tau\rangle}$ and $s = 2\mu \cos^6 \frac{\theta_r}{3}$. As a result, the kinetics of fluorescence deexcitation of the donors in the presence of acceptors, $F_{DA}(t)$, can be expressed for the classical case where both donors and acceptors are localized in the lipid bilayer at the same distance from the lipid–water interface as

$$F_{DA}(t) = G_{intra}(t)G_{inter}(t)F_D(t). \tag{5}$$

Fitting time-resolved fluorescence data by this equation enables to transversely localize fluorescent probes with subnanometre resolution and to accurately determine the surface concentration of acceptors [23].

3 FRET Between Heterogeneously Distributed Donors and Acceptors

3.1 *Heterogeneous Probe Distributions Induced by Lipid Nanodomain Formation or Protein Oligomerization*

Heterogeneous probe distribution can arise in lipid bilayers in two different ways: (1) either due to formation of membrane lipid nanodomains to which donors and/or acceptors have different affinity in comparison with the surrounding nondomain phase (Sects. 3.2 and 3.3) [16] or (2) because of co-clustering of the donors with the acceptors (Sect. 3.4). Both cases result in each donor having a different radial distribution of acceptors around itself, which makes it difficult to derive precise mathematical formulas describing FRET in such a heterogeneous probe environment. As will be shown below, this limitation can be effectively circumvented by analysing FRET using MC simulations, giving rise to a new analytical approach called MC-FRET. In this approach, MC simulations are used to mimic FRET in the heterogeneous probe environment and to generate in-silico TRF decay curves that are compared to the experimental ones in an iterative fashion (Sect. 3.2.1), enabling to study formation of membrane nanodomains [8, 24] (Sects. 3.2 and 3.3), or in-membrane oligomerization of proteins (Sect. 3.4).

3.2 *The Estimation of Lipid Nanodomain Sizes by MC-FRET*

To characterize membrane lipid nanodomains by MC-FRET, the used donor and acceptor probes need to exhibit a distinct affinity to the nanodomains and the region outside of them. In such case, the shape of TRF decays of the donors in the presence of acceptors will be influenced by the size, surface density and inter-leaflet organization of the nanodomains. All of these parameters can be estimated by MC-FRET [17].

To characterize the probe affinity towards nanodomains it is convenient to introduce a distribution constant for donors, $K_D(D)$, or acceptors, $K_D(A)$, as $K_D(D/A) = C(D/A \text{ inside of a domain})/C(D/A \text{ outside of a domain})$ between the nanodomains and the region outside of them with the help of donor/acceptor surface concentrations inside (C(D/A inside)) or outside (C(D/A inside)) of the nanodomains, respectively. The following situations may arise: Case 0): $K_D(D) = 1$ and $K_D(A) = 1$. Consequently, the probes are distributed homogenously across the entire lipid bilayer regardless of the presence of nanodomains (Fig. 1a). Such a situation does not allow for the detection of nanodomains using MC-FRET. Case I) $K_D(D) > 1$ and $K_D(A) > 1$, implying that both donors and acceptors accumulate inside of the nanodomains (Fig. 1b). Such a spatial accumulation of the probes decreases the average distance between the probes and leads to enhanced FRET efficiency and faster relaxation of donors into the ground state (Fig. 1e) as compared to the homogeneous probe distribution (Case 0, Fig. 1a). Case II) $K_D(D) > 1$ and $K_D(A) < 1$ or $K_D(D) < 1$ and $K_D(A) > 1$, which results in the accumulation of donors and acceptors in distinct bilayer regions, and consequently spatial separation of donors from acceptors (Fig. 1c). Such a probe distribution yields a lower FRET efficiency and slower relaxation kinetics in comparison with Case 0 (Fig. 1e). And finally, Case III) $K_D(D) < 1$ and $K_D(A) < 1$. Since both donors and acceptors are excluded from nanodomains in this case, the efficiency of FRET will be increased, and the relaxation kinetics accelerated in comparison with homogenous probe distribution.

3.2.1 Workflow of MC-FRET for the Nanoscopic Characterization of Nanodomains

The MC-FRET analysis is based on fitting experimental TRF decays by the decays generated by MC simulations and follows the basic scheme depicted in Fig. 2. The entire procedure starts by generation of a defined number of nanodomains on the bilayer surface. The nanodomains are assumed to be circular in shape, uniform in size, perfectly registered across the two bilayer leaflets and are characterized by the nanodomain radius, $\langle R \rangle$, and their surface density, $\langle A \rangle$. In the next step, donors and acceptors are distributed between the nanodomains and the remaining bilayer part according to their $K_D(D)$ and $K_D(A)$. The total number of acceptors that enters the simulation is estimated on the same sample prior to nanodomain formation by FRET (see the Baumann-Fayer model (B.-F.) presented in Sect. 2). Then, a donor is randomly excited, and the time Δt_i at which energy transfer takes place calculated according to $\Delta t_i = -\ln \gamma / \Omega_i$, where γ is a randomly generated number between 0–1 and Ω_i is the overall energy transfer rate.

In the case of a randomly excited donor in the presence of multiple acceptors, Ω_i can be expressed as the sum of the rates between the donor and each of the acceptors:

$$\Omega_i = \frac{1}{\langle \tau_D \rangle} \sum_j \left(\frac{R_0}{R_j} \right)^6. \tag{6}$$

Here, j goes through all acceptors in the system and R_0 stands for the Förster radius. For the sake of simplicity, the distance R_j between the i-th selected donor and the acceptors is evaluated only for the acceptors that are closer than the cut-off distance R_c ($10R_0$); the acceptors behind that cut-off are treated as a continuum:

$$\Omega_i = \frac{1}{\langle \tau_D \rangle} \left(\sum_j \left(\frac{R_0}{R_j} \right)^6 + 0.5 \times C_2 \times \left(\frac{R_0}{R_c} \right)^4 \right), \text{for } j, \text{ where } |R_i - R_j| \leq 10R_0. \tag{7}$$

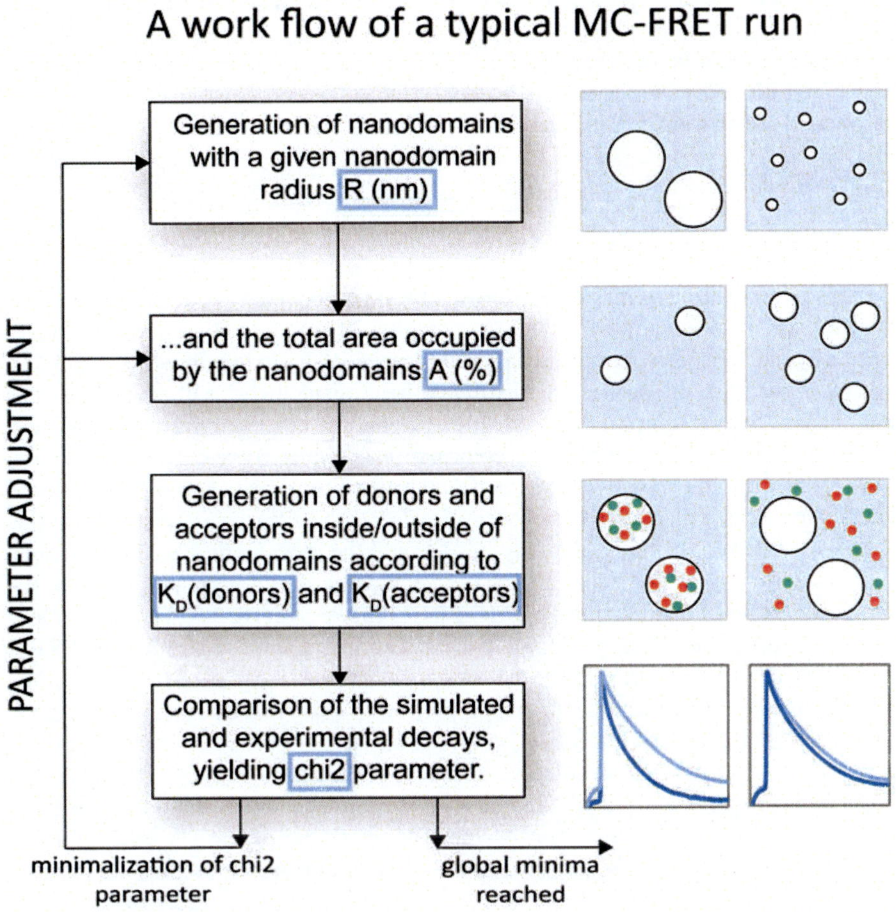

Fig. 2 A workflow of an MC-FRET simulation yielding the optimized nanodomain radius, the total area occupied by the nanodomains and optionally the distribution constants $K_D(D)$, $K_D(A)$

The outcome of each simulation step is the time interval Δt_i between the excitation and energy transfer event. By constructing a histogram of Δt_i intervals, the total survival probability function $G_{TOT}(t)$ is obtained, and the simulated decay of the donors quenched by the acceptors $F_{DA}(t)$ calculated $F_{DA}(t) = G_{TOT}(t)F_D(t)$. The simulated decay is fitted to the experimental one by varying the input simulation parameters, i.e., the domain radius $\langle R \rangle$, the area fraction occupied by the domains $\langle A \rangle$, and K_D(D/A).

3.2.2 What Nanodomain Sizes Can Be Resolved by MC-FRET? [18]

Nanodomains can be detected by MC-FRET only if their formation influences the shape of TRF decays of the donors quenched by acceptors. Thus, to characterize the resolution of MC-FRET it is convenient to introduce a parameter *RES*, defined as

$$\mathrm{RES} = \frac{1}{n}\sum_{i=1}^{n}\frac{\mathrm{abs}\left(F_{\mathrm{DA,homo}}(t_i) - F_{\mathrm{DA,hetero}}(t_i)\right)}{F_{\mathrm{DA,homo}}(t_i)}100. \tag{8}$$

This parameter quantifies the difference between the TRF decay recorded in the absence of nanodomains, $F_{\mathrm{DA,\ homo}}(t_i)$, where probes are distributed homogeneously, and the decay recorded in the presence of nanodomains, $F_{\mathrm{DA,\ hetero}}(t_i)$, where, in contrast, the probes are distributed heterogeneously. With the help of this parameter, the resolution of MC-FRET can be characterized as follows: (1) $RES \leq 10\%$ (red colour on the colormap bar of Fig. 3), yielding very similar TRF decays for the homogeneous and heterogeneous probe distributions. At the same time, this parameter value corresponds to the change in the steady-state intensity of donors, $\langle F_{\mathrm{DA,\ hetero}}\rangle/\langle F_{\mathrm{DA,\ homo}}\rangle$, and the average fluorescence lifetime, $\langle \tau_{\mathrm{DA,\ hetero}}\rangle/\langle \tau_{\mathrm{DA,\ homo}}\rangle$, of less than 10%. Such conditions are unfavourable for the characterization of nanodomains by MC-FRET. (2) $RES \in (10;20\rangle\%$ (yellow colour on the colormap bar of Fig. 3), enabling the detection of nanodomains by MC-FRET. This parameter value is accompanied by the change in $\langle F_{\mathrm{DA,\ hetero}}\rangle/\langle F_{\mathrm{DA,\ homo}}\rangle$ by more than 10% and the change in $\langle \tau_{\mathrm{DA,\ hetero}}\rangle/\langle \tau_{\mathrm{DA,\ homo}}\rangle$ that does not exceed 10%. (3) $RES \in (20;30\rangle\%$ (green colour on the colormap bar of Fig. 3), yielding clearly distinct TRF decays. This parameter value results in the change of $\langle F_{\mathrm{DA,\ hetero}}\rangle/\langle F_{\mathrm{DA,\ homo}}\rangle$ of up to 20% and the change in $\langle \tau_{\mathrm{DA,\ hetero}}\rangle/\langle \tau_{\mathrm{DA,\ homo}}\rangle$ that is above 10%. (4) $RES \in (30;60\rangle\%$ (cyan colour on the colourmap bar of Fig. 3), leading to the change in $\langle F_{\mathrm{DA,\ hetero}}\rangle/\langle F_{\mathrm{DA,\ homo}}\rangle$ of up to 40% and the change in $\langle \tau_{\mathrm{DA,\ hetero}}\rangle/\langle \tau_{\mathrm{DA,\ homo}}\rangle$ of more 10%. (5) Finally, if $RES > 60\%$ (blue colour on the colormap bar of Fig. 3), $\langle F_{\mathrm{DA,\ hetero}}\rangle/\langle F_{\mathrm{DA,\ homo}}\rangle > 30\%$ and $\langle \tau_{\mathrm{DA,\ hetero}}\rangle/\langle \tau_{\mathrm{DA,\ homo}}\rangle > 20\%$.

To calculate the parameter *RES* by Eq. (8), we generated the TRF decays for the homogeneous, $F_{\mathrm{DA,\ homo}}(t_i)$, and heterogeneous, $F_{\mathrm{DA,\ hetero}}(t_i)$, probe distributions assuming various $\langle R\rangle$, $\langle A\rangle$, K_{D}(D) and K_{D}(A). The parameter *RES* is plotted for Cases I, II and III as a function of $\langle R\rangle$ and $\langle A\rangle$ in Fig. 3. A careful inspection of the figure reveals that in case of an extremely high probe affinity to either the nanodomains or the region outside of them (characterized by K_{D}(D/A) = 1000 or K_{D}(D/A) = 1/1000) the resolution of MC-FRET is excellent for most combinations of $\langle R\rangle$ and $\langle A\rangle$. An exception represents a relatively narrow region defined by $\langle R\rangle \in \langle 5;20\rangle$ nm and $\langle A\rangle \in \langle 1;15\rangle$ % for K_{D}(D/A) = 0.001. To identify the Case that provides the best resolution, we generated the diagrams for K_{D}(D/A) = 10 or K_{D}(D/A) = 1/10, respectively (Fig. 3). Based on this comparison, Case II (K_{D}(D) > 1 and K_{D}(A) < 1) performs the best, followed by Case I (K_{D}(D) > 1 and K_{D}(A) > 1), whereas Case II (K_{D}(D) < 1 and K_{D}(A) < 1) performs the worst. Most important of all, even such low and physically realistic K_{D}(D/A) values provide sufficient

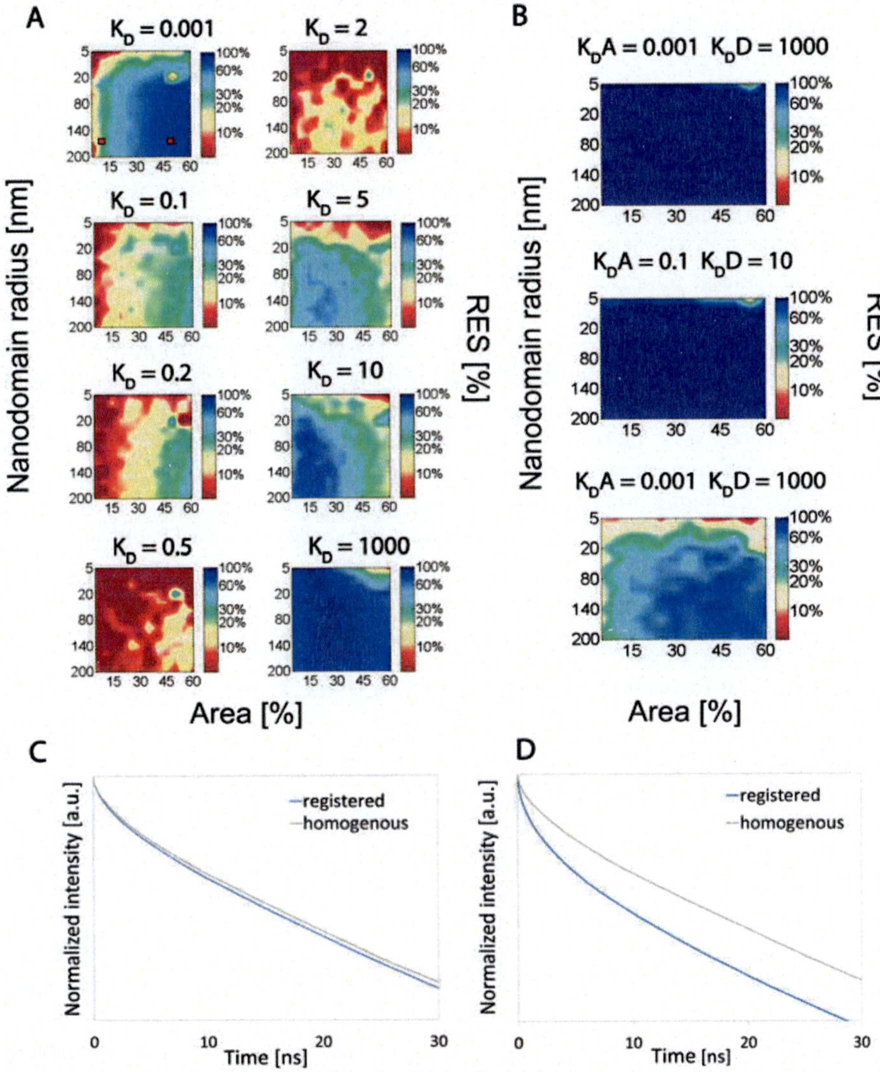

Fig. 3 The resolution of MC-FRET in the detection of membrane nanodomains as a function of the nanodomain radius and the fractional area occupied by the nanodomains shown for various K_Ds. The resolution is characterized by the parameter RES defined in the text. (**a**) $K_D(D) = K_D(A)$, (**b**) $K_D(D) \geq 1$ and $K_D(A) \leq 1$. (**c, d**) Time-resolved fluorescence decays for a homogeneous bilayer and the bilayer containing registered nanodomains characterized by RES = 19.5% (**c**) and RES = 83.8% (**d**). The decays were extracted from the positions in the resolution diagram depicted by solid red squares

resolution for all cases. Although by approaching the K_D(D/A) values even closer to one (K_D(D/A) $= 2$ or K_D(D/A) $= 1/2$) the resolution of MC-FRET continues to decline significantly (Fig. 3), Case III still offers a satisfactory resolution as opposed to Cases I and II.

Overall, the results presented so far show that choosing the right donor–acceptor pair is essential to achieve the best possible resolution. Therefore, we compiled a table that summarizes such donor–acceptor pairs, which, to the best of our knowledge, have so far been used in the detection and characterization of both micro- and nanodomains of different features (Table 1). It follows from the table that although there exist enough fluorescent probes that spontaneously localize both inside and outside the domains, fluorescent probes having a considerable affinity for the domains ($K_D \geq 10$) are essentially non-existent. According to our knowledge, the highest affinity exhibits headgroup-labelled gangliosides ($K_D \geq 20$) frequently used in our laboratory [8, 17, 25] or tryptophan labelled peptide GWALP23 ($K_D = 13$) [26]. On the contrary, plenty of probes excluded from the domains are available (Table 1). Most importantly, probes that have a modest affinity for the domains ($K_D \approx 10$) or the region outside them ($K_D \approx 0.1$) are sufficient for MC-FRET, and a satisfactory number of such probes have been characterized so far (Table 1).

3.3 *Resolving Inter-Leaflet Coupled from Inter-Leaflet Independent Nanodomains by MC-FRET*

Thanks to energy transfer that occurs not only within one bilayer leaflet but also from one leaflet to the other one, the measured FRET efficiency will also depend on the mutual organization of the nanodomains in the lower and upper leaflets [36]. Thus, MC-FRET, in contrast to other fluorescence approaches, offers excellent axial resolution and can be used to study inter-leaflet organization of nanodomains [17]. In principle, the following scenarios may arise (Fig. 4): (1) Nanodomains are perfectly registered across the bilayer leaflets (Fig. 4a), (2) the nanodomains in the upper and lower leaflets are independent from each other (Fig. 4b) and (3) nanodomains are anti-registered (Fig. 4c). In this case, the nanodomains in the two leaflets avoid each other, and thus, the nanodomains in the upper leaflet cannot occupy the lateral positions that have already been taken by the nanodomains in the lower leaflet, and vice versa. To identify the most likely scenario, the experimental TRF decays are fitted by the decays generated for the scenarios 1–3. As the most probable is classified the scenario that provides the best fit and the lowest value of the chi2 parameter. Since the resolution of MC-FRET depends significantly on the extent to which the decays for scenarios 1–3 differ, it can again be characterized by means of the parameter RES. The parameter, however, is now defined as

Table 1 The affinity of selected donor–acceptor pairs to various types of membrane lipid domains characterized by their K_D's, defined as K_D(D/A) = C(D/A inside of a domain)/C(D/A outside of a domain)

D	A	K_D(D)	K_D(A)	Lipid composition	R_0 (nm)	Domain phase	Reference
NBD-DLPE	DiIC12(3)	0.21	0.11	DLPC/DSPC (60:40)	5.2	Gel phase nanodomains	[27]
NBD-DLPE	DiIC12(3)	0.41	0.14	DLPC/DSPC (40:60)	5.2		
NBD-DLPE	DiIC18(3)	0.21	0.15	DLPC/DSPC (60:40)	5.2		
NBD-DLPE	DiIC18(3)	0.41	0.15	DLPC/DSPC (40:60)	5.2		
DHE	DiOC18 saturated	0.61± 0.02	7.4± 0.2	DLPC/DPPC (25:75)	5	Nanodomains of fluid and solid phases	[28]
DiOC18 saturated	DiIC18 unsaturated	7.4± 0.2	0.22± 0.01	DLPC/DPPC (25:75)	4.2 [29]		
DiOC18 saturated	DiIC18 saturated	7.4± 0.2	7.5± 0.3	DLPC/DPPC (25:75)	4.2 [29]		
DiOC18 unsaturated	DiIC18 saturated	0.23± 0.1	7.5± 0.3	DLPC/DPPC (25:75)	4.2 [29]		
NBD-DPPE	Rh-DOPE	3.68 ± 0.55	0.28± 0.08	POPC/PSM/Chol (33:33:33)	2.5/ 2.9Ld/ lo	Nanodomains with L_o character	[30]
TOE	DHE	0.11± 0.025	2.56	DSPC/DOPC/Chol (40/40/20)	2.4	Microdomains with L_o and L_d characters	[31]
TOE	DHE	0.1± 0.01	3.7	bSM/DOPC/Chol (40/40/20)	2.4		
DHE	Bodipy PC	2.56	0.09± 0.015	DSPC/DOPC/Chol (40/40/20)	2.8		
DHE	Bodipy PC	3.7	0.16± 0.025	bSM/DOPC/POPC/Chol (40/40/ 20)	2.8		
TOE	Bodipy PC	0.14± 0.025	0.1± 0.01	DSPC/DOPC/Chol (40/40/20)	2.5		

(continued)

Table 1 (continued)

D	A	K_D(D)	K_D(A)	Lipid composition	R_0 (nm)	Domain phase	Reference
TOE	Bodipy PC	0.12± 0.015	0.11± 0.025	bSM/DOPC/Chol (40/40/20)	2.5		
g-GM_1	r-GM_1	≥20	≥20	DOPC/Chol/SM (65–70/25/5–10)	5.87	Nanodomains with L_d character	[8]
g-GM_1	r-GM_1	10	10	DOPC/SM (85–90/10–15)	5.87		
CF-PEG-DSPE	Rh-PEG-DSPE	5	5	DOPC/Chol/SM (65/25/10)	6.63 [17]		
NBD-PE	Rh-PE	4.3	0.37	POPC/SM/Chol (33:33:33)	5	Nanodomains with L_o character	[18]
BODIPY-PC	Fast-DiI	0.1	0.1	DOPC/DSPC/Chol (30/30/40)	6.5		
DHPE-FITC	DHPE-TR	0.28	<<1	DPhPC/DPPC/Chol (40:35:25)		Nanodomains with L_o character	[32]
DHPE-OG488	DHPE-TR	2.1	<<1				
DHPE-NBD	DHPE-TR	2.2	<<1				
CTxB Alexa 488	DiD	6 ± 3	0.004 ± 0.002	DOPC/SM/Chol/DOPG/GM_1 (23–68/0–45/25/5/2)	5.4	Nanodomains with L_o character	[33]
NBD-PE	Rh-PE	4.3 ± 1.2	0.37 ± 0.06	PSM/POPC/Chol (33:33:33)	6.41	Nanodomains with L_o character	[34]
GWALP23 tryptophan	DHE	13 ± 2	Cca 1	DSPC/DOPC/Chol	6	Nanodomains with L_o and L_d characters	[35]
GWALP23 tryptophan	DHE	8 ± 1	Cca 1	bSM/DOPC/Chol	6		
NBD-cholesterol	Octadecyl rh-B	0.32	0.057	DMPC/Chol (80/20)	4.48	Microdomains with L_o and L_d characters	[26]

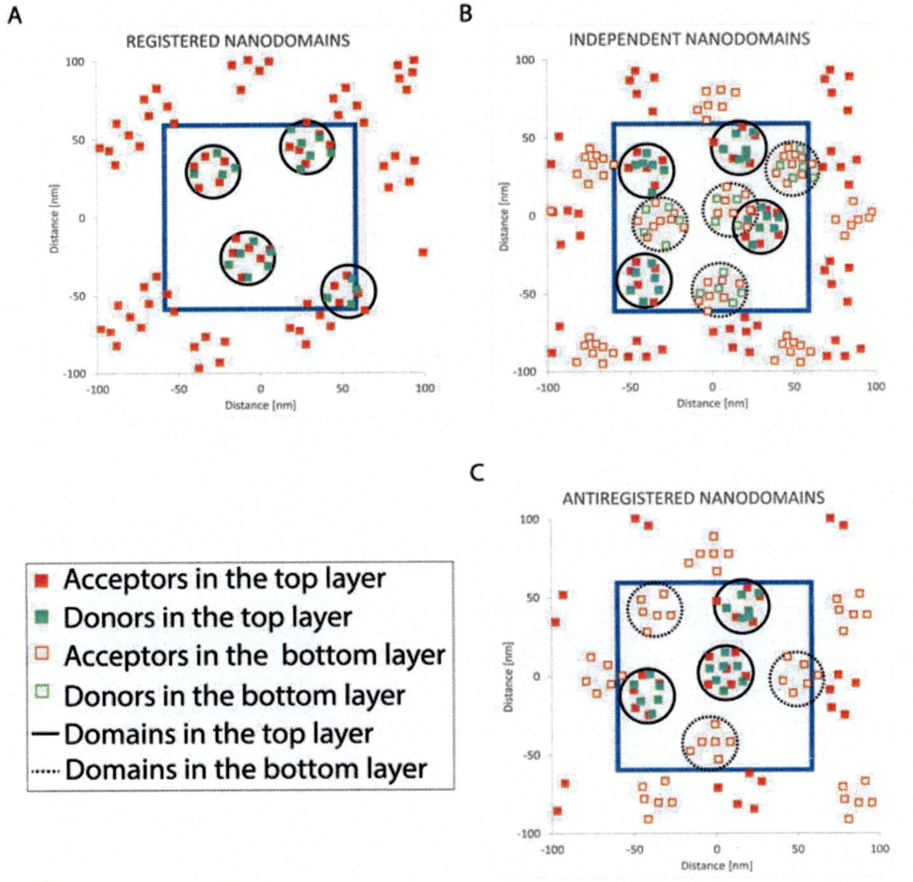

Fig. 4 Schematic planar projection of the simulated lipid bilayer containing: (**a**) perfectly registered; (**b**) independent and (**c**) anti-registered nanodomains. $K_D(D)$, $K_D(A) = 1{,}000$. The blue line indicates the boundaries of the basic simulation box

$$\mathrm{RES} = \frac{1}{n}\sum_{i=1}^{n} \frac{\mathrm{abs}\left(F_{\mathrm{DA,REG}}(t_i) - F_{\mathrm{DA,ANTI/INDEP}}(t_i)\right)}{F_{\mathrm{DA,REG}}(t_i)} 100 \tag{9}$$

Thus, RES expresses the difference between the TRF decay for the registered, $F_{\mathrm{DA,\ REG}}(t)$, and either the anti-registered, $F_{\mathrm{DA,\ ANTI}}(t)$, or independent, $F_{\mathrm{DA,\ INDEP}}(t)$, nanodomains.

In analogy to the cases discussed above, the MC-FRET resolution improves, i.e. the RES values increase, as the probe affinity to one of the phases increases (Fig. 5). Case II ($K_D(D) > 1$ and $K_D(A) < 1$) appears again to be the most robust one, exhibiting reasonable resolution for a wide range of K_D(D/A) values. If $K_D(D) = K_D(A)$ (Cases I and III), a higher probe affinity is generally required to

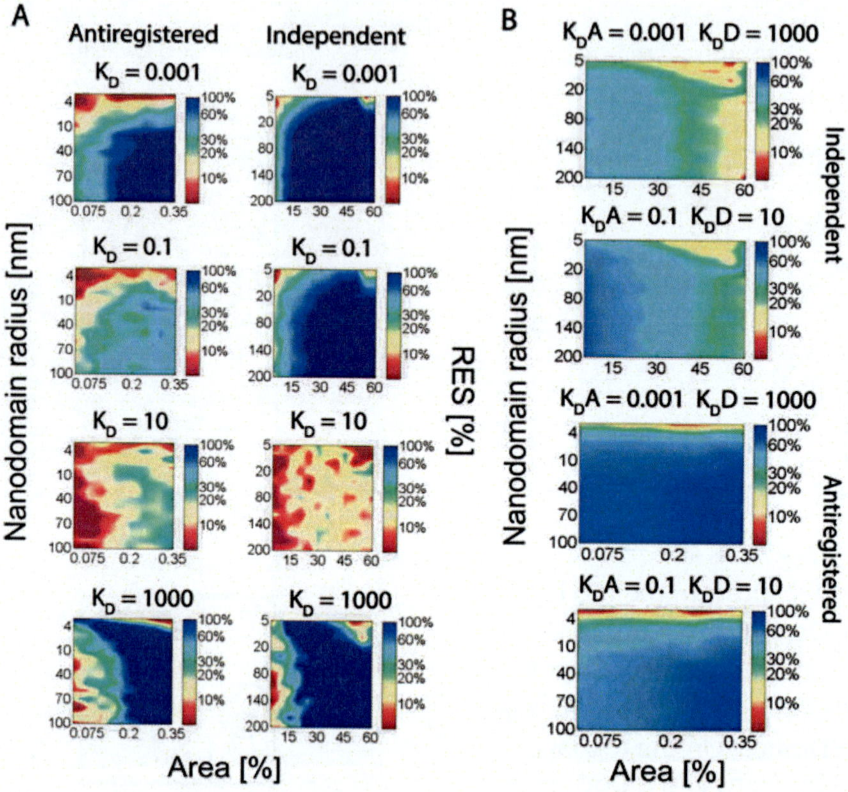

Fig. 5 The potential of MC-FRET to resolve anti-registered/independent from registered nanodomains as a function of the nanodomain radius and the area occupied by the nanodomains. **(a)** $K_D(D) = K_D(A)$, **(b)** $K_D(D) \geq 1$ and $K_D(A) \leq 1$

achieve a similar resolution as for Case II. Both Case I and III achieve the worst resolution for the nanodomains with $\langle R \rangle \leq 7$ nm and $\langle A \rangle \leq 10\%$, which is compensated by a reasonable resolution for the remaining $\langle R \rangle$ and $\langle A \rangle$ if K_D(D, A) ≥ 10 or K_D(D, A) $\leq 1/10$. Overall, the performed analysis identifies Case I as the worst one, yielding at relatively high K_D(D/A) $= 10$ a relatively low $RES \leq 20$ followed by Case III and Case II. Nevertheless, even the least favourable Case I can be used successfully. In Vinklárek et al. [17] we used headgroup-labelled gangliosides GM_1 (K_D(D, A) ≥ 20) to provide the first experimental evidence that lipid nanodomains of the sizes between 10 and 160 nm formed at various lipid ratios of 1,2-dioleoyl-sn-glycero-3-phosphocholine (DOPC), cholesterol, sphingomyelin (SM) and oxidized lipids (1-palmitoyl-2-(5′-oxo-valeroyl)-sn-glycero-3-phosphocholine (POVPC) and 1-palmitoyl-2-glutaryl-sn-glycero-3-phosphocholine (PGPC)) are inter-leaflet coupled. Recently, we used the same fluorescent probes to show that GM_1, GM_2 and GM_3 ganglioside nanodomains ($\langle R \rangle \in \langle 7; 120 \rangle$ nm and $\langle A \rangle \in \langle 38; 61 \rangle$ %), with the actual $\langle R \rangle$ and $\langle A \rangle$ parameters depending on the membrane lipid composition, are also inter-leaflet coupled [25]. Despite these few pieces of evidence, it remains largely unclear to what extent the registration of nanodomains is universal.

3.4 *Quantifying Protein Dimerization by MC-FRET*

Despite numerous advances in fluorescence lifetime microscopy, characterization of protein interactions in the membranes of living cells by FRET remains challenging. Consequently, affinity parameters such as dissociation constants for membrane protein dimerization are usually obtained only for purified recombinant proteins. They are usually in the form of detergent micelles or at best reconstituted to a simple model membranes. Both the systems are far from the native environment that may also participate in the dimerization.

The ability of FRET to deliver quantitative information is restricted to the bilayers that are perfectly planar, allowing for lateral protein surface concentration to be determined accurately. One of such systems are Giant Plasma Membrane Vesicles (GPMVs). Intrinsically, GPMVs represent a planar membrane system containing varying concentrations of various membrane proteins at the conditions closest to live cells. Quantitative FRET has been used on GPMVs for the first time by Chen and Sarabipour [37, 38]. Below, we are going to show how MC-FRET can be used to quantify protein dimerization in the membranes of GMPVs.

3.4.1 Workflow of MC-FRET for the Quantification of Protein Dimerization

The workflow of an MC-FRET simulation addressing protein–protein interactions is similar to that one previously described for nanodomains (see Sect. 3.2.2). In brief, the protocol requires the information about the concentrations of donor and acceptor labelled proteins, which, as shown later in Sect. 3.4.2, can be obtained directly from an MC-FRET experiment by an initial calibration. In the next step, the initial number of protein monomers and dimers in the simulation box (with the size of $100R_0 \times 100R_0$) is calculated. This is achieved by assuming a starting value of the dissociation constant K_d ($K_d = C^2_{\text{monomer}}/C_{\text{dimer}}$), where C stands for surface concentrations and the closest protein–protein distance L_0. Afterwards, monomers and dimers are generated on the bilayer surface, and by following the previously described working scheme (see Fig. 2 and Sect. 3.2.2), the simulated decay is computed. K_d and L_0 are optimized until the global minimum in the difference between the measured and simulated TRF decay has been reached. Alternatively, instead of the TRF decays, FRET efficiencies η can be used to optimize K_d and L_0. In this case, η depends on the quantum yields of donors in the presence, Q_{DA}, and absence Q_D of acceptors:

$$\eta = 1 - \frac{Q_{\mathrm{DA}}}{Q_{\mathrm{D}}}, \tag{10}$$

And thus in combination with Eqs. (6) and (7), we get:

$$\frac{Q_{\mathrm{DA}}}{Q_{\mathrm{D}}} = \frac{1}{M} \lim_{M \to \infty} \left(1 + \sum\nolimits_j \left(\frac{R_0}{R_j}\right)^6\right)^{-1}, \tag{11}$$

And

$$\frac{Q_{\mathrm{DA}}}{Q_{\mathrm{D}}} = \frac{1}{M} \lim_{M \to \infty} \left(1 + \sum\nolimits_j \left(\frac{R_0}{R_j}\right)^6 + 0.5 \times C_2 \times \left(\frac{R_0}{R_c}\right)^4\right)^{-1}, \text{for } j, \text{ where } |R_i - R_j| \leq 10R_0. \tag{12}$$

where M runs over all random excitations and over all configurations of the given number of monomers and dimers.

In the case of the TRF decay analysis, measurement on a single GPMV provides us with enough data for the comparison with the simulation. When using efficiencies, number of GPMVs needs to be examined to obtain enough data to be simulated. (How many depends on the precision of the FRET efficiency determination.)

3.4.2 Determination of Donor and Acceptor Surface Concentrations

The success of MC-FRET analysis relies fundamentally on precise determination of the surface concentration of proteins C. More specifically, the analysis requires the following parameters to be determined accurately: (1) the number of membrane proteins in the defined area and (2) the size of that area (to calculate C = number of membrane proteins in the area/the size of that area). In the native cellular membranes, it is almost impossible to determine the latter parameter accurately by fluorescence microscopy, as the membrane forms many inter-membrane contacts, lamellae, etc., which are far beyond optical resolution. Therefore, the work on GPMVs, composition-wise, the closest model system of a native biological membrane that are flat on the micrometre scale, appears inevitable [38]. Yet, it still represents a challenge. In our laboratory, we use fluorescence correlation spectroscopy (FCS) in the way depicted in Fig. 6.

1. *Determination of the number of proteins in the confocal spot*

   FCS allows for very precise determination of the number of molecules in the area illuminated by a tightly focussed laser beam [39]. This can be achieved by measuring temporal autocorrelation functions (ACFs) of the fluctuations in fluorescence intensity, which contain information about the number of fluorescent

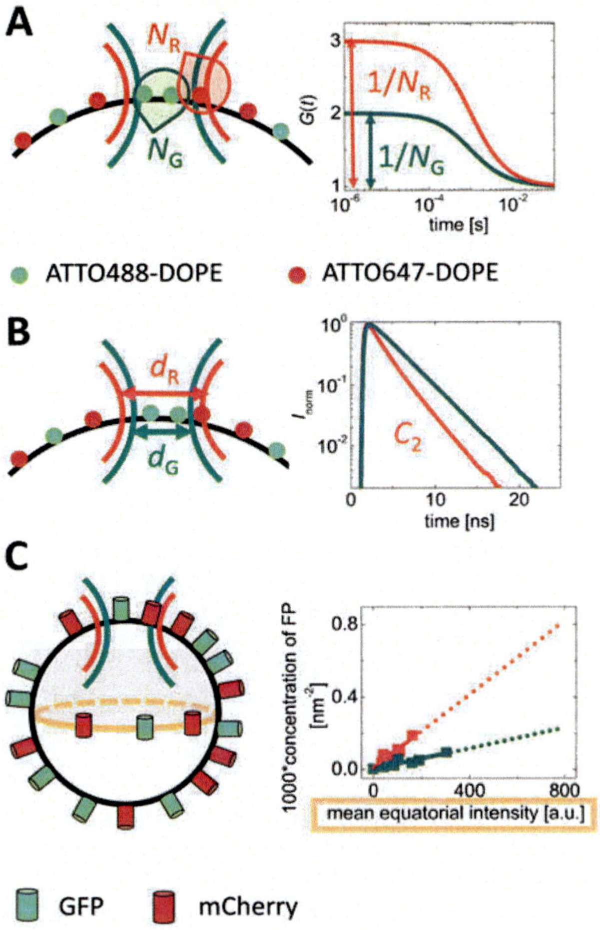

Fig. 6 Determination of the protein surface concentration. (**a**) Determination of the number of fluorescently labelled lipids within the confocal spot by FCS. (**b**) Determination of the surface concentration of homogeneously distributed lipids by FRET. (**c**) Relating mean equatorial fluorescence intensity of fluorescently labelled proteins to their surface concentration

particles (either FP-fused proteins or labelled lipids) and their dynamics in the confocal spot. As shown in Fig. 6a, the amplitudes of ACFs are inversely proportional to the number of diffusing fluorescent particles.

2. *Determination of the size of the confocal spot*

   The profile of the laser light projected on a planar membrane has a 2D Gaussian shape. To estimate the effective size of the beam which FCS counts fluorescent molecules from, one can make use of FRET and Baumann-Fayer model (introduced earlier in Sect. 2) applied to the FRET data acquired in the identical membrane as FCS (ideally the FCS and FRET analysis can be done on the identical dataset). The B.-F. model provides information about the surface concentration of homogeneously distributed, non-interacting acceptors. This concentration can be converted into the FCS-relevant beam size by using information about the number of acceptors in that spot (see step i)). The B.-F. model can be used only for the homogeneously distributed donors and acceptors that in addition fulfil the requirements on dynamics and isotropy. As this is not the case of FPs, in our laboratory, we determined the effective beam size using a pair of non-interacting headgroup-labelled lipids (ATTO488-DOPE and ATTO647N-DOPE) in a fully artificial system of giant unilamellar vesicles (GUVs) [21]. Overall, this calibration yields the effective diameter of the illuminated area: $d = 0.61 * \frac{\lambda}{N_A}$, where λ stays for the wavelength of the excitation light and N_A is the numerical aperture of the objective (Fig. 6b), and represents a powerful approach that provides geometrical characteristics being fully independent of the used fluorescence moiety.

3. *Determination of protein surface concentrations from equatorial intensity*

   MC-FRET experiments are often performed at high loads of acceptor labelled proteins where FCS cannot be applied. Therefore, as shown on Fig. 6c, it is beneficial to construct a calibration curve for the dependence of the mean equatorial fluorescence intensity on the protein surface concentration. By linear extrapolation of this dependence to higher surface concentrations, one can easily calculate (just by measuring the equatorial intensity without the need to employ FCS for each individual GUV) the protein surface concentration of an examined GPMV. Such a calibration dependence can be acquired either directly on GPMVs by varying the amount of plasmid DNA that encodes the protein of interest used for the transfection of cells, or, alternatively, isolated FPs (without the protein of interest) can be attached to the surface of GUVs doped with DGS-NTA(Ni) lipids via His-tag in large range of surface concentrations.

3.4.3 Dealing with Kappa Squared (κ^2)

The dynamic limit conditions considered so far are based on the assumption that donor and acceptor molecules rotate fast as compared to the donor fluorescence lifetime and their respective transition dipoles can sample the entire rotational space. It is, however, unclear to which extent this stringent assumption holds for barrel-shaped fluorescence proteins attached to the membrane of GPMVs (Fig. 7a, b). The theory of FRET accounting for slow and hindered dipole reorientations is fundamentally different as the distribution of distinct dipole orientations have to be taken into account [22, 40]. Recently, we have shown that, even in this case, the formalism introduced for the dynamic and isotropic regime can be used successfully [21].

Under the dynamic limit conditions, the value of κ^2 equals 2/3. In our work, we optimized the value of κ^2 by correcting the value of experimentally determined R_0 for the used donor–acceptor pair. R_0 depends on κ^2 according to $R_0 \sim \kappa^{\frac{1}{3}}$. Thus, the similarity between the experimentally determined and corrected value ($R_{0,\mathrm{corr}}$) of R_0 reports on to what extent our assumptions about dynamic and isotropic chromophore reorientations hold.

More specifically, we prepared GUVs consisting of POPC and 5 mol% of DGS-NTA(Ni) and attached to them fluorescently labelled fusion proteins containing a His-tag (Fig. 7a, b). We systematically varied the concentration of

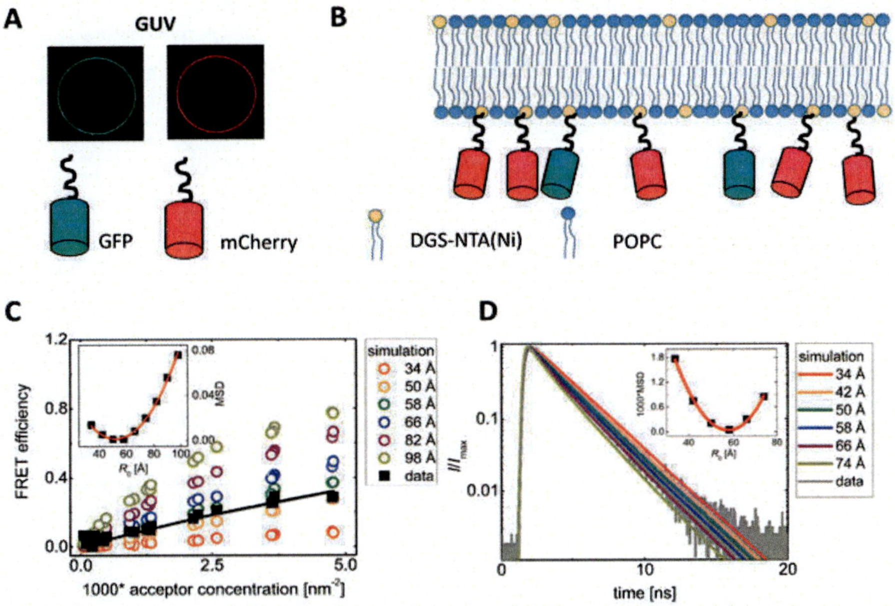

Fig. 7 Determination of the corrected Förster radius, $R_{0,\text{corr}}$. (**a**) Membranes of GUVs with reconstituted GFP (green) and mCherry (red) fluorescent proteins. (**b**) The non-interacting fluorescent proteins are attached to the membrane surface by the bond between DGS-NTA(Ni) (tag in the membrane) and His-tag (tag in the protein). (**c, d**) Determination of $R_{0,\text{corr}}$ by measuring (**c**) the efficiency of FRET or (**d**) TRF decays (grey line), respectively, for various concentrations of acceptors. The MC-FRET simulations were carried out for a set of distinct R_0 values and the resulting dependences were compared with the experimental data (in black) by minimizing the mean square displacement (MSD) values (inset)

acceptor proteins while keeping the concentration of donor proteins constant, and we recorded both the TRF decays and the fluorescence intensity in the equatorial plain of GUVs in both fluorescence channels. In this way, we could estimate the surface concentration of donor and acceptor proteins and the efficiency of FRET for each GUV. The efficiency of FRET is estimated by fitting the TFR decays by a sum of exponentials and using the following relationship

$$\eta = 1 - \sum_i A_{\text{DA}i}\tau_{\text{DA}i} / \sum_i A_{\text{D}i}\tau_{\text{D}i}. \tag{13}$$

Here, ($A_{\text{D}i}$ and $\tau_{\text{D}i}$) or ($A_{\text{DA}i}$ and $\tau_{\text{Da}i}$) are the preexponential factors and individual lifetime components in the absence or presence of acceptors [41].

As shown in Fig. 7c, d, we employed MC-FRET to generate both the FRET efficiencies η and the corresponding donor fluorescence decays F_{DA} for a set of different R_0 and compared those with the experimental data by minimizing the mean square displacement MSD. Whereas in the case of η generation MSD $= (1/N)\sum_{i=1}^{N}(\eta_{i\ \text{measured}} - \eta_{i\ \text{simulated}})^2$, and i runs over all GUVs that have been measured, in case of F_{DA} generation $MSD = (1/N)\sum_{i=1}^{N}(F_{DA,i\ measured} - F_{DA,i\ simulated})^2$ and i runs over all TCSPC channels. The best match between the simulated and experimental data was found for $R_{0,corr}$ in the range 50–58 Å. Of note, it was not necessary to optimize the closest distance between the two fluorophores (set to 30 Å) because the range of physically acceptable values was narrow and its impact on the results was insignificant [21].

Interestingly, we performed the experiment for different lengths of the linker between the fusion protein and the His-tag and found that there was no significant dependence of the FRET efficiency on the linker length [21]. The comparison of $R_{0,corr}$ with $R_0 = 52$ Å determined for dynamic isotropic conditions shows that the behaviour of the used donor/acceptor pair is well described by the formalism used. Of course, this does not mean that the chromophores of the investigated proteins behave as fast isotopically rotating dipoles. Instead, it suggests that the herein presented analysis is accurate enough for solving the dimerization issues that are geometrically resembling the system used here. This is of particular importance when protein–protein interactions in the membrane are studied (see the following Sect. 3.4.4).

3.4.4 Dimerizing and Non-dimerizing Membrane Proteins

Figure 8 shows a demonstrative example using two distinct membrane localized proteins: RHBDL2, a human rhomboid protease, which stays in its monomeric form [21, 42] (Fig. 8a), and the human glutamyl carboxypeptidase II (GCPII) that strongly dimerizes in the membrane via its extracellular interface [21, 43] (Fig. 8b). In both the cases, we have co-transfected HeLa cells with eGFP and mCherry fusions to the investigated proteins and formed GPMVs from these cells as described elsewhere [44, 45]. In line with what has been already described in the above paragraphs, we recorded TRF decays and fluorescence intensity in the equatorial plane, estimated the surface concentration of the green and red labelled proteins and performed the MC-FRET analysis.

In this analysis (see also Sect. 3.4.1), either the experimentally determined FRET efficiency measured on an ensemble of GPMVs is fitted to the simulated FRET efficiency (Fig. 8c, e), or, alternatively, an experimental TRF decay is fitted to those generated by MC-FRET (Fig. 8d, f). In both cases, L_0 – the closest protein–protein distance, related to the excluded surface, and K_d – the dissociation constant – were optimized and, importantly, both approaches provided similar results (Fig. 8g, j). While for RHBDL2, we did not see any sign of protein dimerization at the concentrations achievable by overexpression from transfected plasmids, for GCPII, dimers were detected even at the lowest protein surface concentrations used.

Altogether, efforts to characterize membrane proteins by thermodynamic parameters directly in native membranes have been made by us and others. The quantification, however, depends crucially on the determination of lateral protein concentrations. This can be sufficiently done only in limited number of cases as

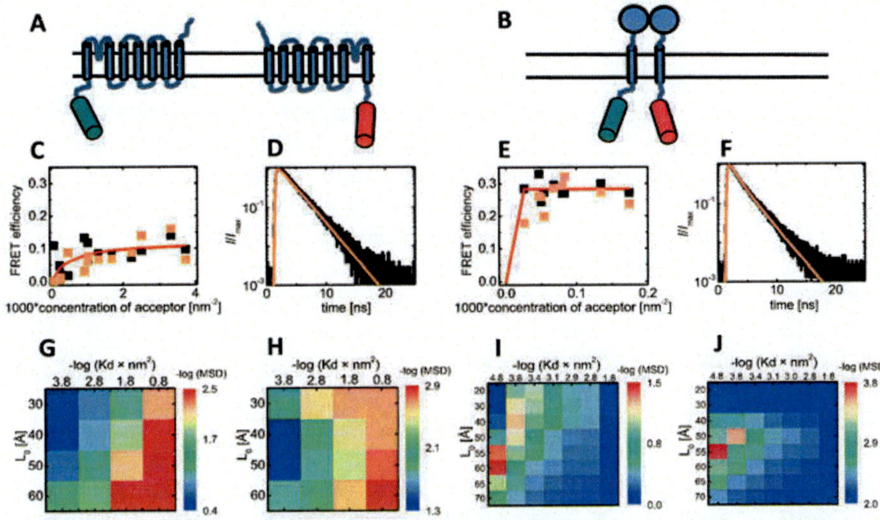

Fig. 8 Quantification of protein dimerization by MC-FRET. (**a**, **b**) A schematic drawing of (**a**) a non-dimerizing protein RHBDL2 and (**b**) a dimerizing protein GCPII. (**c**, **e**) Experimentally determined FRET efficiencies measured on individual GPMVs (black squares) versus FRET efficiencies computed by MC-FRET (orange squares) using as the input simulation parameters the optimal K_d and L_0 values plotted as a function of the acceptor concentration for (**c**) RHBDL2 and (**e**) GCPII. Red line is only intended to guide reader's eyes. (**d**, **f**) An experimental TRF decay (black line) versus a TRF decay generated for the optimal K_d and L_0 parameters (orange line) at given donor/acceptor concentration shown for a selected GPMV containing (**d**) RHBDL2 or (**f**) GCPIIGCPII, respectively. (**g–j**) MSD maps (i.e. 2D plots of MSD as a function K_d and L_0 parameters) generated by MC-FRET for the GPMVs containing (**g**, **h**) RHBDL2 or (**i**, **j**) GCPIIGCPII, respectively. MSD was calculated both by means of the FRET efficiencies (panels **g** and **i**) and TRF decays (panels **h** and **j**; see also the text for more details)

even the plasma membrane turns out to be too ruffled for area determination. Eventually, the experiments had to be carried out in GPMVs, model membranes that are compositionally closest to the native ones. These attempts however revealed important considerations that have to be made when membrane protein oligomerization is addressed even if the quantification is not the main goal. Firstly, as the FRET read-out is almost always present in the 2D systems, a negative control of a similar acceptor expression is essential for drawing conclusions. Secondly, the membrane attached FPs behave as if they were in the isotropic, dynamic regime. Moreover, we do not see any linker length dependence. Therefore, in planning the FRET experiment, there is no need to be overwhelmingly cautious about these aspects.

4 Conclusions

In summary, recent advances in fluorescence microscopy and spectroscopy have enabled to characterize organization of lipids and proteins in plasma membranes and their models with great spatiotemporal resolution. One of these techniques is the recently developed MC-FRET method, which can, by using standard fluorescence lifetime microscopy equipment and analysis tools based on Monte-Carlo simulations, (1) detect nanoscopic heterogeneities formed in lipid bilayers, (2) determine their sizes with the sensitivity down to only a few nanometres in radius, (3) quantify their membrane surface coverage and (4) characterize their mutual inter-leaflet organization. Moreover, it can also be used to study protein dimerization in the membrane by quantifying the dissociation constant for dimerization. Overall, the method reaches a nanometre resolution in all three directions, and thus offers an effective tool in the field of membrane biophysics.

Acknowledgements RŠ and BC acknowledge GAČR Grant 20-01401J. JH would like to acknowledge the support of the Czech Science Foundation by grant number 19-18917S. KS would like to acknowledge the Czech Science Foundation (grant number 18-09556S), and Operational Programme European Regional Development Fund (project no. CZ.02.1.01/0.0/0.0/16_019/0000729). BC would like to acknowledge Project No. SVV 260586 of the Charles University.

References

1. Bernardino de la Serna J, Schütz GJ, Eggeling C, Cebecauer M (2016) There is no simple model of the plasma membrane organization. Front Cell Dev Biol 4(September):1–17
2. Cebecauer M, Amaro M, Jurkiewicz P, Sarmento MJ, Šachl R, Cwiklik L et al (2018) Membrane lipid nanodomains. Chem Rev 118(23):11259–11297
3. Sarmento MJ, Ricardo JC, Amaro M, Šachl R (2020) Organization of gangliosides into membrane nanodomains. FEBS Lett 594(22):3668–3697. https://doi.org/10.1002/1873-3468.13871
4. Shi J, Yang T, Kataoka S, Zhang Y, Diaz AJ, Cremer PS (2007) GM1 clustering inhibits cholera toxin binding in supported phospholipid membranes. J Am Chem Soc 129(18):5954–5961
5. Sarmento MJ, Coutinho A, Fedorov A, Prieto M, Fernandes F (2014) Ca2 + induces PI(4,5)P2 clusters on lipid bilayers at physiological PI(4,5)P2 and Ca2 + concentrations. Biochim Biophys Acta Biomembr 1838(3):822–830
6. Sarmento MJ, Coutinho A, Fedorov A, Prieto M, Fernandes F (2017) Membrane order is a key regulator of divalent cation-induced clustering of PI(3,5)P2 and PI(4,5)P2. Langmuir 33(43): 12463–12477
7. Sjöholm J, Bergstrand J, Nilsson T, Šachl R, Von Ballmoos C, Widengren J et al (2017) The lateral distance between a proton pump and ATP synthase determines the ATP-synthesis rate. Sci Rep 7(April):1–12
8. Koukalová A, Amaro M, Aydogan G, Gröbner G, Williamson PTF, Mikhalyov I et al (2017) Lipid driven nanodomains in giant lipid vesicles are fluid and disordered. Sci Rep 7(1):5460. http://www.nature.com/articles/s41598-017-05539-y
9. Betzig E, Patterson GH, Sougrat R, Lindwasser OW, Olenych S, Bonifacino JS et al (2006) Imaging intracellular fluorescent proteins at nanometer resolution. Science 313(5793): 1642–1645

10. Rust MJ, Bates M, Zhuang X (2006) Sub-diffraction-limit imaging by stochastic optical reconstruction microscopy (STORM). Nat Methods 3(10):793–795
11. Hell SW, Wichmann J (1994) Breaking the diffraction resolution limit by stimulated emission: stimulated-emission-depletion fluorescence microscopy. Opt Lett 19(11):780
12. Piliarik M, Sandoghdar V (2014) Direct optical sensing of single unlabelled proteins and super-resolution imaging of their binding sites. Nat Commun 5(1):1–8
13. Young G, Hundt N, Cole D, Fineberg A, Andrecka J, Tyler A et al (2018) Quantitative mass imaging of single biological macromolecules. Science 360(6387):423–427
14. Owen DM, Magenau A, Williamson D, Gaus K (2012) The lipid raft hypothesis revisited - new insights on raft composition and function from super-resolution fluorescence microscopy. BioEssays 34(9):739–747
15. Cheng X, Smith JC (2019) Biological membrane organization and cellular signaling. Chem Rev 119(9):5849–5880
16. Šachl R, Johansson LB, Hof M (2012) Förster resonance energy transfer (FRET) between heterogeneously distributed probes: application to lipid nanodomains and pores. Int J Mol Sci 13(12):16141–16156
17. Vinklárek IS, Vel'As L, Riegerová P, Skála K, Mikhalyov I, Gretskaya N et al (2019) Experimental evidence of the existence of interleaflet coupled nanodomains: an MC-FRET study. J Phys Chem Lett 10(9):2024–2030
18. Šachl R, Humpolíčková J, Štefl M, Johansson LB, Hof M (2011) Limitations of electronic energy transfer in the determination of lipid nanodomain sizes. Biophys J 101(11):L60-2
19. King C, Sarabipour S, Byrne P, Leahy DJ, Hristova K (2014) The FRET signatures of noninteracting proteins in membranes: simulations and experiments. Biophys J 106(6): 1309–1317. https://doi.org/10.1016/j.bpj.2014.01.039
20. King C, Raicu V, Hristova K (2017) Understanding the FRET signatures of interacting membrane proteins. J Biol Chem 292(13):5291–5310
21. Škerle J, Humpolíčková J, Johnson N, Rampírová P, Poláchová E, Fliegl M et al (2020) Membrane protein dimerization in cell-derived lipid membranes measured by FRET with MC simulations. Biophys J 118(8):1861–1875
22. Baumann J, Fayer MD (1986) Excitation transfer in disordered two-dimensional and anisotropic 3-dimensional systems - effects of spatial geometry on time-resolved observables. J Chem Phys 85(7):4087–4107
23. Šachl R, Boldyrev I, Johansson LB (2010) Localisation of BODIPY-labelled phosphatidylcholines in lipid bilayers. Phys Chem Chem Phys 12(23):6027–6034
24. Amaro M, Šachl R, Aydogan G, Mikhalyov II, Vácha R, Hof M (2016) GM1 ganglioside inhibits b-amyloid oligomerization induced by sphingomyelin. Angew Chemie 55:9411–9415
25. Sarmento MJ, Owen MC, Ricardo JC, Chmelová B, Davidović D (2021) The impact of the glycan headgroup on the nanoscopic segregation of gangliosides. Biophys J 120:5530–5543
26. Loura LMS, Fedorov A, Prieto M (2001) Exclusion of a cholesterol analog from the cholesterol-rich phase in model membranes. Biochim Biophys Acta Biomembr 1511(2):236–243
27. Loura LMS, Fedorov A, Prieto M (2000) Partition of membrane probes in a gel/fluid two-component lipid system: a fluorescence resonance energy transfer study. Biochim Biophys Acta Biomembr 1467(1):101–112
28. Buboltz JT (2007) Steady-state probe-partitioning FRET: a simple and robust tool for the study of membrane phase behavior. Phys Rev E 76:021903
29. Charron DM, Zheng G (2018) Nanomedicine development guided by FRET imaging. Nano Today 18:124–136
30. Silva LC, De Almeida RFM, Castro BM, Fedorov A, Prieto M (2007) Ceramide-domain formation and collapse in lipid rafts: membrane reorganization by an apoptotic lipid. Biophys J 92(2):502–516. https://doi.org/10.1529/biophysj.106.091876
31. Enoki TA, Heberle FA, Feigenson GW (2018) FRET detects the size of nanodomains for coexisting liquid-disordered and liquid-ordered phases. Biophys J 114(8):1921–1935. https://doi.org/10.1016/j.bpj.2018.03.014

32. Bordovsky SS, Wong CS, Bachand GD, Stachowiak JC, Sasaki DY (2016) Engineering lipid structure for recognition of the liquid ordered membrane phase. Langmuir 32(47):12527–12533
33. Štefl M, Šachl R, Humpolíčková J, Cebecauer M, MacHáň R, Kolářová M et al (2012) Dynamics and size of cross-linking-induced lipid nanodomains in model membranes. Biophys J 102(9):2104–2113
34. De Almeida RFM, Loura LMS, Fedorov A, Prieto M (2005) Lipid rafts have different sizes depending on membrane composition: a time-resolved fluorescence resonance energy transfer study. J Mol Biol 346(4):1109–1120
35. Usery RD, Enoki TA, Wickramasinghe SP, Nguyen VP, Ackerman DG, Greathouse DV et al (2018) Membrane bending moduli of coexisting liquid phases containing transmembrane peptide. Biophys J 114(9):2152–2164. https://doi.org/10.1016/j.bpj.2018.03.026
36. Sarmento MJ, Hof M, Šachl R (2020) Interleaflet coupling of lipid nanodomains – insights from in vitro systems. Front Cell Dev Biol 8:284
37. Chen L, Novicky L, Merzlyakov M, Hristov T, Hristova K (2010) Measuring the energetics of membrane protein dimerization in mammalian membranes. J Am Chem Soc 132(10): 3628–3635
38. Sarabipour S, Nuala DP, Hristova K (2015) Characterization of membrane protein interactions in plasma membrane derives vesicles with quantitative imaging fret. Acc Chem Res 48(8): 2262–2269
39. Benda A, Beneš M, Mareček V, Lhotský A, Hermens WT, Hof M (2003) How to determine diffusion coefficients in planar phospholipid systems by confocal fluorescence correlation spectroscopy. Langmuir 19(10):4120–4126
40. Engström S, Lindberg M, Johansson LB (1988) Monte Carlo simulations of electronic energy transfer in three-dimensional systems: a comparison with analytical theories. J Chem Phys 89(1):204–213
41. Becker W (2012) Fluorescence lifetime imaging - techniques and applications. J Microsc 247(2):119–136
42. Kreutzberger AJB, Urban S (2018) Single-molecule analyses reveal rhomboid proteins are strict and functional monomers in the membrane. Biophys J 115(9):1755–1761
43. Schülke N, Varlamova OA, Donovan GP, Ma D, Gardner JP, Morrissey DM et al (2003) The homodimer of prostate-specific membrane antigen is a functional target for cancer therapy. Proc Natl Acad Sci U S A 100(22):12590–12595
44. Del Piccolo N, Placone J, He L, Agudelo SC, Hristova K (2012) Production of plasma membrane vesicles with chloride salts and their utility as a cell membrane mimetic for biophysical characterization of membrane protein interactions. Anal Chem 84(20):8650–8655
45. Sezgin E, Kaiser HJ, Baumgart T, Schwille P, Simons K, Levental I (2012) Elucidating membrane structure and protein behavior using giant plasma membrane vesicles. Nat Protoc 7(6):1042–1051. https://doi.org/10.1038/nprot.2012.059

Part IV
Applications of Fluorescence Spectroscopy to Protein Studies

Single-Molecule Fluorescence Spectroscopy of Intrinsically Disordered Proteins

Hagen Hofmann and Wenwei Zheng

Contents

Abstract The past two decades have seen a substantial leap forward in our understanding of intrinsically disordered proteins, in terms of both thermodynamics and dynamics, but also in terms of structural ensembles. From understanding the principles and biological importance of their solvent pliability up to characterizing their dynamics including an identification of the molecular origins of internal friction, single-molecule FRET experiments have been an important driver of this progress. By now, the methods and analysis tools in single-molecule FRET have grown to an extensive toolbox that allows a straightforward comparison of experiments with analytical theories and results of molecular simulations. This chapter summarizes the technologies behind single-molecule FRET experiments and molecular simulations together with the key findings on intrinsically disordered proteins.

H. Hofmann (✉)
Department of Chemical and Structural Biology, Weizmann Institute of Science, Rehovot, Israel
e-mail: hagen.hofmann@weizmann.ac.il

W. Zheng
College of Integrative Sciences and Arts, Arizona State University, Mesa, AZ, USA
e-mail: wenweizheng@asu.edu

R. Šachl, M. Amaro (eds.), *Fluorescence Spectroscopy and Microscopy in Biology*, Springer Ser Fluoresc (2023) 20: 403–442, https://doi.org/10.1007/4243_2022_38,
Published online: 3 January 2023

Keywords Intrinsically disordered proteins · Molecular simulations · Protein dynamics · Single-molecule FRET

1 Introduction

It often takes time for discoveries to settle in. After the first findings that proteins do not necessarily fold for function, intrinsically disordered proteins (IDPs) and intrinsically disordered regions (IDRs) changed our structure-based view on how proteins interact [1]. While initially being identified as "nasty" flexible extrusions of proteins that notoriously escaped X-ray structure determinations, we now know that IDRs and IDPs longer than 30 amino acids compose nearly 30–44% of the human proteome [1, 2] with functions as diverse as those of classical proteins. As fascinating this discovery has been, our understanding of the "why" and "how" is only slowly emerging. Why have IDPs evolved? How do they combine specificity and affinity without forming structured complexes? What evolutionary advantage do IDPs offer compared to folded proteins? In fact, prokaryotes that are nearly void of IDPs also populated our planet successfully. Answers to these questions are pending and it seems questionable whether they will ever be obtained in detail. IDPs (here synonymously also for IDRs) form molecular complexes upon binding to other proteins, DNA, RNA, or small molecules, but contrary to most folded proteins, IDPs are often multi-functional, which enables them to form hubs in protein interaction networks [1, 3, 4]. Similar to folded proteins, they are often chemically modified via post-translational modifications and their degradation rates in cells are substantially increased compared to folded proteins [5]. While some of these features, e.g., high turnover rates or efficient post-translational modifications, are explained by the lack of stable structures, their ability to interact with a multitude of different ligands [1, 3] is more difficult to understand. Although some IDPs indeed fold upon binding to ligands, thus forming specific binding interfaces in folding-coupled binding reactions, folding is not necessary for binding and specificity [6]. For instance, an extremely flexible protein complex bare of structure is formed between the two IDPs Prothymosin α and Histone H1 [7]. Other examples of high-affinity complexes that retain a large degree of disorder include the RNA-chaperone NCD and nucleic acids [8], Sic1 and Cdc4 [9], the Na^+/H^+ exchanger tail and ERK2 [10], nucleoporin tails and nuclear transport receptors [11], and the cytoplasmic tail of E-cadherin and β-catenin [12].

Clearly, structure determinations based on X-ray or cryogenic electron microscopy (cryo-EM) do not provide detailed information about this enormous flexibility such that mainly four methods have been used in the past to characterize the behavior of IDPs: NMR (nuclear magnetic resonance), SAXS (small-angle X-ray scattering), SANS (small-angle neutron scattering), and fluorescence-based methods, particularly single-molecule Förster resonance energy transfer (smFRET). In fact, the past two decades have seen an overwhelming number of studies in which smFRET approaches provided insights into the physical principles that dictate the behavior of IDPs and unfolded proteins such as their environmental sensitivity [13–32] and their dynamics [23, 33–37].

Here, we discuss some of the common smFRET methods used to characterize the size and the dynamics of IDPs. Our particular goal is to discuss the principles behind dynamic smFRET approaches and how they can be linked to analytical and numerical models of IDPs and unfolded proteins. Given the excellent introduction into the principles of FRET in Chapter "Laboratory instrumentation", we will not repeat the basic theory to avoid redundancy. We start by introducing technical aspects of smFRET experiments in Sect. 2, followed by a discussion of findings, theories, and simulation approaches to characterize IDPs using smFRET experiments (Sect. 3). In Sect. 4, we discuss smFRET methods and analytical theories used to study the dynamics of IDPs.

2 Technical Aspects of smFRET Experiments

2.1 *Confocal smFRET Experiments*

Before we discuss how exactly smFRET experiments help to characterize IDPs, we would like to remind the reader of a few technical aspects of single-molecule spectroscopic techniques. Based on the instrumentation, single-molecule experiments can roughly be divided into two categories: camera-based and confocal experiments. Camera-based imaging experiments offer the advantage to observe many surface-tethered molecules simultaneously. Yet, the time-resolution of such experiments is in the order of 1–10 ms, depending on the camera used (sCMOS or EM-CCD), which is insufficient to capture the very fast sub-microsecond dynamics of disordered proteins. Hence, most insights into IDPs and their behavior were obtained with confocal microscopes that offer a much higher time-resolution. With the development of detectors with high detection efficiencies (SPADs – single-photon avalanche detectors) and single-photon counting devices [38, 39], the time-resolution of confocal smFRET experiments is only limited by the counting precision of a few picoseconds. The design of a confocal microscope is rather simple (Fig. 1a). An excitation laser is focused through an objective of high numerical aperture (*NA*) to a diffraction limited spot whose diameter is given by $d = \lambda/2NA$. Since typical numerical apertures are in the order of 1.35–1.45, the confocal spot size is in the order of several hundreds of nanometers, i.e., small compared to many cellular structures, but very large compared to the size of proteins (2–5 nm) (Fig. 1b). The fluorescence of an emitter in the confocal spot is collected via the same objective, imaged on a pinhole to remove out-of-focus fluorescence, before it is directed onto a SPAD. Clearly, smFRET experiments with two colors (donor and acceptor) require two detectors and often even four detectors are being used to obtain information about the rotational degrees of freedom of protein-attached dyes in fluorescence anisotropy single-molecule measurements.

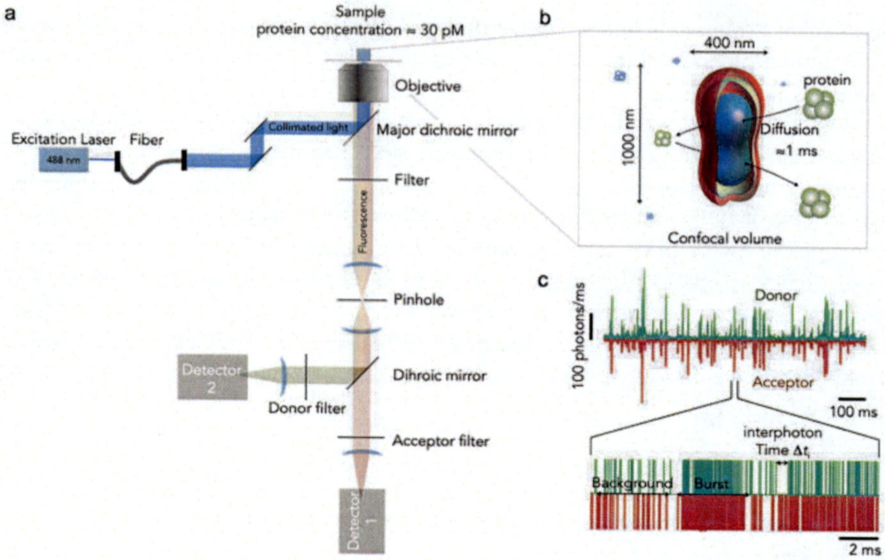

Fig. 1 The principle of confocal single-molecule FRET experiments. (**a**) Scheme of a confocal microscope. Excitation light (blue) from a laser is focused via an objective into the sample solution. Fluorescence photons from the sample are collected via the same objective, filtered, imaged on a pinhole, and finally recorded with single-photon avalanche detectors (SPADs). (**b**) The confocal volume created by the objective has a higher intensity in the center (blue) and lower intensity at the periphery [40]. The diffusion time of molecules through this volume is in the order of a millisecond. (**c**) Example of the raw data of a smFRET experiment. Molecules diffusing in and out of the confocal volume cause bursts of donor and acceptor photons in the time trace, which is shown here at a low resolution of 1 ms (upper panel). At higher time-resolution (0.1 μs), individual photons can be distinguished in the time trace (lower panel). The identification of bursts is based on a threshold of inter-photon times

In the past, most smFRET experiments on IDPs and unfolded proteins were performed with freely diffusing molecules. These types of experiments are relatively easy to realize and typically capture timescales from hundreds of picoseconds (the rotational motions of dyes attached to a protein) up to several milliseconds. The raw data of these experiments are photon traces that contain the arrival times of photons on the detectors (Fig. 1c). The construction of FRET histograms therefore requires an identification of the photons that were emitted by a molecule while it was diffusing through the confocal spot, thus resulting in a burst of photons. In addition, these raw photon counts have to be corrected for a number of experimental imperfections. Importantly, state-of-the-art burst-search algorithms already take these corrections into account [30, 41], which minimizes the risk of biasing the distribution of FRET efficiencies.

2.2 Correction Parameters in smFRET Experiments

In the following, we describe the correction of FRET values for two-color FRET experiments. However, the correction methods are general and also applicable to multi-color experiments. To determine correct FRET efficiencies (E) in a smFRET experiment with the channels 1 (acceptor) and 2 (donor), the raw photon counts (n_1 and n_2) of an individual molecule have to be corrected by the differences in brightness of the two dyes, cross-talk between the channels, background photons, and direct excitation of the acceptor at the wavelength used to excite the donor. Two main approaches are currently being used for these corrections: the determination of correction factors in separate experiments [30, 42, 43] and the determination within the same experiment [44].

Let us start with the separate-sample method. Here, the correction parameters are determined from two separate samples of the donor and acceptor dyes in which their concentrations are adjusted such that both samples have an absorbance of 0.1 at the wavelength of the donor excitation laser [42]. The fluorescence signal of both samples after donor excitation will then be measured in channel 1 and 2, and three correction factors can be determined: γ which accounts for the different detection probabilities of photons from the two dyes, β_{21} and β_{12}, the leakage of donor photons from channel 2 into the acceptor channel 1, and the leakage of acceptor photons (channel 1) into the donor channel 2, respectively. In addition, the probability α to directly excite the acceptor dye at the excitation wavelength specific for the donor can be computed from the extinction coefficients of the dyes at the donor excitation wavelength (see below). If n_1 and n_2 are the detected raw photon counts for a molecule in channel 1 and 2, respectively, and b_1 and b_2 are the background rates in both channels (in units of photon/s), the corrected photon counts for acceptor and donor (n'_D, n'_A) of this molecule are given by

$$\begin{pmatrix} n_A \\ n'_D \end{pmatrix} = \begin{pmatrix} 1 & -\beta_{21} \\ -\beta_{12} & \gamma \end{pmatrix} \begin{pmatrix} n_1 - b_1 T \\ n_2 - b_2 T \end{pmatrix} \text{ and } n'_A = n_A - \alpha\left(n'_D + n_A\right) \tag{1}$$

Here, T is the duration of a burst, which typically is in the order of a millisecond. As an example, for the frequently used FRET-pair Alexa Fluor 488 as donor and Alexa Fluor 594 as acceptor, our lab obtained $\gamma = 1.12 \pm 0.09$, $\beta_{21} = 0.050 \pm 0.003$, and $\beta_{12} = 0.0021 \pm 0.0004$ over 5 years with 21 measurements of these correction factors. As mentioned above, the probability of directly exciting the acceptor dye at the donor excitation wavelength is given by $\alpha = \epsilon_A/(\epsilon_A + \epsilon_D)$ where ϵ_A and ϵ_D are the extinction coefficients of the dyes at the donor excitation wavelength. For the dye pair mentioned above, one finds $\alpha = 0.049$. Hence, the correct FRET efficiency of an identified molecule is given by

$$E = \frac{n'_A}{n'_A + n'_D}. \quad (2)$$

Importantly, to prevent a bias in sampling molecules with different FRET efficiency, the corrections (Eq. 1) have to be taken into account during the identification of molecules (bursts) in the raw photon trace. Commonly, bursts are identified from the measured photon traces following Eggeling et al. [41] and Hoffmann et al. [30]. If $\Delta t_i = t_i - t_{i-1}$ is the inter-photon time of the i-th photon (Fig. 1c), the photon is retained if $\Delta t_i \leq \gamma_j \Delta t_{\max}$, in which $\Delta t_{\max}$ is an inter-photon time threshold set by the researcher, typically ~100 μs, and γ_j is the correction factor of the ith photon detected in channel $j = [1, 2]$, i.e., $\gamma_1 = 1$ and $\gamma_2 = 1.12$ with the definition in Eq. 1. The algorithm then proceeds to the next photon $i + 1$, stops after n photons once $\Delta t_{i+n} > \gamma_j \Delta t_{\max}$, and provides the total length of the burst by $T = t_{n-1} - t_{i-1}$. The photon number in the resulting photon string is now corrected via Eq. 1 using estimated background rates b_1 and b_2. The initial guess of b_1 and b_2 is given by all detected photons in channel 1 and 2, respectively, divided by the total measurement time. A burst is finally identified if $n_A + n_D > n_T$ where again, n_T is a photon threshold that has to be set. Based on our experience, the photon threshold to identify a burst (molecule) is in the order of 30–100 photons in most smFRET experiments. In the next iteration, the photons belonging to each identified burst are removed from the photon trace and a new guess for b_1 and b_2 is computed based on the remaining photons. Subsequently, the burst search is performed again with updated background rates. This procedure converges after three iterations to the correct background rates and a constant number of identified bursts.

The disadvantage of the separate-sample method is that experiments with different samples (proteins) might require different correction factors since the quantum yield of the dyes and therefore γ also depend on the chemical environment. Similarly, experiments at different solvent conditions might change the brightness of donor and acceptor differently, thus leading to errors in the correct determination of FRET efficiency if a single set of correction factors is being used. An alternative method is based on ALEX (alternating laser excitation) [45] or PIE (pulsed-interleaved excitation) [46] excitation schemes. Here, the donor and the acceptor are excited in an alternating manner: a period of donor excitation is followed by a period of acceptor excitation with a second laser (Fig. 2a). If switching between both excitation sources is much faster than the diffusion of a molecule through the confocal spot, important additional information can be obtained. For example, by exciting the acceptor dye directly in a period, molecules that do not contain an active acceptor due to photo-bleaching or imperfect labeling can be identified and removed from the further analysis. To this end, a second parameter, the stoichiometry ratio, defined via

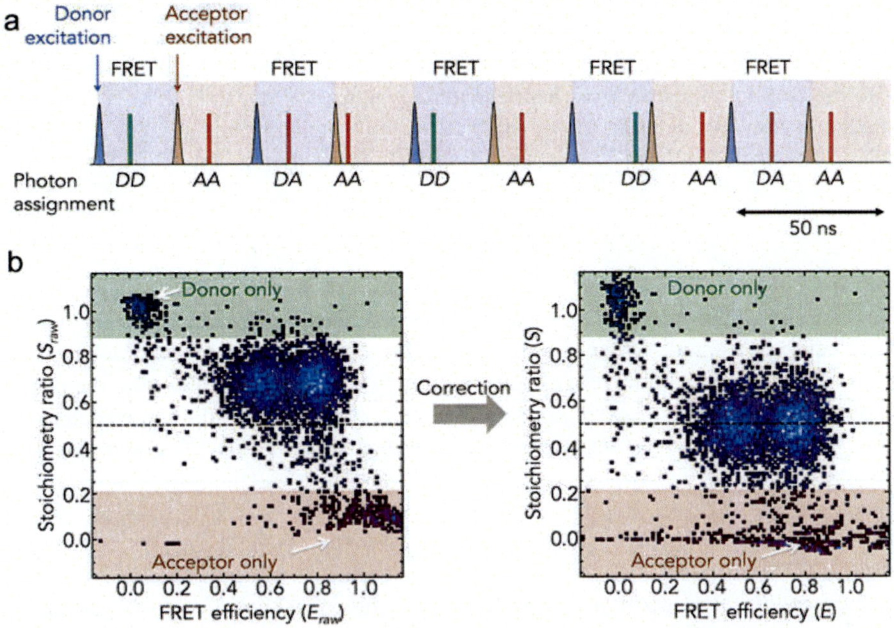

Fig. 2 Pulsed-interleaved single-molecule detection (PIE). (**a**) Excitation scheme of a PIE experiment. Excitation lasers for donor (blue peaks) and acceptor (red peaks) alternate during the experiment. Photons after donor excitation (*DD* or *DA*) are used to compute FRET efficiencies whereas photons after acceptor excitation (AA) are used to compute stoichiometry ratios. (**b**) Stoichiometry versus FRET maps for a sample containing four populations of molecules: donor-only molecules (green area), acceptor-only molecules (red area), and two FRET populations (white area). The maps are shown before (*left*) and after (*right*) correction of the photon counts, the FRET population is at $S = 0.5$

$$S = \frac{n'_{DD} + n'_{DA}}{n'_{DD} + n'_{DA} + \gamma_{PIE} n_{AA}}, \tag{3}$$

is computed for each molecule. Here, the first subscript indicates the excitation and the second subscript indicates the emission, i.e., n'_{DA} indicates the fully corrected acceptor photons after donor excitation. The parameter γ_{PIE} is a correction factor to account for the different excitation intensities for donor and acceptor. A two-dimensional plot of S versus E for all identified molecules now provides valuable information (Fig. 2b). For example, subpopulations of molecules with different FRET efficiency should show the same stoichiometry ratio if properly corrected. How is the correction performed? A detailed discussion of the method can be found in Hellenkamp et al. [44], which summarizes the work by the labs of Shimon Weiss [45, 47, 48] and Don Lamb [46]. In a first step, molecules are identified without any correction. A two-dimensional $E - S$ plot of the uncorrected FRET efficiencies and stoichiometry ratios immediately reveals the acceptor direct excitation probability and the cross-talk probability. For example, the population of molecules without an acceptor dye, often called donor-only molecules, should show a true mean FRET efficiency $\langle E \rangle = 0$. Yet, due to spectral cross-talk, some donor photons leak into the acceptor channel and the uncorrected mean FRET efficiency of the donor-only populations $\langle E_{\mathrm{raw}} \rangle$ (only corrected by background) will be located at $\langle E_{\mathrm{raw}} \rangle > 0$ (Fig. 2b, left). The cross-talk probability is then given by

$$\beta = \frac{\langle E_{\mathrm{raw}} \rangle}{1 - \langle E_{\mathrm{raw}} \rangle}. \tag{4}$$

Similarly, the effect of direct excitation of the acceptor by the donor excitation laser can be obtained. Acceptor-only molecules should have a mean stoichiometry $\langle S \rangle = 0$. Yet, due to a non-zero chance to directly excite the acceptor with the donor excitation laser, the uncorrected stoichiometry of acceptor-only molecules is at $\langle S_{\mathrm{raw}} \rangle > 0$ (Fig. 2b, left). The acceptor direct excitation coefficient is then given by

$$\mathrm{d} = \frac{\langle S_{\mathrm{raw}} \rangle}{1 - \langle S_{\mathrm{raw}} \rangle}. \tag{5}$$

Importantly, the quantity d is different from the quantity α used in the separate-sample method in Eq. 1. However, the two are directly related via

$$\mathrm{d} = \frac{\alpha}{1 - \alpha} \gamma_{PIE}. \tag{6}$$

The correction for cross-talk and acceptor direct excitation then gives a partially corrected acceptor photon number n_{DA} via

$$n_{DA} = n_{D1} - \beta n_{D2} - \mathrm{d} n_{A1}. \tag{7}$$

Again, the first subscript indicates the excitation source (D-excitation or A-excitation) and the second subscript indicates the emission, i.e., channel 1 or 2 with channel 1 being the acceptor channel. With the definitions $n_{DD} = n_{D2}$ and $n_{AA} = n_{A1}$, we can now compute stoichiometries and FRET efficiencies corrected for background, cross-talk, and acceptor direct excitation for each molecule. These partially corrected stoichiometries (S_p) and FRET efficiencies (E_p) depend on the true FRET efficiency (E) and the correction factors γ and γ_{PIE} via

$$E_p = \frac{n_{DA}}{n_{DA} + n_{DD}} = \frac{E}{E + (1 - E)/\gamma} \text{ and } S_p = \frac{n_{DA} + n_{DD}}{n_{DA} + n_{DD} + n_{AA}}$$
$$= \frac{E\gamma + (1 - E)}{E\gamma + (1 - E) + \gamma/\gamma_{PIE}}. \tag{8}$$

From the resulting two-dimensional map (E_p, S_p), we obtain the two missing parameters γ and γ_{PIE} by combining Eq. 8 as

$$\langle S_p \rangle = \frac{1}{1 + \beta_{PIE}(\langle E_p \rangle + \gamma - \gamma \langle E_p \rangle)} \text{ with } \beta_{PIE} = \frac{1}{\gamma_{PIE}}. \tag{9}$$

Here, the bracket $\langle \ldots \rangle$ indicates an average over a subpopulation of molecules. Eq. 9 defines a curve in the $E_p - S_p$ map that intersects the mean position of a subpopulation of molecules. Yet, Eq. 9 has two unknown parameters (γ and γ_{PIE}), which means that their determination requires two positions (molecule populations) in the (E_p, S_p) map. The correction method is therefore only applicable for multimodal FRET histograms, i.e., histograms with at least two subpopulations of different FRET efficiencies (Fig. 2b). If this is not the case, a separate sample, e.g., a mixture of double-stranded DNA, labeled with donor–acceptor at different positions can be used as calibration sample to apply this procedure [44]. Once all correction parameters have been obtained, the fully corrected photon counts for each molecule can be obtained using Eq. 1.

3 Single-Molecule FRET: A Blurry Window into Molecular Disorder

3.1 Compaction and Expansion of Unfolded and Disordered Proteins Probed with smFRET

Ensemble fluorescence techniques, particularly when combined with fast mixing such as in a stopped-flow apparatus, provided the first hint that unfolded proteins compact after quickly diluting out chemical denaturants such as guanidinium chloride (GdmCl) or urea. Yet, at the time, it was unclear whether these so-called burst phases in protein folding experiments resembled a non-cooperative compaction of an unfolded protein or whether it was rather indicative of the fast formation of a structured, i.e., partially folded intermediate. The first clear indication that the dimension of an unfolded protein, measured in terms of the distance between chemically attached FRET donor (D) and acceptor (A) fluorophores, came in 2000 by the group of Shimon Weiss [13]. Using a confocal microscope, the authors monitored D-A labeled chymotrypsin inhibitor molecules while the molecules were randomly diffusing through the confocal volume of their microscope. The experiments revealed a clear bimodal distribution of FRET efficiencies $P(E)$ that included a peak resulting from folded molecules at high FRET and a peak from unfolded molecules at lower FRET efficiency. With the FRET efficiency being a proper molecular reaction coordinate (collective variable) in the sense that it is related to the distance between donor and acceptor dyes according to

$$E(r) = \frac{R_0^6}{R_0^6 + r^6}. \tag{10}$$

where R_0 is the Förster distance and r is the distance between D and A, the experiment provided the distribution of folded and unfolded molecules directly, without requiring fast mixing technologies. Importantly, a change in the concentration of GdmCl caused a change in the average FRET position of the unfolded peak, which indicated that the dimension of unfolded polypeptide chains responds to changing solvent conditions.

Given this result, one is tempted to equate the measured FRET histogram $P(E)$ with a Boltzmann distribution that reflects a potential of mean force $V(E)$ via $V(E) = -\ln P(E)$. Yet, nothing could be further from the truth. First, the average time it takes for a nanometer-sized molecule in water to diffuse through the confocal volume (burst time) is in the order of a millisecond. This means that conformational rearrangements faster than 1 ms will be averaged such that the resulting FRET efficiency E for a single molecule is in fact a time-averaged quantity. Second, the numbers of D- and A-photons (n'_D and n'_A) are only in the order of ~100, which causes a substantial uncertainty in the determination of FRET values of individual molecules. Hence, the width of a FRET-peak is not necessarily related to the width of the underlying distribution of DA distances. Instead, it is in large parts determined by shot noise. The expected width of the FRET histogram that results from shot-noise σ_{noise} can be estimated [49] via

$$\sigma_{\text{noise}}^2 = \langle E^2 \rangle - \langle E \rangle^2 = \langle E \rangle (1 - \langle E \rangle) \langle {n'}_0^{-1} \rangle \leq \langle E \rangle (1 - \langle E \rangle) n_T^{-1}. \tag{11}$$

Here, $n'_0 = n'_D + n'_A$ is the total number of photons obtained for a molecule, the bracket $\langle \ldots \rangle$ indicates the average over a population of molecules, and n_T is the photon threshold to identify a molecule. For example, with a typical threshold of 50 photons and a FRET-peak centered at $\langle E \rangle = 0.5$, one finds $\sigma_{\text{noise}} \leq 0.07$, which is a substantial contribution to the width of a FRET histogram. In the extreme case in which the interconversion of different conformations is much faster than the diffusion time of a molecule through the confocal spot, such as for IDPs and unfolded proteins, shot noise is often the dominating contributor to the width in the FRET histogram (Fig. 3a).

Contrary to the width, the average position of the FRET distribution $\langle E \rangle$ contains distance information, which has been used to quantify the extreme solvent-sensitivity of unfolded proteins. The first well-studied example of this sensitivity was given by the Eaton-lab, who demonstrated that the unfolded state of the cold shock protein (Csp*Tm*) compacts dramatically with decreasing concentrations of GdmCl [14, 50]. These experiments were confirmed with many other unfolded and disordered proteins in the following years [15–18, 28], indicating that an expansion or compaction of unfolded proteins with changing solvent conditions is a rather generic polymer effect. A quantification of this change in dimension in terms of average donor–acceptor distances is possible by expressing the mean FRET

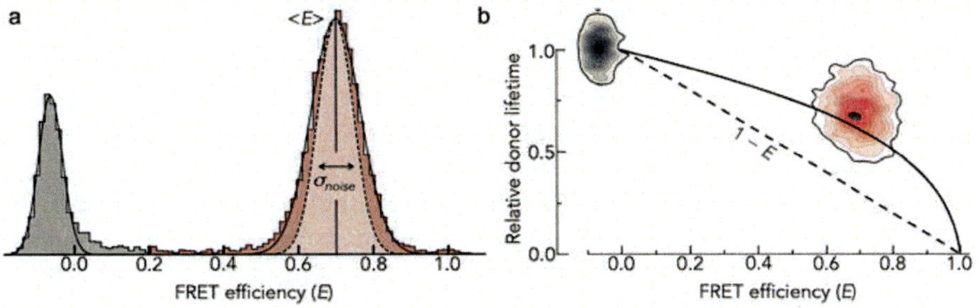

Fig. 3 Identifying conformational heterogeneity in IDPs. (**a**) SmFRET histogram of the intrinsically disordered DNA-binding domain of c-Myc. The histogram shows a single population of molecules (red) whose width is mainly determined by shot noise (dashed line). Molecules lacking an active acceptor are shown in gray. (**b**) 2D-map of the relative donor fluorescence lifetime τ_{DA}/τ_D and FRET efficiency for the data shown in (**a**). The dashed line is the prediction from Förster theory for the case in which all molecules have the same donor–acceptor distance (Eq. 18). The solid line is obtained for the case of a Gaussian distance distribution that is sampled at timescales much slower than the fluorescence lifetime (Eq. 19). Although the FRET distribution of c-Myc molecules is very narrow (**a**), conformational heterogeneity is identified in the lifetime-FRET map (**b**)

efficiency $\langle E \rangle$ as an average over the conformational distribution $P(r)$ of a polypeptide chain according to

$$\langle E \rangle = \int_a^L P(r)E(r)\mathrm{d}r. \tag{12}$$

The integration ranges from the closest possible donor–acceptor distance a to the maximum distance, which for a disordered polypeptide chain is given by its contour length $L = Nb$. Here, N is the number of peptide bonds and b is the distance between two consecutive C_α-atoms ($b = 0.38$ nm). Clearly, the distribution $P(r)$ cannot be recovered solely from the mean FRET efficiency $\langle E \rangle$ such that suitable analytical models for $P(r)$ are required to extract distance information. A variety of models have been used in the past [23], the simplest of it being the Gaussian chain model given by

$$P(r) = \frac{4}{R}\left(\frac{3}{2\pi}\right)^{1/2}\left(\frac{r}{R}\right)^2 \exp\left[-\left(\frac{r}{R}\right)^2\right]. \tag{13}$$

Here, $R = \sqrt{2/3}b_K N_K^{1/2} = \sqrt{2/3}\langle r^2 \rangle^{1/2}$ is the most likely donor–acceptor distance, a parameter that can be obtained from $\langle E \rangle$ by solving Eqs. 12 and 13 numerically. The parameters b_K and N_K are the length and number of virtual Kuhn segments that are related to the real bond length b and amino acid number N via

$$b_K = 2l_p \text{ and } N_K = \frac{b}{b_K} N, \tag{14}$$

with l_p being the persistence length (stiffness) of the chain. Persistence lengths determined from smFRET experiments using Eqs. 12, 13, and 14 are ~0.4 nm at physiological condition [30], which roughly agrees with results from other methods, e.g., see [20] and references therein. Importantly, more recent experiments and simulations [22, 25, 51] found that the Gaussian chain model (Eq. 13) overestimates chain dimensions from smFRET experiments when used in Eq. 12 and a more realistic self-avoiding random walk (SAW) model

$$P(r) = A \frac{4\pi}{R} \left(\frac{r}{R}\right)^{2+g} \exp\left[-B\left(\frac{r}{R}\right)^{\delta}\right], \tag{15}$$

is better suited to extract mean donor–acceptor distances from smFRET experiments on IDPs and unfolded proteins. In Eq. 15, the parameters A and B are obtained from the conditions $1 = \int_0^\infty P(r)\mathrm{d}r$ and $R^2 = \int_0^\infty r^2 P(r)\mathrm{d}r$, and the exponents are given by $g \approx 1/6\nu$ and $\delta = \frac{1}{1-\nu}$ [51–53]. The quantity $\nu = (\ln R - \ln b_K)/\ln N_K$ is the length-scaling exponent of the chain, a parameter that we will discuss in more detail in the following section. Although Eq. 15 is strictly correct only for expanded homopolymers (good solvent) [54], a direct comparison with molecular simulations indicated that it is also a good approximation for more compact chains [51].

Although smFRET histograms do not contain direct information on the width of the distribution $P(r)$ of donor–acceptor distances r in a disordered chain, the fluorescence lifetime of the donor contains this information. By combining confocal single-molecule detection with time-correlated single-photon counting (TCSPC) using pulsed excitation sources, fluorescence lifetimes of both donor and acceptor can easily be obtained. For a distribution of distances $P(r)$, the fluorescence lifetime decay of the donor is given by

$$I(t) = I_0 \int_0^\infty P(r)\, e^{-[k_D + k_T(r)]t} \mathrm{d}r, \tag{16}$$

where $\tau_D = 1/k_D$ is the fluorescence lifetime of the donor in the absence of an acceptor, $k_T(r) = k_D(R_0/r)^6$ is the rate of transferring energy to the acceptor. For the hypothetical case in which all molecules have the same donor–acceptor distance R, i.e., $P(r) = \delta(r - R)$ and $E(R) = \langle E \rangle$,[1] Eq. 16 gives

$$I(t) = I_0\, e^{-k_D\left[1+(R_0/R)^6\right]t} \tag{17}$$

and the average donor fluorescence lifetime is given by

[1] Here, δ indicates the Dirac delta function.

$$\langle \tau_{DA} \rangle = \int_0^\infty t\, I(t)dt / \int_0^\infty I(t)dt = k_D^{-1}\left[1 + (R_0/R)^6\right]^{-1} = \tau_D(1 - \langle E \rangle). \quad (18)$$

Hence, the donor fluorescence lifetime depends linearly on the FRET efficiency, which is the classical prediction from Förster theory (Fig. 3b). Importantly, the same result is obtained for more complicated distance distributions, as long as the dynamics, i.e., the sampling of this distribution and therefore the fluctuations of the energy transfer rate k_T, are much faster than the intrinsic fluorescence lifetime of the donor τ_D [55]. Yet, this is not the case for IDPs and unfolded proteins that typically reconfigure at timescales of tens to hundreds of nanoseconds (Sect. 4), i.e., significantly slower than the fluorescence lifetimes of typical dyes (1–4 ns). In case of such slow sampling, we can compute the static average over the distance distribution $P(r)$ and obtain

$$\langle \tau_{DA} \rangle = \int_0^\infty t\, I(t)\mathrm{d}t / \int_0^\infty I(t)\mathrm{d}t = \int_0^\infty P(r)\, \tau_{DA}(r)^2 \mathrm{d}r / \int_0^\infty P(r)\tau_{DA}(r)\mathrm{d}r \quad (19)$$

with

$$\tau_{DA}(r) = k_D^{-1}\left[1 + (R_0/r)^6\right]^{-1}.$$

A plot of the average donor fluorescence lifetime $\langle \tau_{DA} \rangle$ as a function of the mean FRET efficiency $\langle E \rangle$ will not follow a simple linear scaling (Fig. 3b). In fact, substituting $\tau_{DA}(r) = \tau_D[1 - E(r)]$, we can express Eq. 19 in different form

$$\langle \tau_{DA} \rangle = \tau_D(1 - \langle E \rangle) + \tau_D \frac{\sigma_E^2}{1 - \langle E \rangle}, \quad (20)$$

with $\sigma_E^2 = \langle E^2 \rangle - \langle E \rangle^2$ and $\langle E^2 \rangle = \int E(r)^2 P(r)\mathrm{d}r$, given that the distance distribution is properly normalized such that $\int P(r)\mathrm{d}r = 1$. Notably, the term σ_E^2 should not be confused with the width of the FRET histogram, which is largely affected by shot noise as explained above. It rather is the width of the "true" FRET distribution, which is typically inaccessible in smFRET experiments. Hence, a model for $P(r)$ is required to obtain information about the width of the distance distribution. Since the second term in Eq. 20 is always positive, distance fluctuations quantified by $P(r)$ will increase the average donor fluorescence lifetime compared to the prediction given by Förster theory (Eq. 18, Fig. 3b). Strong deviations from the classical Förster prediction have indeed been observed for a multitude of disordered and unfolded proteins [23, 30, 33] such that a comparison of experimental donor lifetimes with the Förster prediction in Eq. 18 can generally be used to identify distance heterogeneity in proteins [55, 56]. Importantly, these relationships only hold under the condition that the orientation of donor and acceptor dye dipoles averages sufficiently at timescales faster than the fluorescence lifetimes of the dyes. If this is not the case, the relationship between donor–acceptor distances and measured FRET efficiencies is more complicated. For example, the FRET efficiency will now depend on both, the distance and the orientation of the dyes via [57]

$$E(r,\kappa^2)=\left[1+\frac{2}{3\kappa^2}\left(\frac{r}{R_0}\right)^6\right]^{-1} \text{ with } \kappa^2=(\cos\theta_T-3\cos\theta_D\cos\theta_A)^2. \tag{21}$$

Here, θ_T is the angle between donor and acceptor dipole, whereas θ_D and θ_A are the angles between these dipoles and the vector that connects both dipoles. While the value of κ^2 can range from 0 to 4, it will fluctuate in general such that several averaging regimes have to be considered. Assuming an isotropic orientation of the dye dipoles, the distribution of κ^2 is given by [57, 58]

$$p(\kappa^2)=\begin{cases}\frac{1}{2\sqrt{3\kappa^2}}\ln\left(2+\sqrt{3}\right) & 0\leq\kappa^2\leq 1\\ \frac{1}{2\sqrt{3\kappa^2}}\ln\left(\frac{2+\sqrt{3}}{\sqrt{\kappa^2}+\sqrt{\kappa^2-1}}\right) & 1<\kappa^2\leq 4\end{cases}. \tag{22}$$

In the most commonly used limit, the fluctuation of κ^2 is much faster than the fluorescence lifetime of the dyes and an average value for κ^2 obtained from Eq. 22 via $\langle\kappa^2\rangle = \int \kappa^2 p(\kappa^2)d\kappa^2 = 2/3$ can be used. In this limit, the FRET efficiency (Eq. 21) simplifies to Eq. 10. Yet, exceptions have been found for cases in which the dyes interact with extended protein surfaces [59, 60], thus hampering the fast averaging of dipole orientations of the dyes. Fluorescence anisotropy experiments, ideally in a time-resolved manner, should be used to check the applicability of the $\langle\kappa^2\rangle = 2/3$ limit. For intrinsically disordered and unfolded proteins, the fast averaging limit is often fulfilled due to the absence of extended protein surfaces that would facilitate dye–protein interactions.

In summary, the combination of mean FRET efficiencies and fluorescence lifetimes does not only provide information on average chain dimensions, but it also allows an estimate of the width of distance distributions. Yet, analytical models of these distributions such as the SAW model (Eq. 15) are still required to retrieve this information from smFRET experiments. Similarly, experiments alone are insufficient to understand another important property of IDPs, their extreme sensitivity toward changes in external conditions such as denaturants [13–18, 28], temperature [21, 32], crowding [31], or ionic strength [7, 18, 19, 26]. Clearly, a change in any of these conditions will unambiguously affect the balance of attractive and repulsive interactions within the chain, thus resulting in altered chain dimensions. To obtain a more quantitative understanding of the interactions that drive such compactions or expansions, we will extend the Gaussian chain model (Eq. 13) to generate a rather general mean-field homopolymer model.

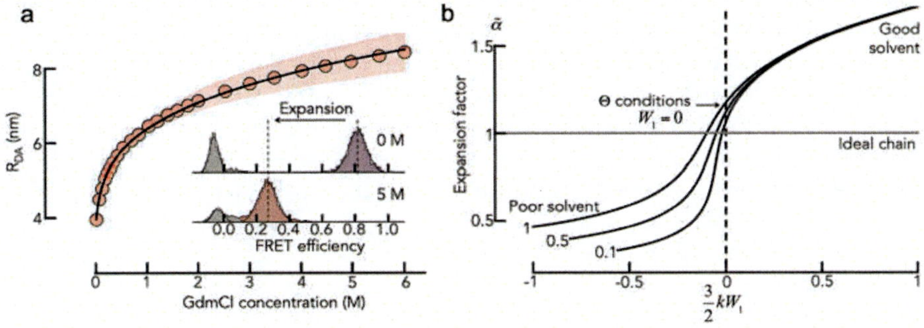

Fig. 4 Solvent-induced expansion of an IDP. (**a**) Average donor–acceptor distances (R_{DA}) of the disordered Max protein as function of the GdmCl concentration. The solid line is a fit with a polyampholyte theory [26]. *Inset:* SmFRET histograms of Max at two concentrations of the denaturant GdmCl (indicated). The FRET population shifts to lower FRET values, indicating an expansion of the chain. (**b**) Coil-to-globule transition as predicted by the mean-field theory of de Gennes (Eq. 25) for $N_K = 100$ and three different values (indicated) of the three-body interaction parameter ($3k^2W_2$)

3.2 Mean-Field Homopolymer Theory

The fact that polymers expand and contract in a solvent-dependent manner is not a new observation. In fact, plenty of examples have been studied in the twentieth century, the most well-known being poly-styrene in cyclohexane [61]. Although chemically tremendously more complex than homopolymers, IDPs exhibit a similar pliability to solution changes (Fig. 4a) and homopolymer theories have therefore been used in the past to model them. Clearly, the results should be interpreted with care. However, it is undeniable that homopolymer models have greatly advanced our understanding of disordered and unfolded proteins not only by providing simplified parameters to describe the molecular processes behind chain expansions or collapse but also by spotting deviations from homopolymer properties.

To model the expansion and compaction of polymers, let us start with the Gaussian chain model in Eq. 13. To simplify notation, we first express Eq. 13 in terms of a reduced distance coordinate $\alpha^2 = \left(\frac{r}{R}\right)^2$, where α is also known as the expansion factor. Keep in mind that R is the most likely donor–acceptor distance of a Gaussian chain, i.e., of an ideal mathematical chain without interactions and volume. Neglecting constant prefactors, Eq. 13 can be re-written as

$$P(\alpha) \propto \alpha^2 \exp\left(-\alpha^2\right). \qquad (23)$$

As any molecular property that underlies the statistical laws of thermodynamics, the distance distribution is associated to a free energy profile ($P(\alpha) \propto e^{-F(\alpha)}$), which is given by $F(\alpha) = -\ln P(\alpha) \propto \alpha^2 - 2\ln \alpha$. Since our chain model at this point is just a mathematical chain of points connected by vectors, i.e., it lacks volume or any type of interactions, the free energy $F(\alpha)$ only describes the elastic entropy of the chain that results from the multiplicity of random conformations of this ideal chain. To introduce interactions of any type between monomers, repulsive or attractive, we will use the mean-field approximation in which the number of interactions within a chain is proportional to the density (volume fraction) $\phi = d^3 N_K/\alpha^3 R^3$ of monomers in the chain. Here, d is the size (radius) of a single monomer in the chain. Thus, a reasonable extension of the free energy of an ideal chain by interactions would be given by a virial expansion of the type

$$F(\alpha) \propto \alpha^2 - 2\ln\alpha + N_K\left(W_1\phi + W_2\phi^2 + \ldots\right), \tag{24}$$

with the virial coefficients W_1 for two-body interactions and W_2 for three-body interactions. To simplify the expression, we will only consider the first two terms of the expansion, which already provide a rather general model that has first been proposed (in slightly different form)[2] by de Gennes in 1975 [62]. Finding the free energy minimum $dF/d\alpha = 0$ gives

$$\tilde{\alpha}^5 - \tilde{\alpha}^3 - \frac{3k^2 W_2}{\tilde{\alpha}^3} = \frac{3}{2} k W_1 \sqrt{N_K} \text{ with } k = (d/b_K)^3. \tag{25}$$

Here, $\tilde{\alpha}$ indicates the expansion factor at the free energy minimum. Notably, Eq. 25 can also be obtained using variational approaches [26, 63, 64]. A plot of $\tilde{\alpha}$ as function of the two-body interaction energy clearly uncovers a substantial compaction of the chain with decreasing (attractive) W_1 and positive (repulsive) W_2 (Fig. 4b). Contrary to folding/unfolding transitions, this compaction is non-cooperative [62, 65] and rather resembles a higher-order phase transition compared to the cooperative first-order like folding–unfolding transition of proteins. However, as pointed out by de Gennes [62], Eq. 25 also predicts a first-order coil-to-globule transition for homopolymers for sufficiently small values of W_2, a finding that has so far not been confirmed experimentally and that is likely an artifact of the theory. It is instructive to analyze the individual energetic contributions to the two-body interaction term W_1, often called χ, which can be expressed as a Flory-Huggins interaction parameter [66]

$$W_1 \propto w_{sp} - \left(w_{ss} + w_{pp}\right)/2, \tag{26}$$

[2] De Gennes derived his model for the radius of gyration (r_G) of the polymer. Since the distribution of radii of gyration for a Gaussian chain is not known in closed analytic form, he used the approximation $P(\alpha) \propto \alpha^3 \exp(-3\alpha^2/2)$ where $\alpha = r_G/R_{G,\text{ ideal}}$. Notably, compared to donor–acceptor distances measured with smFRET, the radius of gyration is by far the better quantity to construct a mean-field theory due to its direct link to the monomer-density of a chain. The long-known fact has more recently gained renewed attention in the so-called FRET-SAXS controversy [see Refs. 22 and 77].

where the energy changes w_{pp}, w_{ss}, and w_{sp} are due to polymer–polymer (*pp*), solvent–solvent (*ss*), and solvent–polymer contacts (*sp*), respectively. Notably, Eq. 26 had been derived for lattice-polymer models and processes such as the adsorption of denaturant molecules on the polypeptide chain might not be adequately captured by it. Hence, an awareness of these pitfalls is required when interpreting experimental results in terms of w_{pp}, w_{ss}, and w_{sp}. The exact molecular effects of GdmCl and urea on polypeptide chains, which might affect multiple interactions simultaneously, have been debated much in the past [67]. A combination of polar interactions, hydrogen bonding, and stacking to non-polar surfaces has been suggested. Yet, it is unclear whether any of these contributions really dominate the action of GdmCl and urea. Based on experimental transfer free energies (Δg_T) of amino acids from water to solutions of these denaturants, it is clear that both denaturants assist in the solvation of non-polar side chains [67, 68]. In fact, the molecular transfer model developed by Thirumalai and co-workers [69, 70] explicitly describes the change in dimension of unfolded proteins and the thermodynamic stability of folded proteins based on the experimentally determined transfer free energies of amino acids. In addition, smFRET experiments on a broad variety of IDPs and unfolded proteins suggest a clear relationship between Δg_T and W_1 [20, 21, 71, 72]. Whereas the approximate values of W_1 between high molar concentrations of denaturants and water are known to be in the order of a few k_BT [20, 26], absolute values for the three-body interaction term W_2 in Eq. 25 are more difficult to obtain. Recent results suggest values in the order of 4–30 k_BT for the term $3k^2W_2$ [26]. Yet, care has to be taken as these energies also depend on the precise numerical factors used in Eq. 25 and mean-field theories alike. In fact, a more independent parameter to quantify the compaction of disordered proteins (and polymers in general) is the length-scaling exponent ν.

For sufficiently long chains, the dimension scales with its length according to $R_{DA} \propto N_K^{\nu}$, where $R_{DA} = \tilde{\alpha} R$ is the equilibrium distance between donor and acceptor that can be determined in smFRET experiments (see Eqs. 12, 13, 14, and 15). The length-scaling exponent ν can be obtained from Eq. 25 for three limits: (1) swollen chains in good solvent, i.e., W_1 is repulsive ($W_1 > 0$) and three-body interactions can be neglected ($W_1 \ll W_2$), (2) compact chains in poor solvents, i.e., W_1 is strongly attractive ($W_1 < 0$) and W_2 is repulsive ($W_2 > 0$), and (3) conditions in which $W_1 \approx 0$ and $W_2 > 0$ such that the chain behaves like an ideal chain, a condition that is also called Θ-condition (Fig. 4b). Let us start with an expanded chain in good solvent. At such conditions, the effect of three-body interactions is substantially reduced and we can safely assume $W_2 \approx 0$. Equation 25 then simplifies to the well-known Flory result [73]

$$\tilde{\alpha}^5 - \tilde{\alpha}^3 = \frac{3}{2} k W_1 \sqrt{N_K}. \tag{27}$$

In the extreme limit of expanded chains ($\tilde{\alpha}^5 \gg \tilde{\alpha}^3$) and with the definitions for $\tilde{\alpha}$ and R, we obtain

$$R_{DA} = C_e b_K N_K^{3/5} \propto N_K^{\frac{3}{5}} \text{ with } C_e = \left(\frac{2}{3}\right)^{\frac{1}{2}} \left(\frac{3kW_1}{2}\right)^{\frac{1}{5}}. \tag{28}$$

Hence, polymers in good solvent should exhibit a length-scaling exponent of $\nu = 3/5$. In fact, the result is surprisingly close to the exact value from renormalization-group field theory of $\nu = 0.588 \pm 0.001$ [74]. Notably, this success of Flory theory is based on a fortuitous cancelation of errors: the term $W_1 N_K \phi$ in Eq. 24 (first term in brackets) overestimates the monomer interaction energy because correlations between monomers along the chain are neglected [75] and the term $\alpha^2 - 2 \ln \alpha$ (Eq. 24) of the Gaussian chain model overestimates the elastic energy of the chain.

A number of experimental studies using a diverse set of methods from SAXS [22, 27, 76, 77], smFRET [20, 22, 27], over NMR [78] to FCS (fluorescence correlation spectroscopy) [20, 22], and molecular simulations [22, 27, 51] demonstrated that IDPs and unfolded proteins at high concentration of denaturants (GdmCl and urea) indeed exhibit a length-scaling exponent of ~3/5. Hence, sequences and amino acid compositions are less relevant for chain dimensions under these conditions. This is particularly the case for proteins in GdmCl, which is not only a denaturant but also a salt, thus capable of effectively screening charge–charge interactions in unfolded proteins and IDPs as shown with smFRET on a number of systems [18, 19, 26].

The second limit that can be obtained from Eq. 25 is that of a strongly collapsed chain in poor solvent. Under these conditions ($\tilde{\alpha}^5 << \tilde{\alpha}^3 >> 1$, and $W_1 < 0$), Eq. 25 simplifies to

$$\tilde{\alpha}^3 = \frac{2kW_2}{|W_1|} N_K^{-1/2}, \tag{29}$$

which gives

$$R_{DA} = C_g b_K N_K^{1/3} \propto N_K^{\frac{1}{3}} \text{ with } C_g = \left(\frac{2}{3}\right)^{\frac{1}{2}} \left(\frac{2kW_2}{|W_1|}\right)^{\frac{1}{3}}. \tag{30}$$

Hence, if the chain forms a highly collapsed globule due to strong polymer–polymer contacts, the growth in dimension with chain length is tremendously reduced compared to the excluded volume limit (Eq. 28). It should be mentioned that although Eq. 30 follows naturally from Eq. 25, it is only valid for mean-field theories constructed in terms of radii of gyration. For a distance such as that measured with smFRET (R_{DA}), the volume fraction of a chain ϕ is ill-defined in the limit of compact chains and the chain volume cannot be reliably estimated from the end-to-end distance. In fact, the de-coupling of distances R_{DA} inferred from smFRET and radii of gyrations R_G for compact chains has been studied in detail [27] and shows that chains with similar R_G can obey largely different FRET efficiencies. Hence, scaling exponents of extremely compact chains cannot be extracted from donor–acceptor distance measurements alone but require either the additional help of molecular simulations or the direct determination of better quantities for chain dimensions such as the radius of gyration [77, 79] or the hydrodynamic radius of a chain [80, 81]. So far, only a few cases of disordered polypeptide chains with scaling exponents close to 1/3 have been published, one being poly-glutamine peptides [80]. Non-equilibrium SAXS experiments using fast continuous-flow mixing also identified a scaling exponent close to 1/3 for early folding intermediates of seven proteins [82]. It is unclear however to which extent the low scaling exponent is affected by structure formation in those intermediates after being diluted from high concentrations of denaturant.

Finally, we discuss the limit at which w_{pp}, w_{ss}, and w_{sp} balance such that $W_1 = 0$. At this Θ-condition, the chain behaves like an ideal chain and only three-body interactions (W_2) and chain elasticity, i.e., configurational entropy, contribute to the chain dimension. Importantly, since $W_2 > 0$ at Θ-conditions, the absolute donor–acceptor distance will not be identical to that of an ideal chain ($\tilde{\alpha} \neq 1$) but rather be close to it ($\tilde{\alpha} \sim 1$). To see this, we start with Eq. 25 and set $W_1 = 0$, which gives

$$\tilde{\alpha}^8 - \tilde{\alpha}^6 = 3k^2 W_2. \tag{31}$$

To obtain an approximate solution, we expand the polynomial (LHS) around $\tilde{\alpha} = 1$, which leads to the series $2(\tilde{\alpha} - 1) + 13(\tilde{\alpha} - 1)^2 + 36(\tilde{\alpha} - 1)^3 + \ldots$. Since $\tilde{\alpha} - 1 < 1$,we only keep the leading term, which admittedly is a brave assumption given the significant pre-factor of the second term, resulting in

$$R_{DA} = C_\Theta b_K N_K^{1/2} \propto N_K^{\frac{1}{2}} \text{ with } C_\Theta = \left(\frac{3}{2}\right)^{\frac{1}{2}}\left(1 + \frac{3}{2}k^2 W_2\right), \tag{32}$$

which exceeds the distance of the ideal chain that would be given by $R = \sqrt{2/3} b_K N_K^{1/2}$. Indeed, a large number of smFRET and SAXS experiments have demonstrated that the scaling exponents of IDPs and unfolded proteins under physiological conditions (absence of denaturants) cluster around $\nu = 1/2$ [20, 22, 29, 77]. These results indicated that polypeptide chains in water are surprisingly close to Θ-conditions, i.e., conditions at which attractive and repulsive interactions in the chain roughly balance. Given the rather small number of proteins tested so far, it is unclear whether this finding is pure coincidence or indeed the result of evolution. In fact, it has been speculated that the balance of attractive and repulsive interactions at Θ-conditions could be advantageous for protein folding reactions, e.g., by allowing a more efficient sampling of intra-chain contacts during folding [20, 83]. Clearly this interpretation is difficult to draw for IDPs that do not necessarily fold, yet, Θ-conditions might still be advantageous for liquid–liquid phase separation processes. Although the scaling exponents for many polypeptide chains are close to 1/2, differences in sequence composition, but even in the precise sequence such as patterning of charges [84–86] or hydrophobic and aromatic residues [87] can cause significant variations in the size of polymers [88, 89]. For example, a nearly balanced number of positively and negatively charged amino acids can strongly compact IDPs [26]. On the opposite, high net-charges cause strong electrostatic repulsions that lead to extremely expanded chains [19]. Naturally, such charge-driven interactions are sensitively affected by ionic strengths, and even small variation within the physiological regime (100–300 mM) can cause large changes in chain dimensions and scaling exponents. Many studies over the past decades explicitly incorporated the effect of charge interactions in mean-field theories [90–93] and their application in interpreting salt-induced compaction of highly charged IDPs and the expansion of nearly charge-balanced IDPs found with smFRET has been enormously successful [19, 26]. Yet, a drawback of these theories is their inability to account for specific charge patterns along the sequence. In fact, mean-field polyelectrolyte and polyampholyte theories are only applicable for well-mixed sequences in which it can be safely assumed that amino acid charges are essentially smeared out across the chain. Although attempts have been made to develop analytical heteropolymer theories for quenched sequences [94], modern approaches to account for sequence specificity in the polymer behavior of IDPs and unfolded proteins are mainly based on molecular simulations.

3.3 More Accurate Polymer Models: Combining smFRET with Molecular Simulations

Mean-field theory is a convenient first option to understand smFRET data of IDPs. However, local secondary structure preferences and long-range interactions caused by either charged amino acids or hydrophobic patches in the sequence might have significant populations and lifetimes. In such cases, homopolymer theories have limited capabilities of faithfully interpreting FRET-based distance information.

To combine smFRET data with molecular simulations, one often relies on the accuracy of the simulation model, which is described by a force field that accounts for interactions between amino acids. Since force fields were often parameterized using data from a variety of IDPs, i.e., not necessarily for the IDP studied in a particular smFRET experiment, there is no guarantee that a simulation will reproduce a FRET signal out of the box, thus complicating an interpretation of physical mechanisms with the model. To ease mechanistic interpretations of smFRET experiments therefore requires a molecular ensemble that can reproduce the smFRET data. One strategy has been to reweight conformations generated with an existing simulation model such that the calculated observables (e.g., FRET efficiency) best match the experimental values [22]. A tremendous advantage of ensemble reweighting is its low computational cost compared to the strategy of adapting parameters using iterative simulations to match experimental outcomes. However, reweighting also has its limits. For instance, if the initial ensemble obtained from a simulation deviates significantly from the experimental measurement, or in other words, if important conformations are not sampled in the simulation, reweighting will not provide a realistic description of the IDP ensemble. Moreover, the outcome from ensemble reweighting methods could be biased by the initial ensemble, which depends on the quality of the force field [95]. In such cases one has to downgrade the expectation from bottom-up simulation models with predictive power and instead either bias the simulations using experimental data [96] or adapt the simulation parameters such that the simulated trajectory can still interpret the experimental data [97]. Further cross-validation using experimental inputs other than smFRET, such as solvation free energy, SAXS, and NMR, is then necessary to verify the refined model. In the following, we will therefore mostly focus on those molecular simulation methods that can be easily integrated with smFRET experiments and do not discuss bottom-up simulation methods.

Depending on the spatial resolution of the model, all-atom implicit/explicit-solvent simulations [97, 98] and coarse-grained simulations [99] have been used in the past to describe smFRET data. Clearly, all-atom explicit-solvent models have the highest resolution but can be limited by force field accuracy and sampling efficiency. Taking advantage of the increasing amount of experimental measurements on IDPs, significant efforts have been undertaken in recent years to improve all-atom force fields [97, 100, 101] such that modern force fields will provide meaningful IDP ensembles. In terms of sampling, most software packages [102–105] have now taken advantage of the rapidly improving GPU resources such that microseconds simulations of an IDP with less than 100 amino acids can be achieved within few days in standard high-performance computing clusters. In addition, the development of specialized supercomputers for molecular dynamics such as "Anton" even enabled IDP-simulations of 100 μs in length within few days [106], which even allows investigations of IDP–ligand interactions [107].

Using all-atom simulations to interpret smFRET experiments requires modeling of the dyes. In fact, commonly used dyes such as Alexa Fluor 488 and Alexa Fluor 594 have significant sizes of a few amino acids (Fig. 5a). The simulation has to either model dyes explicitly or specific correction factors have to be included to calculate FRET efficiencies from simulations without dyes. Using a carefully parameterized force field of FRET dyes, Best et al. showed that the orientational factor $\langle \kappa^2 \rangle$ is indeed close to 2/3, as estimated in many experiments [108] and FRET efficiency calculations can be simplified using Eq. 10 (see Sect. 3.1). However, to improve the sampling efficiency, dyes are not commonly included. In such cases, only distances between the C_α-atoms of the labeled residues are accessible from the simulations. These distances should then be rescaled by a factor of $[(N + N_{\text{linker}})/N]^{\nu}$, where N is the number of peptide bonds between the labeled amino acids and ν is the length-scaling exponent which can be determined by the scaling of internal chain distances with sequence separation in the simulation [109] or from experiments with multiple labeling positions [20]. Importantly, N_{linker} is a free parameter that represents the effective length of both dye linkers in terms of an equivalent number of peptide bonds. It has experimentally been estimated to be ~9 [29, 30], which agrees well with estimates obtained from all-atom simulations with and without dyes [110] (Fig. 5b).

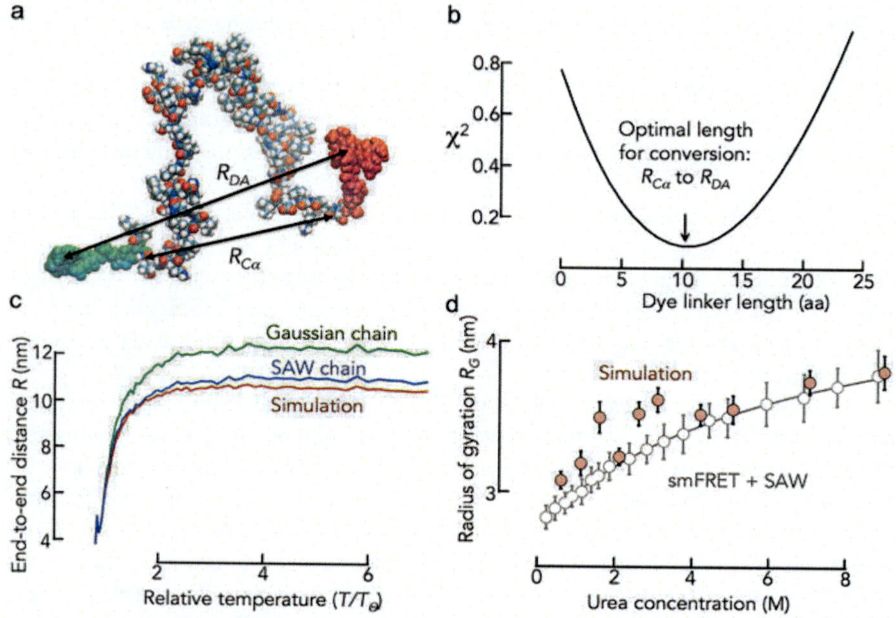

Fig. 5 Comparing smFRET experiments with molecular simulations. (**a**) One representative conformation of a simulation with the IDP ACTR, including the dyes Alexa Fluor 488 and Alexa Fluor 594, is shown. Two different ways of calculating donor–acceptor distances for comparison with smFRET experiments are depicted: R_{DA} between the donor and acceptor for simulations that explicitly include the dyes and $R_{C\alpha}$ between the C_α atoms of the labeled residues for simulations without dyes. (**b**) Deviation between the FRET efficiencies calculated using R_{DA} and $R_{C\alpha}$ as a function of the linker length N_{linker} (in amino acids) used for correcting FRET values from simulations without dyes [110]. (**c**) End-to-end distances from all-atom simulations (red) are compared with the distances obtained from mean FRET values that were computed from the same simulation. Two polymer models, the Gaussian chain (Eq. 13) and the SAW model (Eq. 15) were used to compute the distances from FRET using Eq. 12. (**d**) The radius of gyration R_G obtained from smFRET using SAW model is compared to an all-atom ensemble reweighted using both smFRET and SAXS data

Importantly, the heterogeneous ensemble of IDP conformations is better characterized by different experimental techniques rather than smFRET alone and all-atom simulations have become essential in integrating them, not only to generate realistic IDP ensembles but also to identify and reconcile discrepancies between findings with different experimental methods. For instance, while smFRET experiments showed that unfolded and disordered polypeptide chains expanded substantially upon addition of denaturants, SAXS experiments suggested that expansions were much less pronounced [111]. All-atom simulations were then used as benchmarks to test empirical ways of using polymer models for analyzing the smFRET and SAXS data (see Sect. 3.1) by comparing artificial experimental signals, e.g., FRET efficiencies and SAXS scattering curves together with the correct "answers," i.e., donor–acceptor distance and radius of gyration (R_G) that could be directly calculated from the simulation. This comparison showed that the Gaussian chain model overestimates distances in smFRET experiments whereas Guinier analysis underestimates R_G when the chain deviates from a random coil [22]. A SAW model (Eq. 15), in which the distance distribution is adjusted according to the scaling exponent, has been found to be a better choice for interpreting smFRET and SAXS data for ν larger than 0.5 [51, 112] (Fig. 5c). In addition, simulations [51] also verified that a size-dependent factor λ [113] is necessary to convert the average donor–acceptor distance R_{DA} from smFRET into R_G

$$\lambda = \frac{R_{DA}^2}{R_g^2} = \frac{2(\gamma + 2\nu)(\gamma + 2\nu + 1)}{\gamma(\gamma + 1)}. \tag{33}$$

Here, $\gamma = 1 + g/\nu \approx 7/6$ as in Eq. 15. Equation 33 shows that the estimated conversion factor ranges between 4.1 and 6.3, depending on the chain compaction specified by the length-scaling exponent ν, which added to the deviations found between smFRET and SAXS measurements [27]. With both corrections, the radius of gyration estimated from smFRET experiments was comparable with that estimated from an ensemble using both FRET and SAXS measurement for the unfolded R17 protein at different denaturant concentrations [22, 51] (Fig. 5d). Integrated with molecular simulations, smFRET measurements have also been found to complement other experimental methods such as NMR and PRE (Paramagnetic Relaxation Enhancement) [114], thus showing that all-atom simulations provide an effective way of integrating multiple sources of experimental data.

Although all-atom simulations certainly provide the most accurate representation of molecular processes, particularly when combined with explicit water models, sampling can be time consuming and resource demanding for IDPs with more than 100 residues. It is often informative to use a coarser representation of the molecular complexity at substantially lower computational cost. An ideal "interpolation" between all-atom simulations and simple polymer models are low-resolution coarse-grained models. Coarse-grained models contain about the same number of free parameters as analytical polymer theories but additionally include specific sequence details of IDPs such as the patterning of charges and hydrophobic sequence patches that are absent in mean-field theories. The fast sampling of coarse-grained models even allows a quick adjustment to experimental data and can be used to generate multiple IDP ensembles at a variety of solvent conditions. For instance, in a recent study of the disordered cytoplasmic tail of E-cadherin [12], smFRET experiments of differently labeled variants of E-cadherin showed that a mean-field polyampholyte theory [90] failed to predict the salt-dependent conformational change of all variants. A description using a coarse-grained model instead reproduced the salt-dependence of all variants with just one global fitting parameter and identified the reason for the failure of the mean-field description. The segregation of oppositely charged residues along the sequence, i.e., charge patterning, was found to be essential for the dimension of this IDP, an effect that cannot be captured with mean-field theories. Finally, systems involving IDP assemblies such as in liquid–liquid phase separation (LLPS) processes [115] or already only those that include one IDP in complex with another macromolecular ligand [116] typically exceed the sampling-capabilities of all-atom simulations, thus making coarse-grained models unavoidable.

Clearly, the level of coarse-graining depends on the specific task and the balance between the computational efficiency and modeling accuracy. Coarse-graining levels range from representing individual residues by multiple beads over one bead per residue down to one bead for several residues. Yet, the need of calculating distances between labeled residues in describing smFRET experiments clearly favors one bead per residue models. Coarse-graining at the single-residue level is particularly important if sequence-specific effects play an important role in the system at hand. To capture electrostatic interactions at different ionic strength, which is a relevant factor considering physiological salt concentrations of 100–300 mM [117], the Debye-Hückel approximation is commonly used [118]. The computational efficiency of such a screening potential in contrast to an explicit ion model often outweighs the disadvantage of neglecting the true radial distribution of ions around the charged amino acids in studies of IDPs [119, 120]. In addition, other interactions between every pair of amino acids also add up to significant factors that can sensitively affect the dimension of an IDP. In a recently developed coarse-grained model (HPS model) [99] such pairwise interactions were introduced using the amino acid hydropathy in the Ashbaugh-Hatch functional form [121]. It was shown that a tuning of the strength of this amino acid hydropathy described the experimental results of a variety of IDPs with high accuracy [99, 122–125]. Further tuning the pairwise interaction strengths to match smFRET experiments of a specific IDP is then a straightforward way to understand the main sequence contribution to the overall dimension [12], which is information that cannot be obtained from analytical polymer theories.

The enormous flexibility of coarse-grained models can even be used to study the behavior of IDPs that interact with another folded protein. Since such complexes will also involve specific contacts between both partners, a structure of the complex, obtained experimentally or via computational methods, is a necessary starting point to identify specific contacts. Using this structural information, the stability of the folded binding partner is ensured either via strong harmonic constraints between native residue pairs or by treating the folded partner as a rigid body. In fact, the latter strategy is preferred since it saves computational resources and a number of molecular dynamics packages such as LAMMPs [126] and HOOMD-Blue [127] already provide this option. Specific inter-molecular contacts between IDPs and a folded partner can then be introduced using tunable harmonic or Lennard-Jones (LJ) potentials. Here, harmonic potentials strongly restrain the complex in its bound state such that the IDP will not dissociate from its folded partner and an LJ potential might be a more reasonable choice to sample both bound and unbound states. By tuning the interaction strengths of the potential, experimental smFRET values can be matched and a description of conformational flexibility of an IDP in complex with a folded protein can be obtained. This strategy has recently been used to quantify the conformational ensemble of disordered E-cadherin in complex with the folded protein β-catenin [12].

In summary, both all-atom explicit-solvent and coarse-grained models have their merits and limits. All-atom models provide a more accurate representation of the protein and also describe dyes explicitly, which simplifies the interpretation of smFRET experiments. However, all-atom models are limited by both sampling and force field accuracy. Even though force fields are being continuously improved with an increasing amount of experimental data, the quadratic scaling between timescales and system size will continue to limit sampling. Coarse-grained models are therefore a cost-effective solution. In fact, they can be considered as an extension of analytical polymer models. Yet, coarse-grained models average over many degrees of freedom and are less transferable among different systems. For instance, the strengths of interactions between amino acids often depend on the local sequence context. While charge–charge interactions are typically predicted well by coarse-grained models due to their substantial interaction energies, the combined effect of many weak interactions is much more difficult to catch. Hence, parameterizing (tuning) coarse-grained models with experimental data on the specific system at hand is very helpful for interpreting experimental data in terms of a molecular model. Clearly, specific interactions might be missed and the timescales of motions will be unrealistically fast such that all-atom simulation can be used as a complementation.

4 Probing and Modeling Sub-microsecond Dynamics of Disordered Proteins

4.1 Nanosecond Fluorescence Correlation Spectroscopy (nsFCS) Coupled with FRET

So far, we discussed the average dimension of disordered and unfolded proteins and its susceptibility toward changes in solution conditions. Our starting point for this discussion was the distribution of donor–acceptor distances that we expect for the extremely heterogeneous conformational ensemble of IDPs. In this section, we focus on the timescales at which such disordered ensembles are sampled, i.e., the reconfiguration time. The first attempts to identify this timescale were based on an analysis of the width in FRET histograms [14] and resulted in an upper limit of ≤25 μs for the unfolded state of Csp*Tm*. However, even extremely slow transitions at the timescale of seconds have been reported for surface-immobilized and unfolded RNase H [15]. Yet, already at that time, estimates from ensemble experiments such as contact-quenching, pioneered by Eaton and Hofrichter [128] and triplet–triplet energy transfer [129], pioneered by Kiefhaber and colleagues, suggested reconfiguration timescales <1 μs for flexible polypeptides. Indeed, the first rigorous measurement of donor–acceptor distance fluctuations in an unfolded protein identified dynamics in the order of 50 ns [34]. Technically, these smFRET measurements required a Hanbury Brown and Twiss detection scheme [130] to circumvent dead times of detectors and counting electronics[3] and the inter-photon times between photon pairs were measured with picosecond time-resolution (256–304 ps). Since then, the development of powerful counting electronics with a time-resolution down to 8 ps [38, 39] has greatly simplified smFRET-based nsFCS experiments [7, 23–25, 32, 33, 35–37, 131–135] such that they can even be used to study the dynamics of IDPs inside living cells [136, 137]. In their essentials, smFRET-based nsFCS experiments do not differ much from classical FCS experiments, which retrieve the timescale of fluorescence fluctuations, e.g., due to the diffusion of molecules in and out of the confocal volume, thus allowing the determination of translational diffusion coefficients and Stokes radii as a measure of the size of the diffusing particle. Differences are twofold: first, instead of one color, two colors are monitored in smFRET-based nsFCS experiments, and second, the timescale at which fluctuations are observed is much shorter (ns). Admittedly, the latter factor is of more of technical origin and simply requires fast counting electronics that allows the storage of photon arrival times with high, i.e., picosecond precision.

What is the principle of nsFCS? Due to the flexibility of IDPs, the distance between donor and acceptor dyes attached to them fluctuates thus giving rise to fluctuations in the rate of photons ($\tilde{n} = n/T$) from donor $\tilde{n}_D(t)$ and acceptor $\tilde{n}_A(t)$. The timescale of these fluctuations can be quantified with four correlation functions (g_{DD}, g_{AA}, g_{DA}, g_{AD}) defined by

$$g_{ij}(\tau) = \frac{\langle \tilde{n}_i(t)\tilde{n}_j(t+\tau)\rangle}{\langle \tilde{n}_i(t)\rangle\langle \tilde{n}_j(t)\rangle} \text{ with } i = \{D, A\} \text{ and } j = \{D, A\}. \tag{34}$$

For distance dynamics slower than the fluorescence lifetime of the dyes, the shape of these correlation functions differs characteristically between auto- (g_{DD}, g_{AA}) and cross-correlation functions (g_{DA}, g_{AD}) in the presence of FRET between donor and acceptor (Fig. 6a). Whereas the autocorrelation functions decay (positive amplitude), the cross-correlation functions should increase (negative amplitude) due to the anti-correlated change in photon rates from donor and acceptor. Since only distance changes coupled with FRET can cause such anti-correlated behavior, an increasing cross-correlation function unambiguously indicates the presence of distance dynamics. Conversely, an absence of this increase does not exclude distance fluctuations because other effects, e.g., static quenching of the dyes by aromatic amino acids such as tryptophan or tyrosine [12, 133], can mask the anti-correlated fluctuations of donor and acceptor emission. At very short timescales in the order of the fluorescence lifetimes of the dyes, all correlation functions increase due to photon anti-bunching. Organic dyes in smFRET experiments are single-photon emitters, i.e., after a photon has been emitted, the emission of a second photon takes time since the dye has to be excited again. Thinking of correlation functions as probability of observing two photons separated by the lag-time τ, this probability should ideally be

[3] Here, photons from each color are randomly distributed to two detectors.

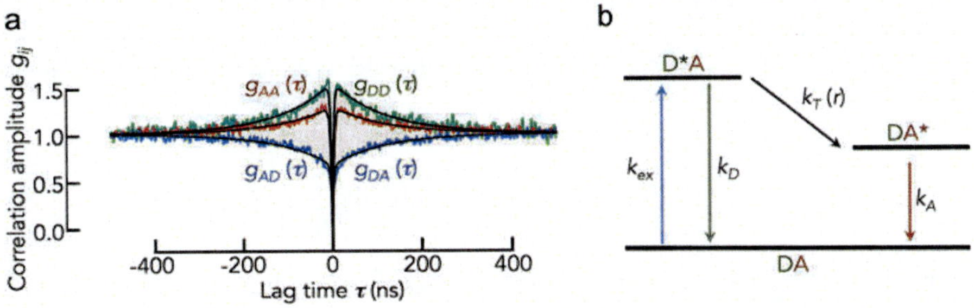

Fig. 6 Probing IDP dynamics with nsFCS. (**a**) Correlation functions from nsFCS for the intrinsically disordered DNA-binding domain of the protein c-Myc together with a global fit with multi-exponential decays as given by Eq. 35 (black lines). (**b**) Simple photo-physical scheme of a smFRET experiment (see text). The rate matrix for this scheme is given by Eq. 37

zero at $\tau = 0$, i.e., $g_{ij}(0) = 0$. Notably, this is rarely the case for cross-correlation functions because donor and acceptor can sporadically populate their excited states at the same time,[4] which can cause the simultaneous emission of a donor and an acceptor photon.

Initially, the practical goal of nsFCS experiments in the characterization of IDP dynamics was to determine the decay time of the correlation functions (Eq. 34). Importantly, the decay of the autocorrelation functions and the rise in the cross-correlation functions should provide the same relaxation time if these decays result from distance dynamics. Hence, the strategy of globally fitting the four correlation functions using exponentials such as

$$g_{ij}(\tau) = \left(1 - a_{ij}e^{-|\tau|/\tau_a^{ij}}\right)\left(1 \pm b_{ij}e^{-|\tau|/\tau_b}\right)\left(1 + c_{ij}e^{-|\tau|/\tau_c^{ij}}\right)/N_{ij} \tag{35}$$

is frequently used [33, 34, 131]. Here, the first factor (RHS) accounts for antibunching with a correlation-function-specific amplitude (a_{ij}) and decay time (τ_a^{ij}), the second factor accounts for FRET-based distance dynamics (bunching) with a correlation-function-specific amplitude (b_{ij}) and a global decay time of distance fluctuations (τ_b), and the third factor accounts for the photo-physical triplet dynamics of the dyes that typically take place at timescales of several microseconds with a correlation-function-specific amplitude (c_{ij}) and decay time (τ_c^{ij}). The factor N_{ij}^{-1} is related to the average number of molecules in the confocal volume. The ± sign in the second factor is for autocorrelations (+) and cross-correlations (−). The global correlation time τ_b (global for all four correlation functions) then characterizes the timescale of distance fluctuation between donor and acceptor. Yet, caution has to be taken. The highly nonlinear distance dependence of FRET (Eq. 10) acts as a filter that flattens photon rate fluctuations much below and above the Förster distance R_0,

[4] For example, after FRET from donor to acceptor, the acceptor is in the excited state. If the donor is re-exited before the acceptor relaxes to the ground state, both dyes will be in the excited state.

which affects the correlation time. To retrieve the correct timescale of distance fluctuations, i.e., the reconfiguration time of the disordered chain, this filter has to be taken into account. Gopich et al. showed that the correlation time is given by [138]

$$\tau_b = D^{-1}\int_a^L P(r)^{-1}\left[\int_a^r \delta\tilde{n}(\rho)P(\rho)\mathrm{d}\rho\right]^2 \mathrm{d}r \Big/ \int_a^L \delta\tilde{n}(r)^2 P(r)\mathrm{d}r. \tag{36}$$

As in Eq. 12, $P(r)$ is again the donor–acceptor distance distribution and D is the intra-chain diffusion coefficient that characterizes the timescale of distance fluctuations. Knowing D and $P(r)$ then fully characterizes the dynamics of an IDP or an unfolded protein. The goal is therefore to compute D assuming a suitable model of the distance distribution. Since τ_b is identical for all correlation functions, it suffices to only consider the photon rates $\tilde{n}$ of one of the dyes, the donor in our case ($\tilde{n} = \tilde{n}_D$). The term $\delta\tilde{n}(r) = \tilde{n}(r) - \langle\tilde{n}\rangle$ can then be computed from the kinetic photo-physical scheme of excited states and ground states in a two-color FRET system (Fig. 6b). Neglecting the possibility that donor and acceptor can simultaneously populate their excited states and using the base (DA, D*A, DA*) where D and A indicate the ground state of donor and acceptor, respectively, and the asterisk indicates the excited states of the dyes, the populations of these photo-physical states expressed in the vector $\mathbf{p} = (p_{DA}\ p_{D^*A}\ p_{DA^*})$ are given by a linear and homogeneous differential equation system

$$\dot{\mathbf{p}} = \mathbf{K}\mathbf{p}$$

with the rate matrix

$$\mathbf{K} = \begin{pmatrix} -k_{ex} & k_D & k_A \\ k_{ex} & -[k_D + k_T(r)] & 0 \\ 0 & k_T(r) & -k_A \end{pmatrix}. \tag{37}$$

Here, $k_{ex} = \sigma I/h\nu$ is the excitation rate of the donor that depends on the laser intensity (I) and the absorption cross-section of the donor (σ) at the excitation wavelength λ ($\nu = c/\lambda$ where c is the speed of light), k_D and k_A are the decay rates of the excited states of donor and acceptor, respectively, and $k_T(r)$ is the distance-dependent rate of energy transfer from donor to acceptor (see Eq. 16). The photon rate of the donor as function of the distance is then given by $\tilde{n}(r) = \phi_D k_D p_{D^*A}^{ss}$ where the superscript ss indicates the steady-state population that is obtained from $\mathbf{0} = \mathbf{K}\mathbf{p}$ and ϕ_D is the quantum yield of the donor dye. As usual, the average donor photon rate is given by $\langle\tilde{n}\rangle = \int \tilde{n}(r)P(r)\mathrm{d}r$ such that the term $\delta\tilde{n}(r)$ is directly accessible. Typical values of these parameters for the dye pair Alexa Fluor 488 and Alexa Fluor 594 are $k_{ex} \approx 0.02\ \mathrm{ns}^{-1}$, $k_D \sim k_A \approx 0.25\ \mathrm{ns}^{-1}$, and $\phi_D > 0.9$. Thus, Eqs. 36 and 37 together with the measured correlation time and a model for the distance distribution $P(r)$ can now be used to compute the intra-chain diffusion coefficient D, i.e., the diffusion coefficient characterizing the motion of donor and acceptor dyes relative to each other.[5] Once D is known, also the reconfiguration time of the chain τ_r can be calculated using a similar approach, i.e., by replacing $\tilde{n}(r)$ with r in Eq. 36. Hence, reconfiguration times obtained using FRET-coupled nsFCS experiments typically depend on the model $P(r)$.

Importantly, Gopich and Szabo derived an elegant formalism with which auto- and cross-correlation functions can be directly computed for any kinetic model [139]. In contrast to the semi-empirical approach of determining correlation times using fits with exponential functions (Eq. 35) followed by a conversion to intra-chain diffusion coefficients and reconfiguration times using Eqs. 36 and 37, the method also effectively utilizes the amplitude information of the correlation functions. With the definitions in Eq. 37, the correlation function is given by

$$g_{ij}(\tau) = \frac{\mathbf{1}^T \mathbf{V}_j e^{\mathbf{K}\tau} \mathbf{V}_i \mathbf{p}_{ss}}{\left(\mathbf{1}^T \mathbf{V}_j \mathbf{p}_{ss}\right)\left(\mathbf{1}^T \mathbf{V}_j \mathbf{p}_{ss}\right)}. \tag{38}$$

Here, $\mathbf{1}^T = (1\ 1\ldots)$ is the transposed unit vector, $e^{\mathbf{K}\tau}$ is the matrix exponential of $\mathbf{K}\tau$, $\mathbf{p}_{ss}$ is the steady-state solution of Eq. 37, and $\mathbf{V}_j$ and $\mathbf{V}_i$ are detection matrices for channel i and j that indicate which transition is being monitored. For donor and acceptor detection, the matrices in our particular case are given by

$$\mathbf{V}_D = \xi_D k_D Q_D \begin{pmatrix} 0 & 1 & 0 \\ 0 & 0 & 0 \\ 0 & 0 & 0 \end{pmatrix} \text{and } \mathbf{V}_A = \xi_A k_A Q_A \begin{pmatrix} 0 & 0 & 1 \\ 0 & 0 & 0 \\ 0 & 0 & 0 \end{pmatrix}. \tag{39}$$

Here, Q_D and Q_A are the quantum yields of donor and acceptor, and ξ_D, ξ_A are the detection efficiencies. Also, leakage (cross-talk) from the donor to the acceptor channel can easily be implemented by correcting the acceptor detection matrix via $\mathbf{V'}_A = \mathbf{V}_A + \beta \mathbf{V}_D$. Using Eqs. 38 and 39 to directly fit the correlation functions of IDPs and unfolded proteins that sample a distance distribution $P(r)$ requires a combination of photophysics with the distance dynamics of the chain [34]. To this end, the dynamics of the chain are described as a one-dimensional diffusive process in the potential of mean force (PMF) given by the distribution $P(r)$. This approach essentially pictures the molecule as a dumbbell with donor and acceptor on opposite ends, linked by a spring whose potential is given by the PMF. The rate matrix then becomes a combination of two parts, intra-chain diffusion and photophysics

$$\mathbf{K}_{\text{diff}} = D\partial/\partial r P(r) \partial/\partial r\, P(r)^{-1}\, \mathbf{I} + \mathbf{K}(r) \tag{40}$$

where $\mathbf{K}(r)$ describes the distance-dependent photophysics and the first term describes diffusion in a PMF. Here, $\mathbf{I}$ is a 3 × 3 unit matrix and D is the diffusion coefficient. To use Eq. 40 for computing correlation functions with Eq. 38, the

[5] Importantly, the intrinsic diffusion of donor and acceptor relative to each other should not be confused with the translational diffusion of the whole molecule.

diffusion operator (first term) has to be discretized, a procedure detailed in Nettels et al. [34]. For a practical perspective, fits of correlation functions with empirical exponentials (Eq. 35) or with an appropriate model (Eqs. 37–40) are very similar in quality. However, the number of free parameters is significantly reduced in the model-dependent fit since the excitation and emission rates of the dyes can be determined in advance, thus leaving only two free parameters, the diffusion D and the average donor–acceptor distance $\langle r^2\rangle^{1/2} = R_{DA}$ as fitting parameters, the latter of which being additionally constrained by the average FRET efficiency $\langle E\rangle$ (Eq. 12).

4.2 Polymer Models for IDP Dynamics and Their Limitations

Which microscopic molecular processes determine the reconfiguration timescales of IDPs? Clearly, the complex arrangement of thousands of atoms in IDPs and an astronomical number of chain conformations poses challenges to analytical theories. Yet, rather simple analytically solvable bead-and-spring models, such as the Rouse model and derived models, turned out to be surprisingly successful [33, 36, 37, 140, 141]. The concept of these model goes back to P. E. Rouse who aimed at describing the viscoelastic properties of homopolymer solutions [140]. In the Rouse model [140], an IDP is composed of $n = N_K + 1$ beads that are connected by N_K segment bonds. As described in Sect. 3.1, these Kuhn segments consist of several bonds such that the chain obeys Gaussian statistics, resulting in a “phantom chain” in which beads are linked via harmonic springs [142]. The length of a segment b_K is chosen such that the mean square end-to-end distance of the chain can be written as $\langle r^2\rangle = b_K^2 N_K$ (Eq. 14). The force-balance of a bead at position k will now be given by three terms: (1) an elastic spring force $\mathbf{f}_{ek}$, (2) a friction force due to the solvent $\mathbf{f}_{sk}$, and (3) a random force $\mathbf{f}_{bk}$ (Fig. 7a):

$$\mathbf{f}_{sk} + \mathbf{f}_{ek} = \mathbf{f}_{bk} \tag{41}$$

Assuming that solvent friction acts on individual beads, $\mathbf{f}_{sk}$ is given by Stokes’ law

$$\mathbf{f}_{sk} = \zeta_s \dot{\mathbf{r}}_k \text{ with } \zeta_s = 6\pi\eta a. \tag{42}$$

Here, ζ_s is the solvent friction coefficient for an individual bead, which is determined by its hydrodynamic radius a and by the solvent viscosity η, and $\mathbf{r}_k$ is the vector that defines the position of bead k. The second term in Eq. 41, i.e., the elastic spring force, can be derived from Hooke’s law. However, a bead in position k experiences an elastic force from two neighbors, namely at $k + 1$ and $k - 1$ such that

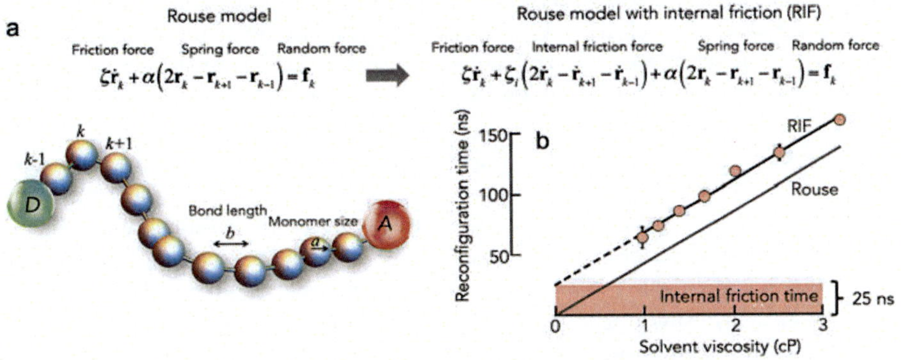

Fig. 7 Polymer dynamics models and their comparison with nsFCS. (**a**) A scheme of a coarse-grained homopolymer and a comparison of the equations of motion (Langevin equations) of the Rouse model without (*left*) and with (*right*) internal friction. (**b**) Reconfiguration times for the IDP (ACTR) obtained with nsFCS (circles) and a fit with the Rouse model with internal friction (RIF, black line). The internal friction component is indicated as red area. A Rouse model with the same monomer size (gray line) extrapolates to zero reconfiguration time at vanishing viscosity

$$\mathbf{f}_{ek} = \alpha(2\mathbf{r}_k - \mathbf{r}_{k+1} - \mathbf{r}_{k-1}) \text{ with } \alpha = 3k_BTN_K/\langle r^2 \rangle \tag{43}$$

and the spring constant α describes the stiffness of the harmonic potentials between neighbored beads. With Eqs. 41–43, we can write the Langevin equation for the x-coordinate of a bead k as

$$\zeta_s\dot{x}(k, t) + \alpha[2x(k, t) - x(k+1, t) - x(k-1, t)] = X(k, t) \tag{44}$$

Here, $X(k,t)$ is the x-component of the random force that satisfies the usual fluctuation dissipation theorem of Gaussian-distributed white noise according to

$$\langle X(k, t)X(k, t')\rangle = 2k_BT\zeta_s\delta(t - t')\Delta(k - k').$$

Here, Δ indicates the Kronecker-delta. The solution has been described in the past [36, 141] and we will not re-iterate it here. It suffices to say that a spectrum of relaxation times is obtained, as expected for a coupled harmonic oscillator, and the mode-dependent relaxation time τ_R of the Rouse model is given by

$$\tau_R \approx \zeta_s/\alpha q^2 \propto \eta \tag{45}$$

with $q = \pi z/N_K$ with $z = \{1, 2, 3, \ldots\}$ being the mode number. Eq. 45 implies that the dynamics of the chain will slow down with increasing solvent viscosity and the relaxation of the chain is faster for higher modes, i.e., for larger z. Notably, Eq. 45 predicts $\tau_r \propto N_K^2$, which is different from the correct scaling $\tau_r \propto N_K^{3\nu}$ given by the Zimm model that also accounts for hydrodynamic interactions [63, 75].

Numerous nsFCS experiments on IDPs have shown that Eq. 45 does not correctly capture the dynamics of IDPs [33, 131] (Fig. 7b). Although the proportionality between reconfiguration time and solvent viscosity has indeed been found, a solvent-independent friction component of the type [143, 144]

$$\tau_r \approx \zeta_{\text{eff}}/\alpha q^2 \text{ with } \zeta_{\text{eff}} = \zeta_s + q^2 \zeta_i \tag{46}$$

had to be invoked to satisfactorily fit the experimental chain reconfiguration time τ_r (Fig. 7a). Here, ζ_i describes an additional friction process independent of the solvent viscosity that comes from processes internal to the protein. Eq. 46 can also be written as $\tau_r = \tau_R + \tau_i$ where τ_i is a viscosity-independent internal friction timescale (Fig. 7b). How can an additional friction process be justified? At high solvent friction such as in aqueous solutions, Kramers' reaction rate theory predicts that the reaction time τ is proportional to the friction coefficient of a particle [145]. Assuming continuum hydrodynamics, the reaction time should therefore scale with the solvent viscosity ($\tau \propto \eta$) without an offset, which has indeed been observed for slow reactions such as the barrier-dominated millisecond folding dynamics of two-state folders [146, 147]. However, substantial deviations were found for comparably fast reactions [148–150] where a linear extrapolation of the reaction time to zero viscosity also identified a limiting time scale τ_i [148]. An alternative to a viscosity dependence with an offset (as in Eq. 46) can be found if the scaling between reaction time and solvent viscosity has a weaker dependence such as $\tau \propto \eta^\beta$ with $\beta < 1$ [151, 152]. Indeed, such fractional viscosity dependencies have been identified for the diffusion of small molecular compounds [153–157] and physical interpretations range from a viscosity-dependent change of the hydrodynamic coupling between solvent and particle [158, 159], over a breakdown of the continuum hydrodynamics due to the granularity of the solvent at small length scales [153, 160], up to a breakdown of Kramers' theory due to memory effects caused by solvent relaxation [161, 162].

Yet, support for the presence of a limiting internal friction timescale (τ_i) came from recent nsFCS experiments of unfolded and intrinsically disordered proteins (IDPs) [33, 131, 163]. Though these experiments suggested that internal friction results from a rather local process that acts on individual Kuhn segments (typically around 4 peptide bonds) [33], its molecular origin remained elusive and several possibilities such as bond rotations [143, 164], mode-coupling [165], or steric interference of chain segments [166] are in discussion [36]. Identifying the molecular events that cause internal friction in unstructured polypeptides might be a crucial step toward understanding the elementary processes in the folding of proteins and IDPs. Recent simulations and experiments point toward a combination of effects, including dihedral angle flipping, intra-chain interactions, and solvent memory effects [37, 167–169]. Importantly, some of these effects have already been incorporated in theories by Bazùa and Williams [143], Allegra and Ganazzoli [164], and de Gennes [166]. Other authors even addressed more essential simplifications in Rouse-type models such as the assumption of harmonic springs in bead-and-spring models [165]. However, a direct comparison of this variety of theories with experimental results demonstrated that all these theories are capable of describing experimental results on the disordered protein ACTR [36].

It might not be surprising that analytical theories capture internal friction effects with an additional fitting parameter yet, an identification of the molecular processes that give rise to this parameter remains elusive based on experiment and theory alone. All-atom simulations with explicit solvent therefore became an ideal tool to disentangle the molecular contributions that define reconfiguration times, intra-chain diffusion coefficients, and internal friction effects. However, precise estimates from all-atom simulations require an optimized water model. In fact, viscosities of different water models vary tremendously [170], which strongly alters intra-chain dynamics. TIP4P/2005 has been found to correctly capture the viscosity of water [170, 171] and therefore force fields using this water model have been shown to provide a faithful estimate of the intra-chain dynamics measured with smFRET-nsFCS experiment in the order of hundred nanoseconds [22, 97]. Most importantly, simulations allow a direct analysis of the molecular origins of internal friction effects in IDPs. Also in simulations, the reconfiguration time of IDPs deviates from the prediction of the simple Rouse model (Eqs. 40–43). By modifying the mass of water to mimic changes in solvent viscosity, internal friction behavior was reproduced in all-atom explicit-solvent simulations down to viscosities that are typically inaccessible experimentally ($\eta \gg 1cP$) [168]. These simulations showed that internal friction effects were primarily associated with backbone dihedral transitions [167, 169] as suspected by Bazùa and Williams four decades ago [143], but also intra-chain interactions contributed to the nonlinear viscosity dependence of IDPs, thus showing that they are sequence-dependent [172].

Acknowledgments H.H. thanks the Israel Science Foundation (ISF 2253/18) and the European Research Council (ERC-CoG 864578). W.Z. acknowledges the support from the National Science Foundation (MCB-2015030) and the National Institutes of Health (R35GM146814).

References

1. van der Lee R et al (2014) Classification of intrinsically disordered regions and proteins. Chem Rev 114:6589–6631
2. Uversky VN, Dunker AK (2008) Biochemistry. Controlled chaos. Science 322(5906): 1340–1341
3. Wright PE, Dyson HJ (2014) Intrinsically disordered proteins in cellular signalling and regulation. Nat Rev Mol Cell Biol 16(1):18–29
4. Ferreon ACM, Ferreon JC, Wright PE, Deniz AA (2013) Modulation of allostery by protein intrinsic disorder. Nature 498(7454):390–394
5. Gsponer J, Futschik ME, Teichmann SA (2008) Tight regulation of unstructured proteins: from transcript synthesis to protein degradation. Science 322:1365–1368
6. Schuler B et al (2020) Binding without folding - the biomolecular function of disordered polyelectrolyte complexes. Curr Opin Struct Biol 60:66–76

7. Borgia A et al (2018) Extreme disorder in an ultrahigh-affinity protein complex. Nature 555(7694):61–66
8. Holmstrom ED, Liu Z, Nettels D, Best RB, Schuler B (2019) Disordered RNA chaperones can enhance nucleic acid folding via local charge screening. Nat Comms 10(1):2453–2411
9. Mittag T et al (2008) Dynamic equilibrium engagement of a polyvalent ligand with a single-site receptor. Proc Natl Acad Sci U S A 105(46):17772–17777
10. Hendus-Altenburger R et al (2016) The human Na + /H + exchanger 1 is a membrane scaffold protein for extracellular signal-regulated kinase 2. BMC Biol 14(1):1–17
11. Milles S et al (2015) Plasticity of an ultrafast interaction between nucleoporins and nuclear transport receptors. Cell 163(3):734–745
12. Wiggers F et al (2021) Diffusion of a disordered protein on its folded ligand. Proc Natl Acad Sci U S A 118(37):e2106690118
13. Deniz AA et al (2000) Single-molecule protein folding: diffusion fluorescence resonance energy transfer studies of the denaturation of chymotrypsin inhibitor 2. Proc Natl Acad Sci U S A 97(10):5179–5184
14. Schuler B, Lipman E, Eaton W (2002) Probing the free-energy surface for protein folding with single-molecule fluorescence spectroscopy. Nature 419(6908):743–747
15. Kuzmenkina E, Heyes C, Nienhaus G (2005) Single-molecule Forster resonance energy transfer study of protein dynamics under denaturing conditions. Proc Natl Acad Sci U S A 102(43):15471–15476
16. Sherman E, Haran G (2006) Coil-globule transition in the denatured state of a small protein. Proc Natl Acad Sci U S A 103(31):11539–11543
17. Merchant K, Best R, Louis J, Gopich I, Eaton W (2007) Characterizing the unfolded states of proteins using single-molecule FRET spectroscopy and molecular simulations. Proc Natl Acad Sci U S A 104(5):1528–1533
18. Hofmann H, Golbik R, Ott M, Hübner C, Ulbrich-Hofmann R (2008) Coulomb forces control the density of the collapsed unfolded state of barstar. J Mol Biol 376(2):597–605
19. Müller-Späth S et al (2010) Charge interactions can dominate the dimensions of intrinsically disordered proteins. Proc Natl Acad Sci U S A 107:14609–14614
20. Hofmann H et al (2012) Polymer scaling laws of unfolded and intrinsically disordered proteins quantified with single-molecule spectroscopy. Proc Natl Acad Sci U S A 109(40): 16155–16160
21. Wuttke R et al (2014) Temperature-dependent solvation modulates the dimensions of disordered proteins. Proc Natl Acad Sci U S A 111(14):5213–5218
22. Borgia A et al (2016) Consistent view of polypeptide chain expansion in chemical denaturants from multiple experimental methods. J Am Chem Soc 138:11714
23. Schuler B, Soranno A, Hofmann H, Nettels D (2016) Single-molecule FRET spectroscopy and the polymer physics of unfolded and intrinsically disordered proteins. Annu Rev Biophys 45: 207–231
24. Grossman-Haham I, Rosenblum G, Namani T, Hofmann H (2018) Slow domain reconfiguration causes power-law kinetics in a two-state enzyme. Proc Natl Acad Sci U S A 115(3):513–518
25. Vancraenenbroeck R, Hofmann H (2018) Occupancies in the DNA-binding pathways of intrinsically disordered helix-loop-helix leucine-zipper proteins. J Phys Chem B 122:11460–11467
26. Vancraenenbroeck R, Harel YS, Zheng W, Hofmann H (2019) Polymer effects modulate binding affinities in disordered proteins. Proc Natl Acad Sci U S A 116(39):19506–19512
27. Fuertes G et al (2017) Decoupling of size and shape fluctuations in heteropolymeric sequences reconciles discrepancies in SAXS vs. FRET measurements. Proc Natl Acad Sci U S A 114(31):E6342–E6351
28. Mukhopadhyay S, Krishnan R, Lemke E, Lindquist S, Deniz A (2007) A natively unfolded yeast prion monomer adopts an ensemble of collapsed and rapidly fluctuating structures. Proc Natl Acad Sci U S A 104(8):2649–2654

29. Aznauryan M et al (2016) Comprehensive structural and dynamical view of an unfolded protein from the combination of single-molecule FRET, NMR, and SAXS. Proc Natl Acad Sci U S A 113(37):E5389–E5398
30. Hoffmann A et al (2007) Mapping protein collapse with single-molecule fluorescence and kinetic synchrotron radiation circular dichroism spectroscopy. Proc Natl Acad Sci U S A 104(1):105–110
31. Soranno A et al (2014) Single-molecule spectroscopy reveals polymer effects of disordered proteins in crowded environments. Proc Natl Acad Sci U S A 111(13):4874–4879
32. Nettels D et al (2009) Single-molecule spectroscopy of the temperature-induced collapse of unfolded proteins. Proc Natl Acad Sci U S A 106:20740–20745
33. Soranno A et al (2012) Quantifying internal friction in unfolded and intrinsically disordered proteins with single molecule spectroscopy. Proc Natl Acad Sci U S A 109:17800–17806
34. Nettels D, Gopich I, Hoffmann A, Schuler B (2007) Ultrafast dynamics of protein collapse from single-molecule photon statistics. Proc Natl Acad Sci U S A 104(8):2655–2660
35. Zosel F, Haenni D, Soranno A, Nettels D, Schuler B (2017) Combining short- and long-range fluorescence reporters with simulations to explore the intramolecular dynamics of an intrinsically disordered protein. J Chem Phys 147:152708
36. Soranno A, Zosel F, Hofmann H (2018) Internal friction in an intrinsically disordered protein-comparing Rouse-like models with experiments. J Chem Phys 148(12):123326
37. Soranno A et al (2017) Integrated view of internal friction in unfolded proteins from single-molecule FRET, contact quenching, theory, and simulations. Proc Natl Acad Sci U S A 114: E1833–E1839
38. Wahl M et al (2008) Scalable time-correlated photon counting system with multiple independent input channels. Rev Sci Instrum 79(12):123113
39. Wahl M, Rahn H-J, Gregor I, Erdmann R, Enderlein J (2007) Dead-time optimized time-correlated photon counting instrument with synchronized, independent timing channels. Rev Sci Instrum 78(3):033106
40. Dertinger T et al (2007) Two-focus fluorescence correlation spectroscopy: a new tool for accurate and absolute diffusion measurements. ChemPhysChem 8(3):433–443
41. Eggeling C et al (2001) Data registration and selective single-molecule analysis using multi-parameter fluorescence detection. J Biotechnol 86(3):163–180
42. Schuler B (2007) Application of single molecule Förster resonance energy transfer to protein folding. Methods Mol Biol 350:115–138
43. Rosenblum G et al (2021) Allostery through DNA drives phenotype switching. Nat Commun 12(1):2967–2912
44. Hellenkamp B et al (2018) Precision and accuracy of single-molecule FRET measurements-a multi-laboratory benchmark study. Nat Methods 15(9):669–676
45. Kapanidis AN et al (2004) Fluorescence-aided molecule sorting: analysis of structure and interactions by alternating-laser excitation of single molecules. Proc Natl Acad Sci U S A 101(24):8936–8941
46. Müller BK, Zaychikov E, Bräuchle C, Lamb DC (2005) Pulsed interleaved excitation. Biophys J 89(5):3508–3522
47. Kapanidis AN et al (2005) Alternating-laser excitation of single molecules. Acc Chem Res 38(7):523–533
48. Laurence T, Kong X, Jäger M, Weiss S (2005) Probing structural heterogeneities and fluctuations of nucleic acids and denatured proteins. Proc Natl Acad Sci U S A 102(48): 17348–17353
49. Gopich I, Szabo A (2005) Theory of photon statistics in single-molecule Förster resonance energy transfer. J Chem Phys 122(1):14707
50. Lipman E, Schuler B, Bakajin O, Eaton W (2003) Single-molecule measurement of protein folding kinetics. Science 301(5637):1233–1235
51. Zheng W et al (2018) Inferring properties of disordered chains from FRET transfer efficiencies. J Chem Phys 148(12):123329

52. Fisher ME (1966) Shape of a self-avoiding walk or polymer chain. J Chem Phys 44(616)
53. Des Cloizeaux J (1974) Lagrangian theory for a self-avoiding random chain. Phys Rev A 10(5):1665–1669
54. O'Brien EP, Morrison G, Brooks BR, Thirumalai D (2009) How accurate are polymer models in the analysis of Förster resonance energy transfer experiments on proteins? J Chem Phys 130(12):124903
55. Gopich IV, Szabo A (2012) Theory of the energy transfer efficiency and fluorescence lifetime distribution in single-molecule FRET. Proc Natl Acad Sci U S A 109(20):7747–7752
56. Kalinin S, Valeri A, Antonik M, Felekyan S, Seidel CAM (2010) Detection of structural dynamics by FRET: a photon distribution and fluorescence lifetime analysis of systems with multiple states. J Phys Chem B 114:7983–7995
57. Dale RE, Eisinger J, Blumberg B (1979) Orientational freedom of molecular probes. Biophys J 26:161–193
58. Schuler B, Lipman EA, Steinbach PJ, Kumke M, Eaton WA (2005) Polyproline and the "spectroscopic ruler" revisited with single-molecule fluorescence. Proc Natl Acad Sci U S A 102(8):2754–2759
59. Hillger F et al (2008) Probing protein-chaperone interactions with single-molecule fluorescence spectroscopy. Angew Chem Int Ed Engl 47(33):6184–6188
60. Hofmann H et al (2010) Single-molecule spectroscopy of protein folding in a chaperonin cage. Proc Natl Acad Sci U S A 107(26):11793–11798
61. Sun S, Nishio I, Swislow G, Tanaka T (1980) The coil-globule transition - radius of gyration of polystyrene in cyclohexane. J Chem Phys 73(12):5971–5975
62. de Gennes P (1975) Collapse of a polymer chain in poor solvents. J Phys Lett 3:L55–L57
63. Doi M, Edwards S (1988) The theory of polymer dynamics. Oxford University Press, New York
64. Dua A, Vilgis TA (2005) Self-consistent variational theory for globules. EPL 71(1):49
65. Sanchez I (1979) Phase transition behavior of the isolated polymer chain. Macromolecules 12: 980–988
66. Hofmann H (2016) Understanding disordered and unfolded proteins using single-molecule FRET and polymer theory. Methods Appl Fluoresc 4(4):042003
67. England J, Haran G (2011) Role of solvation effects in protein denaturation: from thermodynamics to single molecules and back. Annu Rev Phys Chem 62:257–277
68. Nozaki Y, Tanford C (1970) The solubility of amino acids, diglycine, and triglycine in aqueous guanidine hydrochloride solutions. J Biol Chem 245(7):1648–1652
69. Thirumalai D, Liu Z, O'Brien EP, Reddy G (2012) Protein folding: from theory to practice. Curr Opin Struct Biol 23:1–8
70. O'Brien E, Ziv G, Haran G, Brooks B, Thirumalai D (2008) Effects of denaturants and osmolytes on proteins are accurately predicted by the molecular transfer model. Proc Natl Acad Sci U S A 105:13403–13408
71. Ziv G, Haran G (2009) Protein folding, protein collapse, and Tanford's transfer model: lessons from single-molecule FRET. J Am Chem Soc 131(8):2942–2947
72. Ziv G, Thirumalai D, Haran G (2009) Collapse transition in proteins. Phys Chem Chem Phys 11(1):83–93
73. Flory P (1949) The configuration of real polymer chains. J Chem Phys 17(3):303
74. Le Guillou JC, Zinn-Justin J (1977) Critical exponents for the n-vector model in three dimensions from field theory. Phys Rev Lett 39:95–98
75. Rubinstein M, Colby RH (2012) Polymer physics. Oxford University Press, Oxford, p 440
76. Kohn J et al (2004) Random-coil behavior and the dimensions of chemically unfolded proteins. Proc Natl Acad Sci U S A 101(34):12491–12496
77. Riback JA et al (2017) Innovative scattering analysis shows that hydrophobic disordered proteins are expanded in water. Science 358(6360):238–241
78. Wilkins D et al (1999) Hydrodynamic radii of native and denatured proteins measured by pulse field gradient NMR techniques. Biochemistry 38(50):16424–16431

79. Grosberg A, Kuznetsov D (1992) Quantitative theory of the globule-to-coil transition 1. Link density distribution in a globule and its radius of gyration. Macromolecules 25(7):1970–1979
80. Crick SL, Jayaraman M, Frieden C, Wetzel R, Pappu RV (2006) Fluorescence correlation spectroscopy shows that monomeric polyglutamine molecules form collapsed structures in aqueous solutions. Proc Natl Acad Sci U S A 103(45):16764–16769
81. Grosberg A, Kuznetsov D (1992) Quantitative theory of the globule-to-coil transition. 2. Density-density correlation in a globule and the hydrodynamic radius of a macromolecule. Macromolecules 25:1980–1990
82. Uzawa T et al (2006) Time-resolved small-angle X-ray scattering investigation of the folding dynamics of heme oxygenase: implication of the scaling relationship for the submillisecond intermediates of protein folding. J Mol Biol 357(3):997–1008
83. Camacho C, Thirumalai D (1993) Kinetics and thermodynamics of folding in model proteins. Proc Natl Acad Sci U S A 90:6369–6372
84. Mao AH, Crick SL, Vitalis A, Chicoine CL, Pappu RV (2010) Net charge per residue modulates conformational ensembles of intrinsically disordered proteins. Proc Natl Acad Sci U S A 107(18):8183–8188
85. Sherry KP, Das RK, Pappu RV, Barrick D (2017) Control of transcriptional activity by design of charge patterning in the intrinsically disordered RAM region of the Notch receptor. Proc Natl Acad Sci U S A 114(44):E9243–E9252
86. Peran I et al (2019) Unfolded states under folding conditions accommodate sequence-specific conformational preferences with random coil-like dimensions. Proc Natl Acad Sci U S A 116(25):12301–12310
87. Martin EW et al (2020) Valence and patterning of aromatic residues determine the phase behavior of prion-like domains. Science 367(6478):694–699
88. Holm C et al (2004) Polyelectrolyte theory. Springer, Berlin
89. Dobrynin AV, Rubinstein M (2005) Theory of polyelectrolytes in solutions and at surfaces. Prog Polym Sci 30(11):1049–1118
90. Higgs PG, Joanny J-F (1991) Theory of polyampholyte solutions. J Chem Phys 94(2): 1543–1554
91. Ha DT (1992) Conformations of a polyelectrolyte chain. Phys Rev A 46(6):R3012–R3015
92. Kundagrami A, Muthukumar M (2010) Effective charge and Coll-Globule transition of a polyelectrolyte chain. Macromolecules 43(5):2574–2581
93. Bhattacharjee A, Kundu P, Dua A (2011) Self-consisten theory of structures and transitions in weak polyampholytes. Macromol Theory Simul 20:75–84
94. Shakhnovich EI, Gutin AM (1989) Formation of unique structure in polypeptide chains. Theoretical investigation with the aid of a replica approach. Biophys Chem 34(3):187–199
95. Gomes G-NW, Namini A, Gradinaru CC (2022) Integrative conformational ensembles of Sic1 using different initial pools and optimization methods. Front Mol Biosci 9:910956
96. Lindorff-Larsen K, Best RB, Depristo MA, Dobson CM, Vendruscolo M (2005) Simultaneous determination of protein structure and dynamics. Nature 433:128–132
97. Best RB, Zheng W, Mittal J (2014) Balanced protein-water interactions improve properties of disordered proteins and non-specific protein association. J Chem Theory Comput 10(11): 5113–5124
98. Vitalis A, Pappu RV (2008) ABSINTH: a new continuum solvation model for simulations of polypeptides in aqueous solutions. J Comput Chem 30:673–699
99. Dignon GL, Zheng WW, Kim YC, Best RB, Mittal J (2018) Sequence determinants of protein phase behavior from a coarse-grained model. PLoS Comput Biol 14(1):e1005941
100. Robustelli P, Piana S, Shaw DE (2018) Developing a molecular dynamics force field for both folded and disordered protein states. Proc Natl Acad Sci U S A 115(21):E4758–E4766
101. Huang J et al (2017) CHARMM36m: an improved force field for folded and intrinsically disordered proteins. Nat Methods 14(1):71–73

102. Hess B, Kutzner C, Van der Spoel D, Lindahl E (2008) GROMACS4: algorithms for highly efficient, load-balanced, and scalable molecular simulation. J Chem Theor Comput 4(3): 435–447
103. Case DA et al (2005) The Amber biomolecular simulation programs. J Comput Chem 26(16): 1668–1688
104. Eastman P et al (2017) OpenMM 7: rapid development of high performance algorithms for molecular dynamics. PLoS Comput Biol 13(7):e1005659
105. Phillips JC et al (2005) Scalable molecular dynamics with NAMD. J Comput Chem 26:1781–1802
106. Shaw DE et al (2007) Anton, a special-purpose machine for molecular dynamics simulation. In: ISCA'07: 34th annual international symposium on computer architecture, conference proceedings, conference proceedings – annual international symposium on computer architecture. Association for Computing Machinery, New York, pp 1–12
107. Robustelli P, Piana S, Shaw DE (2020) Mechanism of coupled folding-upon-binding of an intrinsically disordered protein. J Am Chem Soc 142(25):11092–11101
108. Best RB, Hofmann H, Nettels D, Schuler B (2015) Quantitative interpretation of FRET experiments via molecular simulation: force field and validation. Biophys J 108:2721–2731
109. Das RK, Pappu RV (2013) Conformations of intrinsically disordered proteins are influenced by linear sequence distributions of oppositely charged residues. Proc Natl Acad Sci U S A 110(33):13392–13397
110. Zheng W et al (2016) Probing the action of chemical denaturant on an intrinsically disordered protein by simulation and experiment. J Am Chem Soc 138(36):11702–11713
111. Yoo TY et al (2012) Small-angle X-ray scattering and single-molecule FRET spectroscopy produce highly divergent views of the low-denaturant unfolded state. J Mol Biol 418:226–236
112. Zheng W, Best R (2018) An extended Guinier analysis for intrinsically disordered proteins. J Mol Biol 430:2540–2553
113. Witten TA, Schäfer L (1978) Two critical ratios in polymer solutions. J Phys A 11(9): 1843–1854
114. Gomes GW et al (2020) Conformational ensembles of an intrinsically disordered protein consistent with NMR, SAXS, and single-molecule FRET. J Am Chem Soc 142(37): 15697–15710
115. Brangwynne CP et al (2009) Germline P granules are liquid droplets that localize by controlled dissolution/condensation. Science 324(5935):1729–1732
116. Dyson HJ, Wright PE (2002) Coupling of folding and binding for unstructured proteins. Curr Opin Struct Biol 12(1):54–60
117. Theillet F-X et al (2014) Physicochemical properties of cells and their effects on intrinsically disordered proteins (IDPs). Chem Rev 114(13):6661–6714
118. Debye P, Hückel E (1923) Zur Theorie der Elektrolyte: I. Gefrierpunktserniedrigung und verwandte Erscheinungen. Phys Z 24(9):185–206
119. Hooper HH, Blanch HW, Prausnitz JM (1990) Configurational properties of partially ionized polyelectrolytes from Monte Carlo simulation. Macromolecules 23(22):4820–4829
120. English AE, Tanaka T, Edelman ER (1998) Polyampholytic hydrogel swelling transitions: limitations of the Debye–Hückel law. Macromolecules 31(6):1989–1995
121. Ashbaugh HS, Hatch HW (2008) Natively unfolded protein stability as a coil-to-globule transition in charge/hydropathy space. J Am Chem Soc 130(29):9536–9542
122. Latham AP, Zhang B (2019) Improving coarse-grained protein force fields with small-angle X-ray scattering data. J Phys Chem B 123(5):1026–1034
123. Dannenhoffer-Lafage T, Best RB (2021) A data-driven hydrophobicity scale for predicting liquid-liquid phase separation of proteins. J Phys Chem B 125(16):4046–4056
124. Regy RM, Thompson J, Kim YC, Mittal J (2021) Improved coarse-grained model for studying sequence dependent phase separation of disordered proteins. Protein Sci 30(7):1371–1379

125. Tesei G, Schulze TK, Crehuet R, Lindorff-Larsen K (2021) Accurate model of liquid-liquid phase behavior of intrinsically disordered proteins from optimization of single-chain properties. Proc Natl Acad Sci U S A 118(44)
126. Plimpton S (1995) Fast parallel algorithms for short-range molecular dynamics. J Comput Phys 117(1):1–19
127. Anderson JA, Lorenz CD, Travesset A (2008) General purpose molecular dynamics simulations fully implemented on graphics processing units. J Comput Phys 227(10):5342–5359
128. Lapidus LJ, Eaton WA, Hofrichter J (2000) Measuring the rate of intramolecular contact formation in polypeptides. Proc Natl Acad Sci U S A 97(13):7220–7225
129. Bieri O et al (1999) The speed limit for protein folding measured by triplet-triplet energy transfer. Proc Natl Acad Sci U S A 96(17):9597–9601
130. Hanbury Brown R, Twiss RQ (1956) Correlation between photons in two coherent beams of light. Nature 177(4497):27–29
131. Borgia A et al (2012) Localizing internal friction along the reaction coordinate of protein folding by combining ensemble and single-molecule fluorescence spectroscopy. Nat Commun 3:1195
132. Schuler B, Müller-Späth S, Soranno A, Nettels D (2012) Application of confocal single-molecule FRET to intrinsically disordered proteins. Methods Mol Biol 896:21–45
133. Haenni D, Zosel F, Reymond L, Nettels D, Schuler B (2013) Intramolecular distances and dynamics from the combined photon statistics of single-molecule FRET and Photoinduced electron transfer. J Phys Chem B 117(42):13015–13028
134. Schuler B, Hofmann H (2013) Single-molecule spectroscopy of protein folding dynamics-expanding scope and timescales. Curr Opin Struct Biol 23:1–12
135. Cubuk J et al (2021) The SARS-CoV-2 nucleocapsid protein is dynamic, disordered, and phase separates with RNA. Nat Commun 12(1):1936–1917
136. König I et al (2015) Single-molecule spectroscopy of protein conformational dynamics in live eukaryotic cells. Nat Methods 12(8):773–779
137. Schuler B, König I, Soranno A, Nettels D (2021) Impact of in-cell and in-vitro crowding on the conformations and dynamics of an intrinsically disordered protein. Angew Chem Int Ed Engl 133:2
138. Gopich IV, Nettels D, Schuler B, Szabo A (2009) Protein dynamics from single-molecule fluorescence intensity correlation functions. J Chem Phys 131(9):095102
139. Gopich IV, Szabo A (2008) Theory of photon counting in single molecule spectroscopy. World Scientific Publishing Co. Pte. Ltd., Singapore
140. Rouse PE (1953) A theory of the linear viscoelastic properties of dilute solutions of coiling polymers. J Chem Phys 21:1272–1280
141. Makarov DE (2010) Spatiotemporal correlations in denatured proteins: the dependence of fluorescence resonance energy transfer (FRET)-derived protein reconfiguration times on the location of the FRET probes. J Chem Phys 132(3):035104
142. de Gennes P-G (1979) Scaling concepts in polymer physics. Cornell University Press, Ithaca
143. Bazúa ER, Williams MC (1973) Molecular formulation of the internal viscosity in polymer dynamics, and stress symmetry. J Chem Phys 59:2858–2868
144. Khatri BS, McLeish TCB (2007) Rouse model with internal friction: a coarse grained framework for single biopolymer dynamics. Macromolecules 40(18):6770–6777
145. Kramers H (1940) Brownian motion in a field of force and the diffusion model of chemical reactions. Physica 7:284–304
146. Plaxco KW, Baker D (1998) Limited internal friction in the rate-limiting step of a two-state protein folding reaction. Proc Natl Acad Sci U S A 95(23):13591–13596
147. Jacob M, Geeves M, Holtermann G, Schmid FX (1999) Diffusional barrier crossing in a two-state protein folding reaction. Nat Struct Biol 6(10):923–926
148. Ansari A, Jones CM, Henry ER, Hofrichter J, Eaton WA (1992) The role of solvent viscosity in the dynamics of protein conformational changes. Science 256(5065):1796–1798

149. Pabit SA, Roder H, Hagen SJ (2004) Internal friction controls the speed of protein folding from a compact configuration. Biochemistry 43(39):12532–12538
150. Hagen SJ, Qiu L, Pabit SA (2005) Diffusional limits to the speed of protein folding: fact or friction? J Phys Condens Matter 17(18):S1503–S1514
151. Chung HS, Eaton WA (2013) Single-molecule fluorescence probes dynamics of barrier crossing. Nature 502(7473):685–688
152. Frauenfelder H, Fenimore PW, Chen G, McMahon BH (2006) Protein folding is slaved to solvent motions. Proc Natl Acad Sci U S A 103(42):15469–15472
153. Evans DF, Tominaga T, Davis HT (1981) Tracer diffusion in polyatomic liquids. J Chem Phys 74(2):1298
154. Evans DF, Tominaga T, Chan C (1979) Diffusion of symmetrical and spherical solutes in protic, aprotic, and hydrocarbon solvents. J Solution Chem 8:461–478
155. Pollack GL, Enyeart JJ (1985) Atomic test of the stokes-Einstein law. II. Diffusion of Xe through liquid hydrocarbons. Phys Rev A 31:980–984
156. Hiss TG, Cussler EL (1973) Diffusion in high viscosity liquids. AIChE 19:698–703
157. Ellerton HD, Mulcahy DE, Dunlop PJ, Reinfelds G (1964) Mutual frictional coefficients of several amino acids in aqueous solution at 25. J Phys Chem 68(2):403–408
158. Harris KR (2009) The fractional Stokes-Einstein equation: application to Lennard-Jones, molecular, and ionic liquids. J Chem Phys 131(5):054503
159. Zwanzig R, Harrison AK (1985) Modifications of the Stokes-Einstein formula. J Chem Phys 83(11):5861–5862
160. Bhattacharyya S, Bagchi B (1997) Anomalous diffusion of small particles in dense liquids. J Chem Phys 106(5):1757–1763
161. Grote RF, Hynes JT (1981) Reactive modes in condensed phase reactions. J Chem Phys 74: 4465–4475
162. Grote RF, Vanderzwan G, Hynes JT (1984) Frequency-dependent friction and solution reaction-rates. J Phys Chem 88(20):4676–4684
163. Neuweiler H, Johnson C, Fersht A (2009) Direct observation of ultrafast folding and denatured state dynamics in single protein molecules. Proc Natl Acad Sci U S A 106(44):18569–18574
164. Allegra G, Ganazzoli F (1981) Configurations and dynamics of real chains 2. Internal viscosity. Macromolecules 14(4):1110–1119
165. Adelman SA, Freed KF (1977) Microscopic theory of polymer internal viscosity: mode coupling approximation for the Rouse model. J Chem Phys 67(4):1380
166. de Gennes PG (1977) Origin of internal viscosities in dilute polymer solutions. J Chem Phys 66(12):5825–5826
167. Echeverria I, Makarov DE, Papoian GA (2014) Concerted dihedral rotations give rise to internal friction in unfolded proteins. J Am Chem Soc 136:8708–8713
168. Schulz JCF, Schmidt L, Best RB, Dzubiella J, Netz RR (2012) Peptide chain dynamics in light and heavy water: zooming in on internal friction. J Am Chem Soc 134(14):6273–6279
169. de Sancho D, Sirur A, Best RB (2014) Molecular origins of internal friction effects on protein-folding rates. Nat Commun 5:4307
170. Gonzalez MA, Abascal JL (2010) The shear viscosity of rigid water models. J Chem Phys 132(9):096101
171. Abascal JLF, Vega C (2005) A general purpose model for the condensed phases of water: TIP4P/2005. J Chem Phys 123:234505
172. Zheng W, Hofmann H, Schuler B, Best RB (2018) Origin of internal friction in disordered proteins depends on solvent quality. J Phys Chem B 122(49):11478–11487

Insights into the Conformational Dynamics of Potassium Channels Using Homo-FRET Approaches

Ana Coutinho, Clara Díaz-García, Ana Marcela Giudici, and María Lourdes Renart

Contents

A. Coutinho
iBB, Institute for Bioengineering and Biosciences, Instituto Superior Técnico, Universidade de Lisboa, Lisbon, Portugal

Associate Laboratory i4HB – Institute for Health and Bioeconomy at Instituto Superior Técnico, Universidade de Lisboa, Lisbon, Portugal

Dep. de Química e Bioquímica, Faculdade de Ciências, Universidade de Lisboa, Lisbon, Portugal
e-mail: ana.coutinho@tecnico.ulisboa.pt

C. Díaz-García
iBB, Institute for Bioengineering and Biosciences, Instituto Superior Técnico, Universidade de Lisboa, Lisbon, Portugal

Associate Laboratory i4HB – Institute for Health and Bioeconomy at Instituto Superior Técnico, Universidade de Lisboa, Lisbon, Portugal
e-mail: clara.diaz@tecnico.ulisboa.pt

A. M. Giudici and M. L. Renart (✉)
Instituto de Investigación, Desarrollo e Innovación en Biotecnología Sanitaria de Elche (IDiBE), Universidad Miguel Hernández, Elche, Alicante, Spain
e-mail: marcela@umh.es; lrenart@umh.es

R. Šachl, M. Amaro (eds.), *Fluorescence Spectroscopy and Microscopy in Biology*, Springer Ser Fluoresc (2023) 20: 443–478, https://doi.org/10.1007/4243_2022_24,
Published online: 11 August 2022

Abstract Potassium channels are integral membrane proteins that are ubiquitously found among prokaryotic and eukaryotic organisms. They play an essential role in diverse vital physiological functions, and the elucidation of the molecular mechanisms underlying their functional properties is critical due to their impact on numerous diseases. Even before the resolution of the first crystallographic structure of a potassium channel in 1998, fluorescence techniques were used to shed light on the structure–function relationship of these macromolecules. In this chapter, we first revise the main hetero- and homo-FRET strategies that have been implemented to elucidate the conformational dynamics of potassium channels, which complement the information provided by other biophysical techniques. We then use the prokaryotic potassium channel KcsA as a case study to illustrate how homo-FRET based approaches helped uncovering the conformational plasticity of its selectivity filter and the importance of the occupancy of the ion binding site S2 on the allosteric coupling between its inner and outer gates.

Keywords Conformational plasticity · Fluorescence spectroscopy · FRET · Homo-FRET · Intrinsic protein fluorescence · Ion channels · KcsA · Membrane protein dynamics · Potassium channels · Time-resolved fluorescence

1 Introduction

Biological membranes are intrinsically impermeable to ions and polar molecules, even though these compounds play a critical role in the cellular function and homeostasis. Selective permeability is therefore conferred by different types of membrane proteins, including pumps and ion channels. Particularly, ion channels are oligomeric pore-forming membrane proteins that facilitate nearly diffusion-limited rates of ions (~10^7–10^8 ions/s) across the membrane down to their electrochemical gradients by shifting from a closed to an open conformation upon an external stimulus. These biomolecules are widely distributed among prokaryotic and eukaryotic organisms, playing multiple roles such as the propagation of the action potential, the coordination of the heartbeat, the insulin secretion or even the orchestration of the immune response, among others [1–5]. The dysfunction of ion channels, triggered by different mutations or by autoimmune disorders, has been associated with numerous pathological conditions such as generalized epilepsy, episodic ataxia, long QT syndrome, cystic fibrosis, neonatal diabetes mellitus and many others [6–8]. In this way, the research on ion channel's structure and function is the cornerstone for the rational development of new therapies against these "channelopathies".

As there are a vast number of different functional and structural studies on ion channels, in this chapter we will focus on the potassium (K^+) channels superfamily. Nevertheless, almost all the biophysical approaches described here can be translated to the study of any type of channel. Potassium channels (as any other ion channel) are dynamical entities, playing their function through an ensemble of diverse conformations instead of a single static structure. They present a high selectivity for K^+ over Na^+ (the physiological competitor) and can also permeate other monovalent cations such as Rb^+, Cs^+, Tl^+ and NH_4^+ [1, 9]. The regulation of the functional properties of these channels, i.e., selectivity, high-rate K^+ permeation, blockade and inactivation, is orchestrated by multiple factors, including ion–protein, protein–protein and lipid–protein interactions [10–13].

The understanding of the functional properties of ion channels, in general, and K^+ channels, in particular, requires the ability to precisely resolve the structure and dynamics of the macromolecule in different experimental conditions, such as in the presence of a specific ligand or embedded in a lipid bilayer. Most of the high-resolution structures of ion channels typically represent snapshots of a particular state of the protein in a crystallized or frozen form, solved by X-ray crystallography or single-particle cryo-electron microscopy (cryo-EM). Other techniques, such as nuclear magnetic resonance (NMR) and electron paramagnetic resonance (EPR) spectroscopies (including double electron-electron resonance (DEER) measurements), small-angle X-ray scattering (SAXS) and small-angle neutron scattering (SANS), or cross-linking mass spectrometry (XL-MS), give information on the ensemble-average of all contributing conformations, even though some advances have been made in recent years to capture transient intermediates [14]. Due to the harsh experimental conditions often used by some of the above-mentioned biophysical techniques, their validity must be confirmed in milder and more physiological-like environments. Steady-state and time-resolved fluorescence spectroscopy often fulfil these requisites [15].

In this chapter, we will illustrate how Förster resonance energy transfer (FRET) is a powerful tool that can be used to map conformational transitions in homo-multimeric membrane proteins like the K^+ channels. Briefly, FRET is a nonradiative long-range dipole–dipole coupling mechanism in which energy is transferred from an excited donor fluorophore (D) to a nearby ground-state acceptor (A) with an absorption spectrum overlapping the emission spectrum of the donor. The energy transfer efficiency, E, depends on the inverse of the sixth power of the distance, R, between D and A [16, 17]. This steep sensitivity of the FRET signal to the donor–acceptor distance is the basis for its wide application as a "spectroscopic ruler", a term first coined by Haugland and Stryer [18]. Usually, the donor and acceptor pair consists of two chemically distinct molecules (hetero-FRET), but FRET can also occur between identical molecules (homo-FRET or donor–donor energy migration) when they display a small Stokes shift [16, 17]. The fundamentals and methodological background for performing hetero- and homo-FRET measurements have already been presented throughout this book. In the following sections, we will first summarize how FRET measurements can probe the conformational changes related to potassium channel's function, with an emphasis on the implementation of homo-FRET based methodologies to the study of homo-multimeric symmetric proteins. We will then use the prototypical bacterial KcsA potassium channel as a case study to illustrate how these homo-FRET based approaches can be applied to answer specific mechanistic questions about this system, namely the influence of the degree of ion occupancy on the structural rearrangements undergone by the selectivity filter (SF) and its allosteric coupling to the activation gate.

2 FRET Toolbox for Studying Multimeric Symmetric Proteins

2.1 *Hetero-FRET Measurements*

2.1.1 Single Donor–Acceptor Pair

The Förster theory of resonance energy transfer for a single donor–acceptor pair has been presented previously in this book, so only a brief summary is given here to introduce the notation to be used subsequently in this chapter. According to Förster, the energy transfer rate constant, k_T, between a single D and A pair separated by a distance R is given by:

$$k_T = \frac{1}{\tau_D}\left(\frac{R_0}{R}\right)^6 \tag{1}$$

where τ_D is the fluorescence lifetime of the donor in the absence of the acceptor, and R_0 is the Förster or critical radius [16]. The magnitude of the Förster radius depends on the spectral features of the donor and acceptor molecules and can be experimentally evaluated (in Å) according to:

$$R_0 = 0.2108 \cdot \sqrt[6]{\kappa^2 \cdot n^{-4} \cdot \Phi_D \cdot J(\lambda)} \tag{2}$$

where κ^2 is the orientational factor related to the relative orientations of the transition dipoles of the donor and acceptor, which is usually set to 2/3 (isotropic dynamic regime), n is the refractive index of the medium and Φ_D is the fluorescence quantum yield of the donor. $J(\lambda)$ (in M^{-1} cm^{-1} nm^4) is the spectral overlap integral between the donor and acceptor:

$$J(\lambda) = \int_0^\infty F_D(\lambda) \cdot \varepsilon_A(\lambda) \cdot \lambda^4 \cdot d\lambda \Big/ \int_0^\infty F_D(\lambda) \cdot d\lambda \tag{3}$$

where $F_D(\lambda)$ is the emission spectrum of the donor and $\varepsilon_A(\lambda)$ is the absorption spectra of the acceptor in M^{-1} cm^{-1}.

Usually, energy transfer is described by its efficiency, E, defined as the probability that an excited donor will return to the ground-state by giving its energy to a nearby acceptor. This can be expressed in terms of rate constants:

$$E = \frac{k_T}{k_f + k_{nr} + k_T} = \frac{k_T}{{\tau_D}^{-1} + k_T} \tag{4}$$

where k_f and k_{nr} are the radiative and nonradiative rate constants of the D, respectively. The combination of Eqs. (1) and (4) yields:

$$E = \frac{R_0^6}{R_0^6 + R^6} \tag{5}$$

The characteristic distance, R_0, can then be interpreted as the distance at which the energy transfer efficiency between an isolated D and A pair is 50%. This expression also reveals that the FRET efficiency is extremely sensitive to the donor–acceptor distance, R, when it falls within the range $0.5 \cdot R_0 < R < 1.5 \cdot R_0$.

Since FRET is a photophysical mechanism that provides an additional deactivation pathway to the excited fluorophores (Eq. 4), both the fluorescence intensity, I_D, and lifetime, τ_D, of the donor decrease in the presence of a nearby acceptor to I_{DA} and τ_{DA}, respectively. The energy transfer efficiency, E, can then be computed experimentally using either fluorescence parameter:

$$E = 1 - \frac{I_{DA}}{I_D} = 1 - \frac{\tau_{DA}}{\tau_D} \tag{6}$$

Alternatively, when A is a fluorescent molecule, a concomitant increase in the sensitized emission of the acceptor can also be detected. Once R_0 is determined spectroscopically (Eq. 2), the separating D–A distance can be calculated by introducing the experimentally determined energy transfer efficiency in Eq. (5) [16, 17]. The FRET technique typically measures distances in the 30–80 Å range, and therefore is particularly useful to evaluate changes in distances associated with conformational rearrangements in proteins.

2.1.2 Homo-Tetrameric Proteins

The application of FRET measurements to evaluate the conformational landscape of an ion channel under different experimental conditions often relies on fluorescent labelling the protein by using a natural or an engineered cysteine residue in the amino acid sequence by site-directed mutagenesis. These natural/engineered cysteines are the targets for covalently attaching an extrinsic label – donor/acceptor fluorophore – with an adequate highly reactive moiety, like a maleimide or iodoacetamide group [19–21].

Most of the potassium ion channels are homo-tetrameric proteins, and therefore their position-specific labelling with two different probes (D and A) is challenging, particularly when they react with the same group as the free sulfhydryl of cysteine residues. In fact, random labelling a protein with a mixture of the donor/acceptor pair results in a heterogeneous population of labelled proteins, even when there is a single cysteine residue per subunit [22–25]. This means that the interpretation of the measured energy transfer efficiencies is complicated by the geometric relationship among the subunits as FRET can occur both between the laterally adjacent (with $R_1 = R$) and the diagonally opposed subunits (with $R_2 = \sqrt{2}\ R$, and therefore $k_{T2} = k_{T1}/8$) (Table 1). In addition, the combinatorially conjugation of both the donor and acceptor among the four subunits of the protein must be considered [23, 25, 26]. Although both these factors preclude the exact determination of distances in a macroscopic hetero-FRET approach, nevertheless the performance of a cysteine-scan mutagenesis along the protein sequence allows mapping the regions where there is a systematic difference in energy transfer efficiencies between distinct functional states of the channels. This, in turn, contributes to the identification of the regions involved in the structural rearrangements undergone by the protein during its gating cycle or upon the addition of an activation or inhibitory ligand [23, 24].

Three main experimental strategies have been implemented to overcome these limitations. The first one entails mapping the conformational transitions of ion channels in vitro or in living cells using luminescence resonance energy transfer (LRET) [15, 27–30]. Briefly, whereas in FRET, energy transfer occurs between two organic fluorophores, in LRET, energy transfer usually takes place between a long-lived Tb^{+3}-chelate donor and a short-lived organic dye acceptor. Although LRET shares the same fundamental mechanism as conventional hetero-FRET, it offers some advantages since the use of chelated lanthanides as donors enhances the selectivity of the signal even in the presence of donor-only or acceptor-only labelled proteins [31, 32]. In particular, the lanthanide donors like Tb^{3+} have very narrow transitions, allowing for a clear separation between Tb donor emission and acceptor sensitized emission. Moreover, the long (ms) excited-state lifetime of the lanthanide donors allows for the collection of the sensitized acceptor emission signal (which only arises form donor–acceptor pairs) after a time delay sufficiently long so that the fluorescence from the directly excited acceptors decays to zero, thereby avoiding interference from direct excitation of the acceptor label. In this way, energy transfer efficiencies can be accurately calculated using the ratio between the sensitized acceptor lifetime and the donor-only lifetime [31, 32]. However, the long lifetimes of the luminescent lanthanides also allow for significant structural movements of the protein during their excited state, and therefore LRET can potentially skew the

Table 1 Time-resolved and steady-state fluorescence anisotropy ($r_i(t)$ *and* $<r>_i$, respectively) computed for a set of *i* fluorophores undergoing fully reversible isotropic homo-FRET within a cyclic molecular or supramolecular organization [39]. The systems correspond to a single protein with multiple identical fluorophores (intramolecular homo-FRET) or a homo-multimer (intermolecular homo-FRET). In this last case, only one fluorophore/subunit is considered. In all cases, fluorescence depolarization due to the overall rotation of the system was neglected

Topological model	Analytical solutions	Equation
Pair of fluorophores ($i = 2$)		
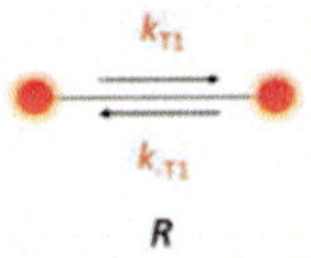	• $r_2(t) = \frac{r_0}{2}\left(1 + e^{-2k_{T1}t}\right)$ • $< r>_2 = \frac{r_0}{2}\left[1 + \frac{1}{1+2(R_0/R)^6}\right]$	(T1.1) (T1.2)
Equilateral triangle ($i = 3$)		
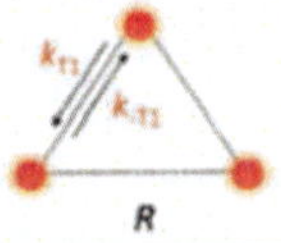	• $r_3(t) = \frac{r_0}{3}\left(1 + 2e^{-3k_{T1}t}\right)$ • $< r>_3 = \frac{r_0}{3}\left[1 + \frac{2}{1+3(R_0/R)^6}\right]$	(T2.1) (T2.2)
Equilateral square ($i = 4$)		
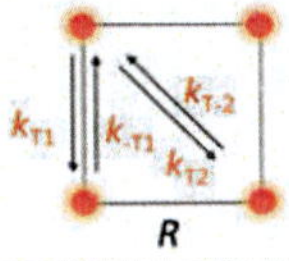	• $r_4(t) = \frac{r_0}{4}\left(1 + e^{-4k_{T1}t} + 2e^{-9/4 k_{T1}t}\right)$ • $< r>_4 = \frac{r_0}{4}\left[1 + \frac{1}{1+4(R_0/R)^6} + \frac{2}{1+9/4(R_0/R)^6}\right]$	(T3.1) (T3.2)

measurements towards the distances of closest approach, instead of giving an average value [15].

The second approach involves the concatenation of subunits, i.e. the preparation of a tandem of dimers. In this case, only one protomer contains the cysteine mutation and so there are only two instead of four cysteine residues within a tetrameric channel. Due to the asymmetric structures (the subunits are linked by their intracellular C- and N-terminal ends), these two dimers can only assemble anti-parallel, originating a single distance between the diagonally opposed subunits. This approach can be combined with the first one, e.g. as demonstrated by Fauré et al. 2012 [33]. Recently, the group of Nichols has successfully implemented this strategy to study the membrane-embedded inward rectifier KirBac1.1 channel using single-molecule FRET measurements [34–36]. This technique, which has already been described in this book, allows to uncover the presence of sub-conformational substates of the fluorescently labelled proteins with distinct average FRET efficiencies, which are often masked in the more traditional ensemble-average FRET measurements. In addition, the possible modulation of the interconversion kinetics between these substates (dynamic heterogeneity) by a perturbation can also be studied in detail, allowing to further investigate the structure–function relationship of the potassium channels.

Thirdly, novel biorthogonal labelling strategies can be developed, as recently demonstrated by Gupta et al. [37]. The authors used the incorporation of a non-canonical methionine analogue containing an azido group and cysteine mutagenesis to label the Shaker voltage-gated potassium channel (K_v) site – specifically with two different fluorophores simultaneously. Using voltage-clamp fluorometry, Gupta et al. were then able to monitor voltage-dependent conformational changes of this membrane protein in live cells [37].

In the next section, we describe a fourth alternative that relies on performing homo-FRET measurements.

2.2 *Homo-FRET Measurements*

As already mentioned, FRET can also occur between identical fluorophores when their absorption and emission spectra display a small Stokes shift [17, 38]. Considering a pair of fluorophores separated by a distance *R*, energy transfer between these two molecules is now reversible if the fluorophores are in the same environment ($k_{T1} = k_{-T1} = k_T$ (Eq. 1)), and the traditional readouts of a hetero-FRET experiment (fluorescence intensity/lifetime) can no longer be used because they become invariant. As such, homo-FRET can only be detected by carrying out polarized fluorescence measurements. In fact, each energy transfer step within a cluster of identical fluorophores leads to a FRET-induced angular displacement (depolarization) of the emitted fluorescence which can be monitored by measuring the time-resolved or steady-state fluorescence anisotropy of the sample [17, 38]. Berberan-Santos and co-workers have derived analytical solutions for the time evolution of fluorescence depolarization among *i* identical fluorophores (energy migration) in cyclic multichromophoric systems [39–41]. As an introduction to the general features of homo-FRET within an intra- or inter-molecular cyclic system, we will discuss next the specific cases of fluorophores arranged in an equilateral triangle or square, in addition to the case of an isolated pair of fluorophores (Table 1 and Fig. 1).

Table 1 summarizes the fluorescence anisotropy decays, $r_i(t)$, computed for a triad ($i = 3$, three-fold symmetry) and a quartet ($i = 4$, four-fold symmetry) of fluorophores undergoing homo-FRET within a cyclic regular geometry. The case corresponding to a pair of fluorophores ($i = 2$) is also included for completeness. The analytical solutions presented in Table 1 were obtained considering isotropic homo-FRET interactions among the fluorophores and taking into consideration that the time-resolved fluorescence anisotropy, $r(t)$, of each system can be described as:

$$r(t) = \sum_{i=1}^{i} \rho_i(t) \cdot r_i \tag{7}$$

where $\rho_1(t)$ and r_1 are the survival probability and anisotropy of the initially excited fluorophore, and $\rho_i(t)$ and r_i for $i > 1$ are the survival probabilities and anisotropies of

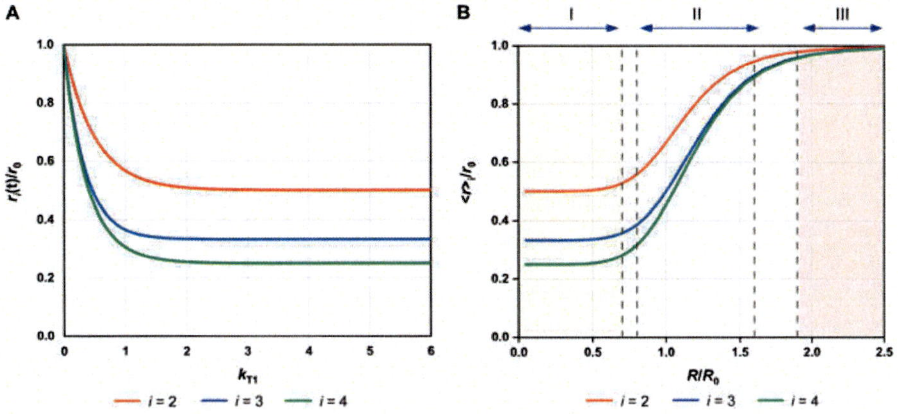

Fig. 1 The fluorescence anisotropy of a set of *i* fluorophores undergoing fully reversible isotropic homo-FRET within a cyclic molecular or supramolecular organization is determined by the specific geometry of the system. (**a**) Reduced fluorescence anisotropy decay, $r(t)/r_0$, and (**b**) steady-state fluorescence anisotropy, $<r>/r_0$, for a pair of fluorophores ($i = 2$), and a triad ($i = 3$) or a quartet ($i = 4$) of fluorophores arranged in an equilateral triangle or square geometry, respectively. r_0 is the initial (fundamental) anisotropy of the fluorophore. Regions I ($R/R_0 < 0.7$), II ($0.8 < R/R_0 < 1.6$), and III ($R/R_0 > 1.9$) are described in the text. The overall rotational tumbling of each system in solution was neglected

the indirectly excited ones (i.e. via energy transfer), respectively. In the absence of any overall rotational diffusion of the system, r_1 equals r_0, the initial (fundamental) anisotropy of the fluorophore [39].

As can be seen in Fig. 1a, the reduced anisotropy $r(t)/r_0$ decreases from 1 over time until a plateau value is reached, corresponding to the reduced limiting anisotropy, r_∞/r_0. In other words:

$$\lim_{t\to\infty} r_i(t) = r_\infty = \frac{r_0}{i} \tag{8}$$

This result is a consequence of that in a fully reversible system, the excitation becomes evenly distributed among all the fluorophores at sufficiently long times, i.e., the probability of fluorescence emission becomes identical for all the fluorophores, irrespective of the one that was initially excited. Since it has been shown that for isotropically oriented dipoles, the anisotropy after one energy transfer step is 0.016 [42], the contribution from the indirectly excited fluorophores to the anisotropy of the system under study can be disregarded ($r_i(i > 1) \sim 0$), and thus the limiting anisotropy is inversely proportional to the number of fluorophores *i* under consideration.

The steady-state fluorescence anisotropy for each system, $<r>_i$, is obtained by integration:

$$< r>_i = \frac{\int_0^{\infty} I(t) \cdot r_i(t) \cdot dt}{\int_0^{\infty} I(t) \cdot dt} \tag{9}$$

where *I*(*t*) is the fluorescence intensity decay of the fluorophore, which for simplicity is assumed to be mono-exponential. In Fig. 1b, the variation of the reduced steady-state fluorescence anisotropy, $<r>_i/r_0$, is represented as a function of the relative lateral inter-fluorophore distance, R/R_0, for each of the systems depicted in Table 1. Three distinct regions can be clearly identified: (*i*) **Region I: for $R/R_0 < 0.7$**, energy migration among the fluorophores included in each system is so efficient that the emission anisotropy becomes independent of *R*, reaching a limiting plateau value of $<r>_i/r_0 = 1/i$ (or $<r>_i = r_0/i$). If a monomeric protein containing a single fluorophore self-assembles in solution, then under this limit the steady-state anisotropy of the system allows evaluating the stoichiometry *i* of the homo-multimers (oligomers) formed, as previously described by Runnels and Scarlata [42]. (*ii*) **Region II: for $0.8 < R/R_0 < 1.6$**, the steady-state anisotropy of the fluorophores is extremely responsive to changes in the lateral inter-fluorophore distance, *R*. This regime is ideally suited to detect the occurrence of intra- or inter-molecular conformational changes within each system studied. Finally, (*iii*) **Region III: for $R/R_0 > 1.9$**, the lateral inter-fluorophore distances are much larger than the Förster radius, and so the fluorescence emission arises primarily from the initially excited fluorophore due to the absence of homo-FRET. In this case, the emission anisotropy of the system converges to the value presented by an isolated fluorophore, i.e. its initial (fundamental) anisotropy, r_0. It should be noted that although Eqs. (T1.2) and (T2.2) are equivalent to Eq. (18) given in Runnels and Scarlata for $N = 2$ and $N = 3$, respectively [42], Eq. (T3.2) deviates from their solution for $N = 4$, because the authors deemed that all molecules in a cluster of *N* molecules have the same separating distance. This is equivalent to consider that the four fluorophores are arranged tetrahedrally at variance with the square geometry considered in Table 1 [41].

As the number of fluorophores *i* included in each system increases, the nearest neighbour interactions become dominant. Cardoso and Berberan-Santos have shown that the probability of energy transfer to acceptors other than the nearest for $i \geq 4$ is always residual, attaining a maximum value of only ~6% for $i = 4$ [39]. This means that, e.g., the contribution of direct energy transfer between the fluorophores located at diagonally opposed vertices of a square is minor and mainly affects the long-time behaviour of the anisotropy decay [41].

2.3 General Applications of Homo-FRET Measurements

Briefly, homo-FRET measurements under Region I ($R/R_0 < 0.7$) have been extensively used in the literature to evaluate the average number of fluorescently labelled peptides/proteins in a cluster [42–44] and the degree of oligomerization of proteins involved in a cooperative partition equilibrium [45]. These measurements are particularly useful for detecting small oligomers since the decrease in the overall anisotropy of the system becomes less pronounced as the degree of oligomerization increases, as exemplified in Fig. 1b [46]. More recently, these studies have been extended to live cells through the implementation of fluorescence anisotropy imaging microscopy (FAIM) measurements [47]. These allow for the detection and quantification of the molecular self-assembly of proteins and lipid clustering in vivo [43, 46, 48].

Region II ($0.8 < R/R_0 < 1.6$) can also been exploited in homo-FRET studies to measure inter-fluorophore distances and detect conformational changes in proteins. The pioneering work of Johansson and collaborators established that time-resolved anisotropy measurements can be used to determine distances between a pair of identical fluorescent dyes covalently attached to a protein that undergo energy homo-transfer [38, 49–51]. In these works, the reorientational motions of the donor molecules were taken explicitly into consideration. Later, Zou et al. [52] used T4 lysozyme, with a known crystal structure, to confirm that the steady-state anisotropies of different fluorescein-labelled mutants could be used to obtain distances when regime II applies. These authors found a similar distance trend upon labelling the same enzyme mutants with spin and fluorescent labels. This result further validates the use of homo-FRET measurements to determine distances since EPR are less affected than fluorescence measurements by orientational effects of the probes. More recently, several genetically expressed biosensors based on intramolecular homo-FRET have also been developed, allowing to detect signalling events in vivo [53–55].

3 Homo-FRET Studies on the Potassium Channel KcsA: A Practical Guide from a Symmetric Tetrameric Membrane Protein

3.1 Functional and Structural Characteristics of the KcsA Channel

KcsA is a homo-tetrameric proton-activated, voltage-modulated potassium channel cloned from the soil bacteria *Streptomyces lividans* [56]. Its high-resolution structure was solved by X-ray crystallography, being the first ion channel to be crystallized [57]. Since then, KcsA has been used as a prototypical protein on the biophysical studies of K^+ channels due to its homology to the eukaryotic counterparts and the simplicity of its structure: each one of the four identical monomers is constituted by

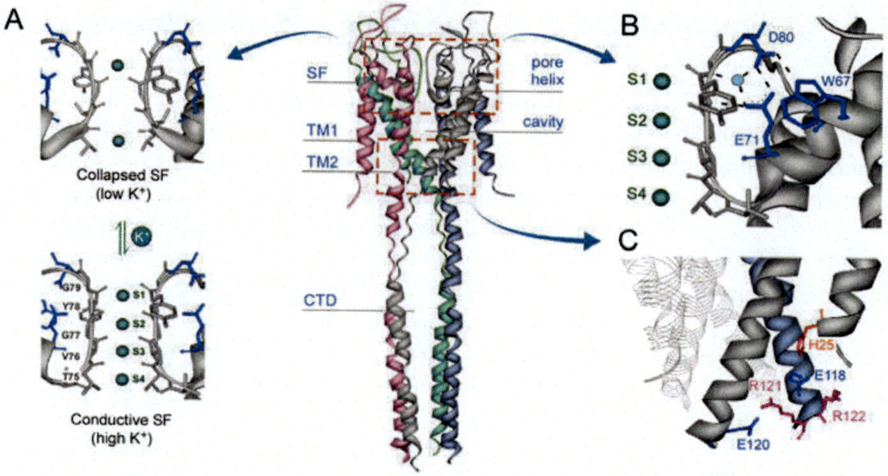

Fig. 2 X-ray structure of the full-length KcsA channel in the closed state (PDB: 3EFF, central panel) (TM1 and TM2: transmembrane helices 1 and 2, respectively; SF: selectivity filter; CTD: C-terminal domain) (**a**) Schematic representation of the equilibrium between the collapsed and conductive conformations of the selectivity filter. The collapsed form (PDB: 1K4D) is found in the presence of either low K^+ or high Na^+ concentration. The conductive form (PDB: 1K4C) is found in the presence of high K^+ concentration, as well as of other permeant cations, such as Tl^+, Rb^+ or Cs^+ (even though no Rb^+ or Cs^+ is detected in the S2 binding site). (**b**) Zoom-in of the network of interactions between W67, E71 and D80 in the inactivation triad. The H-bonds are mediated by a molecule of H_2O (cyan). (**c**) Constituent residues of the pH sensor at the end of the TM2 helices. E120 can interact with either R121 or R122 from the same subunit, whereas an H-bond is found between H25 and E118 of neighbour monomers

two transmembrane helices (TM1 and TM2), the amphipathic N-terminal domain (not solved by X-ray crystallography) and the intracellular C-terminal domain. The TM1 and TM2 sections are connected by a loop that contains a tilted short-helix (pore-helix) and the SF (Fig. 2) [57]. The SF signature sequence $T_{75}VGYG_{79}$ is highly conserved among the K^+ channels superfamily and delimits four putative K^+ binding sites (Fig. 2a). Its conformational plasticity has a great impact on the high-rate of ion permeation, selectivity and inactivation properties of the channel, positioning the SF as a gate (outer gate) of these membrane proteins [58]. Figure 2a illustrates the high-resolution structures of two typical conformations of the SF: the conductive form, observed in the presence of high concentrations of permeant cations (K^+, Rb^+, Cs^+ or Tl^+), and the non-conductive or collapsed form, found at low concentrations of permeant cations or high concentrations of blocking species such as Na^+ or Li^+, being Na^+ the relevant physiological blocker of the K^+ channels [1, 59]. In the conductive conformation, the four binding sites are present, even though only K^+ and Tl^+ can bind to all of them with equal probability (Rb^+ and Cs^+ are never found at the S2 binding site) [57, 60–62]. It was postulated that the different ion occupancy of the SF by the permeant cations could affect the ion flux (conductivity) and inactivation rates of K^+ channels [63, 64].

The activation gate (inner gate) of KcsA is located at the end of the TM2 segments, formed by a bundle-crossing of these helices. The amino acids involved in the pH sensor are connected through several salt bridges and H-bond interactions, being sensitive to the local amount of H^+ [65] and thus regulating the K^+ flux by changing the conformation of the intracellular half of the channel [66, 67].

Even though KcsA does not present the typical 6-TM topology of the voltage-gated K^+ channels (K_v), its functional properties are also modulated by voltage [9] due to the presence of the E71 residue at the pore-helix [68]. This negatively charged amino acid interacts through a H-bond network with D80 (at the end of the signature sequence) and W67 (at the pore-helix) residues to form the so-called inactivation triad [68, 69] (Fig. 2b). Different types of mutations incorporated at the E71 residue showed its connection to the SF, especially at the level of T76 and G77, therefore influencing the rate of C-type inactivation process, as explained below [63, 70, 71].

In the presence of the proper stimuli, i.e., pH ~ 4.0, the gating of KcsA, as well as of other K^+ channels, involve conformational rearrangements at the inner and outer gates [57, 58, 63, 68, 69, 71–83]. Macroscopic currents measurements performed on asolectin-reconstituted KcsA showed that after gating at acidic pH a peak of maximal current is detected, followed by an exponential decrease of the flux. The maintenance of the gating stimulus therefore induces a slow inactivation process, characterized by a time constant of 1–3 s, that consequently leads to a steady-state with a very low open probability of the wild-type (WT) channel [63]. So, based on the conformation of the activation gate (closed/open) and the SF or outer gate (conductive/inactivated), the gating cycle of KcsA is defined by at least four kinetic states (Fig. 3): *closed/conductive (pH 7, K^+), open/conductive (transient state found in K^+ after the pH 4 gating), open/inactivated (pH 4, K^+, at steady state) and closed/inactivated (pH 7, K^+, immediately after increasing the pH from 4 to 7)* [84]. Thus, this gating cycle clearly pictures the allosteric coupling between the opening movement at the inner gate and the changes in SF conformation, where the slow inactivation process takes place. The structure of this inactivated form and the molecular details of the inactivation process remain elusive. In the case of KcsA, X-ray diffraction studies suggested that the inactivated SF is almost identical to the collapsed structure observed at low K^+ amounts; this hypothesis was supported by complementary experimental approaches which suggested that only slight modifications on the SF filter accompanied the inactivation of these channels [85–88]. Still, the SF rearrangements that are coupled to C-type inactivation remain a matter of debate about whether they involve pinching, dilation, or other more subtle SF changes [64, 89].

3.2 Fluorescence Properties of the KcsA Channel

In general terms, the aromatic residues (Trp, Tyr and Phe) found in the structure of K^+ channels are frequently distributed between the intracellular and extracellular mouths, acting as anchors during protein insertion into the lipid bilayer. Interestingly, two of these aromatic residues (tyrosine and tryptophan) are often present at the pore-helix, being highly conserved among the K^+ channels superfamily and frequently involved

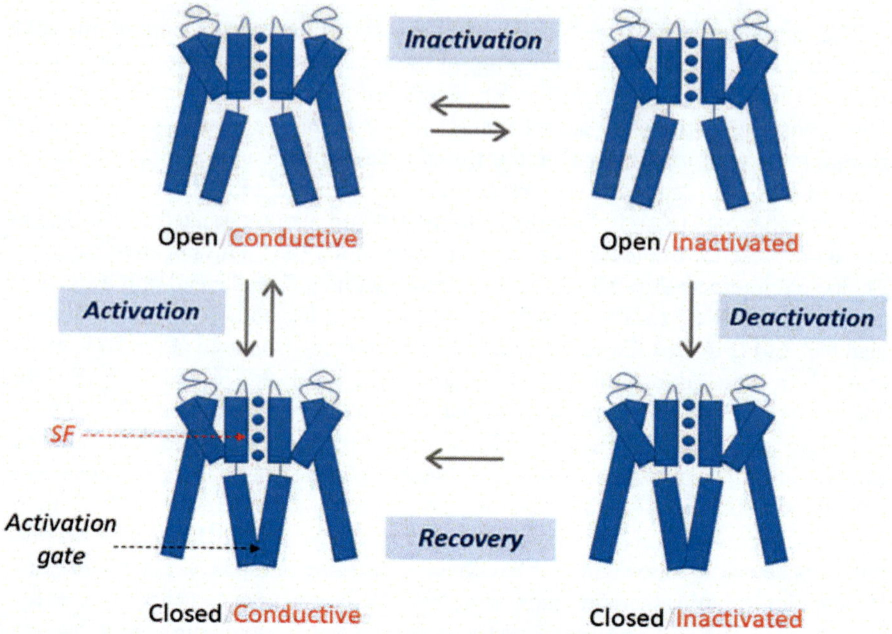

Fig. 3 The gating cycle of the prototypical K^+ channel KcsA. The four-state gating model is controlled by the presence of two gates: the activation gate (inner gate) at the intracellular entrance of the pore and the selectivity filter (outer gate). Each channel state is defined by the combination of the conformational state of the inner gate (closed or open) and the outer gate (conductive or inactivated)

in the modulation of the SF conformation [90]. In the case of the KcsA channel, its intrinsic fluorescence spectrum results from the combination of the emission of five W residues per monomer: W26 and W113, located at the inner mouth of the transmembrane section and W67, W68 and W87, at the opposite end of the protein (Fig. 4a) [10]. Both the W67 and W68 residues at the pore-helix were found to sense the type and concentration of cations bound at the SF, and the changes on their emission spectra could be used to estimate the channel's affinity for permeant and blocking cations [10, 91]. W67 was selected as a fluorescent reporter to study the conformational dynamics of the SF due to its main role on the regulation of the C-type inactivation process in KcsA as a participant in the inactivation triad [69] (Fig. 2b). Control patch-clamp recordings of this single-tryptophan channel W67 KcsA (W26,68,87,113F KcsA) reconstituted in asolectin vesicles demonstrated that this quadruple mutant functions as a wild-type like protein [41].

W67 KcsA remains a highly fluorescent protein, retaining 40–45% of the WT overall florescence intensity, due to the shielded position of this residue (protected from the aqueous solution) (Figs. 2b and 4a, b). Its emission spectra still respond to the presence of permeant and blocking cations within the filter (Fig. 4c), presenting a blue-shifted spectrum in the presence of K^+, Rb^+ or Cs^+ (maximum emission wavelength circa 325 nm) and a high quantum yield ($\Phi \sim 0.3$), yielding a Förster

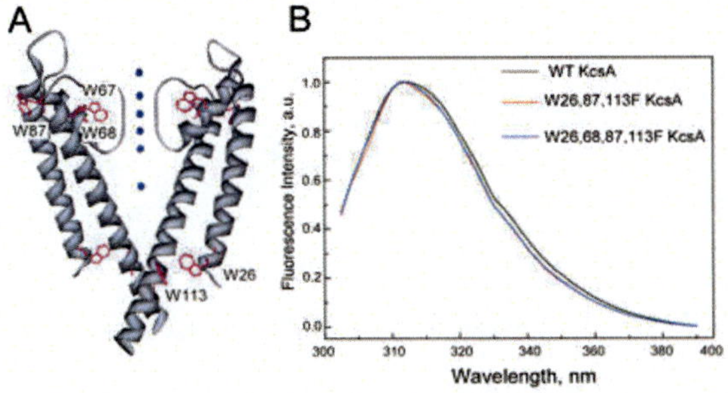

Fig. 4 The tryptophan residues, particularly W67, are good fluorescent reporters of K^+ binding to KcsA. (**a**) The intrinsic fluorescence properties of the KcsA channel mainly arise from five W residues: W26 and W113 at the intracellular mouth, W87 facing the lipid membrane, and W67 and 68 just behind the SF. (**b**) The fluorescence emission spectra ($\lambda_{exc} = 295$ nm) are only slightly shifted to lower wavelengths when the most intracellular Trp residues (26, 87 and 113) of KcsA are mutated to Phe. (**c**) The centre-of-mass of the emission spectra, $<\lambda>$, is a useful parameter to examine the binding of K^+ to the SF (stabilization of the conductive form). In fact, the largest change on $<\lambda>$ was detected when only Trp67 is monitored (single-tryptophan mutant W26,68,87,113F or W67 KcsA)

radius of $R_0 = 12$ Å at 200 mM KCl, pH 7.0 [41], close to the highest values described in the literature for tryptophan–tryptophan Förster pairs [16]. Direct inspection of several high-resolution crystal structures of KcsA obtained under different experimental conditions revealed a proximity between their W67 residues, ranging from 15 to 18 Å. Consequently, such distances fall within the working Region II of Fig. 1b ($0.8 < R/R_o < 1.6$), and therefore their determination should be possible using homo-FRET measurements. The fluorescence anisotropy of W67 can then potentially monitor the conformational changes undergone by the pore-helices of this mutant KcsA channel, which are linked to the SF through a network of hydrogen bonds (Fig. 2b). This possibility was confirmed experimentally by performing steady-state and time-resolved fluorescence anisotropy measurements of the detergent-solubilized W67 KcsA mutant channel in the presence of different SF modulators. An analytical framework describing the occurrence of homo-FRET within a symmetric tetrameric protein was then used to obtain W67–W67 intersubunit (lateral) distances as detailed in the next section.

3.3 *Time-Resolved Anisotropy Measurements and Homo-FRET Analytical Framework to Characterize the Conformational Dynamics of the Tetrameric W67 KcsA Channel*

Considering that the time-resolved anisotropy decays of W67 KcsA were carried out in a fluid solution, the fluorescence anisotropy decays of this protein were well described by:

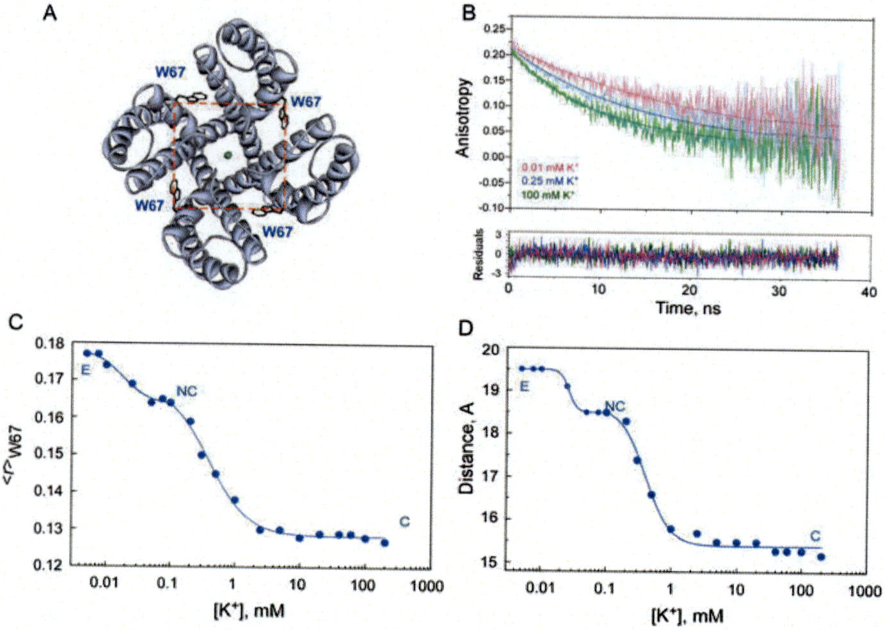

Fig. 5 The homo-FRET process within W67 KcsA is sensitive to the number of K^+ ions in the SF. (**a**) Top view of the KcsA channel (PDB: 1K4C), highlighting in blue sticks the symmetric position of the W67 residues (square geometry), one in each of subunit of the homo-tetrameric channel. (**b**) Time-resolved anisotropy decays of the W67 KcsA channel (6 μM, monomer based) in 20 mM Hepes, 10 mM Na^+, 5 mM DDM, pH 7.0 buffer supplemented with 0.01 mM, 0.25 mM and 100 mM KCl. The samples were excited at 300 nm as the fundamental anisotropy of tryptophan reaches its highest value ($r_0 \sim 0.3$) at this wavelength [93], thus allowing to maximize the dynamical range of the time-resolved measurements. The initial anisotropies of the decays, $r_{(0)} \sim$ 0.20-0.23, were always significantly lower than the fundamental anisotropy of tryptophan, $r_0 = 0.3$, due to very rapid rotational motions of W67 residues on the subnanosecond timescale, which were beyond the time resolution of the instrumentation. The solid lines are the best fit of Eq. (10) to the decays using $\phi_g = 43$ ns fixed. (**c**) The steady-state anisotropy, $<r>_{W67}$, and (**d**) the lateral W67–W67 intersubunit distances calculated from the time-resolved anisotropy decays of W67 KcsA (as described in Sect. 3.3) along its titration with K^+ show two consecutive bindings events. According to [41, 91], this behaviour corresponds to (*1*) the stabilization of the non-conductive (NC) state from an "empty" filter (E), and (2) the subsequent stabilization of the conductive form of the SF (C), where the four K^+ binding sites can be occupied by the permeant cation

$$r(t) = r_4(t) \cdot \exp\left(-t/\phi_g\right) = \frac{r_{(0)}}{4}\left(1 + e^{-4k_{T1}t} + 2e^{-9/4 k_{T1}t}\right) \cdot \exp\left(-t/\phi_g\right) \qquad (10)$$

The first term of Eq. (10), $r_4(t)$, accounts for the fluorescence depolarization due to the intramolecular homo-FRET process occurring among the four W67 residues of KcsA, a tetrameric protein with four-fold symmetry (Fig. 5a and Table 1). In turn, the second term considers the additional depolarization caused by the overall rotational diffusion of the detergent-solubilized protein in solution. The global rotational correlation time, ϕ_g, was determined independently using a L90C KcsA mutant fluorescently labelled with pyrene, a long-lived fluorescent probe, and was subsequently fixed in all the analyses ($\phi_g = 43 \pm 3$ ns [41]). Equation (10) was therefore fitted to the fluorescence anisotropy decays of W67 KcsA obtained under different experimental conditions using only two free fitting parameters: the experimental initial anisotropy, $r(0)$, and the lateral energy transfer rate constant, k_{T1}, respectively. For each experimental condition tested, the ensemble-averaged W67–W67 lateral distance R within each channel could then be retrieved from the fitted energy transfer rate constant, k_{T1}, using Eq. (1) [41, 92].

The three main assumptions used in the derivation of the formalism described by $r_4(t)$ (Eq. T3.1) were (*1*) pure homo-FRET, (*2*) isotropic FRET and (*3*) unpolarized emission of the indirectly excited fluorophores [41]. Pure homo-FRET implies that the process taking place among the W67 residues is fully reversible. Considering that these residues are shielded from the aqueous environment, all four fluorophores are indeed essentially spectroscopically equivalent. Regarding the second approximation, the orientational dependence of FRET is expressed by the orientational factor, κ^2. By using $\kappa^2 = 2/3$ in Eq. (2), the dynamic regime (fast isotropic rotational motion of the partners) was assumed, which does not strictly apply here. However, the mixed polarization of the absorption spectrum of tryptophan (^1L_a and ^1L_b transitions with nearly orthogonal transition moments [93]) is equivalent to rotational randomization of the orientations [94]. Additionally, the initial anisotropy measured for W67 KcsA, $r(0) \sim 0.20$–0.23, was always less than the fundamental anisotropy of the tryptophan residues under our experimental conditions, $r_0 \sim 0.3$. This means that the W67 residues have some fast local rotational dynamics that helps depolarizing the emitted fluorescence, which in turn contributes to relax the orientational constraints of the system. Taken together, these last two factors support the third approximation that the emission from the indirectly excited fluorophores is largely unpolarized.

In the more general case, the energy transfer efficiency may be influenced by the mutual orientation of the fluorophores, and so it may not be adequate to assume complete depolarization for the emission of the indirectly excited fluorophores [38, 40, 43, 95]. As an example, Cardoso and Berberan-Santos [39] showed using Monte Carlo simulations that the decrease of the fluorescence anisotropy with R_0/R is less marked for the case of anisotropic compared to isotropic orientations in a heptagon, due to a significant contribution of unfavourable relative orientations of the transition dipole moments of each chromophore in the first case.

Finally, although the KcsA-DDM complex is not strictly spherical, the difference between the short and long axis of the prolate ellipsoid is not as pronounced as suggested by Fig. 2 when the rim of detergent molecules surrounding the transmembrane section of the channel is considered [96]. The assumptions made can also be justified a posteriori given that the fluorescence anisotropy decays for the detergent-solubilized W67 KcsA channels were always adequately fitted using only two fitting parameters and a fixed long rotational correlation time of 43 ns, yielding consistent results under a wide range of experimental conditions (pH and variable ionic composition of the buffer).

3.4 An Efficient Homo-FRET Process Among W67 Residues at the Pore-Helices Detects the Conformational Plasticity of the SF at pH 7.0

K^+ ions, which are the relevant physiological permeant cations of KcsA, can modulate the SF conformation of the protein. The X-ray structures of this channel in the closed state showed that its SF can undergo a transition from a non-conductive/collapsed form to a conductive structure by simply increasing its concentration (Fig. 2a). Accordingly, the steady-state anisotropy of W67 KcsA, $\langle r \rangle_{W67}$, showed a strong dependence on the concentration of K^+ present in the buffer solution at pH 7 (Fig. 5c). The dose-response plot clearly described two consecutive binding events with apparent dissociation constants, $K_{D1} \sim 0.045$ mM and $K_{D2} \sim 0.40$ mM, respectively, in agreement with previous thermodynamic studies performed with the WT KcsA channel [91, 97]. The first binding event corresponds to the conversion of an "empty" SF (state E, without K^+, but in 10 mM Na^+ to avoid the dissociation of the tetramer into monomers) to a non-conductive or collapsed state (state NC), which should correspond to the crystallographic structure observed at low K^+ concentration (PDB: 1K4D). Here, the S1 and S4 binding sites are occupied by K^+, while the S2 and S3 binding sites are not detected. Upon further increasing the concentration, K^+ ions self-induce a conformational change in the SF upon which the S2 and S3 binding sites are formed and filled, and the conductive structure (state C) of the SF is stabilized. Therefore, the anisotropy data strongly suggest that the K^+ ions induce local conformational changes in the KcsA mutant channel at the SF level, bringing together the W67 residues located at the short pore-helix behind the SF of each subunit. This local rearrangement increases the efficiency of tryptophan–tryptophan energy migration (homo-FRET), and consequently produces a stronger depolarization of the fluorescence emitted by the mutant protein which, in turn, causes a decrease in its steady-state anisotropy.

The performance of time-resolved anisotropy measurements confirmed this interpretation. As described in Sect. 3.3, the four-fold axis of symmetry of the KcsA channel, and consequently the square geometry delimited by the W67 residues (Fig. 5a), allowed deriving a formalism describing the fluorescence anisotropy decay of the W67 KcsA mutant-DDM complex (Eq. 10) [41]. Figure 5b clearly shows that the anisotropy decays of W67 KcsA became faster at higher K^+ concentrations due to a more efficient intramolecular homo-FRET component. As explained above (Sect. 3.3), the average intramolecular energy transfer rate constant, k_{T1}, can be obtained by fitting Eq. (10) to the anisotropy decays of W67 KcsA; in turn, W67–W67 intersubunit lateral distances can then be computed using Eq. (1). In this way, the first K^+ binding event was found to elicit an approximation of ~1-1.5 Å between the pore-helices, whereas the W67–W67 intersubunit distances were further narrowed down by ~3 Å during the stabilization of the conductive form of the SF (second binding event) (Fig. 5d).

The time-resolved anisotropy decays of W67 KcsA were also measured in the presence of a saturating concentration of other modulators of the SF of KcsA,

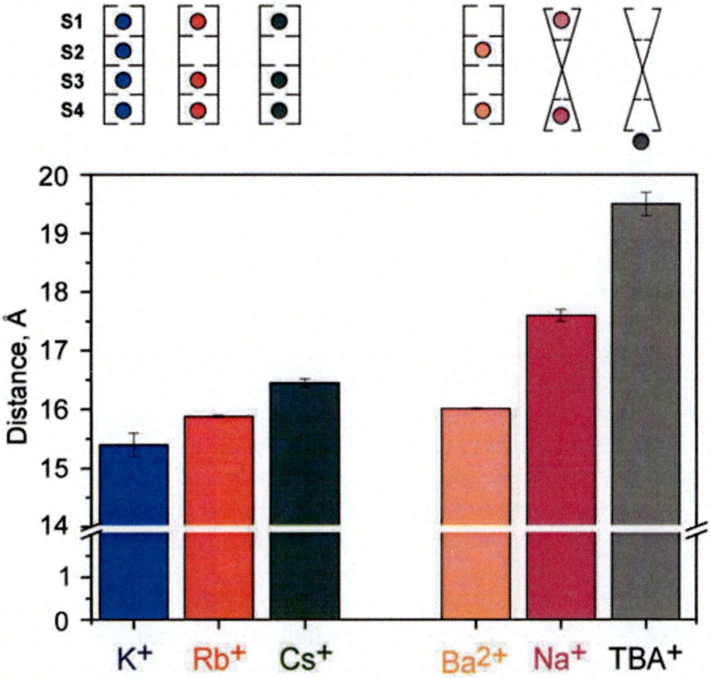

Fig. 6 The W67–W67 intersubunit distances critical depend on the degree of ion occupancy of the SF from W67 KcsA at pH 7.0 (closed state). These lateral distances were calculated from fitting the time-resolved anisotropy decays of each sample with Eq. (10). The first group (left) corresponds to the permeant cations, where the selectivity filter is able to adopt a conductive conformation. The second group (right) encompasses three blocking agents with different degree of interactions with the SF. The measurements were carried out with W67 KcsA solubilized in 5 mM DDM detergent micelles

namely different permeant and non-permeant cations and blockers, and the W67–W67 intersubunit distances were calculated using the derived homo-FRET formalism. As shown in Fig. 6, the largest intersubunit distance at the pore-helix level was found in the presence of the pore blocker tetrabutylammonium ion, TBA^+ ($R \sim$ 19–20 Å), which is perfectly correlated to its location on the ion pathway, interacting with some residues on the central cavity and leaving the SF almost empty [98]. This distance is close to the one determined at the beginning of W67 KcsA titration, when the SF is essentially depleted of ions (initial state E, Fig. 5d). An "intermediate" distance was then found when 200 mM of the non-permeant cations Na^+ was present in the buffer ($R \sim$ 17–18 Å, Fig. 6)). It was previously reported that these non-permeant cations are only able to interact with the SF of KcsA at the most external sites of the filter (S1 and S4, average SF ion occupancy of 1) with an mM affinity [91, 99, 100] and thus inducing the collapsed, non-conductive structure. The intermediate plateau level detected in the K^+ dose-response curve of W67 KcsA (intermediate state NC, Fig. 5d) is again in good agreement with these distances.

Regarding the permeant cations tested, the efficiency of the homo-FRET was high in the presence 200 mM of K^+, Rb^+ and Cs^+, and much shorter W67–W67 lateral distances were measured ($R \sim 15.5$–16.5 Å, Fig. 6), scaling with the ionic radii of the cations ($K^+ < Rb^+ < Cs^+$). These distances correspond to the final conductive conformation of the SF (final state C, Fig. 5d). According to the crystallographic data available so far, whereas K^+ is able to interact with equal probability with all the four binding sites, Rb^+ and Cs^+ are excluded from the S2 site [61, 62]. Interestingly, when the blocker Ba^{2+} was used in the sample buffer, the W67–W67 intersubunit distances calculated from the time-resolved anisotropy decays ($R \sim 16$ Å, Fig. 6) revealed that the SF has a conductive-like conformation rather than a Na^+-like collapsed structure. This is consistent with the ability of Ba^{2+} to interact with both S2 and S4 binding sites within the filter, leaving both positions S1 and S3 empty [99, 101, 102]. Altogether, these data reveal that the SF of KcsA has a high degree of conformational plasticity, and that its cation-dependent structural changes are controlled by the number of ions in the filter. The SF loop undergoes two consecutive cation-induced disorder-order transitions as its backbone carbonyls progressively coordinate the ions, bridging together its adjacent pore-helices, and eliciting a very high homo-FRET efficiency among the strategically positioned W67 residues. The first transition, where only the outermost S1 and S4 sites of the SF are occupied, produces a modest $\Delta R \sim 1$–1.5 Å. In turn, upon further occupancy of the innermost binding sites S2 and/or S3, the canonical conductive conformation of the SF is formed at the end of the second transition, which is accompanied by a much more pronounced ΔR of ~3 Å, emphasizing the role of the central SF-bound ions in stabilizing the filter.

Finally, the impact of the E71A mutation in the W67 KcsA context on the conformational dynamics of the SF was also tested, since this mutation is known to abolish the inactivation process by "freezing" the conductive conformation and weakening the connection with the inactivation triad [69]. As expected, the W67–W67 intersubunit distances calculated from the time-resolved anisotropy decays of this non-inactivating mutant channel remained constant at ~15 Å not only in the presence of high concentrations of permeant cations (K^+, Rb^+ and Cs^+), but also of the non-permeant Na^+ [41]. This decrease in the conformational plasticity of the SF has also been previously related to a decrease in the selectivity of E71A KcsA [103].

3.5 *The Conformational Dynamics of the Outer Vestibule Is Hindered by the Presence of a Monoclonal Fab Fragment*

The final test of the derived homo-FRET formalism requires a comparison between the inter-tryptophan distances obtained at saturating cation concentrations and the available crystallographic data. Overall, the W67–W67 intersubunit distances are in fair agreement with the W67–W67 Cδ2–Cε2 distances calculated from the crystallographic structures of KcsA that were obtained using a truncated channel (Δ125

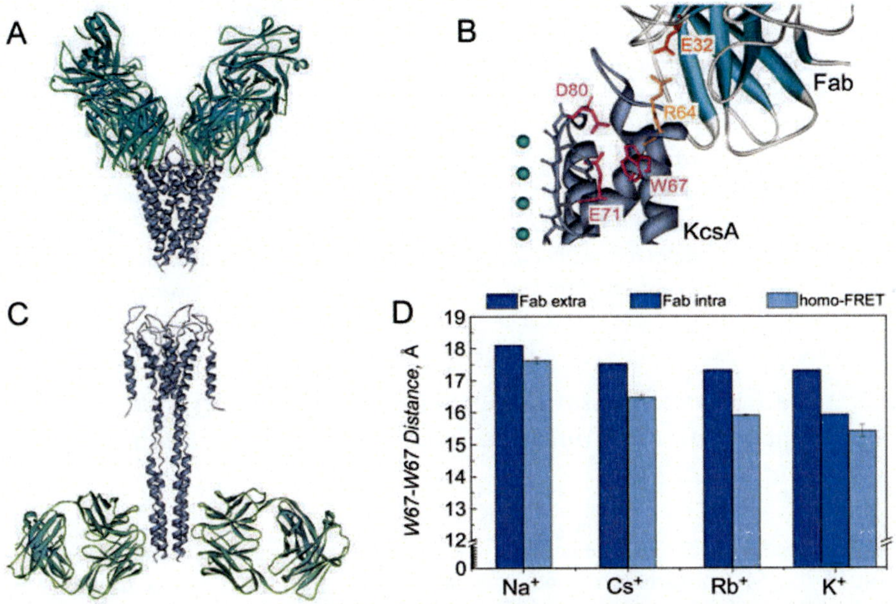

Fig. 7 The co-adjuvant crystallization Fab fragments hampers the conformational flexibility of the KcsA SF. (**a** and **b**) X-ray crystallographic structures of (**a**) the C-terminal truncated KcsA channel bound to an extracellular Fab fragment (PDB: 1K4C) and (**c**) the full-length protein bound to an intracellular Fab fragment (PDB: 3EFF). Both structures are considered to represent a closed state of the KcsA channel. (**b**) Detailed illustration of the interaction area between the extracellular antibody fragment and KcsA. A salt bridge between the E32 residue from the Fab and R64 from KcsA was postulated as the main factor responsible for the changes on the C-type inactivation rate decrease induced by this antibody fragment [69]. As clearly shown, R64 is near the inactivation triad of the channel (W67–D80–E71) (PDB: 1K4C). (**d**) The homo-FRET W67–W67 inter-subunit lateral distances can be compared to those from X-ray crystallographic data (Cδ2-Cε2 distances). The distances recovered from the X-ray structures solved in the presence of the extracellular Fab fragment are always larger than those derived from the time-resolved anisotropy decays of W67 KcsA. On the other hand, the homo-FRET distance measured in the presence of a saturating concentration of K^+ is in fair agreement with the one determined in the presence of the intracellular Fab fragment bound to the full-length protein

KcsA). These structures were solved in the presence of Fab fragments that target the turret region (pore-helix and external loop) [61, 104], being the interaction between the E32 residue from the antibody and the R64 from KcsA a key on the epitope–paratope interaction [69] (Fig. 7a, b, d). The Fab fragments act as a crystallization co-adjuvant by restricting the mobility of certain protein regions, and hence improving the resolution of the crystals. However, these W67–W67 Cδ2–Cε2 distances show only small differences in the presence of different conducting and blocking cations ($R \sim 17$–18 Å, and therefore $\Delta R_{max} < 1$ Å), whereas a more pronounced $\Delta R_{max} \sim 3$ Å was detected within the homo-FRET data (Fig. 7d). Importantly, a much closer agreement was obtained between the inter-fluorophore distance measured in the presence of a high concentration of K^+ and the one calculated from the crystallographic data produced with the full-length channel bound to an intracellular Fab fragment [105] (Fig. 7b, d). This is a clear evidence that the addition of an external Fab fragment critically affected the structure of the outer mouth of the channel, imposing restrictions to its structural plasticity at the pore-helix level that somehow modified its connection to the SF conformation (Fig. 7c). Indeed, electrophysiological recordings showed that these external Fab fragments were able to induce a remarkable decrease on the C-type inactivation rate of the channel, almost identical to the effect seen by E71A or R64A mutations [69]. On the other hand, a milder effect was detected with the intracellular monoclonal Fab fragment which only induced a two-fold reduction in the rate of entry into the inactivated state.

3.6 The Conformational Dynamics of the Selectivity Filter Depends on the Conformational State of the Activation Gate of KcsA

As described in Sect. 3.1, the gating cycle of KcsA involves an allosteric modification of the SF conformation once the opening of the inner gate is triggered at acidic pH: after a peak of maximal K^+ current, the channel enters into a non-conductive inactivated state, which acts as regulator of the permeation process [63, 68, 69, 78]. During the last years, attempts have been made to characterize the inactivated state of this K^+ channel [64]. The prevailing view obtained from several X-ray crystallographic studies is that the structure of the SF in the open-inactivated state reminisces that detected in the closed-collapsed form at low $[K^+]$ [60, 64, 66, 84]. To check the effect of the opening of the inner gate on the conformational landscape of its SF, the potassium-induced changes in the steady-state and time-resolved fluorescence anisotropy of W67 KcsA were again used to monitor the conformational equilibrium of the SF but now at pH 4.0. The homo-FRET approach failed to detect the stabilization of the intermediate collapsed structure, since at low concentrations of K^+ the anisotropy and W67–W67 intersubunit distances remained fairly constant ($\langle r \rangle \sim 0.162$ and $R \sim 18$ Å, respectively) (Fig. 8a). Nevertheless, increasing the concentration of K^+ from 10–20 mM to 200–300 mM self-induced again a notable change on these two parameters, leading to a final structure which is very much alike to the closed channel at pH 7.0 ($R \sim 15.3$ Å). Interestingly, this transition displays a one-order of magnitude higher K_D than in the closed state, therefore revealing an important change on the protein–ion interaction at the SF in the open-inactivated state.

Several open structures of the KcsA channel were obtained by X-ray crystallography, using a constitutively open mutant channel [66, 106]. It was observed that the W67–W67 lateral distances calculated from the anisotropy decays for the open/inactivated W67 KcsA channel at high K^+ concentrations and pH 4.0 ($R = 15.3 \pm 0.1$ Å) are almost identical to those W67–W67 (Cδ2–Cε2) distances determined from X-ray data obtained in the presence of an intracellular Fab fragment

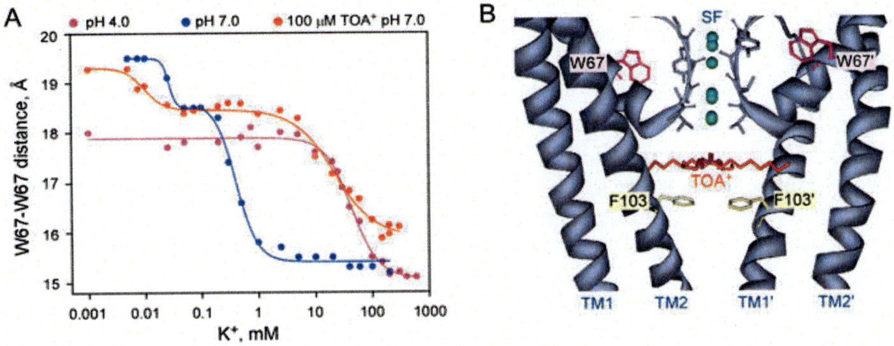

Fig. 8 K^+ binding behaviour of W67 KcsA in the closed (pH 7.0), open (pH 4.0) and TOA^+-blocked states. (**a**) The W67–W67 intersubunit distances calculated from the time-resolved anisotropy decays of the W67 KcsA channel detect two consecutive K^+ binding events at pH 7.0, whereas the opening of the inner gate at pH 4.0 triggered a loss of conformational plasticity at the W67 level at low K^+ concentrations and a decrease on the affinity for the permeant cation. Despite these differences, the structure of the pore-helix at saturating concentrations of this cation, i.e., at the conductive or inactivated conformation of the SF, is almost identical ($R \sim 15.3$ Å). The effect of the pH 4.0 condition is partially emulated by the addition of 100 μM TOA^+ to the samples at pH 7.0, particularly the decrease in the SF affinity for K^+ (second binding event). In the presence of this alkylammonium salt, the W67–W67 intersubunit distances are larger than in the control. (**b**) This is related to binding of the long octyl chains of TOA^+ at the central cavity and the modification of the rotameric state of the F103 residue, just below the blocker [107]

($R = 15.2$ Å; PDB 3PJS), but not when an extracellular Fab was used to stabilize the crystals ($R = 17.7$ Å; PDB 3F5W). These results reinforce the conclusion that the binding of the extracellular Fab fragment alters the conformation of the extracellular loop and the SF dynamics, since the same Fab fragment has been described to have a profound effect on KcsA inactivation [69]. Moreover, a comparison between the distances determined from the homo-FRET approach for the closed/conductive (pH 7.0, 200 mM K^+, $R = 15.4$ Å), closed/non-conductive collapsed (pH 7.0, 0.1 mM K^+, $R = 18.5$ Å) and open/inactivated (pH 4.0, 200 mM K^+, $R = 15.3$ Å) states clearly rule out the hypothesis that the inactivated state is structurally similar to its collapsed state, in agreement with more recent studies [85–87].

Finally, regarding the modulation of the conformational equilibria of the SF, the alkylammonium salts have been broadly used to characterize the functional properties of K^+ channels [108–112]. Particularly, the tetraoctylammonium ion (TOA^+) was described to allosterically induce the C-type inactivated form of the SF after its binding at the central cavity of the KcsA [107]. Regarding our fluorescence anisotropy studies, the addition of 100 μM of TOA^+ to the W67 KcsA channel at pH 7.0 induced two main effects on the K^+ binding to the SF: (1) a general increase in the W67–W67 intersubunit distances for most of the concentrations of K^+ tested; (2) an increase of the K_{D2} value (second binding event) by an order of magnitude (from 0.45 mM to 4.5 mM) (Fig. 8a), suggesting that similarly to the acidic pH, TOA^+ by itself causes inactivation of the channel SF at pH 7.0. According to the X-ray crystallographic data, this effect is mediated by a change on the rotameric state of the sidechain of the F103 residue (Fig. 8b) [107]. This phenylalanine had been described as a main effector on the allosteric communication between the inner and outer gates of the KcsA channel, since the F103A mutation profoundly reduced the slow inactivation rate [113]. In this respect, it should be noted that the available X-ray data on the TOA^+–KcsA complex was obtained in the presence of the extracellular Fab fragment [107]. Based on such data, the authors concluded that the SF conformation in the TOA^+–bound channel is an inactivated structure, similar to that detected in KcsA alone at low K^+ concentrations [60] (PDB 1K4D), where the innermost sites S2 and S3 are absent. In contrast to that conclusion, the results obtained from the time-resolved anisotropy decays of W67 KcsA revealed that the stack of K^+ binding sites remains accessible in the TOA^+–KcsA complex, although with a lower affinity, and that the W67–W67 intersubunit distances are close to those found in the resting channel in the absence of TOA^+. These structural characteristics are shared by several inactivated models of KcsA [86, 97, 114] and it seems, therefore, a general feature of the inactivated SF of this potassium channel.

3.7 Examining the Bidirectional Crosstalk Between the Inner and Outer Gates of KcsA Through Homo-FRET Monitoring of Two Independent Fluorescent Reporters

Several biophysical approaches, mainly using NMR and EPR techniques, have also reported that the communication between the inner and outer gates of KcsA is not unidirectional since changes in the SF conformation (related to its K^+ occupancy) can also modify the open-closed equilibrium of the inner gate [115–117]. As in KcsA, it is presumed that most other K^+ channels are controlled through an allosteric coupling between the gates at both ends of the pore, even though the molecular basis of such regulatory mechanisms is still unclear [64, 117, 118]. The monitoring of the dynamic behaviour of the inner gate of KcsA required labelling the protein with an extrinsic fluorophore. Francisco Bezanilla and collaborators [79, 119] were the first to use time-resolved fluorescence measurements to investigate the opening dynamics of KcsA by tracking the changes on the fluorescence lifetime measurements (in the frequency domain) of the tetrametylrhodamine-5-maleimide (TMR-M) covalently attached to the G116C KcsA mutant channel. Here, the same G116C mutation was incorporated in the context of the quadruple mutant W26, 68, 87,113F KcsA, where there is only a single Trp residue (W67) in each monomer. This new mutant protein, G116C W67 KcsA, was also labelled with tetrametylrhodamine-5-maleimide (TMR-M).

3.8 *Consequences of Incomplete Labelling of G116C W67 KcsA with an Extrinsic Probe*

In contrast to the hetero-FRET measurements, the use of a single extrinsic fluorophore greatly simplifies the labelling of the samples. However, one must bear in mind that the incomplete random labelling of a multimeric protein, even with only one labelling site/subunit, always results in various subpopulations of fluorescently labelled channels. In the case of a tetrameric protein, these populations correspond to unlabelled, single, double, triple and fully labelled proteins (Fig. 9a). The relative proportion of each subpopulation is controlled by the final average degree of protein labelling (dye-to-protein molar ratio, D:P) (Fig. 9b, c) which, in turn, is critically dependent on the initial D:P molar ratio used in the label reaction. Therefore, the application of steady-state and time-resolved fluorescence measurements to the characterization of these systems requires performing a comparative study of two samples, prepared with a low and a high D:P, respectively. The first case is used to characterize the fluorescence properties of a predominantly singly labelled sample, i.e. when essentially no homo-FRET occurs. On the other hand, samples prepared with a higher degree of labelling are enriched in multiply labelled species (Fig. 9b, c) – these are then used to characterize the changes in the efficiency of the homo-FRET process within the fluorescently labelled proteins that accompany the inner gating of the channel [120].

3.9 *Allosteric Coupling Between the Activation Gate and the SF*

The simultaneous presence of two independent fluorescent reporters, one at each gate of the channel (W67 near to the SF and TMR at the inner gate (Fig. 10a), respectively, allowed to investigate the allosteric crosstalk between the inner gate and the SF in the same sample [120]. In this case, the location of W67 at the pore-helix, very far away from the TMR at the C116 position (≈36 Å, G116-Cα – W67-Cα distance) prevented any contact quenching or hetero-FRET process. Whereas W67 is a genetically encoded fluorescent reporter and hence is present in all G116C W67 KcsA subunits, the covalent attachment of the TMR is a random process, resulting in a heterogeneous population with distinct subpopulations of fluorescently labelled channels with a variable number of covalently linked probes [23], as described above. The number of TMR dyes attached to the protein could be controlled by adjusting the labelling conditions, ranging from 5 to 50% of labelled tetramer (on average) of the G116C W67 KcsA channel. An under labelled sample, predominantly consisting of singly labelled proteins, allowed for the characterization of the rotational dynamics of the conjugated dye/protein and to study the influence of pH on its fluorescence properties. On the other hand, samples prepared with a higher degree of labelling, which are enriched in multiply labelled protein species, gave

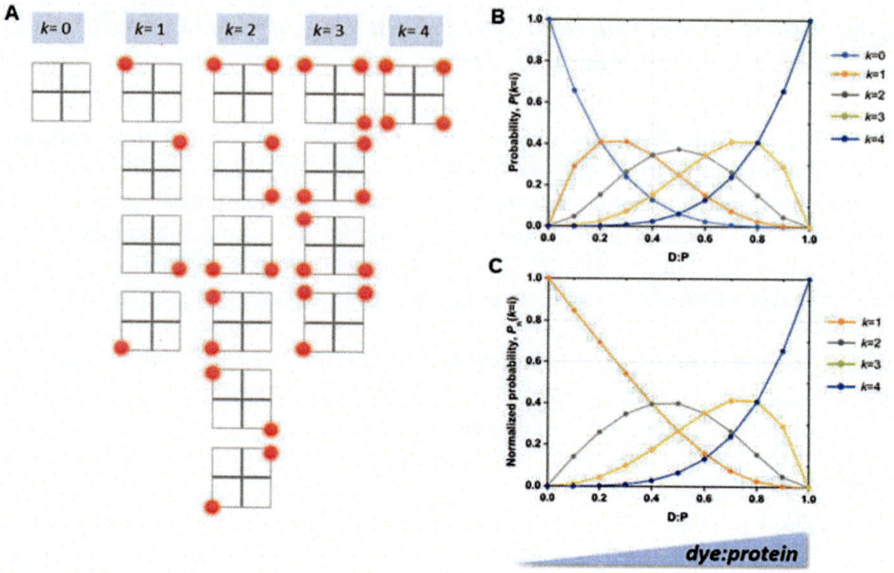

Fig. 9 The average degree of labelling, D:P, of a tetrameric protein controls the final proportion of the different fluorescently labelled species present in solution. (**a**) The random labelling of a tetrameric protein with a single labelling site per subunit produces a heterogeneous population of fluorescently labelled species with a variable number of extrinsic dyes, *k*, covalently attached to each protein. (**b**) The probabilities of finding unlabelled ($k = 0$), single ($k = 1$), double ($k = 2$), triple ($k = 3$) and fully ($k = 4$) labelled proteins can be calculated for a given D:P using a binomial distribution. (**c**) The fluorescence anisotropy measurements only detect emitting species, and so the probability distributions need to be re-normalized by excluding the unlabelled proteins since these are non-fluorescent

information about the extent of the homo-FRET process among the fluorescently labelled subunits within each tetrameric channel (Fig. 10b, c). Particularly, when working at high D:P ratios (>0.30), the close proximity of the TMR dyes in the closed state of the channel (pH 7.0) resulted in a highly efficient homo-FRET process, monitored by low $<r>_{TMR}$ values (Fig. 11) and fast time-resolved anisotropy decays (Fig. 10b). This efficiency was reduced when the opening of the protein was triggered at acidic pH, leading to higher $<r>_{TMR}$ values (Fig. 11) and slower anisotropy decays (Fig. 10c). This way, a new dual homo-FRET based sensor was obtained, capable of detecting the structural rearrangements of the inner gate after the pH-induced activation and the SF conformation in the same sample. In this case, the intersubunit distances at the inner gate level were not calculated due to the concomitant formation of H- and J-type dimers of conjugated TMR at pH 7.0 and 4.0, respectively [120].

In addition, the ability to detect the conformational state of the inner and outer gates in the same sample allowed for the characterization of their allosteric communication in the KcsA channel. A typical experiment to study the pH-induced activation of the inner gate consisted of a pH titration, where a stock of

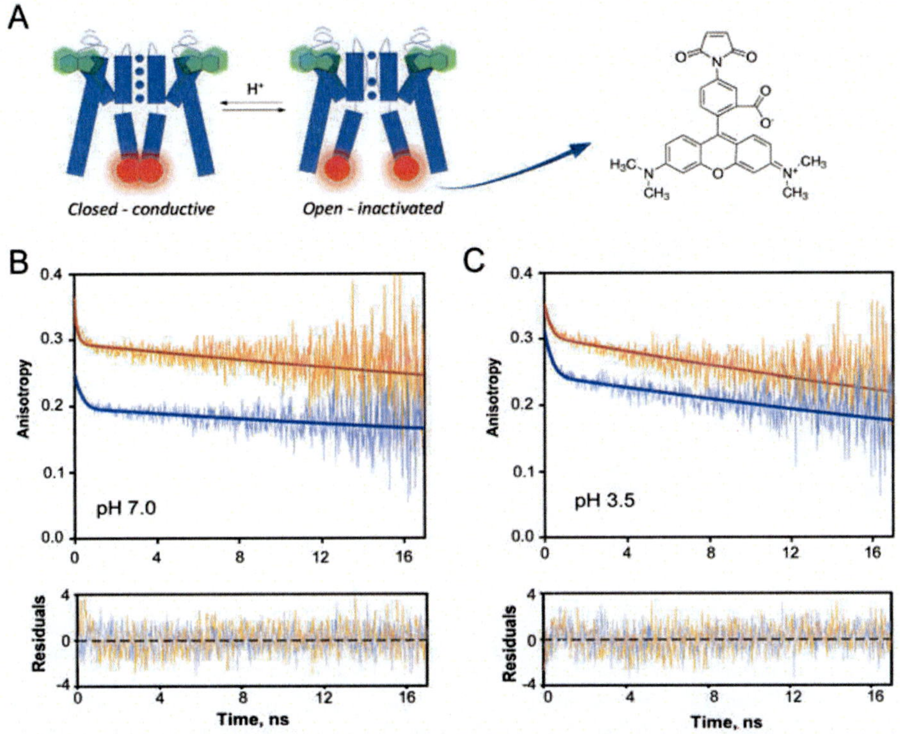

Fig. 10 Probing the conformational dynamics of the inner gate of KcsA using homo-FRET measurements. (**a**) The conformational equilibrium between the closed and open states of G116C W67 KcsA can be monitored by the covalent attachment of tetramethylrhodamine-5-maleimide (TMR) dye at the C116 position. (**b**) The labelling ratio (dye:protein ratio, D:P) has a great impact on the time-resolved anisotropy decays of the covalently linked TMR, as illustrated by the representative decays of the TMR conjugated to G116C W67 KcsA protein at (**b**) pH 7.0 and (**c**) pH 3.5. The D:P of the samples was 0.12 (orange) and 0.45 (blue). The solid lines are the best fits of a double exponential function to the experimental data. Bottom panels: the weighted residuals (W. residuals) are randomly distributed around zero showing the goodness of the fits. The measuring conditions were $\lambda_{ex} = 488$ nm $\lambda_{em} = 575$ nm. The buffers contained 5 mM DDM and 200 mM KCl ($T = 25°C$)

TMR-labelled protein stored in 20 mM Hepes, 5 mM DDM, 200 mM NaCl, pH 7.0 was divided into small aliquots and subsequently dialyzed to different pH and ionic conditions. The steady-state ($<r>_{TMR}$, Fig. 11) and time-resolved anisotropy decays (Fig. 10) of the TMR dyes bound to the G116C W67 KcsA presented a notable dependence on the pH and ionic environment only in the multiply labelled channels (high D:P). Concomitantly, the average lifetime of these samples remained essentially constant, reinforcing the conclusion that intramolecular homo-FRET is the major factor responsible for the large changes detected in $<r>_{TMR}$ within the 3.5 to 7.0 pH range. The changes detected on the $<r>_{TMR}$ values according to the pH of

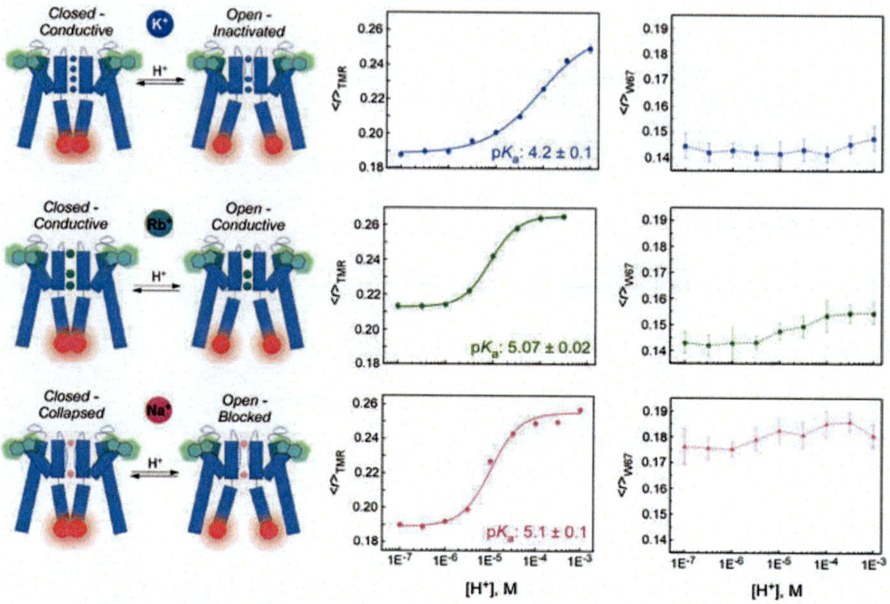

Fig. 11 The ion occupancy at the S2 site of the SF contributes to controlling the allosteric crosstalk between the inner and outer gates of the KcsA potassium channel. The left panels display a schematic representation of the allosteric communication between the inner and outer gates of the KcsA channel, reported by fluorescence anisotropy of the covalently bound TMR at position C116 (red) and the intrinsic fluorophore W67 (green), respectively. The conformation of the SF (outer gate), reported by the fluorescence anisotropy of W67, $<r>_{W67}$, in the TMR-labelled G116C W67 KcsA channels (right panels) is sensitive to the different ion occupancy presented by K^+ (blue), Rb^+ (green) and Na^+ (pink). In the case of K^+, its ability to bind with equal probability to the four binding sites at the SF (when the inner gate is in the closed conformation) leads to closer W67–W67 inter-subunit distances, and hence lower anisotropy levels. The activation pK_a, calculated from the variation of the fluorescence anisotropy of covalently linked TMR, $<r>_{TMR}$, with pH (central panels) was ~4.0. On the other hand, the inner gate opened at lower [H^+] (pK_a of ~ 5.0) when Na^+ or Rb^+ is bound to the SF at S1-S4 or S1-S3-S4 positions, respectively, i.e., when there was a loss of ion occupancy at the S2 site. The inability of both cations to interact with the S2 site at the outer gate is also characterized by higher W67 anisotropies. The final D:P of the samples were 0.45 in K^+ and Na^+ and 0.38 in Rb^+

the buffer, from the low-anisotropy closed state to the high-anisotropy open state, presented a sigmoidal behaviour, allowing for the calculation of an activation pK_a of the channel in different experimental conditions. Figure 11 (central panels) compares the results obtained in the presence of 200 mM K^+, Rb^+ or Na^+. The calculated values were p$K_a = 4.2 \pm 0.1$, p$K_a = 5.07 \pm 0.02$ and p$K_a = 5.1 \pm 0.1$ for K^+, Rb^+ an Na^+, respectively. The simultaneously monitoring of the $<r>_{W67}$ allowed to evaluate the conformational state of the SF in each case (Fig. 11, right panels). The anisotropy values of W67 determined at pH 7.0 and 4.0 agree with those previously measured for W67 KcsA indicating that the SF of the G116C W67 KcsA mutant channel labelled with TMR is still discriminating between permeant and blocking cations. Therefore, the lower activation pK_a obtained for the KcsA inner gate in 200 mM KCl compared to RbCl and NaCl seems to be conditioned primarily by its average degree of ion occupancy. In fact, although both permeant ions K^+ and Rb^+ induce a conductive SF at neutral pH, the SF displays a diminished ion occupancy in the second case: the crystal structures of KcsA solved in the presence of Rb^+ ions characteristically miss an ion at the second K^+ binding site [61, 62] in contrast to the SF in the presence of K^+, which displays an average occupancy of 2 ions, either in the S1-S3 or the S2-S4 positions [61]. Hence, the closed conformation of the inner gate is more stable only when the SF is in the presence of high K^+ amounts (i.e., when there is an equal occupancy of the four binding sites). On the other hand, a lower average ion occupancy of the SF (200 mM Rb^+ or Na^+) destabilizes the outer gate which, in turn, results in a higher activation pK_a, thus providing definitive evidence for the existence of an allosteric coupling between the two gates of the KcsA channel as previously reported [121].

4 Concluding Remarks

Ion channels in general, and K^+ channels, in particular, are highly dynamic entities where the conformational plasticity of the transmembrane pore and its allosteric interplay with the regulatory domains play a central role in their functional behaviour, namely the high-rate ion flux, selectivity and inactivation processes. Over the last decades, FRET studies (including time-resolved approaches) have shown to be a powerful tool to gain insight into the molecular basis of the structure–function relationship of such membrane proteins since they are performed in physiological-like experimental conditions (i.e., at low concentration in aqueous solution at room temperature), offering good-quality structural information (even at the Å level).

Due to the high level of symmetry often found in the K^+ channels family, the analysis of homo-FRET processes has also emerged as a valuable instrument to characterize the conformational plasticity of these proteins. Particularly, the homo-FRET formalism derived for the homo-tetrameric KcsA channel has been shown to be a powerful tool that allowed to quantitatively measure the W67–W67 intersubunit distances through polarized fluorescence measurements of a single-tryptophan mutant of KcsA under mild experimental conditions. Particularly, the measurements were performed in the absence of bound antibody fragments that, as shown here, may hinder the pore-helix dynamics putatively associated with the conversion of KcsA into an inactivated state. Hence, the high-resolution structures provided by protein crystallography may not always reflect the predominant structure displayed by the protein in less extreme experimental conditions and these studies should be complemented by the implementation of other biophysical techniques, including steady-state and time-resolved hetero-FRET or homo-FRET approaches.

In more general terms, a significant advantage of homo-transfer over heterotransfer experiments is that it greatly simplifies protein labelling with extrinsic probes. However, the accuracy of the distances recovered from applying this homo-

FRET approach, as all FRET measurements in general, is constrained by orientational effects of the probes. In the case of using extrinsic fluorophores, it is useful to covalently label the proteins with dyes that have short linkers. In this way, the depolarization of the emitted fluorescence due to local motions of the probe relatively to the protein backbone is minimized, thereby maximizing the dynamical range of the fluorescence anisotropy measurements. However, this, in turn, compromises the approximation that the indirectly excited fluorophores in a cyclic arrangement have a very low anisotropy. Still, the use of fluorophores that have a mixed polarization, like the tryptophan residues, allows to overcome these problems as shown here.

Acknowledgements This work was supported by the grant PGC2018-093505-B-I00 from the Spanish "Ministerio de Ciencia e Innovación"/FEDER, UE and national funds from FCT-Fundação para a Ciência ea. Tecnologia, I.P., under the scope of the project UIDB/04565/2020 and UIDP/04565/2020 of the Research Unit Institute for Bioengineering and Biosciences – iBB and the project LA/P/0140/2020 of the Associate Laboratory Institute for Health and Bioeconomy – i4HB.

C.D.-G. acknowledges support from the Medical Biochemistry and Biophysics Doctoral Programme (M2B-PhD) from FCT (ref.: SFRH/PD/BD/135154/2017).

References

1. Hille B (2001) Ion channels of excitable membranes.3rd edn. Sinauer Associates, Sunderland
2. Ashcroft FM (1999) Ion channels and disease.1st edn. Academic Press, San Diego
3. Doyle DA (2004) Molecular insights into ion channel function. Mol Membr Biol 21:221–225
4. Miller C (2000) An overview of the potassium channel family. Genome Biol 1:reviews0004. https://doi.org/10.1186/gb-2000-1-4-reviews0004
5. Catterall WA (1995) Structure and function of voltage-gated ion channels. Annu Rev Biochem 64:493–531
6. Kass RS (2005) The channelopathies: novel insights into molecular and genetic mechanisms of human disease. J Clin Invest 115:1986–1989
7. Kullmann DM (2002) The neuronal channelopathies. Brain 125:1177–1195
8. Kumar P, Kumar D, Jha SK et al (2016) Ion channels in neurological disorders. In: Donev R (ed) Advances in protein chemistry and structural biology. Academic Press, pp 97–136
9. LeMasurier M, Heginbotham L, Miller C (2001) KcsA: it's a potassium channel. J Gen Physiol 118:303–314
10. Renart ML, Barrera FN, Molina ML et al (2006) Effects of conducting and blocking ions on the structure and stability of the potassium channel KcsA. J Biol Chem 281:29905–29915. https://doi.org/10.1074/jbc.M602636200
11. Renart ML, Giudici AM, Díaz-García C et al (2020) Modulation of function, structure and clustering of K^+ channels by lipids: lessons learnt from KcsA. Int J Mol Sci 21:2554
12. Krishnan MN, Bingham JP, Lee SH et al (2005) Functional role and affinity of inorganic cations in stabilizing the tetrameric structure of the KcsA K^+ channel. J Gen Physiol 126:271–283
13. Molina ML, Barrera FN, Fernández AM et al (2006) Clustering and coupled gating modulate the activity in KcsA, a potassium channel model. J Biol Chem 281:18837–18848. https://doi.org/10.1074/jbc.M600342200

14. Lerner E, Barth A, Hendrix J et al (2021) FRET-based dynamic structural biology: challenges, perspectives and an appeal for open-science practices. eLife 10:e60416. https://doi.org/10.7554/eLife.60416
15. Taraska JW, Zagotta WN (2010) Fluorescence applications in molecular neurobiology. Neuron 66:170–189
16. Lakowicz JR (2007) Principles of fluorescence spectroscopy.3rd edn. Springer, New York
17. Valeur B, Berberan-Santos MN (2012) Molecular fluorescence.2nd edn. Wiley-VCH, Weinheim
18. Stryer L, Hauglandt RP (1967) Energy transfer: a spectroscopic ruler. Proc Natl Acad Sci U S A 58:719–726. https://doi.org/10.1073/pnas.58.2.719
19. Ahern CA, Pless SA (2015) Novel chemical tools to study ion channel biology.1st edn. Springer, New York
20. Nanda JS, Lorsch JR (2014) Labeling of a protein with fluorophores using maleimide derivatization. In: Methods in enzymology. Elsevier, pp 79–86
21. Akabas MH (2015) Cysteine modification: probing channel structure, function and conformational change. In: Ahern C, Pless S (eds) Advances in experimental medicine and biology1st edn. Springer, New York, pp 25–54
22. Glauner KS, Mannuzzu LM, Gandhi CS, Isacoff EY (1999) Spectroscopic mapping of voltage sensor movement in the Shaker potassium channel. Nature 402:813–817. https://doi.org/10.1038/45561
23. Wang S, Lee SJ, Heyman S et al (2012) Structural rearrangements underlying ligand-gating in Kir channels. Nat Commun 3:617. https://doi.org/10.1038/ncomms1625
24. Sadler EE, Kapanidis AN, Tucker SJ (2016) Solution-based single-molecule FRET studies of K^+ channel gating in a lipid bilayer. Biophys J 110:2663–2670. https://doi.org/10.1016/j.bpj.2016.05.020
25. Corry B, Jayatilaka D, Rigby P (2005) A flexible approach to the calculation of resonance energy transfer efficiency between multiple donors and acceptors in complex geometries. Biophys J 89:3822–3836. https://doi.org/10.1529/biophysj.105.069351
26. Cheng W, Yang F, Takanishi CL, Zheng J (2007) Thermosensitive TRPV channel subunits coassemble into heteromeric channels with intermediate conductance and gating properties. J Gen Phys 129:191–207. https://doi.org/10.1085/jgp.200709731
27. Heyduk T (2002) Measuring protein conformational changes by FRET/LRET. Curr Opin Biotechnol 13:292–296
28. Posson DJ, Ge P, Miller C et al (2005) Small vertical movement of a K^+ channel voltage sensor measured with luminescence energy transfer. Nature 436:848–851. https://doi.org/10.1038/nature03819
29. Richardson J, Blunck R, Ge P et al (2006) Distance measurements reveal a common topology of prokaryotic voltage-gated ion channels in the lipid bilayer. Proc Natl Acad Sci U S A 103:15865–15870. https://doi.org/10.1073/pnas.0607532103
30. Cha A, Snyder GE, Selvin PR, Bezanilla F (1999) Atomic scale movement of the voltage-sensing region in a potassium channel measured via spectroscopy. Nature 402:809–813. https://doi.org/10.1038/45552
31. Selvin PR (2002) Principles and biophysical applications of lanthanide-based probes. Annu Rev Biophys Biomol Struct 31:275–302. https://doi.org/10.1146/annurev.biophys.31.101101.140927
32. Selvin PR, Hearst JE (1994) Luminescence energy transfer using a terbium chelate: improvements on fluorescence energy transfer. Proc Natl Acad Sci U S A 91:10024–10028. https://doi.org/10.1073/pnas.91.21.10024
33. Faure É, Starek G, McGuire H et al (2012) A limited 4 Å radial displacement of the S4-S5 linker is sufficient for internal gate closing in Kv channels. J Biol Chem 287:40091–40098. https://doi.org/10.1074/jbc.M112.415497

34. Wang S, Vafabakhsh R, Borschel WF et al (2015) Structural dynamics of potassium-channel gating revealed by single-molecule FRET. Nat Struct Mol Biol 23:31–36. https://doi.org/10.1038/nsmb.3138
35. Wang S, Lee SJ, Maksaev G et al (2019) Potassium channel selectivity filter dynamics revealed by single-molecule FRET. Nat Chem Biol 15:377–383. https://doi.org/10.1038/s41589-019-0240-7
36. Matamoros M, Nichols CG (2021) Pore-forming transmembrane domains control ion selectivity and selectivity filter conformation in the kirbac1.1 potassium channel. J Gen Physiol 153:e202012683. https://doi.org/10.1085/jgp.202012683
37. Gupta K, Toombes GE, Swartz KJ (2019) Exploring structural dynamics of a membrane protein by combining bioorthogonal chemistry and cysteine mutagenesis. eLife 8:e50776. https://doi.org/10.7554/eLife.50776
38. Kalinin S, Johansson LBA (2004) Utility and considerations of donor-donor energy migration as a fluorescence method for exploring protein structure-function. J Fluoresc 14:681–691. https://doi.org/10.1023/B:JOFL.0000047218.51768.59
39. Cardoso S, Berberan-Santos MN (2021) Reversible electronic energy transfer (homo-FRET) in cyclic molecular and supramolecular systems: fluorescence anisotropy decays for the isotropic interaction. J Phys Chem A 125:8476–8481. https://doi.org/10.1021/acs.jpca.1c04975
40. Berberan-Santos MN, Choppinet P, Fedorov A et al (1999) Multichromophoric cyclodextrins. 6. Investigation of excitation energy hopping by Monte-Carlo simulations and time-resolved fluorescence anisotropy. J Am Chem Soc 121:2526–2533. https://doi.org/10.1021/ja983601n
41. Renart ML, Giudici AM, Poveda JA et al (2019) Conformational plasticity in the KcsA potassium channel pore helix revealed by homo-FRET studies. Sci Rep 9:6215. https://doi.org/10.1038/s41598-019-42405-5
42. Runnels LW, Scarlata SF (1995) Theory and application of fluorescence homotransfer to melittin oligomerization. Biophys J 69:1569–1583. https://doi.org/10.1016/S0006-3495(95)80030-5
43. Bader AN, Hoetzl S, Hofman EG et al (2011) Homo-FRET imaging as a tool to quantify protein and lipid clustering. ChemPhysChem 12:475–483. https://doi.org/10.1002/cphc.201000801
44. Blackman SM, Piston DW, Beth AH (1998) Oligomeric state of human erythrocyte band 3 measured by fluorescence resonance energy homotransfer. Biophys J 75:1117–1130. https://doi.org/10.1016/S0006-3495(98)77601-5
45. Melo AM, Fedorov A, Prieto M, Coutinho A (2014) Exploring homo-FRET to quantify the oligomer stoichiometry of membrane-bound proteins involved in a cooperative partition equilibrium. Phys Chem Chem Phys 16:18105–18117. https://doi.org/10.1039/C4CP00060A
46. Chan FTS, Kaminski CF, Kaminski Schierle GS (2011) HomoFRET fluorescence anisotropy imaging as a tool to study molecular self-assembly in live cells. ChemPhysChem 12:500–509. https://doi.org/10.1002/cphc.201000833
47. Tramier M, Coppey-Moisan M (2008) Fluorescence anisotropy imaging microscopy for homo-FRET in living cells. Methods Cell Biol 85:395–414. https://doi.org/10.1016/S0091-679X(08)85017-0
48. Pastor-Flores D, Talwar D, Pedre B, Dick TP (2020) Real-time monitoring of peroxiredoxin oligomerization dynamics in living cells. Proc Natl Acad Sci U S A 117:16313–16323. https://doi.org/10.1073/pnas.1915275117
49. Bergström F, Hägglöf P, Karolin J et al (1999) The use of site-directed fluorophore labeling and donor-donor energy migration to investigate solution structure and dynamics in proteins. Proc Natl Acad Sci U S A 96:12477–12481. https://doi.org/10.1073/pnas.96.22.12477
50. Karolin J, Fa M, Wilczynska M et al (1998) Donor-donor energy migration for determining intramolecular distances in proteins: I. Application of a model to the latent plasminogen activator inhibitor-1 (PAI-1). Biophys J 74:11–21. https://doi.org/10.1016/S0006-3495(98)77762-8

51. Fa M, Bergström F, Hägglöf P et al (2000) The structure of a serpin-protease complex revealed by intramolecular distance measurements using donor-donor energy migration and mapping of interaction sites. Structure 8:397–405. https://doi.org/10.1016/S0969-2126(00)00121-0
52. Zou P, Surendhran K, Mchaourab HS (2007) Distance measurements by fluorescence energy homotransfer: evaluation in T4 lysozyme and correlation with dipolar coupling between spin labels. Biophys J 92:L27–L29. https://doi.org/10.1529/biophysj.106.098913
53. Ross BL, Tenner B, Markwardt ML et al (2018) Single-color, ratiometric biosensors for detecting signaling activities in live cells. eLife 7:e35458. https://doi.org/10.7554/elife.35458
54. Cameron WD, Bui CV, Hutchinson A et al (2016) Apollo-NADP+: a spectrally tunable family of genetically encoded sensors for NADP+. Nat Methods 13:352–358. https://doi.org/10.1038/nmeth.3764
55. Warren SC, Margineanu A, Katan M et al (2015) Homo-FRET based biosensors and their application to multiplexed imaging of signalling events in live cells. Int J Mol Sci 16:14695–14716. https://doi.org/10.3390/ijms160714695
56. Schrempf H, Schmidt O, Kummerlen R et al (1995) A prokaryotic potassium ion channel with two predicted transmembrane segments from streptomyces lividans. EMBO J 14:5170–5178
57. Doyle DA, Morais CJ, Pfuetzner RA et al (1998) The structure of the potassium channel: molecular basis of K^+ conduction and selectivity. Science 280:69–77
58. Berneche S, Roux B (2005) A gate in the selectivity filter of potassium channels. Structure 13:591–600
59. Nimigean CM, Miller C (2002) Na+ block and permeation in a K^+ channel of known structure. J Gen Physiol 120:323–335
60. Zhou Y, Morais-Cabral JH, Kaufman A, MacKinnon R (2001) Chemistry of ion coordination and hydration revealed by a K^+ channel-fab complex at 2.0 A resolution. Nature 414:43–48
61. Morais-Cabral JH, Zhou Y, MacKinnon R (2001) Energetic optimization of ion conduction rate by the K^+ selectivity filter. Nature 414:37–42
62. Zhou Y, MacKinnon R (2003) The occupancy of ions in the K^+ selectivity filter: charge balance and coupling of ion binding to a protein conformational change underlie high conduction rates. J Mol Biol 333:965–975
63. Chakrapani S, Cordero-Morales JF, Perozo E (2007) A quantitative description of KcsA gating I: macroscopic currents. J Gen Physiol 130:465–478
64. Xu Y, McDermott AE (2019) Inactivation in the Potassium Channel KcsA. J Struct Biol X 3:100009. https://doi.org/10.1016/j.yjsbx.2019.100009
65. Thompson AN, Posson DJ, Parsa PV, Nimigean CM (2008) Molecular mechanism of pH sensing in KcsA potassium channels. Proc Natl Acad Sci U S A 105:6900–6905
66. Cuello LG, Jogini V, Cortes DM, Perozo E (2010) Structural mechanism of C-type inactivation in K(+) channels. Nature 466:203–208
67. Cuello LG, Cortes DM, Jogini V et al (2010) A molecular mechanism for proton-dependent gating in KcsA. FEBS Lett 584:1126–1132
68. Cordero-Morales JF, Cuello LG, Perozo E (2006) Voltage-dependent gating at the KcsA selectivity filter. Nat Struct Mol Biol 13:319–322
69. Cordero-Morales JF, Cuello LG, Zhao Y et al (2006) Molecular determinants of gating at the potassium-channel selectivity filter. Nat Struct Mol Biol 13:311–318
70. Jekhmane S, Medeiros-Silva J, Li J et al (2019) Shifts in the selectivity filter dynamics cause modal gating in K^+ channels. Nat Commun 10:123. https://doi.org/10.1038/s41467-018-07973-6
71. Chakrapani S, Cordero-Morales JF, Perozo E (2007) A quantitative description of KcsA gating II: single-channel currents. J Gen Physiol 130:479–496
72. Liu Y, Holmgren M, Jurman ME, Yellen G (1997) Gated access to the pore of a voltage-dependent K^+ channel. Neuron 19:175–184
73. Perozo E, Cortes DM, Cuello LG (1998) Three-dimensional architecture and gating mechanism of a K^+ channel studied by EPR spectroscopy. Nat Struct Biol 5:459–469

74. Jiang Y, Lee A, Chen J et al (2002) The open pore conformation of potassium channels. Nature 417:523–526
75. Kelly BL, Gross A (2003) Potassium channel gating observed with site-directed mass tagging. Nat Struct Biol 10:280–284
76. Liu Y, Jurman ME, Yellen G (1996) Dynamic rearrangement of the outer mouth of a K^+ channel during gating. Neuron 16:859–867
77. Ogielska EM, Aldrich W (1999) Functional consequences of a decreased potassium affinity in a potassium channel pore. Ion interactions and C-type inactivation. J Gen Physiol 113:347–358. https://doi.org/10.1085/jgp.113.2.347
78. Gao L, Mi X, Paajanen V et al (2005) Activation-coupled inactivation in the bacterial potassium channel KcsA. Proc Natl Acad Sci U S A 102:17630–17635
79. Blunck R, Cordero-Morales JF, Cuello LG et al (2006) Detection of the opening of the bundle crossing in KcsA with fluorescence lifetime spectroscopy reveals the existence of two gates for ion conduction. J Gen Physiol 128:569–581
80. Ader C, Schneider R, Hornig S et al (2008) A structural link between inactivation and block of a K^+ channel. Nat Struct Mol Biol 15:605–612. https://doi.org/10.1038/nsmb.1430
81. Kiss L, Korn SJ (1998) Modulation of C-type inactivation by K^+ at the potassium channel selectivity filter. Biophys J 74:1840–1849. https://doi.org/10.1016/S0006-3495(98)77894-4
82. Kurata HT, Fedida D (2006) A structural interpretation of voltage-gated potassium channel inactivation. Prog Biophys Mol Biol 92:185–208
83. Zheng J, Sigworth FJ (1997) Selectivity changes during activation of mutant Shaker potassium channels. J Gen Physiol 110:101–117. https://doi.org/10.1085/jgp.110.2.101
84. Cuello LG, Cortes DM, Perozo E (2017) The gating cycle of a K^+ channel at atomic resolution. elife 6:e28032. https://doi.org/10.7554/eLife.28032
85. Devaraneni PK, Komarov AG, Costantino CA et al (2013) Semisynthetic K^+ channels show that the constricted conformation of the selectivity filter is not the C-type inactivated state. Proc Natl Acad Sci U S A 110:15698–15703
86. Matulef K, Komarov AG, Costantino CA, Valiyaveetil FI (2013) Using protein backbone mutagenesis to dissect the link between ion occupancy and C-type inactivation in K^+ channels. Proc Natl Acad Sci U S A 110:17886–17891
87. Kratochvil HT, Maj M, Matulef K et al (2017) Probing the effects of gating on the ion occupancy of the K^+ channel selectivity filter using two-dimensional infrared spectroscopy. J Am Chem Soc 139:8837–8845. https://doi.org/10.1021/jacs.7b01594
88. Liu S, Focke PJ, Matulef K et al (2015) Ion-binding properties of a K^+ channel selectivity filter in different conformations. Proc Natl Acad Sci U S A 112:15096–15100. https://doi.org/10.1073/pnas.1510526112
89. Hoshi T, Armstrong CM (2013) C-type inactivation of voltage-gated K^+ channels: pore constriction or dilation? J Gen Physiol 141:151–160
90. Shealy RT, Murphy AD, Ramarathnam R et al (2003) Sequence-function analysis of the K^+-selective family of ion channels using a comprehensive alignment and the KcsA channel structure. Biophys J 84:2929–2942
91. Renart ML, Triano I, Poveda JA et al (2010) Ion binding to KcsA: implications in ion selectivity and channel gating. Biochemistry 49:9480–9487. https://doi.org/10.1021/bi101235v
92. Giudici AM, Díaz-García C, Renart ML et al (2021) Tetraoctylammonium, a long chain quaternary ammonium blocker, promotes a noncollapsed, resting-like inactivated state in KcsA. Int J Mol Sci 22:490. https://doi.org/10.3390/ijms22020490
93. Valeur B, Weber G (1977) Resolution of the fluorescence excitation spectrum of indole into the 1La and 1Lb excitation bands. Photochem Photobiol 25:441–444. https://doi.org/10.1111/j.1751-1097.1977.tb09168.x
94. Haas E, Ephraim-Katchalski-Katzir, Steinberg IZ (1978) Effect of the orientation of donor and acceptor on the probability of energy transfer involving electronic transitions of mixed polarization. Biochemistry 17:5064–5070. https://doi.org/10.1021/bi00616a032

95. Teijeiro-Gonzalez Y, Crnjar A, Beavil AJ et al (2021) Time-resolved fluorescence anisotropy and molecular dynamics analysis of a novel GFP homo-FRET dimer. Biophys J 120:254–269. https://doi.org/10.1016/j.bpj.2020.11.2275
96. Calcutta A, Jessen CM, Behrens MA et al (2012) Mapping of unfolding states of integral helical membrane proteins by GPS-NMR and scattering techniques: TFE-induced unfolding of KcsA in DDM surfactant. Biochim Biophys Acta Biomembr 1818(9):2290–2301. https://doi.org/10.1016/j.bbamem.2012.04.005
97. Giudici AM, Renart ML, Díaz-García C et al (2019) Accessibility of cations to the selectivity filter of KcsA in the inactivated state: an equilibrium binding study. Int J Mol Sci 20:689. https://doi.org/10.3390/ijms20030689
98. Yohannan S, Hu Y, Zhou Y (2007) Crystallographic study of the tetrabutylammonium block to the KcsA K^+ channel. J Mol Biol 366:806–814. https://doi.org/10.1016/j.jmb.2006.11.081
99. Lockless SW, Zhou M, MacKinnon R (2007) Structural and thermodynamic properties of selective ion binding in a K^+ channel. PLoS Biol 5:e121
100. Thompson AN, Kim I, Panosian TD et al (2009) Mechanism of potassium-channel selectivity revealed by Na+ and Li+ binding sites within the KcsA pore. Nat Struct Mol Biol 16:1317–1324. https://doi.org/10.1038/nsmb.1703
101. Piasta KN, Theobald DL, Miller C (2011) Potassium-selective block of barium permeation through single KcsA channels. J Gen Physiol 138:421–436
102. Rohaim A, Gong LD, Li J et al (2020) Barium blockade of the KcsA channel in open and closed conformation datasets. Data Brief 32:106135. https://doi.org/10.1016/j.dib.2020.106135
103. Cheng WW, McCoy JG, Thompson AN et al (2011) Mechanism for selectivity-inactivation coupling in KcsA potassium channels. Proc Natl Acad Sci U S A 108:5272–5277
104. Zhou Y, MacKinnon R (2001) Fab-mediated crystallization of a potassium channel. NSLS Act Rep 2001:25–27
105. Uysal S, Vasquez V, Tereshko V et al (2009) Crystal structure of full-length KcsA in its closed conformation. Proc Natl Acad Sci U S A 106:6644–6649
106. Tilegenova C, Cortes DM, Cuello LG (2017) Hysteresis of KcsA potassium channel's activation– deactivation gating is caused by structural changes at the channel's selectivity filter. Proc Natl Acad Sci U S A 114:3234–3239. https://doi.org/10.1073/pnas.1618101114
107. Lenaeus MJ, Burdette D, Wagner T et al (2014) Structures of KcsA in complex with symmetrical quaternary ammonium compounds reveal a hydrophobic binding site. Biochemistry 53:5365–5373
108. Armstrong CM, Binstock L (1965) Anomalous rectification in the squid giant axon injected with tetraethylammonium chloride. J Gen Physiol 48:859–872
109. Armstrong CM (1969) Inactivation of the potassium conductance and related phenomena caused by quaternary ammonium ion injection in squid axons. J Gen Physiol 54:553–575. https://doi.org/10.1085/jgp.54.5.553
110. Hille B (1967) The selective inhibition of delayed potassium currents in nerve by tetraethylammonium ion. J Gen Physiol 50:1287–1302
111. Choi KL, Aldrich RW, Yellen G (1991) Tetraethylammonium blockade distinguishes two inactivation mechanisms in voltage-activated K^+ channels. Proc Natl Acad Sci U S A 88: 5092–5095. https://doi.org/10.1073/pnas.88.12.5092
112. Heginbotham L, MacKinnon R (1992) The aromatic binding site for tetraethylammonium ion on potassium channels. Neuron 8:483–491. https://doi.org/10.1016/0896-6273(92)90276-J
113. Cuello LG, Jogini V, Cortes DM et al (2010) Structural basis for the coupling between activation and inactivation gates in K(+) channels. Nature 466:272–275
114. Matulef K, Annen AW, Nix JC, Valiyaveetil FI (2016) Individual ion binding sites in the K(+) channel play distinct roles in C-type inactivation and in recovery from inactivation. Structure 24:750–761
115. Ader C, Schneider R, Hornig S et al (2009) Coupling of activation and inactivation gate in a K^+-channel: potassium and ligand sensitivity. EMBO J 28:2825–2834

116. Wylie BJ, Bhate MP, McDermott AE (2014) Transmembrane allosteric coupling of the gates in a potassium channel. Proc Natl Acad Sci U S A 111:185–190. https://doi.org/10.1073/pnas.1319577110
117. Xu Y, Bhate MP, McDermott AE (2017) Transmembrane allosteric energetics characterization for strong coupling between proton and potassium ion binding in the KcsA channel. Proc Natl Acad Sci U S A 114:8788–8793. https://doi.org/10.1073/pnas.1701330114
118. Mironenko A, Zachariae U, de Groot BL, Kopec W (2021) The persistent question of potassium channel permeation mechanisms. J Mol Biol 433:167002
119. Blunck R, McGuire H, Hyde HC, Bezanilla F (2008) Fluorescence detection of the movement of single KcsA subunits reveals cooperativity. Proc Natl Acad Sci U S A 105:20263–20268. https://doi.org/10.1073/pnas.0807056106
120. Díaz-García C, Renart ML, Poveda JA et al (2021) Probing the structural dynamics of the activation gate of KcsA using homo-FRET measurements. Int J Mol Sci 22:19954
121. Medintz I, Hildebrandt N (2013) FRET – Förster resonance energy transfer: from theory to applications.1st edn. Wiley-Vch, Weinheim

Intrinsic Fluorescence Kinetics in Proteins

Olaf Rolinski and Rhona Muir

Contents

Abstract Detecting changes in the intrinsic fluorescence of proteins enables non-invasive monitoring of their biological activities. Here, we present the applications of the fluorescence intensity decay measurements of the intrinsic amino acids to study three proteins involved in different processes: the temperature-induced phase transition of human serum albumin, Alzheimer's beta-amyloid aggregation, and collagen glycation. The complexity of the intrinsic fluorescence is addressed by using the non-Debye model of fluorescence kinetics and/or by applying the time-resolved emission spectra (TRES) technique.

Keywords Beta-amyloid aggregation · Collagen glycation · Fluorescence kinetics · HSA phase transition · Intrinsic protein fluorescence · Time-resolved emission spectra

O. Rolinski (✉) and R. Muir
Department of Physics, University of Strathclyde, Glasgow, UK
e-mail: o.j.rolinski@strath.ac.uk

R. Šachl, M. Amaro (eds.), *Fluorescence Spectroscopy and Microscopy in Biology*, Springer Ser Fluoresc (2023) 20: 479–508, https://doi.org/10.1007/4243_2022_39,
Published online: 31 January 2023

1 Introduction

Protein structure (conformation, monomeric or oligomeric form, and complexing with other molecules) is usually related to its function and determines its performance in its natural biological habitat. Methods of determining protein structures and monitoring their activities are in high demand in biomedical research, and numerous analytical techniques have been introduced in recent decades. They include X-ray crystallography [1], NMR spectroscopy [2] (offering better spatial resolution), cryo-electron microscopy [3], or electronic spectroscopies [4–6]. Among the latter, circular dichroism [4] and fluorescence spectroscopy [5, 6] have become very popular in protein studies due to their ease and speed of operation. Florescence spectroscopy is a multi-dimensional technique; i.e. it allows measurement of several main observables (steady-state spectra, fluorescence intensity decays, and steady-state and time-resolved fluorescence anisotropy) at a variety of experimental conditions (composition of the sample, excitation and detection wavelength and polarisation, and temperature). When applied to protein, the multi-dimensionality of fluorescence spectroscopy may enable recognition of the functionally different protein forms and monitoring of their changes triggered by chemical and physical factors. This is due to high sensitivity of the fluorescence of excited fluorophore to the local nm-scale environment.

Typical applications of fluorescence spectroscopy to protein studies include tagging the protein with the fluorophore with the absorption and fluorescence spectra in the visible to infrared region, which allows its selective excitation, and separation of the fluorophore fluorescence from the protein's intrinsic fluorescence. The variety of commercially available fluorescent tags for proteins [7, 8] enables sensing of many features of protein assemblies, like conformational stages, formation of α- and β-sheets or helix structures, molecular separations, exposure to the solvent, access of the quencher molecules, and local pH.

Although effective in many applications, this approach may fail when the presence of the fluorescent tag modifies the protein's performance. In this case, the collected data report on the protein/fluorophore system, rather than on the protein in its natural environment. Note that minimising the tagged proteins to proteins ratio in the sample is not helpful because the fluorescence signal comes from the protein/tag complexes only. Indeed, the fluorescent tag can even stop the investigated process, for example the aggregation of the Alzheimer's beta-amyloids [9].

The intrinsic fluorescence, i.e. the fluorescence of three amino acids: tryptophan, tyrosine, and phenylalanine, and the fluorescence coming from the formation of fluorescent residues during activities of some proteins, offers an alternative approach to fluorescence-based protein sensing. As intrinsic fluorescence comes from the residues that are natural components of the protein or are formed during their natural activities, sensing based on intrinsic fluorophores is non-invasive and can be applied when the tag-based technique disturbs the studied process too much. However, the intrinsic fluorophores were not created by nature with the sensing application in mind; thus, their fluorescence responses to the changes in the local environment are usually complex and data interpretation is more demanding than in the case of the specialised bespoke sensors. Moreover, it is not always the case that the investigated process makes any measurable impact on the intrinsic fluorescence; therefore, experimental verification is needed in each case. However, if a substantial change in fluorescence signal is detected, a sensitive and non-invasive sensing procedure can be usually developed.

Fluorescence intensity and anisotropy decays are main observables providing information on protein behaviour in biological media. Intensity decay reflects the kinetics of protein interactions with surrounding molecules, with processes like collisional quenching or FRET impacting the depopulation of the excited state. Anisotropy reports the rotational freedom and rigidity of the protein environment [10] and has been used in the studies of many key proteins like HSA [11] and green fluorescent protein (GFP) [12].

In this chapter, we present the studies of protein assemblies based on the measurements of the intrinsic fluorescence intensity decays by the time-correlated single photon counting [13, 14] (TCSPC) technique. We discuss the alternative data collection and analysis methods and then focus on the technique exploring the time-resolved emission spectra (TRES). Several examples of fluorescence kinetics in protein of medical importance are presented.

1.1 Fluorescent Residues in Protein as Potential Fluorescence Sensors

Among the 20 amino acids which constitute the building bricks of proteins, tryptophan (Trp), tyrosine (Tyr), and phenylalanine (Phe) are fluorescent, which makes them candidates for the intrinsic, non-invasive sensors (Fig. 1).

The overlapping absorption spectra of all three amino acids require using excitation sources of specific wavelengths if a selective excitation is needed. Exciting only one amino acid in the protein may not be always possible, but the problem can be partially addressed by using the commercially available nanoLEDs [15–17], offering excitation wavelengths of 295 nm [15] (for selective excitation of Trp in all proteins), 279 nm [16] (for selective excitation of Tyr in the proteins containing no Trp), and 265 nm [17] (for Phe, if Trp and Tyr are not present).

Table 1 presents the key spectroscopic parameters of the fluorescent amino acids. According to the rotamer model [18–21], individual amino acids exist in one of three different conformational forms (rotamers), each exhibiting an individual fluorescence lifetime, determined by local environment. The anticipated orientations of the rotamers in respect of the protein backbone are shown in Fig. 2.

Relatively high peak molar absorption coefficient of Trp, 5,600 M^{-1} cm^{-1} at 295 nm, combined with its good quantum yield, 0.13, results in a reasonable effective brightness and makes Trp the most used intrinsic fluorescence sensor in proteins. For example, exciting a single Trp in HSA with the excitation source

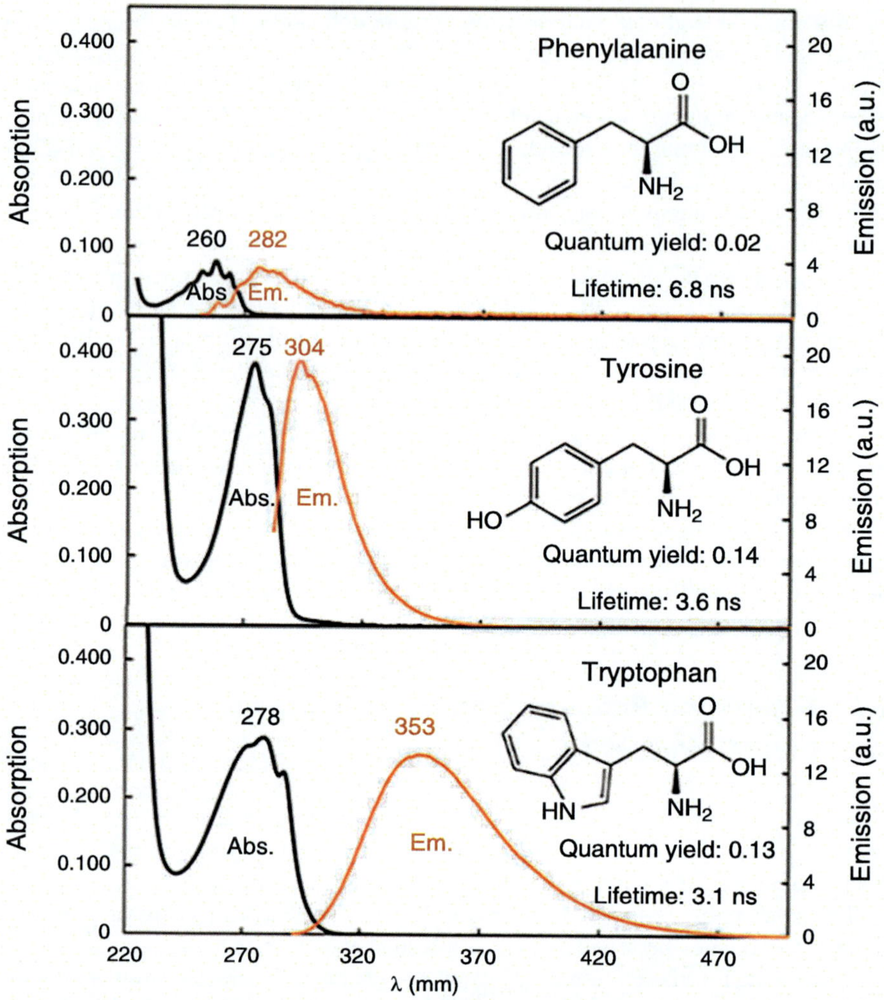

Fig. 1 Absorption (black) and emission (orange) spectrum of intrinsic fluorophores: Phe (50 μM), Tyr (50 μM), and Trp (50 μM) in water shown with the quantum yield, fluorescence lifetime, and chemical structure of each fluorophore. (Reproduced from: A.Alghamdi, PhD Thesis, University of Strathclyde, 2021)

Table 1 Spectroscopic parameters of fluorescent amino acids

	$\varepsilon(\lambda_{max})$ M^{-1} cm^{-1}	Φ	Effective brightness	Rotamer population
Trp (indole)	5,579(295)	0.13	725	R1>>R2>R3
Tyr (phenol)	1,405(279)	0.14	197	R1>R2≅R3
Phe	195(258)	0.024	5	R1>R2≅R3

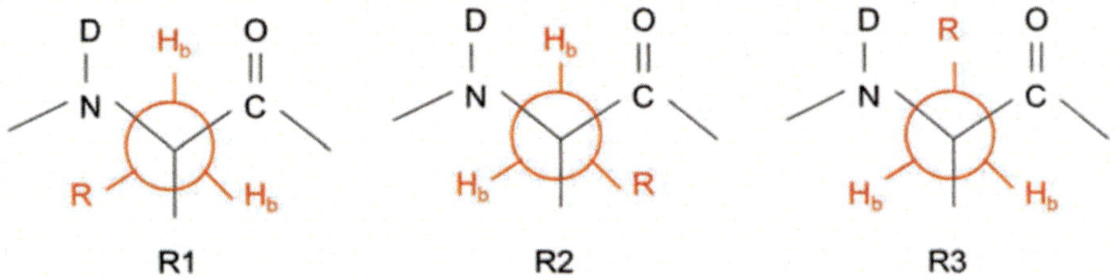

Fig. 2 Rotamers orientations

295 nm allows detailed studies of Trp excited-state kinetics, because Tyr and Phe are not excited and their fluorescence does not interfere with Trp fluorescence [22].

If there is no Trp in the protein (e.g. vasopressin, beta amyloid (Aβ)), Tyr fluorescence may be used effectively as an intrinsic sensor reporting protein activity [23, 24].

In addition to the fluorescence of individual amino acids, proteins also form specific fluorescent residues. Since the isolation of GFP from *Aequorea Victori*a [25], where the fluorescent residue is the sequence of three amino acids serine–tyrosine–glycine, producing green fluorescence, a number of fluorescent proteins have been developed. They enabled labelling and visualising proteins in living cells in thousands of biological applications [26–28].

2 Modelling the Kinetics of Intrinsic Fluorescence

In this section, we discuss the potential mathematical models of the fluorescence kinetics which may be used for representing the experimental fluorescence intensity decays collected in a TCSPC experiment. The relationship between the sought-after model fluorescence intensity decay *I(t)* and the measurable fluorescence response *F (t)* and the prompt *L(t)* curves [29, 30] is given by the convolution integral

$$F(t) = \int_0^t L(t - t')I(t')\mathrm{d}t' \tag{1}$$

Both *F(t)* and *L(t)* have digital representation (they are the contents of the data stored in the multi-channel analyser of TCSPC instrument after collecting signal from the fluorescent sample and the scattering sample, respectively), while *I(t)* is a formula representing the fluorescence intensity decay following a δ-pulse excitation. This is a predicted solution of the differential equation, or the set of differential equations describing the anticipated fluorescence kinetics. In practical applications, the experimental decays are fitted to the extended formula. This also takes into account the additional technical effects of TCSPC instrumentation, namely

$$F(t) = a + bL(t + \Delta) + c\int_0^t L(t + \Delta - t')I(t')\mathrm{d}t' \tag{2}$$

where *a, b*, and *c* are the background level, the contribution of the scattered light, and the scaling parameter, respectively, and *Δ* is the time shift between the prompt and decay curves due to the colour effect of the light detector and finite channel width of the time axis. The decay *I(t)* is dependent on the kinetic model parameters. The analysis of the TCSPC data requires two stages: firstly, determining whether the assumed *I(t)* fits well to the experimental data and determining the values of fitted parameters and, secondly, checking if the parameters recovered from a series of the decays measured at different experimental settings are consistent with the character of the assumed model of the kinetics (e.g. whether the decays measured at increasing temperatures give shorter fluorescence lifetimes due to expected effect of collisional quenching).

2.1 Exponential and Multi-exponential Decays

In the simplest case, the *I(t)* is the solution of the two-level model $dI(t)/dt = -kI(t)$ (Fig. 3a), where *k* is the rate of the decay, $k = 1/\tau$, and τ is the fluorescence lifetime. The resulting exponential decay, $I(t) = I_0\exp[-t/\tau]$, is rarely the case in proteins. For example, single Tyr in a buffer solution has a single-exponential decay, but in a peptide its decay is already more complex. A single Trp, even only in a buffer, has the decay more complex than exponential.

The analytical solutions of the more complex models of the kinetics are known only in a very few cases. For example, if the measured fluorescence is the sum of fluorescence emitted by a few independent residues, each showing its own exponential decay, the solution has a form of the multi-exponential function

$$I(t) = \sum\nolimits_{n=1}^{N} b_n \exp\left(-\frac{t}{\tau_n}\right) \tag{3}$$

where τ_n are the fluorescence lifetimes of the individual residues and b_n are related to the contributions of these residues c_n to the decay ($c_n = b_n\tau_n / \sum\nolimits_{i=1}^{N} b_i\tau_i$). A multi-

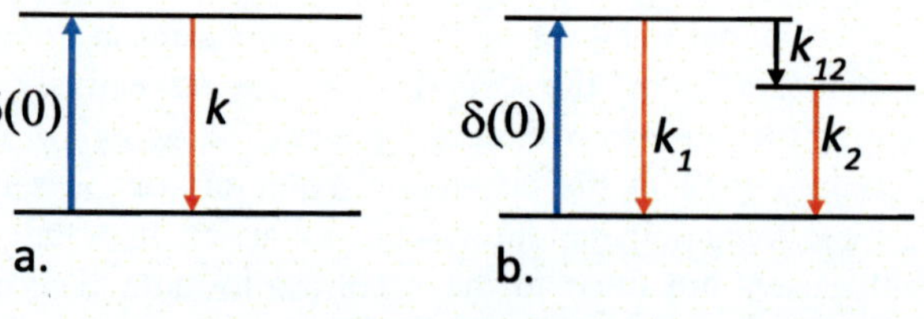

Fig. 3 Examples of known analytical forms of intensity decays *I(t)* at different models of the kinetics: a simple two-level model (**a**) and the case of the excited-state reaction (**b**)

exponential expression for $I(t)$ is also obtained in the case of a few other simple formal models of the kinetics, e.g. the fluorescence of the product of excited-state reaction (Fig. 3b). In the case of more complex underlying kinetics however, the resulting fluorescence decay usually cannot be obtained in the form of analytical formula.

Nevertheless, fitting a multi-exponential function to the experimental data became a common procedure, as the goodness of fit can be almost always achieved by selecting a large enough number of exponential components in a fitted function. This practice has led to two dominating models of protein kinetics, both of which offer the explanation of multi-exponentiality of the decay: the rotamer model and the model of dielectric relaxation. In the rotamer model [18–21], the discrete ground-state conformations (rotamers), usually two or three, show characteristic fluorescence decay times, resulting in a multi-exponential decay of the whole ensemble. In the dielectric relaxation model [29, 31], a single conformation is assumed, but its fluorescence spectrum shifts towards longer wavelengths due to the solvent response to the excited-state dipole moment of the fluorophore. If this relaxation occurs at the same time scale as fluorescence, then even if the excited state decays exponentially, the experimental decay (measured at the fixed detection wavelength) is non-exponential but may be represented by a multi-exponential function.

The generic feature of protein fluorescence is the lifetime–wavelength correlation; i.e. the mean decay lifetimes are increasing with the detection wavelength. This generic observation supports the dielectric relaxation model, and because the rotamer model does not explain this effect, it is used as evidence against this model. On the other hand, the lack of negative amplitudes in protein decay functions challenges the relaxation model. The difficulty associated with the interpretation of multi-exponential decays has led to developing alternative approaches to lifetime data analysis, and here we present two alternative strategies.

In the first strategy, the effects of heterogeneous environment and the dielectric relaxation are modelled together and lead to so-called non-Debye kinetics [32]. This approach reveals more detailed information on the studied kinetics but is limited to relatively simple protein fluorescent systems. The example of non-Debye model of Trp kinetics in HSA [32, 33] is presented in Sect. 2.2.

The second strategy does not assume any specific model of the kinetics but explores the advantage of multi-exponential functions being good mathematical representations of the experimental decays. Fitting the multi-exponential model allows for extracting the fluorescence intensity decays $I_\lambda(t)$ from the convolution integral (1) for a series of the decays measured at different detection wavelengths λ, and constructing the time-resolved emission spectra (TRES) $I_t(\lambda)$ [34, 35]. TRES-based analysis provides unbiased information on the unknown model of the kinetics in multi-component fluorescent systems. The examples of applying TRES to study protein oligomerisation and complexation (glycation) are presented in Sect. 2.3.

2.2 Modelling Microheterogeneity and Dielectric Relaxation in Protein Fluorescence Kinetics

In the simple exponential kinetics of the fluorescence decay, the probability that the fluorophore is still in the excited state at time t after excitation is $P(t) = \exp[-kt]$, where k is the rate of the decay. Here, we assume a more general case, where this probability decays exponentially until the maximum quenching time t_c, and then remains constant on the level $\exp[-t/t_c]$. In other words, $P(t) = \exp[-k \min(t,t_c)]$. In the non-homogeneous system of many fluorophores of decay rate values k, and maximum quenching times t_c characterised by certain density distributions, the fluorescence intensity decay of the whole ensemble will be given by [32, 33]

$$I_{\alpha,\kappa}(t) = \exp\left[-\frac{1}{\kappa}\int_0^{\kappa\left(\frac{t}{\tau_F}\right)^{\alpha}}\left(1-\exp\left[1-\frac{1}{x}\right]\right)\mathrm{d}x\right] \tag{4}$$

The formula (4), as a generalisation of the exponential decay, is more flexible to represent responses of fluorescent amino acids in heterogeneous biological environment. For the special case of $\alpha = 1$ and $\kappa = 0$, it becomes the exponential decay $I_{1,0}(t) = \exp[-t/\tau_F]$. For $\alpha < 1$ and $\kappa > 0$, the reduced index of stability α impacts the whole decay, while κ only the tail (Fig. 4a).

For $\kappa = 0$ the expression (4) becomes the well-known stretched-exponential function [36] $I_{\alpha,0}(t) = \exp[\left(-\frac{t}{\tau_F}\right)^{\alpha}]$ (Fig. 4b), and for $\alpha = 1$, we receive the Becquerel function [37], $I_{1,\kappa}(t) \sim (1 + \kappa(t/\tau_F))^{-1/\kappa}$ (Fig. 4c). Both stretched-exponential and Becquerel's functions have been used for describing fluorescence kinetics in multiple applications [36, 37].

Microheterogeneity of the fluorescent molecular system represented by Eq. (4) can be combined with the dielectric relaxation, offering a more realistic model of protein kinetics. In the original paper by Toptygin and Brand [38], the wavenumber- and time-dependent fluorescence intensity decay $I_\nu(\nu,t)$ is given by

$$I_\nu(\nu, t) = \nu^3 S(\nu - \nu_c(t))D(t) \tag{5}$$

where $S(\nu)$ is the vibrational envelope representing the fluorescence transitions, $D(t)$ is the decay of the population of the excited state, and $\nu_c(t)$ is the time evolution of the spectral centroid. In the original paper, $D(t)$ and $\nu_c(t)$ are modelled by linear combinations of exponentials and S(t) is expanded in Taylor series. Consequently, Eq. (5) is approximated by a multi-exponential function. Although this function is likely to fit well to the experimental data, the approximation used here may lead to the loss of information on protein system or misinterpretation of the data. Indeed, the model supporting multi-exponentiality, i.e. the existence of subsystems of distinct lifetimes, is unlikely to be adequate for the biological media, where the distribution of lifetimes seems to be more relevant. Therefore, in the approach we proposed [33], $D(t)$ is represented by Eq. (4) and $\nu_c(t)$ by $\sim\exp[-(t/\tau_s)^{\alpha_s}]$, namely

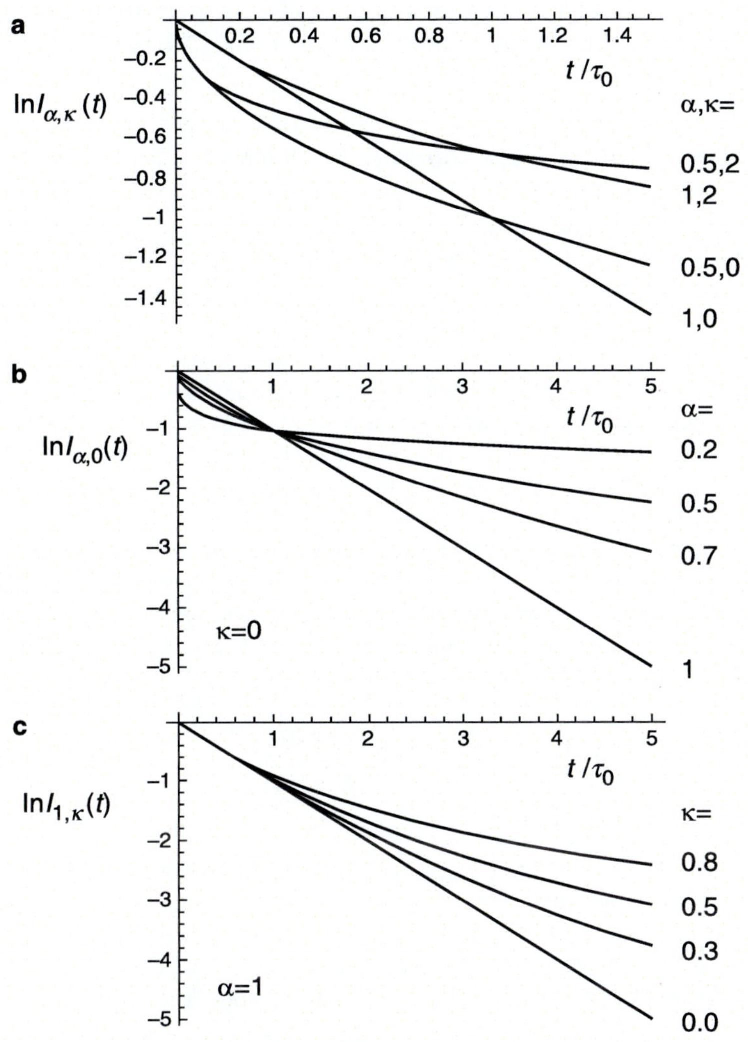

Fig. 4 The lifetime distributions $g(\tau)$ recovered from the MEM method applied to the synthetic decay simulated for the $I_{\alpha,\kappa}(t)$ model. The parameters used are $\tau_0 = 2$ ns, $\kappa = 0.0001$, and $\alpha = 1$, and then for the same τ_0, $\kappa = 0.1$ and $\alpha = 1.0$–0.8 (**a**), and $\tau_0 = 2$ ns, $\alpha = 0.9$, and $\kappa = 0.05$–0.3 (**b**). (Reproduced from [32] with permission of IoP Publishing)

$$I_\nu(\nu, t) = \nu^3 \, S(\nu - \nu_\infty - (\nu_0 - \nu_\infty) \exp\left[-\left(\frac{t}{\tau_s}\right)^{\alpha_s}\right]) I_{\alpha,\kappa}(t) \tag{6}$$

This model function depends on the decay parameters, namely τ, α, and κ and on the dielectric relaxation parameters ν_0, ν_∞, τ_s,and α_S. Determining these parameters requires gathering a series of experimental decays measured at different detection wavelengths and applying a global data analysis. In Sect. 3.1, an example of applying this approach to fluorescence kinetics of a single Trp in HSA is presented.

2.3 Time-Resolved Emission Spectra (TRES)

If the analytical formula for the fluorescence intensity decay in complex system cannot be anticipated due to its complexity, determination of TRES may be an efficient way towards revealing the actual kinetics. Indeed, determining the exact model of the excited-state fluorescence kinetics in many protein systems is difficult for two main reasons: a variety of molecular-level processes can occur simultaneously, and there is a lack of analytical formulae for fluorescence intensity decays in more complex kinetics. In such cases, it is practical to use the model-free TRES approach, which exploits the ability of multi-exponential functions to serve as flexible analytical representations of the complex (and, in general, not multi-exponential) decays.

In this approach, the matrix of wavenumber- and time-dependent fluorescence intensity decays $I_\nu(\nu,t)$ (Eq. 5) is obtained experimentally by measuring the fluorescence intensity decay $I_\lambda(t)$ at a series of detection wavelengths λ. The raw data are then fit to multi-exponential functions, and the TRES $I_t(\lambda)$ are obtained from the equation

$$I_t(\lambda) = I_\lambda(t) \times \frac{S(\lambda)}{\int_0^\infty I_\lambda(t)\mathrm{d}t} \tag{7}$$

where $S(\lambda)$ is the steady-state fluorescence spectrum, and the denominator is proportional to the total number of emitted photons. The obtained spectra are then converted from the wavelength $I_t(\lambda)$ to the wavenumber scale according to $I_t(\nu) = \lambda^2 I_t(\lambda)$. In the next stage, multiple normalised Toptygin-type distributions are used [39–43] to represent the shape of spectral components.

$$F(\nu) = \sum_{i=1}^{M} \frac{C_i(t)}{\sqrt{2\pi}\sigma_i(t)\nu_i(t)\left(\nu_i(t)^2 + 3\sigma_i(t)^2\right)} \nu^3\, exp\left(\frac{-(\nu-\nu_i(t))^2}{2\sigma_i(t)^2}\right) \tag{8}$$

Here, M is the number of spectral components, ν_i is the peak position, and σ_i is the half-width of the distribution. It should be noted that the $1/\left(\sqrt{2\pi}\sigma_i(t)\nu_i(t)\left(\nu_i(t)^2 + 3\sigma_i(t)^2\right)\right)$ term is the normalisation factor, which implies that $C_i(t)$ has the meaning of the fluorescence contributions of individual components. The revealed evolutions of $C_i(t)$, $\nu_i(t)$, and $\sigma_i(t)$ in ns time scale help to understand the mechanisms of the underlying kinetics.

In Sects. 3.2 and 3.3, we present several cases of protein studies, where TRES does not determine the exact excited-state kinetics, but reveals main features of the monitored protein activities, which narrows down the range of the potential kinetic models. The examples include the beta-amyloid aggregation in a buffer and in the presence of several compounds impacting aggregation, which is one of key mechanisms of Alzheimer's disease, and the process of glycation, i.e. binding glucose to certain proteins, which is related to the complications of diabetes.

3 Applications of Intrinsic Fluorescence Studies of Protein

3.1 *Non-Debye Kinetics of Trp in HSA*

Applying the approach from Sect. 2.2 allows resolving the microheterogeneity and dielectric relaxation of Trp fluorescence kinetics [32, 33]. The TCSPC-based measurements were performed using the FluoroCube fluorescence lifetime system (Horiba Jobin Yvon IBH, Glasgow, UK) and the AnAlGaN type of pulse light emitting diode (pulse width ~600 ps) working at 295 nm, and 1 MHz repetition rate was used for excitation. The measurements of the 3×10^{-5}M HSA solution in 0.01 M phosphate buffer at pH 7.4 were performed in the range of wavelengths from 330 nm to 360 nm with 5 nm increments. The temperature was changed from 10°C to 60°C in 10° steps. The model decay (6) for the $I_{\alpha,\kappa}(t)$ given by Eq. (4) has been investigated by global deconvolution analysis, where a number of decays measured for the different detection wavelengths were fitted simultaneously. Figure 5 shows the example of global analysis of the series of decays measured at 50°C.

The results of fitting, obtained for the samples at all temperatures, are shown in Fig. 6. The parameters characterising the decay of the excited-state population, τ_F and α_F, decrease slowly with temperature, resulting from increased collisional quenching and a broader range of transition rates, respectively, while the κ_F is negligibly small. Good values of χ^2 and random distributions of residuals demonstrate that the relaxation of the excited states of Trp in HSA can be represented by a single stretched-exponential function. At the same time, the changes in parameters τ_S and α_S are more significant, and the substantial drop in τ_S between 30°C and 40°C suggests a change in HSA conformation. This affects the rate of protein's dielectric relaxation, while the Trp local microenvironment is not disturbed (no substantial changes in τ_F and α_F). In general, the above spectrally and time-resolved fluorescence model allows separation of two processes in proteins: the excited-state decay and dielectric relaxation. This approach paves the way for the dual relaxation-based intrinsic fluorescence sensing of protein systems.

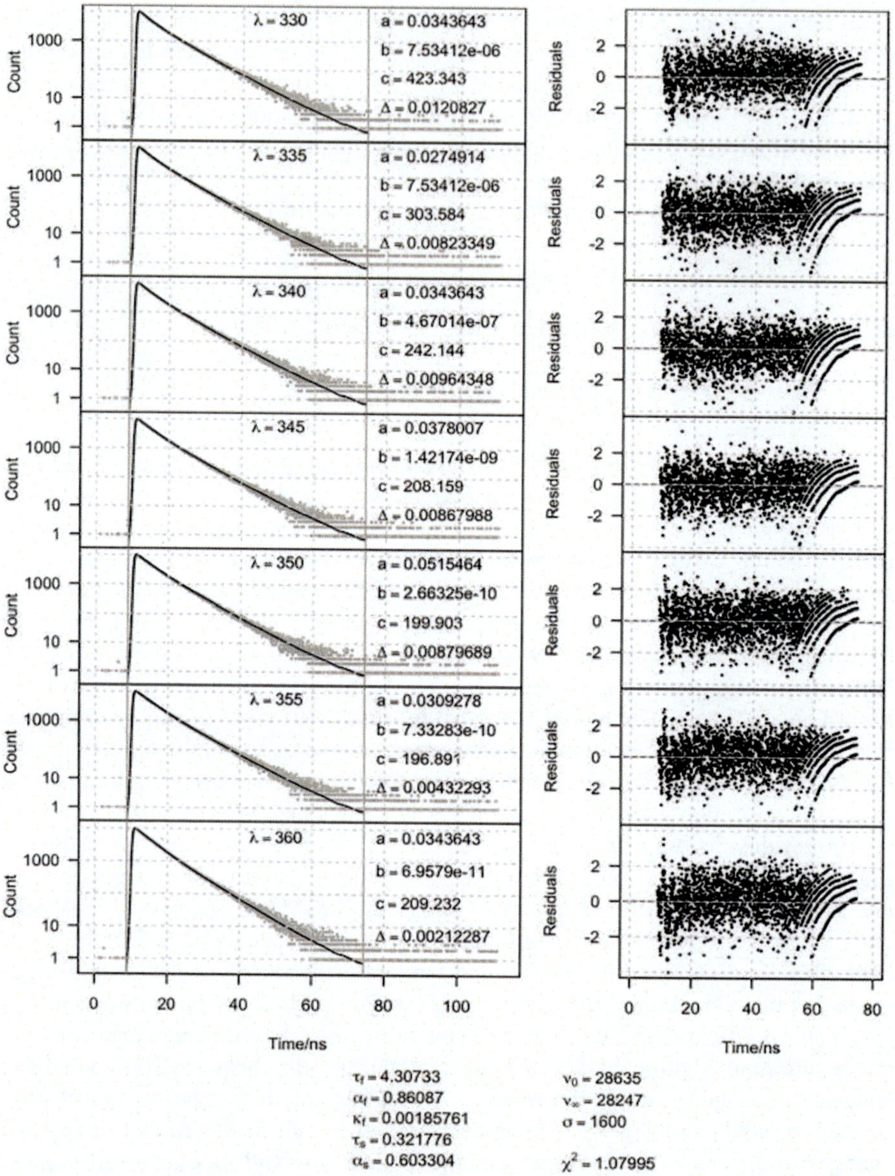

Fig. 5 The raw decays of HSA fluorescence at 50°C measured at different wavelengths, the fitted functions, and the distribution of residuals, together with the parameters of the model defined by Eq. (6) recovered in the global analysis. Note that the model parameters are common for all sets of decays, and only technical parameters (*a*–*c*) and Δ are specific to each decay. (Reproduced from [33] with permission of IoP Publishing)

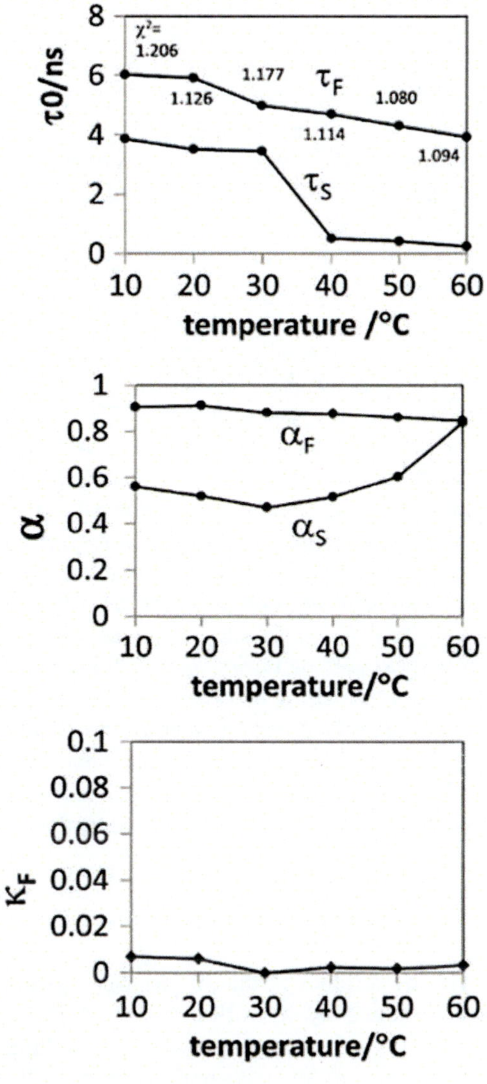

Fig. 6 Temperature dependence of the parameters τ_F and τ_S (**a**), α_F and α_S (**b**), and κ_F (**c**) obtained for the double-relaxation model (6) of fluorescence kinetics, from the global analysis of the series of decays measured at the different detection wavelengths. (Reproduced from [33] with permission of IoP Publishing)

3.2 *Oligomerisation/Glycation of Beta-Amyloid $A\beta_{1\text{-}40}$*

Beta-amyloids (Aβ) are fragments of an integral membrane amyloid precursor protein (APP). The processes responsible for the metabolism of APP to Aβ are reasonably well understood; however, the physiological role of Aβ is not clear. Aβ monomers appear in physiological fluids at nM concentrations. At higher concentrations monomers start to aggregate, initially into small permeable oligomers, and, at later stages, into plaques and fibrils that are hallmarks of Alzheimer's disease (AD). According to current research on AD [44], the toxic action of Aβ starts well before the fibrils are formed. This is due to small oligomers that interfere with the neural cell membranes forming Ca^{2+}-permeable pores. Consequently, the disturbed calcium balance is the key mechanism of the cell death and the mechanism of AD. Preventing aggregation of Aβ peptides into small oligomers is therefore central in developing therapeutic drugs targeting AD. Aβ oligomers-triggered intracellular overdose of Ca^{2+} produces a series of further neurotoxic events like oxidative stress, tau phosphorylation, and neuronal loss. Currently, there is no method allowing monitoring Aβ aggregation on early stages.

Sensitivity of fluorescence to sub-nm spatial and ns time scale of molecular processes makes it a perfect tool for Aβ aggregation studies, but the extrinsic fluorescent labels (Thioflavin T, Congo Red, Atto 488A, HL647) are known to modify the Aβ smallest oligomers [9, 45]; thus, their feasibility for monitoring the earliest stages of Ab oligomerisation is questionable.

To enable detection of Aβ aggregation starting from the earliest monomer→dimer stage, we have developed an approach using Aβ's intrinsic fluorophore tyrosine for a non-invasive interaction with the peptides during their aggregation. Tyrosine-based studies of Aβ have demonstrated that early aggregation impacts the intrinsic fluorescence responses significantly, contrary to the fluorescence of extrinsic labels which show the lag phase at the early stage [46]. Measuring the time-resolved emission spectra (TRES) of the aggregating peptides allows separation of the simple monomer's fluorescence and more complex responses from the (evolving during aggregation) distributions of di- to oligomers [39]. Moreover, the data show that the presence of compounds like copper ions [40], glucose [41], or the flavonoid quercetin dramatically alter the fluorescence of aggregates, which constitutes the experimental foundation of the approach.

Presented below are studies of free Aβ aggregation and the effects of different compounds (copper ions and glucose) that impact the performance of aggregation.

3.2.1 Free $A\beta_{1\text{-}40}$ in a Buffer Solution

Tyrosine residues in the 50 μM $A\beta_{1-40}$ sample (in HEPES buffer of pH 7.4) [39] were excited by a Horiba NanoLed pulsed source at the wavelength 279 nm [16]. This wavelength allowed for excluding excitation of phenylalanines, also present in the $A\beta_{1\text{-}40}$ peptide. The TCSPC measurements were conducted on a Horiba Scientific DeltaFlex fluorometer (Horiba Jobin Yvon IBH Ltd., Glasgow, UK). For every measurement, 12 fluorescence intensity decays at detection wavelengths ranging from 294 to 327 nm with 3 nm increments were collected.

The decay parameters were recovered from fitting to a multi-exponential decay (2), see Fig. 7a, b, and then used to calculate TRES using Eq. 7. Plots were then converted to the wavenumber scale (see examples for Day 1 and Day 8 in Fig. 7c).

Solid lines in Fig. 7c are the plots of the Toptygin-type distributions (8) for the parameters obtained from their fitting to the experimental TRES points. The two-component distributions ($N = 2$) were required to satisfy the goodness-of-fit

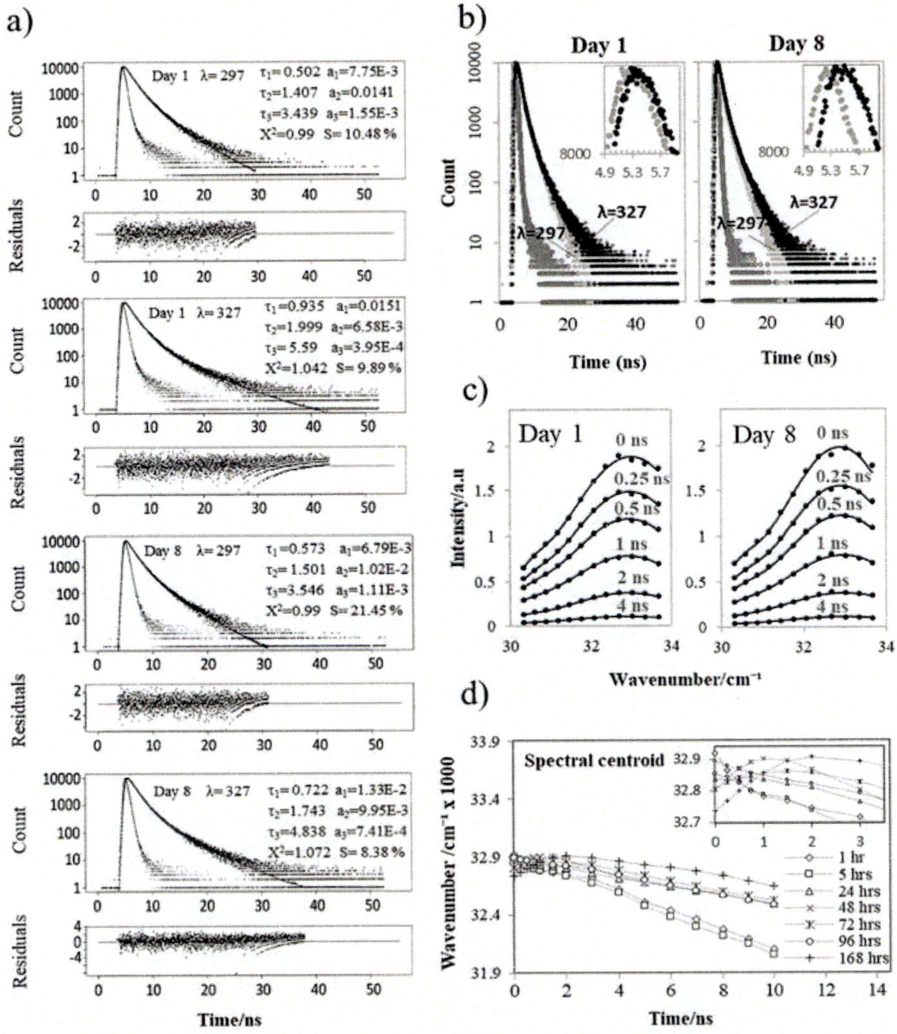

Fig. 7 (**a**) Fluorescence decay data of beta-amyloid at the detection wavelengths 297 and 327 nm together with the excitation pulse, fitted three-exponential function, the scattered light term (S), and the distribution of residuals. (**b**) The time–wavelength correlation observed in two decays of beta-amyloid at detection wavelengths 297 and 327 nm and at two different stages of aggregation: Day 1 and Day 8. The insets show the peak areas of the experimental curve. (**c**) Time-resolved emission spectra (TRES) obtained at two different stages of aggregation (Day 1 and Day 8) fitted to a two -Gaussian profile. (Reproduced from [39] with permission of IoP Publishing)

criteria, suggesting the existence of two fluorescent forms. The evolution of these parameters at different times after excitation in the ns time scale, measured in different days after sample preparation, is shown in Fig. 8.

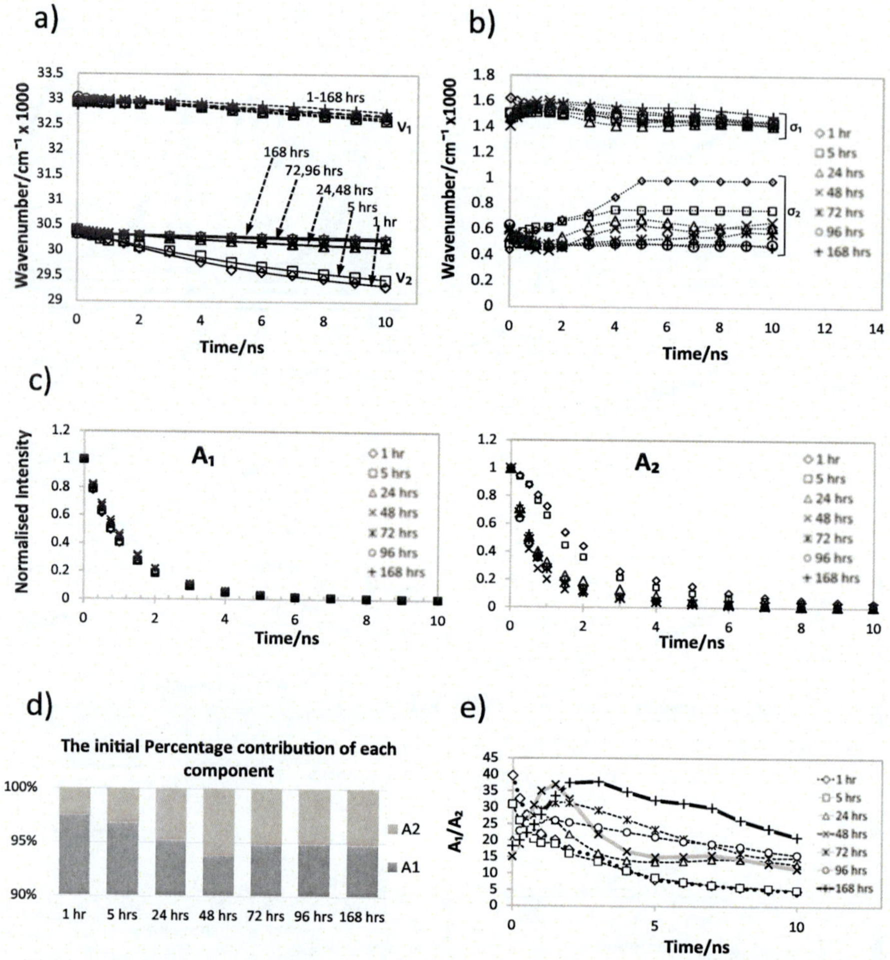

Fig. 8 **(a)** Evolution of the peak positions $\nu_1(t)$ and $\nu_2(t)$ with time. **(b)** Standard deviation of both peaks $\sigma_1(t)$ and $\sigma_2(t)$ at different stages of aggregation. **(c)** Fluorescence intensity decay of each component (monomers C_1, oligomers C_2). **(d)** The initial percentage contribution of each component $C_1(0)$ and $C_2(0)$ at different stages of aggregation. **(e)** The ratio of the monomer to oligomer contribution $C_1(t)/C_2(t)$ plotted against time. (Reproduced from [39] with permission of IoP Publishing)

We can observe two components showing initial peak positions (Fig. 8a) at 33,000 cm^{-1}, which can be attributed to A$\beta_{1\text{-}40}$ monomers, and at 30,500 cm^{-1}, which corresponds to oligomers. Both peaks have different characteristic values of standard deviations (Fig. 8b). The plots $C_1(t)$ and $C_2(t)$ (Fig. 8c) represent fluorescence intensity decays of the monomers and oligomers, respectively. We note that the character of the monomer's decays does not change during the period of experiment, while the decays of the oligomers evolve due to changes in Tyr's environment when oligomers are getting bigger. The location of the monomers peak does not show significant changes over time, which indicates fast dielectric relaxation that is practically completed before the fluorescence occurs. During the first few hours after sample preparation, monomer's fluorescence contribution reaches about 98% of all fluorescence (Fig. 8d, e), but its value gradually decreases in the following days, which is consistent with formation of oligomers which are fluorescent at longer wavelengths. Moreover, the initial oligomers show clearly dielectric relaxation manifested in the drop in $\nu_2(t)$ in the first day (Fig. 8a), while in the following days the shift in the oligomer's spectral peak is negligible, indicating formation of the larger and thus more rigid oligomers.

3.2.2 A$\beta_{1\text{-}40}$ with Cu^{+2} Ions

High concentrations of metal ions like copper Cu^{2+} and zinc Zn^{2+} are detected within senile plagues of Alzheimer's brain tissue. It has been demonstrated in experiments in vitro [47, 48] that these ions can bind to the A$\beta_{1\text{-}40}$ directly and modulate the aggregation process. Moreover, the A$\beta_{1\text{-}40}$–copper complexes may be involved in the formation of reactive oxygen species and contribute to the oxidative stress observed in the disease. Therefore, we have studied [40] the interactions of Cu^{2+} with A$\beta_{1\text{-}40}$ by detecting the changes in TRES measured in the solution of A$\beta_{1\text{-}40}$ containing copper ions.

Presence of copper in the solution can potentially cause two types of effect on the fluorescence kinetics of intrinsic tyrosine: the direct effect – quenching of tyrosine, and the indirect effect – influencing the aggregation process. Indeed, adding mM concentrations of Cu^{2+} to Tyr's solution reduces its fluorescence lifetime (Fig. 9a), which can be explained by collisional quenching resulting in the linear Stern–Volmer plot (Fig. 9b). However, at the μM concentration range this effect is negligible (Fig. 9c). Fluorescence decays of Tyr in NAYA are two-exponential, but the effects of μM levels of copper are still insignificant (Fig. 9d, e). However, in the case of beta-amyloid fragment, A$\beta_{1\text{-}16}$, due to increasing complexity of the local environment of Tyr, a three-exponential model of the decay is needed to obtain a good fit. Moreover, this is the smallest beta-amyloid fragment we have investigated, for which the fluorescence lifetimes and their contributions are responding to the change of copper concentration in μM region. This effect is likely to be due to the A$\beta_{1\text{-}16}$–Cu^{2+} complex formation, as His6 and His13 (or His14) are expected to form a copper ion binding site [49].

Preliminary measurements of the copper ion effect on the aggregation of the full A$\beta_{1\text{-}40}$ Alzheimer's peptide have demonstrated that copper ions substantially complicate the fluorescence intensity decays, suggesting accelerated aggregation and formation of different aggregates than in the case of free amyloids. TRES measurements performed for the A$\beta_{1\text{-}40}$ sample containing 15 μM Cu^{2+} have shown clearly (Fig. 10) that the presence of copper alters the fluorescence kinetics of Tyr from the very beginning. Measurements were taken at several stages of aggregation: 1, 5, 24, 48, 72, 96, and 168 h after sample preparation. The times taken by individual measurements were longer at later days due to decreasing fluorescence yield during

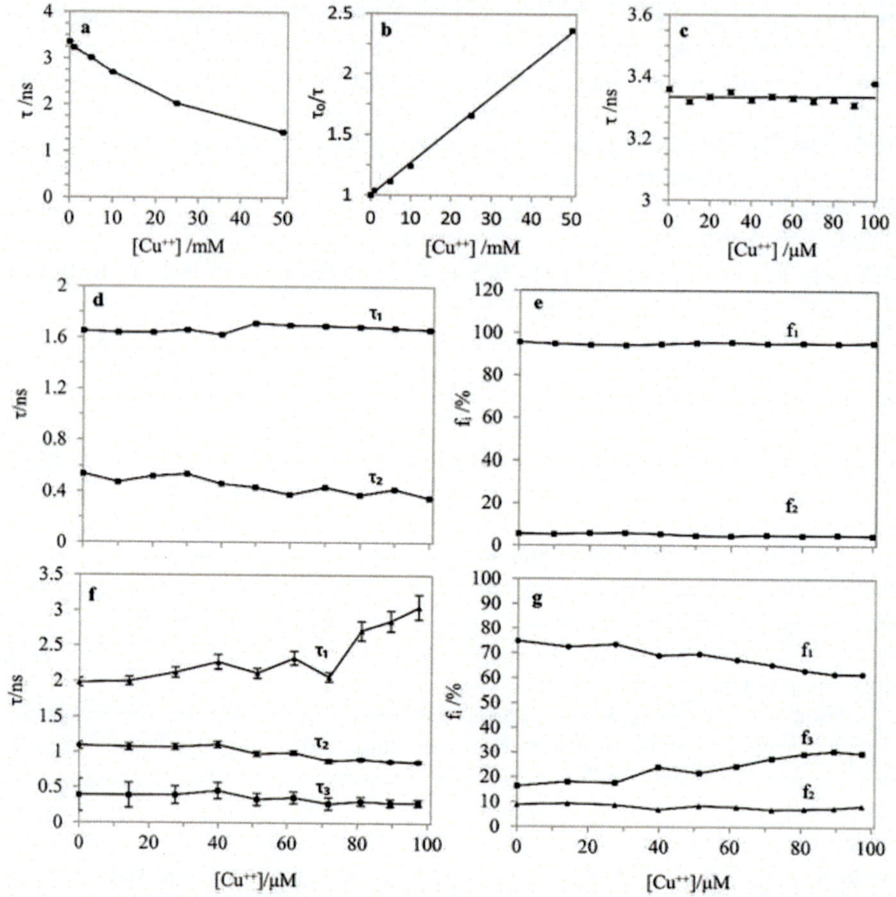

Fig. 9 Plots showing the correlation between Tyr's fluorescence lifetime τ and the concentration of Cu^{2+} in mM (**a**) and μM (**c**) scales; the Stern–Volmer plot of fluorescence quenching of Tyr by Cu^{2+} ions (**b**); plot showing the effect of Cu^{2+} on parameters obtained from fitting NAYA's fluorescence decay to a two-exponential model: decay times τ_1 and τ_2 (**d**), and the percentage contributions f_1 and f_2 (**e**); parameters obtained from fitting Tyr's fluorescence decay in a 50 μM $A\beta_{1-16}$ solution to a three-exponential decay model plotted against Cu^{2+} concentration: Tyr fluorescence decay times τ_1, τ_2, and τ_3 (**f**), percentage contributions f_1, f_2, and f_3 (**g**). Error bars represent the standard deviation. (Reproduced from [40] with permission of John Wiley and Sons)

aggregation (the numbers indicating the age of the sample are the times when the first wavelength has been started). Shown are the examples of TRES for the free $A\beta_{1-40}$ sample and the sample with 15 μM Cu^{2+}, at 1 h and 168 h after their preparation. The substantial differences in the shapes of the TRES seen in Fig. 10 are reflected in the performance of their individual parameters determined by fitting experimental TRES to the model (8) [40].

Figure 11 shows the evolution of two peak positions identified for the amyloid–copper sample (b). These data are compared with the previously obtained [39] peaks

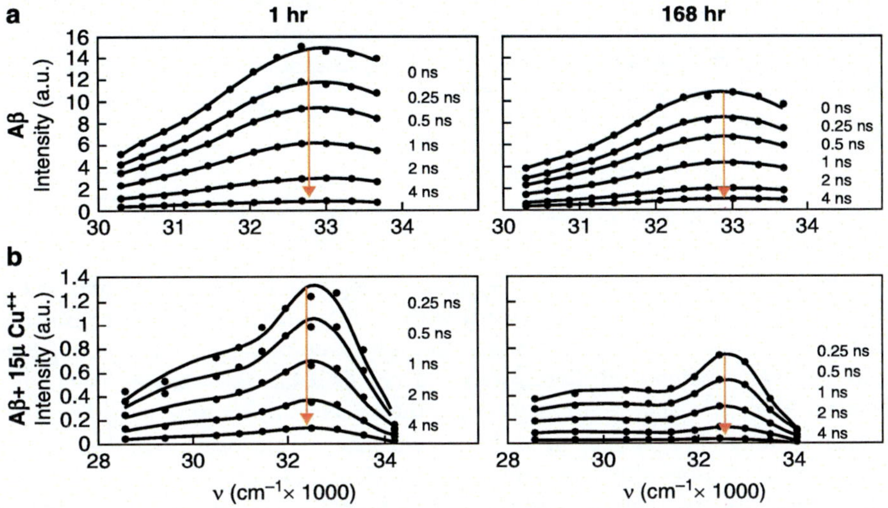

Fig. 10 Time-resolved emission spectra (TRES) obtained for 50 μM $A\beta_{1-40}$ (**a**) and 50 μM $A\beta_{1-40}$ with 15 μM Cu^{2+} (**b**) at two stages of aggregation (1 h and 168 h). The solid lines represent the two-Toptygin-type function fits. Reproduced from [40] with permission of John Wiley and Sons)

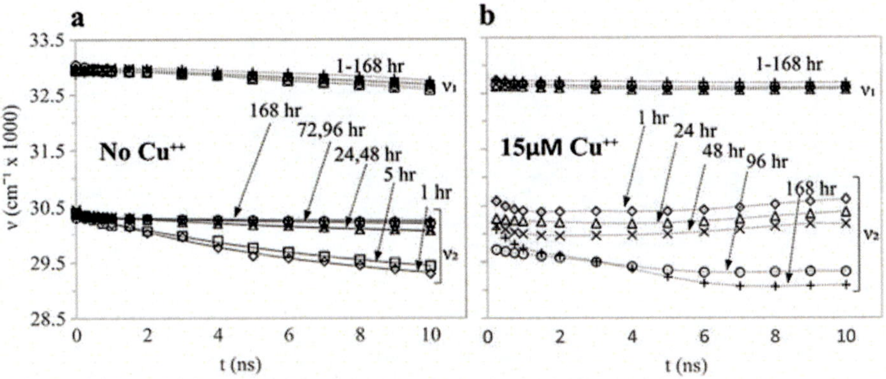

Fig. 11 Evolution of the peak positions $\nu_1(t)$ and $\nu_2(t)$ for free $A\beta_{1\text{-}40}$ (**a**) and Aβ1−40 with 15 μM Cu^{2+} (**b**) at different stages of aggregation (◇ 1 h, □ 5 h, Δ 24 h, × 48 h, ∗ 72 h, ○ 96 h, and + 168 h). (Reproduced from [40] with permission of John Wiley and Sons)

for the free amyloid (a). The position of the higher wavenumber peaks $\nu_1(t)$, related to monomers, remains very similar to that observed for the free $A\beta_{1\text{-}40}$ sample and again indicates rapid dielectric relaxation occurring faster than fluorescence decay. This very similar behaviour of monomers is expected and confirms credibility of our approach. Simultaneously, the entirely differentperformance of the $\nu_2(t)$ dependence representing oligomers is characterised by a slight decrease during the first 6 ns, followed by an increase. This observation is likely to be caused by the presence of at least two types, or even a distribution of oligomers, each characterised by a different peak evolution and different relaxation time. In the samples older than 96 h, $\nu_2(t)$ has no increasing stage which suggests only one fluorescent compound, with the other fluorescence being reduced due to reabsorption or precipitation of aggregates to larger fibrils.

3.2.3 Aβ_{1-40} with Glucose

The potential influence of glucose on Aβ_{1-40} aggregation is essential not only in the context of molecular mechanisms of Alzheimer's disease but can also be considered as one of the potential long-term side effects observed in diabetes. The correlation between the occurrence of diabetes and Alzheimer's has been proven in clinical research [50]. Our approach based on the fluorescence of tyrosine in aggregating Aβ_{1-40} peptides [41] has confirmed the influence of glucose molecules on the formation of amyloid oligomers and their glycation.

Examples of fluorescence spectra of 50 μM Aβ_{1-40} in PBS, measured in the absence and in presence of 50 mM glucose at different excitation wavelengths, and after two different times of incubation (1 and 384 h), are shown in Fig. 12. Spectra observed at excitation 279 nm are typical for Tyr. At excitation wavelength

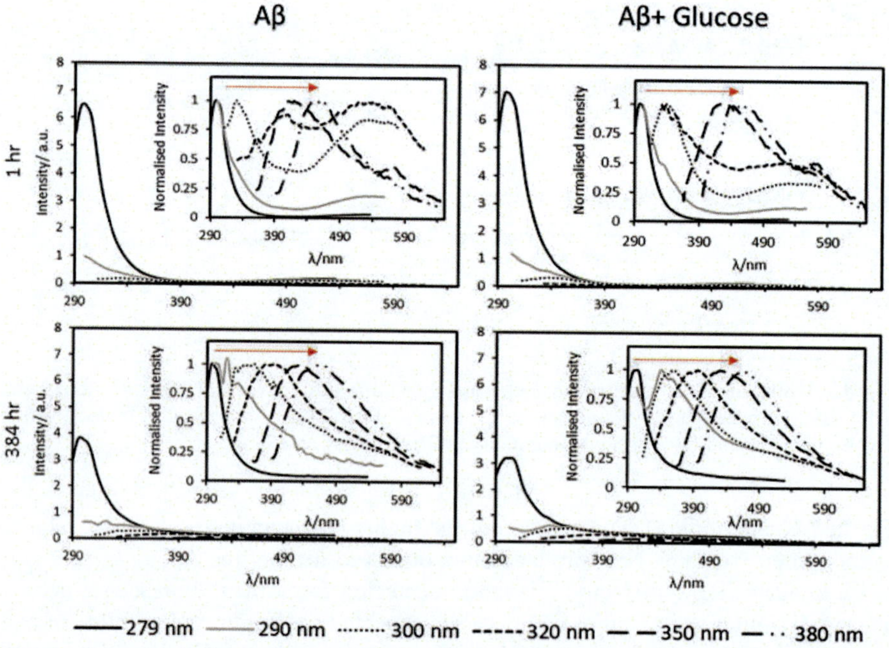

Fig. 12 Steady-state measurements of 50 μM Aβ1-40 in PBS buffer (10 mM, pH 7.4) in the absence (left) and presence (right) of 50 mM glucose at excitation wavelengths 279 nm (solid black), 290 nm (solid grey), 300 nm (dotted), 320 nm (short dash), 350 nm (long dash), and 380 nm (dash-dot-dot) after incubation for 1 h (top) and 384 h (bottom). Insets show the normalised emission spectra. (Reproduced from [41] with permission of Elsevier)

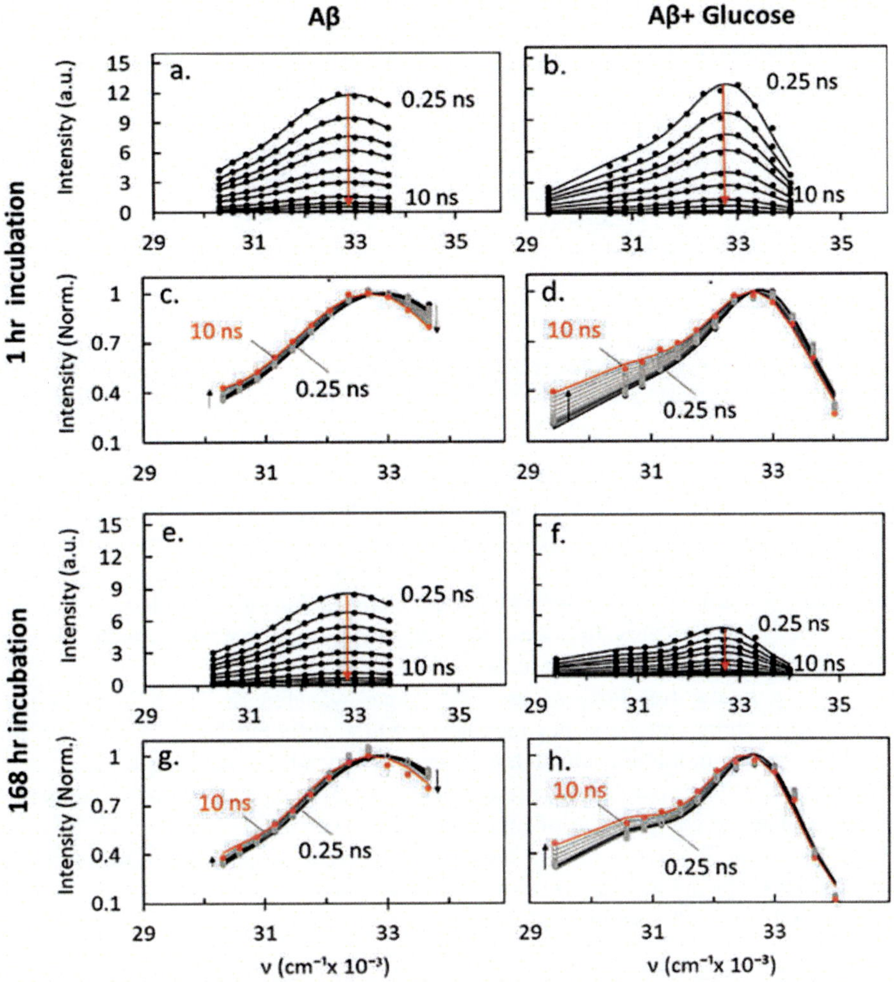

Fig. 13 Time-resolved emission spectra (TRES) obtained for 50 μM Aβ1-40 in HEPES buffer (0.1 μM, pH 7.4) after 1 h of incubation (**a**) and 168 h of incubation (**e**). TRES obtained for 50 μM Aβ1-40 in the presence of 50 mM glucose after 1 h of incubation (**b**) and 168 h of incubation (**f**). Normalised TRES for Aβ1-40 in the absence and presence of glucose after 1 h of incubation (**c, d**) and 168 h of incubation (**g, h**). The solid lines represent the two-Toptygin-type function fits. (Reproduced from [41] with permission of Elsevier)

greater than 300 nm, the positions of the fluorescence peaks shift towards longer wavelengths, suggesting formation of fluorescent oligomers of the broad distribution of sizes, as evidenced by the excitation–emission peaks correlation.

TRES measurements were performed at several stages of aggregation over a week. Figure 13 shows the initial (1 h) and the final (168 h) set of TRES obtained from the samples without and with (50 mM) glucose.

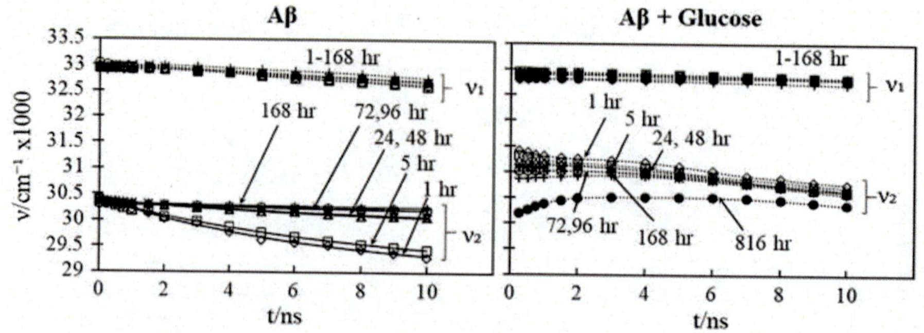

Fig. 14 Peak's position $\nu 1$ (*t*) and $\nu 2$ (*t*), obtained from fitting $A\beta_{1\text{-}40}$ TRES to the Eq. (8), plotted against time in nanoseconds at different stages of aggregation (1 h, 5 h, 24 h, 48 h, 72 h, 96 h, and 168 h) for sample with and without glucose. (Reproduced from [41] with permission of Elsevier)

The raw TRES data were analysed by fitting to them Eq. (8). Similarly to free $A\beta_{1\text{-}40}$, the sample containing glucose also demonstrated the presence of two components ($N = 2$) in (8). Figure 14 shows the positions of the two peaks $\nu_1(t)$ and $\nu_2(t)$. Initial positions of the $\nu_1(t)$ peaks in glucose-containing sample at all ages are around 33,000 cm^{-1} and show almost no changes during 10 ns after excitation. Again, this suggests very fast dielectric relaxation of monomers, which are not affected by the presence of glucose.

Contrary to the free $A\beta_{1\text{-}40}$ case, the longer wavelength TRES peak in the glucose-containing sample is not monotonic and is different for different ages of sample. The non-monotonic character of the $\nu_2(t)$ dependence suggests fluorescence of more than one form of fluorescent oligomers. As the $\nu_2(t)$ evolves as the sample is getting older, the composition of these oligomers (i.e. the distribution of sizes) is likely to change. The complexity of the observed $\nu_2(t)$ at different sample ages might be a result of $A\beta_{1\text{-}40}$ binding to glucose (glycation) and/or alterations in the amyloid aggregation due to the presence of glucose.

3.3 *Glycation of Collagen*

Collagen is one of the most abundant fibrous proteins in mammals [51], where it is found in the skin, tendons, blood vessels, bones, and teeth. Collagen has a long biological half-life, in skin approximately 5 years [52], and so it is highly susceptible to glycation and the formation of advanced glycation end products (AGEs). This is a process that occurs in response to high blood glucose levels, such as those seen in diabetes, and the presence of these AGEs is related to diabetes complications [52, 53]. On collagen, glucose binds to a free lysine residue [54] to form a Schiff base and follows the classical glycation pathway.

Collagen AGEs can accumulate in skin, and it has been established that the risk of diabetes complications can be predicted by identifying the presence of AGEs in dermal tissue. The studies have found [55] that long-term treatment of hyperglycaemia reduced the levels of AGEs in skin collagen. This suggests that skin AGEs, including collagen AGEs, can be used as a biomarker for diabetes complications [56].

We study the potential for using the time-resolved intrinsic fluorescence of collagen as a biomarker of its glycation. Collagen's fluorescence in human skin [57] originates from two sources: tyrosine (Tyr) [58, 59] and collagen cross-links. As there are no tryptophan [60] or phenylalanine [61] residues in collagen, they do not interfere with Tyr. It can be excited at approximately 275 nm and has peak fluorescence emission at 305 nm [59].

There are two types of collagen cross-links [57]: pepsin-digestible and collagenase-digestible. The former are excited at approximately 340 nm and emit at 400 nm, and the latter are excited at 360 nm with fluorescence emission peak at 440 nm [62]. AGE accumulation drives more cross-linking within collagen fibres, and so impacts its intrinsic fluorescence [63], making fluorescence emission spectra a valuable tool for monitoring AGE accumulation [42, 64].

Previous studies have used skin auto-fluorescence as a means of studying glycation-induced changes both in diabetic patients and in healthy controls [64, 65], using steady-state fluorescence. This, however, is not able to fully explain the mechanisms of glycation and so helps in the search for anti-glycation factors that could prevent AGE formation. We explore the time-resolved emission spectra (TRES) and aim to verify whether the alterations in TRES triggered by collagen glycation are sufficient to help in understanding the actual mechanisms of glycation, and thus potential pathways of its prevention.

The time-resolved responses of the pepsin-digestible cross-links in collagen were studied [42] using two samples: free collagen and collagen-glucose solutions. The collagen sample had a concentration of 20 μM and the collagen−glucose sample contained 20 μM collagen and 40 mM glucose.

Figure 15 shows how the steady-state fluorescence spectrum of each sample evolves over time. Free collagen initially has the fluorescence peak at~405 nm. The intensity of the spectrum then increases, mostly during the first 7 days, and undergoes a red spectral shift up to~425 nm on Day 35. The fluorescence spectrum of the collagen−glucose sample increases steadily over the 35 days. It reaches a higher value and is broader than the spectrum of the free collagen sample.

Figure 15b allows for the comparison of the fluorescence emission of free collagen and collagen−glucose samples over 35 days, with both spectra normalised. From Day 0, the sample with glucose shows a broader spectrum, and there is also a difference in the wavelength of peak fluorescence: ~405 nm for free collagen, which moves gradually to 425, and ~425 nm for collagen–glucose, which remains stable. We attribute the emission in the 400–430 nm range to the formation of dityrosine from the neighbouring tyrosine residues [66], and this process occurs in both the free collagen and glucose−collagen samples. However, the further increase in the fluorescence intensity and the shift of the spectra observed in the collagen−glucose sample only is likely to be an indication of multiple cross-links formed between glycated collagen molecules [56, 67].

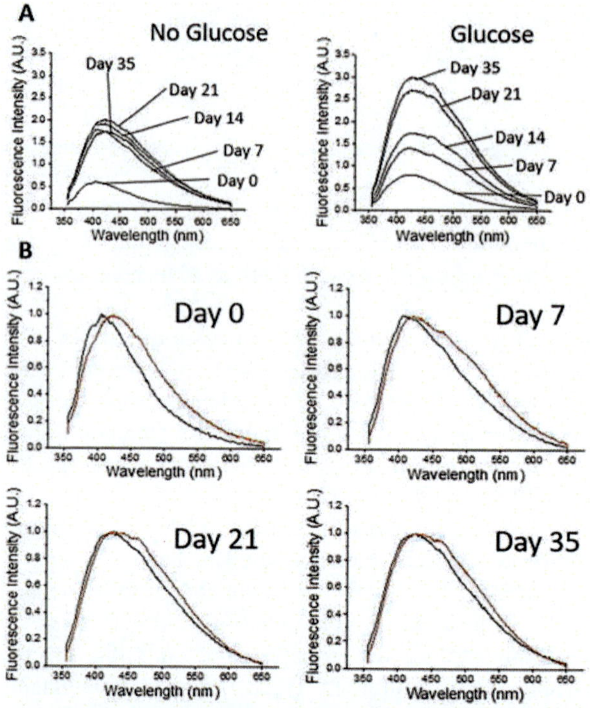

Fig. 15 Fluorescence emission spectra for free collagen and collagen glucose when excited at 340 nm. (**a**) shows fluorescence changes over time for collagen and collagen-glucose at days 0, 7, 14, 21, and 35. (**b**) shows collagen (black) and collagen–glucose (red) on days 0, 7, 14, 21, and 35, with both samples' spectra normalised to peak fluorescence. (Reproduced from [42] with permission of ACS Publications)

TRES measurements were performed at various time points after sample preparation to gain insights into the collagen glycation. The fluorescence decays were recorded from 390 to 500 nm, in increments of 10 nm, and the decay parameters were recovered from fitting these decays to a multi-exponential model (2), where three components were required. These parameters were then used to calculate the TRES using Eq. 7, which was then converted to a function of wavenumber. The absolute and normalised TRES for collagen and collagen–glucose are shown in Fig. 16.

The TRES of free collagen (black curves) on Day 0 show no change in their shape during the decay (see the plot of normalised spectra, black curves in Fig. 16e), indicating no ns time scale processes. The broad peak of the spectrum between ~24,700 and~23,300 cm^{-1} (405–429 nm) is consistent with the steady-state spectra obtained. The measurements performed at later days show a peak at ~21,100 cm^{-1} (475 nm) and the gradual transition of the fluorescence peak towards the red on a nanosecond time scale.

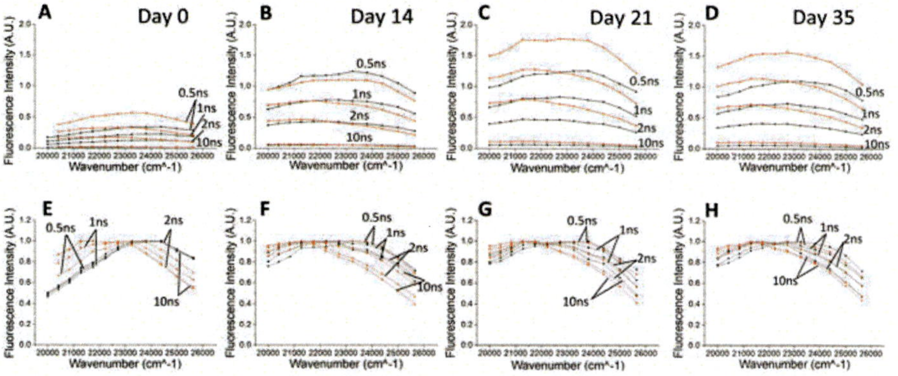

Fig. 16 Examples of TRES spectra of free collagen (black) and collagen-glucose (red) on Day 0 (**a**), Day 14 (**b**), Day 21 (**c**), and Day 35 (**d**). Plots (**e**), (**f**), (**g**), and (**h**) show the same spectra after normalisation. (Reproduced from [30] with permission of ACS Publications)

In the sample containing glucose (red spectra), TRES data show a similar transition, but the emission peak at 475 nm is already apparent at Day 0. This may suggest that glucose accelerates the formation of this fluorescent residue.

A Toptygin model (8) for M = 1 was fitted to the TRES and produced the evolution of the parameters shown in Fig. 17. The initial (Day 0) peak position v(t) is ~22,500 cm^{-1} for the free collagen sample (Fig. 17a) and ~22,000 cm^{-1} for collagen−glucose (Fig. 17c), and both samples behave differently; see Fig. 17e for comparison. There is no shift in $\nu(t)$ following excitation for free collagen at Day 0; however, the measurements taken on the later days show a substantial drop of the initial $\nu(t)$ to ~20,900 cm^{-1}. All $\nu(t)$ dependencies show a further red shift during the first 2 ns, and then the $\nu(t)$ plots start to increase (which is most visible for the data obtained on day 7). As previously observed in Aβ, this increase contradicts the homogeneous dielectric relaxation process. It can only be explained by the fluorescence of several species of individual fluorescence lifetimes, dielectric relaxation rates, etc.

In the case of collagen−glucose sample, Fig. 17e shows the $\nu(t)$ on Day 0 drops from 22,000 to ~20,500 cm^{-1} during the first 2 ns. In the case of older samples, the starting value of $\nu(t)$ is~20,600 cm^{-1}, which is also followed by a similar red shift within the first 2 ns. This behaviour of $\nu(t)$ demonstrates that in the presence of glucose, the formation of cross-links occurs differently, leading to only one type of the fluorescent cross-link.

The *log(C(t))* curves (Fig. 7b, d) seem to confirm this observation. The changes in the rates of decay on different days in the pure collagen sample (Fig. 7b) are not monotonic. Indeed, the decay in Day 7 is faster than in Day 0, and then it slows down again at the later days. This suggests the presence of multiple fluorescent cross-links whose contributions change over time. The decays observed in the glycated sample (Fig. 7d), however, show a stable trend of decreasing decay rates, which is likely to be observed when there is a single or a set of uniform fluorescent residues.

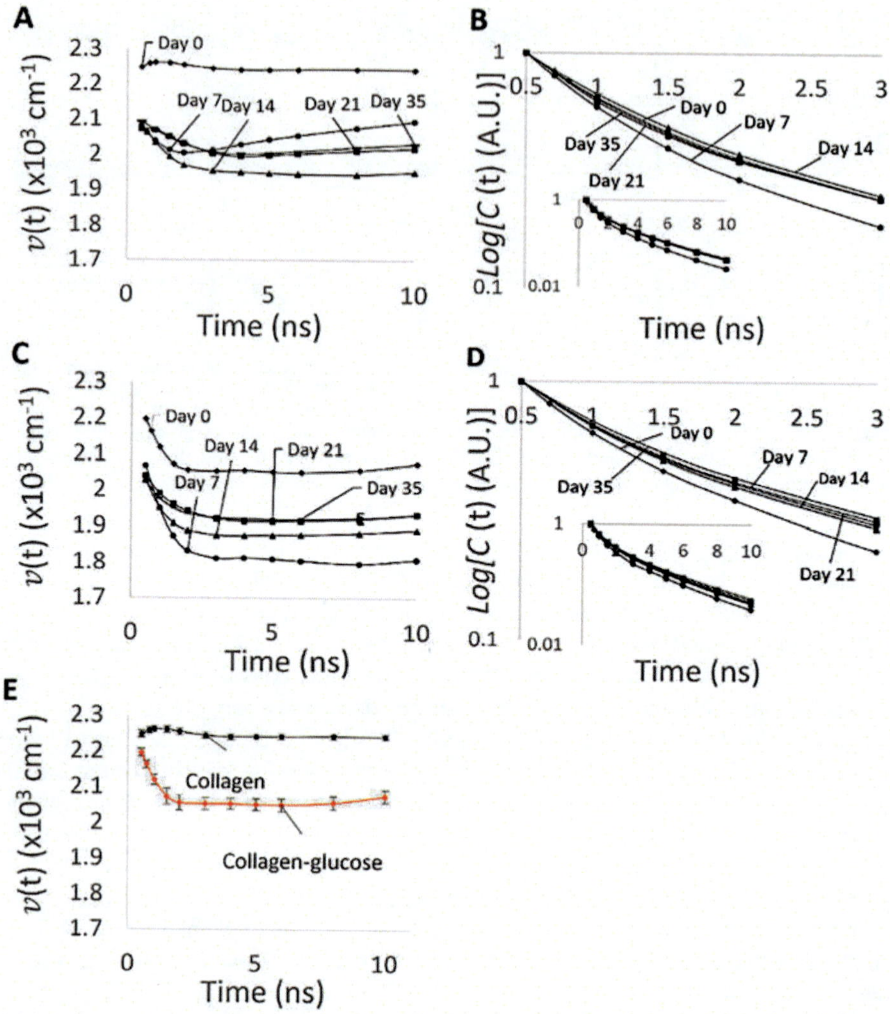

Fig. 17 Plots **a** and **c** show the time evolution of the peak position $\nu(t)$ of the emitting fluorophores for collagen and collagen–glucose, respectively, excited at 340 nm. The normalised time evolution of fluorescence intensity C(t) is shown for collagen and collagen–glucose in plots **b** and **d**. The insert shows the same data over a longer time range. The five lines show Day 0, Day 7, Day 14, Day 21, and Day 35. Part E shows $\nu(t)$ for collagen and collagen – glucose on Day 0 for comparison, with the error bars showing the 90% confidence intervals for the fitted values. (Reproduced from [42] with permission of ACS Publications)

To summarise, the TRES analysis has shown that in a pure collagen sample, more than one type of fluorescent cross-link is formed in the 35-day period, while in the collagen−glucose sample, a single fluorescent form is created. The TRES show a substantial impact of glucose on the formation of pepsin–digestible collagen cross-links and on the formation pathways of new cross-links and protein aggregates.

4 Conclusion

Gathering molecular-level information from the intrinsic fluorescence of biomolecules involved has an advantage of non-invasiveness, but not always offers a detailed information of the investigated process. Indeed, each of the described cases of TRES applications does not fully explain the underlying kinetics; however, it narrows down the selection of the potential models. Determining the exact model needs to be supported by the complementary experimental methods and by theoretical approaches like molecular dynamics (MD) simulations. The latter has the capacity to predict the possible molecular mechanisms (quenching, FRET, exposure to water, etc.) that may impact the fluorescence intensity decay. In the next step, the (analytical or numerical) solution of the system of differential equations representing the expected kinetics needs to be fitted the experimental decays. It is likely that the multiple experimental decays gathered at different sample conditions will be required to confirm validity of more complex models (e.g. the model presented in Sect. 3.1, which combines the non-Debye kinetics and dielectric relaxation).

References

1. Kreutzer AG, Samdin TD, Guaglianone G, Spencer RK, Nowick JS (2020) X-ray crystallography reveals parallel and antiparallel β-sheet dimers of a β-hairpin derived from Aβ16–36that assemble to form different tetramers. ACS Chem Nerosci 11(15):2340–2347
2. Phyo P, Zhao X, Templeton AC, Xu W, Cheung JK, Su Y (2021) Understanding molecular mechanisms of biologics drug delivery and stability from NMR spectroscopy. Adv Drug Deliv Rev 174:1–29
3. Raimondi V, Grinzato A (2022) A basic introduction to single particles cryo-electron microscopy. AIMS Biophysics 9(1):5–20
4. Miles AJ, Janes RW, Wallace BA (2021) Tools and methods for circular dichroism spectroscopy of proteins: a tutorial review. Chem Soc Rev 50(15):8400–8413
5. Yang Z, Xu H, Wang J, Chen W, Zhao M (2021) Single-molecule fluorescence techniques for membrane protein dynamics analysis. Appl Spectrosc 75(5):491–505
6. Mirdha L, Chakraborty H (2021) Fluorescence-based techniques for the detection of the oligomeric status of proteins: implication in amyloidogenic diseases. Eur Biophys J 50(5): 671–685
7. Deng H, Yan S, Huang Y, Lei C, Nie Z (2020) Design strategies for fluorescent proteins/mimics and their applications in biosensing and bioimaging. Trends Anal Chem 122:115757
8. Perfilov MM, Gavrikov AS, Lukyanov KA, Mishin AS (2021) Transient fluorescence labeling: low affinity – high benefits. Int J Mol Sci 22(21):11799
9. Amaro M, Wellbrock T, Birch D, Rolinski O (2014) Inhibition of beta-amyloid aggregation by fluorescent dye labels. Appl Phys Lett 104
10. Vinegoni C, Feruglio PF, Gryczynski I, Mazitschek R, Weissleder R (2019) Fluorescence anisotropy imaging in drug discovery. Adv Drug Deliv Rev 151-152:262–288
11. Gordon F, Casamayou-Boucau Y, Ryder AG (2022) Evaluating the interaction of human serum albumin (HSA) and 1,2-dimyristoyl-sn-glycero-3-phosphocholine (DMPC) liposomes in different aqueous environments using anisotropy resolved multi-dimensional emission spectroscopy (ARMES). Colloids Surf B Biointerfaces 211:112310

12. Kaur H, Nguyen K, Kumar P (2022) Pressure and temperature dependence of fluorescence anisotropy of green fluorescent protein. RSC Adv 12(14):8647–8655
13. Hirvonen LM, Suhling K (2017) Wide-field TCSPC: methods and applications. Meas Sci Technol 28(1)
14. Madonini F, Villa F (2021) Single photon avalanche diode arrays for time-resolved Raman spectroscopy. Sensors 21(13):4287
15. McGuinness CD, Sagoo K, McLoskey D, Birch DJS (2005) Selective excitation of tryptophan fluorescence decay in proteins using a subnanosecond 295nm light-emitting diode and time-correlated single-photon counting. Appl Phys Lett 86(26):261911
16. McGuinness CD, Sagoo K, McLoskey D, Birch DJS (2004) A new sub-nanosecond LED at 280 nm: application to protein fluorescence. Meas Sci Technol 15(11):L19–L22
17. McGuinness CD, Macmillan AM, Sagoo K, McLoskey D, Birch DJS (2006) Excitation of fluorescence decay using a 265nm pulsed light-emitting diode: evidence for aqueous phenylalanine rotamers. Appl Phys Lett 89(6):063901
18. Donzel B, Gauduchon P, Wahl P (1974) Conformation in the excited state of two tryptophanyl diketopiperazines. J Am Chem Soc 96(3):801–808
19. Szabo AG, Rayner DM (1980) Fluorescence decay of tryptophan conformers in aqueous solution. J Am Chem Soc 102(2):554–563
20. Adams PD, Chen Y, Ma K, Zagorski MG, Sönnichsen FD, McLaughlin ML, Barkley MD (2002) Intramolecular quenching of tryptophan fluorescence by the peptide bond in cyclic hexapeptides. J Am Chem Soc 124(31):9278–9286
21. Gasymov OK, Abduragimov AR, Glasgow BJ (2012) Tryptophan Rotamer distribution revealed for the α-helix in tear lipocalin by site-directed tryptophan fluorescence. J Phys Chem B 116(45):13381–13388
22. Starosta R, Santos FC, De Almeida RFM (2020) Human and bovine serum albumin time-resolved fluorescence: tryptophan and tyrosine contributions, effect of DMSO and rotational diffusion. J Mol Struct 1221:128805
23. Rousslang KW, Reid PJ, Holloway DM, Haynes DR, Dragavon J, Ross JBA (2002) Time-resolved phosphorescence of tyrosine, tyrosine analogs, and tyrosyl residues in oxytocin and small peptides. J Protein Chem 21(8):547–555
24. Dai Q, Wang X, Liu C, Feng Z, Ge B, Ma X, Zhang Z, Yu J, Wang X, Huang F (2021) Detection of tyrosinase in living cells using an enteromorpha prolifera based fluorescent probe. Anal Chim Acta 1169:338605
25. Shimomura O, Johnson FH, Saiga Y (1962) Extraction, purification and properties of aequorin, a bioluminescent protein from the luminous hydromedusan, aequorea. J Cell Comp Physiol 59: 223–239
26. Telford WG, Hawley T, Subach F, Verkhusha V, Hawley RG (2012) Flow cytometry of fluorescent proteins. Methods 57(3):318–330
27. Giepmans BN, Adams SR, Ellisman MH, Tsien RY (2006) The fluorescent toolbox for assessing protein location and function. Science 312(5771):217–224
28. Vizcaino-Caston I, Wyre C, Overton TW (2012) Fluorescent proteins in microbial biotechnology – new proteins and new applications. Biotechnol Lett 34(2):175–186
29. Lakowicz JR (2006) Principles of fluorescence spectroscopy.3rd edn. Springer, New York
30. Valeur B (2005) Pulse and phase fluorometries: an objective comparison. In: Hof M, Hutterer R, Fidler V (eds) Fluorescence spectroscopy in biology, vol 3. Springer, pp 30–48
31. Lakowicz JR (2000) On spectral relaxation in proteins. Photochem Photobiol 72(4):421
32. Rolinski OJ, Vyshemirsky V (2016) Fluorescence kinetics of tryptophan in a heterogeneous environment. Methods Appl Fluoresc:4
33. Rolinski OJ, McLaughlin D, Birch DJS, Vyshemirsky V (2016) Resolving environmental microheterogeneity and dielectric relaxation in fluorescence kinetics of protein. Methods Appl Fluoresc 4(2):024001

34. Chang M-F, Li L, Cao X-D, Jia M-H, Zhou J-S, Chen J-Q, Xu J-H (2017) Fluorescence dynamics of LicT protein by time-resolved spectroscopy. Acta Phys Chim Sin 33(5): 1065–1070
35. Alghamadi A, Chung LHC, Rolinski OJ (2021) Monitoring the assembly and aggregation of polypeptide materials by time-resolved emission spectra. In: Ryadnov MG (ed) Methods in molecular biology, vol 2208, pp 167–177
36. Panigrahi SK, Mishra AK (2017) Novel use of UV broad-band excitation and stretched exponential function in the analysis of fluorescent dissolved organic matter: study of interaction between protein and humic-like components. Methods Appl Fluoresc 5(3):035001
37. Menezes F, Fedorov A, Baleizão C, Valeur B, Berberan-Santos MN (2013) Methods for the analysis of complex fluorescence decays: sum of Becquerel functions versus sum of exponentials. Methods Appl Fluoresc 1(1):015002
38. Toptygin D, Brand L (2000) Spectrally- and time-resolved fluorescence emission of indole during solvent relaxation: a quantitative model. Chem Phys Lett 322(6):496–502
39. Alghamdi A, Vyshemirsky V, Birch DJS, Rolinski OJ (2018) Detecting beta-amyloid aggregation from time-resolved emission spectra. Methods Appl Fluoresc 6(2):024002
40. Alghamdi A, Wellbrock T, Birch DJS, Vyshemirsky V, Rolinski OJ (2019) Cu 2+ effects on beta-amyloid oligomerisation monitored by the fluorescence of intrinsic tyrosine. ChemPhysChem 20(23):3181–3185
41. Alghamdi A, Forbes S, Birch DJS, Vyshemirsky V, Rolinski OJ (2021) Detecting beta-amyloid glycation by intrinsic fluorescence – understanding the link between diabetes and Alzheimer's disease. Arch Biochem Biophys 704:108886
42. Muir R, Forbes S, Birch DJS, Vyshemirsky V, Rolinski OJ (2021) Collagen glycation detected by its intrinsic fluorescence. J Phys Chem B 125(39):11058–11066
43. Muir R, Forbes S, Birch DJS, Vyshemirsky V, Rolinksi OJ. NADH glycation in vitro monitored by intrinsic fluorescence of the nicotinamide ring. Methods Appl Fluoresc. Submitted
44. Fantini J, Chahinian H, Yahi N (2020) Progress toward Alzheimer's disease treatment: leveraging the Achilles' heel of Aβ oligomers? Protein Sci 29(8):1748–1759
45. Zhu BY, Cheng Y, Li GB, Yang SY, Zhang ZR (2016) Synthesis and evaluation of styrylpyran fluorophores for noninvasive detection of cerebral β-amyloid deposits. Bioorg Med Chem 24(4):827–834
46. Rolinski OJ, Amaro M, Birch DJ (2010) Early detection of amyloid aggregation using intrinsic fluorescence. Biosens Bioelectron 25(10):2249–2252
47. Sasanian N, Bernson D, Horvath I, Wittung-Stafshede P, Esbjörner EK (2020) Redox-dependent copper ion modulation of amyloid-β (1-42) aggregation in vitro. Biomol Ther 10(6):924
48. Coskuner O (2016) Divalent copper ion bound amyloid-β(40) and amyloid-β(42) alloforms are less preferred than divalent zinc ion bound amyloid-β(40) and amyloid-β(42) alloforms. J Biol Inorg Chem 21(8):957–973
49. Yugay D, Goronzy DP, Kawakami LM, Claridge SA, Song T-B, Yan Z, Xie Y-H, Gilles J, Yang Y, Weiss PS (2016) Copper ion binding site in β-amyloid peptide. Nano Lett 16(10): 6282–6289
50. Jash K, Gondaliya P, Kirave P, Kulkarni B, Sunkaria A, Kalia K (2020) Cognitive dysfunction: a growing link between diabetes and Alzheimer's disease. Drug Dev Res 81(2):144–164
51. Fratzl P (2008) Collagen: structure and mechanics, an introduction. In: Collagen. Springer, Boston, pp 1–13
52. Avery NC, Bailey AJ (2006) The effects of the Maillard reaction on the physical properties and cell interactions of collagen. Pathol Biol 54(7):387–395
53. Verzijl N, DeGroot J, Oldehinkel E, Bank RA, Thorpe SR, Baynes JW, Bayliss MT, Bijlsma JW, Lafeber FP, Tekoppele JM (2000) Age-related accumulation of Maillard reaction products in human articular cartilage collagen. Biochem J 350(Pt 2):381–387
54. Reigle KL, Di Lullo G, Turner KR, Last JA, Chervoneva I, Birk DE, Funderburgh JL, Elrod E, Germann MW, Surber C, Sanderson RD, San Antonio JD (2008) Non-enzymatic glycation of

type I collagen diminishes collagen-proteoglycan binding and weakens cell adhesion. J Cell Biochem 104(5):1684–1698
55. Genuth S, Sun W, Cleary P, Sell DR, Dahms W, Malone J, Sivitz W, Monnier VM (2005) Glycation and carboxymethyllysine levels in skin collagen predict the risk of future 10-year progression of diabetic retinopathy and nephropathy in the diabetes control and complications trial and epidemiology of diabetes interventions and complications P. Diabetes 54(11): 3103–3111
56. Fokkens BT, Smit AJ (2016) Skin fluorescence as a clinical tool for non-invasive assessment of advanced glycation and long-term complications of diabetes. Glycoconj J 33(4):527–535
57. Kollias N, Gillies R, Moran M, Kochevar IE, Anderson RR (1998) Endogenous skin fluorescence includes bands that may serve as quantitative markers of aging and photoaging. J Invest Dermatol 111(5):776–780
58. Nguyen TT, Gobinet C, Feru J, Pasco SB, Manfait M, Piot O (2012) Characterization of type I and IV collagens by Raman microspectroscopy: identification of spectral markers of the dermo-epidermal junction. Spectrosc Int J 27:421–427
59. Teale FWJ, Weber G (1957) Ultraviolet fluorescence of the aromatic amino acids. Biochem J 65(3):476–482
60. Nuytinck L (2000) Glycine to tryptophan substitution in type I collagen in a patient with OI type III: a unique collagen mutation. J Med Genet 37(5):371–375
61. Shen Y, Zhu D, Lu W, Liu B, Li Y, Cao S (2018) The characteristics of intrinsic fluorescence of type I collagen influenced by collagenase I. Appl Sci 8(10):1947
62. Kollias N, Gillies R, Moran M, Kochevar IE, Anderson RR (1998) Endogenous skin fluorescence in vivo on human skin. J Investig Dermatol 111(5):776–780
63. Na R, Stender IM, Henriksen M, Wulf HC (2001) Autofluorescence of human skin is age-related after correction for skin pigmentation and redness. J Invest Dermatol 116(4): 536–540
64. Meerwaldt R, Graaff R, Oomen PHN, Links TP, Jager JJ, Alderson NL, Thorpe SR, Baynes JW, Gans ROB, Smit AJ (2004) Simple non-invasive assessment of advanced glycation endproduct accumulation. Diabetologia 47(7):1324–1330
65. Meerwaldt R, Hartog JWL, Graaff R, Huisman RJ, Links TP, Den Hollander NC, Thorpe SR, Baynes JW, Navis G, Gans ROB, Smit AJ (2005) Skin autofluorescence, a measure of cumulative metabolic stress and advanced glycation end products, predicts mortality in hemodialysis patients. J Am Soc Nephrol 16(12):3687–3693
66. Menter JM (2006) Temperature dependence of collagen fluorescence. Photochem Photobiol Sci 5(4):403
67. Brennan M (1989) Changes in solubility, non-enzymatic glycation, and fluorescence of collagen in tail tendons from diabetic rats. J Biol Chem 264(35):20947–20952

Dynamics and Hydration of Proteins Viewed by Fluorescence Methods: Investigations for Protein Engineering and Synthetic Biology

Jan Sykora, Zbynek Prokop, Jiri Damborsky, Abhinav, Martin Hof, and Mariana Amaro

Contents

Abstract Protein engineering and synthetic biology are currently very active areas of research and development. In the pursuit of engineering proteins with specific capabilities, it has become evident that the scrutiny of structural and geometrical properties does not suffice to achieve the proposed goals. The dynamics and hydration of specific protein areas seem to be of higher influence than it has been once thought. This chapter introduces three different fluorescence spectroscopy techniques (time-dependent fluorescent shift, HMC hydration assay based on unnatural amino acid fluorescence, and photoinduced electron transfer–fluorescence correlation spectroscopy) that allow for assessing the dynamics and hydration of

J. Sykora, Abhinav, M. Hof, and M. Amaro (✉)
J. Heyrovský Institute of Physical Chemistry of the CAS, Prague, Czech Republic
e-mail: amaro@jh-inst.cas.cz

Z. Prokop and J. Damborsky
Loschmidt Laboratories, Department of Experimental Biology and RECETOX, Faculty of Science, Masaryk University, Brno, Czech Republic

International Clinical Research Center, St. Anne's University Hospital Brno, Brno, Czech Republic

R. Šachl, M. Amaro (eds.), *Fluorescence Spectroscopy and Microscopy in Biology*, Springer Ser Fluoresc (2023) 20: 509–532, https://doi.org/10.1007/4243_2022_40, Published online: 3 January 2023

proteins in a site-specific fashion and showcase their usefulness in advancing the design of more efficient enzymes. Systematic application of these techniques to various biomolecular systems will allow a thorough description of these important protein properties, which are rarely taken into account during protein engineering and synthetic biology projects.

Keywords Access tunnels · Catalytic activity · Catalytic efficiency · Enzyme · Fluorescence correlation spectroscopy · Fluorescence spectroscopy · HMC hydration assay · Molecular gating · Photoinduced electron transfer · Protein design · Protein dynamics · Solvation · Substrate specificity · Time-dependent fluorescence shift · Unnatural amino acid fluorescence

1 Introduction

The research area of protein engineering and synthetic biology has undergone a huge outburst in recent years encompassing the developments of a wide range of new systems (e.g. from the redesigning of biomolecules to genetic engineering of DNA to create new organisms) employing a manifold of different methodologies and disciplines [1, 2]. To present a focused view on this convoluted topic, this chapter discusses the biomolecular design and engineering side of synthetic biology research. Specifically, the design of novel catalytic protein structures with improved functionality over existing proteins lies in the scope.

Enzymes are natural catalysts that can find broad applications in biotechnology and medicine due to the very specific and effective manner in which they transform chemical compounds [3]. Improvements in the functionality of enzymes and creation of new functions are highly sought-after goals to achieve the current standards and demands of the industry [4]. The development of novel enzymes nowadays is typically guided by structural information and combines the use of de novo protein design [5, 6], focused directed evolution, and in silico methods [7, 8]. Nonetheless, despite several recent advances [9, 10], the majority of de novo designed enzymes do not compete with the natural versions in terms of specificity and catalytic rates [11]. The causes of such difficulties are still an open question and show that our understanding of enzymatic catalysis is incomplete. Clearly, structural considerations alone are not sufficient enough in guiding protein design [12, 13]. Two aspects often overlooked in the past that have an important influence on the function of the catalyst are protein dynamics and hydration [14–16]. In this chapter, we will spotlight our use of fluorescence methods to explore the dynamics and hydration of model proteins belonging to the haloalkane dehalogenase enzyme family and show the importance of these factors in enzyme engineering.

Haloalkane dehalogenases (HLDs), which belong to the α/β-hydrolase superfamily, are enzymes that catalyse the cleavage of the carbon–halogen bond in halogenated compounds by an S_N2 mechanism (Fig. 1a). Their active site is buried inside

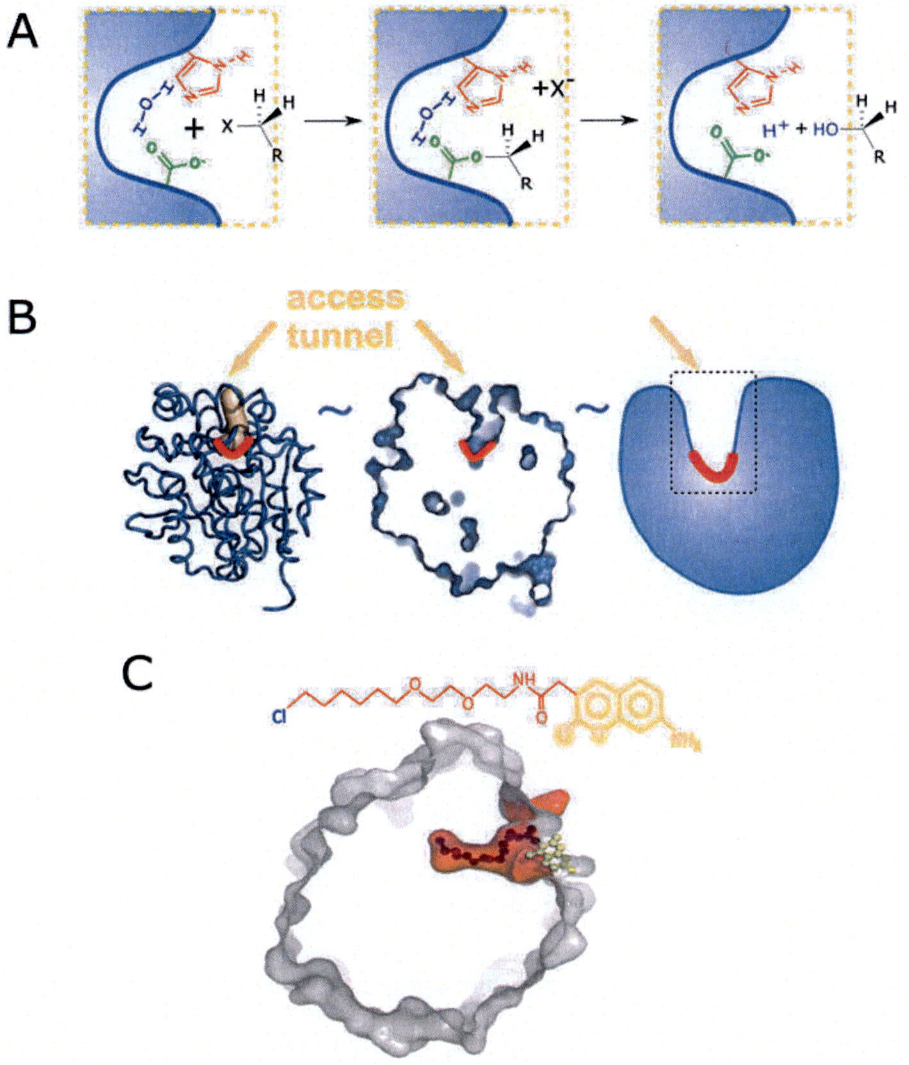

Fig. 1 Simplified scheme for the reaction catalysed in the active site of HLDs including the role of a nucleophilic aspartate (in green) and histidine (in red) involved in the nucleophilic substitution (S_N2) and following alkyl-enzyme intermediate hydrolysis, respectively (**a**), schematic structure of HLDs with the active site highlighted in red (**b**), chemical formula of Promega HaloTag Coumarin (the fluorescence reporter in yellow, linker in red), and the structure of a covalent adduct of Promega HaloTag Coumarin located in the tunnel of HLD (with the catalytic histidine mutated out) (**c**)

the protein core, in a predominantly hydrophobic cavity at the interface of the α/β-hydrolase core domain and the helical cap domain. The active site is connected to the bulk solvent via different transport tunnels (Fig. 1b) [17–21]. HLDs catalyse the hydrolysis of a wide range of halogenated substrates, i.e. (chlorinated, brominated, iodinated) alkanes, alkenes, cycloalkanes, alcohols, carboxylic acids, esters, ethers, epoxides, amides, and nitriles [22]. Several substrates are serious environmental pollutants (e.g. 1,2-dichloroethane, 1,2,3-trichloropropane, 1-chlorobutane, hexachlorocyclohexane), and thus, HLDs can find use in applications in bioremediation [23] and detoxification [24]. They can also be of use in industrial biocatalysis [25, 26], biosensing [27], and cell imaging [28].

HLDs present a thoroughly understood and studied enzyme family. The profound knowledge of the catalytic mechanisms of HLDs and their structures provides a framework for developing and testing different strategies of protein engineering or de novo protein design [29]. The main goal of these experiments is to find all conceivable factors that have a remarkable influence on the activity, substrate specificity, and enantioselectivity of HLDs. These findings could be then potentially generalized to other enzyme families and proteins. Herein, the applied strategy resides in combining kinetics data and molecular dynamics simulations obtained for newly designed mutants with results of specifically tailored fluorescence techniques. Even though there are a plethora of different approaches available for mapping various attributes of protein hydration and dynamics (e.g. neutron scattering, NMR approaches, optical Kerr effect, extended depolarized light scattering (EDLS), and terahertz and two-dimensional (2D)-IR spectroscopies [30–33]), fluorescence methods are worth considering for several reasons:

1. They can provide site-specific information, which does not apply to techniques monitoring collective motions. The ongoing advances in labelling strategies of proteins and purification methods enable highly efficient and specific positioning of the fluorescence reporters within the protein scaffold.
2. Low concentration of the protein sample (typically nM–μM) is sufficient for the measurements, which is usually below the precipitation threshold of most biomolecules.
3. The instrumentation both for steady-state and time-resolved fluorescence is nowadays accessible and relatively easy to handle.

On the other hand, the obvious disadvantage of fluorescence techniques is the usage of an external reporter, which can affect the behaviour and activity of the enzymes. To partly purge the gained experimental findings from undesired artefacts caused by fluorescent labelling, collecting data for more probes with different chemical structures and with multiple fluorescence approaches is a useful strategy.

The use of fluorescence spectroscopy methods to study the role of protein hydration and dynamics in controlling enzyme activity and enantioselectivity is the subject of this chapter. Carefully designed fluorescence experiments combined with molecular dynamics (MD) simulations and enzyme kinetics can reveal the qualitative importance of these protein characteristics for enzyme functioning and thus unravel their relevance in the special area of de novo enzyme design and protein engineering.

2 Hydration and Mobility Monitored by Time-Dependent Fluorescence Shift

Time-dependent fluorescence shift (TDFS) is an experimental method originally developed to follow the solvation dynamics of neat solvents [34]. It takes advantage of the local perturbation of the system provoked by the change in the dipole moment of the fluorophore after its excitation. The microenvironment of the fluorophore, aligned to its dipole moment in the ground state, is suddenly exposed to energetically non-favourable conditions. As a result, a continuous relaxation process occurs to re-establish the equilibrium (i.e. energetic minimum) between the excited fluorophore and surrounding molecules (Fig. 2a). This results in a gradual decrease in the energy of the solvated fluorophore that can be monitored by recording time-resolved emission spectra (TRES). TRES show the time-dependent red shift caused by the relaxation dynamics of the dye's microenvironment (Fig. 2b). The overall spectral shift between the initial Franck–Condon state and the fully relaxed state (overall TDFS, also called dynamic Stokes shift and denoted as $\Delta\nu$) increases with the solvent polarity [34], while the TDFS kinetics characterized by the relaxation time τ_r is primarily proportional to the viscosity of the solvent (Fig. 2c) [35].

When following the TDFS of a fluorescent dye attached to a protein, the situation is analogous yet more precarious compared to neat solvents. The environment is more heterogeneous being composed of water molecules and differently hydrated amino acid segments. This immense heterogeneity makes the characterization of protein hydration tricky not only in the case of TDFS but in general. Indeed, protein hydration has been reported to occur on largely diverse time and length scales, often depending on the selected experimental and/or theoretical approaches [30–32, 36]. The literature data on this topic are being built up since the 1970s, which has brought a huge set of diverse conclusions and hypotheses making the topic of protein hydration rather fuzzy. Recently, extrapolation of the collected data and advances both in experimental and in theoretical methods appear to converge to a general consensus on the interpretation of this complex phenomenon. In brief, the translational and rotational motions of individual water molecules are only subtly retarded upon interaction with the protein, approximately by a factor of 3–5 compared to the bulk solvent [37]. This corresponds to time scales faster than 5 ps. Motions occurring on longer time scales ranging from tens of picoseconds up to nanoseconds and even microseconds are caused by collective coupled motions of protein segments and water molecules [32]. In this respect, hydrated charged, polar, and/or polarizable groups in side chains and the backbone of the protein are dominant contributors that cause the heterogeneous character of protein hydration dynamics.

In connection to the anomalously retarded solvation observed in biological systems, the concept of so-called biological water [36, 38, 39] was coined to refer to a "special" state bestowing the water a role decisive for protein function. The unique anomalous arrangements of water molecules hydrating the biomolecule were believed to be the driving force in protein activity. However, considering that similar anomalous disturbance of the water structure also happens at non-biological

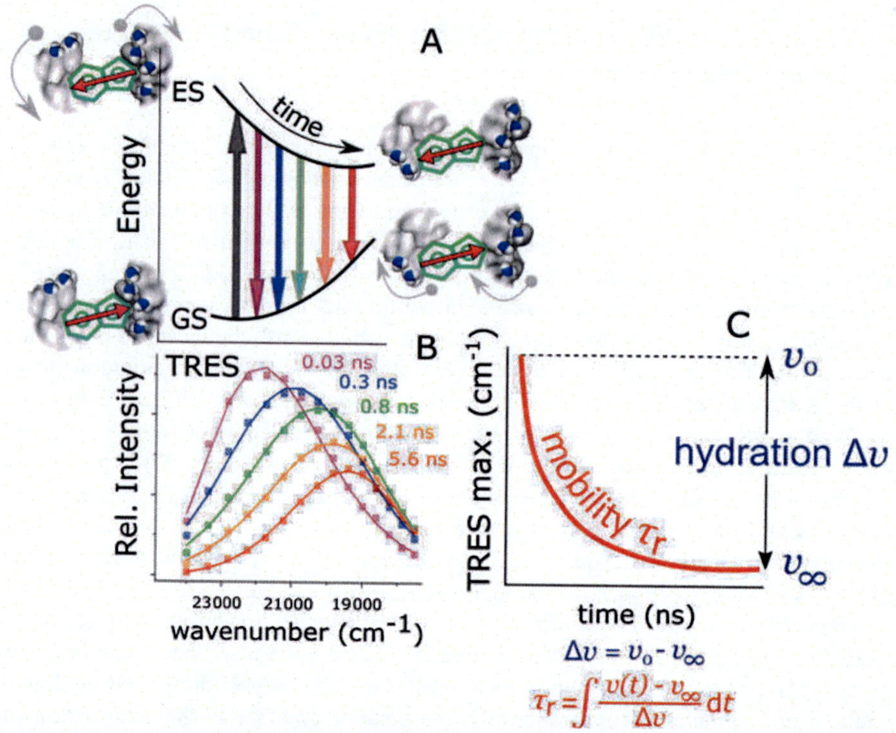

Fig. 2 Scheme of the TDFS phenomenon: the probe microenvironment formed by the hydrated protein segments responds to the sudden change in the probe dipole moment (red arrow) upon excitation by continuously rearranging in time (GS and ES denote the ground and excited state, respectively) (**a**), this relaxation process leads to the decrease in the energy of the system which is projected to the time-dependent shift of time-resolved emission spectra (TRES) (**b**), two parameters containing information on the probe microenvironment are gained: the dynamic Stokes shift denoted as $\Delta\nu$ reflects the hydration of the protein segments, while the TDFS kinetics characterized by relaxation time τ_r is sensitive to their mobility. These parameters are calculated using the shown formulas (**c**). The contribution from the relaxation of bulk water is minimal on the time range longer than tens of ps, which is the typical operation time scale of most commercial spectrometers

interfaces, we do not see any need for introducing this term. Its usage might be misleading as it often serves to exaggerate the time and length scales of a water action in the protein's vicinity. Recent contributions have confirmed that the first solvation layer of a protein bears substantially different properties compared to bulk water, while modifications in the second and more remote solvation layers are hardly detectable [40].

2.1 *Role of Protein Hydration and Mobility at the Tunnel Mouth of HLDs*

2.1.1 TDFS as a Detector of Site-Specific Hydration and Mobility

TDFS is a technique with the potential to follow coupled motions of water and protein segments and provide information that can be correlated to local protein properties, i.e. the extent of hydration (parameter $\Delta\nu$) and mobility (parameter τ_r) of the dye's microenvironment. This TDFS capability can be demonstrated using the systematic study of DbjA [41] and DhaA [19], two HLDs that possess different tunnel morphology and different enantioselectivity (Fig. 3b) [22, 26]. The tunnel mouth region of HLDs is known to be important for their function affecting both the transport of substrates towards the active site and the rate of product release [42, 43].

The first prerequisite for a successful TDFS application, including HLD studies, is the selective labelling of the protein at the region of interest. To do so one can make use of the general attribute of enzymes that offer the labelling option of employing a fluorescent dye as the substrate for the enzymatic reaction. To create a stable covalent adduct with the fluorescent reporter, a single point mutation in the

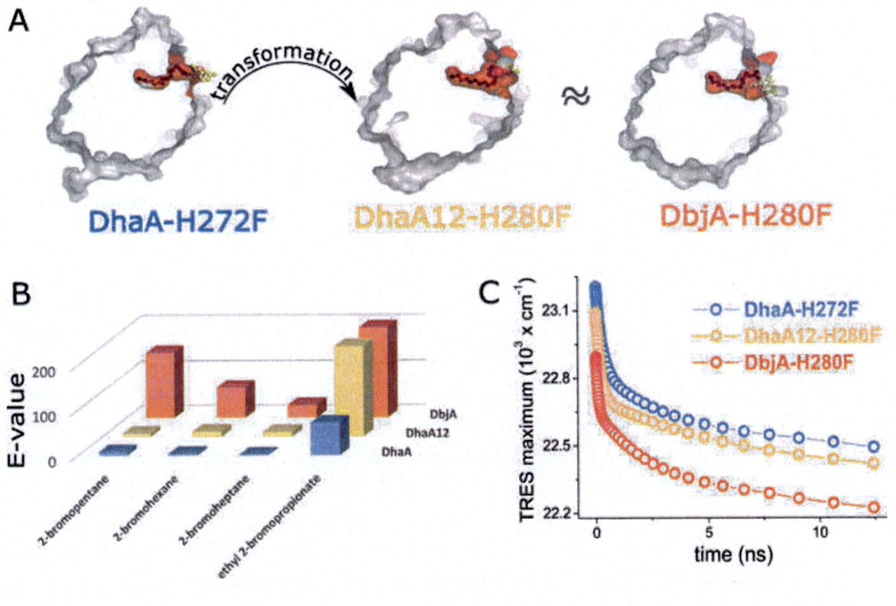

Fig. 3 Tunnel transplantation from DbjA to DhaA results in the DhaA12 mutant (DhaA transformation). Cross-section of the investigated HLDs labelled with *HT-coumarin* illustrating their tunnel architecture (in red) and the fluorophore positioning (in yellow) showing the similar tunnel architectures of DhaA12-H280F and DbjA-H280F (**a**), E-values reflecting enantioselectivity of DhaA, DhaA12, and DbjA to various substrates (**b**), time course of TRES shift for the investigated HLDs labelled with *HT-coumarin* (**c**)

Table 1 Parameters gained from molecular dynamics (MD) simulations and TDFS measurements for the tunnel mouth regions of selected enzymes using HT-coumarin dye [44, 46]. The overall extent of hydration in the vicinity of the fluorophore is proportional to the measured dynamic Stokes shift $\Delta\nu$ and to the "density of water" value calculated by MD. The mobility of the dye's microenvironment is inversely related to the B-factor obtained by MD (which quantifies the flexibility of the protein's residues) and the TDFS's relaxation time, τ_r. The intrinsic errors for $\Delta\nu$ and τ_r are 50cm^{-1} and 0.05ns, respectively. Finally, the amino acid residues that are in most frequent contact with the dye identified by MD are listed. This is an important criterion when comparing the data among different HLDs. Only those mutants with similar amino acid surroundings should be compared quantitatively

	Enzyme			
Parameter	DhaA-H280F	DbjA-H272F	DhaA12-H280F	Interpretation
Density of water (5 Å around dye) [kg m^{-3}][a] Dynamic stokes shift $\Delta\nu$ [cm^{-1}]	2.3 ± 0.7 950	6.0 ± 0.3 1,300	3.3 ± 0.7 1,000	The extent of tunnel mouth hydration
Mobility expressed by B-factor of tunnel mouth near dye [Å^2][a] Relaxation time τ_r [ns]	112 4.1	378 2.8	158 3.8	Mobility of hydrated protein segments within the tunnel mouth
Residues forming the most van der Waals contacts with the dye[a]	Phe144 Lys175	Gly183 Arg179	Gly183 Arg179	

[a] Parameters gained from quantum mechanics/molecular mechanics stochastic boundary molecular dynamics (QM/MM SBMD) simulations

catalytic pentad of the HLD (typically histidine to phenylalanine) is performed yielding the mutants DbjA-H280F and DhaA-H272F [44]. Histidine catalyses the hydrolysis of the intermediate; thus, its replacement halts the progress of the reaction resulting in a covalently bound alkyl-enzyme intermediate (Fig. 1a, c). In particular for HLDs, one can use fluorescent substrates consisting of the fluorophore on one side and a linker terminated by a halogen on the opposite side, where the linker length can be tailored. In this particular study, the Promega HaloTag® Coumarin (*HT-coumarin,* composed of environment-sensitive Coumarin 120 with an attached linker terminated by Cl) has been used as the substrate (Figs. 1c and 3a). Due to the long linker, the fluorescence reporter is positioned at the tunnel mouth of the labelled enzyme [44]. To prove that the labelling procedure is successful and selective, mass spectroscopy combined with fluorescence quenching can be applied [44].

The TDFS for the HLD mutants DbjA-H280F and DhaA-H272F labelled with *HT-coumarin* are shown in Fig. 3c, and the data are summarized in Table 1. The detected mobility parameter τ_r reaches nanosecond time scales for both enzymes even though the dye is positioned in a rather exterior part of their tunnel mouths. Such slow dynamics are more prolonged than the TDFS responses usually observed at protein surfaces [45]. Moreover, the mobility of the *HT-coumarin* microenvironment in DhaA-H272F is slowed down to a higher extent compared to that of DbjA-H280F. This finding agrees with diffraction and MD simulation data [44], which predict higher mobility in the wider tunnel mouth of DbjA-H280F. The observed higher degree of hydration for the more open tunnel mouth of DbjA-H280F is seen by means of both MD and TDFS (Table 1) [44]. The presented data show intrinsic differences in the environment of the tunnel mouths of DhaA and DbjA enzymes supporting an important role of this region in discriminating between various substrates, and illustrating that TDFS serves as a powerful tool to characterize the architecture of particular protein regions, especially when comparing structurally related proteins.

As TDFS is not a label-free technique, the effect of the chemical structure of the probe on the detected TDFS response in proteins has to be considered when interpreting the recorded data. To investigate this effect, the above-described HLD mutants DhaA-H272F and DbjA-H280F have also been labelled with probes containing the dimethylaminonaphthalene fluorescent group, which are denoted as MUC dyes. The TDFS data have been then compared to the results obtained with *HT-coumarin* (Table 2) [47]. The MUC probes are designed similarly to the *HT-coumarin*; i.e. they contain the fluorescent reporter at one end and the -Cl atom at the other end, connected by a linker identical to *HT-coumarin* (MUC6) or one-carbon longer (MUC7) [47].

The MUC data showed remarkably slower TDFS relaxation times compared to *HT-coumarin* for both investigated enzymes. The relaxation process was even extended beyond the lifetime of the MUC dyes for the DhaA variant. Increasing the temperature from 10°C to 45°C did not accelerate the TDFS response sufficiently to be fully observable on the time scales determined by the lifetime of the MUC dye, allowing only 37% of the process to be captured. Nonetheless, it enabled a better comparison between the TDFS gained for DbjA-H280F and DhaA-H272F. The latter enzyme, which possesses a narrower tunnel mouth, shows a lower degree of hydration and a decreased mobility of the MUC dye's microenvironment compared to DbjA-H280F, which is in agreement with the TDFS results gained for *HT-coumarin* and diffraction and simulation data mentioned above [44, 46]. The narrower tunnel of DhaA promotes specific interactions between the dye and the protein that modulate the TDFS response. The particular low hydrophilicity and dipole moment of the MUC dye may give rise to hydrophobic or electrostatic interactions of the chromophore with the protein matrix. We can thus conclude that the chemical nature of the dye can alter the TDFS behaviour; nonetheless, findings gained for probes positioned at a similar site within the enzyme structure report qualitatively comparable information and correlate with the differences in mobility and hydration of those specific regions of the examined proteins.

2.1.2 TDFS and Structure–Function Relationship of HLDs: Enantioselectivity Study

The de novo protein design can succeed fully only when embracing and comprising all conceivable factors that can affect the activity and/or enantioselectivity of the enzyme. Apart from the proper geometrical arrangement of the active site and its close environment, the broader protein context has been suggested as a potentially essential factor contributing to the catalytic potency. To refine this concept, the

Table 2 TDFS data for the tunnel mouth regions of DbjA-H280F and DhaA-H272F enzymes using HT-coumarin and MUC6 dyes [47]. The dynamic Stokes shift, $\Delta\nu$, is proportional to the overall extent of hydration in the vicinity of the fluorophore, the relaxation time, τ_r, is inversely related to the mobility of the dye's microenvironment, and the observed TDFS reports on the fraction of the TDFS that is captured within the experimental time window. The intrinsic errors for $\Delta\nu$ and τ_r are $50 cm^{-1}$ and 0.05ns, respectively

	Dynamic Stokes shift $\Delta\nu$ [cm^{-1}]			Relaxation time τ_r [ns]			Observed TDFS [%]		
Dye enzyme	HT-coumarin	MUC6	MUC6 (45°C)	HT-coumarin	MUC6	MUC6 (45°C)	HT-coumarin	MUC6	MUC6 (45°C)
DhaA-H280F	950	1,050[a]	1,300[a]	4.1	0.1[a]	1.7[a]	90	20	37
DbjA-H272F	1,300	1,420		2.8	6.6		70	75	

[a] A large part of the reorientation of the microenvironment is not complete during the fluorescence lifetime of the MUC dye; i.e. the process is slower than the experimental nanosecond time window

above-discussed HLD mutants of DbjA and DhaA (DbjA-H280F and DhaA-H272F) have been used as model systems. DbjA shows high enantioselectivity towards the β-bromoalkanes in contrast to DhaA (Fig. 3b) [26]. By applying advanced techniques of protein engineering, stepwise mutations in the DhaA enzyme resulted in a newly designed protein DhaA12-H280F (DhaA12) [46] whose architecture of the access tunnel and geometry of the active site are identical to those of DbjA (Fig. 3a). Indeed, the active site transplantation from DbjA to DhaA has been realized with Ångström precision as confirmed by X-ray crystallography [46]. If the active site is the dominant factor determining HLD enantioselectivity towards β-bromoalkanes, the newly designed HLD variant DhaA12 is to possess remarkably higher enantioselectivity compared to DhaA and ideally reach the level of DbjA. However, as apparent from Fig. 3b, the E-values recorded for DhaA12 fall behind those of DbjA considerably, being more comparable to its original template DhaA. Thus, factors other than the 3D arrangement of the active site also have to be considered to fully reconstitute the exceptional capabilities of the DbjA enzyme. Hydration and mobility at the access path including the tunnel mouth could be factors of relevance. The tunnel mouth's degree of mobility and hydration may assist in the discrimination of the respective enantiomers. To test the hypothesis, TDFS experiments have been conducted for DhaA12-H280F labelled with *HT-coumarin* and compared to the above-presented data obtained for DbjA-H280F and DhaA-H272F (Fig. 3c, Table 1) [46]. The time evolution of TRES maxima for DbjA-H280F is remarkably different from those of DhaA12-H280F and DhaA-H272F. This is confirmed by inspecting the derived hydration and mobility parameters ($\Delta\nu$ and τ_r, respectively), which clearly show that DbjA-H272F has a significantly higher extent of hydration and of mobility at the tunnel mouth region in comparison to the other variants (Table 1). In fact, the newly designed DhaA12 shows a very small improvement in environmental mobility over that of the parent DhaA. This scenario is analogous to that shown by the enantioselectivity data (Fig. 3b), which demonstrate insufficient improvement in DhaA12's action regarding β-bromoalkanes despite its 3D geometry and arrangement of the active site being identical to that of DbjA. This finding brings strong evidence that geometrical factors do not guarantee full functionality of the enzymes and that the broader protein context, including the hydration and mobility of regions relatively remote to the active site, needs to be considered in protein design.

3 Hydration Assay Based on Steady-State Fluorescence of HMC

Although TDFS presents a powerful tool for investigating protein hydration, it cannot be applied universally. Besides the need for time-resolved instrumentation, not all fluorescent reporters have the necessary photophysical properties that allow for TDFS experiments [48]. For instance, the *HT-coumarin* mentioned above is a probe suitable for TDFS, but note that not all coumarin dyes are of this kind. The 7-hydroxy-4-methylcoumarin (4-Methylumbelliferone, 7H4MC, HMC) is one of such probes whose photophysics is too complex for a straightforward TDFS interpretation. Specifically, HMC can adopt four different forms in the excited state: neutral, anionic, tautomeric, and complex (Fig. 4a). Therefore, there are four possible fluorescence emitting states, each with its own excited-state kinetics, which renders the interpretation of the dynamic shift data practically impossible. Each of these forms exhibits a different emission maximum, making them spectrally separable, and importantly their population is driven by the local hydration [49, 50]. In the absence of water, only the neutral and the complex forms can be generated, while the anionic and tautomeric forms require an aqueous environment. Thus, the integral fluorescence from the anionic and tautomeric forms reflects the extent of hydration of the microenvironment surrounding the HMC dye. This information can be accessed from the steady-state fluorescence emission spectra of the dye, and the so-called hydration parameter ($A^A + A^T$) is defined as the sum of the spectral contributions of these water-conditioned forms (Fig. 4c) [51]. Interestingly enough, the formation of the tautomer demands the presence of water molecules with residence times longer than the fluorescence lifetime of HMC (typically $\sim$ 6 ns), i.e. the type of hydration often referred to as "structured water". In summary, the photophysics of HMC is ruled by the hydration state of the dye and has the potential to provide information not only on the extent of hydration (analogous to the dynamic Stokes shift $\Delta\nu$ from TDFS data) but also on the presence of structured water, which is not accessible by TDFS. On the other hand, this "HMC hydration assay" does not report on the mobility of the dye's microenvironment.

3.1 *Hydration at the Tunnel Mouth of HLDs: Site-Specific HMC Hydration Assay in Proteins Using Unnatural Amino Acid*

Insertion of the HMC chromophore into a specific position within the protein backbone can be carried out via site-directed mutagenesis using the concept of an unnatural amino acid (UAA). The UAA, which is a chemically synthesized L-(7-hydroxycoumarin-4-yl) ethylglycine in this case, is encoded by a nonsense codon whose translation is mediated through a specially modified tRNA/aminoacyl–tRNA synthetase pair [52]. This methodology has allowed the creation of two HLD mutants of DhaA and DbjA, namely the DhaA:C176UAA and DbjA:G183UAA enzymes, which bear the HMC fluorophore at their tunnel mouth (Fig. 4b) [51]. The hydration parameter ($A^A + A^T$) determined for DhaA and DbjA corresponds to what can be anticipated by the MD simulations and TDFS measurements that have been shown in the previous sections of this chapter. The tunnel opening of DbjA: G183UAA shows a significantly higher degree of hydration compared to the DhaA variant as illustrated by its elevated hydration parameter ($A^A + A^T$)

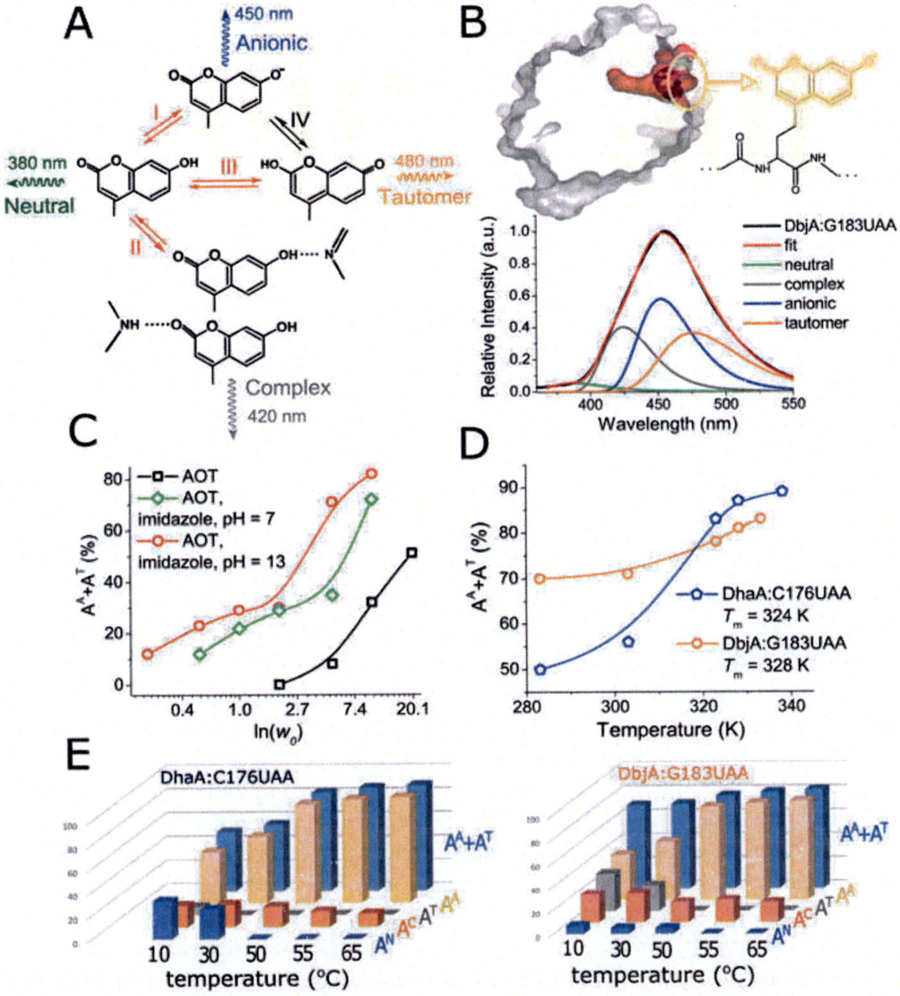

Fig. 4 Scheme of HMC photophysics in the excited state. The complex is a hydrogen bonded form of the dye; for example, in proteins an adjacent amino group can be involved (**a**), approximate location of the UAA fluorophore in the investigated HLDs, and example of an emission spectrum of HMC decomposed into the spectral contributions of each form of HMC described in panel (**a**) (**b**), data obtained in reverse AOT micelles at defined hydration degrees (w_O) proving the correlation of the hydration parameter ($A^A + A^T$) with the water content (**c**), "hydration parameter" ($A^T + A^A$) for DhaA and DbjA mutants obtained at various temperatures to follow the effect of thermal denaturation of the proteins (**d**), areas of the emission spectra of the different forms of the HMC fluorophore in DhaA and DbjA mutants at various temperatures (**e**)

(Fig. 4d). MD simulations of these specific HLDs containing the UAA support that the level of hydration for the DbjA variant is substantially larger than for the DhaA one. The number of water molecules within 5 ångströms of the fluorophore is calculated to be 9 ± 2 for DhaA:C176UAA and 21 ± 5 or 28 ± 4 for DbjA: G183UAA (for two dominant chromophore locations). For the wild-type DhaA and DbjA, the number of water molecules in the same region is calculated to be 16 ± 4 and 27 ± 4, respectively. Further validation of the HMC hydration assay comes from experiments performed at higher temperatures. Increasing the temperature of the system till past the thermal denaturation point shows a progressive elevation of the hydration parameter (Fig. 4d). Upon denaturation, the hydration level sensed by the UAA reflects the high exposure to the aqueous environment of the originally buried chromophore. Moreover, the increasing temperature results in an increase in the contribution of the anionic form at the expense of the other forms as is to be expected when the chromophore is exposed to the bulk aqueous medium (Fig. 4e). These findings demonstrate the applicability of the HMC assay as an indicator of site-specific hydration in proteins that only requires steady-state fluorimetry.

Combining the HMC hydration assay with MD simulations can bring to light even more information about the proteins in question. Curiously, the DbjA: G183UAA mutant is characterized by a significantly higher content of the tautomeric form compared to DhaA:C176UAA (Fig. 4e). This finding points to the presence of structured water, and MD simulations not only support this conclusion but also elucidate the origin of the experimentally observed phenomenon. Seemingly, the wider tunnel mouth of the DbjA variant allows the UAA to exist in an additional conformation that cannot exist in DhaA:C176UAA according to MD. This conformation has the UAA buried into the tunnel mouth entry causing itself to lock several water molecules within the active site pocket for a time substantially longer than the lifetime of HMC (up to 60 ns) [51]. This was not observed for DhaA:C176UAA. MD simulations also help to understand why DbjA and DhaA mutants exhibited fluorescence from the complexed form of the chromophore. MD indicates that the dye can often form hydrogen bonds with neighbouring amino acid residues in either of the proteins, which is likely the promoter of the formation of the complexed form [51]. Of note, the conclusions on the extent of hydration gained by HMC hydration assay for DbjA:G183UAA and DhaA: C176UAA fully agree with those of the TDFS experiments performed on the DbjA-H280F and DhaA-H272F mutants, which demonstrates the robustness and consistency of the herein applied different fluorescence techniques.

4 Protein Dynamics Probed by Photoinduced Electron Transfer: Fluorescence Correlation Spectroscopy (PET-FCS)

Up to this point, the fluorescence methods that have been presented report on the mobility and hydration of the environment within the intimate proximity of the fluorophore. To follow the complex conformational mobility of proteins on longer length scales, other fluorescence techniques can be used [53]. The combination of photoinduced electron transfer (PET) quenching with the single-molecule sensitivity of fluorescence correlation spectroscopy (FCS) results in one of such approaches. PET-FCS can provide information on protein dynamics that occur on time scales ranging from nano- to milliseconds [54].

PET is an ultrafast (femto- to picoseconds) photochemical quenching process of charge transfer between the fluorophore and a quencher molecule that occurs at distances typically shorter than 1 nm (van der Waals interaction). For the case of proteins, it is notable that tryptophan is the only residue capable of acting as an efficient electron donor for a PET reaction with a fluorophore [54]. Moreover, the efficiency of PET between a suitable chromophore and tryptophan diminishes significantly beyond 0.5–1 nm [55]. This serendipitously renders PET a fantastic site-specific phenomenon of use in protein studies. In a PET experiment, a single fluorophore must be placed in a strategic region of interest and the quencher tryptophan may either be a native or a mutated residue. Conformational changes of the protein that result in periodic approaching and receding of the dye and tryptophan (on the appropriate length scale) will cause off/on switching of the single-molecule fluorescence signal (Fig. 5a). These fluorescence intensity fluctuations can be followed by fluorescence correlation spectroscopy (FCS), which is a method primarily used for following the diffusion of fluorescent species in solution [56]. However, in principle, FCS can be used to study any process that results in fluorescence intensity fluctuations [57]. Briefly, in FCS one records the time trace of a fluorescence signal that arises when fluorescent molecules pass through (i.e. diffuse in and out of) a small well-defined detection volume, usually achieved with a confocal microscope set-up. Statistical analysis of the fluorescence fluctuations present in the recorded trace allows the generation of a correlation function (typically an autocorrelation function, ACF) that reveals the diffusion characteristics of the molecules in study. When processes other than the particle's diffusion cause recurrent changes in brightness (i.e. blinking), additional fluorescence fluctuations are generated and further decays are observed in the ACF (Fig. 5b). Such additional fluctuations can be caused by numerous phenomena in inter- or intra-molecular interactions, e.g. protein–ligand binding or intersystem crossing to a triplet state. Of interest here is that the fluorescent molecules, namely proteins, can undergo conformational transitions that result in blinking due to the tailored PET process while diffusing through the detection volume. FCS analysis of the ACF of PET-active systems can report on the rates of the conformational transitions on times ranging from nano- to milliseconds (Fig. 5b). Hence, PET-FCS is the technique of choice to investigate a very special HLD enzyme named LinB86.

4.1 *Real-Time Capture of Molecular Gating in HLDs*

LinB86 is an HLD enzyme product of protein engineering that remarkably surpasses its wild-type ancestor LinB in catalytic activity [58]. It has been constructed by inserting two modifications to the wild-type LinB scaffold: one is the mutation of the

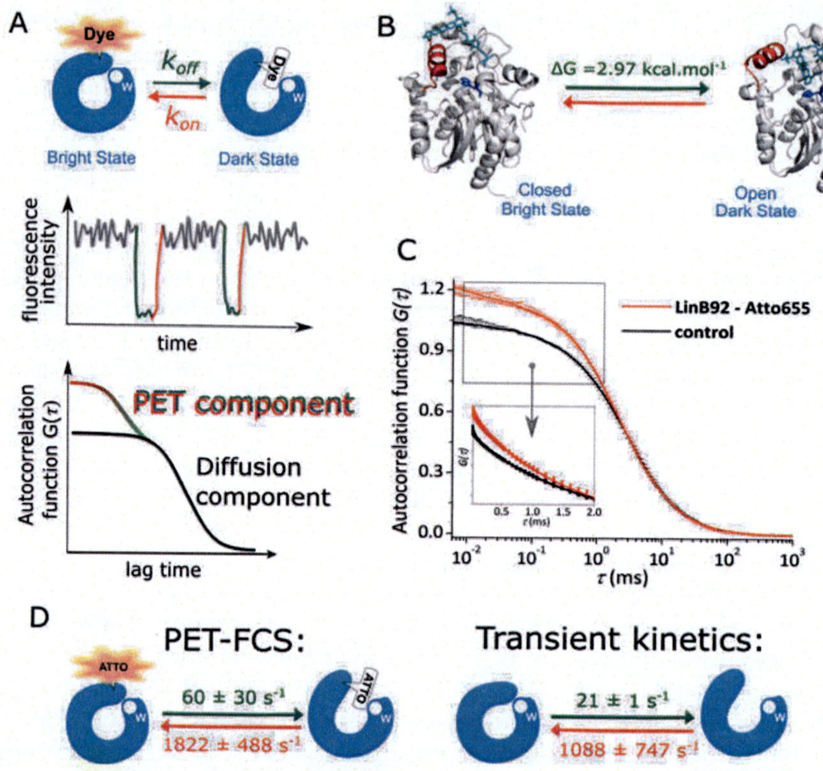

Fig. 5 Principle of the PET-FCS technique: the conformational transition of the protein causes switching between a bright state and a dark "PET-quenched" state. These transitions are the source of fluctuations in the fluorescence intensity time trace. If such a "blinking" protein is monitored by means of FCS, an additional "PET" component reflecting the conformational transition appears in the autocorrelation function (ACF) (**a**), results gained by MD simulations illustrate the equilibrium between two different conformations of LinB92: the closed bright state with Atto655 (in cyan) and Trp177 (in blue) apart, and the open dark state with Atto655 and Trp177 in contact. Note the large relocation of the cap domain (in red) upon the conformational transition (**b**), comparison of ACFs obtained for LinB92 and control mutant LinB99 with back-mutated Trp177. While the control ACF contains only a diffusion component, LinB92 shows an additional faster PET component (**c**), comparison of the k_{on} and k_{off} constants obtained by PET-FCS for the LinB92 labelled with Atto655 and those gained by the label-free "transient kinetics" approach (**d**)

alanine at the main access tunnel to the bulkier tryptophan (L177W), and the other is the mutation of three amino acid residues for less bulky ones in a strategic position found through focused directed evolution (W140A, F143L, I211L). The first modification results in blocking the main tunnel and increases the rigidity of the access site surroundings, which improves the S_N2 chemical step of the catalytic cycle (Fig. 1a) [58, 59], while the latter one improves the chemical step of hydrolysis and, importantly, introduces additional flexibility to the enzyme molecule facilitating the transient formation of an open conformation [14, 58]. This new flexibility of

LinB86 significantly accelerates the transport rates of both substrate and product via the access tunnel (note that product release is the rate-limiting step) [14, 58]. In summary, LinB86 has accelerated chemical steps of catalysis as well as substrate/ product transport to/out of the active site in comparison to the parent LinB. Transient kinetics data suggest the existence of a dynamic equilibrium between a closed and an open conformation of LinB86 and that these modifications of conformational dynamics are the mechanism for the enhanced catalytic efficiency of the LinB86 mutant [14, 58].

The single-molecule PET-FCS approach has the potential to test the above hypothesis, but the "on–off switch" must be properly designed, meaning that a couple of prerequisites must be fulfilled to perform a successful experiment. The first prerequisite is to choose a dye that can undergo the tryptophan-induced PET and can also be inserted into a specific site of the protein scaffold. The cysteine reactive Atto-oxazine fluorophores (e.g. Atto655-maleimide (Atto655) and AttoOxa12-maleimide (AttoOxa12)) fulfil both of these criteria [55]. Appending the fluorophore to the protein scaffold requires introducing a cysteine residue, which specifically reacts with the maleimide in the dye molecule. The second essential aspect is the positioning of the dye within the protein scaffold. The fluorophore must be placed in the protein region that is involved in the conformational change of interest, and, simultaneously, it has to be placed in the vicinity of the tryptophan (W177) to facilitate the close contact necessary for PET to occur. In this particular case, the search for the right spot for the fluorophore has meant to run a computational generation of a group of cysteine-mutated LinB86 variants that have been evaluated in terms of backbone flexibility, solvent accessibility, and protein stability. From these, the four seemingly most promising LinB86 mutants have been constructed and experimentally characterized, and the LinB86 variant displaying the highest stability has been chosen for the PET-FCS assay. The chosen variant has the single point mutation Q172C and is henceforth denoted as LinB92. To exclude non-specific contributions to the PET-FCS data, a control variant of LinB92 has been prepared with a back-mutated W177 (W177L, denoted as LinB99) that renders the dominant PET pathway inactive [14].

The Atto dyes attached to the LinB92 variant showed remarkably lower brightness values compared to the control variant LinB99 (Table 3). This indicates that the designed "PET-switch" works correctly; i.e. Atto fluorescence is specifically quenched by W177 via the PET mechanism in the LinB92 enzyme. The ACFs recorded for LinB92 and LinB99, both labelled with Atto655, are shown in Fig. 5c. Firstly, it is clearly visible that the ACFs recorded for LinB92 decay faster compared to LinB99. This is due to the additional (sub)-millisecond "non-diffusional" component caused by the repetitive PET quenching of the dye by W177, which has been removed in LinB99 [14]. This PET component of the ACF emerges from the fluctuations in fluorescence intensity caused by a periodic conformational transition of the enzyme that leads to an alternation between a dark state (PET-quenched, fluorescence *off*) where W177 is in the close vicinity of the fluorophore and a bright state (unquenched, fluorescence *on*) where W177 is distant from the fluorophore (Fig. 5b). The same behaviours are observed for LinB92 and LinB99 labelled with

Table 3 Brightness values for the labelled HLD LinB86 mutants. The brightness of a molecule is expressed as the number of photons emitted per second and is relative to the brightness of the standard solution of Atto655-maleimide in a glycine buffer, pH = 8.6 at 20°C. The lower brightness of LinB92 compared to the control LinB99 for both fluorescent labels evidences the specific quenching of the fluorophore via PET with tryptophan 177, meaning a successful design of the "PET-switch" [60]

	Brightness (kHz)	
	Atto655	AttoOxa12
LinB92-	12.6 ± 1.8	11.6 ± 2.5
LinB99-	18.4 ± 3.6	16.8 ± 1.9

the AttoOxa12 fluorophore [14]. MD simulations of the LinB92 variant have identified the two stable conformations that exist in a dynamic equilibrium. The PET-quenched state of LinB92 is an open conformation where the fluorophore can move into the mouth of the access tunnel and come into close contact with W177. The fluorescent, unquenched, state corresponds to a closed conformation of LinB92 where W177 and Atto655 are significantly separated (Fig. 5b). Importantly, the simulations show identical openings of LinB86 and LinB92 that involve a large movement of one of the cap-domain helices; thus, the fluorophore does not seem to disturb the gating process.

Secondly, the on and off rates describing the conformational equilibria can be deduced from the measured FCS data and compared with those obtained by transient kinetic data showing a strong correlation (Fig. 5d). Additionally, analysis of the MD data also provides information on the equilibrium probability of the two states and unambiguously identified the closed conformation as the more stable one (Fig. 5b). The strong agreement between the dynamical information gained by the different experimental approaches and also by MD provides a solid scaffold to support the theory that the tailored gating dynamics of LinB86 is responsible for the unique catalytic performance of this rationally designed enzyme. Furthermore, the transient kinetics study of the LinB variants using bromide binding, which showed a decrease in the observed rates at elevated bromide concentrations, was paralleled in the PET-FCS study showing a similar effect, i.e. slow-down of k_{on} with increasing Br^- concentration [14]. The fact that the PET-FCS experiment is capable of showing the changes in gating dynamics induced by bromide solidifies that this method is capturing the real-time gating dynamics of the LinB86 enzyme at a molecular level. While PET-FCS has been used to study protein conformational dynamics and folding/denaturation pathways [54, 61, 62], this work exemplifies how PET-FCS can be used to obtain real-time dynamic information on protein gating phenomena, or dynamical movements in general, which are increasingly seen as vital elements of protein functioning [63–65].

5 Conclusions and Perspectives

In this chapter, we present a thorough overview of various fluorescence techniques suitable for analysis of protein solvation and dynamics. We describe strategies of fluorescence labelling that yield probes positioned at function-relevant regions of the HLD enzymes. Moreover, we stress the importance of the knowledge of probe photophysics, which is essential for the proper choice of the experimental approach and interpretation of results. We show that probes with plain environment-sensitive photophysics may be used for monitoring TDFS in proteins in a time-resolved manner, revealing the mobility and the degree of hydration within the probe micro-environment. We demonstrate how complex photophysics comprising hydration-driven excited-state reactions are utilized to reflect the hydration of the probe's microenvironment in the case of the steady-state fluorescence "HMC hydration assay". Finally, we evidence how coupling a protein's conformational transition to the switching between a bright and a dark state of a fluorophore can give an estimate on conformational kinetics and aid in mapping conformational dynamics. Each of the presented methods provides slightly different information on dynamics and hydration in proteins. This chapter documents how their combination delivers a comprehensive insight into the hydration and dynamics of HLDs that can be correlated with enzyme kinetics data to identify factors important for HLD activity. Although these techniques present a powerful toolbox, the retrieved information is ensemble-averaged. Capturing the inner diversity of enzyme action would require single-molecule techniques. For instance, single-molecule TIRF microscopy could follow changes in brightness caused by conformational transitions of enzymes, substrate binding, or product release.

Microbial enzyme HLDs served as a model system for testing the strengths and weaknesses of fluorescence methods for analysing protein solvation and dynamics [16, 44, 47, 51, 60]. Why are these proteins attractive as the working horses for detailed mechanistic studies? HLDs fulfil multiple criteria, which makes them particularly suitable for analysing effects of solvation and dynamics on protein function: (1) HLDs are hydrolases and employ water as co-substrate, (2) active sites are buried in the protein core and protein dynamics assist the process of substrate (halogenated alkanes) binding and products (primary alcohols and halide ions) release via multiple access tunnels, (3) the structures of several dozen family members have been solved to high resolution using protein crystallography, (4) the reaction mechanism is known from crystallographic analysis, enzyme kinetic studies, and quantum chemistry calculations, (5) transient kinetics has been investigated for multiple family members and rate-limiting (physical) steps have been confirmed for specific substrates, (6) proteins are made of only two domains, do not require co-factors, and are generally easy to produce, and (7) hundreds of mutants have been constructed and characterized during three decades of research [66]. We envision that more research groups will use HLDs as a model system for mechanistic studies employing fluorescence methods and will contribute to a better understanding of the relationships between structure and protein solvation/dynamics.

Protein solvation and dynamics are particularly important for the physical steps of the enzymatic catalytic cycle. These include substrate binding and product release. For some reaction mechanisms, water serves as the co-substrate and repetitive solvation of the active site is essential for proper catalytic functioning. Protein dynamics can assist physical steps of the catalytic cycle in proteins with buried active sites and/or with tightly bound hydrophobic substrates/products that prefer to escape from the solvent environment. Conformational changes are typically needed for efficient association and dissociation of enzyme–substrate and enzyme–product complexes. These conformational changes or gating can involve flexible loops [67], but also secondary elements or even whole domains [68]. Very often, they can be rate-limiting and need to be engineered for better catalytic efficiency. However, engineering proteins dynamics is very challenging often due to the lack of mechanistic information and directed evolution approaches must be combined with rational protein design [67]. Fluorescence methods are perfectly positioned for collecting high-quality quantitative data even at a single molecule level to guide the rational design of protein dynamics. Appropriate software tools are then needed for structure-based protein design of dynamical elements.

Acknowledgements The authors would like to express their thanks to the Czech Ministry of Education (INBIO CZ.02.1.01/0.0/0.0/16_026/0008451, ENOCH CZ.02.1.01/0.0/0.0/16_019/0000868, RECETOX RI LM2018121 and EXCELES Neuro LX22NPO5107). This project has received funding from the European Union's Horizon 2020 research and innovation programme under grant agreement TEAMING No. 857560. The article reflects the author's view and the agency is not responsible for any use that may be made of the information it contains. This project has received funding from the European Union's Horizon 2020 research and innovation programme under the Marie Skłodowska-Curie grant agreement No. 860592.

References

1. El Karoui M, Hoyos-Flight M, Fletcher L (2019) Future trends in synthetic biology – a report. Front Bioeng Biotechnol 7
2. (2017) Synthetic biology: volume 1, Royal Society of Chemistry, London
3. Robinson PK Enzymes: principles and biotechnological applications. Essays Biochem 59(2015):1–41
4. Wiltschi B, Cernava T, Dennig A, Casas MG, Geier M, Gruber S, Haberbauer M, Heidinger P, Acero EH, Kratzer R, Luley-Goedl C, Muller CA, Pitzer J, Ribitsch D, Sauer M, Schmolzer K, Schnitzhofer W, Sensen CW, Soh J, Steiner K, Winkler CK, Winkler M, Wriessnegger T (2020) Enzymes revolutionize the bioproduction of value-added compounds: from enzyme discovery to special applications. Biotechnol Adv 40
5. Pan XJ, Kortemme T (2021) Recent advances in de novo protein design: principles, methods, and applications. J Biol Chem 296
6. Zanghellini A (2014) de novo computational enzyme design. Curr Opin Biotechnol 29:132–138
7. Damborsky J, Brezovsky J (2014) Computational tools for designing and engineering enzymes. Curr Opin Chem Biol 19:8–16
8. Kries H, Blomberg R, Hilvert D (2013) De novo enzymes by computational design. Curr Opin Chem Biol 17:221–228

9. Grisewood MJ, Hernandez-Lozada NJ, Thoden JB, Gifford NP, Mendez-Perez D, Schoenberger HA, Allan MF, Floy ME, Lai RY, Holden HM, Pfleger BF, Maranas CD (2017) Computational redesign of acyl-ACP Thioesterase with improved selectivity toward medium-chain-length fatty acids. ACS Catal 7:3837–3849
10. Lapidoth G, Khersonsky O, Lipsh R, Dym O, Albeck S, Rogotner S, Fleishman SJ (2018) Highly active enzymes by automated combinatorial backbone assembly and sequence design. Nat Commun 9
11. Baker D (2010) An exciting but challenging road ahead for computational enzyme design. Protein Sci 19:1817–1819
12. Giger L, Caner S, Obexer R, Kast P, Baker D, Ban N, Hilvert D (2013) Evolution of a designed retro-aldolase leads to complete active site remodeling. Nat Chem Biol 9:494–U449
13. Welborn VV, Head-Gordon T (2019) Computational design of synthetic enzymes. Chem Rev 119:6613–6630
14. Kokkonen P, Sykora J, Prokop Z, Ghose A, Bednar D, Amaro M, Beerens K, Bidmanova S, Slanska M, Brezovsky J, Damborsky J, Hof M (2018) Molecular gating of an engineered enzyme captured in real time. J Am Chem Soc 140:17999–18008
15. Schafer JW, Zoi I, Antoniou D, Schwartz SD (2019) Optimization of the turnover in artificial enzymes via directed evolution results in the coupling of protein dynamics to chemistry. J Am Chem Soc 141:10431–10439
16. Sykora J, Brezovsky J, Koudelakova T, Lahoda M, Fortova A, Chernovets T, Chaloupkova R, Stepankova V, Prokop Z, Smatanova IK, Hof M, Damborsky J (2014) Dynamics and hydration explain failed functional transformation in dehalogenase design. Nat Chem Biol 10:428–430
17. Marek J, Vevodova J, Smatanova IK, Nagata Y, Svensson LA, Newman J, Takagi M, Damborsky J (2000) Crystal structure of the haloalkane dehalogenase from Sphingomonas paucimobilis UT26. Biochemistry 39:14082–14086
18. Mazumdar PA, Hulecki JC, Cherney MM, Garen CR, James MNG (2008) X-ray crystal structure of mycobacterium tuberculosis haloalkane dehalogenase Rv2579. Biochim Biophys Acta 1784:351–362
19. Newman J, Peat TS, Richard R, Kan L, Swanson PE, Affholter JA, Holmes IH, Schindler JF, Unkefer CJ, Terwilliger TC (1999) Haloalkane dehalogenases: structure of a Rhodococcus enzyme. Biochemistry 38:16105–16114
20. Tratsiak K, Prudnikova T, Drienovska I, Chrast L, Damborsky J, Rezacova P, Kuty M, Chaloupkova R, Smatanova IK (2014) Crystal structure of the novel haloalkane dehalogenases. Acta Crystallogr A Found Adv 70:C1678
21. Verschueren KHG, Seljee F, Rozeboom HJ, Kalk KH, Dijkstra BW (1993) Crystallographic analysis of the catalytic mechanism of haloalkane dehalogenase. Nature 363:693–698
22. Koudelakova T, Chovancova E, Brezovsky J, Monincova M, Fortova A, Jarkovsky J, Damborsky J (2011) Substrate specificity of haloalkane dehalogenases. Biochem J 435:345–354
23. Jugder BE, Ertan H, Bohl S, Lee M, Marquis CP, Manefield M (2016) Organohalide respiring bacteria and reductive dehalogenases: key tools in organohalide bioremediation. Front Microbiol 7
24. Coleman NV, Mattes TE, Gossett JM, Spain JC (2002) Phylogenetic and kinetic diversity of aerobic vinyl chloride-assimilating bacteria from contaminated sites. Appl Environ Microbiol 68:6162–6171
25. Swanson PE (1999) Dehalogenases applied to industrial-scale biocatalysis. Curr Opin Biotechnol 10:365–369
26. Prokop Z, Sato Y, Brezovsky J, Mozga T, Chaloupkova R, Koudelakova T, Jerabek P, Stepankova V, Natsume R, van Leeuwen JGE, Janssen DB, Florian J, Nagata Y, Senda T, Damborsky J (2010) Enantioselectivity of haloalkane dehalogenases and its modulation by surface loop engineering. Angew Chem Int Ed 49:6111–6115

27. Bidmanova S, Chaloupkova R, Damborsky J, Prokop Z (2010) Development of an enzymatic fiber-optic biosensor for detection of halogenated hydrocarbons. Anal Bioanal Chem 398:1891–1898
28. Los GV, Encell LP, McDougall MG, Hartzell DD, Karassina N, Zimprich C, Wood MG, Learish R, Ohane RF, Urh M, Simpson D, Mendez J, Zimmerman K, Otto P, Vidugiris G, Zhu J, Darzins A, Klaubert DH, Bulleit RF, Wood KV (2008) HatoTag: a novel protein labeling technology for cell imaging and protein analysis. ACS Chem Biol 3:373–382
29. Damborsky J, Chaloupkova R, Pavlova M, Chovancova E, Brezovsky J (2010) Structure–function relationships and engineering of haloalkane dehalogenases. In: Timmis KN (ed) Handbook of hydrocarbon and lipid microbiology. Springer, Berlin, pp 1081–1098
30. Biedermannova L, Schneider B (2016) Hydration of proteins and nucleic acids: advances in experiment and theory. A review. Biochim Biophys Acta 1860:1821–1835
31. Laage D, Elsaesser T, Hynes JT (2017) Water dynamics in the hydration shells of biomolecules. Chem Rev 117:10694–10725
32. Mondal S, Mukherjee S, Bagchi B (2017) Protein hydration dynamics: much ado about nothing? J Phys Chem Lett 8:4878–4882
33. Smith JC, Merzel F, Bondar AN, Tournier A, Fischer S (2004) Structure, dynamics and reactions of protein hydration water. Philos Trans R Soc B Biol Sci 359:1181–1189
34. Horng ML, Gardecki JA, Papazyan A, Maroncelli M (1995) Subpicosecond measurements of polar solvation dynamics – Coumarin-153 revisited. J Phys Chem 99:17311–17337
35. Ito N, Arzhantsev S, Heitz M, Maroncelli M (2004) Solvation dynamics and rotation of coumarin 153 in alkylphosphonium ionic liquids. J Phys Chem B 108:5771–5777
36. Bagchi B (2005) Water dynamics in the hydration layer around proteins and micelles. Chem Rev 105:3197–3219
37. Mattea C, Qvist J, Halle B (2008) Dynamics at the protein-water interface from O-17 spin relaxation in deeply supercooled solutions. Biophys J 95:2951–2963
38. Pal SK, Peon J, Zewail AH (2002) Biological water at the protein surface: dynamical solvation probed directly with femtosecond resolution. Proc Natl Acad Sci U S A 99:1763–1768
39. Bagchi B (2013) Biological water. In: Water in biological and chemical processes: from structure and dynamics to function. Cambridge University Press, Cambridge, pp 81–96
40. Persson F, Soderhjelm P, Halle B (2018) The spatial range of protein hydration. J Chem Phys 148
41. Sato Y, Natsume R, Tsuda M, Damborsky J, Nagata Y, Senda T (2007) Crystallization and preliminary crystallographic analysis of a haloalkane dehalogenase, DbjA, from Bradyrhizobium japonicum USDA110. Acta Crystallogr Sect F Struct Biol Commun 63:294–296
42. Chaloupkova R, Sykorova J, Prokop Z, Jesenska A, Monincovaa M, Pavlova M, Tsuda M, Nagata Y, Damborsky J (2003) Modification of activity and specificity of haloalkane dehalogenase from Sphingomonas paucimobilis UT26 by engineering of its entrance tunnel. J Biol Chem 278:52622–52628
43. Bosma T, Pikkemaat MG, Kingma J, Dijk J, Janssen DB (2003) Steady-state and pre-steady-state kinetic analysis of halopropane conversion by a Rhodococcus haloalkane dehalogenase. Biochemistry 42:8047–8053
44. Jesenska A, Sykora J, Olzynska A, Brezovsky J, Zdrahal Z, Damborsky J, Hof M (2009) Nanosecond time-dependent stokes shift at the tunnel mouth of Haloalkane dehalogenases. J Am Chem Soc 131:494–501
45. Qiu WH, Kao YT, Zhang LY, Yang Y, Wang LJ, Stites WE, Zhong DP, Zewail AH (2006) Protein surface hydration mapped by site-specific mutations. Proc Natl Acad Sci U S A 103: 13979–13984
46. Sykora J, Brezovsky J, Koudelakova T, Lahoda M, Fortova A, Chernovets T, Chaloupkova R, Stepankova V, Prokop Z, Kuta Smatanova I, Hof M, Damborsky J (2014) Dynamics and hydration explain failed functional transformation in dehalogenase design. Nat Chem Biol 10: 428–430

47. Amaro M, Brezovsky J, Kovacova S, Maier L, Chaoupkova R, Sykora J, Paruch K, Damborsky J, Hof M (2013) Are time-dependent fluorescence shifts at the tunnel mouth of Haloalkane dehalogenase enzymes dependent on the choice of the chromophore? J Phys Chem B 117:7898–7906
48. Scollo F, Evci H, Amaro M, Jurkiewicz P, Sykora J, Hof M (2021) What does time-dependent fluorescence shift (TDFS) in biomembranes (and proteins) report on? Front Chem 9
49. Choudhury SD, Nath S, Pal H (2008) Excited-state proton transfer behavior of 7-hydroxy-4-methylcoumarin in AOT reverse micelles. J Phys Chem B 112:7748–7753
50. Choudhury SD, Pal H (2009) Modulation of excited-state proton-transfer reactions of 7-Hydroxy-4-methylcoumarin in ionic and nonionic reverse micelles. J Phys Chem B 113: 6736–6744
51. Amaro M, Brezovsky J, Kovacova S, Sykora J, Bednar D, Nemec V, Liskova V, Kurumbang NP, Beerens K, Chaloupkova R, Paruch K, Hof M, Damborsky J (2015) Site-specific analysis of protein hydration based on unnatural amino acid fluorescence. J Am Chem Soc 137:4988–4992
52. Mills JH, Lee HS, Liu CC, Wang JY, Schultz PG (2009) A genetically encoded direct sensor of antibody-antigen interactions. ChemBioChem 10:2162–2164
53. Gordon SE, Munari M, Zagotta WN (2018) Visualizing conformational dynamics of proteins in solution and at the cell membrane. eLife 7:e37248
54. Sauer M, Neuweiler H (2014) PET-FCS: probing rapid structural fluctuations of proteins and nucleic acids by single-molecule fluorescence quenching. In: Engelborghs Y, Visser AJWG (eds) Fluorescence spectroscopy and microscopy, vol 1076. Humana Press, pp 597–615
55. Doose S, Neuweiler H, Sauer M (2005) A close look at fluorescence quenching of organic dyes by tryptophan. ChemPhysChem 6:2277–2285
56. Elson EL, Magde D (1974) Fluorescence correlation spectroscopy. 1. Conceptual basis and theory. Biopolymers 13:1–27
57. Hess ST, Huang SH, Heikal AA, Webb WW (2002) Biological and chemical applications of fluorescence correlation spectroscopy: a review. Biochemistry 41:697–705
58. Brezovsky J, Babkova P, Degtjarik O, Fortova A, Gora A, Iermak I, Rezacova P, Dvorak P, Smatanova IK, Prokop Z, Chaloupkova R, Damborsky J (2016) Engineering a de novo transport tunnel. ACS Catal 6:7597–7610
59. Biedermannova L, Prokop Z, Gora A, Chovancova E, Kovacs M, Damborsky J, Wade RC (2012) A single mutation in a tunnel to the active site changes the mechanism and kinetics of product release in haloalkane dehalogenase LinB. J Biol Chem 287:29062–29074
60. Kokkonen P, Bednar D, Dockalova V, Prokop Z, Damborsky J (2018) Conformational changes allow processing of bulky substrates by a haloalkane dehalogenase with a small and buried active site. J Biol Chem 293:11505–11512
61. Neuweiler H, Johnson CM, Fersht AR (2009) Direct observation of ultrafast folding and denatured state dynamics in single protein molecules. Proc Natl Acad Sci U S A 106:18569–18574
62. Ries J, Schwarze S, Johnson CM, Neuweiler H (2014) Microsecond folding and domain motions of a spider silk protein structural switch. J Am Chem Soc 136:17136–17144
63. Merski M, Moreira C, Abreu RMV, Ramos MJ, Fernandes PA, Martins LM, Pereira PJB, Macedo-Ribeiro S (2017) Molecular motion regulates the activity of the mitochondrial serine protease HtrA2. Cell Death Dis 8
64. Agarwal PK, Bernard DN, Bafna K, Doucet N (2020) Enzyme dynamics: looking beyond a single structure. ChemCatChem 12:4704–4720

65. Gardner JM, Biler M, Risso VA, Sanchez-Ruiz JM, Kamerlin SCL (2020) Manipulating conformational dynamics to repurpose ancient proteins for modern catalytic functions. ACS Catal 10:4863–4870
66. Koudelakova T, Bidmanova S, Dvorak P, Pavelka A, Chaloupkova R, Prokop Z, Damborsky J (2013) Haloalkane dehalogenases: biotechnological applications. Biotechnol J 8:32
67. Schenkmayerova A, Pinto GP, Toul M, Marek M, Hernychova L, Planas-Iglesias J, Liskova VD, Pluskal D, Vasina M, Emond S, Doerr M, Chaloupkova R, Bednar D, Prokop Z, Hollfelder F, Bornscheuer UT, Damborsky J (2021) Engineering the protein dynamics of an ancestral luciferase. Nat Commun 12
68. Gora A, Brezovsky J, Damborsky J (2013) Gates of enzymes. Chem Rev 113:5871–5923